Ornamental Geophytes

From Basic Science to Sustainable Production

Edited by
Rina Kamenetsky
Hiroshi Okubo

Ornamental Geophytes
From Basic Science to Sustainable Production

CRC Press
Taylor & Francis Group
Boca Raton   London   New York

CRC Press is an imprint of the
Taylor & Francis Group, an **informa** business

Cover photos by William B. Miller (USA) and Sedat Kiran (Turkey).

CRC Press
Taylor & Francis Group
6000 Broken Sound Parkway NW, Suite 300
Boca Raton, FL 33487-2742

First issued in paperback 2016

© 2013 by Taylor & Francis Group, LLC
CRC Press is an imprint of Taylor & Francis Group, an Informa business

No claim to original U.S. Government works

Version Date: 20120716

ISBN 13: 978-1-138-19861-6 (pbk)
ISBN 13: 978-1-4398-4924-8 (hbk)

Library of Congress Cataloging-in-Publication Data

Ornamental geophytes : from basic science to sustainable production / editors: Rina Kamenetsky and Hiroshi Okubo.
   p. cm.
  Includes bibliographical references and index.
  ISBN 978-1-4398-4924-8 (alk. paper)
  1. Bulbs (Plants) 2. Flowers. 3. Plants, Ornamental. I. Kamenetsky, Rina. II. Okubo, H. (Hiroshi)

SB425.O76 2012
635.9'1--dc23   2012009136

Visit the Taylor & Francis Web site at
http://www.taylorandfrancis.com

and the CRC Press Web site at
http://www.crcpress.com

Rina Kamenetsky dedicates this book to her father, Dr. Josef Malyar, who inspired and encouraged her scientific career.

Hiroshi Okubo dedicates this book to his mentor, Dr. Shunpei Uemoto, who stimulated and supported his scientific work.

Contents

Foreword

This book is a timely and welcome publication in the field of flower bulb production and research. The first book on this topic was published in 1972 by Alun R. Rees and was titled *The Growth of Bulbs*. It was followed up by a second publication, *Ornamental Bulbs, Corms, and Tubers*, in 1992. Subsequently, our book *The Physiology of Flower Bulbs* was published in 1993 and the literature search for it concluded in 1992. Since then, there have been many technological advances in areas such as taxonomy, plant breeding, biotechnology, and physiology. In addition, there has been a worldwide "Environmental Awakening" that has required significant changes in commercial bulb and flower production and their utilization. Concurrently, there has been an increased globalization of flower bulbs. In addition, it is well known that research is essential for any viable industry and it must be continuous. Therefore, researchers and the leaders of the flower bulb industry must address the new challenges that have and will continue to develop. Strategic and long-term planning is essential for everyone. Very importantly, traditional sources of research and educational funds are rapidly declining, not only in The Netherlands but also worldwide. Researchers will have to be creative in obtaining and maintaining funds (see Chapter 1). This is especially true for long-term programs like tulip breeding (see Chapter 6), which, at present, takes about 15–20 years to bring a new cultivar into mainstream production.

Thus, this book, which highlights the recent advances in this scientific field, is very essential. It will benefit not only researchers who have been engaged for years but also new researchers who must meet and challenge the existing dogmas. Last, information will be valuable to the industry to bring the value of flower bulbs to the worldwide spectrum of consumers. In the end, these are the most important and last link in the chain of utilization and profitability of all flower bulb products.

The flower bulb industry and researchers are indebted to Dr. Rina Kamenetsky and Dr. Hiroshi Okubo for encouraging and assembling this needed book on ornamental geophytes. It will be invaluable in meeting the challenges for the twenty-first century and beyond.

August A. De Hertogh
North Carolina State University, Raleigh, NC

Marcel Le Nard
Institut National de la Recherche Agrononique, France

Editors

Rina Kamenetsky was born in Almaty, Kazakhstan (in the former USSR). She received her BSc and MSc with distinction (1979) from the Kazakh State University and her PhD in plant physiology from the Main Botanical Garden of the Kazakh National Academy of Sciences at Almaty (1984). The early part of her professional career was in Kazakhstan, where she held research positions at the Laboratory of Medicinal and Edible Plants of the Main Botanical Garden, and at the Laboratory of Plant Ecology in the Institute of Botany. Dr. Kamenetsky immigrated to Israel in 1990 and held the position of senior scientist at Ben-Gurion University of the Negev. In 1994, she joined the Department of Ornamental Horticulture at the Agricultural Research Organization, the Volcani Center in Bet Dagan. She served as chair of this department in 2004–2007. In 2006 she was appointed as adjunct professor of the Hebrew University of Jerusalem and part-time professor of the Henan University of Science and Technology, China.

Currently, her research interests focus in three major directions: (1) the mechanisms of internal and environmental control of flowering and dormancy in geophytes and herbaceous perennial plants; (2) strategies and technologies for the development and production of ornamental and edible crops; (3) plant biodiversity, the introduction of new ornamental plants, and their cultivation in warm regions. She developed an academic course on the biology and production of geophytes, and has been teaching at the Hebrew University of Jerusalem for the last 10 years. She also supervises PhD and MSc students and hosts postdoctoral fellows and foreign trainees in her lab.

The results of her research have been published in more than 180 papers, reported at international meetings and symposia, and summarized in 20 scientific reviews and book chapters. Dr. Kamenetsky serves as a chairperson of Working Group Flowerbulbs and Herbaceous Perennials and Israel's representative to the Council of the International Society of Horticulture Science (ISHS). She is a reviewer for 20 scientific journals, and a member of several scientific editorial boards. Dr. Kamenetsky is an active participant in international scientific teams, and received research grants in collaboration with colleagues from Holland, France, Germany, Italy, the United States, Canada, and Central Asia. She has also been invited to take part in numerous international missions, consultations, and meetings in Chile, Argentina, China, Japan, Kazakhstan, and Uzbekistan, and to serve as a member of the organizing and scientific committees of 10 international symposia. The results of her research on the flowering physiology of ornamentals such as peony, tulips, *Allium*, *Eremurus*, and others are being applied in commercial floriculture.

During her scientific carrier, Rina was continuously supported by her close family—her husband Moshe Goldstein and her parents Nelly and Josef Malyar. Her little grandson Liam Sela also contributed to her ambition to edit and publish this book.

Hiroshi Okubo was born in Fukuoka, Japan. He received his BS, MS, and PhD in horticultural science from Kyushu University, Fukuoka, Japan. He joined the Laboratory of Horticultural Science, Faculty of Agriculture, Kyushu University as an assistant professor in 1981. He was promoted to associate professor in 1990 and to full professor in 2000. He is the director of the University Farm of Kyushu University, and is concurrently professor at Nanjing Forestry University, Nanjing, China from 2009.

His research interests are focused on (1) the physiology of growth and development in horticultural crops, particularly in ornamental geophytes;

(2) the introduction and breeding of underutilized plants, particularly in tropical regions; and (3) species and varietal differentiation of horticultural crops. In addition to bulbous plants, his research includes camellias, rhododendrons, orchids, asparagus, onions, tropical legumes, solanums, gourds, and so on. He has been involved in many research trips to tropical Asian countries. This experience has helped him develop his hypotheses of dormancy, not only in geophytes, but also in plants as a whole. Dr. Okubo is the author of more than 200 professional papers, book chapters, technical publications, and popular press articles. He is the recipient of an award of the Japanese Society for Horticultural Science (1988) and the outstanding paper award of the Botanical Society of Japan (2009).

Professor Okubo has served as a scientific committee member and as an invited speaker at many international symposia. He has also presented invited lectures in universities and other institutions in China, Poland, Vietnam, Indonesia, and others. He serves as a reviewer for a number of international and domestic scientific journals.

Acknowledgments

The preparation of this book took almost three years. More than 40 leading scientists from all over the world were involved in the writing and reviewing, as well as in fruitful discussions of its contents. We would like to thank all the contributors and coauthors for their willingness to share their experience and knowledge. We also wish to thank the hundreds of growers, research fellows, students, technicians, administrators, extension officers, and other colleagues for their constant support of our research, which made possible the accumulation of this great amount of information on geophytes science and production.

We are grateful to our mentors I. O. Baitulin, I. R. Rakhimbaev, Y. Gutterman, S. Uemoto, and H. Imanishi for the love of geophytes and horticulture that we learned from them. We also wish to thank Dr. A. A. De Hertogh who encouraged us to initiate this project, and constantly supported its realization during these three years.

Special thanks to our families who were patient and helpful during the years of preparation of this book. This was a challenging endeavor, and we always felt the support and encouragement at home.

Introduction

Rina Kamenetsky and Hiroshi Okubo

Ornamental geophytes, also called "flower bulbs," belong to more than 800 different botanical genera and exhibit great diversity in their developmental biology and physiological responses to environmental factors. Morphologically, these plants are characterized by renewal buds located in an underground storage organ, for example, rhizome, tuber, corm, or bulb. Ornamental geophytes play a significant role in the global flower industry, and are utilized for the commercial production of cut flowers, potted plants, and propagation materials, as well as in landscaping and gardening. The most important and well-known crops are tulip, lily, gladiolus, narcissus, iris, hyacinth, and crocus.

A comprehensive treatise *The Physiology of Flower Bulbs*, consisting of 40 chapters on the general aspects of geophyte biology and on particular genera, was published by A. A. De Hertogh and M. Le Nard in 1993. Since then, four international symposia on flower bulbs (Lilien-Kipnis et al. 1997; Littlejohn et al. 2002; Okubo et al. 2005; Van den Ende et al. 2010) and two special symposia on the genus *Lilium* (Lee and Roh 1996; Grassotti and Burchi 2011) have taken place under the aegis of the International Society of Horticultural Sciences (ISHS). Numerous scientific papers and trade articles, as well as several excellent reviews and books (e.g., De Hertogh 1996; Hanks 2002), have also been published. Although a great deal of modern research on ornamental geophytes has been conducted since the beginning of the 1990s, up-to-date information has not been comprehensively summarized and presented and is, therefore, not always available to researchers and horticulturalists.

The main objective of this book is to summarize the recent advances in the biology and global production of ornamental geophytes. Twenty chapters, compiled by leading experts, provide up-to-date reviews on geophyte taxonomy, physiology, genetics, production, plant protection, and postharvest biology. In general, the latest advances in geophyte science reflect three main changes in plant science and horticulture: (1) a demand for sustainable and environment-friendly production, (2) widespread employment of new molecular technologies, and (3) the globalization of the production and marketing chains.

The growing public concern over the potential impact on the environment from the use of pesticides and other chemicals, the increasing costs of energy and the urbanization of many of the bulb-producing areas have had a major effect on the production of ornamental geophytes. The development of sustainable production systems requires the integration of basic and applied research that examines the whole agro-ecosystem, and a multidisciplinary approach to addressing research needs. Progress in the utilization of renewable resources, environment-friendly integrated management, and postharvest biology of ornamental geophytes is summarized in several of the chapters of this book.

Since new research technologies and molecular skills have developed dramatically over the last 20 years, much of the data on the molecular regulation of plant development, disease resistance, or the bioengineering of geophytes are completely new. Taxonomy, phylogeny, and the breeding for useful traits are also employing novel sophisticated techniques, resulting in tremendous progress in these fields. Molecular technologies have increased our ability to detect and identify pests and pathogens, which is of much major benefit to clean plant programs.

For many years, the cultivation of flower bulbs has been the prerogative of developed ornamental horticulture in countries with moderate climates. Four centuries of bulb production, breeding, the development of new products, and the bulb trade led to the clear leadership of The

Netherlands in this domain. Significant production of ornamental geophytes takes place also in the United Kingdom, France, Japan, and the United States. However, major shifts in the global economy are leading to remarkable changes in the floriculture industry. Increased globalization, the liberalization of markets, and the development of new products play an important role in shaping market trends and building consumer demand for goods and services, including those for ornamental geophytes. In recent decades, the production of flowers and propagating material of high quality has expanded in countries of the Southern Hemisphere and in regions with warm climates. This development has allowed the industry to take advantage of counter seasonal markets and enables the constant supply of fresh flowers to European, Japanese, and North American markets during winter.

In this book, the current status of research and the bulb industry is reviewed for a few countries and regions. These regions can, to a certain extent, be seen as models for different approaches to the geophyte industry. Although four of these new production regions are located in the Southern Hemisphere, their perspectives on the bulb industry and markets are different. Thus, Brazil produces mainly introduced popular bulbous crops for the domestic market, while only a small share of the propagation material is exported. On the other hand, Chile exports tulip and lily bulbs, produced on a large commercial scale, in partnership with international companies. Both Chile and South Africa are extremely rich in native geophyte species, which can be developed as new and original ornamental crops. Although botanical research has been done on many new and promising species, the local flower industries still tend to concentrate on the major popular geophytes crops, while commercial production of the local ornamental geophytes remains very modest. New Zealand does not possess natural resources of ornamental geophytes. Geophytes research in this country is market oriented and dedicated to the most efficient breeding and production methods for propagation material and cut flowers. Since the domestic market is very small, the main efforts are directed at the export of ornamental products.

Turkey represents a very special case in floriculture. The history of trade in ornamental geophytes from this country is quite dramatic. For many years, only bulbs collected in the wild were exported, but, after 1990, the Turkish government decided to limit bulb collecting from native populations, and now a significant share of bulb export is generated from commercial production. Turkey is very rich in natural resources of ornamental geophytes, which can be developed as new ornamental crops in the future.

Very rapid changes in horticulture in general and floriculture in particular are taking place in South-East Asia. In this book, the production of and research in ornamental geophytes in China, Korea, Vietnam, Thailand, and Taiwan are reviewed. Accelerated economic development during the last two decades has led to an increase in the standard of living in Asia, resulting in a swift increase in the demand for and subsequent supply of flowers. It is assumed that, in the near future, flower production and consumption in this region will determine the direction of global ornamental horticulture, as well as the continuing success of the world flower market. Unfortunately, it is rather difficult to estimate geophyte production in the two largest countries of Asia—China and India, since statistics are neither always available nor reliable. It is known that India is one of the largest flower producers in the world (http://www.apeda.gov; http://floriculturetoday.in), but most of the production is directed at the domestic market. Geophyte flowers are probably produced in the northern regions of India, but precise details are not available. In recent years, research on geophyte species has been reported from Iran, one of the major centers of geophyte diversity. Although data on the flower bulb industry in Iran are not accessible, it is known, for example, that research is being conducted on the endemic lily species (Fataei 2011). Ornamental geophytes are also widely distributed in Central Asia (Uzbekistan, Kazakhstan, Kirgizstan), but no data on commercial production in these countries are available.

In recent years, we have seen increased participation of researchers from the new production regions at international symposia on floriculture and ornamental geophytes. For example, among the 134 participants at the *VI ISHS Symposium on Flower Bulbs* in Skierniewice, Poland, in 1992

(Saniewski et al. 1992), only 13 participants came from emerging production regions, whereas at the *X Symposium* in Lisse, The Netherlands, in 2008 (Van den Ende et al. 2010), more than 50 of the 155 participants represented new production countries. Similarly, only 11 countries/regions participated in the *First International Symposium on the Genus* Lilium in Korea in 1994 (Lee and Roh 1996), whereas the *Second* Lilium *Symposium* in 2010 in Italy (Grassotti and Burchi 2011) brought together researchers from 21 countries, with 24 of the 74 participants representing Argentina, Chile, China, Colombia, Eastern Europe, Iran, Korea, Mexico, and Thailand.

New production centers are located in different climates and, although they can take advantage of the vast experience accumulated in the traditional production areas, they must still develop local production systems and evaluate potential crop assortment, as well as postharvest technologies and marketing policies. This relates not only to agrotechniques and protocols, but also to the general approach to market-oriented research, sustainable production systems, the careful use of pesticides, and the balance between natural resources and commercial production. This experience, together with the transfer of knowledge and international scientific collaboration, will help establish profitable and environment-friendly production chains in different countries and continents.

This treatise does not aim to provide technical support or production protocols. The scientific and technological information is supplemented by specific examples on seasonality, fertilization, temperatures regimes, or plant protection practices, but this information relates only to specific experiments or climatic conditions. In addition, the use of specific chemicals and pesticides and their concentrations is not approved in some countries. Therefore, the information included in this book only serves as a reference for current and future practice of geophytes production and consumption. Although the morphological structures and life cycles of the storage organs of geophytes are variable, the term "flower bulb" is liberally used by the authors. In addition to its botanical meaning as an underground storage organ, the terms "bulb" and "bulbing" are also used in horticulture as a generic term for different underground storage organs.

We hope that this book will serve in future improvement of our knowledge about geophytes—one of the most fascinating and diverse plant groups, as well as for future progress in the global ornamental industry.

REFERENCES

De Hertogh, A. 1996. *Holland Bulb Forcer's Guide*, 5th Ed. Hillegom, The Netherlands: The International Flower Bulb Centre and The Dutch Bulb Exporters Association.

De Hertogh, A., and M. Le Nard. 1993. *The Physiology of Flower Bulbs*. Amsterdam: Elsevier.

Fataei, E. 2011. Recognition of new habitats of *Lilium ledebourii* Baker in Northwest Ardabil Province of Iran using GIS technology. *Acta Hort.* 900:65–70.

Grassotti, A., and G. Burchi. 2011. *Acta Horticulturae 900*. Leuven, Belgium: International Society for Horticultural Science.

Hanks, G. R. 2002. *Narcissus and Daffodil: The Genus* Narcissus. London: Taylor & Francis.

Lee, J. S., and M. S. Roh. 1996. *Acta Horticulturae 414*. Leuven, Belgium: International Society for Horticultural Science.

Lilien-Kipnis, H., A. Borochov, and A. H. Halevy. 1997. *Acta Horticulturae 430*. Leuven, Belgium: International Society for Horticultural Science.

Littlejohn, G., R. Venter, and C. Lombard. 2002. *Acta Horticulturae 570*. Leuven, Belgium: International Society for Horticultural Science.

Okubo, H., W. B. Miller, and G. A. Chastagner. 2005. *Acta Hortiiculturae 673*. Leuven, Belgium: International Society for Horticultural Science.

Saniewski, M., J. C. M. Beijersbergen, and W. Bogatko. 1992. *Acta Horticulturae 325*. Leuven, Belgium: International Society for Horticultural Science.

Van den Ende, J. E., A. T. Krikke, and A. P. M. den Nijs. 2010. *Acta Horticulturae 886*. Leuven, Belgium: International Society for Horticultural Science.

Contributors

Paul Arens
Plant Breeding
Wageningen University and Research Centre
Wageningen, The Netherlands

Anna Bach
Department of Ornamentals
University of Agriculture in Krakow
Krakow, Poland

Ibrahim Baktir
Department of Horticulture
Akdeniz University
Antalya, Turkey

Mark P. Bridgen
Department of Horticulture
Cornell University
Riverhead, New York

Gary A. Chastagner
Department of Plant Pathology
Research and Extension Center
Washington State University
Puyallup, Washington

Joo Bee Chuah
The New Zealand Institute for Plant and
    Food Research Ltd.
Palmeston, New Zealand

Glenn E. Clark
The New Zealand Institute for Plant and
    Food Research Ltd.
Pukekohe Research Centre
Pukekohe, New Zealand

Márcia da S. Peetz
Hórtica Consultoria e Treinamento
São Paulo, Brazil

Margery L. Daughtrey
Department of Plant Pathology and
    Plant-Microbe Biology
Cornell University
Riverhead, New York

August A. De Hertogh
Department of Horticultural Science
North Carolina State University
Raleigh, North Carolina

Graham D. Duncan
Kirstenbosch National Botanical Garden
South African National Biodiversity Institute
Cape Town, South Africa

Moshe A. Flaishman
Department of Fruit Tree Sciences
Agricultural Research Organization
The Volcani Center
Bet Dagan, Israel

Seiichi Fukai
Department of Horticultural Science
Kagawa University
Miki-cho, Kagawa, Japan

Keith A. Funnell
The New Zealand Institute for Plant and
    Food Research Ltd.
Palmeston, New Zealand

Gordon R. Hanks
Warwick Crop Centre
The University of Warwick
Wellesbourne, Warwick, United Kingdom

Cai-Zhong Jiang
Crops Pathology and Genetic Research Unit
    USDA-ARS
Davis, California

Antonio H. Junqueira
Hórtica Consultoria e Treinamento
São Paulo, Brazil

Rina Kamenetsky
Department of Ornamental Horticulture
Agricultural Research Organization
The Volcani Center
Bet Dagan, Israel

Kathryn K. Kamo
Floral and Nursery Plants Research Unit
U.S. Department of Agriculture
Beltsville, Maryland

Frans A. Krens
Plant Breeding
Wageningen University and Research Centre
Wageningen, The Netherlands

Marcel Le Nard
Institut National de la Recherche Agronomique
Ploudaniel, France

Agnieszka Marasek-Ciołakowska
Department of Physiology and Biochemistry
Research Institute of Horticulture
Skierniewice, Poland

Alan W. Meerow
USDA-ARS-SHRS
National Germplasm Repository
Miami, Florida

Timothy W. Miller
Department of Crop and Soil Sciences
Washington State University
Mount Vernon, Washington

William B. Miller
Department of Horticulture
Cornell University
Ithaca, New York

Ed R. Morgan
The New Zealand Institute for Plant and
    Food Research Ltd.
Palmeston, New Zealand

Hiroshi Okubo
Faculty of Agriculture
Kyushu University
Fukuoka, Japan

Eduardo A. Olate
Department of Plant Science
Pontificia Universidad Catolica de Chile
Santiago, Chile

Hanu R. Pappu
Department of Plant Pathology
Washington State University
Pullman, Washington

Michael S. Reid
Department of Plant Sciences
University of California
Davis, California

Flavia Schiappacasse
Facultad de Ciencias Agrarias
Universidad de Talca
Talca, Chile

Dariusz Sochacki
Floriculture Division
Research Institute of Horticulture
Skierniewice, Poland

Giulio C. Stancato
Instituto Agronomico (IAC)
Campinas, São Paulo, Brazil

Antonio F. C. Tombolato
Instituto Agronomico (IAC)
Campinas, São Paulo, Brazil

Roberta P. Uzzo
Instituto Agronomico (IAC)
Campinas, São Paulo, Brazil

Johan van Scheepen
KAVB
Hillegom, The Netherlands

Jaap M. van Tuyl
Plant Breeding
Wageningen University and
    Research Centre
Wageningen, The Netherlands

Iris Yedidia
Department of Ornamental Horticulture
Agricultural Research Organization
The Volcani Center
Bet Dagan, Israel

Michele Zaccai
Department of Life Sciences
Ben Gurion University
Beer Sheva, Israel

Meira Ziv
The R. H. Smith Institute of Plant Science
    and Genetics
Hebrew University of Jerusalem
Rehovot, Israel

1 Globalization of the Flower Bulb Industry

*August A. De Hertogh, Johan van Scheepen, Marcel Le Nard,
Hiroshi Okubo, and Rina Kamenetsky*

CONTENTS

I. INTRODUCTION

It is well known that ornamental geophytes exhibit great diversity in their morphology, growth and developmental cycles, and physiological responses to environmental factors (De Hertogh and Le Nard 1993). These diverse ornamental crops represent a significant segment of the global flower industry and are primarily utilized for: (1) commercial bulb production; (2) flower production, including outdoor and forced fresh cut flowers; (3) growing as potted plants; and (4) commercial landscaping and private gardens. Like all other plants, ornamental geophytes significantly contribute to the concept of "Quality to Life". They add simple pleasures to everyday living and make people feel better, which is critical when they are under stress or in need of a sign of appreciation.

Most of the ornamental geophytes are cultivated in temperate-climate regions of the world (Table 1.1) with The Netherlands being the leading producer. It is important to note that although ornamental geophytes exist in over 800 different genera (Bryan 1995), the industry is dominated by seven genera: *Tulipa*, *Lilium*, *Narcissus*, *Gladiolus*, *Hyacinthus*, *Crocus*, and *Iris* (Table 1.2, Chapter 4). In addition, several crops, for example, *Freesia*, *Alstroemeria*, *Hippeastrum*, *Zantedeschia*, *Anemone*, and *Ranunculus*, are of significant economical importance in the cut flower and pot plant industries (www.vbn.nl).

As the global demand for ornamental geophytes increases, it is obvious that innovative production and marketing efforts are needed. The contribution of research to the development of the global flower bulb industry was recently reviewed by Benschop et al. (2010). In this chapter, historical perspectives and the most important aspects of globalization of production and utilization of ornamental geophytes will be reviewed. A complete analysis of historical developments and the current status of the research and industry is important for the establishment of future priorities and needs. These aspects will also be covered in this book.

TABLE 1.1

Estimated World Production of Flower Bulb as Reported in 1972, 1992, and 2005

Country	Production Area Reported (ha)		
	1972	1992	2005
The Netherlands	10,303	16,031	20,921
UK	6016	4300	4660
France	70	1285	1289
China	—	—	1281
USA	2196	4449	995
Japan	1469	1622	883
Israel	10	427	456
Poland	—	990	335
New Zealand	—	15	258
Canada	202	400	—
Chile	—	—	240
South Africa	—	425	200
Brazil	—	433	200
Others	2103	2218	482
Total	22,369	32,595	32,200

Source: Adapted from Rees, A. R. 1972. *The Growth of Bulbs. Applied Aspects of the Physiology of Ornamental Bulbous Crop Plants.* London: Academic Press; De Hertogh, A., and M. Le Nard. 1993. *The Physiology of Flower Bulbs.* Amsterdam: Elsevier; Benschop, M. et al. 2010. *Hort. Rev.* 36:1–115.

TABLE 1.2

Production of Dutch-Grown Flower Bulbs in 2003/2004, 2004/2005, and 2007/2008

Taxa	Production Area Reported (ha)		
	2003–2004	2004–2005	2007–2008
Tulipa	10,982	10,034	9885
Lilium	3212	3275	3699
Narcissus	1796	1721	1687
Gladiolus	1151	1060	1019
Hyacinthus	1121	1140	854
Crocus	668	566	463
Iris (bulbous)	481	464	360
Total	19,411	18,260	17,967

Source: PT/BKD. 2008. Bloembollen, Voorjaarsbloeiers, beplante oppervlakten, seizoen 2007/2008.

II. FLOWER BULBS IN A HISTORICAL PERSPECTIVE

A. From Ancient Times to the Seventeenth Century

Ornamental geophytes have been admired by many people and civilizations and are frequently mentioned in ancient history and mythology (Reynolds and Meachem 1967; Doerflinger 1973;

Genders 1973; Scheider 1981; Bryan 1989, 2002). Early inhabitants were attracted and influenced by horticultural characteristics such as fragrance, color, and time of flowering—the same characteristics that are important for the current global flower bulb industry. The Greek philosopher Theophrastus (371–287 B.C.), an associate and later the successor of Aristotle at the Lyceum of Athens, wrote that some species were valued as spices and dyes (*Crocus sativus*, "The Saffron Crocus"), medicinal plants (colchicums and scillas) or used as food (gladioli, scillas, and *Asphodelus*) (Negbi 1989).

Bulb flowers, for example, the anemone, hyacinth, lily, narcissus, and crocus, have adorned various forms of art for centuries and are associated with stories or customs from various cultures (http://www.mythencyclopedia.com).

Many famous Greek myths were inspired by spring bulb flowers (Margaris 2000). The first example is *Lilium candidum* ("The Madonna Lily"), which is native to the Eastern Mediterranean (Bryan 1989, 1995, 2002). There is a myth that this lily was born from the milk of the goddess Hera. In early Greece, *L. candidum* was the flower of the Goddesses—Aphrodite (Venus) and Diana. Today, white lilies symbolize pure virginal love in the Christian world. *L. candidum* is, in all probability, the original "Easter Lily", since it was also associated with the "Virgin Mother" and was a symbol of motherhood in the Semitic world. In addition, the lily was the holy flower of the ancient Assyrians and dates back to 700 B.C. (Doerflinger 1973). Until the sixteenth century, the "Madonna Lily" was known only as a garden plant. Thus, the "Lilies of the Field", mentioned in the Bible (St. Matthew 6:28), are believed to be this lily.

Currently, the "Easter Lily" that is forced for this holiday is *L. longiflorum*, which originates from the Ryukyu Islands of Southern Japan (Kiplinger and Langhans 1967). In the Orient, Feng Shui believers admire the lily as an emblem of summer and abundance. Also, in Japan, lilies have been appreciated since ancient times. They are mentioned in poems and can be seen in paintings and fabrics. Fifteen of the 100 known *Lilium* species grow wild in Japan and about half of them are endemic (Shimizu 1971). These species have extensively contributed to the modern Asiatic and Oriental hybrid lily cultivars. At least 39 *Lilium* species are endemic to China (Liang 1980). In A.D. 520, the first book was published, in which the lily was mentioned for herbal/medical uses. In the period A.D. 900–960, an agricultural treatise was published and the lily was described as one of the crops. It was indicated that propagation of the lily was carried out by scaling and was advised to use chicken manure as a fertilizer (Haw 1986).

In Greek mythology, Hyacinth was a beautiful youth who was loved by Apollo (Doerflinger 1973). He was killed accidentally by a discus thrown by Apollo. According to another legend, the wind god Zephyr, out of jealousy, blew the discus to kill Hyacinth. From his blood, a flower grew and it was named after him.

In the Iliad, Homer compares the color of the sky at sunrise with that of the "Saffron Crocus". The dried stigmas of this crocus produce a yellow color used to dye fabrics not only in antiquity but also on rugs in recent times. Many people are familiar with the wall paintings of the crocus gatherers on the island of Thera (Santorini). The crocus was a favorite motif among ancient painters, as witnessed by the twelfth century B.C. pot base from Kea, an island near Athens.

Iris was the goddess of the rainbow and a messenger of the Olympian gods in Greek mythology. She was later described as a handmaiden and personal messenger of the goddess Hera. Iris was also a goddess of the sea and sky. Her father, Thaumas ("The Wondrous"), was a sea god and her mother Elektra ("The Amber")—a shining cloud goddess.

Also, in Greek mythology, a young man named Narcissus fell hopelessly in love with his image, reflected in a pool of water (Doerflinger 1973). He was punished by the goddess Nemesis for rejecting the nymph, Echo. As a result, while he sat, Narcissus faded away and was turned into a flower of the same name. A second version says that he killed himself when he discovered that touching the water blurred his image. It has been reported that Narcissi were used in the temples of ancient Greece (Jefferson-Brown 1969) and they were associated with the goddess Persephone, the daughter of Zeus and Demeter, who was kidnapped by the god Hades while picking the narcissus flowers.

Narcissus pseudonarcissus is native to the United Kingdom, where it is known as the "Lent Lily" (Tompsett 2006). Many of the currently available daffodil cultivars are related to this species. *Narcissus tazetta* var. *tazetta* (*chinensis*), the "Chinese Sacred Lily", does not originate from the Orient (Bryan 1989, 2002). This species, which originated from the Mediterranean, was transported to China by traders in ancient times and was then used for medicinal purposes. It was introduced into Japan between A.D. 794 and 1185 (Tsukamoto 1985).

Ornamental geophytes are a prominent part of the floral imagery of the Taj Mahal—the pinnacle of Indo-Mughal art. Built in the seventeenth century, this mausoleum in Agra, India, includes one of the most beautiful monuments in the world and gardens in the Persian style. The external and internal walls are profusely ornamented with bas-reliefs and stone inlays of flowers. The main ornamental geophytes include *Anemones*, *Ranunculus*, *Fritillaria*, iris, lilies, *Pancratium*, and *Gloriosa* (Janick et al. 2010).

Many flower bulb species are native to North, South, and Central America (Bryan 1989, 2002) and were domesticated by American Indians long before Europeans arrived. Some of the flower bulbs of the Americas are: (1) the Aztec Lily (*Sprekelia formosissima*), also called "St. James lily" or "Jacobean lily", native to Mexico and Guatemala (Doerflinger 1973); (2) Canna (*Canna* spp) which is native to tropical America and as far north as the State of South Carolina; (3) Dahlias (*Dahlia* spp) which were domesticated by Pre-Columbian Indians and is native to Mexico, Central America, and Columbia (Bryan 1989, 2002); and (4) the "Tuberose" (*Polianthes tuberosa*) that was domesticated by Pre-Columbian Indians of Mexico. Its precise origin has been lost (Benschop 1993).

Since 1180, "Fleur de Lys" has adorned the coat of arms of French Kings (Correvon and Massé 1905; Ward 1999). However, this flower was not a lily, as the name implies, but the yellow flowering *Iris pseudoacorus* (Correvon and Massé 1905). Louis XI incorporated the motif into the coat of arms of the Italian Medici family. Subsequently, it was included in the coat of arms of Florence and Tuscany.

Tulips were known in Persia as early as in the twelfth century (Lodewijk 1979; Lythberg 2010) and were praised by the poet Omar Khayyam (Segal 1992). In the fifteenth century, tulips were used in city gardens and those of the Sultans of Turkey. Solms-Laubach (1899) has covered the history of "field" and "garden" tulips in Western Europe, their introduction into the different countries, and the descriptions of the tulips known at that time. Lodewijk (1979), Segal (1992), and Pavord (1999) have also traced the introduction of the tulip from Turkey to various European countries. They reported that in the eighteenth century, Ahmed III of Turkey imported tulips grown in The Netherlands and, thus, the cycle was complete.

These examples demonstrate that flower bulbs have been appreciated and cultivated for thousands of years, long before they were widely grown commercially or extensively studied and researched. Clearly, this was due to the diversity of their horticultural characteristics, for example, flower colors, shapes, heights, and flowering periods.

B.   From the Seventeenth to the Twenty-First Century

Without question, the introduction of the tulip from Turkey, in the middle of the sixteenth century, was the main factor leading to the development of a flower bulb industry in Europe (Lodewijk 1979; Van der Sloot 1994; Pavord 1999). In fact, with the exception of *Narcissus*, all the major genera that have contributed to the development of the flower bulb industry in The Netherlands have been introduced into Europe. Thus, the establishment and development of the flower bulb industry in The Netherlands over the past 400 years were initiated by the introduction and adaptation of nonindigenous species of flower bulbs.

The diversification of the tulip occurred early in its culture. Three centuries before their introduction into Europe, tulips exhibited floral diversity in Persian gardens (Botschantzeva 1982). In Turkey, hybridization was carried out in the sixteenth century and continued during the seventeenth and eighteenth centuries (Pavord 1999). In Western Europe, tulip hybridization started shortly after its introduction in the seventeenth century and by the end of the seventeenth century hundreds of cultivars were available in France and in The Netherlands (De La Quintinye 1697; Pavord 1999). It must be noted that some cultivars bred in the eighteenth or nineteenth century, for example, 'Keizerskroon' (1750) and 'Couleur Cardinal' (1845), are still being cultivated (Van Scheepen 1996; PT/BKD 2008).

Hyacinths were introduced to northern Europe in the middle of the sixteenth century and active hybridization started in The Netherlands after 1700 (Doorenbos 1954). By the end of the eighteenth century, about 2000 cultivars were being grown. The major goal of the early hybridizations was to satisfy the requests of wealthy patrons who promoted bulb production and utilization. These individuals, through their influence on fashion trends, were able to affect the development of new ornamental crops. For example, during the seventeenth century, the tulip dominated the hyacinth (Doorenbos 1954). However, after 1700, the hyacinth came into fashion and it was the queen of the bulb flowers until 1890, when the tulip regained its prominence.

Breeding of flower bulbs continued without scientific support up to the nineteenth century, when breeding programs for narcissi (Wylie 1952) and gladioli (Fairchild 1953) were initiated. These programs were based not only on the heredity characteristics, proposed by Gregor Mendel in 1886, but also on the roles of the chromosomes. The extensive use of interspecific hybridization led to the release of highly improved cultivars. The same was true for *Lilium* (Rockwell et al. 1961) and bulbous *Iris* (Dix 1974) that were actively hybridized after the beginning of the twentieth century. These hybridizations and the increase in cultivar diversification contributed significantly to the global development of the flower bulb industry. This was not only due to commercial breeders but also due to dedicated hobbyists. Many of the latter were or are members of various flower bulb associations and societies, like the American Daffodil Society, the North American Gladiolus Council, and the British Daffodil Society (see Section V).

The selected characteristics of the various bulb cultivars were directly influenced by their projected uses and by the economic/social situations. In addition, the evolution of scientific knowledge established possibilities for significant technical advances. For example, in the seventeenth to nineteenth centuries, tulips were marketed as single bulbs (Doorenbos 1954). However, as the economies of the countries grew, bulbs were widely used in gardens and marketed in groups, for example, packages of 12 or more. Later, they were used as fresh cut flowers and potted plants. Although forcing of the tulip 'Duc van Tol' was already reported in 1760 (Pavord 1999), commercial forcing increased significantly when Nicolaas Dames, a Dutch bulb grower, developed the basic techniques for hyacinths in 1910 (Doorenbos 1954). It is worth noting that these techniques are still being used today. Subsequently, forcing procedures have been developed for many flower bulbs (De Hertogh and Le Nard 1993; De Hertogh 1996). Thus, mass utilization of flower bulbs has progressed as the primary focus changed from gardening to a large-scale commercial flower bulb and fresh cut flower industry. Pavord (1999) and Franks (1958) indicated that this tendency was evident as early as in the end of the eighteenth century. The position of The Netherlands as the dominant country in this field has continued, and at the end of the twentieth century they controlled about 92% of the total world bulb trade (Schubach 2010).

III. WORLDWIDE PRODUCTION AND UTILIZATION IN THE TWENTIETH AND TWENTY-FIRST CENTURIES

It is projected that the annual global value of the flower bulb industry at the beginning of the twenty-first century exceeded US$1 billion (Benschop et al. 2010). The estimated data on worldwide flower bulb production from 1972 to 2005, without question, undervalued the number of hectares of production (Table 1.1). The Netherlands has the most comprehensive reporting system and reliable data on bulb production (www.bkd.eu; PT/BKD 2008), and the Dutch industry dominates bulb production, especially tulips and lilies, which comprise about 75% of the total Dutch production (Table 1.2). In the recent past, there was a significant increase in lily bulb production in The Netherlands and, at present, lilies are the second leading flower bulb produced in this country (Table 1.2). Concurrently, there has been a 25% decline in hyacinth and iris production and a 10% decline in tulip, gladiolus and narcissi bulb production in The Netherlands. The data illustrate two significant trends. First, a marked increase in world bulb production was recorded from 1972 to 1992. Second, since 1992, total world bulb production has not changed significantly. One of the reasons might be the absence of comprehensive global records. For example, acreage from the Southern Hemisphere and new production areas are not reported. At the same time, the production in The Netherlands was reduced to 17,763 hectares in 2009/2010 (Schubach 2010). This is a significant decline. In contrast, China has reported an increase to 4680 hectares (Schubach 2010). Clearly, during the period from 1992 to 2005 (Table 1.1) The Netherlands and the United Kingdom have maintained their position as the leading bulb production countries. The United States, Japan, and France have declined, while China, Chile, Israel, and New Zealand have increased their bulb production. In spite of the fact that the changes in the global industry over the last few decades have been slow, there has been a significant increase in world population. Therefore, it is obvious that the industry must plan for pending changes. For example, the Southern Hemisphere production of lily and tulip bulbs was started by a few Dutch bulb companies in cooperation with Chilean and New Zealand growers in order to reduce the length of the bulb storage. As a result, flower quality is enhanced when the bulbs are forced in the fall and winter in the Northern Hemisphere.

A.   THE GLOBAL SUPPLY CHAIN OF ORNAMENTAL GEOPHYTES

Major shifts in the world economy, societies, and technology are leading to dramatic changes in the floriculture industry. The increased globalization, liberalization of markets, and development of new products play an important role in shaping market trends and building consumer demand for goods and services, including those of ornamental plants. There has been a movement from a production-driven to a customer-driven strategy. The major driving forces in the ornamental sector are: (1) consumer demand for new products and varieties; (2) the growing awareness of environmental issues; (3) the increasing popularity of gardening as a hobby; and (4) the creation of new working places. Improving the availability, quality, and marketing of products will present additional opportunities to expand this sector. Therefore, new ornamental products must be developed by researchers and breeders in collaboration with efficient producers and satisfied consumers, linked together in mutually beneficial ways.

The global floricultural sector has, and will continue to, experience changes. In addition to the traditional flower bulb-producing and -consuming countries (Tables 1.1 through 1.3), globalization and increased competition have led to the development of new production centers. For example, floricultural production in Latin America, Africa, and Asia continues to increase. In addition, China, India, Malaysia, Pakistan, Taiwan, Thailand, Singapore, Sri Lanka, and Vietnam are emerging as new production centers. It is anticipated that the north–south axis will be important in the export markets. Africa will increase flower exports to Europe, and South America to the United States and Canada. Within Asia, there will be a growing interregional trade with emerging countries such as Malaysia, Thailand, and the Philippines. Australia and New Zealand have the potential to enter the niche market in Asia with high-quality flower products (De Groot 1999; Seideman 2004).

During the last few decades, constant increase in the competition for existing flower bulb markets enhanced the demand for high-quality bulbs and bulb flowers (Benschop et al. 2010). In addition, the globalization of the horticultural trade has led to advances in the transfer of knowledge and economic progress in developing countries. Thus, bulb production is no longer limited to countries with temperate climates. The production of bulbs and bulb flowers of high quality in regions with warm climates has become significant during the last few decades (Kamenetsky 2005). This has been promoted by relatively inexpensive land, low labor costs, and the expansion of international trade.

TABLE 1.3
Value of Marketable Ornamental Geophytes Produced in The Netherlands and the Ratio of Landscaping to Forcing Usage in Each Export Country

Country	Value in Millions of U.S. Dollars			Ratio (Landscaping:Forcing)
	1996–1997	1999–2000	2005–2006	
USA	115	147	179	2:1
Japan	114	102	102	1:3
Germany	95	90	104	2:1
UK	51	65	97	3:1
Italy	53	61	56	1:4
France	55	56	65	2:1
Sweden	24	24	28	1:2
Canada	15	20	29	1:1

Source: Van der Veer, A. 2006. Export Bloembollen naar Afzetkannaal 2004/2005. Rept. PT2006-02. Zoetermeer, The Netherlands: Productschap Tuinbouw. Afdeling Marktinformatie en Marktonderzoek.

Worldwide, the major markets for forcing of flowering bulbs as cut flowers or potted plants and their use in landscaping and garden have remained unchanged. The only "new" market in recent years has been the use of bulbs by landscape planters, which has increased significantly. Van der Veer (2006) and Benschop et al. (2010) pointed out that the use of flower bulbs for forcing and/or landscaping varied in each country (Table 1.3). For example, in the United Kingdom, the ratio is 3:1 in favor of landscaping, while in Italy it is 1:4 in favor of forcing. This utilization pattern must be taken into account as production and marketing strategies are developed.

Historically, commercial ornamental production was carried out by many small and medium growers and marketed through a large number of retail nurseries and garden centers. This trend has changed in recent years. Currently, products are being sold through "big box" systems, for example, supermarkets, home centers, and specialized wholesale systems for garden plants. As with any industrial product, flower bulbs must meet consumer expectations, and provide a fair compensation for the money expended. However, unlike fruits and vegetables, many ornamental products depend on fashion and season. For example, flowers with yellow and orange colors are popular in fall, while perennial bulbs are desired for sustainable gardening.

B. Role of Research in the Development of New Products and New Technologies

Scientific research programs on ornamental geophytes deal with a large spectrum of plant biology. They range from cellular and molecular biology to adaptability to the environment and the regulation of plant life cycle or propagation. However, even in-depth biological knowledge is not sufficient for the successful creation of new commercial products and new market niches. Therefore, multidisciplinary approaches are required for the improvement of existing crops and potentially useful species and their development into new commercial crops and novel products. As the industry looks to the future, it must take into account possible changes in global climatic changes and shifts. Increased attention will have to be paid not only to research on stress physiology and genetics (Palta 2010) but also to the evaluation of the sources and costs of the energy inputs for bulb production, forcing, and transportation.

TABLE 1.4

"Chain of Profitability" for Flower Bulb Research and Bulb Industry Based on Consumer Satisfaction

Informational Area	Examples of Specific Needs
General aspects	Research—The key link for industry profitability. The focus must be on "Problem Prevention", bulb and plant quality, industry sustainability, and consumer satisfaction.
	Education—To attract and support the brightest individuals to the industry and support other important educational programs.
Plant materials	Genetic improvement through plant breeding of traditional and new crops.
	Biotechnology must be supported to produce improved crops.
Planting materials for forcing	Containers should focus on biodegradable types.
	Planting media must include renewable resources and the suppression of pests.
Pest control	Systems must prevent the development of pests. Prediction systems must be developed for the outbreak of pests.
	Proper utilization of pesticides, when needed.
Production protocols	Optimal production management systems for all crops that will enhance the post-production life of the plants and flowers.
Post-production	Optimal handling systems for fresh cut flowers, potted plant products, and bulbs used for landscaping and gardening.

In addition to scientific publications, the results of the research are either a superior product or enhanced knowledge that will aid the industry and the consumer. Therefore, the major focal points of the research, education, and technological developments should be based on a proposed "Chain of Profitability" (Table 1.4). Consumer demand is the most important link in this chain. On the other hand, this approach focuses on profitability, since the consumer ultimately pays for the product. All participants in the value chain, from the producer to the retailer, must make a profit. Applied research programs provide direct support to the production and marketing segments of the "Chain of Profitability" and affect it directly. Therefore, public systems for the transfer of technology must be developed. Some examples of such technologies or products are: extended post-production vase and plant life, high tolerance to physiological stress conditions, improved resistance to diseases, insects, or animals (deer, rabbits, and voles). Evaluation standards must be established for the products of flower bulbs. As an example, the different uses of tulips require specific characteristics and developmental traits for garden plants (Table 1.5), cut flowers (Table 1.6), or potted plants (Table 1.7).

TABLE 1.5

Goals and Desirable Characteristics for Tulips Utilized in Landscapes and Gardens

Goal No.	Characteristics
1	A wide range of colors, flowering types, and plant heights for the entire flowering season is essential.
2	All cultivars should perennialize, especially in warm climatic zones, for example, USDA zones 8, 9, and 10.
3	The flower life should be greater than 10 days under a wide range of temperature conditions.
4	The cultivars should be resistant to *Fusarium* and *Botrytis*.
5	Stem strength must be excellent.
6	Variegated foliage is desirable.
7	Cultivars should not be susceptible to flower abortion, bulb necrosis ("Kernrot"), or "Stem Topple".

Source: Adapted from De Hertogh, A. A. 1990. Basic criteria for selecting flower bulbs for North American markets. *N. C. Hort. Res. Series No. 85.*

TABLE 1.6

Goals and Desirable Characteristics for Tulips Forced as Fresh Cut Flowers

Goal No.	Characteristics
1	There must be a wide range of colors, flower types, and plant heights for the entire forcing season.
2	Should force uniformly in <30 days in a 13–16°C greenhouse ("Standard Forced").
3	Should be single flowered.
4	Should not be susceptible to flower abortion, "Kernrot", or "Stem Topple".
5	Should have a large (>5 cm) floral bud.
6	Flowers must color above the foliage.
7	There must be a minimum of pigment changes as the flower senesces.
8	Flowers must retain basic flower shape and not be excessively thermonastic.
9	Flowers should have an appealing fragrance.
10	At the optimal cutting stage, they must be at least >40 cm (16 in.) in length for the wholesale market. Cash and carry markets can be shorter.
11	Flowers must have excellent stem strength.
12	Leaves must be strong, upright, and at least 3, in number. The first leaf should be well above the bulb nose. Cultivars with variegation are also desirable.
13	After cutting, internodal growth must be minimal.
14	Vase-life must be in excess of 6 days when placed at 20°C.
15	Cut tulip cultivars should have a low water uptake.

Source: Adapted from De Hertogh, A. A. 1990. Basic criteria for selecting flower bulbs for North American markets. *N. C. Hort. Res. Series No. 85.*

TABLE 1.7

Goals and Desirable Characteristics for Tulips Forced as Flowering Potted Plants

Goal No.	Characteristics
1	A wide range of flower colors that are available for the entire forcing season.
2	Single, multi-flowered, and semi-double cultivars are desired.
3	Should have a large (>5 cm) floral bud when coloring.
4	Must be forced uniformly (<21 days) at 16–17°C ("Standard Forced").
5	Must have a flowering height of 18–25 cm (10 cm pots) or 25–30 cm (15 cm pots) so PGR's are not required. Various plant heights are required for large containers.
6	Variegated foliage types are desirable.
7	Must have excellent stem and foliage strength.
8	Must flower above the foliage.
9	Flowers should have an appealing fragrance.
10	At marketing stage, the plants must have excellent storage capabilities.
11	Post-greenhouse internodal growth must be minimal.
12	Must have a "Full Pot Plant Look", thus the leaves are important for this effect.
13	Must not be susceptible to flower abortion, "Stem Topple", or bud necrosis ("Kernrot").

Source: Adapted from De Hertogh, A. A. 1990. Basic criteria for selecting flower bulbs for North American markets. *N. C. Hort. Res. Series No. 85.*

IV. FLOWER BULB ORGANIZATIONS

Numerous commercial, semi-government, and government agencies are associated with the world-wide flower bulb industry (Table 1.8). The history, specific roles, and organizational structures of these agencies were reviewed by Benschop et al. (2010). In this chapter, we shall primarily focus on the major organizations and their impact on consumer uses and products, research support, and on the future of the bulb production industry.

TABLE 1.8

Acronyms and Organizations of the Flower Bulb (Geophyte) Industry

Acronym	Name of the Organization and the Affiliated Country
AFE	American Floral Endowment—United States of America
AGREXCO	AGRicultural Export COmpany—Israel
ANTHOS	Same as BGBB—The Netherlands
ASFG	Association of Specialty Cut Flower Growers—United States of America
BGBB	Bond voor de Groothandel in Bloembollen en Boomkwekerijproducten (Association for the export of flower bulbs and trees and shrubs)—The Netherlands
BKD	Bloembollenkeuringsdienst (Flower Bulb Inspection Service)—The Netherlands
BOND	Bond voor Bloembollenhandelaren (now part of the BGBB, see above)—The Netherlands
BRC	Bulb Research Centre (see also LBO)—The Netherlands
CBC	Centrale Bloembollen Comite (Central Flower Bulb Committee, and later the IBC, see below. It was also known as the Associated Bulb Growers of Holland)—The Netherlands
CNB	Cooperatieve Nederlandse Bloembollencentrale—(A Dutch Flower Bulb Cooperative, a member owned auction)—The Netherlands
CPRO-DLO	Centrum voor Plantenveredelinsen Reproduktie Onderzoek-DLO (Centre for Plant Breeding and Reproduction Research. Together with its partners AB-DLO and IPO-DLO, CPRO-DLO formed Plant Research International, part of Wageningen UR)—The Netherlands
FIA	Foundacion Para la Innovacion Agraria—Chile
HOBAHO	Flower bulb auction started by: Homan-Bader-Hogewoning (currently owned by the Flower Auction—Aalsmeer)—The Netherlands
IBC	International Bloembollen Centrum (International Flower Bulb Centre)—The Netherlands
IBS	International Bulb Society—United States of America
INIA	Instituto de Investigaciones Agropecuarias—Chile
INRA	Institut National de la Recherche Agronomique—France
IVT	Instituut voor de Veredeling van Tuinbouwgewassen (Institute for Horticultural Plant Breeding—later the CPRO-DLO and PRI)—The Netherlands
JFTA	Japan Flower Trader's Association—Japan
KAVB	Koninklijke Algemeene Vereeninging voor Bloembollencultuur (Royal General Bulbgrowers' Association)—The Netherlands
KBGBB	Koninklijke Bond voor de Groothandel in Bloembollen en Boomkwekerijproducten—The Netherlands (see ANTHOS and BOND)
LBO	Laboratorium voor Bloembollen Onderzoek (Laboratory for Flower Bulb Research or Bulb Research Centre—BRC)—The Netherlands
MSU	Michigan State University at East Lansing, Michigan—United States of America
NAFWA	North American Flowerbulb Wholesalers Association—United States of America
NBA	Niigata Flower Bulb Growers Cooperatives Association—Japan
NCSU	North Carolina State University at Raleigh, NC—United States of America
NFI	Netherlands Flowerbulb Institute (formerly the New York Office of the BOND)—United States of America
NFIC	Netherlands Flowerbulb Information Center (formally part of the IBC and successor to the NFI)—United States of America
NWBGA	NorthWest Bulb Growers Association—United States of America
PBG	Proefstation voor Bloemisterij en Glasgroente in Aalsmeer and Naaldwijk (Research Station for Floriculture and Glasshouse Vegetables at Aalsmeer and Naaldwijk)—The Netherlands
PD	Plantenziektenkundige Dienst (Plant Protection Service)—The Netherlands
PPO	Plantaardig PraktijkOnderzoek—The Netherlands
PRI	Plant Research International, Wageningen—The Netherlands
PT	Productschap Tuinbouw (Horticultural Marketing Board)—The Netherlands
PVS	Produktschap voor Siergewassen (Ornamental Marketing Board)—The Netherlands
RHS	Royal Horticultural Society—London, England
SBO	Stiching Bloembollen Onderzoek—The Netherlands
TBGA	Toyama Bulb Growers Association—Japan
USDA/APHIS	United States Department of Agriculture/Animal Plant Health Inspection Service—United States of America
USDA/ARS	United States Department of Agriculture/Agricultural Research Service—United States of America
USDA/RI	United States Department of Agriculture/Research Initiatives—United States of America
WSU	Washington State University—United States of America

Note: The acronyms are derived from the first letter of the names of the organizations.

In The Netherlands, the major organizations are

1. HOBAHO (www.hobaho.nl), which has a research center in Hillegom, and was established in 1920 as a private company. In 1988, the "Testcentrum voor Siergewassen" was founded with the goal of carrying out hybridization and studies on the physiology of bulbous crops and to cover the entire production chain from hybridization to the sale of the flowers. The hybridization studies have focused on disease resistance, polyploidy, flower colors and shapes, quality maintenance, and the ability to be forced worldwide. Since 2006, the HOBAHO has been a subsidiary of Flora Holland.
2. The CNB (www.cnb.nl) was founded in 1975. They have representatives in all flower bulb-growing areas in The Netherlands and coordinate commercial bulb transactions between exporters and growers.
3. The KAVB (www.kavb.nl) was established in 1860 by the flower bulb growers, exporters, and traders (Franken 1931; Krelage 1946; Dwarswaard and Langeslag 2009). Initially, the major goal was the organization of flower exhibitions to promote the sale of flower bulbs. Subsequently, the KAVB became a bulb growers' organization, but exporters are still members. The KAVB is active in political, economic and environmental issues, and is responsible for the national and international registration of almost all commercially produced flower bulbs. The KAVB regularly publishes official classified lists of tulips, hyacinths, and "specialty bulbs" and maintains the most extensive library of the flower bulb literature in the world.
4. ANTHOS (www.anthos.org) represents the companies that trade in flower bulbs and nursery stock products. It is divided into five country groups. The North American group is the oldest and has been a leader in the support of many exporter-related needs. Since 1965, ANTHOS has supported research programs at Michigan State University, North Carolina State University (De Hertogh and Le Nard 1993; De Hertogh 1996), and Cornell University (www.flowerbulbs.cornell.edu). In addition, ANTHOS coordinates and funds the USDA/APHIS and Canadian Agricultural Inspection for the Pre-shipping Bulb Inspection System.
5. BKD (www.bloembollenkeuringsdienst.nl) is the Flowerbulb Inspection Service, started in 1923. Its inspections are based on EU- and national legislation. The main mission of the BKD is the monitoring and improvement of bulb quality and to provide an international service regarding the health and quality of ornamental geophytes.
6. The International Bulb Centre (IBC, www.bulbsonline.org) was terminated in December 2011. It had several responsibilities, for example, market research, public relations, advertising, technical advice and product information, and promotional and educational activity. It is expected that other organizations such as ANTHOS will assume these functions and they will be administered by specific market groups.

In Japan, the Niigata Flower Bulb Growers Cooperatives Association (NBA, founded in 1953) and the Toyama Bulb Growers Association (TBGA, founded in 1948) are local growers' associations in Niigata and Toyama prefectures, respectively. Their activities consist of consulting on bulb production, providing temperature treatments, and importing bulbs for the local growers. The Japan Flower Bulb Trader's Association (JFTA) was founded in 1991 in Yokohama, Japan to contribute to the development of the flower industry. The association maintains relationships with the Ministry of Agriculture, Forestry and Fisheries, Japan (MAFF), the Embassy of The Netherlands in Japan, and Plant Protection Stations of MAFF to promote smooth import business for bulb traders.

In the United States, the various branches of the USDA, AFE, ASCFG, the Gloeckner Foundation, and the "Team Concept" of the Specialty Crops Programs of the USDA Agricultural Food Research Institute (AFRI), a part of the National Institute of Food and Agriculture (NIFA, which previously was CSREES) are active in the development, marketing, and research on ornamental geophytes. The North American Flowerbulb Wholesalers Association (NAFWA, www.nafwa.com) was organized in 1983. The objectives of NAFWA are to establish and support the marketing and research needs of North America, to support educational activity, to assist with the flower bulb inspection programs and to promote and distribute research and marketing information.

Unfortunately, countries such as France and the United Kingdom still need to form new organizations to support research and market programs in ornamental geophytes. Clearly, China must also be included in research and development programs, along with additional organizations in Japan and other Asian countries. Since the global flower bulb industry is dominated by many small and very diverse businesses, the programs of all the organizations must be integrated to be effective worldwide.

In the past, the majority of flower bulb organizations (Table 1.8) that have supported research and educational programs belonged to government or semi-government agencies, industry, or hobbyists. However, in most countries, financial support has been discontinued or reduced and links between the industry and research have weakened. To meet its future needs in the competitive global market, the flower bulb industry must produce and promote superior bulb products, accompanied by up-to-date knowledge (Tables 1.5 through 1.7). This requires continuous and substantial amounts of financial support for research in order to sustain and train researchers and educators. At present, this approach is being government supported in all the developed countries. The government agencies also set the regulations and the industry has to develop products and/or systems that satisfy the regulations and/or needs.

V. FLOWER BULB EXHIBITIONS

Flower exhibitions that are open to the public are an excellent forum in which to stimulate consumers to purchase and use flower bulbs not only for gardens and landscaping but also as forced products, for example, fresh cut flowers and potted plants. The first exhibition in The Netherlands was organized in Haarlem in 1818 and the first exhibition outside The Netherlands took place in Belgium in 1839 (Krelage 1946). Thus, there is a long history of exhibiting flower bulbs in these settings.

Without question, the most extensive and well-known flower bulb garden in the world is the Keukenhof Garden (www.keukenhof.nl) located in Lisse, The Netherlands. Eleven bulb growers established the garden in 1949. They wanted to show their bulbs primarily to other growers and exporters. Wisely, they also included the public and, in the spring of 1950, their first flowering season, over 200,000 visitors visited the garden. Thus, the primary orientation was immediately changed from an industry perspective to the demands of the general public and consumers. In 1950, the first greenhouse was constructed for indoor exhibitions and, in 1956, the windmill, which still exists, was brought from the Province of Groningen. Currently, the garden comprises over 32 hectares and annually attracts over 800,000 people from all over the world. There is a need, however, to increase the educational programs in the garden. A wide variety of programs could assist in stimulating sales and satisfying consumers (Table 1.4).

In addition to the Keukenhof, Hortus Bulborum (www.hortus-bulborum.nl), which is located in Limmen, North Holland, The Netherlands, there is also a special bulb garden that is open to the public. It was started in 1928 (Leijenhorst 2004) and contains collections of many old cultivars of tulips, hyacinths, daffodils, and specialty bulbs, such as crocuses and fritillarias. When in flower, many of the bulbs are very unique. They are used as germplasm for the breeding of new cultivars. Among the 300 annual/outdoor flower bulb exhibitions in The Netherlands (www.dutchflowerlink.nl), the primary ones are: (1) The Midwinterflora (www.midwinterflora.nl), held in Lisse; (2) Creilerflora (www.creilerflora.nl) held in Creil; (3) Westerkoggeflora in de Goorn in January; (4) Driebanflora in Venhuizen; (5) Lenteweelde held in Obdam; (6) Holland Flowers

Festival (www.hollandflowersfestival.nl) held in Bovenkarspel; and (7) Lentetuin Breezand (www.lentetuin.nl) held in Breezand.

There are several other flower exhibitions or events in The Netherlands that feature flower bulbs. First, there is the annual "Flower Parade" (www.bloemencorsobloembollenstreek.nl) that was organized in 1948 and is held in mid- to late-April. Normally, the parade starts in Noordwijk and ends in Haarlem. It attracts over a million viewers along this route. Also, each year there are local competitions of floral mosaics in villages such as Anna Paulowna, Breezand, and Limmen in the Province of North Holland. The mosaics have a special theme and are primarily constructed with hyacinths florets. Every 10 years, the Dutch ornamentals industry organizes a "Floriade" in a selected town. The Floriades were started in 1960 in Rotterdam and are normally open from April to October. The next Floriade was organized in Venlo in 2012. Lastly, each fall, the HortiFair is held in Amsterdam and attracts about 40,000 participants and visitors. Flower bulbs are also featured in this exhibition.

In England, two major gardens feature flower bulbs as part of their programs. They are: (1) Springfield's Festival Gardens (www.gardenvisit.com) in Spalding, Lincolnshire, which was remodeled in 2003/2004, and (2) The Easton Walled Gardens (www.eastonwalledgardens.co.uk) in Grantham, Lincolnshire. In addition, Hampton Court, which is located outside London, normally has large flower bulb displays. The Chelsea Flower Show is held annually in May and is sponsored by the Royal Horticultural Society (www.rhs.org.uk). Forced bulbs are an integral part of each of these exhibitions.

In British Columbia, Canada, Butchart Gardens (www.butchartgardens.com) has a large display of spring and summer flowering bulbs. The garden is located near Victoria on Vancouver Island. Also, in Ottawa, there is an annual display of tulips that are provided by the Dutch Royal Family in appreciation of Canada's assistance during World War II.

In the United States, Longwood Gardens (www.longwoodgardens.org) in Kennett Square, Pennsylvania features bulbs not only in gardens but also in their large conservatories and greenhouses. In May 2010 and 2011, they hosted the "Lilytopia", a "Lily Show", which was linked to the one held annually in Keukenhof. Hopefully, this show will become an annual event. In the state of Washington, the Puyallup Valley Daffodil Festival and the Skagit Valley Tulip Festival are held in April. They attract about 500,000 visitors annually.

In Japan, the Tonami Tulip Fair in Tonami City, Toyama Prefecture, features about 450 cultivars during late April to early May. The unique World Lily Gardens of Yurigahara Park is located in Sapporo, Hokkaido. It is open year-round and is free of charge. About 100 lily species and cultivars are grown outdoors, and the flowering season extends from early June to late September.

Because of the large indigenous population of flower bulbs in South Africa, the National Botanic Gardens at Kirstenbosch always has bulbs in flower either in its gardens or in greenhouses (Du Plessis and Duncan 1989; Goldblatt and Manning 2000).

VI. ORNAMENTAL GEOPHYTE SOCIETIES

Over the centuries, many flower bulb societies have been founded which have contributed greatly to the global industry (Doerflinger 1973; Table 1.9). These societies publish a variety of newsletters, yearbooks, bulletins, and books (Koenig and Crowley 1972; Fairchild 1979; Howie 1984). They cover plant collections, the identification and taxonomy of indigenous species, and the breeding and release of new cultivars and species. In addition, most have Annual Meetings, at which they exchange information and plant materials, show new cultivars, and present special awards. One worldwide society that is dedicated to the dissemination of knowledge and the conservation of flower bulbs is the International Bulb Society (IBS) (www.bulbsociety.org) that publishes the journal *Herbertia* and annually awards the Herbert Medal to an individual making a significant contribution to the knowledge and importance of ornamental geophytes.

TABLE 1.9
Websites of Selected Flower Bulb Societies

Taxa	Society	Website
All ornamental geophytes	International Bulb Society	www.bulbsociety.org
	Pacific Bulb Society	www.pacificbulbsociety.org
	Royal Dutch Bulb Growers' Society (KAVB)	www.kavb.nl
	Royal Horticultural Society	http://www.rhs.org.uk/index.asp
	Indigenous Bulb Association of South Africa	www.safrican.org.za
Aroids	The Aroid Society	www.aroid.org
Clivia	The Clivia Society	www.cliviasociety.org
Cyclamen	The Cyclamen Society	www.cyclmen.org
Dahlia	American Dahlia Society	www.dahlia.org
	Nederlands Dahlia Vereniging	www.nederlandsedahliavereniging.nl
Gladiolus	North American Gladiolus Council	www.gladworld.org
Iris	American Iris Society	www.iris.org
	British Iris Society	www.britishirissociety.org.uk
Lily	North American Lily Society	www.lilies.org
Narcissus	American Daffodil Society	www.daffodilusa.org
	The Daffodil Society of The United Kingdom	www.daffsoc.freeserve.co.uk
Peony	American Peony Society	www.americanpeonysociety.org
	British Peony Society	http://www.peonysociety.org.uk
	Canadian Peony Society	http://www.peony.ca
	German Peony society	http://www.paeonia.de
	Danish Peony Society	http://www.danskpaeonselskab.dk
Tulip	Wakefield and North of England Tulip Society	http://www.tulipsociety.co.uk

Source: Adapted from Benschop, M. et al. 2010. *Hort. Rev.* 36:1–115.

CHAPTER CONTRIBUTORS

August A. De Hertogh is professor emeritus at North Carolina State University, Raleigh, North Carolina. He was born in Chicago, Illinois, to Belgian immigrants and earned his BS and MS from North Carolina State University and a PhD from Oregon State University, Corvallis, Oregon. Dr. De Hertogh has held positions at the Boyce Thompson Institute, Michigan State University, and North Carolina State University. His major research areas have been the development of environmental and plant growth regulator technology to control flowering during the forcing and postharvest phases of potted plants and fresh cut flowers. He is the author of over 200 publications on ornamental geophytes, with the most significant being *The Holland Bulb Forcer's Guide* and *The Physiology of Flower Bulbs*, coedited with Dr. Marcel Le Nard. Dr. De Hertogh is a recipient of the Mansholt Medal, the Nicolaas Dames Medal, and the Herbert Medal and has been inducted into the SAF's Floriculture Hall of Fame.

Johan van Scheepen is a taxonomist and librarian at the Royal General Bulbgrowers' Association, Hillegom, The Netherlands. He grew up in the western part of The Netherlands, and studied biology (taxonomy) at Rijksuniversiteit Leiden, The Netherlands and at the Paris Museum of Natural History. He has held the position of taxonomist at Reading University, United Kingdom. His major research areas are cultivar registration and classification in flower bulbs.

Marcel Le Nard is retired from the Institut National de la Recherche Agronomique (INRA), France. His entire career was spent at INRA. He led research programs on tulip, bulbous iris, and other bulbous crops. His major areas of research are the physiology of flower bulbs, production and postharvest handling, and the genetics and breeding of ornamental geophytes. Dr. Le Nard is a recipient of the Xavier Bernard scientific award of the Académie d'Agriculture de France (1983) and the Herbert Medal (2002).

Hiroshi Okubo is a professor of horticultural science at Kyushu University, Fukuoka, Japan. He was born in Japan and received his BS, MS, and PhD in horticultural science from Kyushu University. His major expertise includes the physiology of growth and development in horticultural crops, particularly in ornamental geophytes. He also conducts research on tropical plant species. Dr. Okubo is the author of more than 200 professional papers, book chapters, technical publications, and popular press articles. He is the director of the University Farm of Kyushu University and a recipient of an award of the Japanese Society for Horticultural Science (1988).

Rina Kamenetsky is a senior researcher at the Agricultural Research Organization (ARO, The Volcani Center) and a professor at the Hebrew University of Jerusalem in Israel. She was born in Almaty, Kazakhstan, and earned her BSc and MSc from Kazakh State University and a PhD from the National Academy of Science of Kazakhstan. Dr. Kamenetsky has held positions in the Main Botanical Garden and Institute of Botany in Kazakhstan and the J. Blaustein Institute for Desert Research, Ben-Gurion University of the Negev, Israel. Her main research areas include the mechanisms of the internal and environmental regulation of flowering and dormancy in geophytes, biodiversity and the introduction of new ornamental plants and their cultivation in warm regions. She is the author of over 150 publications on geophytes.

REFERENCES

Benschop, M. 1993. Polianthes. In *The Physiology of Flower Bulbs*, eds. A. De Hertogh, and M. Le Nard, 589–601. Amsterdam: Elsevier.

Benschop, M., R. Kamenetsky, M. Le Nard, H. Okubo, and A. De Hertogh. 2010. The global flower bulb industry: Production, utilization, research. *Hort. Rev.* 36:1–115.

Botschantzeva, Z. P. 1982. *Tulips: Taxonomy, Morphology, Cytology, Phytogeography and Physiology*. Rotterdam, The Netherlands: A. A. Balkema. (Translated and edited by H. Q. Varenkamp)

Bryan, J. E. 1989. *Bulbs*. Portland, OR: Timber Press.

Bryan, J. E. 1995. *Manual of Bulbs*. Portland, OR: Timber Press.

Bryan, J. E. 2002. *Bulbs* (Revised edition). Portland, OR: Timber Press.

Correvon, H., and H. Massé. 1905. *Les iris dans les jardins*. Paris: Librairie Horticole.

De Groot, N. S. P. 1999. Floriculture worldwide trade and consumption patterns. *Acta Hort.* 495:101–122.

De Hertogh, A. A. 1990. Basic criteria for selecting flower bulbs for North American markets. *N. C. Hort. Res. Series No. 85*.

De Hertogh, A. 1996. *Holland Bulb Forcer's Guide*. 5th ed. Hillegom, The Netherlands: International Flower Bulb Centre and The Dutch Bulb Exporters Association.

De Hertogh, A., and M. Le Nard. 1993. *The Physiology of Flower Bulbs*. Amsterdam: Elsevier.

De la Quintinye, J. B. 1697. Instructions pour les Jardins Fruitiers et Potagers, Tome 1. In *Nouvelle Instruction pour la Culture des fleurs: Content la manière de les Cultiver, et les Ouvrage qu'il faut Faire Chaque Mois de L'Annee Selon Leurs Différent especes. 3-ème ed.*, 119–139. Amsterdam: Henri Desbordes.

Dix, J. F.-Ch. 1974. *Het Geslacht Iris. Bloembollencultuur* 84:760–761, 790–791, 821, 846–847, 869–870, 952–954.

Doerflinger, F. 1973. *The Bulb Book*. Newton Abbot, Devon, UK: David and Charles.

Doorenbos, J. 1954. Notes on the history of bulb breeding in The Netherlands. *Euphytica* 3:1–11.

Du Plessis, N., and G. Duncan. 1989. *Bulbous Plants of Southern Africa—A Guide to their Cultivation and Propagation*. Cape Town: Tafelberg.

Dwarswaard, A., and S. Langslag. 2009. KAVB: Almost 150 years active. *Flor. Int.* 19:46–47.

Fairchild, L. 1979. *How to Grow Glorious Gladiolus. Bull. 139*. Ypsilanti, MI: North American Gladiolus Council.

Fairchild, L. M. 1953. *The Complete Book of the Gladiolus*. New York: Farrar, Straus and Young.

Franken, T. 1931. *Het Bloembollenboek*. Amsterdam: Andries Blitz.

Franks, H. G. 1958. *Home Truths about Bulbs—The Story of Holland's Bulbland*. Harlem, The Netherlands: Planeta.

Genders, R. 1973. *Bulbs, a Complete Handbook of Bulbs, Corms and Tubers*. London: Robert Hale.

Goldblatt, P., and J. Manning. 2000. *Wildflowers of the Fairest Cape*. Cape Town, South Africa: ABC Press.

Haw, S. G. 1986. *The Lilies of China*. London: B. T. Batsford.

Howie, V. 1984. *Let's Grow Lilies*. Waksee, IA: North American Lily Society.

Janick, J., R. Kamenetsky, and S. H. Puttaswamy. 2010. Horticulture of the Taj Mahal: Gardens of the imagination. *Chronica Hort*. 50(3):34–37.

Jefferson-Brown, M. J. 1969. *Daffodils and Narcissi—A Complete Guide to the Narcissus Family*. London: Faber & Faber.

Kamenetsky, R. 2005. Production of flower bulbs in regions with warm climates. *Acta Hort*. 673:59–66.

Kiplinger, D. C., and R. W. Langhans. 1967. *Easter Lilies. The Culture, Diseases, Insects and Economics of Easter Lilies*. Ithaca, NY: Cornell University.

Koenig, N., and W. Crowley. 1972. *The World of Gladiolus*. Edgewood, MD: The North American Gladiolus Council, Engelwood Press.

Krelage, E. H. 1946. *Drie Eeuwen Bloembollenexport—De Geschiedenis van den Bloembollenhandel en der Hollandsche Bloembollen tot 1938's*. Gravenhage: Rijksuitgeverij, Dienst van de Nederlandsche Staatscourant.

Leijenhorst, L. 2004. *Hortus Bulborum—Treasury of Historical Bulbs*. Stichting Uitgeverij Noord-Holland: Wormer.

Liang, S. Y. 1980. *Lilium* L. In *Flora reipublicae popularis sinicae, Vol. 14*, eds. F.-T. Wang, and T. Tang, 116–157. Beijing: Science Press (In Chinese).

Lodewijk, T. 1979. *The Book of Tulips*. New York: Vendome Press.

Lythberg, B. 2010. *The Tulip Anthology*. San Francisco: Chronicle Books.

Margaris, N. S. 2000. Flowers in Greek mythology. *Acta Hort*. 541:23–29.

Negbi, M. 1989. Theophrastus on geophytes. *Bot. J. Linn. Soc.* 100:15–43.

Palta, J. 2010. Developing strategies for sustainable production in a changing global climatic scenario. *ASHS Newsletter* 26(5):1, 8.

Pavord, A. 1999. *The Tulip*. London: Bloomsbury Publ.

PT/BKD. 2008. Bloembollen, Voorjaarsbloeiers, beplante oppervlakten, seizoen 2007/2008.

Rees, A. R. 1972. *The Growth of Bulbs. Applied Aspects of the Physiology of Ornamental Bulbous Crop Plants*. London: Academic Press.

Reynolds, M., and W. L. Meachem. 1967. *The Garden Bulbs of Spring*. New York: Funk and Wagnalls.

Rockwell, F. F., E. C. Grayson, and J. de Graff. 1961. *The Complete Book of Lilies*. New York: Doubleday and Co.

Scheider, A. F. 1981. *Park's Success with Bulbs*. Greenwood, SC: George W. Park Seed Co.

Schubach, A. 2010. *International Statistics Flowers and Plants, Vol. 58*. Voorhout, The Netherlands: International Assoc. Hort. Producers (AIPH).

Segal, S. 1992. *The Tulip Trade in Holland in the 17th Century—Tulips Portrayed*. Lisse, The Netherlands: Museum voor de Bloembollenstreek.

Seideman, T. 2004. *Despite Globalization Traumas, Flower Industry Blooms. World Trade*. http://www.worldtrademag.com/Articles/Feature_Article/4e61cf5149af7010VgnVCM100000f932a8c0.

Shimizu, M. 1971. *The Lilies of Japan*. Tokyo: Seibundo Shinkosha (In Japanese).

Solms-Laubach, H. Grafen zu. 1899. *Weizen und Tulpen und deren Geschichte*. Leipzig, Germany: Verlag von Arthur Felix.

Tompsett, A. 2006. *Golden Harvest—The Story of Daffodil Growing in Cornwall and the Isles of Scilly*. Cornwall, UK: Alison Hodge.

Tsukamoto, Y. 1985. *My Flower Museum*. Tokyo: Asahi Shinbunsha (In Japanese).

Van der Sloot, H. 1994. *Tulp 400 Jaar*. Rijswijk, The Netherlands: Elmar, B.V.

Van der Veer, A. 2006. Export Bloembollen naar Afzetkannaal 2004/2005. Rept. PT2006-02. Zoetermeer, The Netherlands: Productschap Tuinbouw. Afdeling Marktinformatie en Marktonderzoek.

Van Scheepen, J. 1996. *Classified List and International Register of Tulip Names*. Hillegom, The Netherlands: Royal General Bulbgrowers' Association (KAVB).

Ward, B. J. 1999. *A Contemplation upon Flowers. Garden Plants in Myth and Literature*. Portland, OR: Timber Press.

Wylie, A. P. 1952. The history of the garden narcissi. *Heredity* 6:137–156.

2 Taxonomy and Phylogeny

Alan W. Meerow

CONTENTS

I. INTRODUCTION

Our understanding of angiosperm phylogeny has undergone a revolution over the last three decades, largely due to two spectacular advances in the science of systematic botany (Judd et al. 2007). With the advent of polymerase chain reaction (PCR) technology (Saiki et al. 1988), direct comparison of the nucleotide sequences of organismal DNA became possible. Second, phylogenetic analysis (cladistics) has become the standard methodology for testing hypotheses of phylogeny among organisms in systematic biology (Wiley 1981; Felsenstein 2004) based upon principles formally enumerated by Hennig (1966). The main principle of cladistics defines any inclusive group of organisms (a clade), regardless of taxonomic rank, by the presence of one or more shared, derived character states (synapomorphies). Such a group is described as being monophyletic. To accept a taxonomic grouping based on shared primitive character states (symplesiomorphies) is not acceptable, and results in polyphyletic (taxonomic groups with multiple evolutionary origins) or paraphyletic groups (groups from which one or more members of common descent are excluded). Clades in a phylogenetic tree that share immediate common ancestry are often referred to as "sister" groups. A further principle of parsimony, still the most widely used approach in cladistics, states that the shortest possible phylogenetic tree (or cladogram), that is, the one that requires the least number of steps (character state changes), is the most accurate. One or more outgroups outside of the taxa of immediate interest (the ingroup) are included to polarize the character state changes (base substitutions in the instance of DNA sequences) at the outset of the cladistic analysis. Several confidence tests of a particular phylogenetic resolution are employed by systematists, the most widely used being the bootstrap[*] (Felsenstein 1985, 1988; Sanderson 1989; Hillis and Bull 1993). A high bootstrap value for a particular clade is a sign of robustness; a low value means that the clade is not well supported. Two other approaches used in phylogenetic reconstruction are maximum likelihood (Huelsenbeck and Crandall 1997) and Bayesian analysis (Beaumont 2010). The combination of phylogenetic analysis with DNA sequence data, or the field of molecular systematics, has in many

---

[*] Bootstrap is a computer-based resampling method for assigning confidence measures to phylogenetic tree resolution.

cases transformed our concepts of the relationships among the major groups of flowering plants (Judd et al. 2007).

Most of the geophytic ornamental plants are concentrated within the monocotyledonous orders. While the geophytic habit occurs in the eudicots, it is not a common life form, and is rarely predominant within any eudicot family. True bulbs (versus corms and tubers) are, for the most part, absent from eudicot families with the exception of *Oxalis* (Oxalidaceae; Oberlander et al. 2009). The following discussion will primarily focus on the taxonomy and phylogeny of the currently accepted monocotyledous orders, families, and, ultimately, genera of horticultural importance (e.g., *Allium*, *Alstroemeria*, *Lilium*, *Narcissus*).

II. OVERVIEW OF CURRENT ANGIOSPERM PHYLOGENY

A consortium of plant systematists calling itself the Angiosperm Phylogeny Group (APG) has distilled the vast body of both chloroplast and nuclear DNA sequence data accumulated over the last 20+ years into a formal phylogenetic classification of the flowering plants that has undergone three revisions (APG 1998; APG II 2003; APG III 2009). While there is yet no universal acceptance of all recognized taxa, the APG classification will likely remain the foundation upon which future classification systems will build, for better or worse. The Angiosperm Tree of Life project (http://www.flmnh.ufl.edu/angiospermATOL/index.html) maintains a dynamic database of phylogenetic information and the most current cladogram based on a multiple gene sequence data set can be viewed there.

It is generally accepted that the New Caledonian endemic family Amborellaceae is sister to all other angiosperms (Hansen et al. 2007; Jansen et al. 2007; Moore et al. 2007; Soltis et al. 2011), and is treated at the ordinal level in APG III (APG 2009) as Amborellales. Two other orders form a grade at the base of the angiosperm tree-of-life (Figure 2.1) after the branching of Amborellales, the Nymphaeales (three families) and Austrobaileyales (four families). Together, these three orders are sometimes collectively referred to as the "basal angiosperms" (Moore et al. 2007). The next branch is a clade comprised of the Chloranthales in a sister relationship to the magnoliids (Magnoliales +

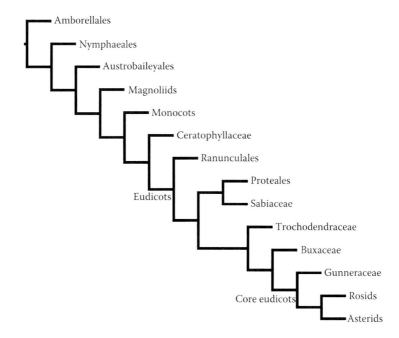

FIGURE 2.1   Major lineages of the angiosperms as supported by large-scale DNA sequence analyses (e.g., Soltis et al. 2011) and adapted by the Angiosperm Phylogeny Group (APG III 2009). (Adapted from Soltis, D. E. et al. 2011. *Am. J. Bot.* 98:704–730; APG III. 2009. *Bot. J. Linn. Soc.* 161:105–121.)

Laurales, Piperales + Canellales). The monocotyledons (*Monocotyledoneae*) form the next clade of the tree. A grade of basal eudicot clades follows successively: Ranunculales, Proteales + Sabiaceae, Trochodendraceae, and Buxaceae. The remainder of the tree comprises the "core eudicots", the *Eudicotyledoneae* (Cantino et al. 2007), which consists of two large groups (*Pentapetalae* clade) after the branching of Gunnerales: the rosids (Wang et al. 2009) and the asterids (Moore et al. 2010), each of which is further subdivided into clades of related families. Soltis et al. (2011) designate these two divisions as the *Superrosidae* and *Superasteridae* clades. *Superrosidae* consists of Saxifragales as sister to the *Rosidae*; *Rosidae* comprises Vitaceae as sister to *Malvidae + Fabidae*. The *Superasteridae* clade embraces Berberidopsidales, Santalales, Caryophyllales, *Asteridae*, and possibly Dilleniaceae. The latter has variably been resolved as sister to *Superrosidae* (Moore et al. 2010), or sister to both *Superrosidae + Superasteridae* (Moore et al. 2011). Cornales is the first branch within the *Asteridae*, followed by Ericales. The rest of the *Asteridae* clade consists of two subclades: *Lamiidae* and *Campanulidae* (Olmstead et al. 1992, 2000; Bremer et al. 2002).

III. MONOCOT PHYLOGENY

The most recent synthesis of molecular data (Chase et al. 2006) recognizes 11 orders within the monocotyledons, the first seven of which form a grade (Figure 2.2). The genus *Acorus* (Acorales) is sister to all other monocots. The next branch is the primarily aquatic Alismatales (but including Araceae, the aroids), followed by Petrosaviales (mycotrophic herbs), a clade of Dioscoreales/ Pandanales, then Liliales, followed by Asparagales. Liliales and Asparagales embrace the majority of important ornamental geophytes. Asparagales is sister to the only monophyletic group composed of more than two orders, collectively known as the commelinids, comprising Arecaceae, the palms (first branch), followed by a clade of Commelinales/Zingiberales, and terminated by Poales, the grasses and sedges, but including Bromeliaceae.

A. NEW PHYLOGENY OF THE LILIOID MONOCOTYLEDONS

Huber (1969) highlighted the heterogeneity present in many traditional monocot families, especially Liliaceae. Much of this work was refined and placed into the phylogenetic context by Dahlgren and

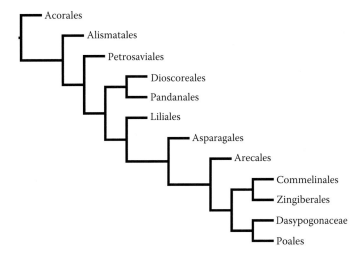

FIGURE 2.2 Phylogenetic relationships between the orders of the monocotyledons. (Based on Chase, M. W. et al. 2006. In *Monocots: Comparative Biology and Evolution (Vol. 1, excluding Poales)*, eds. J. T. Columbus, E. A. Friar, C. W. Hamilton, J. M. Porter, L. M. Prince, and M. G. Simpson, 63–75. Claremont: Rancho Santa Ana Botanic Garden.)

coworkers (Dahlgren and Clifford 1982; Dahlgren and Rasmussen 1983; Dahlgren et al. 1985). To date, phylogenetic analyses of the monocotyledons, based on both morphological and gene sequence matrices, have supported this classification with some amendment (Chase et al. 2006).

In a study by Dahlgren et al. (1985), based on Huber's (1969) seminal work, the families of monocots rich in geophytes are classified into two orders, Asparagales and Liliales, that have evolved many traits in parallel. Dahlgren et al. (1985) listed 16 characters that differentiated Liliales and Asparagales, but most do not occur in all taxa and several at least are plesiomorphic states. The important and consistent characters that separate the two orders are the presence of phytomelan in the seed coat of Asparagales (Huber 1969), the universal absence of septal nectaries in Liliales (Rudall et al. 2000), and simultaneous microsporogenesis in Asparagales, although the latter has reverted to successive in Amaryllidaceae, Asparagaceae, Hypoxidaceae, and Xanthorrhoeaceae (Rudall 2002a,b). Roots of Asparagales also contain a hypodermal layer (Kauff et al. 2000). An inferior ovary is also a synapomorphy for Asparagales, but reversals to a superior ovary have occurred in five families (Kocyan and Endress 2001; Rudall 2002a,b; Rudall and Bateman 2002, 2004). As Dahlgren et al. (1985) and Goldblatt (1995) point out, the boundaries between the two orders are difficult to define on morphological grounds alone, although multiple gene sequences support these two orders as monophyletic groups (Chase et al. 2000b, 2006).

1. Asparagales

Asparagales are the largest order of monocots (Chase et al. 1995a,b, 2000b, 2006). The order is estimated to have diverged from the rest of the monocotyledons between 120 and 133 million years before the present (MYBP; Janssen and Bremer 2004; Magallón and Castillo 2009). Thirty-one families were included in Asparagales by Dahlgren et al. (1985). Analyses of *rbc*L sequence data (Chase et al. 1995a,b) resulted in the transfer of Orchidaceae and Iridaceae from Liliales (Dahlgren et al. 1985) to Asparagales. Conversely, several families treated by Dahlgren et al. (1985) within Asparagales have been moved to Liliales. APG (1998) recognized 26 families in the order, reduced ultimately to 14 by APG III (2009).

Asparagales has consistently formed two groups in molecular phylogenetic analyses (Figure 2.3), a clade of what has been termed as the "lower" asparagoids (characterized by a predominance of simultaneous microsporogenesis and frequently inferior ovaries) and a clade of "higher" asparagoids with a uniformly successive microsporogenesis and frequent occurrence of superior ovaries (Rudall et al. 1997; Nadot et al. 2006). Relationships between the families within each group presented problems in early single- or few-gene phylogenies (Chase et al. 1995a,b), and the macromorphological synapomorphies for many of the families are not apparent (Pires et al. 2006). No morphological synapomorphies unite the "astelioid" clade (Figure 2.3), but three of the constituent families (Asteliaceae, Hypoxidaceae, and Lanariaceae) share branched hairs and mucilage canals (Rudall et al. 1998). Pseudoumbellate inflorescences are found in Amaryllidaceae, and in Asparagaceae subfam. Brodioideae (Themidaceae). Zygomorphic flowers occur variously in at least four clades: Orchidaceae (Rudall and Bateman 2002, 2004), Iridaceae (Rudall and Goldblatt 2001), Tecophilaeaceae (Simpson and Rudall 1998), and some Amaryllidaceae (Meerow and Snijman 1998; Rudall et al. 2002; Meerow 2010). Certain chemical characters are synapomorphic for particular clades, for example, anthraquinones for Xanthorrhoeaceae (Kite et al. 2000) and alkaloids and allyl sulfide compounds in Amaryllidaceae (subfamilies Amarylloideae and Allioideae, respectively).

The more recent multi-gene analyses summarized in Chase et al. (2006) confirm Orchidaceae as sister to the rest of the Asparagales. The next branch terminates a largely Southern Hemisphere clade of families, Asteliaceae, Blandfordiaceae, Hypoxidaceae, and Lanariaceae, which have been proposed (but not yet accepted) for merger as Hypoxidaceae (Soltis et al. 2005). Boryaceae (two woody genera, *Borya* and *Alania*, endemic to Australia) is supported weakly as sister to this group in Graham et al. (2006) and Pires et al. (2006), both using only plastid genes (Figure 2.3a). Chase et al.'s (2006) data, which included nuclear and mitochondrial DNA sequences as well as plastid, resolve a clade with a basal grade of Boryaceae, followed by Tecophilaeaceae, then Doryanthaceae (Figure 2.3b). Next is a

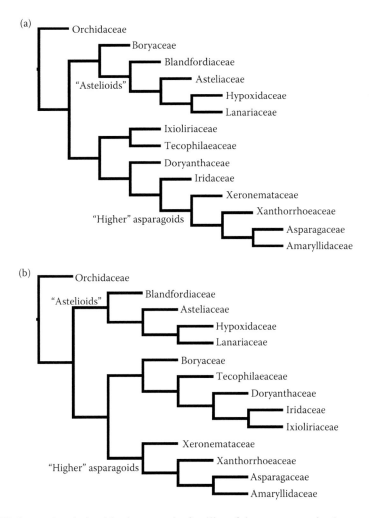

FIGURE 2.3 Phylogenetic relationships between the families of the monocot order Asparagales. (a) (Based on Graham, S. W. et al. 2006. In *Monocots: Comparative Biology and Evolution (Vol. 1, excluding Poales)*, eds. J. T. Columbus, E. A. Friar, J. M. Porter, L. M. Prince, and M. G. Simpson, 3–21. Claremont: Rancho Santa Ana Botanical Garden; Pires, J. C. et al. 2006. In *Monocots: Comparative Biology and Evolution (Vol. 1, excluding Poales)*, eds. J. T. Columbus, E. A. Friar, J. M. Porter, L. M. Prince, and M. G. Simpson, 287–304. Claremont: Rancho Santa Ana Botanic Garden; with only plastid sequence data.) (b) (Based on Chase, M. W. et al. 2006. In *Monocots: Comparative Biology and Evolution (Vol. 1, excluding Poales)*, eds. J. T. Columbus, E. A. Friar, C. W. Hamilton, J. M. Porter, L. M. Prince, and M. G. Simpson, 63–75. Claremont: Rancho Santa Ana Botanic Garden; with combined plastid, nuclear, and mitochrondrial sequence data.)

clade with a sister relationship between Ixioliriaceae, long a problematic taxon, and Iridaceae (Figure 2.3b). Pires et al. (2006) and Graham et al. (2006) using plastid genes only resolve Ixioliriaceae and Tecophilaeaceae as sister families (Figure 2.3a). It is hoped that the Monocot Tree of Life Project (http://www.botany.wisc.edu/monatol/) will add clarity to these conflicting phylogenies.

Among the higher asparagoids, the recently recognized Xeronemataceae (Chase et al. 2000a), a relict family consisting of a single genus of two species from New Caledonia and several small islands offshore from New Zealand, is basal to Xanthorrhoeaceae (Devey et al. 2006), submerging previously accepted families Asphodelaceae and Hemerocallidaceae. The terminal clade consists of two sister families as recognized by APG III (2009): Amaryllidaceae (comprising three subfamilies, Agapanthoideae, Allioideae, and Amaryllloideae), and a broadly delimited Asparagaceae (comprising the former families Agavaceae, Anthericaceae, Aphyllanthaceae, Hyacinthaceae, Laxmanniaceae, Ruscaceae, and Themidaceae). Seven subfamilies are recognized: subfam. Agavoideae (former Agavaceae and Hesperocallidaceae), subfam. Aphyllanthoideae, subfamily Asparagoideae, subfam. Brodiaeoideae (Themidaceae), subfam. Lomandroideae (Laxmanniaceae), subfam. Nolinoideae (Ruscaceae, including Convallariaceae), subfam. Scilloideae (Hyacinthaceae). The broad circumscription of Asparagaceae has generated controversy in the taxonomic community, but no published rebuttals have yet been appeared. This author finds the APG III concept untenable for this clade, and will refer to previous family designations [cf. APG II (2003)] in the rest of this chapter. Only selected families of this clade that contain ornamental geophytic genera will be reviewed below.

a. *Iridaceae*

The Iridaceae are a medium-sized family of 77 genera and 1750 species (Goldblatt and Manning 2008) mostly found in the Southern Hemisphere. Africa is the center of diversity of the family, and the majority of species are concentrated in the temperate and Mediterranean regions in the southern part of the continent (Goldblatt and Manning 2008), with secondary diversity in tropical and subtropical America. Many genera of Iridaceae are economically important because of their ornamental value.

Dahlgren et al. (1985) classified the iris family in Liliales near Colchicaceae on the basis of their extrorse anthers, nonphytomelanous seeds, mottled tepals, perigonal nectaries, and nuclear endosperm development. Perigonal nectaries are now known to represent an independent, derived state in Iridaceae, as are mottled tepals, and septal nectaries are the ancestral state for the family (Goldblatt 1998). The more ancestral Iridaceae are characterized by helobial endosperm formation. Despite the lack of clear-cut morphological links to Asparagales, multiple gene sequence analyses place Iridaceae well within this order (Chase et al. 2006; Graham et al. 2006; Pires et al. 2006). The family is typically characterized by the possession of isobilateral, equitant leaves, styloid crystals, inferior ovaries, and flowers with three stamens (Goldblatt 1998).

The exact relationships of the family to the rest of the Asparagales are still uncertain (Figure 2.3); plastid genes alone resolve a sister relationship to the higher asparagoids (Figure 2.3a; Graham et al. 2006; Pires et al. 2006), while a combined plastid, nuclear and mitochondrial sequence data set supports a sister relationship to Ixioliriaceae (Figure 2.3b; Chase et al. 2006).

Dahlgren et al. (1985) divided Iridaceae into five subfamilies (Isophysoideae, Aristeoideae, Sisyrinchioideae, Iridoideae, and Ixioideae). Goldblatt (1990) first recognized four different subfamilies (Isophysidoideae (one genus, Tasmania), Nivenioideae (seven genera, Australia, South Africa, and Madagascar, three with woody stems), Iridoideae (27 genera, cosmopolitan), and Ixioideae (27 mostly African genera), on the basis of cladistic analysis of morphological characters. These classifications differ mainly in the position of *Patersonia* (subfam. Nivenioideae according to Goldblatt); and subfam. Sisyrinchioideae *sensu* Dahlgren (1985), which was combined in Iridoideae by Goldblatt (1990); and in the inclusion of the achlorophyllous mycoparasite *Geosiris* in the family. Iridoideae comprised tribes Mariceae, Tigrideae, Iridineae, and Sisyrinchieae, and subfamily Ixioideae comprised tribes Pillansieae, Watsonieae, and Ixieae.

Rudall (1994) analyzed 33 characters, many of them anatomical, across the family. This analysis recognized the same four subfamilies and seven tribes as Goldblatt (1990), but the relationships among the subfamilies were incongruent. In Goldblatt's (1990) scheme, Ixioideae were the most derived clade, whereas Rudall's (1994) analysis resolved them as sister to the rest of the family. Goldblatt (1990) placed the Tasmanian *Isophysis* as sister to the rest of the family, whereas Rudall placed it sister to Nivenioideae. *Isophysis* together with Nivenioideae then formed the most derived clade.

Using molecular data derived from the plastid gene *rps4*, Souza-Chies et al. (1997) resolved *Isophysis* as sister to the rest of the family. Subfamily Ixioideae formed a well-supported clade, albeit with little internal resolution, while subfamily Nivenioideae formed a paraphyletic grade with Ixioideae as the terminal clade. There were insufficient data to resolve the monophyly of subfamily Iridoideae. Reeves et al. (2001) used sequences of four plastid genes across the family, and confirmed the monophyly of all four subfamilies (Ixioideae is now correctly known as Crocoideae) except for Nivenioideae, which resolved similar to the results reported by Souza-Chies et al. (1997). Achlorophyllous *Geosiris* fell within the Nivenioideae. Most of the tribes were monophyletic, and *Isophysis* sister to the rest of the family. Goldblatt et al. (2008) resolved the paraphyly of Nivenioideae by recognizing three additional monophyletic subfamilies, Aristeoideae (previously maintained by Dahlgren et al. 1985) and the new Geosiridaceae and Patersonioideae (Figure 2.4). They hypothesized that Iridaceae differentiated in the late Cretaceous and originated in Antarctica-Australasia. Janssen and Bremer (2004) estimate an age for the family at 96 MYBP.

It is within subfamilies Crocoideae and Iridoideae where the greatest generic diversity of the family is concentrated and where the genera most important to horticulture can be found. Goldblatt et al. (2006) used multiple plastid sequences to develop a phylogeny of the primarily African Crocoideae. Within the Crocoideae, *Crocus* and *Gladiolus* are the most important genera horticulturally. *Moraea* is another very large genus in the family into which Goldblatt et al. (2002) transferred many previously recognized small genera of South African origin.

In the Iridoideae, the endemic American genera are classified within three tribes: Tigrideae, Trimezieae, and Sisyrinchieae (Goldblatt and Manning 2008). Rodriguez and Sytsma (2006) showed that *Tigridia*, as currently circumscribed, was not monophyletic. On that basis, Goldblatt and Manning (2008) formally transferred several small genera into *Tigridia*.

A few horticulturally important genera are described below.

The genus *Crocus* consists of 88 diminutive, cormous species. Mathew (1982) recognized 81 species, but seven additional species have since been described (Petersen et al. 2008). The genus ranges from Central and Southern Europe to North Africa and from Southwest Asia to western China. The center of diversity is in Turkey and the Balkans (Mathew 1982). With morphological characters alone, Mathew (1982) divided the genus into two subgenera, subgenus *Crocus* with extrorse anthers, and subgenus *Crociris* with introrse anthers. The latter subgenus includes only *C. banaticus*. Subgenus *Crocus* was further subdivided into two sections, *Crocus* and *Nudiscapus*, with six and nine series, respectively. Petersen et al.'s (2008) analysis of multiple plastid and coding and noncoding regions conflicted with the morphologically based classification. The strikingly divergent *C. banaticus* (subgenus *Crociris*) was not resolved as sister to subgenus *Crocus* but was nested within it. Of the 15 series recognized by Mathew (1982), eight were resolved as monophyletic and two were rendered nonmonophyletic by only a single species. The largest two sections, series *Reticulati* and series *Biflori*, were not monophyletic.

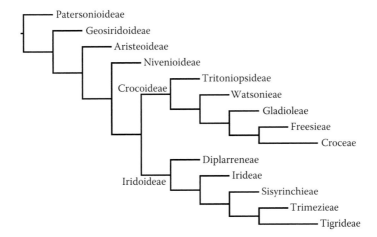

FIGURE 2.4 Phylogenetic relationships at the subfamilial and tribal levels within Iridaceae. (Based on Reeves, G. et al. 2001. *Am. J. Bot.* 88:2074–2087; Goldblatt, P. et al. 2006. In *Monocots: Comparative Biology and Evolution* (*Vol. 1, excluding Poales*), eds. J. T. Columbus, E. A. Friar, J. M. Porter, L. M. Prince, and M. G. Simpson, 399–411. Claremont: Rancho Santa Ana Botanical Garden; Goldblatt, P. et al. 2008. *Syst. Bot.* 33:495–508.)

According to the most recent phylogenetic analysis of subfamily Crocoideae (Goldblatt et al. 2006), the South African genus *Syringodea* is sister to *Crocus*. *Romulea* is resolved as the sister group to these two genera.

Gladiolus is second only to *Iris* in the number of species within Iridaceae (Goldblatt and Manning 2008), and is also only one of a few genera in the subfamily Crocoideae to diversify in areas other than the Cape region of South Africa (Goldblatt 1996). The most closely related genus to *Gladiolus* is *Melasphaerula*, which consists of a single species in the western Cape (Goldblatt and Manning 2008). The genus has been sumptuously monographed by Goldblatt (1996, 1998), who recognized a total of 255 species, and the reader is referred to them for the taxonomic history of the genus. Goldblatt and Manning (1998) dispensed with Goldblatt's (1996) previous recognition of two subgenera in *Gladiolus*, and classified the 163 southern African species into seven sections and 27 series. Section *Gladiolus* is now restricted to the 10 species found in Eurasia. The tropical African species are split between sections *Densiflorus* and *Linearifolius* (Goldblatt 1998), which also contain species in southern Africa.

Iris is the largest genus in Iridaceae. It is almost exclusively a temperate, Northern Hemisphere floristic element, with more than 260 species, the greatest number in Eurasia to East Asia. While the genus is diverse in ecology, the majority are found in dry, semi-desert, or rocky montaneous habitats. The most recent classification of the genus by Mathew (1989), which is built on previous work by Dykes (1913), Lawrence (1953), and Rodionenko (1987), divided *Iris* into six subgenera and 12 sections. Section *Limniris* in subg. *Limniris* was further classified into 16 series. Rodionenko's (1987) recognition of the bulbous subgenera *Hermodactyloides*, *Scorpiris*, and *Xiphium* as distinct genera (*Iriodictyum*, *Juno*, and *Xiphium*, respectively), was not accepted by Mathew (1989), who did, however, recognize the genus *Hermodactylus* for *I. tuberosa*. Rodionenko (1987) also recognized only two sections in subg. *Iris*: *Iris*, comprised of species with nonarilate seeds, and *Hexapogon*, with arils. Taylor (1976) argued against grouping all arilate species in section *Hexapogon*.

Tillie et al. (2001) used data from the plastid *trnL-trnF* spacer and gene *rps4* across a sample of all subgenera and sections except *Hexapogon*, *Monolepis*, and *Micropogon*. Resolution was poor with these data, but *Belamcanda chinensis* was resolved as sister to *Iris dichotoma* (which had sometimes been segregated as the genus *Pardanthopsis*) at the base of the subgenus *Iris* clade, with no bootstrap support, however. Wilson (2004) used two different plastid regions across 46 species of *Iris*, representing all subgenera and all sections except *Regelia*, *Brevituba*, and *Monolepis*. She also resolved *B. chinensis* as embedded within the genus. Goldblatt and Mabberley (2005) later made the formal transfer into *Iris* as *I. domestica*. The two largest subgenera, *Iris* and *Limniris*, both appeared polyphyletic. When section *Hexapogon* was removed from the data matrix, subgenus *Iris* was weakly monophyletic. Subgenus *Limniris* as circumscribed by Mathew (1989) had eight independent origins in the sequence topology. Subgenus *Scorpiris*, series *Spuria* (subgenus *Limniris* section *Limniris*), and a clade of section *Limniris* species from North America and Asia were identified as potentially monophyletic. Wilson (2006) then mapped various morphological charcters onto a *matK* phylogeny of the genus and determined that rhizomes are an ancestral state for *Iris* and that bulbs have evolved three times in the genus. The characters of sepal beards and crests, and arillate seeds all showed extensive homoplasy (parallelism).

Wilson (2009) analyzed the phylogenetic relationships among the 16 series within *Iris* section *Limniris* and demonstrated its paraphyly. Series *Spuriae*, *Foetidissimae*, *Syriacae*, *Longipetalae*, *Vernae*, *Unguiculares*, and some parts of *Tenuifoliae* were placed outside of the core group of *Limniris* species. The largest clade of *Limniris* species, which she called the "core *Limniris* clade", encompassed series *Californicae*, *Sibiricae*, *Laevigatae*, *Tripetalae*, *Prismaticae*, *Ensatae*, *Ruthenicae*, *Hexagonae*, *Chinenses*, and *I. songarica*. Series *Sibiricae*, *Tripetalae*, and *Chinenses* were polyphyletic and *Laevigatae* was paraphyletic. Later, Wilson (2011) expanded the number of taxa and gene regions (plastid *matK*, *ndhF*, and *trnK*) and to more fully test the subgeneric classification of Mathew (1989), as well as the status of the former genera *Belamcanda*, *Hermodactylus*, and *Pardanthopsis*. *Iris* subg. *Nepalensis* and subg. *Xiphium* were resolved as monophyletic, while subg. *Hermodactyloides*, subg. *Limniris*, subg. *Iris* and subg. *Scorpiris* were not. *Iris tuberosa* was supported as a member of *Iris* subg. *Hermodactyloides*, which Wilson (2011) combined with subg. *Xiphium*. Subg. *Pardanthopsis*, was resurrected for *I. domestica* and *I. dichotoma* (*Belamcanda* and *Pardanthopsis*, respectively). In total, Wilson (2011) recognizes eight subgenera in *Iris*, the remaining six of which are: *Crossiris*, *Limniris*, *Lophiris*, *Nepalensis*, *Siphonostylis*, and *Xyridion*. Her "Limniris III" clade was not yet given a formal name because it was resolved without bootstrap support.

b.   *Amaryllidaceae*

The precise relationship of Amaryllidaceae to other Asparagales remained elusive until Fay and Chase (1996) used the plastid gene rubisco (*rbcL*) to argue that *Agapanthus*, Alliaceae, and Amaryllidaceae form a monophyletic group (also evident in Chase et al. 1995a, b) and that together they are related most closely to Hyacinthaceae *sensu stricto (s.s.)* and the resurrected family Themidaceae (the former tribe Brodiaeeae of Alliaceae). They recircumscribed Amaryllidaceae to include *Agapanthus*, previously included in Alliaceae, as subfamily Agapanthoideae. Subsequent analyses of multiple DNA sequences from both the chloroplast and nuclear genomes have shown quite strongly that *Agapanthus*, Amaryllidaceae, and Alliaceae represent a distinct lineage within the monocot order Asparagales (Meerow et al. 1999; Fay et al. 2000), but the exact relationships among the three groups have been difficult to resolve with finality (Graham et al. 2006). APG II (APG 2003) recommended treating all three as a single family, Alliaceae (which had nomenclatural priority at that time), and more emphatically in APG III (APG 2009), but as Amaryllidaceae, reflecting the successful proposal for superconservation of the name (Meerow et al. 2007).

On the basis of the cladistic relationships of chloroplast DNA sequences (Meerow et al. 1999), all three subfamilies originated in Africa and infrafamilial relationships are resolved along biogeographic lines (Figure 2.5). Subfamily Amarylloideae, the largest in the number of genera, has colonized all continents except Antarctica. Janssen and Bremer (2004) estimate the age of the family at 87 MYBP.

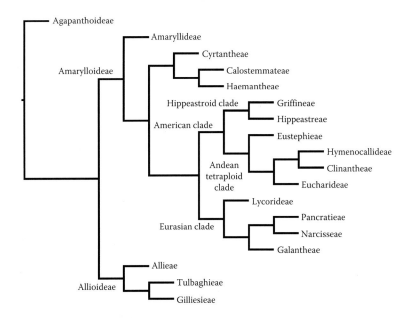

FIGURE 2.5   Subfamilial and tribal level phylogeny of the Amaryllidaceae *sensu* APG III (2009). (Based on Meerow, A. W. 2010. In *Diversity, Phylogeny and Evolution in the Monocotyledons*, eds. O. Seberg, G. Petersen, A. S. Barfod, and J. I. Davis, 145–168. Aarhus, Denmark: Aarhus University Press; Meerow, A. W. et al. 1999. *Am. J. Bot.* 86:1325–1345; Meerow, A. W. et al. 2000a. In *Monocots: Systematics and Evolution*, eds. K. L. Wilson, and D. A. Morrison, 368–382. Collingwood, Australia: CSIRO Publ; Meerow, A. W. et al. 2000b. *Syst. Bot.* 25:708–726.)

A few horticulturally important taxa are described below.

The genus *Agapanthus* (subfamily Agapanthoideae) is restricted to South Africa and consists of 6–10 species of rhizomatous, evergreen or deciduous perennials, most with blue flowers (Snoeijer 2004). The flowers have superior ovaries, and the genus contains sapoidal saponins.

Allioideae is represented in Africa by the South African endemic genus *Tulbaghia*, and a single species of *Allium*, but is most diverse generically in southern South America (Chile and Argentina). Three tribes can be recognized (Chase et al. 2009): Allieae, Gilliesieae, and Tulbaghiae, of which the first and third consist of only a single genus, *Allium* and *Tulbaghia*, respectively. *Tulbaghia* is endemic to South Africa and is sister to the South American Gillieseae. The subfamily is characterized by superior ovaries and the unique allyl sulfide chemistry that gives many members their characteristic garlic odor. Allieae is the largest tribe, notably due to the speciose genus *Allium* (Li et al. 2010).

Allium contains over 800 species (Fritsch et al. 2010; Li et al. 2010), and is one of the largest genera of monocots known. More than 50 species are used as edible, medicinal and ornamental crops. Variable morphologically as well as ecologically, it has spread across the Holarctic region, inhabiting dry subtropics to boreal vegetation. Only a single species of *Allium* occurs outside the Holarctic zone, *A. dregeanum*, native to South Africa (De Wilde-Duyfjes 1976). An Old World center of diversity encompasses the Mediterranean Basin to Central and Eastern Asia, with a second smaller one in western North America. Both Li et al. (2010) and Friesen et al. (2006) review the infrageneric taxonomic history of this complex genus. Molecular studies have either addressed the phylogenetic relationships of the entire genus (Mes et al. 1997; Dubouzet and Shinoda 1999; He et al. 2000; Fritsch and Friesen 2002; Friesen et al. 2006; Li et al. 2010) or specific subgenera and sections (*Amerallium*: Samoylov et al. 1995, 1999; *Melanocrommyum*: Dubouzet and Shinoda 1998; Mes et al. 1999; Gurushidze et al. 2008, 2010; Fritsch et al. 2010; *Rhizirideum*: Dubouzet et al. 1997; section Cepa: Gurushidze et al. 2007; origins of *A. ampeloprasum* horticultural races and section *Allium*: Hirschegger et al. 2010). Other molecular phylogenetic investigations have been concerned with the origins of economically important *Allium* crops (e.g., Friesen and Klaas 1998; Friesen et al. 1999; Blattner and Friesen 2006). Nguyen et al. (2008) examined the phylogeny of the western North American species and their adaptation to serpentine soils.

Friesen et al.'s (2006) analysis of 195 species of *Allium* using the ITS region of nrDNA presented a new subgeneric classification consisting of 15 monophyletic subgenera. Earlier, Friesen et al. (2000) showed that the anomalous *Milula* with a spicate inflorescence was nested within the Himalayan species of *Allium*. *Nectaroscordum* and *Caloscordum* are also retained within *Allium*. The most extensive study to date is that of Li et al. (2010) using ITS sequences along with the intron of the plastid gene *rps16* across over 300 taxa in the genus, and included a biogeographical analysis of the genus. Large genetic distances among the subgeneric groups suggest an ancient origin for the genus, with substantial differentiation by the early Tertiary. Three major clades are consistently resolved (Fritsch 2001; Fritsch and Friesen 2002; Friesen et al. 2006; Li et al. 2010). Subgenera *Amerallium*, *Anguinum*, *Vvedenskya*, *Porphyroprason*, and *Melanocrommyum* originated in eastern Asia. The putatively oldest lineage consists of only bulbous plants (subgenera *Nectaroscordum*, *Microscordum*, and *Amerallium*) that only rarely produce a rhizome (Fritsch and Friesen 2002). The second clade includes subgenera *Caloscordum*, *Anguinum*, *Vvedenskya*, *Porphyroprason*, and *Melanocrommyum*, and the third subgenera *Butomissa*, *Cyathophora*, *Rhizirideum*, *Allium*, *Cepa*, *Reticulatobulbosa*, and *Polyprason*. The latter two contain both rhizomatous and bulbous species. The third lineage is the most poorly resolved in these analyses, and includes a number of nonmonophyletic subgenera (Li et al. 2010). A scenario of rapid radiation was proposed for this clade. The first two clades contain both Old and New World species; almost all of the western North American species are classified in subgenus *Amerallium* (Nguyen et al. 2008), which has sparingly extended to central and eastern North America. The only other North American species are members of subg. *Anguinum* (Li et al. 2010).

The tribe Gilliesieae is entirely restricted to the American continents and is most diverse in southern South America, especially Argentina and Chile, and includes such established minor ornamental bulb crops as *Ipheion* and *Leucocoryne*. Only *Nothoscordum* extends outside of that region, with species in North America. The sister relationship of the tribe to *Tulbaghia* (Fay and Chase 1996; Fay et al. 2006) suggests an austral entry into South America, perhaps via Antarctica, as has been suggested for many groups showing a similar biogeographic scenario (Raven and Axelrod 1974). Unfortunately, generic limits within the group remain problematic, with many species variously treated as members of diverse genera (Rahn 1998; Zöllner and Arriagada 1998; Rudall et al. 2002; Fay et al. 2006). Two groups can be distinguished with the tribe (Rudall et al. 2002), one with actinomorphic and one with zygomorphic flowers. Fay et al. (2006) represents the most recent attempt to resolve the generic relationships of the group, but is marred by the lack of representation of several rare and poorly known genera, and low sampling of others. A combination of plastid and ribosomal DNA sequences robustly supports the two floral morphological clades (Fay et al. 2006). However, within the actinomorphic clade, both *Ipheion* and *Nothoscordum* resolve as biphyletic. Interestingly, the zygomorphic flowers of *Gilliesia* are hypothesized to be insect mimics with a pseudocopulatory pollination syndrome (Rudall et al. 2002).

Outside of Allioideae, the largest subfamily of Amaryllidaceae is Amarylloideae (Figure 2.5). This subfamily is also economically the most important, albeit for its large number of ornamental bulbs rather than any food value. It is characterized by an inferior ovary, a unique group of alkaloidal compounds, many with bioactive properties (Meerow and Snijman 1998), and base chromosome number of $x = 11$ (Meerow and Snijman 1998). Tribe Amaryllideae, entirely southern African with the exception of pantropical *Crinum*, were sister to the rest of Amaryllidaceae with very high bootstrap support in Meerow et al.'s (1999) analysis of plastid genes (Figure 2.5). The remaining two African tribes of the family, Haemantheae (including Gethyllideae) and Cyrtantheae (consisting of only *Cyrtanthus*), were well supported, but their position relative to the Australasian Calostemmateae and a large clade comprising the Eurasian and American genera was not clear. Most surprising, the Eurasian and American elements of the family were each monophyletic sister clades. Ito et al. (1999) resolved a very similar topology for a more limited sampling of Amaryllidaceae and related asparagoids using plastid *matK* sequences. Plastid *ndhF* sequences (Meerow and Snijman 2006) resolved Cyrtantheae as sister to a clade of Calostemmataeae and Haemantheae.

Much generic diversity of the tribe Amaryllideae is confined to South Africa (Snijman and Linder 1996). Compared to other tribes in Amaryllidaceae, Amaryllideae is marked by a large number of synapomorphies (Snijman and Linder 1996; Meerow and Snijman 1998): extensible fibers in the bulb tunics, bisulculate pollen with spinulose exine, scapes with a sclerenchymatous sheath, unitegmic or ategmic ovules, and nondormant, water-rich, nonphytomelanous seeds with chlorophyllous embryos. A few of the genera extend outside of South Africa proper, but only *Crinum*, with seeds well suited to oceanic dispersal (Koshimizu 1930), ranges through Asia, Australia, and America. Snijman and Linder's (1996) phylogenetic analysis of the tribe based on morphological, floral and seed anatomical, and cytological data resulted in the recognition of two monophyletic subtribes: Crininae (*Boophone*, *Crinum*, *Ammocharis*, and *Cybistetes*) and Amaryllidinae (*Amaryllis*, *Nerine*, *Brunsvigia*, *Crossyne*, *Hessea*, *Strumaria*, and *Carpolyza*). Meerow et al.'s (1999) incomplete sampling of this tribe for three plastid sequences resolved *Amaryllis* as sister to the rest of the tribe. Weichhardt-Kulessa et al. (2000) presented an analysis of internal transcribed spacer (ITS) sequences for a part of the tribe (subtribe Strumariinae *sensu* Müller-Doblies and Müller-Doblies [1985, 1996]). Meerow and Snijman (2001) analyzed morphology and ITS sequences across the entire tribe. *Amaryllis* is sister to the remaining genera, followed by *Boophone*. All other genera were included in two clades conforming to Snijman and Linder's (1996) subtribes Amaryllidinae (less *Amaryllis*, thus now Strumariinae) and Crininae (less *Boophone*), and *Carpolyza* was transferred into *Strumaria*.

Meerow et al. (2003) presented phylogenetic and biogeographical analyses of nrDNA ITS and plastid *trnL-F* sequences for all continental groups of the genus *Crinum* and related genera. Their results indicated that *C. baumii* is more closely related to *Ammocharis* and *Cybistetes* than to *Crinum s.s.* Three clades are resolved in *Crinum s.s.* The first one unites a monophyletic American group with tropical and North African species. Meerow et al. (2003) hypothesized that emergent aquatic tropical African species with actinomorphic perianths were likely a sister group to the American species, which was shown to be the case by Kwembeya et al. (2007). The second clade included all southern African species and the Australian endemic *C. flaccidum*. The third includes monophyletic Madagascar, Australasian, and Sino-Himalayan clades, with southern African species. The salverform, actinomorphic perianths of subg. *Crinum* appear to have evolved several times in the genus from ancestors with zygomorphic perianths (subg. *Codonocrinum*); thus neither subgenus is monophyletic. Biogeographical analyses place the origin of *Crinum* in southern Africa. The genus underwent three major waves of radiation corresponding to the three main clades resolved in the trees. Two entries into Australia or the genus were indicated, as were separate Sino-Himalayan and Australasian dispersal events. These results were confirmed by Kwembeya et al. (2007).

The three tribes Calostemmateae, Cyrtantheae, and Haemantheae form a clade that is sister to the American and Eurasian tribes of the subfamily (Figure 2.5; Meerow et al. 1999; Meerow and Snijman 2006), although their exact relationships with each other remain ambiguous (Meerow and Snijman 2006; Bay-Smidt et al. 2011). Calostemmateae consist of two Australasian genera (*Proiphys*, forest understory herbs of Malaysia, Indonesia, the Philippines, and tropical Australia; and *Calostemma*, endemic to Australia). The indehiscent capsules of both genera are similar in appearance to the unripe berry-fruits of *Scadoxus* and *Haemanthus* (Haemantheae), but early in the development of the seed, the embryo germinates precociously, and a bulbil forms within the capsule and functions as the mature propagule (Rendle 1901). A reasonable hypothesis is that it represents an early entry into Australia directly from Africa.

Cyrtantheae consists of a single genus. *Cyrtanthus* is endemic to sub-Saharan Africa, with well over 90% of its species concentrated in South Africa (Dyer 1939; Reid and Dyer 1984). With about 55 species it is the largest genus of southern Africa's Amaryllidaceae (Snijman and Archer 2003) and one of the largest in the family overall (Snijman and Meerow 2010). The genus exhibits a high level of floral morphological diversity which is unparalleled in any other genus of the family. Conversely, the genus shows great consistency in chromosome number, with 2n = 16 characteristic of most, if not all, of the species (Wilsenach 1963; Ising 1970; Strydom et al. 2007). It is also the only African genus with flattened, winged, phytomelanous seed, so common in the American clade of the family (Meerow and Snijman 1998). Snijman and Meerow (2010) explored the phylogeny of the genus in the context of floral and ecological adaptation using plastid *ndhF* and nuclear ribosomal DNA.

Haemantheae are the only group of Amaryllidaceae that have evolved a baccate fruit (Meerow et al. 1999; Meerow and Clayton 2004). It is entirely African, and like *Cyrtanthus*, most of its diversity is in South Africa (Meerow and Snijman 1998). Meerow and Clayton (2004) analyzed plastid *trnL-F* and nrDNA ITS sequences across the tribe. Two main clades are resolved, one comprising the monophyletic rhizomatous genera *Clivia* and *Cryptostephanus*, and a larger clade that unites *Haemanthus* and *Scadoxus* as sister genera to an *Apodolirion/Gethyllis* subclade.

In the American clade, the relationships of the endemic American genera (the entry of *Crinum* onto the continent is considered a separate event) were well resolved using the spacer regions of nuclear ribosomal DNA (Meerow et al. 2000a), and the major relationships have also been supported by plastid genes and introns (Meerow et al. 1999, 2000b; Meerow and Snijman 2006; Meerow 2010). The American genera of the family form two major clades (Figure 2.5). The first, or hippeastroid clade, are diploid (n = 11), primarily the extra-Andean element of the family (although several of the genera do have Andean representatives), comprising the Brazilian endemic tribe Griffineae (*Griffinia* and *Worsleya*) sister to genera treated as the tribe Hippeastreae in most recent classifications (Dahlgren et al. 1985; Müller-Doblies and Müller-Doblies 1996; Meerow and Snijman 1998). Several genera within the hippeastroid clade resolve as polyphyletic (*Rhodophiala*, *Zephyranthes*) and the possibility of reticulate evolution (i.e., early hybridization) in these lineages was hypothesized (Meerow 2010). A resolute plastid phylogeny, which would shed light on this issue because of its uniparental inheritance, is still elusive. The second clade constitutes the tetraploid-derived (n = 23) Andean-centered tribes (Figure 2.5). The Andean clade is characterized by three consistent deletions, two in the ITS1 and one in the ITS2 regions (Meerow et al. 2000a). The tribes Hymenocallideae (Meerow et al. 2002) and its sister tribe Clinantheae were recognized. A petiolate-leaved Andean subclade, containing elements of both Eucharideae and Stenomesseae (tribe Eucharideae), was also resolved. Monographs of several genera in this clade are available (*Eucharis*, Meerow 1989; *Eucrosia*, Meerow 1987) and a regional treatment of the family was produced for the Flora of Ecuador (Meerow 1990), which includes most species in the genus *Phaedranassa*. Interestingly, in both the American subclades there is a small tribe that is sister to the rest of the group, the Eustephieae in the Andean group and the Griffineae in the hippeastroid clade (Figure 2.5). These two small tribes may represent either ancestral or merely very isolated elements of their respective clades.

The most economically important genus of the American Amaryllidaceae subfam. Amaryllloideae, *Hippeastrum*, is still not very well understood taxonomically. It did not help matters that from 1938 to 1984, controversy ensued over the type specimen for the name *Amaryllis belladonna*, and thus the correct application of the generic name. Goldblatt (1984) and Meerow et al. (1997) review the history of the controversy in detail. Despite prior general consensus that *A. belladonna* applies to the species from South Africa known as the Cape belladonna, a vociferous and unyielding cohort (Uphoff 1938, 1939; Traub and Moldenke 1949; Traub 1954, 1983; Tjaden 1979, 1981) launched the contrary argument that the Linnean binomial must be applied to the neotropical *Hippeastrum equestre* (=*H. puniceum*). Goldblatt (1984) sought to end the ongoing controversy with a nomenclatural proposal that *Amaryllis* be conserved with, as its ultimate type, a specimen in the Clifford Herbarium (now at the British Museum) clearly assignable to the Cape belladonna from South Africa. This proposal was accepted at the *Fourteenth International Botanical Congress* in 1987.

Hippeastrum consists of 50–60 entirely New World species, although one species, *H. reginae*, appears to have been introduced into Africa. No modern revision of the genus has appeared since that of Traub and Moldenke (1949). The species are concentrated in two main areas of diversity, one in eastern Brazil and the other in the central southern Andes of Peru, Bolivia, and Argentina, on the eastern slopes and adjacent foothills. A few species extend north to Mexico and the West Indies. Meerow et al. (2000a) included seven species in their molecular phylogenetic analysis of the American genera of subfam. Amaryllloideae, representative of the biogeographic range of the genus. Their results suggested that the genus is robustly monophyletic and originated in Brazil. *Hippeastrum reticulatum*, with unusual fruit and seed morphology, was sister to all other species. The low rates of base substitution in both plastid and nrDNA sequences, and the consistent interfertility of species— well-mined by bulb breeders (Meerow 2009)—suggest that the genus underwent a relatively recent radiation. Many of the species seem to intergrade with one another. Traub and Moldenke (1949) attempted a formal subgeneric classification of the genus (as *Amaryllis*) based on floral morphology, but most of their infrageneric taxa do not appear to be monophyletic (Meerow and Snijman 1998).

The Eurasian clade of the Amaryllidaceae (Figure 2.5) contains the members of the family that have adapted to the highest latitudes in the Northern Hemisphere, and also those with the greatest economic value as spring flowering temperate zone garden plants (*Narcissus, Galanthus, Leucojum*). The clade was only recently recognized as a monophyletic group, resolved as sister to the endemic American genera by plastid DNA sequences (Ito et al. 1999; Meerow et al. 1999; Lledó et al. 2004). The Eurasian clade encompasses four tribes that were previously recognized (Meerow and Snijman 1998): Galantheae, Lycorideae, Narcisseae, and Pancratieae, the overall relationships of which were obscured by their diversity of chromosome number and morphology (Traub 1963). Müller-Doblies and Müller-Doblies (1978a) earlier observed similarities between the internal bulb morphology of *Ungernia* (Lycorideae) and *Sternbergia* (Narcisseae). With the exception of the Central and East Asian Lycorideae, the clade is centered within the Mediterranean region (Meerow and Snijman 1998; Lledó et al. 2004). There are 11 genera in the clade, comprising ca. 120 spp., with *Lycoris* (ca. 20 spp.) and *Narcissus* (40 spp.) the largest genera (Meerow and Snijman 1998).

Lledó et al. (2004) presented a cladistic analysis of the clade that focused on the relationships of *Leucojum* and *Galanthus* using plastid *matK*, nuclear ribosomal ITS sequences, and morphology. *Leucojum* was revealed as paraphyletic, and the genus *Acis* was resurrected to accommodate the linear-leaved Mediterranean *Leucojum* species with solid scapes. While their sampling within these three genera was extensive, only a single species each of the genera *Pancratium*, *Sternbergia*, *Narcissus*, and *Vagaria*, along with the monotypic *Lapiedra*, were used as outgroups. *Hannonia* was not included. Consequently, the phylogenetic relationships of the entire clade were not explicitly examined in their analyses. A similar case holds for Graham and Barrett's (2004) study of floral evolution in *Narcissus* using plastid *ndhF* and *trnL-F* sequences, which included only *Lapiedra* and one species each of *Galanthus*, *Leucojum*, and *Sternbergia* as outgroups in their analyses.

Meerow et al. (2006) analyzed the clade using plastid *ndhF* and rDNA ITS sequences for 33 and 29 taxa, respectively; all genera were represented by at least one species. Both sequence matrices resolve the Central and East Asian tribe Lycorideae as sister to the Mediterranean-centered genera of the clade, and two large subclades were recognized solved within the greater Mediterranean region: Galantheae, consisting of *Acis*, *Galanthus*, and *Leucojum*; and Narcisseae (sister genera *Narcissus* and *Sternbergia*, and *Pancratium*). However, there were areas of incongruence between the two markers, which disappeared when three predominantly monotypic genera, *Hannonia*, *Lapiedra*, and *Vagaria*, centered in North Africa, were removed from the alignments. The authors hypothesized that incomplete lineage sorting took place after the divergence of Galantheae and Narcisseae/*Pancratium* from a common ancestor, with the three small or monotypic genera retaining a mosaic of the ancestral haplotypes. After the vicariant divergence of the Asian Lycorideae, North Africa and the Iberian Peninsula are the most likely areas of origin for the rest of the clade (Meerow et al. 2006).

Narcissus is the most important genus of temperate zone spring flowering bulbs in the Amaryllidaceae. The genus is taxonomically very complex (Fernandes 1968a; Webb 1980; Mathew 2002), no doubt in part due to its propensity to hybridize in nature (Marques 2010), and the many horticultural hybrids and selections (Mathew 2002). Consequently, the number of species varies considerably in different studies. For example, Webb (1980) recognized 26 species; Fernandes (1968a) accepted 63. Blanchard (1990) favored Fernandes' (1968a) treatment. The genus is most speciose in the Western Mediterranean area, particularly the Iberian Peninsula and NW Africa. The genus is also fascinating biologically due to the occurrence of all four major classes of heterostyly in the genus, from stylar monomorphism, stigma-height dimorphism, distyly, to tristyly (reviewed in Barrett and Harder 2005). It is the only heterostylous genus of Amaryllidaceae.

Fernandes (1968a) divided *Narcissus* into two subgenera, *Hermione* with base chromosome number $x = 5$ and *Narcissus* with $x = 7$. He recognized 10 sections (*Apodanthae* [as *Apodanthi*], *Aurelia*, *Bulbocodium*, *Ganymedes*, *Jonquilla*, *Narcissus*, *Pseudonarcissus*, *Serotini*, *Tapeinanthus*, and *Tazettae*) based on his decades of karyotypic studies in the genus (summarized in Fernandes 1967, 1968a,b, 1975). Pérez et al. (2003) used the short plastid intergenic spacer between *trnL* and *trnF* across a small sampling of *Narcissus* species and did not get much resolution beyond the two recognized subgenera. Graham and Barrett (2004) provided a phylogenetic analysis of the plastid *trnL-F* and *ndhF* regions sequenced from 32 *Narcissus* species representing all 10 sections recognized by Fernandes (1975) and Blanchard (1990). This report strongly supported monophyletic subgenera *Hermione* and *Narcissus*, but not of all sections. Only section *Apodanthae* was clearly monophyletic, but several clades corresponded approximately to recognized sections (Graham and Barrett 2004). The most robust study is that of Marques (2010) who utilized plastid, mitochondrial, and nrDNA (ITS) across a large sampling of species, most with multiple accessions. She uncovered a striking incongruence between trees supported by the cytoplasmic versus the nuclear sequences, which she attributed to widescale hybridization throughout the evolutionary history of the genus. Tests for the recombination in the ITS alignments supported this hypothesis. Again, only a few of Fernandes' (1968a) sections were found to be monophyletic.

The relationship between *Galanthus* and *Leucojum sensu lato (s.l.)* has long been recognized, as has their relationship to *Narcissus* and *Sternbergia* (Müller-Doblies and Müller-Doblies 1978b;

Davis 1999, 2001). Both genera share pendulous, predominantly white flowers, similar internal bulb morphology and poricidal anthers (Müller-Doblies and Müller-Doblies 1978b). Unlike *Narcissus*, both lack a floral tube or a paraperigone (corona). *Galanthus* is marked by the striking length differences between the inner and outer tepal series, which are only subequal in *Leucojum* and *Acis* (Meerow and Snijman 1998).

Galanthus consists of 18 species, mostly distributed in Europe, Asia Minor, and the Near East (Davis 1999, 2001). Stern (1956) recognized three series in *Galanthus*, erected primarily on the basis of leaf vernation: *Nivales* Beck (leaves flat), *Plicati* (leaves plicate) and *Latifolii* (leaves convolute). Davis (1999) combined series *Nivales* and *Plicati* into series *Galanthus* and divided series *Latifolii* into two subseries: *Glaucaefolii* and *Viridifolii*. Molecular phylogenetic studies (Lledó et al. 2004; Larsen et al. 2010) indicate that the two subseries are not monophyletic.

Leucojum s.l. originally contained 10 species (Stern 1956), mostly occurring in the western Mediterranean area, from the Atlantic coast of Portugal and Morocco to the northern Balkans and Crimea, but today the genus comprises only two species: *L. vernum* and *L. aestivum* (Lledó et al. 2004; Meerow et al. 2006; Larsen et al. 2010), both broadly distributed in central and northern Europe, Turkey, and the Caucasus. *Leucojum* is characterized by hollow scapes, broad leaves, and clavate styles. Both species have a base chromosome number of $x = 11$. *L. vernum*, the type of the genus, is widespread in central and northern Europe. Its seeds have a pale outer testa and elaiosomes. *L. aestivum* is found throughout the Mediterranean and central Europe to Turkey and eastern Caucasus. It differs from *L. vernum* by its water-dispersed seed with a dark testa and lack of elaiosomes.

The remaining *Leucojum* species are now classified in the genus *Acis*, divided into subgenus *Acis* or *Ruminia* (Lledó et al. 2004; Meerow et al. 2006; Larsen et al. 2010), characterized by solid scapes, narrow leaves, and filiform styles. The subgenera *Acis* and *Ruminia* are differentiated by the morphology of the epigynous staminal disc, six-lobed in *A.* subgenus *Ruminia*, and unlobed in *Acis* subgenus *Acis*. *Acis* subg. *Acis* is the larger of the two subgenera with five species.

c. Hyacinthaceae

The Hyacinthaceae is classified by APG III (2009) as belonging to the Asparagaceae subfam. Scilloideae (Figure 2.3). Earlier, this family was recognized as a natural group within the classic (and polyphyletic) concept of Liliaceae *s.l.* on the basis of anatomical (Fuchsig 1911) and embryological (Schnarf 1929; Wunderlich 1937; Buchner 1948) characters. A more complete taxonomic history of the family can be found in Pfosser and Speta (1999). Speta (1998a) recognized 67 genera and ca. 900 species in the family, which he subdivided into five subfamilies. Four of these were well supported by molecular data (Pfosser and Speta 1999). Important horticultural genera include *Eucomis*, *Hyacinthus*, *Lachenalia*, *Muscari*, *Ornithogalum*, *Scilla*, and *Veltheimia*. Hyacinthaceae is allied with Themidaceae (Asparagaceae subfam. Brodieaoideae *sensu* APG III (2009) and Aphyllanthaceae (Fay and Chase 1996; Fay et al. 2000).

Hyacinthaceae is largely found through Africa and the Mediterranean, but occurs in Northern Europe and Asia as well. *Oziroë* represents the sole New World genus, with five species from South America (Speta 1998a; Guaglianone and Arroyo-Leuenberger 2002). The family's diversity center is in sub-Saharan Africa (Williams 2000). Molecular and phytochemical data support the monophyly of the family (Pfosser and Speta 1999; Chase et al. 2000b), and the presence of genera from South Africa in the four major clades. Four major clades are treated as subfamilies Oziroëoideae, Ornithogaloideae, Urgineoideae, and Hyacinthoideae (Speta 1998a; Pfosser and Speta 1999; Manning et al. 2003). A fifth previously recognized subfamily, Chlorogaloideae (Speta 1998a,b) is more closely related to the families Anthericaceae, Funkiaceae, and Agavaceae (Pfosser and Speta 1999, 2001), all treated as part of a polymorphic Asparagaceae by APG III (APG 2009).

Circumscription of genera within the Old World subfamilies has remained problematic (Pfosser and Speta 1999; Stedje 2001a,b; Manning et al. 2003, 2009), particularly as regards the "core" genus in each (i.e., *Scilla*, *Ornithogalum*, and *Urginea*). Phylogenetic analyses (Pfosser and Speta 1999;

Manning et al. 2003, 2009; Martínez-Azorín et al. 2011) make it painfully obvious that these genera represented paraphyletic or polyphyletic constructs. Speta (1998a, 2001) delimited genera very narrowly in an attempt to recognize only monophyletic groups, which resulted in an explosion of generic names for the family, a few of which could be distinguished in terms of morphological characters (Stedje 2001a,b; Manning et al. 2003). Manning et al. (2003) adopted a combined approach wherein genera were recognized by both clear morphological synapomorphies and molecular-based monophyly. This treatment lowered the number of genera recognized in sub-Saharan Africa, Madagascar and India to 15 from Speta's (1998a,b, 2001) 45.

Manning et al.'s (2003) reduction of all genera within subfamily Ornithogaloideae to a single genus *Ornithogalum* was not widely accepted and a later treatment (Manning et al. 2009) reestablished *Albuca*, *Dipcadi*, and *Pseudogaltonia*. Martínez-Azorín et al. (2011), combining plastid and nrDNA ITS sequences, choose to recognize 19 monophyletic genera within Ornithogaloideae, arrayed in the three previously recognized monophyletic tribes Albuceae, Dipcadieae, and Ornithogaleae. Manning et al. (2007) also revised *Ornithogalum* subgenus *Aspasia* section Aspasia.

Subfamily Hyacinthoideae is the largest of the four subfamilies, and is yet to receive the degree of attention that has been afforded the phylogeny of Ornithogaloideae. Speta (1998a) recognized two tribes, Massonieae, with 30–45 primarily sub-Saharan African genera, and the smaller Hyacintheae of Eurasian and northern Africa. He adopted a fairly narrow concept for the genus *Scilla*, recognizing a number of previously established segregates from that genus, such as Müller-Doblies and Müller-Doblies (1997) published a partial revision of Massonieae, recognizing 15 genera. As with Ornithogaloideae, generic limits have been in a frequent state of flux (Speta 1998a,b; Stedje 1998; Pfosser and Speta 1999; Goldblatt and Manning 2000; Hamatani et al. 2008), most particulary in terms of the sub-Saharan African taxa (Stedje 2001a; Pfosser et al. 2003). Within Hyacintheae, there have not been any extensive rearrangements proposed to Speta's (1998a) treatment, although Stedje (2001a) was highly critical of many of his segregate genera, especially the monotypic ones. The most economically important genera in this tribe are *Hyacinthus*, *Muscari*, and *Scilla*. Speta (1998a) recognizes 3, 50, and 30 species in each, respectively, and no recent revisions of these genera have appeared. A molecular analysis and revisionary treatment of the largely Mediterranean *Hyacinthoides*, once part of the large and polyphyletic Linnean genus *Scilla*, appeared recently (Grundmann et al. 2010).

d.   Themidaceae

The resurrected plant family Themidaceae (Fay and Chase 1996; Pires et al. 2001; Pires and Sytsma 2002) is treated by APG III (2009) as Asparagaceae subfam. Brodieaoideae. It comprises a clade of 12 genera mostly found in western North America. Formerly recognized as tribe Brodiaeae of the Alliaceae (e.g., Dahlgren et al. 1985), these genera have closer affinities to Hyacinthaceae and other families in the higher asparagoids of the Asparagales (Fay and Chase 1996; Fay et al. 2000; Pires et al. 2001). The exact sister relationships of the family to other clades of the higher Asparagles have never been consistently resolved with high support, one of the reasons why APG III (2009) argues for maintaining a broadly circumscribed Asparagaceae (Chase et al. 2009).

Two principal groups have been recognized in the family: the poorly known *Milla* and its allies (*Bessera*, *Dandya*, *Milla*, and *Petronymphe*), primarily Mexican endemics, and the more familiar *Brodiaea* complex, centered in the western United States. The Mexican genera have showy flowers and often terete leaves. Several new species and a new genus have recently been described (Delgadillo 1992; Lopez-Ferrari and Espejo 1992; Turner 1993; Howard 1999). There have also been segregates recognized within the *Brodiaea* complex (e.g., *Androstephium*, *Bloomeria*, *Muilla*, and *Triteleiopsis*) that have not been well understood (Ingraham 1953; Shevock 1984; White et al. 1996). *Brodiaea s.l.* (*Brodiaea*, *Dichelostemma*, *Triteleia*) represents an evolutionary radiation characterized by polyploidy and marked floral diversity (Moore 1953; Keator 1967, 1989; Niehaus 1971, 1980). Pires and Sytsma (2002) analyzed the entire family with multiple plastid DNA sequences. Four major clades were identified: (1) the *Milla* complex; (2) *Brodiaea*, *Dichelostemma*, and *Triteleiopsis*; (3) *Triteleia*, *Bloomeria*, and *Muilla clevelandii*; and (4) *Androstephium* and the other species of *Muilla*. They concluded that the morphological characters previously used to circumscribe the genera within the *Brodiaea* complex evolved at least twice. Moreover, common distribution patterns appeared to be independent radiations.

2. Liliales

Dahlgren et al. (1985) originally recognized 10 families in Liliales: Alstroemeriaceae, Colchicaceae, Uvulariaceae, Calachortaceae, Liliaceae, Geosiridaceae, Iridaceae, Apostasiaceae, Cypripediaceae, and Orchidaceae. Plastid DNA sequences have since resulted in Iridaceae (including Geosidridaceae) and Orchidaceae (including Apostasiaceae and Cypripediaceae) being transferred to Asparagales (Chase et al. 1995a,b). Further, cladistic analyses of combined plastid genes *rbc*L and *trn*L-F resolved four main lineages within the Liliales (Rudall et al. 2000): (1) Liliaceae (including Calachortaceae and some former members of Uvulariaceae), Philesiaceae, and Smilacaceae; (2) Campynemataceae; (3) the colchicoid lilies (Colchicaceae including *Petermannia* and *Uvularia*), Alstroemeriaceae and *Luzuriaga*; and (4) Melanthiaceae (including Trilliaceae). The relationships between these lineages were not well resolved. A cladistic analysis using morphological characters provides much less resolution among and within these groups (Rudall et al. 2000), while a combined analysis yielded a tree topology similar to the molecular data alone, with the exception of the position of *Calochortus*. Patterson and Givnish (2002) using plastid genes *rbc*L and *ndh*F supported the treatment of Tamura (1998), recognizing Calochortaceae as a distinct family. Fay et al. (2006) combined five plastid regions and one mitochondrial gene across Liliales and segregated Petermanniaceae from Colchicaceae and tentatively included the achlorophyllous Corsiaceae in the order (Figure 2.6). With the expanded data set they argued for including Uvulariaceae in Colchicaceae and Calochortaceae in Liliaceae.

APG (1998) and II (2003) recognized nine families: Alstroemeriaceae, Campynemataceae, Colchicaceae, Liliaceae, Luzuriagaceae, Melanthiaceae, Philesiaceae Dumort, Ripogonaceae, and Smilacaceae. Rudall et al. (2000) suggested combining Philesiaceae and Ripogonaceae with Smilacaceae. APG III (2009) adds Corsiaceae and Petermanniaceae, and submerged Luzuriagaceae into Alstroemeriaceae, for a total of 10 families (Figure 2.6). Bremer (2000) estimated that Liliales diverged from their sister group (Figure 2.2) in the Early Cretaceous, a minimum of 100 MYBP,

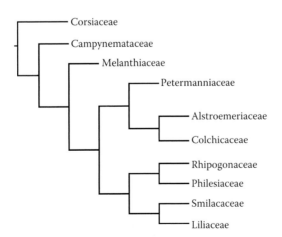

FIGURE 2.6 Phylogenetic relationships between the families of the monocot order Liliales. (Based on Rudall, P. J. et al. 2000. In *Monocots: Systematics and Evolution*, eds. K. L. Wilson, and D. A. Morrison, 347–359. Collingwood, Australia: CSIRO Publ.; Fay, M. F., P. J. Rudall, and M. W. Chase. 2006. In *Monocots: Comparative Biology and Evolution (Vol. 1, excluding Poales)*, eds. J. T. Columbus, E. A. Friar, C. W. Hamilton, J. M. Porter, L. M. Prince, and M. G. Simpson, 365–371. Claremont: Rancho Santa Ana Botanic Garden.)

with the modern lineages of the order diverging from each other ca. 82 MYBP. Vinnersten and Bremer (2001) hypothesize that the four major clades of the order *sensu* Rudall et al. (2000) date back to ca. 65 MYBP.

a. Liliaceae

In Rudall et al.'s (2000) synthesis, Liliaceae *s.s.* appears to consist of two main groups. The larger clade, based on plastid sequences, is made up of three subclades (all genera not listed): (1) a *Clintonia–Gagea* clade; (2) the core Liliaceae (*Lilium*, *Fritillaria*, *Nomocharis*, *Cardiocrinum*), and (3) a *Tulipa–Erythronium* group. The smaller main clade represents part of what Dahlgren et al. (1985) treated as Uvulariaceae (*Tricyrtis* and allies). *Calochortus* is sister to Liliaceae in the *rbc*L/*trn*L-F trees, but is embedded between the two main clades of the family in the combined analyses presented by Rudall et al. (2000; Figure 2.7). Patterson and Givnish (2002), using the more rapidly evolving chloroplast gene *ndhF*, resolved *Calochortus* as sister to *Tricyrtis*. Tamura (1998) recognized Calochortaceae, isolating *Calochortus* in the monogeneric bulbous tribe Calochorteae. The remaining four rhizomatous genera (including *Tricyrtis*) were placed in the tribe Tricyrtideae. Liliaceae, as narrowly circumscribed, is a predominantly holarctic family (Tamura 1998). Hayashi and Kawano's (2000) study with plastid *rbcL* and *matk* supported Tamura's (1998) concept. Patterson and Givnish's (2002) combined *rbcL* and *ndhF* analysis of the "core" Liliales supported a monophyletic Liliaceae, including one clade (*Lilium*, *Fritillaria*, *Nomocharis*, *Cardiocrinum*, *Notholirion*) that they assessed as having diversified in the Himalayas ca. 12 million years ago and another (*Erythronium*, *Tulipa*, *Gagea*, *Lloydia*) that originated in East Asia more or less contemporaneously. The genera *Medeola* and *Clintonia* were sister to Liliaceae *s.s.* and can be characterized by their rhizomatous stems, inconspicuous flowers, fleshy animal-dispersed fruits, and broad reticulate-veined leaves (Hayashi et al. 2001). *Calochortus* was resolved as sister to *Tricyrtis*. Interestingly, *Tricyrtis* and the clade of Uvulariaceae *pro parte* (*sensu* Dahlgren et al. 1985), *Prosartes–Streptopus–Scoliopus*, have morphological traits in common with *Medeola-Clintonia*.

Rønsted et al. (2005), using plastid maturase sequences (*matK*), resolved monophyletic tribes Lilieae [*Fritillaria*, *Lilium*, *Nomocharis* (nested in *Lilium*), *Cardiocrinum*, and *Notholirion*], Tulipeae (*Amana*, *Erythronium*, *Tulipa*, *Gagea*, and *Lloydia*), and Medeoleae (*Clintonia* and *Medeola*), all with strong bootstrap support (Figure 2.7). They also obtained a weakly supported clade including *Prosartes*, *Scoliopus*, *Streptopus*, and *Calochortus* as sister to all other Liliaceae

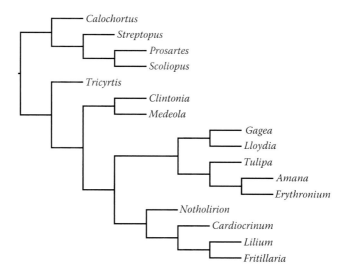

FIGURE 2.7 Phylogenetic relationships between the genera of the Liliaceae. (Based on Rønsted, N. et al. 2005. *Mol. Phylogen. Evol.* 35:509–527.)

including *Tricyrtis*, which indicated that Calochortaceae *sensu* Tamura (1998) was paraphyletic (Figure 2.7). Rønsted et al. (2005) indicate a sister relationship of Tulipeae to Lilieae (*Cardiocrinum*, *Fritillaria*, *Lilium*, *Notholirion*; Figure 2.7). *Fritillaria* and *Lilium* are sister genera (Figure 2.7).

Wikström et al. (2001) estimated the divergence of Liliaceae from Smilacaceae at ca. 65 MYBP, and the age of the crown clade to ca. 50 MYBP. Vinnersten and Bremer (2001) dated the Liliaceae/Smilacaceae divergence at ca. 55 MYBP, and the origin of modern diversity in the family 10–15 MY later. Vinnersten and Bremer (2001) also suggested a North American birthplace for the family, with dispersal to Eurasia 30–40 MYBP, the area of origin for subfamily Lilioideae. If so, *Erythronium* represents a secondary return to western America where the genus is most speciose. *Lloydia*, mostly Eurasian, is also represented in North America (Tamura 1998).

Horticulturally, *Lilium* is one of the most important genera of this family. There are ca. 100 spp. of *Lilium* found throughout the temperate and boreal regions of the Northern Hemisphere (McRae 1998). Southwestern and Himalayan Asia-China (ca. 70 species) are considered the center of diversity for the genus (Lighty 1968; Baranova 1969). Only two species extend into the tropics. Endlicher (1836) classified the genus into five sections: *Amblirion*, *Martagon*, *Pseudolirium*, *Eulirion*, and *Cardiocrinum*, the latter two now recognized as part of *Fritillaria* and the genus *Cardiocrinum*, respectively. Wilson (1925) divided the genus into four subgenera, with all true lilies belonging to *Eulirion*, further divided into four sections, *Leucolirion*, *Archelirion*, *Pseudolirium*, and *Martagon*. Comber (1949), using 13 morphological characteristics and two seed germination types, classified *Lilium* into seven sections and nine subsections: section *Martagon*, typified by *L. martagon*; section *Pseudolirium* (*L. philadelphicum*); section *Liriotypus* (*L. candidum*); section *Archelirion* (*L. auratum*); section *Sinomartagon* (*L. davidii*); section *Leucolirion* (*L. longiflorum*); and section *Daurolirio* (*L. dauricum*). Sections *Pseudolirium*, *Sinomartagon*, and *Leucolirion* were divided into four, three, and two subsections, respectively. Although generally considered the most authoritative classification for many years, Haw (1986) noted points of disputation. The difficulty of relying on morphological characteristics alone is due to convergence for some floral characters (Nishikawa et al. 1999).

Molecular studies have helped clarify taxonomic disparities in lily classification. Nishikawa et al. (1999) conducted the first comprehensive molecular phylogenetic analysis of genus *Lilium* using the nrDNA ITS across 55 species. Their results indicated that only three of seven sections, *Archelirion*, *Pseudolirium*, and *Martagon*, were monophyletic. The largest section, *Sinomartagon* (ca. 30 species), was highly polyphyletic, and section *Daurolirion* was paraphyletic. Internal support for most clades was low.

Hayashi and Kawano (2000) resolved three well-supported major clades in *Lilium* with plastid *matK* sequences that were not congruent with classification based on morphological characters, and also showed that the genus *Nomocharis* was nested within *Lilium*. The first clade represented species distributed from Japan to Burma, southwestern China and the Himalayas, with the notable exception of *L. philadelphicum* from eastern North America. The second united species from Asia, the Mediterranean and central Europe. The third included only North American species. Thus, two entries into North America for the genus were indicated.

Nishikawa et al. (2001) further evaluated phylogenetic relationships of section *Sinomartagon* based on the ITS using 64 species. The section was highly polyphyletic and resolved into five smaller groups. They also found that the addition of a single species into the analysis (*L. brownii*) rendered section *Archelirion* no longer monophyletic. Only two of seven sections, *Pseudolirium* (26% bootstrap support) and *Martagon* (97% bootstrap support), were monophyletic.

The most comprehensive study of *Lilium* phylogeny is that of Lee et al. (2011), analyzing ITS sequences from 83 species and 14 varieties. Only section *Martagon* was found to be monophyletic. Sections *Archelirion*, *Liriotypus*, and *Pseudolirium* were polyphyletic because two, one, and two species in each section, respectively, were resolved in other clades. Two major lineages of section *Leucolirion* were resolved, and section *Sinomartagon* was the most highly polyphyletic of Comber's (1949) groups.

Several regional studies have also been conducted in the genus. Dubouzet and Shinoda (1999) used ITS sequences to study phylogenetic relationships among 16 Japanese species. Ikinci et al. (2006) and Muratovic et al. (2010) evaluated the phylogeny of European spp., all but one of which (*L. martagon* in section *Martagon*) were classified in section *Liriotypus* by Comber (1949). Ikinci et al. (2006) showed that section *Liriotypus* is not monophyletic, and that *L. bulbiferum* is more closely related to other sections. Muratovic et al. (2010) resolved three major clades within the European lilies: *L. martagon* group (BP = 100%), *L. bulbiferum* (BP = 100%), and a third comprising the rest of Comber's (1949) section *Liriotypus* (BP = 100%). As yet, no one has proposed a revised phylogenetic classification for the genus.

For *Fritillaria*, a number of subgeneric classifications have been proposed. Baker (1874) recognized 10 subgenera based on the characters of the bulb, style, nectary, and capsule. Bentham and Hooker (1883) reduced these to five sections and transferred Baker's subgenus *Notholirion* to *Lilium*. For *Flora Orientalis*, Boissier (1882) subdivided section *Fritillaria* into two subsections delimited by an entire or trifid style: subsections *Trichostyleae* (including Baker's subgenera *Fritillaria* and *Monocodon*) and *Olostyleae* (including Baker's subgenera *Rhinopetalum* and *Amblirion* except *F. pudica*). Subgenera *Liliorhiza* and *Goniocarpa* do not occur in the area covered by *Flora Orientalis*. Komarov (1935) elevated Baker's (1874) subgenera *Korolkowia* and *Rhinopetalum* to the rank of genus. Turrill and Sealy (1980) divided *Fritillaria* (excluding *Korolkowia*) into four sections, correcting Boissier's subsection *Trichostyleae* to *Fritillaria*. All American species were placed in section *Liliorhiza*. Rix (2001) recognizes eight subgenera: *Davidii* (including only *F. davidii*), *Liliorhiza* (with several series), *Japonica* (including *F. japonica*, *F. amabilis*, and three closely related species), *Fritillaria* (with two sections, *Olostyleae* and *Fritillaria*, subdivided into series), *Rhinopetalum*, *Petilium*, and the monotypic *Theresia* and *Korolkowia*. Rix (1977) considers the center of diversity to be Iran where several biogeographical groups come together. The large nectaries of *Fritillaria* flowers have been important characters for delimiting subgeneric taxa. Bakshi-Khaniki and Persson (1997) reviewed nectary morphology in 31 southwestern Asian *Fritillaria* species. Subgenus *Rhinopetalum* Baker is marked by deeply depressed nectaries with a slit-like orifice bordered by two densely hairy lobes, and argued that this warranted recognition of the segregate genus *Rhinopetalum*, previously suggested by Komarov (1935).

Rønsted et al. (2005) analyzed a combined sequence matrix of ITS and plastid *rpl1* and *matK* sequences across 27 *Fritillaria* species, and their results largely supported the subgeneric classification of Rix (2001). Internal relationships of subgenus *Fritillaria* were not well resolved, however, and Rix's (2001) series were not monophyletic.

Despite the economic importance of the genus *Tulipa*, no molecular phylogenetic study has yet appeared for the genus. The number of species in the genus has varied from more than 100 (Hall 1940; Botschantzeva 1982) to ca. 40 (Stork 1984). Most recently, Van Raamsdonk et al. (1997) recognized 58, while Zonneveld (2009) claims 87. De Reboul (1847) presented the first intergeneric classification of the genus, but mostly only species known in southern Europe were included. Regel (1873) and Baker (1874) made significant advances. Regel (1873) recognized two main groups, *Eriostemones* and *Leiostemones* (Boissier 1882), and Baker (1874) established series within them (Boissier 1882). Van Raamsdonk and De Vries (1995) formally raised the two sections to the rank of subgenus, subg. *Tulipa* (*Leiostemones)* and subg. *Eriostemones*.

Subg. *Tulipa*, ca. 40 species (Marais 1984), occupies the Central Asian center of diversity for the genus. It includes the species most closely identified with cultivated tulips, *T. gesneriana* (Van Eijk et al. 1991). It has been subdivided into five sections (Van Raamsdonk and De Vries 1995): sect. *Clusianae* (south of the mountain ranges Kopeth Dagh, Pamir Alai and western Himalayas), sect. *Kolpakowskianae* (Tien Shan Mountains to western China), sect. *Tulipanum* (central Europe, Asia Minor, Lebanon, Syria, Caucasus, Iran, Iraq, and Afghanistan), sect. *Eichleres* (broadly distributed in Central Asia and Caucasus), and sect. *Tulipa* (Europe and the Middle East). The reader is referred to Van Raamsdonk and De Vries (1995) for full enumeration. The second smaller subgenus, *Eriostemones*, consists of ca. 20 spp. and extends into the Caucasus west to western Europe. The subgenus is divided into three sections (Van Raamsdonk and De Vries 1992, as subsections) based on morphological, crossability, chromosome and chemical studies (Nieuwhof et al. 1990; Van Eijk et al. 1991; Van Raamsdonk and De Vries 1992). Section *Australes* extend from Europe, the Aegean Sea region and Asia Minor. Species of section *Saxatiles* are found in the Near East and Crete. The range of sect. *Biflores* is eastern Europe to Central Asia and western China (Hoog 1973). Both diploids and polyploids occur in subg. *Eriostemones* (Kroon and Jongerius 1986).

Zonneveld (2009) surveyed nuclear genome size 123 taxa of *Tulipa* and proposed a revised classification in conjunction with morphological, geographical, and molecular data. He recognized 87 species in four subgenera, *Clusianae*, *Tulipa*, *Eriostemones*, and *Orithyia*, divided further into 12 sections. Seven of the eight series of section *Eichleres* are now placed in four sections: (1) section *Lanatae*, mainly confined to species from the Pamir-Alay and including series *Lanatae*, (2) section *Multiflorae* (including series *Glabrae*), (3) section *Vinistriatae* (including series *Undulatae*), and (4) section *Spiranthera*.

b. Colchicaceae

This cormous and rhizomatous family includes the horticultural genera *Gloriosa*, *Sandersonia*, *Littonia*, and *Colchicum* and five other genera (Nordenstam 1998), including *Uvularia*, the only North American genus of the family. The 19 genera are distributed in Africa, Asia, Australia, Eurasia, and North America. The pattern suggests an early Gondwanan distribution, but the dating estimation of Vinnersten and Bremer (2001) suggested divergence from a common ancestor with Alstroemeriaceae/Luzuriagaceae at ca. 60 MYBP, and crown age for the family of ca. 34 MYBP.

De Candolle (1805) was the first to use the family name Colchicaceae, in the *Flore Francaise*. He included six genera: *Bulbocodium*, *Colchicum*, and *Merendera*, which are still members of the family, and *Erythronium*, *Tofieldia*, and *Veratrum*, today assigned to Liliaceae, Tofieldiaceae, and Melanthiaceae, respectively. The subsequent taxonomic chronology is complex, and the reader is referred to Nordenstam (1998), Hayashi et al. (1998), Vinnersten and Reeves (2003) and Vinnersten and Manning (2007) for details. Nordenstam (1998) included 19 genera and ca. 225 species in the Colchicaceae, noting the need for a phylogenetically based intrafamilial classification. He suggested dividing the family into two subfamilies, one distinguished by having a corm, parallel sheathing leaves, dry capsules, and the presence of colchicine alkaloids (wurmbaeoid genera), and the other characterized by rhizomes; nonsheathing leaves, occasionally with net venation; dry or fleshy capsules, and alkaloids without a troplone ring (uvularioid genera). Using a sensitive assay, Vinnersten and Larsson (2010) determined that the uvularioid genera do in fact contain colchicine.

The molecular data of Vinnersten and Reeves (2003) did not support the division of the family into two subfamilies. Rather, the Colchicaceae are composed of a basal grade of Australian and American genera followed by a core clade of mainly African genera. The tribe Iphigenieae was highly paraphyletic. Their data also supported merging *Colchicum* and *Androcymbium*, as well as *Littonia* and *Gloriosa*. Vinnersten and Manning (2007) synthesized these data into a phylogenetic classification, recognizing 15 genera in five monophyletic and well-supported tribes, rather than any subfamilies (Figure 2.8). Manning et al. (2007) later transferred all *Androcymbium* sp. into *Colchicum*.

c. Alstroemeriaceae

The Alstroemeriaceae were traditionally allied to the Amaryllidaceae largely because of the inferior ovary (Herbert 1837; Kunth 1850; Baker 1888; Pax 1888; Pax and Hoffmann 1930). Hutchinson (1964) placed Alstroemeriaceae with Petermanniaceae and the Philesiaceae in his order Alstroemeriales. Huber (1969) was the first to explicitly state a relationship to Colchicaceae. Dahlgren et al. (1985) placed Alstroemeriaceae in the Liliales, and the relationships of the family in Liliales have been further refined by Rudall et al. (2000), Chase et al. (1995a,b, 2000b, 2006), and Fay et al. (2006).

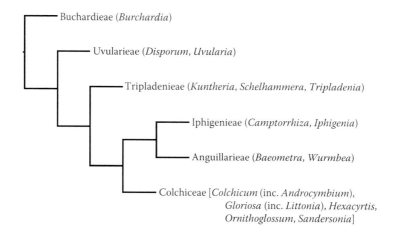

FIGURE 2.8 Phylogenetic relationships between the tribes of Colchicaceae, with member genera shown. (Adapted from Vinnersten, A., and G. Reeves. 2003. *Am. J. Bot.* 90:1455–1462; Vinnersten, A., and J. Manning. 2007. *Taxon* 56:171–178.)

If narrowly treated, this New World endemic family resolves as an isolated lineage most closely related to the Luzuriagaceae [and now combined with it in APG III (2009)], and a monophyletic Colchicaceae (Chase et al. 1995a,b, 2000b; Rudall et al. 2000). Luzuriagaceae, a Southern Hemispheric family, occurs from southern South America to New Zealand and southeastern Australia and consists of two genera, *Luzuriaga*, with four species (Arroyo and Leuenberger 1988), three in temperate southern South America and one in New Zealand, and the second genus, *Drymophila*, with two spp. in Eastern Australia and Tasmania (Conran and Clifford 1998). A broadened Alstroemeriaceae *sensu* APG III (2009) is thus a classic Austral floristic realm family.

Alstroemeriaceae subfamily Alstroemerioideae is a New World group distributed from 23°N in Mexico to 54°S in Argentina and Chile. Bayer (1998) recognized five genera: *Alstroemeria*, *Bomarea*, *Leontochir*, *Schickendantzia*, and *Taltalia*. The latter two genera are segregates from *Alstroemeria* and were not supported by cladistic analyses of chloroplast DNA variation (Aagesen and Sanso 2003). *Bomarea* consists of over 100 species of mostly twining perennials with tuberous roots in primarily high mountain forests of Central and South America, and has been divided into four subgenera (Hofreiter and Tillich 2002), which do not appear to be supported by molecular phylogenetic analysis (Alzate et al. 2008). Few species are cultivated, although many are exquisitely beautiful. In Aageson and Sanso's (2003) analysis, *Bomarea* and the monotypic *Leontochir* are sister genera, although only three spp. of *Bomarea* were included in the analysis. *Leontochir* is characterized by its prostrate habit, the presence of *Fritillaria*-like nectaries, and a unilocular ovary (Wilkin 1997). Aagesen and Sanso (2003) ultimately included it within *Bomarea* on morphological grounds. In Alzate et al.'s (2008) molecular phylogenetic analysis of *Bomarea*, *Leontochir* was embedded within it, strengthening the argument of treating it as part of that genus. Chacon (2010), applying molecular clock models to multiple sequence matrices across the entire family, hypothesized that Alstroemeriaceae originated in Cretaceous East Gondwana, with the diversification of the Alstroemerioideae during the Eocene, ca. 44 MYBP, when a land connection between Australia and South America still existed, and Antarctica supported forest. *Alstroemeria* began to diversify ca. 26 MYBP in southern South America, before the Andes presented significant dispersal barriers, expanding east into Brazil and radiating during the middle Miocene, ca. 13 MYBP.

Alstroemeria has not formally subdivided into subgenera, but two groups are frequently recognized, the Chilean and the Brazilian species (Aker and Healy 1990). In Aagesen and Sanso (2003), the Andean species of *Alstroemeria* were embedded within the Brazilian species. Bayer (1987) monographed the Chilean species, recognizing 31 species, but considered a subgeneric classification impossible until the Brazilian species were better understood. Ravenna (2000) described numerous novel Brazilian species and presented an infrageneric classification that violated the provisions of the International Code for Botanical Nomenclacture (Aageson and Sanso 2003). De Assis (2001) completed a revision of the Brazilian species, placing many of Ravenna's (2000) new taxa into synonymy, and recognized 39 spp. from Brazil.

Han et al. (2000) hypothesized that Chilean and Brazilian *Alstroemeria* species originated independently from *A. aurea*, which occurs on both sides of the Andes. This relationship was not supported by Aagesen and Sanso's (2003) study, but they did suggest that *A. aurea* and the few other trans-Andean species could have shared a common ancestor with the Brazilian group. Plastid *ndhF* sequences (De Assis and Meerow, unpubl. data) also resolved *A. aurea* as sister to the Brazilian species. Chacon (2010) resolves the Brazilian species as monophyletic, but embedded within the Andean species.

d. Melanthiaceae

Batsch (1802) first described the Melanthiaceae (comprising *Melanthium*, *Veratrum*, *Helonias*, and *Narthecium*) and distinguished them from Liliaceae by the apically free carpels or free styles). The taxonomic history of this group of lilies has been problematic, to say the least (Zomlefer et al. 2001), but this group appears to be well defined morphologically by extrorse anthers and three styles (although these characters occur elsewhere in Liliales). The complex history of the family, which includes transfers of genera to completely different orders of the monocotyledons, is detailed in Ambrose (1980), Goldblatt (1995), and Zomlefer (1997a,b,c). For example, the tribe Tofieldieae is now recognized as belonging to the Alismatales (as the family Tofieldiaceae), while part of the Narthecieae is classified within the Dioscoreales (Chase et al. 1995a,b, 2000b; Caddick et al. 2000) as the family Narthecieaceae.

The Melanthiaceae *sensu* APG III (2009) comprises 11–16 genera (154–2011 spp.) of predominately woodland and/or alpine perennial herbs occurring mainly in the temperate to Arctic zones of the Northern Hemisphere (with one species of *Schoenocaulon* extending into South America). Tamura (1998) did not include Trilliaceae (Zomlefer 1996; Tamura 1998) in his treatment. Trilliaceae resolves as embedded within Melanthiaceae in many molecular analyses (Chase et al. 2000b; Soltis et al. 2000; Rudall et al. 2000).

B. Cannaceae, Costaceae, and Zingiberaceae

These three families represent the three primarily geophytic families of the commelinid order Zingiberales (Figure 2.2; Kress 1990, 1995; Kress et al. 2001), which also includes the bananas (Musaceae), lobster claws (Heliconiaceae), and bird-of-paradise (Strelitziaceae). Janssen and Bremer (2004) estimate that the order diverged from the rest of the commelinids ca. 114 MYBP, with crown age for the order at 88 MYBP. Crown age for Zingiberaceae was estimated to be only 26 MYBP.

With 53 genera and over 1200 species, the pantropical Zingiberaceae is the largest of the eight families of the order (Kress 1990). Petersen's (1889) classification of the family, amended by others over time, recognized four tribes (Globbeae, Hedychieae, Alpinieae, and Zingibereae) based on morphological characters, such as the number of locules and placentation in the ovary, staminodia development, modifications of the single fertile anther, foliar orientation. Molecular phylogenetic analyses (ITS and plastid *matK* region) revealed considerable homplasy for these characters and paraphyly for three of the tribes. The African *Siphonochilus* and Bornean *Tamijia* are the first branches in the phylogeny. Kress et al. (2002) proposed a new classification with four subfamilies and four tribes: Siphonochiloideae (Siphonochileae), Tamijioideae (Tamijieae), Alpinioideae (Alpinieae, Riedelieae), and Zingiberoideae (Zingibereae, Globbeae).

Only one genus of Zingiberaceae, *Renealmia*, occurs in tropical America, and is also found in Africa. Three are endemic to Africa (*Aframomum*, *Aulotandra*, and *Siphonochilus*), and the remainder range from east Asia to the South Pacific. The family is still poorly known taxonomically, and molecular studies suggest that numerous monotypic or small genera are embedded within larger genera such as *Curcuma* and *Alpinia* (Kress et al. 2002; Ngamriabsakul et al. 2003).

Phylogenetic studies of horticulturally important genera of the Zingiberaceae include those of Wood et al. (2000) for *Hedychium*; Rangsiruji et al. (2000a,b), and Kress et al. (2005) on *Alpinia*; Ngamriabsakul et al. (2000) for *Roscoea*; Williams et al. (2004) on *Globba*.

In earlier classifications (e.g., Petersen 1889; Schumann 1904) the family Costaceae, which is resolved as the sister to Zingiberaceae *s.s.* (Kress 1990, 1995; Kress et al. 2001), was included in that family. Costaceae have several distinctive features, including the absence of aromatic oils, branching of aerial stems, and a spiral monostichous phyllotaxy (Specht et al. 2001; Specht 2006). Specht and Stevenson (2006) provided a new classification for the family that maintained the small genera *Tapeinochilos*, *Monocostus*, and *Dimerocostus*, but divided the polyphyletic pantropical *Costus* into four genera: *Chamaecostus*, *Cheilocostus*, *Costus*, and *Paracostus*.

Cannaceae comprise the single New World genus *Canna*. The family is most closely related to the Marantaceae (Kress 1990; Kress et al. 2001). Despite their cohesive identity based on morphology (Maas-van de Kamer and Maas 2008), a precise species taxonomy for cannas has remained elusive, in part due to a long history of cultivation by humans (Prince 2010). Tanaka (2001) recognized 19 species; Maas-van de Kamer and Maas (2008) only 10, placing a number of poorly understood species into a broad *C. indica* complex. The center of diversity for the genus is South America. Prince (2010) found incongruence between plastid and nuclear (nrDNA ITS) sequence phylogenies for the genus, indicative of possible interspecific hybridization or incomplete lineage sorting. Chloroplast DNA resolved the North American *C. flaccida* as sister to the rest of the genus.

C.   ARACEAE

Araceae, which belongs to the order Alismatales (Figure 2.2), is, like the palms, a signature tropical family (Grayum 1990; Mayo et al. 1997), and one of the most diverse groups of monocots in its ecology and structure (Mayo et al. 1997). It occurs on all continents except Antarctica, and includes 105 genera and 3300 species (Mayo et al. 1997). The geophytic habit, while not dominant among the aroids, has evolved several times in the family. Two genera, *Zantedeschia* and *Caladium*, stand out as the most economically important of the geophytic aroids, although *Arisaema*, *Amorphophallus*, *Arum*, *Biarum*, and *Sauromatum* also figure in the bulb trade. The aroids are strongly supported as a monophyletic group with the inclusion of the reduced, floating aquatic Lemnaceae (Mayo et al. 1997). Chase et al. (2000b) resolved Araceae within the mostly aquatic monocot order Alismatales, as sister to Tofieldiaceae (formerly part of Liliaceae, then Melanthiaceae), but more current multigene analyses position the aroids as sister to the remainder of the alismatid families (Chase et al. 2006; Soltis et al. 2011). Janssen and Bremer (2004) estimate that Araceae diverged from the rest of Alismatales ca. 128 MYBP, and was extant by 117 MYBP.

French et al. (1995) presented the first molecular phylogeny of the family using plastid genes, and this was followed by a variety of DNA sequence phylogenies, mostly focused on particular aroid tribes or genera (e.g., Renner and Zhang 2004; Renner et al. 2004; Rothwell et al. 2004; Tam et al. 2004; Gonçalves et al. 2007; Cabrera et al. 2008; Gauthier et al. 2008; Wong et al. 2010). The most extensive is that of Cabrera et al. (2008), who unequivocally established the relationships of the duckweeds (the former Lemnaceae, now Araceae subfamily Lemnoideae). Cusimano et al. (2011) integrated the DNA sequence data with morphological and anatomical characters within the family and recognized 44 clades, which they recommend as the basis for a new formal classification.

Mayo et al. (1997) provided the most stable classification for Araceae to date, and the reader is directed there and to Cabrera et al. (2008) and Cusimano et al. (2011) for details of the complex taxonomic history of the group and the details of its intrafamilial classification. Seven subfamilies were recognized, which Mayo et al. (1997) divided into two informal aggregates, the "proto-Araceae" (subfam. Gymnostachydoideae and Orontioideae) and the "true Araceae" (subfam.

Pothoideae, Monsteroideae, Lasioideae, Calloideae, and Aroideae). The major concentration of geophytic taxa in the family is placed within the largest and monecious subfamily Aroideae. Several tribes are universally or primarily characterized by geophytic habit, for example, the neotropical and mostly tuberous Spathicarpeae (Gonçalves et al. 2007); the cormous, broadly distributed Old World Areae (Cusimano et al. 2010), which includes *Arum*, *Biarum*, and *Sauromatum*; Arisaemateae (*Arisaema*), Caladieae, and Colocasieae (Renner and Zhang 2004).

Boyce (1993, 2006) has recently monographed *Arum* and *Biarum* (Boyce 2008). Linz et al. (2010) completed a molecular phylogeny of *Arum*, which largely supported Boyce's most current classification (Boyce 2006). *Arisaema* comprises about 150 species of mostly temperate forest understory herbs from tubers or rhizomes (Gusman and Gusman 2002), the majority ranging from the Himalayas through East and Southeast Asia. The genus is biologically fascinating for its ability to change reproductive gender from year to year (Schlessman 1988). Renner et al. (2004) used plastid sequences to construct a phylogeny of the genus and test biogeographical hypotheses.

The genus *Caladium* belongs to the tribe Caladieae, a largely neotropical group in subfamily Aroideae (Mayo et al. 1997). *Caladium* comprises about 12 species, of which *C. bicolor* has been the major contributor to the horticultural cultivars. Other commonly cultivated species are *C. lindenii*, which was previously classified as a species of *Xanthosoma*, and *C. humboldtii*, which Madison (1981) considered as likely a variant of *C. bicolor*. Only a single morphological character distinguishes these genera effectively: pollen of *Xanthosoma* is shed in tetrads while that of *Caladium* is shed as single grains. The tribe Caladieae consists almost exclusively of New World genera (Grayum 1990; French et al. 1995; Mayo et al. 1997). The only non-neotropical genus included in the Caladieae is *Hapaline*, which is entirely palaeotropical in distribution, but with morphology, cytology, anatomy, and biochemistry indicating a close relationship to neotropical genera (Boyce 1996). Using AFLP, Loh et al. (2000) confirmed that *C. humboldtii* is a distinct species and that *C. lindenii* is a true *Caladium* species.

Zantedeschia, the calla lilies, are classified in their own tribe, the Zantedeschieae, which resolves as sister to a clade comprised of the Spathicarpeae (Cabrera et al. 2008; Cusimano et al. 2011). Eight species are recognized (Singh et al. 1996).

IV.   GEOPHYTIC EUDICOTS

With the exception of *Oxalis*, the geophytic habit among eudicots families is limited to tubers and rhizomes, with the ocassional corm. The Ranunculaceae, as one of the basal eudicots, is fairly rich in geophytic plants, most notably some members of the genera *Anemone*, *Eranthis*, and *Ranunculus*.

Oxalis (Oxalidaceae, in the fabid/rosid order Oxalidales) is the only genus of eudicot with true bulbs, but many species form tubers. The genus is extremely variable morphologically, with most of the diversity in South America (Lourteig 1994, 2000). About 15% (ca. 35 spp. confined to section *Ionoxalis*) of endemic American *Oxalis* species have true bulbs (Lourteig 2000), while all indigenous southern African *Oxalis* (>200 spp.) are bulbous (Oberlander et al. 2009).

The Begoniaceae (fabid/rosid, order Cucurbitales) also has its share of geophytic species, most notably the Andean tuberous species of *Begonia* section *Eupetalum* that have given rise to the *B. × tuberhybrida* cultivars (Tebbitt 2005). For current phylogenetic insights into this large and complex genus, see Forrest and Hollingworth (2003), Forrest et al. (2005), and Goodall-Copestake et al. (2009).

Finally, the asterid eudcidots can claim two horticulturally important geophytic genera, both members of the composite family (Asteraceae, a campanulid asterid eudicot), tuberous *Dahlia* (tribe Coreopsideae), and the cormous *Liatris* (tribe Eupatorieae).

Liatris is classified within subtribe Liatrinae of tribe Eupatorieae (King and Robinson 1987). Its closest relatives are *Carphephorus*, *Garberia*, and *Hartwrightia* (King and Robinson 1987; Nesom 2005). Nesom (2005, 2006) recognizes 37 species of this North American genus, organized into five sections: *Liatris*, *Graminifolium*, *Pilifilis*, *Suprago*, and *Vorago*.

There are 36 recognized species in *Dahlia*, several only recently described (Saar and Sørensen 2000, 2005; Saar 2002; Saar et al. 2002, 2003b; Hansen and Sørensen 2003). The majority are endemic to Mexico but a few range to as far south as northern South America (Saar et al. 2003b). The garden dahlia is usually treated as the cultigens *D. variabilis*, in which thousands of cultivars have been recognized (The International Register of *Dahlia* Names 1969 et seq.), but whose taxonomic status remains occluded (Hansen and Hjerting 1996). Sherff (1955) recognized 18 species, with eight varieties and three forms in three sections, *Pseudodendron*, *Epiphytum*, and *Dahlia*. Sørensen's (1969) revision contained 29 species and six varieties in four sections. He split off sect. *Entemophyllon* from sect. *Dahlia*, and further divided sect. *Dahlia* into two subsections. Giannasi (1975a,b) analyzed flavonoids across 19 species, and was able to characterize two groups based on the presence of flavonols (sections *Dahlia* and *Pseudodendron*) versus 6-methoxy and 6,40-dimethoxy flavones (sect. *Entemophyllon*). On this basis he suggested that only two sections, *Entemophyllon* and *Dahlia*, could be recognized. Ten additional species have since been described (Saar and Sørensen 2000, 2005; Saar 2002; Saar et al. 2002, 2003a; Hansen and Sørensen 2003). Two molecular phylogenetic studies (Gatt et al. 2000; Saar et al. 2003a) have not supported these subdivisions, but the latter (Saar et al. 2003a) resolved a fairly well-supported phylogeny for 35 species that should lead to a phylopgenetically acceptable subgeneric classification in the future.

V. CONCLUDING REMARKS

Our current understanding of the relationships among the angiosperms, especially monocotyledons, represents a quantum leap over the knowledge of just a few decades ago. New data continue to accumulate at an ever increasing rate. Within Asparagales, a precise understanding of the relationships among the basal, "lower" families is still elusive (Chase et al. 2006). Within Liliales, the relationships among the component families appear more resolute (Fay et al. 2006), but the exact relationships among the lilioid orders Dioscoreales, Liliales, and Pandanales (Figure 2.2) are not yet fully resolved. The Monocot Tree of Life Project will no doubt contribute toward completing this fundamental gap in our understanding.

CHAPTER CONTRIBUTOR

Alan W. Meerow is a senior research geneticist, USDA-ARS-SHRS, National Germplasm Repository, Miami, Florida, USA. He was born in the Bronx, New York and received a BS from University of California, Davis, and MS and PhD from the University of Florida. He has held the position of professor at the University of Florida, Fort Lauderdale Research and Education Center 1987–1999, and has been with USDA since 1999. The major topics of his research are plant systematics, phylogeny, population genetics, genomics, and breeding. He has published more than 200 refereed journal articles, books or book chapters, proceedings, extension publications, trade and popular magazine articles. Dr. Meerow is a recipient of the Herbert Medal of the International Bulb Society (1998) and the Peter H. Raven Award of the American Society of Plant Taxonomists (2005).

REFERENCES

Aagesen, L., and A. M. Sanso. 2003. The phylogeny of the Alstroemeriaceae, based on morphology, *rps16* intron, and *rbcL* sequence data. *Syst. Bot.* 28:47–69.

Aker, S., and W. Healy. 1990. The phytogeography of the genus *Alstroemeria. Herbertia* 46:76–87.

Alzate, F., M. E. Mort, and M. Ramirez. 2008. Phylogenetic analyses of *Bomarea* (Alstroemeriaceae) based on combined analyses of nrDNA ITS, *psbA-trnH*, *rpoB-trnC* and *matK* sequences. *Taxon* 57:853–862.

Ambrose, J. D. 1980. A re-evaluation of the Melanthioideae (Liliaceae) using numerical analyses. In *Petaloid monocotyledons*, eds. C. D. Brickell, D. F. Cutler, and M. Gregory, 65–81. London: Academic Press.

APG. 1998. An ordinal classification for the families of flowering plants. *Ann. Missouri Bot. Gard.* 85:531–553.

APG II. 2003. An update of the Angiosperm Phylogeny Group classification for the orders and families of flowering plants: APG II. *Bot. J. Linn. Soc.* 141:399–436.

APG III. 2009. An update of the Angiosperm Phylogeny Group classification for the orders and families of flowering plants: APG III. *Bot. J. Linn. Soc.* 161:105–121.

Arroyo, S. C., and B. E. Leuenberger. 1988. Leaf morphology and taxonomic history of *Luzuriaga* (Philesiaceae). *Willdenowia* 17:159–172.

Baker, J. G. 1874. Revision of the genera and species of Tulipeae. *J. Linn. Soc. London, Botany* 14:211–310.

Baker, J. G. 1888. *Handbook of the Amaryllideae*. London: George Bell & Sons.

Bakshi-Khaniki, G., and K. Persson. 1997. Nectary morphology in South West Asia *Fritillaria* (Liliaceae). *Nord. J. Bot.* 17:579–611.

Baranova, M. V. 1969. The geographical distribution of *Lilium* species in the flora of the USSR. *R.H.S. Lily Year Book* 32:39–55.

Barrett, S. C. H., and L. D. Harder. 2005. The evolution of polymorphic sexual systems in daffodils (*Narcissus*). *New Phytol.* 165:45–53.

Batsch, A. J. 1802. *Tabula affinitatum regni vegetabilis*. Weimar, Germany: Landes-Industrie-Comptior.

Bayer, E. 1987. Die Gattung *Alstroemeria* in Chile. *Mitt. Bot. Staatssamml. München* 24:1–362.

Bayer, E. 1998. Alstroemeriaceae. In *The Families and Genera of Vascular Plants III. Flowering Plants, Monocotyledons: Lilianae (except Orchidaceae)*, ed. K. Kubitzki, 79–83. Berlin: Springer.

Bay-Smidt, M. G. K., A. K. Jäger, K. Krydsfeldt, A. W. Meerow, G. I. Stafford, J. van Staden, and N. Rønsted. 2011. Phylogenetic selection of target species in Amaryllidaceae tribe Haemantheae for acetylcholinesterase inhibition and affinity to the serotonin reuptake transport protein. *S. Afr. J. Bot.* 77:175–183.

Beaumont, M. A. 2010. Approximate Bayesian computation in evolution and ecology. *Annu. Rev. Ecol. Evol. Syst.* 41:379–406.

Bentham, G., and J. D. Hooker. 1883. *Fritillaria. Gen. Pl.* 3:817–818.

Blanchard, J. W. 1990. *Narcissus: A Guide to Wild Daffodils*. Surrey, UK: Alpine Garden Society.

Blattner, F. R., and N. Friesen. 2006. Relationship between Chinese chive (*Allium tuberosum*) and its putative progenitor *A. ramosum* as assessed by random amplified polymorphic DNA (RAPD). In *Documenting Domestication: New Genetic and Archaeological Paradigms*, eds. M. A. Zeder, D. G. Bradley, E. Emshwiller, and B. D. Smith, 134–142. Berkeley: University of California Press.

Boissier, P. E. 1882. *Fritillaria. Flora Orientalis* 5:176–190.

Botschantzeva, Z. P. 1982. *Tulips: Taxonomy, Morphology, Cytology, Phytogeography and Physiology*. Rotterdam, The Netherlands: A. A. Balkema. (Translated and edited by H. Q. Varenkamp)

Boyce, P. C. 1993. *The Genus Arum*. Royal Botanic Gardens, Kew: The Bentham-Moxon Trust.

Boyce, P. C. 1996. The genus *Hapaline* (Araceae: Aroideae: Caladieae). *Kew Bull.* 51:63–82.

Boyce, P. C. 2006. *Arum*: A decade of change. *Aroideana* 29:132–137.

Boyce, P. C. 2008. A monograph of the genus *Biarum*—Special part. *Curtis's Bot. Mag.* 25:1–119.

Bremer, K. 2000. Early Cretaceous lineages of monocot flowering plants. *Proc. Natl. Acad. Sci. USA* 97:4707–4711.

Bremer, B., K. Bremer, N. Heidari, P. Erixon, R. G. Olmstead, A. A. Anderberg, M. Källersjö, and E. Barkhordarian. 2002. Phylogenetics of asterids based on 3 coding and 3 non-coding chloroplast DNA markers and the utility of non-coding DNA at higher taxonomic levels. *Mol. Phylogen. Evol.* 24:274–301.

Buchner, L. 1948. Vergleichende embryologische Studien an Scilloideae. *Österr. Bot. Z.* 95:428–450.

Cabrera, L. I., G. A. Salazar, M. W. Chase, S. J. Mayo, J. Bogner, and P. Dávila. 2008. Phylogenetic relationships of aroids and duckweeds (Araceae) inferred from coding and noncoding plastid DNA. *Am. J. Bot.* 95:1153–1165.

Caddick, L. R., P. J. Rudall, P. Wilkin, and M. W. Chase. 2000. Yams and their allies: Systematics of Dioscoreales. In *Monocots: Systematics and Evolution*, eds. K. L. Wilson, and D. A. Morrison, 475–487. Collingwood, Australia: CSIRO Publ.

Cantino, P. D., J. A. Doyle, S. W. Graham, W. S. Judd, R. G. Olmstead, D. E. Soltis, P. S. Soltis, and M. J. Donoghue. 2007. Towards a phylogenetic nomenclature of *Tracheophyta*. *Taxon* 56:822–846.

Chacon, J. 2010. http://www.sysbot.biologie.uni-muenchen.de/en/people/chacon/.

Chase, M. W., M. R. Duvall, H. G. Hills, J. G. Conran, A. V. Cox, L. E. Eguiarte, J. Hartwell et al. 1995a. Molecular phylogenetics of *Lilianae*. In *Monocotyledons: Systematics and Evolution, Vol. 1*, eds. P. J. Rudall, P. J. Cribb, D. F. Cutler, and C. J. Humphries, 109–137. Kew: Royal Botanic Gardens.

Chase, M. W., M. F. Fay, D. S. Devey, O. Maurin, N. Rønsted, T. J. Davies, Y. Pillon et al. 2006. Multigene analyses of monocot relationships: a summary. In *Monocots: Comparative Biology and Evolution (Vol. 1, excluding Poales)*, eds. J. T. Columbus, E. A. Friar, C. W. Hamilton, J. M. Porter, L. M. Prince, and M. G. Simpson, 63–75. Claremont: Rancho Santa Ana Botanic Garden.

Chase, M. W., J. L. Reveal, and M. F. Fay. 2009. A subfamilial classification for the expanded asparagalean families Amaryllidaceae, Asparagaceae and Xanthorrhoeaceae. *Bot. J. Linn. Soc.* 161:132–136.

Chase, M. W., P. J. Rudall, M. F. Fay, and K. L. Stobart. 2000a. Xeronemataceae, a new family of asparagoid lilies from New Caledonia and New Zealand. *Kew Bull.* 55:865–870.

Chase, M. W., D. E. Soltis, P. S. Soltis, P. J. Rudall, M. F. Fay, W. H. Hahn, S. Sullivan, J. Joseph, T. Givnish, K. J. Sytsma, and J. C. Pires. 2000b. Higher-level systematics of the monocotyledons: An assessment of current knowledge and a new classification. In *Monocots: Systematics and Evolution*, eds. K. L. Wilson, and D. A. Morrison, 3–16. Collingwood, Australia: CSIRO Publ.

Chase, M. W., D. W. Stevenson, P. Wilkin, and P. J. Rudall. 1995b. Monocot systematics: A combined analysis. In *Monocotyledons: Systematics and Evolution, Vol. 1*, eds. P. J. Rudall, P. J. Cribb, D. F. Cutler, and C. J. Humphries, 685–730. Kew: Royal Botanic Gardens.

Comber, H. F. 1949. A new classification of the genus *Lilium*. *R.H.S. Lily Year Book* 13:86–105.

Conran, J. G., and H. T. Clifford. 1998. Luzuriagceae. In *The Families and Genera of Vascular Plants III. Flowering Plants, Monocotyledons: Lilianae (except Orchidaceae)*, ed. K. Kubitzki, 365–369. Berlin: Springer.

Cusimano, N., M. D. Barrett, W. L. A. Hetterscheid, and S. S. Renner. 2010. A phylogeny of the Areae (Araceae) implies that *Typhonium*, *Sauromatum*, and the Australian species of *Typhonium* are distinct clades. *Taxon* 59:439–447.

Cusimano, N., J. Bogner, S. J. Mayo, P. C. Boyce, S. Y. Wong, M. Hesse, W. L. A. Hetterscheid, R. C. Keating, and J. C. French. 2011. Relationships within the Araceae: Comparison of morphological patterns with molecular phylogenies. *Am. J. Bot.* 98:654–668.

Dahlgren, R. M. T., and H. T. Clifford. 1982. *The Monocotyledons: A Comparative Study*. London: Academic Press.

Dahlgren, R. M. T., H. T. Clifford, and P. F. Yeo. 1985. *The Families of the Monocotyledons: Structure, Evolution, and Taxonomy*. Berlin: Springer.

Dahlgren, R. M. T., and F. N. Rasmussen. 1983. Monocotyledon evolution: Characters and phylogenetic analysis. *Evol. Biol.* 16:255–395.

Davis, A. P. 1999. *The Genus* Galanthus: *A Botanical Magazine Monograph*. Portland, OR: Timber Press.

Davis, A. P. 2001. The genus *Galanthus*—Snowdrops in the wild. In *Snowdrops: A Monograph of Cultivated Galanthus*, eds. M. Bishop, A. P. Davis, and J. Grimshaw, 9–63. Cheltenham, UK: Griffin Press.

De Assis, M. C. 2001. *Alstroemeria* L. (Alstroemeriaceae) do Brasil. PhD dissertation, University of São Paulo, São Paulo, Brazil.

De Candolle, A. P. 1805. Colchicaceae. In *Flore francaise 3*, eds. M. M. Lamarck, and A. P. de Candolle, 192–193. Paris: Stoupe.

De Reboul, E. 1847. Sulla divisione del genere *Tulipa* in sezioni naturali. *G. Bot. Ital.* 2:57–61.

De Wilde Duyfjes, B. E. E. 1976. *A revision of the genus Allium L. (Liliaceae) in Africa*. Wageningen: H. Veenman and Zonen B.V.

Delgadillo, R. R. 1992. Una nueva especie de *Bessera* (Liliaceae) del Occidente de Jalisco, México. *Bol. Inst. Bot., Univ. Guadalajara* 1:131–136.

Devey, D. S., I. Leitch, P. J. Rudall, J. C. Pires, Y. Pillon, and M. W. Chase. 2006. Systematics of Xanthorrhoeaceae sensu lato, with an emphasis on *Bulbine*. In *Monocots: Comparative Biology and Evolution (Vol. 1, excluding Poales)*, eds. J. T. Columbus, E. A. Friar, C. W. Hamilton, J. M. Porter, L. M. Prince, and M. G. Simpson, 345–351. Claremont: Rancho Santa Ana Botanic Garden.

Dubouzet, J. G., and K. Shinoda. 1998. Phylogeny of *Allium* L. subg. *Melanocrommyum* (Webb et Berth.) Rouy based on DNA sequence analysis of the internal transcribed spacer region of nrDNA. *Theor. Appl. Genet.* 97:541–549.

Dubouzet, J. G., and K. Shinoda. 1999. Phylogenetic analysis of the internal transcribed spacer region of Japanese *Lilium* species. *Theor. Appl. Genet.* 98:954–960.

Dubouzet, J. G., K. Shinoda, and N. Murata. 1997. Phylogeny of *Allium* L. subgenus *Rhizirideum* (G. Don ex Koch) Wendelbo according to dot blot hybridization with randomly amplified DNA probes. *Theor. Appl. Genet.* 95:1223–1228.

Dyer, R. A. 1939 (published 1940). Description, classification and phylogeny. A review of the genus *Cyrtanthus*. *Herbertia* 6:65–103.

Dykes, W. R. 1913. *The Genus Iris*. Cambridge: The University Press. Photo-offset reprint by Dover Publications, New York, 1974.

Endlicher, S. 1836. *Genera Plantarum Secundum Ordines Naturales Disposita*. Wien: Fr. Back.

Fay, M. F., and M. W. Chase. 1996. Resurrection of Themidaceae for the *Brodiaea* alliance, and recircumscription of Alliaceae, Amaryllidaceae and Agapanthoideae. *Taxon* 45:441–451.

Fay, M. F., P. J. Rudall, and M. W. Chase. 2006. Molecular studies of subfamily Gilliesioiodeae (Alliaceae). In *Monocots: Comparative Biology and Evolution (Vol. 1, excluding Poales)*, eds. J. T. Columbus, E. A. Friar, C. W. Hamilton, J. M. Porter, L. M. Prince, and M. G. Simpson, 365–371. Claremont: Rancho Santa Ana Botanic Garden.

Fay, M. F., P. J., Rudall, S. Sullivan, K. L. Stobart, A. Y. De Bruijn, G. Reeves, F. Qamaruz-Zaman et al. 2000. Phylogenetic studies of Asparagales based on four plastid DNA loci. In *Monocots: Systematics and Evolution, Vol. 1*, eds. K. L. Wilson, and D. A. Morrison, 360–371. Collingwood, Australia: CSIRO Publ.

Felsenstein, J. 1985. Confidence limits of phylogenies: an approach using the bootstrap. *Evolution* 39:783–791.

Felsenstein, J. 1988. Phylogenies from molecular sequences: inference and reliability. *Annu. Rev. Genetics* 22:521–565.

Felsenstein, J. 2004. *Inferring Phylogenies*. Sunderland: Sinaeuer Associates.

Fernandes, A. 1967. Contribution à la connaissance de la biosystématique de quelques espèces du genre *Narcissus* L. *Portugaliae Acta Biol., ser. B. Sistem., Ecol., Biog. Paleontol.* 9:1–44.

Fernandes, A. 1968a. Keys to the identification of native and naturalizaed taxa of the genus *Narcissus* L. *Daffodil Tulip Year Book* 1968:37–66.

Fernandes, A. 1968b. Improvements in the classification of the genus *Narcissus* L. *Plant Life.* 24:51–57.

Fernandes, A. 1975. L'évolution chez le genre *Narcissus* L. *Anales Inst. Bot. A. J. Cavanilles.* 32:843–872.

Forrest, L. L., and P. M. Hollingsworth. 2003. A recircumscription of *Begonia* based on nuclear ribosomal sequences. *Plant Syst. Evol.* 241:193–211.

Forrest, L. L., M. Hughes, and P. M. Hollingsworth. 2005. A phylogeny of *Begonia* using nuclear ribosomal sequence data and morphological characters. *Syst. Bot.* 30:671–682.

French, J. C., M. G. Chung, and Y. K. Hur. 1995. Chloroplast DNA phylogeny of the Ariflorae. In *Monocotyledons: Systematics and Evolution, Vol. 1*, eds. P. J. Rudall, P. J. Cribb, D. F. Cutler, and C. J. Humphries, 255–275. Kew: Royal Botanic Gardens.

Friesen, N., R. M. Fritsch, and F. R. Blattner. 2006. Phylogeny and new intrageneric classification of *Allium* (Alliaceae) based on nuclear ribosomal DNA ITS sequences. In *Monocots: Comparative Biology and Evolution (Vol. 1, excluding Poales)*, eds. J. T. Columbus, E. A. Friar, C. W. Hamilton, J. M. Porter, L. M. Prince, and M. G. Simpson, 372–395. Claremont: Rancho Santa Ana Botanic Garden.

Friesen, N., R. M. Fritsch, S. Pollner, and F. R. Blattner. 2000. Molecular and morphological evidence for an origin of the aberrant genus *Milula* within Himalayan species of *Allium* (Alliaceae). *Mol. Phylogen. Evol.* 17:209–218.

Friesen, N., and M. Klaas. 1998. Origin of some minor vegetatively propagated *Allium* crops studied with RAPD and GISH. *Genet. Res. Crop Evol.* 45:511–523.

Friesen, N., S. Pollner, K. Bachmann, and F. R. Blattner. 1999. RAPDs and noncoding chloroplast DNA reveal a single origin of the cultivated *Allium fistulosum* from *A. altaicum* (Alliaceae). *Am. J. Bot.* 86:554–562.

Fritsch, R. M. 2001. Taxonomy of the genus *Allium* L.: Contribution from IPK Gatersleben. *Herbertia* 56:19–50.

Fritsch, R. M., F. R. Blattner, and M. Gurushidze. 2010. New classification of *Allium* L. subg. *Melanocrommyum* (Webb & Berthel.) Rouy (Alliaceae) based on molecular and morphological characters. *Phyton* 49:145–220.

Fritsch, R. M., and N. Friesen. 2002. Evolution, domestication, and taxonomy. In *Allium Crop Science: Recent Advances*, eds. H. D. Rabinowitch, and L. Currah, 5–30. Wallingford, UK: CABI Publ.

Fuchsig, H. 1911. Vergleichende Anatomie der Vegetationsorgane der Lilioideen. *Sitzungsber. Kais. Akad. Wiss. Wien, Math.-Naturw. Kl.* 120:1–43, I-III.

Gatt, M. K., K. R. W. Hammett, and B. G. Murray. 2000. Molecular phylogeny of the genus *Dahlia* Cav. (Asteraceae, Heliantheae—Coreopsidinae) using sequences derived from the internal transcribed spacers of nuclear ribosomal DNA. *Bot. J. Linn. Soc.* 133:229–239.

Gauthier, M. P. L., D. Barabe, and A. Bruneau. 2008. Molecular phylogeny of the genus *Philodendron* (Araceae): delimitation and infrageneric classification. *Bot. J. Linn. Soc.* 156:13–27.

Giannasi, D. E. 1975a. Flavonoid chemistry and evolution in *Dahlia* (Compositae). *Bull. Torrey Bot. Club* 102:404–412.

Giannasi, D. E. 1975b. The flavonoid systematics of the genus *Dahlia* (Compositae). *Mem. New York Bot. Gard.* 26:1–125.

Goldblatt, P. 1984. (748) Proposal to conserve 1176 *Amaryllis* and typification of *A. belladonna* (Amaryllidaceae). *Taxon* 33:511–516.

Goldblatt, P. 1990. Phylogeny and classification of Iridaceae. *Ann. Missouri Bot. Gard.* 77:607–627.

Goldblatt, P. 1995. The status of R. Dahlgren's orders Liliales and Melanthiales. In *Monocotyledons: Systematics and Evolution, Vol. 1*, eds. P. J. Rudall, P. J. Cribb, D. F. Cutler, and C. J. Humphries, 181–200. Kew: Royal Botanic Gardens.

Goldblatt, P. 1996. Gladiolus *in Tropical Africa: Systematic, Biology and Evolution*. Portland, OR: Timber Press.

Goldblatt, P. 1998. Iridaceae. In *The Families and Genera of Vascular Plants III. Flowering Plants, Monocotyledons: Lilianae (Except Orchidaceae)*, ed. K. Kubitzki, 295–333. Berlin: Springer.

Goldblatt, P., T. J. Davies, J. C. Manning, M. van der Bank, and V. Savolainen. 2006. Phylogeny of Iridaceae subfamily Crocoideae based on combined multigene plastid DNA analysis. In *Monocots: Comparative Biology and Evolution (Vol. 1, excluding Poales)*, eds. J. T. Columbus, E. A. Friar, J. M. Porter, L. M. Prince, and M. G. Simpson, 399–411. Claremont: Rancho Santa Ana Botanical Garden.

Goldblatt, P., and D. J. Mabberley. 2005. *Belamcanda* Adanson included in *Iris* Linnaeus, and the new combination, *I. domestica* (Linnaeus) Goldblatt & Mabberley (Iridaceae: Irideae). *Novon* 15:128–132.

Goldblatt, P., and J. Manning. 1998. *Gladilous in Southern Africa*. Vlaeberg: Fernwood Press.

Goldblatt, P., and J. C. Manning. 2000. Cape plants. A conspectus of the Cape flora of South Africa. *Strelitzia* 9:93–108.

Goldblatt, P., and J. C. Manning. 2008. *The Iris Family: Natural History and Classification*. Portland, OR: Timber Press.

Goldblatt, P., A. Rodriguez, M. P. Powell, T. J. Davies, J. C. Manning, M. van der Bank, and V. Savolainen. 2008. Iridaceae 'out of Australasia'? Phylogeny, biogeography, and divergence time based on plastid DNA sequences. *Syst. Bot.* 33:495–508.

Goldblatt, P., V. Savolainen, O. Porteous, I. Sostaric, M. Powell, G. Reeves, J. C. Manning, T. G. Barraclough, and M. W. Chase. 2002. Radiation in the Cape flora and the phylogeny of peacock irises *Moraea* (Iridaceae) based on four plastid DNA regions. *Mol. Phylogen. Evol.* 25:341–360.

Gonçalves, E. G., S. J. Mayo, M-A. van Sluys, and A. Salatino. 2007. Combined genotypic–phenotypic phylogeny of the tribe Spathicarpeae (Araceae) with reference to independent events of invasion to Andean regions. *Mol. Phylogen. Evol.* 43:1023–1039.

Goodall-Copestake, W. P., D. J. Harris, and P. M. Hollingsworth. 2009. The origin of a mega-diverse genus: Dating *Begonia* (Begoniaceae) using alternative datasets, calibrations and relaxed clock methods. *Bot. J. Linn. Soc.* 159:363–380.

Graham, S. W., and S. C. H. Barrett. 2004. Phylogenetic reconstruction of the evolution of stylar polymorphisms in *Narcissus* (Amaryllidaceae). *Am. J. Bot.* 91:1007–1021.

Graham, S. W., J. M. Zgurski, M. A. McPherson, D. M. Cherniawsky, J. M. Saarela, E. S. C. Horne, S. Y. Smith et al. 2006. Robust inference of monocot deep phylogeny using an expanded multigene plastid data set. In *Monocots: Comparative Biology and Evolution (Vol. 1, excluding Poales)*, eds. J. T. Columbus, E. A. Friar, J. M. Porter, L. M. Prince, and M. G. Simpson, 3–21. Claremont: Rancho Santa Ana Botanical Garden.

Grayum, M. H. 1990. Evolution and phylogeny of the Araceae. *Ann. Missouri Bot. Gard.* 77:628–697.

Grundmann, M., F. J. Rumsey, S. W. Ansell, S. J. Russell, S. C. Darwin, J. C. Vogel, M. Spencer, J. Squirrell, P. M. Hollingsworth, S. Ortiz, and H. Schneider. 2010. Phylogeny and taxonomy of the bluebell genus *Hyacinthoides*, Asparagaceae (Hyacinthaceae). *Taxon* 59:68–82.

Guaglianone, E. R., and S. Arroyo-Leuenberger. 2002. The South American genus *Oziroë* (Hyacinthaceae-Oziroëoideae). *Darwinia* 40:61–76.

Gurushidze, M., R. M. Fritsch, and F. R. Blattner. 2008. Phylogenetic analysis of *Allium* subg. *Melanocrommyum* infers cryptic species and demands a new sectional classification. *Mol. Phylogen. Evol.* 49:997–1007.

Gurushidze, M., R. M. Fritsch, and F. R. Blattner. 2010. Species-level phylogeny of *Allium* subgenus *Melanocrommyum*: Incomplete lineage sorting, hybridization and *trnF* gene duplication. *Taxon* 59:829–840.

Gurushidze, M., S. Mashayekhi, F. R. Blattner, N. Friesen, and R. M. Fritsch. 2007. Phylogenetic relationships of wild and cultivated species of *Allium* section *Cepa* inferred by nuclear rDNA ITS sequence analysis. *Plant Syst. Evol.* 269:259–269.

Gusman, G., and L. Gusman. 2002. *The Genus Arisaema. A Monograph for Botanists and Nature Lovers*. Ruggell: Gantner Verlag.

Hall, A. D. 1940. *The Genus Tulipa*. London: The Royal Horticultural Society.

Hamatani, S., Y. Masuda, K. Kondo, E. Kodaira, and H. Ogawa. 2008. Molecular phylogenetic relationships among *Lachenalia, Massonia* and *Polyxena* (Liliaceae) on the basis of the internal transcribed spacer (ITS) region. *Chromosome Bot.* 3:65–72.

Han, T., M. De Jeu, H. van Eck, and E. Jacobsen. 2000. Genetic diversity of Chilean and Brazilian *Alstroemeria* species assessed by AFLP analysis. *Heredity* 84:564–569.

Hansen, D. R., S. G. Dastidar, Z. Cau, C. Penaflorb, J. V. Kuehlc, J. L. Boorec, and R. K. Jansen. 2007. Phylogenetic and evolutionary implications of complete chloroplast genome sequences of four early-diverging angiosperms: *Buxus* (Buxaceae), *Chloranthus* (Chloranthaceae), *Dioscorea* (Dioscoreaceae), and *Illicium* (Schisandraceae). *Molec. Phylogen. Evol.* 45:547–563.

Hansen, H. V., and J. P. Hjerting. 1996. Observations on chromosome numbers and biosystematics in *Dahlia* (Asteraceae, Heliantheae) with an account on the identity of *D. pinnata, D. rosea,* and *D. coccinea. Nord. J. Bot.* 16:445–455.

Hansen, H. V., and P. D. Sørensen. 2003. A new species of *Dahlia* (Asteraceae, Coreopsideae) from Hidalgo state, Mexico. *Rhodora* 105:101–105.

Haw, S. G. 1986. *The Lilies of China.* Portland, OR: Timber Press.

Hayashi, K., and S. Kawano. 2000. Molecular systematics of *Lilium* and allied genera (Liliaceae): Phylogenetic relationships among *Lilium* and related genera based on *rbcL* and matK gene sequence data. *Plant Spec. Biol.* 15:73–93.

Hayashi, K., S. Yoshida, H. Kato, F. H. Utech, D. F. Whigham, and S. Kawano. 1998. Molecular systematics of the genus *Uvularia* and selected Liliales based upon *matK* and *rbcL* gene sequence data. *Plant Spec. Biol.* 13:129–146.

Hayashi, K., S. Yoshida, F. H. Utech, and S. Kawano. 2001. Molecular systematics in the genus *Clintonia* and related taxa based on *rbcL* and *matK* gene sequence data. *Plant Spec. Biol.* 16:119–137.

He, X. J., S. Ge, J. M. Xu, and D. Y. Hong. 2000. Phylogeny of Chinese *Allium* (Liliaceae) using PCR-RFLP analysis. *Science in China (series C)* 43:454–463.

Hennig, W. 1966. *Phylogentic Systematics.* Urbana: University of Illinois Press.

Herbert, W. 1837. *Amaryllidaceae.* London: James Ridgeway and Sons.

Hillis, D. M., and J. J. Bull. 1993. An empirical test of bootstrapping as a method for assessing confidence in phylogenetic analysis. *Syst. Biol.* 42:182–192.

Hirschegger, P., J. Jakše, P. Trontelj, and B. Bohanec. 2010. Origins of *Allium ampeloprasum* horticultural groups and a molecular phylogeny of the section *Allium* (*Allium*: Alliaceae). *Mol. Phylogen. Evol.* 54:488–497.

Hofreiter, A., and H.-J. Tillich. 2002. The delimitation, infrageneric subdivision, ecology and distribution of *Bomarea* Mirbel (Alstroemeriaceae). *Fed. Rep.* 113:528–544.

Hoog, M. H. 1973. On the origin of *Tulipa.* In *Lilies and Other Liliaceae,* eds. E. Napier, and J. N. O. Platt, 47–64. London: Royal Horticultural Society.

Howard, T. M. 1999. Three new *Milla* species from Mexico. *Herbertia* 54:232–237.

Huber, H. 1969. Die Samenmerkmale und Verwandtschaftsverhältnisse der Liliiflorae. *Mitt. Bot. Staatssamml. München* 8:219–538.

Huelsenbeck, J. P., and K. A. Crandall. 1997. Phylogeny estimation and hypothesis testing using maximum likelihood. *Annu. Rev. Ecol. Syst.* 28:437–466.

Hutchinson, J. 1964. *The Genera of Flowering Plants.* Oxford: Oxford University Press.

Ikinci, N., C. Oberprieler, and A. Güner. 2006. On the origin of European lilies: Phylogenetic analysis of *Lilium* section Liriotypus (Liliaceae) using sequences of the nuclear ribosomal transcribed spacers. *Willdeowia* 36:647–656.

Ingraham, J. A. 1953. A monograph of the genera *Bloomeria* and *Muilla* (Liliaceae). *Madroño* 12:19–27.

Ising, G. 1970. Evolution of karyotypes in *Cyrtanthus. Hereditas* 65:1–28.

Ito, M., A. Kawamoto, Y. Kita, T. Yukawa, and S. Kurita. 1999. Phylogenetic relationships of Amaryllidaceae based on *matK* sequence data. *J. Plant Res.* 112:207–216.

Jansen, R. K., Z. Cai, L. A. Raubeson, H. Daniell, C. W. dePamphilis, J. Leebens-Mack, K. F. Müller et al. 2007. Analysis of 81 genes from 64 plastid genomes resolves relationships in angiosperms and identifies genome-scale evolutionary patterns. *Proc. Natl. Acad. Sci. USA* 104:19369–19374.

Janssen, T., and K. Bremer. 2004. The age of major monocot groups inferred from 800+ *rbcL* sequences. *Bot. J. Linn. Soc.* 146:385–398.

Judd, W. S., C. S. Campbell, E. A. Kellogg, P. F. Stevens, and M. J. Donoghue. 2007. *Plant Systematic: a Phylogenetic Approach, 3rd ed.* Sunderland, MA: Sinauer Associates.

Kauff, F., P. J. Rudall, and J. C. Conran. 2000. Systematic root anatomy of Asparagales and other monocotyledons. *Plant Syst. Evol.* 223:139–154.

Keator, G. 1967. Ecological and taxonomic studies of the genus *Dichelostemma*. PhD dissertation, University of California, Berkeley, California, USA.

Keator, G. 1989. The brodiaeas. *Four Seasons* 8:4–11.

King, R. M., and H. Robinson. 1987. The genera of the Eupatorieae (Asteraceae). *Monogr. Syst. Bot. Missouri Bot. Gard.* 22:1–581.

Kite, G. C., R. J. Grayer, P. J. Rudall, and M. S. J. Simmonds. 2000. The potential for chemical charcters in monocotyledon systematic. In *Monocots: Systematics and Evolution*, eds. K. L. Wilson, and D. A. Morrison, 101–113. Collingwood, Australia: CSIRO Publ.

Kocyan, A., and P. K. Endress. 2001. Floral structure and development and systematic aspects of some 'lower' Asparagales. *Plant Syst. Evol.* 229:187–216.

Komarov, V. L. 1935. *Korolkowia, Rhinopetalum, Fritillaria. Fl. U.S.S.R.* 4:227–246.

Koshimizu, T. 1930. Carpobiological studies of *Crinum asiaticum* L. var. *japonicum* Bak. *Mem. Coll. Sci., Kyoto Imp. Univ., Ser. B., Biology* 5:183–227.

Kress, W. J. 1990. The phylogeny and classification of the Zingiberales. *Ann. Missouri Bot. Gard.* 77:698–721.

Kress, W. J. 1995. Phylogeny of the Zingiberanae: Morphology and molecules. In *Monocotyledons: Systematics and Evolution, Vol. 1*, eds. P. J. Rudall, P. J. Cribb, D. F. Cutler, and C. J. Humphries, 443–460. Kew: Royal Botanic Gardens.

Kress, W. J., A.-Z. Liu, M. Newman, and Q.-J. Li. 2005. The molecular phylogeny of *Alpinia* (Zingiberaceae): A complex and polyphyletic genus of gingers. *Am. J. Bot.* 92:167–178.

Kress, W. J., L. M. Prince, W. J. Hahn, and E. A. Zimmer. 2001. Unraveling the evolutionary radiation of the families of the Zingiberales using morphological and molecular evidence. *Syst. Biol.* 50:926–944.

Kress, W. J., L. M. Prince, and K. J. Williams. 2002. The phylogeny and a new classification of the gingers (Zingiberaceae): Evidence from molecular data. *Am. J. Bot.* 89:1682–1696.

Kroon, G. H., and M. C. Jongerius. 1986. Chromosome numbers of *Tulipa* species and the occurrence of hexaploidy. *Euphytica* 35:73–76.

Kunth, C. S. 1850. *Enumeratio Plantarum, Vol. 5*. Stuttgart: Cotta.

Kwembeya, E. G., C. S. Bjorå, B. Stedje, and I. Nordal. 2007. Phylogenetic relationships in the genus *Crinum* (Amaryllidaceae) with emphasis on tropical African species: Evidence from *trnL-F* and nuclear ITS DNA sequence data. *Taxon* 56:801–810.

Larsen, M. M., A. Adersen, A. P. Davis, M. D. Lledó, A. K. Jäger, and N. Rønsted. 2010. Using a phylogenetic approach to selection of target plants in drug discovery of acetylcholinesterase inhibiting alkaloids in Amaryllidaceae tribe Galantheae. *Biochem. Syst. Ecol.* 38:1026–1034.

Lawrence, G. H. M. 1953. A reclassification of the genus *Iris*. *Gentes Herb.* 8:346–371.

Lee, C. S., S.-C. Kim, S. H. Yeau, and N. S. Lee. 2011. Major lineages of the genus *Lilium* (Liliaceae) based on nrDNA ITS sequences, with special emphasis on the Korean species. *J. Plant Biol.* 54:159–171.

Li, Q.-Q., S.-D. Zhou, X.-J. He, Y. Yu, Y.-C. Zhang, and X.-Q. Wei. 2010. Phylogeny and biogeography of *Allium* (Amaryllidaceae: Allieae) based on nuclear ribosomal internal transcribed spacer and chloroplast *rps16* sequences, focusing on the inclusion of species endemic to China. *Ann. Bot.* 106:709–733.

Lighty, R. W. 1968. Evolutionary trends in lilies. *Lily Yearbook RHS* 31:40–44.

Linz, J., J. Stökl, I. Urru, T. Krügel, M. C. Stensmyr, and B. S. Hansson. 2010. Molecular phylogeny of the genus *Arum* (Araceae) inferred from multi-locus sequence data and AFLPs. *Taxon* 59:405–415.

Lledó, M. D., A. P. Davis, M. B. Crespo, M. W. Chase, and M. F. Fay. 2004. Phylogenetic analysis of *Leucojum* and *Galanthus* (Amaryllidaceae) based on plastid *matK* and nuclear ribosomal spacer (ITS) DNA sequences and morphology. *Plant Syst. Evol.* 246:223–243.

Loh, J. P., R. Kiew, A. Hay, A. Kee, L. H. Gan, and Y.-Y. Gan. 2000. Intergeneric and interspecific relationships in Araceae tribe Caladieae and development of molecular markers using amplified fragment length polymorphism (AFLP). *Ann. Bot.* 85:371–378.

Lopez-Ferrari, A. R., and A. Espejo. 1992. Una nueva especie de *Dandya* (Alliaceae) de la Cuenca del Rio Balsas, México. *Acta Bot. Mex.* 18:11–15.

Lourteig, A. 1994. *Oxalis* L. subgénero *Thamnoxys* (Endl.) Reiche emend. Lourteig. *Bradea* 7:1–199.

Lourteig, A. 2000. *Oxalis* L. subgénero *Monoxalis* (Small) Lourteig, *Oxalis* y *Trifidus* Lourteig. *Bradea* 7:201–629.

Maas-van de Kamer, H., and P. J. M. Maas. 2008. The Cannaceae of the world. *Blumea* 53:247–318.

Madison, M. 1981. Notes on *Caladium* (Araceae) and its allies. *Selbyana* 5:342–377.

Magallón, S., and A. Castillo. 2009. Angiosperm diversification through time. *Am. J. Bot.* 96:349–365.

Manning, J. C., F. Forest, D. S. Devey, M. F. Fay, and P. Goldblatt. 2009. A molecular phylogeny and a revised classification of Ornithogaloideae (Hyacinthaceae) based on an analysis of four plastid DNA regions. *Taxon* 58:77–107.

Manning, J. C., P. Goldblatt, and M. F. Fay. 2003. A revised generic synopsis of Hyacinthaceae in Sub-Saharan Africa, based on molecular evidence, including new combinations and the new tribe Pseudoprospereae. *Edinburgh J. Bot.* 60:533–568.

Manning, J. C., M. Martínez-Azorín, and M. B. Crespo. 2007. A revision of *Ornithogalum* subgenus *Aspasia* section *Aspasia*, the chincherinchees (Hyacinthaceae). *Bothalia* 37:133–164.

Marais, W. 1984. *Tulipa* L. In *Flora of Turkey and the East Aegean Islands 8*, ed. P. H. Davis, 302–311. Edinburgh: Edinburgh University Press.

Marques, I. C. 2010. Evolutionary outcomes of natural hybridization in *Narcissus* (Amaryllidaceae): The case of *N. × perezlarae* s.l. PhD dissertation, University of Lisboa, Lisbon, Portugal.

Martínez-Azorín, M., M. B. Crespo, A. Juan, and M. F. Fay. 2011. Molecular phylogenetics of subfamily Ornithogaloideae (Hyacinthaceae) based on nuclear and plastid DNA regions, including a new taxonomic arrangement. *Ann. Bot.* 107:1–37.

Mathew, B. 1982. *The Crocus: A Revision of the Genus Crocus (Iridaceae)*. Portland, OR: Timber Press.

Mathew, B. 1989. *The Iris*. London: Batsford.

Mathew, B. 2002. Classification of the genus *Narcissus*. In *Narcissus and Daffodil*, ed. G. R. Hanks, 30–52. London: Taylor & Francis.

Mayo, S. J., J. Bogner, and P. C. Boyce, 1997. *The Genera of Araceae*. Kew: Royal Botanic Gardens.

McRae, E. A. 1998. *Lilies: A Guide for Growers and Collectors*. Portland, OR: Timber Press.

Meerow, A. W. 1987. A monograph of *Eucrosia* (Amaryllidaceae). *Syst. Bot.* 12:460–492.

Meerow, A. W. 1989. Systematics of the Amazon lilies, *Eucharis* and *Caliphruria* (Amaryllidaceae). *Ann. Missouri Bot. Gard.* 76:136–220.

Meerow, A. W. 1990. 202, Amaryllidaceae. In *Flora of Ecuador no. 41*. eds. G. Harling, and L. Andersson, 1–52. Göteborg: University of Göteborg.

Meerow, A. W. 2009. Titlting at windmills: 20 years of *Hippeastrum* breeding. *Israel J. Plant Sci.* 57:303–313.

Meerow, A. W. 2010. Convergence or reticulation? Mosaic evolution in the canalized American Amaryllidaceae. In *Diversity, Phylogeny and Evolution in the Monocotyledons*, eds. O. Seberg, G. Petersen, A. S. Barfod, and J. I. Davis, 145–168. Aarhus, Denmark: Aarhus University Press.

Meerow, A. W., and J. R. Clayton. 2004. Generic relationships among the baccate-fruited Amaryllidaceae (tribe Haemantheae) inferred from plastid and nuclear non-coding DNA sequences. *Plant Syst. Evol.* 244:141–155.

Meerow, A. W., M. F. Fay, M. W. Chase, C. L. Guy, Q. Li, D. Snijman, and S.-L. Yang. 2000a. Phylogeny of the Amaryllidaceae: Molecules and morphology. In *Monocots: Systematics and Evolution*, eds. K. L. Wilson, and D. A. Morrison, 368–382. Collingwood, Australia: CSIRO Publ.

Meerow, A. W., M. F. Fay, C. L. Guy, Q.-B. Li, F. Q. Zaman, and M. W. Chase. 1999. Systematics of Amaryllidaceae based on cladistic analysis of plastid *rbcL* and *trnL-F* sequence data. *Am. J. Bot.* 86:1325–1345.

Meerow, A. W., J. Francisco-Ortega, D. N. Kuhn, and R. J. Schnell. 2006. Phylogenetic relationships and biogeography within the Eurasian clade of Amaryllidaceae based on plastid *ndhF* and nrDNA ITS sequences: Lineage sorting in a reticulate area? *Syst. Bot.* 31:42–60.

Meerow, A. W., C. L. Guy, Q.-B. Li, and J. R. Clayton. 2002. Phylogeny of the tribe Hymenocallideae (Amaryllidaceae) based on morphology and molecular characters. *Ann. Missouri Bot. Gard.* 89:400–413.

Meerow, A. W., C. L. Guy, Q-B. Li, and S-Y. Yang. 2000b. Phylogeny of the American Amaryllidaceae based on nrDNA ITS sequences. *Syst. Bot.* 25:708–726.

Meerow, A. W., D. J. Lehmiller, and J. R. Clayton. 2003. Phylogeny and biogeography of *Crinum* L. (Amaryllidaceae) inferred from nuclear and limited plastid non-coding DNA sequences. *Bot. J. Linn. Soc.* 141:349–363.

Meerow, A. W., J. L. Reveal, D. A. Snijman, and J. H. Dutilh. 2007. (1793) Proposal to conserve the name Amaryllidaceae against Alliaceae, a "superconservation" proposal. *Taxon* 56:1299–1300.

Meerow, A. W., and D. A. Snijman. 1998. Amaryllidaceae. In *The Families and Genera of Vascular Plants III. Flowering Plants, Monocotyledons: Lilianae (Except Orchidaceae)*, ed. K. Kubitzki, 83–110. Berlin: Springer.

Meerow, A. W., and D. A. Snijman. 2001. Phylogeny of Amaryllidaceae tribe Amaryllideae based on nrDNA ITS sequences and morphology. *Am. J. Bot.* 88:2321–2330.

Meerow, A. W., and D. A. Snijman. 2006. The never-ending story: Multigene approaches to the phylogeny of Amaryllidaceae, and assessing its familial limits. In *Monocots: Comparative Biology and Evolution, (Vol. 1, excluding Poales)*, eds. J. T. Columbus, E. A. Friar, J. M. Porter, L. M. Prince, and M. G. Simpson, 353–364. Claremont: Rancho Santa Ana Botanical Garden.

Meerow, A. W., J. van Scheepen, and J. H. A. Dutilh. 1997. Transfers from *Amaryllis* to *Hippeastrum*. (Amaryllidaceae). *Taxon* 46:15–19.

Mes, T. H. M., N. Friesen, R. M. Fritsch, M. Klaas, and K. Bachmann. 1997. Criteria for sampling in *Allium* based on chloroplast DNA PCR-RFLP's. *Syst. Bot.* 22:701–712.

Mes, T. H. M., R. M. Fritsch, S. Pollner, and K. Bachmann. 1999. Evolution of the chloroplast genome and polymorphic ITS regions in *Allium* subg. *Melanocrommyum. Genome* 42:237–247.

Moore, H. E. 1953. The genus *Milla* (Amaryllidaceae-Allieae) and its allies. *Gentes Herb.* 8:262–294.

Moore, M. J., C. D. Bell, P. S. Soltis, and D. E. Soltis. 2007. Using plastid genome-scale data to resolve enigmatic relationships among basal angiosperms. *Proc. Natl. Acad. Sci. USA* 104:19363–19368.

Moore, M. J., N. Hassan, M. A. Gitzendanner, R. A. Bruenn, M. Croley, A. Vandeventer, J. W. Horn, A. Dhingra, S. F. Brockington, M. Latvis, J. Ramdial, R. Alexandre, A. Piedrahita, Z. Xi, C. C. Davis, P. S. Soltis, and D. E. Soltis. 2011. Phylogenetic analysis of the plastid inverted repeat for 244 species: Insights into deeper-level angiosperm relationships from a long, slowly evolving sequence region. *Int. J. Plant Sci.* 172:541–558.

Moore, M. J., P. S. Soltis, C. D. Bell, J. G. Burleigh, and D. E. Soltis. 2010. Phylogenetic analysis of 83 plastid genes further resolves the early diversification of eudicots. *Proc. Natl. Acad. Sci. USA* 107:4623–4628.

Müller-Doblies, D. and U. Müller-Doblies. 1978a. Zum Bauplan von *Ungernia*, der einzigen endemischen Amaryllidaceen—Gattung Zentralasiens. *Bot. Jahrb. System.* 99:249–263.

Müller-Doblies D., and U. Müller-Doblies. 1978b. Studies on tribal systematics of Amaryllidoideae 1. The systematic position of *Lapiedra* Lag. *Lagascalia* 8:13–23.

Müller-Doblies, D., and U. Müller-Doblies. 1985. De Liliifloris notulae 2: De taxonomia subtribus Strumariinae (Amaryllidaceae). *Bot. Jahrb. System.* 107:17–47.

Müller-Doblies, D., and U. Müller-Doblies. 1996. Tribes and subtribes and some species combinations in Amaryllidaceae J. St.-Hil. emend. R. Dahlgren & al. 1985. *Fed. Rep.* 107(5–6):S.c. 1–9.

Müller-Doblies, U., and D. Müller-Doblies. 1997. A partial revision of the tribe Massonieae (Hyacinthaceae) 1. Survey, including three novelties from Namibia: A new genus, a second species in the monotypic *Whiteheadia*, and a new combination in *Massonia. Fed. Rep.* 108:49–96.

Muratovic, E., O. Hidalgo, T. Garnatje, S. Siljak-Yakovlev. 2010. Molecular phylogeny and genome size in European lilies (genus *Lilium*, Liliaceae). *Adv. Sci. Lett.* 3:180–189.

Nadot, S., L. Penet, L. L. Dreyer, A. Forchioni, and A. Ressayre. 2006. Aperture pattern and microsporogenesis in Asparagales. In *Monocots: Comparative Biology and Evolution (Vol. 1, excluding Poales)*, eds. J. T. Columbus, E. A. Friar, C. W. Hamilton, J. M. Porter, L. M. Prince, and M. G. Simpson, 97–103. Claremont: Rancho Santa Ana Botanic Garden.

Nesom, G. L. 2005. Infrageneric classification of *Liatris* (Asteraceae: Eupatorieae) *Sida* 21:1305–1321.

Nesom, G. L. 2006. *Liatris*. In *Flora of North America, Vol. 21*, ed. FNA Editorial Committee, 512–534. Oxford: Oxford University Press.

Ngamriabsakul, C., M. F. Newman, and Q. C. B. Cronk. 2000. Phylogeny and disjunction in *Roscoea* (Zingiberaceae). *Edinburgh J. Bot.* 57:39–61.

Ngamriabsakul, C., M. F. Newman, and Q. C. B. Cronk. 2003. The phylogeny of tribe *Zingibereae* (Zingiberaceae) based on ITS (nrDNA) and trnL–F (cpDNA) sequences. *Edinburgh J. Bot.* 60:483–507.

Nguyen, N. H., H. E. Driscoll, and C. D. Specht. 2008. A molecular phylogeny of the wild onions (*Allium*; Alliaceae) with a focus on the western North American center of diversity. *Mol. Phylogen. Evol.* 47:1157–1172.

Niehaus, T. F. 1971. A biosystematic study of the genus *Brodiaea* (Amaryllidaceae). *Univ. Calif. Publ. Bot.* 60:1–66.

Niehaus, T. F. 1980. The *Brodiaea* complex. *Four Seasons* 6:11–21.

Nieuwhof, M., L. W. D. van Raamsdonk, and J. P. van Eijk. 1990. Pigment composition of flowers of *Tulipa* species as a parameter for biosystematic research. *Biochem. Syst. Ecol.* 18:399–404.

Nishikawa, T., K. Okazaki, K. Arakawa, and T. Nagamine. 2001. Phylogenetic analysis of section *Sinomartagon* in genus *Lilium* using sequences of the internal transcribed spacer region in nuclear ribosomal DNA. *Breed. Sci.* 51:39–46.

Nishikawa, T., K. Okazaki, T. Uchino, K. Arakawa, and T. Nagamine. 1999. A molecular phylogeny of *Lilium* in the internal transcribed spacer region of nuclear ribosomal DNA. *J. Mol. Evol.* 49:238–249.

Nordenstam, B. 1998. Colchicaceae. In *The Families and Genera of Vascular Plants III. Flowering Plants, Monocotyledons: Lilianae (Except Orchidaceae)*, ed. K. Kubitzki, 175–185. Berlin: Springer.

Oberlander, K. C., E. Emshwiller, D. U. Bellstedt, and L. L. Dreyer. 2009. A model of bulb evolution in the eudicot genus *Oxalis* (Oxalidaceae). *Mol. Phylogen. Evol.* 51:54–63.

Olmstead, R. G., K.-J. Kim, R. K. Jansen, and S. J. Wagstaff. 2000. The phylogeny of the Asteridae sensu lato based on chloroplast *ndhF* gene sequences. *Mol. Phylogen. Evol.* 16:96–112.

Olmstead, R. G., H. Michaels, K. M. Scott, and J. D. Palmer. 1992. Monophyly of the Asteridae and identification of their major lineages inferred from DNA sequences of *rbcL*. *Ann. Missouri Bot. Gard.* 79:249–265.

Patterson, T. B., and T. J. Givnish. 2002. Phylogeny, concerted convergence, and phylogenetic niche conservatism in the core Liliales: Insights from *rbc*L and *ndh*F sequence data. *Evolution* 56:233–252.

Pax, F., and K. Hoffman. 1930. Amaryllidaceae. In *Die natürlichen Pflanzenfamilien, 2, aufl. 15a*, ed. H. Engler, 424–425. Leipzig: Wilhelm Engelmann.

Pax, F. A. 1888. Amaryllidaceae. In *Die natürlichen Pflanzenfamilien. II, 5. Abteilung*, eds. A. Engler, and K. Prantl, 119–121. Leipzig: Wilhelm Engelmann.

Petersen, G., O. Seberg, S. Thorsøe, T. Jørgensen, and B. Mathew. 2008. A phylogeny of the genus *Crocus* (Iridaceae) based on sequence data from five plastid regions. *Taxon* 57:487–499.

Petersen, O. G. 1889. *Musaceae, Zingiberaceae, Cannaceae, Marantaceae.* In *Die natürlichen Pflanzenfamilien, 2, aufl. 6*, eds. A. Engler, and K. Prantl, 1–43. Leipzig: Wilhelm Engelmann.

Pérez, R., P. Vargas, and J. Arroyo. 2003. Convergent evolution of flower polymorphism in *Narcissus* (Amaryllidaceae). *New Phytol.* 161:235–252.

Pfosser, M., and F. Speta. 1999. Phylogenetics of Hyacinthaceae based on plastid DNA sequences. *Ann. Missouri Bot. Gard.* 86:852–875.

Pfosser, M., and F. Speta. 2001. Bufadienolides and DNA sequences: On lumping and smashing of subfamily Urgineoideae (Hyacinthaceae). *Stapfiadiverse* 75:177–250.

Pfosser, M., W. Wetschnig, S. Ungar, and G. Prenner. 2003. Phylogenetic relationships among genera of Massonieae (Hyacinthaceae) inferred from plastid DNA and seed morphology *J. Plant Res.* 116:115–132.

Pires, J. C., M. F. Fay, W. S. Davis, L. Hufford, J. Rova, M. W. Chase, and K. J. Sytsma. 2001. Molecular and morphological phylogenetic analyses of *Themidaceae* (Asparagales). *Kew Bull.* 56:601–626.

Pires, J. C., I. J. Maureira, T. J. Givnish, K. J. Sytsma, O. Seberg, G. Petersen, J. I. Davis, D. W. Stevenson, P. J. Rudall, M. F. Fay, and M. W. Chase. 2006. Phylogeny, genome size, and chromosome evolution of Asparagales. In *Monocots: Comparative Biology and Evolution (Vol. 1 excluding Poales)*, eds. J. T. Columbus, E. A. Friar, J. M. Porter, L. M. Prince, and M. G. Simpson, 287–304. Claremont: Rancho Santa Ana Botanic Garden.

Pires, J. C., and K. J. Sytsma. 2002. A phylogenetic evaluation of a biosystematic framework: *Brodiaea* and related petaloid monocots (Themidaceae). *Am. J. Bot.* 89:1342–1359.

Prince, L. M. 2010. Phylogenetic relationships and species delimitation in *Canna* (Cannaceae). In *Diversity, Phylogeny, and Evolution in the Monocotyledons*, eds. O. Seberg, G. Petersen, A. S. Barfod, and J. I. Davis, 307–331. Aarhus, Denmark: Aarhus University Press.

Rahn, K. 1998. Alliaceae, Themidaceae. In *The Families and Genera of Vascular Plants III. Flowering Plants, Monocotyledons: Lilianae (Except Orchidaceae)*, ed. K. Kubitzki, 70–78, 436–441. Berlin: Springer.

Rangsiruji, A., M. F. Newman, and Q. C. B. Cronk. 2000a. Origin and relationships of *Alpinia galanga* (Zingiberaceae) based on molecular data. *Edinburgh J. Bot.* 57:9–37.

Rangsiruji, A., M. F. Newman, and Q. C. B. Cronk. 2000b. A study of the infrageneric classification of *Alpinia* (Zingiberaceae) based on the ITS region of nuclear rDNA and the trnL-F spacer of chloroplast DNA. In *Monocots: Systematics and Evolution*, eds. K. L. Wilson, and D. A. Morrison, 695–709. Collingwood, Australia: CSIRO Publ.

Raven, P. H., and D. I. Axelrod. 1974. Angiosperm biogeography and past continental movements. *Ann. Missouri Bot. Gard.* 61:539–673.

Ravenna, P. 2000. New or interesting Alstroemeriaceae—I. *Onira* 4(10):33–46.

Reeves, G., M. W. Chase, P. Goldblatt, P. Rudall, M. F. Fay, A. V. Cox, B. Lejeune, and T. Souza-Chies. 2001. Molecular systematics of Iridaceae: Evidence from four plastid DNA regions. *Am. J. Bot.* 88:2074–2087.

Regel, E. A. 1873. Enumeratio specierum hucusque cognitarum generis Tulipae. *Acta Horti Petrop.* 2:437–457.

Reid, C., and R. A. Dyer. 1984. *A Review of the Southern African Species of* Cyrtanthus. La Jolla, CA: American Plant Life Society.

Rendle, A. B. 1901. The bulbiform seeds of certain Amaryllidaceae. *J. Roy. Hort. Soc.* 26:89–96.

Renner, S. S., and L.-B. Zhang. 2004. Biogeography of the *Pistia* clade (Araceae): Based on chloroplast and mitochondrial DNA sequences and Bayesian divergence time inference. *Syst. Biol.* 53:422–432.

Renner, S. S., L.-B. Zhang, and J. Murata. 2004. A chloroplast phylogeny of *Arisaema* (Araceae) illustrates Tertiary floristic links between Asia, North America, and East Africa. *Am. J. Bot.* 91:881–888.

Rix, E. M. 1977. *Fritillaria* L. (Liliaceae) in Iran. *Iran. J. Bot.* 1:75–95.

Rix, E. M. 2001. *Fritillaria. A Revised Classification.* Edingburgh: The *Fritillaria* Group of the Alpine Garden Society, UK.

Rodionenko, G. I. 1987. *The Genus Iris L. (Questions of Morphology, Biology, Evolution and Systematics).* London: The British Iris Society. (English translation)

Rodriguez, A., and K. J. Sytsma. 2006. Phylogeny of the 'tiger-flower group' (Tigirideae: Iridaceae): Molecular and morphological evidence. In *Monocots: Comparative Biology and Evolution (Vol. 1 excluding Poales)*, eds. J. T. Columbus, E. A. Friar, J. M. Porter, L. M. Prince, and M. G. Simpson, 412–424. Claremont: Rancho Santa Ana Botanical Garden.

Rønsted, N., S. Law, H. Thornton, M. F. Fay, and M. W. Chase. 2005. Molecular phylogenetic evidence for the monophyly of *Fritillaria* and *Lilium* (Liliaceae; Liliales) and the infrageneric classification of *Fritillaria*. *Mol. Phylogen. Evol.* 35:509–527.

Rothwell, G. W., M. R. van Atta, H. W. Ballard Jr., and R. A. Stockey. 2004. Molecular phylogenetic relationships among Lemnaceae and Araceae using the chloroplast *trn*L-*trn*F spacer. *Mol. Phylogen. Evol.* 30:378–385.

Rudall, P. 1994. Anatomy and systematics of Iridaceae. *Bot. J. Linn. Soc.* 114:1–21.

Rudall, P. 2002a. Homologies of inferior ovaries and septal nectarines in monocotyledons. *Int. J. Plant Sci.* 163:261–276.

Rudall, P. J. 2002b. Unique floral structures and iterative evolutionary themes in Asparagales: Insights from a morphological cladistic analysis. *Bot. Rev.* 68:488–509.

Rudall, P. J., and R. M. Bateman. 2002. Roles of synorganisation, zygomorphy and heterotopy in floral evolution: The gynostemium and labellum of orchids and other lilioid monocots. *Biol. Rev. (London)* 77:403–441.

Rudall, P. J., and R. M. Bateman. 2004. Evolution of zygomorphy in monocot flowers: Iterative patterns and developmental constraints. *New Phytol.* 162:25–44.

Rudall, P. J., R. M. Bateman, M. F. Fay, and A. Eastman. 2002. Floral anatomy and systematics of Alliaceae with particular reference to *Gilliesia*, a presumed insect mimic with strongly zygomorphic flowers. *Am. J. Bot.* 89:1867–1883.

Rudall, P. J., M. W. Chase, D. F. Cutler, J. Rusby, and A. Y. de Bruijn. 1998. Anatomical and molecular systematics of Asteliaceae and Hypoxidaceae. *Bot. J. Linn. Soc.* 127:1–42.

Rudall, P. J., C. A. Furness, M. F. Fay, and M. W. Chase. 1997. Microsporogenesis and pollen sulcus type in Asparagales (Lilianae). *Can. J. Bot.* 75:408–430.

Rudall, P. J., and P. Goldblatt. 2001. Floral anatomy and systematic position of *Diplarrhena* (Iridaceae): A new tribe Diplarrheneae. *Ann. Bot. (Rome), n. s.* 2:59–66.

Rudall, P. J., K. L. Stobart, W. P. Hong, J. G. Conran, C. A. Furness, G. C. Kite, and M. W. Chase. 2000. Consider the lilies: Systematics of Liliales. In *Monocots: Systematics and evolution*, eds. K. L. Wilson, and D. A. Morrison, 347–359. Collingwood, Australia: CSIRO Publ.

Saar, D. E. 2002. *Dahlia neglecta* (Asteraceae: Coreopsideae), a new species from Sierra Madre Oriental, Mexico. *Sida* 20:593–596.

Saar, D. E., N. O. Polans, and P. D. Sørensen. 2003a. A phylogenetic analysis of the genus *Dahlia* (Asteraceae) based on internal and external transcribed spacer regions of nuclear ribosomal DNA. *Syst. Bot.* 28:627–639.

Saar, D. E., and P. D. Sørensen. 2000. *Dahlia parvibracteata* (Asteraceae, Coreopsideae), a new species from Guerrero, Mexico. *Novon* 10:407–410.

Saar, D. E., and P. D. Sørensen. 2005. *Dahlia sublignosa* (Asteraceae): A species in its own right. *Sida* 21:2161–2167.

Saar, D. E., P. D. Sørensen, and J. P. Hjerting. 2002. *Dahlia spectabilis* (Asteraceae, Coreopsideae), a new species from San Luis Potosí, Mexico. *Brittonia* 54:116–119.

Saar, D. E., P. D. Sørensen, and J. P. Hjerting. 2003b. *Dahlia campanulata* and *D. cuspidata* (Asteraceae, Coreopsideae): Two new species from Mexico. *Acta Bot. Mex.* 64:19–24.

Saiki, R. K., D. H. Gelfand, S. Stoffel, S. J. Scharf, R. Higuchi, G. T. Horn, K. B. Mullis, and H. A. Erlich. 1988. Primer-directed enzymatic amplification of DNA with a thermostable DNA polymerase. *Science* 239:487–491.

Samoylov, A., N. Friesen, S. Pollner, and P. Hanelt. 1999. Use of chloroplast DNA polymorphisms for the phylogenetic study of *Allium* subgenera *Amerallium* and subgenus *Bromatorrhiza* (Alliaceae) II. *Fed. Rep.* 110:103–109.

Samoylov, A., M. Klaas, and P. Hanelt. 1995. Use of chloroplast polymorphisms for the phylogenetic study of subgenera *Amerallium* and *Bromatorrhiza* (genus *Allium*). *Fed. Rep.* 106:161–167.

Sanderson, M. J. 1989. Confidence limits on phylogenies: The bootstrap revisited. *Cladistics* 5:113–129.

Schlessman, M. A. 1988. Gender diphasy ("sex choice"). In *Plant Reproductive Ecology: Patterns and Strategies*, eds. J. Lovett Doust, and L. Lovett Doust, 139–153. New York: Oxford University Press.

Schnarf, K. 1929. Die Ebryologie der Liliaceae und ihre systematische Bedeutung. *Sitzungsber. Kais. Akad. Wiss. Wien, Math.-Naturw.* 38(1):69–92.

Schumann, K. 1904. Zingiberaceae. In *Das Pflanzenreich IV 46 heft 20*, ed. A. Engler, 1–458. Leipzig: Englemann.

Sherff, E. E. 1955. *Dahlia*. In *North Amer. Flora Part 2*, eds. E. E. Sherff, and E. J. Alexander, 45–59. New York: New York Botanical Garden.

Shevock, J. 1984. Redescription and distribution of *Muilla coronata* (Liliaceae). *Aliso* 10:621–627.

Simpson, M. G., and P. J. Rudall. 1998. Tecophilaeaceae. In *The Families and Genera of Vascular Plants III. Flowering Plants, Monocotyledons: Lilianae (Except Orchidaceae)*, ed. K. Kubitzki, 429–436. Berlin: Springer.

Singh, Y., A. E. van Wyk, and H. Baijnath. 1996. Taxonomic notes on the genus *Zantedeschia* Spreng. (Araceae) in southern Africa. *S. Afr. J. Bot.* 62:321–324.

Snijman, D. A., and R. H. Archer. 2003. Amaryllidaceae. In *Plants of Southern Africa: An Annotated Checklist (Strelitzia 14)*, eds. G. Germishuizen, and N. L. Meyer, 957–967. Pretoria: National Botanical Institute.

Snijman, D. A., and H. P. Linder. 1996. Phylogenetic relationships, seed characters, and dispersal system evolution in Amaryllideae (Amaryllidaceae). *Ann. Missouri Bot. Gard.* 83:362–386.

Snijman, D. A., and A. W. Meerow. 2010. Floral and macroecological evolution within *Cyrtanthus* (Amaryllidaceae): Inferences from combined analyses of plastid *ndhF* and nrDNA ITS sequences. *S. Afr. J. Bot.* 76:217–238.

Snoeijer, W. 2004. Agapanthus: *a Revision of the Genus*. Portland, OR: Timber Press.

Soltis, D. E., S. A. Smith, N. Cellinese, K. J. Wurdack, D. C. Tank, S. F. Brockington, N. F. Refulio-Rodriguez, J. B. Walker, M. J. Moore, B. S. Carlsward, C. D. Bell, M. Latvis, S. Crawley, C. Black, D. Diouf, Z. Xi, C. A. Rushworth, M. A. Gitzendanner, K. J. Sytsma, Y.-L. Qiu, K. W. Hilu, C. C. Davis, M. J. Sanderson, R. S. Beaman, R. G. Olmstead, W. S. Judd, M. J. Donoghue, and P. S. Soltis. 2011. Angiosperm phylogeny: 17 genes, 640 taxa. *Am. J. Bot.* 98:704–730.

Soltis, D. E., P. S. Soltis, M. W. Chase, M. E. Mort, D. C. Albach, M. Zanis, V. Savolainen, W. H. Hahn, S. B. Hoot, M. F. Fay, M. Axtell, S. M. Swensen, L. M. Prince, W. John Kress, K. C. Nixon, and J. S. Farris. 2000. Angiosperm phylogeny inferred from 18S rDNA, *rbcL*, and *atpB* sequences. *Bot. J. Linn. Soc.* 133:381–461.

Soltis, D. E., P. S. Soltis, P. K. Endress, and M. W. Chase. 2005. *Phylogeny and Evolution of Angiosperms*. Sunderland, MA: Sinauer Associates.

Sørensen, P. D. 1969. Revision of the genus *Dahlia* (Compositae, Heliantheae—Coreopsidinae). *Rhodora* 71:309–416.

Souza-Chies, T. T., G. Bittar, S. Nadot, L. Carter, E. Besin, and B. Lejeune. 1997. Phylogenetic analysis of Iridaceae with parsimony and distance methods using the plastid gene *rps*4. *Plant Syst. Evol.* 204:109–123.

Specht, C. D. 2006. Systematics and evolution of the tropical monocot family Costaceae (Zingiberales): A multiple dataset approach. *Syst. Bot.* 31:89–106.

Specht, C. D., W. J. Kress, D. W. Stevenson, and R. DeSalle. 2001. A molecular phylogeny of Costaceae (Zingiberales). *Mol. Phylogen. Evol.* 21:333–345.

Specht, C. D., and D. W. M. Stevenson. 2006. A new phylogeny-based generic classification of Costaceae (Zingiberales). *Taxon* 55:153–163.

Speta, F. 1998a. Hyacinthaceae. In *The Families and Genera of Vascular Plants III. Flowering Plants, Monocotyledons: Lilianae (Except Orchidaceae)*, ed. K. Kubitzki, 261–285. Berlin: Springer.

Speta, F. 1998b. Systematische Analyse der Gattung *Scilla* L. (Hyacinthaceae). *Phyton (Horn, Austria)* 38:1–141.

Speta, F. 2001. Die echte und die falsche Meerzwiebel: *Charybdis* Speta und *Stellarioides* Medicus (Hyacinthaceae), mit Neubeschreibungen und Neukombinationen im Anhang. *Stapfia* 75:139–176.

Stedje, B. 1998. Phylogenetic relationships and generic delimitation of sub-Saharan *Scilla* (Hyacinthaceae) and allied African genera as inferred from morphological and DNA sequence data. *Plant Syst. Evol.* 211:1–11.

Stedje, B. 2001a. Generic delimitation of Hyacinthaceae, with special emphasis on sub-Saharan genera. *Syst. Geog. Plants* 71:449–454.

Stedje, B. 2001b. The generic delimitation within Hyacinthaceae, a comment on works by F. Speta. *Bothalia* 31:192–195.

Stern, F. C. 1956. *Snowdrops and Snowflakes. A Study of the Genera Galanthus and Leucojum*. London: The Royal Horticultural Society.

Stork, A. 1984. *Tulipes sauvages et cultivés. Série documentaire 13*. Geneva: Conservatoire et Jardin Botaniques.

Strydom, A., R. Kleynhans, and J. J. Spies. 2007. Chromosome studies on African plants. 20. Karyotypes of some *Cyrtanthus* species. *Bothalia* 37:103–108.

Tam, S.-M., P. C. Boyce, T. M. Upson, D. Barabé, A. Bruneau, F. Forest, and J. S. Parker. 2004. Intergeneric and infrafamilial phylogeny of subfamily Monsteroideae (Araceae) revealed by chloroplast *trnL-F* sequences. *Am. J. Bot.* 91:490–498.

Tamura, M. N. 1998. Calochortaceae, Liliaceae, Melanthiaceae, Trilliaceae. In *The Families and Genera of Vascular Plants III. Flowering Plants, Monocotyledons: Lilianae (Except Orchidaceae)*, ed. K. Kubitzki, 164–172, 343–353, 369–380, 444–452. Berlin: Springer.

Tanaka, N. 2001. Taxonomic revision of the family Cannaceae in the New World and Asia. *Makinoa New Series* 1:1–74.

Taylor, J. J. 1976. A reclassification of *Iris* species bearing arillate seeds. *Proc. Biol. Soc. Washington* 89:411–420.

Tebbitt, M. C. 2005. *Begonias: Cultivation, Identification, and Natural History*. Portland, OR: Timber Press.

Tillie, N., M. W. Chase, and T. Hall. 2001. Molecular studies in the genus *Iris* L.: A preliminary study. *Ann. Bot.* 1:105–112.

Tjaden, W. 1981. *Amaryllis belladonna* Linn.—An up-to-date summary. *Plant Life* 37:21–26.

Tjaden, W. L. 1979. *Amaryllis belladonna* and the Guernsey lily: an overlooked clue. *J. Soc. Bibliog. Nat. Hist.* 9:251–256.

Traub, H. P. 1954. Typification of *Amaryllis belladonna* L. *Taxon* 3:102–111.

Traub, H. P. 1963. *Genera of the Amaryllidaceae*. La Jolla, CA: American Plant Life Society.

Traub, H. P. 1983. The lectotypification of *Amaryllis belladonna* L. (1753). *Taxon* 32:253–267.

Traub, H. P., and H. N. Moldenke. 1949. *Amaryllidaceae: Tribe Amarylleae*. Stanford: Amaryllis Society.

Turner, B. L. 1993. *Jaimehintonia* (Amaryllidaceae: Allieae), a new genus from northeastern Mexico. *Novon* 3:86–88.

Turrill, W. B., and J. R. Sealy. 1980. Studies in the genus *Fritillaria* (Liliaceae). *Hookers Icones Plantarum* 39:1–280.

Uphoff, J. C. T. 1938. The history of nomenclature—*Amaryllis belladonna* (Linn.) Herb., and *Hippeastrum* (Herb.). *Herbertia* 5:100–111.

Uphoff, J. C. T. 1939. Critical review of Sealy's "*Amaryllis and Hippeastrum*". *Herbertia* 6:163–166.

Van Eijk, J. P., L. W. D. van Raamsdonk, W. Eikelboom, and R. J. Bino. 1991. Interspecific crosses between *Tulipa gesneriana* cultivars and wild *Tulipa* species: A survey. *Sex. Plant Reprod.* 4:1–5.

Van Raamsdonk, L. W. D., and T. de Vries. 1992. Biosystematic studies in *Tulipa* sect. *Eriostemones* (Liliaceae). *Plant Syst. Evol.* 179:27–41.

Van Raamsdonk, L. W. D., and T. de Vries. 1995. Species relationships and taxonomy in *Tulipa* subg. *Tulipa* (Liliaceae). *Plant Syst. Evol.* 195:13–44.

Van Raamsdonk, L. W. D., W. Eikelboom, T. de Vries, and Th. P. Straathof. 1997. The systematics of the genus *Tulipa* L. *Acta Hort.* 430:821–828.

Vinnersten, A., and K. Bremer. 2001. Age and biogeography of major clades in Liliales. *Am. J. Bot.* 88:1695–1703.

Vinnersten, A. and S. Larsson. 2010. Colchicine is still a chemical marker for the expanded Colchicaceae. *Biochem. Syst. Ecol.* 38:1193–1198.

Vinnersten, A., and J. Manning. 2007. A new classification of Colchicaceae. *Taxon* 56:171–178.

Vinnersten, A., and G. Reeves. 2003. Phylogenetic relationships within Colchicaceae. *Am. J. Bot.* 90:1455–1462.

Wang, H., M. J. Moore, P. S. Soltis, C. D. Bell, S. F. Brockington, R. Alexandre, C. C. Davis, M. Latvis, S. R. Manchester, and D. E. Soltis. 2009. Rosid radiation and the rapid rise of angiosperm-dominated forests. *Proc. Natl. Acad. Sci. USA* 106:3853–3858.

Webb, D. A. 1980. *Narcissus* L. In *Flora Europaea 5*, eds. T. G. Tutin, V. H. Heywood, N. A. Burges, D. M. Moore, D. H. Valentine, and S. M. Walters, 78–84. Cambridge: Cambridge University Press.

Weichhardt-Kulessa, K., T. Börner, J. Schmitz, U. Müller-Doblies, and D. Müller-Doblies. 2000. Controversial taxonomy of Strumariinae (Amaryllidaceae) investigated by nuclear rDNA (ITS) sequences. *Plant Syst. Evol.* 223:1–13.

White, S. D., A. C. Sanders, and M. D. Wilcox. 1996. Noteworthy collections. *Madroño* 43:334–338.

Wikström, N., V. Savolainen, and M. W. Chase. 2001. Evolution of the angiosperms: Calibrating the family tree. *Proc. R. Soc. Lond. B* 268:2211–2220.

Wiley, E. O. 1981. *Phylogenetics, the Theory and Practice of Phylogenetic Systematics*. New York: John Wiley and Sons.

Wilkin, P. 1997. *Leontochir ovallei. Kew Magazine* 14(1):7–12, pl. 308.

Williams, K. J., W. J. Kress, and P. S. Manos. 2004. The phylogeny, evolution, and classification of the genus *Globba* and tribe Globbeae (Zingiberaceae): appendages do matter. *Am. J. Bot.* 91:100–114.

Williams, R. 2000. Hyacinthaceae. In *Seed Plants of Southern Africa: Families and Genera. Strelitzia 10*, ed. O. A. Leistner, 610–619. Pretoria: National Botanical Institute.

Wilsenach, R. 1963. A cytotaxonomic study of the genus *Cyrtanthus. Cytologia* 28:170–180.

Wilson, C. A. 2004. Phylogeny of *Iris* based on chloroplast *matK* gene and *trnK* intron sequence data. *Mol. Phylogen. Evol.* 33:402–412.

Wilson, C. A. 2006. Patterns of evolution in characters that define *Iris* subgenera and sections. In *Monocots: Comparative Biology and Evolution (Vol. 1, excluding Poales)*, eds. J. T. Columbus, E. A. Friar, C. W. Hamilton, J. M. Porter, L. M. Prince, and M. G. Simpson, 425–433. Claremont: Rancho Santa Ana Botanic Garden.

Wilson, C. A. 2009. Phylogenetic relationships among the recognized series in *Iris* section *Limniris. Syst. Bot.* 34:277–284.

Wilson, C. A. 2011. Subgeneric classification in *Iris* re-examined using chloroplast sequence data. *Taxon* 60:27–35.

Wilson, E. H. 1925. *The lilies of Eastern Asia*. London: Dulau and Company.

Wong, S. Y., P. C. Boyce, A. S. bin Othman, and C. P. Leaw. 2010. Molecular phylogeny of tribe Schismatoglottideae (Araceae) based on two plastid markers and recognition of a new tribe, Philonotieae, from the neotropics. *Taxon* 59:117–124.

Wood, T. H., W. M. Whitten, and N. H. Williams. 2000. Phylogeny of *Hedychium* and related genera (Zingiberaceae) based on ITS sequence data. *Edinburgh J. Bot.* 57:261–270.

Wunderlich, R. 1937. Zur vergleichenden Embryologie der Liliaceae-Scilloideae. *Flora* 132:48–90.

Zöllner, O., and L. Arriagada. 1998. The tribe Gilliesieae (Alliaceae) in Chile. *Herbertia* 53:104–107.

Zomlefer, W. B. 1996. The Trilliaceae in the southeastern United States. *Harvard Papers in Botany* 1(9):91–120.

Zomlefer, W. B. 1997a. The genera of Melanthiaceae in the southeastern United States. *Harvard Papers in Botany* 2:133–177.

Zomlefer, W. B. 1997b. The genera of Tofieldiaceae in the southeastern United States. *Harvard Papers in Botany* 2:179–194.

Zomlefer, W. B. 1997c. The genera of Nartheciaceae in the southeastern United States. *Harvard Papers in Botany* 2:195–211.

Zomlefer, W. B., N. H. Williams, W. M. Whitten, and W. S. Judd. 2001. Generic circumscription and relationships in the tribe Melanthieae (Liliales, Melanthiaceae), with emphasis on *Zigadenus*: evidence from ITS and *trnL-F* sequence data. *Am. J. Bot.* 88:1657–1669.

Zonneveld, B. J. M. 2009. The systematic value of nuclear genome size for "all" species of *Tulipa* L. (Liliaceae). *Plant Syst. Evol.* 281:217–245.

3 Biodiversity of Geophytes
Phytogeography, Morphology, and Survival Strategies

Rina Kamenetsky

CONTENTS

I. INTRODUCTION

The term "geophyte" was derived from the Greek (from *ge*, earth, land; *phyton*, plant) and was coined by Raunkiær (1934) in his famous system of plant life forms. His categories were characterized by the location of the regenerative buds and the parts shed during the seasons unfavorable to growth (Figure 3.1). In this system, cryptophytes were classified as perennial plants that survive the unfavorable seasons not only by seed but also by specialized underground storage organs. Cryptophytes are divided into three groups: (1) geophytes, with renewal buds resting in dry ground, for example, crocus and tulip; (2) helophytes, resting in marshy grounds, for example, reedmace (*Typha*, Typhaceae) and marsh-marigold (*Caltha*, Ranunculaceae); and (3) hydrophytes, resting by being submerged under water, for example, water lily (*Nymphaea*, Nymphaeaceae) and frogbit (*Hydrocharis*, Hydrocharitaceae) (Figure 3.1).

Geophytic species are found both among monocotyledonous and eudicotyledonous taxa. In general, storage organs facilitate survival during periods of unfavorable weather conditions such as high or low temperatures, drought, or improper light levels. A wide range of natural habitats worldwide are characterized by a short annual period of favorable growth conditions, for example, strong seasonal patterns of temperature in arctic–alpine habitats, the pattern of light

57

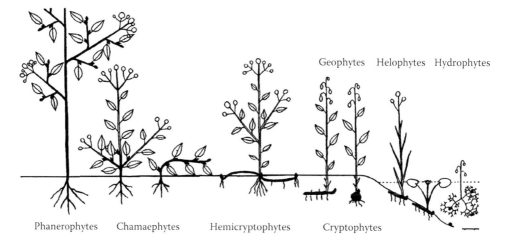

Geophytes   Helophytes   Hydrophytes

Phanerophytes   Chamaephytes   Hemicryptophytes   Cryptophytes

FIGURE 3.1   The Raunkiær system for categorizing plants by life form. (Adapted from Raunkiær, C. 1934. *The Life Forms of Plants and Statistical Geography.* Oxford: Oxford University Press.)

availability in deciduous forests and the pattern of water availability in Mediterranean and (semi) desert habitats (Werger and Huber 2006). Such environments necessitate plant life strategies that make optimal use of the narrow window in time available for growth and reproduction, and allow survival during the relatively long and unfavorable periods. Underground organs are reserves of water, storage materials, and nutrients such as carbohydrates, proteins, and mineral salts. The phase of geophyte development during unfavorable conditions is often referred to as a dormancy period or resting stage; however, such terms are misleading (see also Chapter 9). The storage organ continues to change and constantly senses its environment, and is never physiologically dormant even when aerial growth is halted. This period is also referred to as "intrabulb development" (Kamenetsky 1994). The aboveground organs of the plant (usually annual deciduous shoots or leaves) typically die off during the dry periods of the year or in the winter season, leaving only the storage organs in the soil. When the environmental conditions are appropriate, new foliage emerges aboveground using the food reserves in the underground organs. Thus, in their natural habitats, geophytes are capable of perennial life cycles. Environmental signals, including temperature, moisture, and photoperiod, affect dormancy and flowering of geophytes (Dole 2003; Erwin 2006; Chapters 8 and 9).

This chapter reviews classic and newly acquired data on geographical distribution, ecological adaptations and morphological features of geophytes. These themes have attracted the interest of naturalists and researchers since ancient times, but have become especially valuable with the rapidly changing environment, requirements for ecological restoration, and fast development of ornamental horticulture industries around the world.

II.   NATURAL GROWTH HABIT

In addition to phylogenetic and taxonomic systems (Chase 2004; Chapter 2), geophytes are classified by the morphological structure of the underground organs (e.g., bulbs, corms, rhizomes, or tubers), as well as by their ecological and environmental traits. Plant evolution in climatic areas with marked seasonal changes has led to their adaptation to periods of high and low temperatures and drought and to significant changes in the plants' morphology and annual developmental cycles. Although the direct ancestors of modern geophytes are not known, it was proposed that the geophytes evolved in the tropic and subtropic zones where rainfalls become seasonal. Adaptation to different environments due to an extension of an original habitat or to climate change resulted in the development of new morphological traits. This transition to xero-mesophytic and mesophytic habitats occurred independently in many taxonomic groups, resulting in a large diversity of geophytic morphological structures and life cycles.

In tropical climates, the annual cycle of geophytes is not affected by external factors (e.g., dry periods), but some periodicity occurs owing to suppression of the lateral buds by apical dominance of the terminal shoot and flower. Should the environment change (including dry periods), plants enter resting periods. During this period, terminal buds do not develop new organs, or die and then are replaced by the lateral buds (Rees 1972). One of the reasons could be the death of the root system and failure of the new buds to establish new roots until moist conditions improved. The adaptation for this process includes developing adequate storage tissues to allow survival during unfavorable periods. In cold regions, geophytes die down below the soil level at the onset of winter, since their aboveground parts are not able to produce secondary protective tissues against freezing temperatures. Presumably, after migration from the tropics, diverse adaptations allowed re-growth only after the end of the cold period in the regions with temperate climate, or development of summer dormancy mechanisms, essential for survival in the areas with Mediterranean climate. In addition to protection of renewing buds in response to severe environments, plants mitigate the risk of bud mortality by multiplying buds through clonal growth. Therefore, clonality is an important feature of geophytes, and is particularly widespread in harsh environments such as high latitudes (Klimeš et al. 1997).

According to the life cycle, flowering, and dormancy patterns, the natural growth habit of geophytes can be classified in several ways:

1. *Annual or Perennial Geophytic Organs.* In a strict sense, all geophyte species are botanically perennials and their life cycle from seed to senescence is longer than one (annual) or two (biennial) years. At the same time, geophytic organs of some species are replaced by a new storage unit every year. These annual geophytic organs are found in *Tulipa*, *Gladiolus*, and *Freesia*. Species with perennial storage organs containing parts assembled over several years include *Cyclamen*, *Hyacinthus*, *Narcissus*, *Nerine*, or *Hippeastrum* (Halevy 1990).

2. *Synanthous or Hysteranthous.* Many geophytes are synanthous, that is, leaves are differentiated before the floral parts, and aboveground foliage is produced before flowering, for example, *Hyacinthus*, *Lilium*, *Narcissus*, and *Tulipa*. The synanthous pattern of flowering occurs in spring (De Hertogh and Le Nard 1993). Other geophytes are known as hysteranthous. They flower at the end of summer (September–October) without bearing leaves, and the leaves appear only after the first rains (November–December), after the flowers have already wilted (Dafni et al. 1981a,b). Examples of hysteranthous geophytes are species of *Colchicum*, *Sternbergia*, *Amaryllis*, *Urginea*, *Scilla*, and others. In hysteranthous plants, the photosynthesis rate is very low from flower emergence until anthesis. The bulb reserves are sufficient to allow flower stem elongation and flowering. In contrast, in proteranthous growth the foliage dies down before the flower is produced and photosynthesis takes place before flowering (e.g., *Boophone haemanthoides*, Du Plessis and Duncan 1989).

3. *Evergreen or Deciduous.* This trait depends on climatic conditions in natural habitats. Tropical perennial geophytes do not experience natural growth respites, and are practically evergreen. *Clivia* and some *Agapanthus* species are examples of this type (Halevy 1990). Geophytes, native to temperate and Mediterranean zones with thermo- and photoperiodism, are mostly deciduous. Their growth period is adapted to wet and mild winters (Mediterranean climate) or warm summers (temperate climate), or intermediate seasons—spring and autumn (continental climate).

III. BIODIVERSITY AND PHYTOGEOGRAPHY OF GEOPHYTES IN NATURAL HABITATS

Modern climate classification, based on the work of Wladimir Köppen published in 1884, combines the seasonality of precipitation with the concept that native vegetation is the best expression of climate (Peel et al. 2007). In this system, five primary climatic types—tropical, dry, mild mid-latitude, cold mid-latitude, and polar—are further divided into secondary classifications such as rain forest, monsoon, tropical savanna, humid subtropical, humid continental, oceanic climate, Mediterranean climate, steppe, subarctic climate, tundra, polar ice cap, and desert. While geophyte species constitute ca. 3% of the world flora (Fragman and Shmida 1997), their occurrence in the floras of different climatic areas varies from ca. 40% in the Cape Mediterranean zone of South Africa (Proches et al. 2006) to only a solitary species in the subarctic. In general, several distinct bio-morphological types can be categorized by their occurrence in different climatic zones.

A. TROPICAL AND SUBTROPICAL ZONES

In geophytes originating from these regions, several growth cycles take place and overlap in one plant. Although the growth rate of these species might change during their annual cycle, they have no specific photoperiod or temperature requirements for growth pauses or resumptions. Well-known examples of evergreen geophytes are species of *Clivia*, *Hippeastrum*, *Crinum*, and *Zephyranthes* (Hartsema 1961; Rees 1972, 1992; Du Plessis and Duncan 1989).

The genus *Hippeastrum* is found in tropical and subtropical America from Mexico and the West Indies to Bolivia and southern Brazil (Rees 1972). The *Hippeastrum*'s bulb is composed of leaf bases only, branching sympodially, while the inflorescence differentiates in the apical meristem. The nonperiodic bulb produces new leaf primordia once per month, and inflorescence is differentiated after every fourth leaf. In a uniform climate, it is possible that flowering occurs several times a year, but seasonal pattern is possible under varying temperature. Similarly, *Ornithogalum saundersiae* was found to be a perennial synanthous geophyte without a natural dormant period. This species was able to grow and flower all the year round in Kenya, without any temperature treatment or environmental control (Kariuki and Kako 1999). Another example is the genus *Griffinia*, which consists of about a dozen species of rainforest or dry forest understory bulbous geophytes, endemic to eastern Brazil (Meerow et al. 2002). Most species of *Griffinia* grow in the lush, wet understory of tropical forests, are adapted to low light levels and are at least semievergreen.

The genus *Crinum* is the only pantropical genus of the Amaryllidaceae, with species occurring in Africa, America, Asia, and Australia. *Crinum* seeds are adapted for oceanic dispersal, which probably engendered its broad distribution (Arroyo and Cutler 1984). Nuclear rDNA ITS sequences (Meerow et al. 2003) support a southern African origin of the genus, and indicate three major waves of radiation. It was proposed that Asian and Malagasy *Crinum* species are phylogenetically related, while Sino-Himalayan and Australasian species present two different dispersals. A monophyletic American clade denotes a single dispersal event into the Western Hemisphere, and the American species are allied with tropical and North African species.

B. TEMPERATE ZONES

Species from temperate climates undergo external growth recess as a reaction to short days and low temperatures in autumn. Woodland species in temperate deciduous forests present the most variable group of this bio-morphological type (Whigham 2004). Among woodland herbs, Uemura (1994) identified two groups of evergreen and three groups of deciduous species, as well as wintergreen and achlorophyllous species. The deciduous groups, which accounted for most species, were based on whether the species had heteroptic (i.e., plants with both summer green and overwintering leaves), summer green, or spring green leaves. Kawano (1985) recognized a similar range of phenological patterns in Japan and suggested that the diversity of phenological patterns is a result of adaptive response to woodland habitats with conspicuous periodicity in various physical and biotic regimes. Based on leaf phenology of woodland species, Givnish (1987) categorized spring ephemerals, early summer, late summer, wintergreen, and evergreen plants. In spring ephemerals foliage elongates between 5 and 15 cm above the ground, thus permitting more efficient temperature regulation and efficient use of light during the short period between emergence and development of the tree canopy. Early summer species develop leaves from 10 to 160 cm above the ground and display a variety of umbrella-like structures that minimize shading and maximize light capture with the lowest possible structural costs. Late summer species had indeterminate growth and produce leaves at a greater height (40–160 cm). In evergreen and wintergreen species leaves are close to the ground, resulting in enhanced winter photosynthesis (Minoletti and Boerner 1993; Tissue et al. 1995). The majority of woodland herbs are perennial, and most biomass and nutrients are stored in roots, rhizomes, bulbs, and corms (Kawano 1975; Muller 1979; Kawano et al. 1992). Nutrient cycling is related to patterns of resource allocation. For example, in *Allium tricoccum*, carbon and nutrients, stored underground, are allocated to aboveground biomass early in the growing season, peaking in leaves and reproductive structures in the spring or early summer. As the growing season progresses, nutrients accumulate in underground structures, declining in aboveground organs concurrently with leaves senesce and development of fruits and seeds (Nault and Gagnon 1988). Other *Allium* species from the temperate zones form new leaves throughout the year, and the low winter temperatures only slow down these processes (Kamenetsky and Rabinowitch 2006). During the winter dormancy, however, the apical meristem produces only leaf primordia, while a long photoperiod is required for the flowering of these species.

Spring ephemerals of deciduous forests appear shortly after snow melt and senesce, concurrently with closure of the forest canopy. During this short growth period, they take advantage not only of the high light conditions, but also of the low-temperature regime that favors growth in these species (Lapointe 2001). *Anemone nemorosa* (wood anemone), *Sanguinaria canadensis* (bloodroot), and *Trillium* are examples of this type. Spring ephemerals not only tolerate cold temperatures, but also they actually grow better in a low-temperature regime and produce a larger perennial organ (Nault and Gagnon 1993; Lapointe and Lerat 2006). Air and soil temperatures differently affect growth in spring ephemerals. Thus, low air temperatures enhanced the plant size of *Crocus vernus*, while soil temperature has a greater impact on plant development than air temperature. In the higher-temperature regime, corm size reached a maximum very early, well ahead of the first visual sign of leaf senescence, while the corm continued to grow for much longer in the cooler temperature regime even during leaf senescence (Badri et al. 2007). It was suggested that leaf senescence in spring ephemerals is induced by sink limitation once the perennial organ has been filled with carbohydrates (Lapointe 2001). After flowering, the leaves remain green for another month. Once the tree canopies fill out and reduce the available light, the plants enter summer dormancy. During the summer, the trees use most of the available moisture, thus drying out the surrounding soil.

C. CONTINENTAL THERMOPERIODIC ZONES

This type of climate is found mainly in the Irano-Turanian region, extending eastwards from Anatolia (Asiatic Turkey) to include most of Syria, Iran and north-east Afghanistan, northern Iraq and parts of the Middle East. Outside south-west Asia it extends northwards into Central Asia and eastward to the Tien Shan and Altai Mountain chains. This region is very diverse topographically, including deserts, low plains, high plateaus, and high mountains. Geophytes typical of these regions cease underground growth at relatively high temperatures at the beginning of summer, but require a period of low temperatures during the winter for leaf and stem elongation and successful flowering in spring. The length of their aboveground development is ~3–5 months. Species of *Tulipa*, *Gagea*, *Allium*, and *Eremurus* are good examples of this bio-morphological type. Although tulips experience both summer and winter dormancy, meristems remain active, and are able to produce vegetative and reproductive organs (Botschantzeva 1982; Le Nard and De Hertogh 1993). The geophytic genera of Alliaceae, Liliaceae, Asphodelaceae, and Iridaceae are largely represented in the regions with continental climates. For example, the species spectrum of the South-Western Tien-Shan mountains in Central Asia contains ca. 12% of geophytes (Tojibaev 2010). The primary centers of diversity of many species and forms are located in these regions, which are characterized by a high proportion of geophyte endemism. For example, the Fergana valley in Central Asia is rich in many endemic geophytes, e.g., *Allium alaicum*, *A. viridiflorum*, *Eremurus zenaidae*, *Tulipa ferganica*, *T. rosea*, and *Ungernia ferganica*, to mention only a few species mainly restricted to the lower mountain belt (Hall 2007).

D. SUBALPINE AND ALPINE ZONES

These climatic zones are located above tree line in the high mountains around the world, for example, the Australian Alps, Himalayas, Hengduan Mountains, French Alps, Sierra Nevada, and Rocky Mountains. The species spectra in these zones depend on the location of the zone in the Earth, but in general geophytes are a relatively common element of alpine and subalpine vegetation (Klimeš 2003). Geophytes represent 4.2% of the vascular plants from altitudes of 4180–6000 m a.s.l. in E Ladakh (W Himalayas, NW India) (Klimeš 2003). Bulbous and rhizomatous species are common in the high mountains of Asia and Caucasus, and constitute ca. 6% of the Iranian alpine flora. Some geophytic species are restricted to the alpine area, for example, *Allium capitellatum*, *Gagea alexeenkoana*, *G. glacialis*, *Tulipa humilis*, and *Iris barnumae* (Noroozi et al. 2008). Vegetation of the wetland habitats in the dry slopes of the Alborz Mountains in Iran (1486–3730 m a.s.l.) contains 32.2% of geophytes, with a significant prevalence of the rhizomatous species (27.2% of all species) (Naqinezhad et al. 2010).

From the viewpoint of resource allocation, alpine perennial plants may be expected to have a deeper dormancy than temperate plants because of severe low temperatures during the unfavorable season. It is known that some alpine perennial plants have the ability to tolerate extremely low temperatures in winter and also to resist moderate frosts, even in summer (Körner 1999). In comparison with the other life forms inhabiting the alpine zone (mostly chamaephytes and hemicriptophytes), winter buds of geophytes are the least hardy to freezing. Although the buds of geophytes are protected from severe low temperatures below the ground, breaking the dormancy and activating the buds in autumn would lead to serious frost damage. Therefore, geophytes avoid the activation of buds by exhibiting the deepest dormancy (Yoshie 2008).

Alpine and subalpine geophytes flower in late spring or early summer, and the blooming period is determined by snowmelt and variation in winter snowpack accumulation. In some years, fruit set is diminished or prevented entirely by killing frosts. In the subalpine population of *Erythronium grandiflorum* (Liliaceae) in western Colorado (USA), spatio-temporal mosaic of blooming was affected by drift patterns and formation of early melt holes in spring, with effects on both population–genetic structure and the exposure of different patches to different conditions of weather and pollinator availability (Harder et al. 1985; Thomson 2010).

E. MEDITERRANEAN AND SEMIARID ZONES

Although geophytes are widespread around the world in many habitats, nowhere are they more diverse and abundant than in the five mediterranean-climate ecosystems of the world: the Mediterranean Basin of Eurasia and Northern Africa, much of California, parts of Western and South Australia, southwestern South Africa, isolated sections of Central Asia and parts of central Chile (Doutt 1994; Rundel 1996). Prior to a hot and dry summer, geophytes cease aboveground growth and enter summer dormancy. In autumn, water availability and decreasing temperatures affect dormancy release; the plants sprout and develop leaves and inflorescences during the mild winter. Most geophytes, inhabiting Mediterranean and semiarid areas, do not require cold induction for floral development and stalk elongation (Widmer and Lyons 1985; Kamenetsky and Fritsch 2002).

The Cape Mediterranean zone of South Africa is generally the most abundant in geophytes, constituting up to 40% in some regional floras (Proche\u015f et al. 2006). A large proportion of the ornamental geophytes cultivated worldwide originated from the Cape, such as the genera *Gladiolus*, *Clivia*, and *Freesia*. Indeed the diversity of bulbous, cormous, and tuberous plants growing in the winter–rainfall region of southern Africa is exceptional (Parsons and Hopper 2003; Proche\u015f et al. 2005). Geophyte species diversity and bulb size may be dependent on climatic factors, and especially on rainfall quantity and reliability. Although the Mediterranean climate is characterized by most precipitation occurring in the cooler months, the proportion of winter rainfall varies widely across the Cape region. The amount of winter precipitation also varies, with the north-western areas being extremely arid (<100 mm/year), while parts of the southwest and south have annual totals in excess of 2000 mm. In contrast to the low proportion of annuals, the Cape Region has perhaps the highest proportion of geophytes of any part of the world, and is four to five times richer in this life form than is documented for other Mediterranean floras. At least 1550 species, over 17% of the total, have specialized underground organs including bulbs, corms, rhizomes, or tubers and are seasonally dormant. The overwhelming number of geophytes is monocots, with over 1300 geophytic species, 662 of which belong to one family, Iridaceae. Most of these geophytes are seasonal and lie dormant underground in the dry season, but the few more or less evergreen species with similar underground organs (e.g., *Agapanthus* and *Kniphofia*) can also be included in the geophyte category (Goldblatt and Manning 2002). The advantage of the geophytic habit in these areas is the ability to survive short-term drought and respond rapidly to improved moisture conditions (Dafni et al. 1981a; Rees 1989; De Hertogh and Le Nard 1993). The prolonged dry periods may limit organ replenishment and compromise flowering, ultimately leading to the local extinction of populations. There are two strategies to overcome unpredictable drought: first, leafing and flowering can occur opportunistically, in response to rainfall events (e.g., *Cyrtanthus*, Amaryllidaceae and *Tritonia*, Iridaceae); second, larger storage organs can buffer plants against unpredictable winter rains (e.g., *Eriospermum*, Eriospermaceae and *Oxalis*, Oxalidaceae) (Proche\u015f et al. 2005).

The evolutionary success of geophytes in the Cape Mediterranean zone of South Africa extends well into the arid parts of this region. Geophytes of the semiarid to arid Succulent Karoo (a winter–rainfall desert adjacent to the true Mediterranean zone and incorporating Namaqualand) remain a very important component of the flora, both in terms of abundance and in terms of diversity (Snijman 1984; Goldblatt 1986; Hilton-Taylor 1987; Duncan 1998). In the Karoo Garden Reserve they comprise ~29% of the flora (Perry et al. 1979). Both the species diversity and the growth form diversity of geophytes in the succulent Karoo are remarkable. In addition to typical monocot geophytic growth forms with upright rosettes of basal leaves, there is an unusual growth form in which leaves lie fully pressed against the soil surface, the so-called "prostrate-leaved geophytes" (Esler et al. 1999). Commonly encountered prostrate-leaved geophytes in the succulent Karoo include *Brunsvigia* and *Haemanthus* (Amaryllidaceae); *Lachenalia*, *Massonia*, and *Whiteheadia* (Hyacinthaceae); *Eriospermum* (Eriospermaceae); and *Satyrium* and *Holothrix* (Orchidaceae). While this growth form is found in many geophyte lineages in the succulent Karoo biome and the Cape Mediterranean zone, it occurs infrequently through the summer–rainfall temperate regions of Africa. It is nearly absent in other regions worldwide. The adaptive significance of this growth form is still unclear, and Esler et al. (1999) argue that it might be associated with the avoidance of herbivory, reduction in competition, creation of a CO_2-enriched environment below the leaves, reduction of water loss, precipitation of dew on the leaves and maintenance of optimal leaf temperatures for growth.

Another successful life strategy is presented by the hysteranthous floral rhythm, which may be considered as a character of phylogenetically advanced forms in different families (Dafni et al. 1981a,b). This mechanism contributes to rapid and efficient flowering and optimal pollination and dispersal conditions under seasonal Mediterranean-like climates. Numerous Mediterranean geophytes are hysteranthous, and flower in a leafless stage during autumn. On the basis of their life cycles and other features two types can be recognized: (1) the *Urginea*-type (*Urginea, Scilla, Narcissus,* and *Pancratium*) flowers with supraterranean ovary originating from a perennial storage organ at the end of the annual cycle; seed dispersal and germination follow flowering immediately; (2) the *Crocus*-type (*Crocus, Colchicum, Merendera,* and *Sternbergia*) flowers with subterranean ovary, developing at the beginning of a new reproductive cycle from an annual storage organ, and seed dispersal and germination are delayed to the spring and the next autumn (Table 3.1; Figure 3.2; Dafni et al. 1981b).

The perennial storage organs of the *Urginea* types seem to be connected to unpredictable climates with a very short growing season (Dafni et al. 1981b). Such habitats are typical of the Mediterranean basin and particularly of its border zones toward the deserts where many of the

TABLE 3.1
Characteristics of the Two Types of Hysteranthous Geophytes

Biological Trait	*Urginea* Type	*Crocus* Type
Life span of the storage organ	Perennial	Annual
Flowering stem	Distinct	Reduced
Ovary position	Supraterranean	Subterranean
Flowering onset	Precedes leaf elongation in spring	Delayed until decay of spring foliage
Seed dispersal	Immediately after flowering	In spring and summer, following underground after-ripening during winter
Germination	Without dormancy	After dormancy of at least one year
Origin	Paleotropical	High mountains of the Middle East

Source: Dafni, A. et al. 1981b. *Plant Syst. Evol.* 137:181–193.

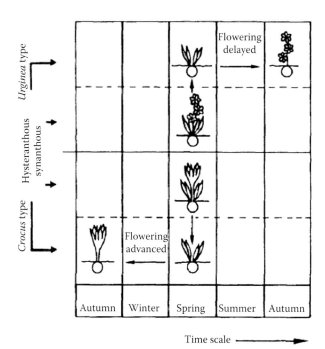

FIGURE 3.2 Scheme of suggested evolution of hysteranthous from synanthous geophytes. (Adapted from Dafni, A. et al. 1981b. *Plant Syst. Evol.* 137:181–193.)

species belonging to this type are growing. The main adaptive advantage of the evolution of the *Urginea* type with its delayed flowering are the immediate germination before the rain period which prevents the risk of long seed exposure to environmental hazards and the better exploitation of pollinators during autumn. Similar examples of the transition from synanthous to hysteranthous forms can be found at the Cape (South Africa) (Dafni et al. 1981b). In the *Crocus* type the annual storage organs are correlated with more predictable climates where the growing season is not too short and where the annual replacement of the storage organ is less risky. The subterranean ovary enables the seeds to escape the harsh winter conditions. This is the main advantage together with the advanced flowering time and reduced competition for pollinators. The *Crocus* type is found within both Mediterranean and Irano-Turanian regions. Shmida (1977) assumes that the plants of this type (e.g., *Merendera*, *Crocus*, and *Sternbergia*) originated under "arid-alpine" conditions, with extremely dry summers and hard winters, and they prevail in the high mountains of the Middle East and extend into NW Africa and the Iberian Peninsula. Most probably, hysteranthous geophytes were derived from synanthous species during their evolution in the Mediterranean areas (Figure 3.2; Dafni et al. 1981b), and intermediate forms of the same species can be found in different habitats. Thus, *Scilla autumnalis* is described as a sub-hysteranthous (McNeill 1980) or hysteranthous species (Dafni et al. 1981a), while *Scilla bifolia* blooms late in winter or spring as a synanthous plant (Zohary 1978; McNeill 1980). *Pancratium maritimum* (Amaryllidaceae) is a hysteranthous or sub-hysteranthous bulbous plant which inhabits the shores of the Mediterranean, the Atlantic, in the Magreb, Iberian peninsula and the Black and Caspian seas (Webb 1980).

F. ARID ZONES

Desert ephemerals have a short growing season corresponding to the timing and amount of precipitation. In general, the vegetation of arid regions is characterized by sparseness of plant cover and the preponderance of a limited number of plant species. The life form distribution is closely correlated with topography and landform (Shaltout et al. 2010). In some cases, prolonged dormancy in the absence of water is difficult to define as "summer dormancy", since it can occur over periods of 2–5 years (Rossa and von Willert 1999). Some examples of prolonged dormancy of perennial geophytes can be found in the Atacama Desert of Chile (Vidiella et al. 1999), and the Namaqualand desert in South Africa (Rossa and von Willert 1999; Desmet 2007). Plant species with underground storage organs (bulbs in *Leucocoryne*, *Rhodophiala*, *Bulbinella*, and *Ornithogalum*; storage roots in *Chloraea*) can survive long periods of drought, and have a short period of vegetative growth and flowering.

IV. MORPHOLOGICAL STRUCTURES OF GEOPHYTES

Underground storage organs are usually classified into bulbs, rhizomes, corms, tubers, and storage roots. Evolution has led to much more variation than these simple categories denote; therefore, in many cases the borders between different types are indistinct. For example, many rhizomatous *Allium* species possess both rhizomes and a "false" bulb, made of leaf sheaths. The rhizomes are the result of consecutive growth of several generations of the bulb axes. Other *Allium* species have typical bulbs (Rees 1972; Kamenetsky and Rabinowitch 2006).

Although geophytes are perennial plants, the absence of a cambium is one of their fundamental features. Holttum (1955) argued that the lack of cambium led to continuous sympodial growth and branching. This type of growth is typical in moist tropical climates. During the colonization of areas with climatic periodicity, the sympodial growth form proved an ecological advantage to the production of dormant organs, thus allowing species lacking a cambial layer to spread to areas with various climatic patterns. Despite the absence of a cambium, geophytes have localized groups of meristematic cells. They occur not only in the plant apex, but also in the axils of the scales and leaf sheaths, on the leaf blade, or just below the flower in the tulip (Rees 1972).

Storage organs of geophytes permit survival during periods of high or low temperatures, drought, or improper light levels. Consequently, the success of a geophytic species depends on growing rapidly when environmental conditions are favorable. Geophytic species respond to many environmental signals that determine when to enter or exit dormancy, including temperature, moisture, and photoperiod (Dole 2003; Chapter 9).

Eco-physiological advantages of the geophyte habit are

1. The food reserves in underground storage organs allow the plants to survive extended periods of restricted growth, such as cold winters and dry, hot summers. This advantage was probably the primary reason for the great geophytic diversity in Mediterranean areas.
2. Differentiation of the monocarpic shoots in the renewal buds prior to the stage of active growth under favorable environmental conditions. This attribute allows growth early in the spring in temperate latitudes where incoming radiation is low, or in areas with cold winters and hot, dry summers, and short springs necessitating rapid growth and flowering (e.g., the Irano-Turanian floristic region).
3. Storage reserves in the underground organs allow separation between vegetative and reproductive stages (e.g., *Colchicum* and *Urginea*), thus providing an ecological advantage in plant survival and competition for pollinators.
4. Geophytic organs serve for vegetative propagation by the formation of lateral shoots in the axils of bulb scales, foliage leaf bases or other parts. Vegetative propagation not only serves for plant distribution in the natural habitat, but also plays an important adaptive role. Under unpredictable environmental conditions, when the flowering and fruiting period can be negatively affected by unfavorable conditions (dryness, low or high temperatures), seed propagation is not always successful. In such cases, vegetative propagation might provide an effective alternative to sexual propagation, even for several consecutive years.

Major morphological structures of the underground storage organs of geophytes are described next.

A. Rhizomes

Rhizomes are modified, elongated underground stems, which function primarily as storage organs, with well-defined nodes. In general, both shoots and adventitious roots arise from the nodes. The plants produce false bulbs or shoots made of leaf sheaths of different thickness (Figure 3.3a). In some *Allium* and *Iris* species, the fleshy rhizomes consist of successive concrescence of the underground stems developing over several seasons, and grow in a horizontal, oblique, or vertical direction (Kamenetsky and Rabinowitch 2006).

Species with rhizomatous underground organs are found in both monocot and dicot taxa. They are common mainly in the temperate zone, mostly in mesophytic habitats, such as meadows or forests, and also in arid areas of the sub-alpine and alpine mountainous zones (Hanelt et al. 1992; De Hertogh and Le Nard 1993; Kamenetsky and Rabinowitch 2006). Some members of the rhizomatous group undergo two to three flowering cycles per summer, and form one to several complete sets of leaves and a few prophylls per flower scape per cycle (Cheremushkina 1992; Kruse 1992).

Rhizomatous geophytes include species with thick and fleshy horizontal, oblique and oblique-vertical rhizomes. *Canna, Alstroemeria, Allium* species from the subgenus *Rhiziridium, Iris* (series *Oncocyclus, Regelia, Regeliocyclus,* and *Aribred*), *Asparagus, Convallaria,* and *Hosta* represent this morphological type. Some species (e.g., the endemic species *Allium caespitosum* from Kazakhstan and China) underwent a very strong specialization to the point that it resembles sedge (*Carex*) rather than a bulbous plant. It produces a long twine-like and highly branched rhizome which anchors the plant in the shifting sands. The plants do not produce bulbs, but long-branched

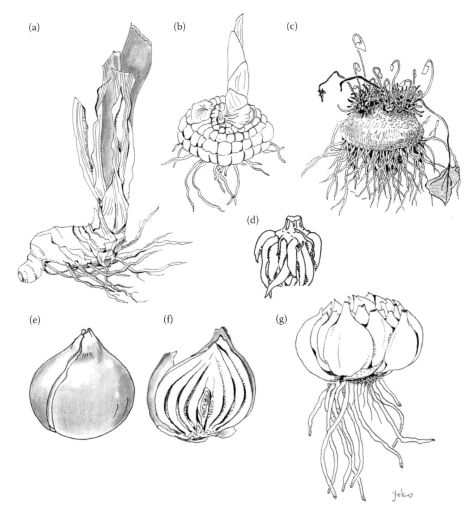

FIGURE 3.3 Morphological structure of the underground storage organs. (a) Rhizome. Horizontal rhizome of *Iris* is metamorphosis of underground stem. Shoots and roots are located in the defined nodes. (b) Corm. Underground corm of *Gladiolus* with vertical axis of growth is metamorphosis of stem; the renewal buds are located in the defined nodes. (c) Tuber of *Cyclamen* is formed from hypocotyl. (d) Tuberous roots of *Ranunculus asiaticus*. (e) Bulb of *Tulipa*. (f) Annual bulb of *Tulipa* is made of specialized scales; a new developing monocarpic shoot is visible within the bulb. (g) Nontunicate perennial bulb of *Lilium*. (Courtesy of Yoko Okubo.)

roots and thin shoots. The latter are composed of two or three leaves which develop in the rhizome's nodes (Kamenetsky and Rabinowitch 2006).

B. Corms

Corms are modified underground stems, typically round, with well-defined nodes, and a vertical axis of growth (Figure 3.3b). *Gladiolus* and *Crocus* (Iridaceae) are good examples of this group. The single internoded corm of *Gynandriris sisyrinchium* (Iridaceae) carries the renewal bud on its apex. The new growth, developing from the renewal bud after the first rains, consists of underground sheath leaves, through the longest of which the scape inflorescence emerges. The axillary replacement corm, already conspicuous at flowering, continues to grow until the season's end (Galil 1981).

C. TUBERS

According to their origin, tubers are separated into three types. Stem tubers such as the potato *Solanum tuberosum*, *Gloriosa*, *Anemone*, and *Zantedeschia* have buds distributed over the entire surface and the primary storage tissue is formed by the stem. Enlarged hypocotyls, such as those of *Cyclamen* or *Gloxinia*, are similar to the stem tubers, but the primary storage tissue has been derived from hypocotyl (Figure 3.3c). The tuberous storage roots are comprised of enlarged root tissue (Figure 3.3d) and do not produce nodes and internodes or leaves. At the proximal end they are attached to an underground crown (tuberous compressed underground stem), while fine roots for water and mineral absorption are developing on the distal end. The renewal buds originate on the crown, although some species (e.g., *Paeonia officinalis*, Kapinos and Dubrov 1993) are able to produce new buds on the storage roots. Tuberous roots are biennial (*Asphodelus*, *Ranunculus*, *Dahlia*, *Astilbe*, and *Eremurus*) or perennial (*Paeonia*).

D. BULBS

In bulbs, the primary storage organ is formed by the swollen leaf bases and/or specialized scales (modified leaves), which are positioned on a compressed short stem called the basal plate (Figure 3.3e–g). Bulbs can be either tunicate, enclosed in dry leaf bases, for example, *Tulipa* and *Hyacinthus*, or nontunicate, without a covering, for example, *Fritillaria* and *Lilium*.

Bulbous geophytes are categorized into several distinct subgroups:

1. Perennial bulbs formed by the foliage leaf sheaths, which are replaced gradually. In one bulb 2–4 generations of leaf sheaths can be found simultaneously. Examples of this

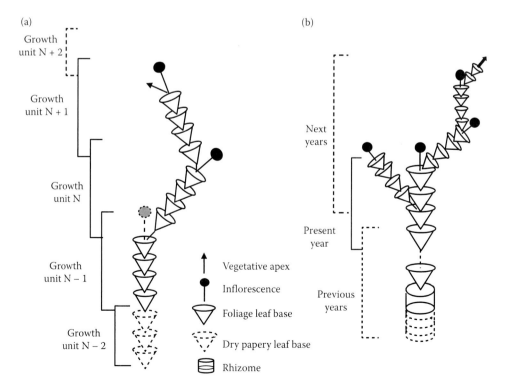

FIGURE 3.4 Diagrammatic representation of the morphological structure of a perennial bulb, formed by the foliage leaf sheaths. (a) *Hippeastrum* plant produces several monocarpic shoots in one bulb. Branching is sympodial, and one bulb contains several generations of flowering units. N, monocarpic shoot flowering now. (b) *Allium nutans*. "False" bulb is formed from the leaf sheaths and is developed on a short vertical rhizome.

subgroup are *Hippeastrum* and *Crinum* (Figure 3.4a). In some species (e.g., *Allium* species of the subgenus *Rhizirideum*, Kamenetsky and Rabinowitch 2006), bulbs are developed on the condensed underground stem, similar to a short rhizome (Figure 3.4b).

2. Perennial bulbs consisting of thick leaf sheaths (false scales) and specialized scales, attached to a condensed stem disk (basal plate) (Figure 3.5). *Narcissus* and *Allium* species have this bulb structure (Kamenetsky and Rabinowitch 2006; Noy-Porat et al. 2009). Bulbs of the Mediterranean hysteranthous plant *Urginea maritima* are made of numerous scales, the outer of which become papery, while new scales are forming inside of the bulb (Dafni et al. 1981a).

3. Annual tunicate bulbs are made only of a basal plate and fleshy, thick specialized scales (Figure 3.6). At the end of the growing season, leaf sheaths dry out and form the enveloping dry tunics, which do not contain any storage material. Roots are frequently ephemeral

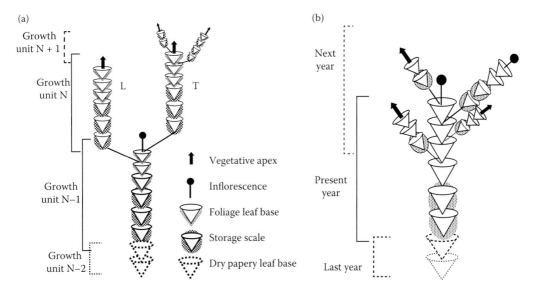

FIGURE 3.5 Diagrammatic representation of morphological structure of perennial bulb made of thick leaf sheaths (false scales) and specialized scales. (a) *Narcissus tazetta*. The bulb is branching sympodially and consists of specialized thick scales and leaf sheaths of the foliage leaves. N, monocarpic shoot flowering now. T, terminal bulb unit; L, lateral bulb unit. (b) *Allium caesium*. Branching occurs in the axes of foliage leaves.

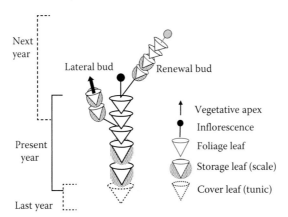

FIGURE 3.6 Diagrammatic representation of the morphological structure of the *Tulipa* annual bulb made of specialized scales.

and allow fast and efficient water absorption during the short vegetation period. *Tulipa* and *Hyacinthus* exemplify bulbs of this group. *Allium* species of this group belong to the subgenus *Melanocrommyum*, which have evolved under arid conditions prevailing in Central Asia and in the Middle East (Fritsch 2001), and includes species such as *Allium altissimum*, *A. karataviense*, *A. rothii*, and others.

V. ORIGIN OF THE CULTIVATED ORNAMENTAL GEOPHYTES

Bryan (1989, 1995, 2002) has reported four highly defined centers of origin that account for many ornamental geophytic species: (1) the Mediterranean basin (Greece, Italy, Northern Africa, Spain, Turkey); (2) Asia (China, Japan, Russia); (3) the mountain regions extending from Chile in South America to the State of Washington in the United States and British Columbia in Canada; and (4) Southern Africa in which the largest total number of genera has been found. High diversity of the native geophytes can also be found in the Irano-Turanian floristic region (Iran, Afghanistan, Central Asia), and in subtropical zones (Figure 3.7).

The Middle East and Irano-Turanian region are the origins of almost all the currently grown "classic" bulbous crops (Benschop et al. 2010), but other potentially useful species can be found in these regions (Avishai et al. 2005). Mediterranean species of *Scilla*, *Pancratium*, *Iris*, and *Fritillaria* have great ornamental potential (Halevy 2000; Avishai et al. 2005; Kamenetsky 2005). Many wild *Allium* species could contribute to the variety of cultivated flower bulbs (Kamenetsky and Fritsch 2002).

The flora of South Africa, which includes over 2700 flower geophyte species, has provided horticulture with the well-known *Gladiolus*, *Freesia*, *Nerine*, and *Zantedeschia* ("calla lily"), but

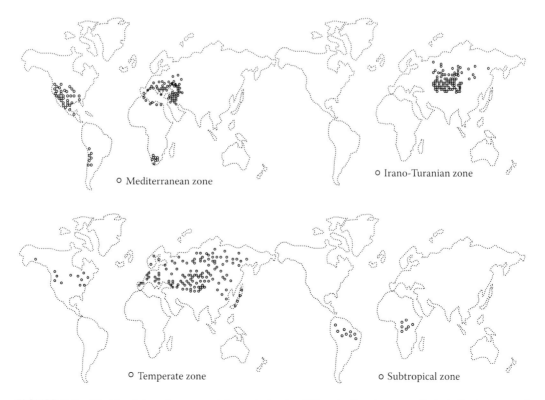

FIGURE 3.7 World origins of ornamental geophytes in different climate zones. Dots indicate areas of origin. (Adapted from Benschop, M. et al. 2010. *Hort. Rev.* 36:1–115.)

there are other species that need to be evaluated, bred, and developed. These species are known as "specialty bulbs", for example, *Ixia*, *Agapanthus*, *Gloriosa*, *Cyrtanthus*, *Lachenalia*, and *Babiana* (Du Plessis and Duncan 1989; Ehlers et al. 2002; Niederwieser et al. 2002).

South America also has a diverse bulbous flora and these resources have not been fully explored. A collection of ornamental *Alstroemeria* and *Hippeastrum* species native to Brazil was started in 1989 (Tombolato et al. 1992; Tombolato and Matthes 1998). It has the goal of creating types of plants and flowers that are uniquely different from existing commercial cultivars. Other genera, for example, *Griffinia* (Amaryllidaceae), *Neomarica* (Iridaceae), and *Gomphrena* (Amaranthaceae), have been collected either because of their showy flowers or unique growth habit (Tombolato and Matthes 1998). New cultivars of *Eucrosia* (Meerow et al. 1992; Roh et al. 1992) have been developed and are being produced commercially.

Research projects on the breeding, propagation, physiology, and production of flowering geophytes from Chile have been in progress for several years (Bridgen et al. 2002). Various interspecific and intraspecific hybrids of *Alstroemeria* species have been bred by means of a combination of traditional and biotechnological techniques. Other species that are being bred and studied include *Conanthera bifolia*, *C. campanulata*, *C. trimaculata*, several *Rhodophiala* spp., *Zephyra elegans*, *Bomarea* (former *Leontochir*) *ovallei*, *Pasithea caerulea*, and *Herbertia lahue*. Genotypic differences have been evaluated for the very unique Chilean species, *Leucocoryne coquimbensis*, *L. coquimbensis* var. *alba*, *L. purpurea*, and *L. ixioides*. All these have potential for increased breeding and development (Bridgen et al. 2002).

The Australian native plant *Anigozanthos* (Haemodoraceae), which is known as the "Kangaroo Paw", was grown mainly outdoors until a few years ago (Goodwin 1993). However, recently introduced high-yielding interspecific hybrids are being greenhouse-grown for year-round production as potted plants. These new hybrids are propagated by *in vitro* tissue culture (Halevy 1999). In Japan, the endangered species, *Arisaema sikokianum*, which is endemic to the Shikoku and Honshu Islands, has been evaluated for commercial use (Fukai et al. 2002).

Introduction of local species into commercial production and their development as new ornamental crops will be reviewed in detail in Chapter 5 and Chapters 14 through 19.

ACKNOWLEDGMENT

The author thanks Dr. Alan W. Meerow, USDA-ARS-SHRS, National Germplasm Repository, Miami, Florida, for the critical review of this chapter.

CHAPTER CONTRIBUTOR

Rina Kamenetsky is a senior researcher at the Agricultural Research Organization (ARO, The Volcani Center) and a professor at the Hebrew University of Jerusalem in Israel. She was born in Almaty, Kazakhstan, and obtained her BSc and MSc from Kazakh State University and PhD from the National Academy of Science of Kazakhstan. Dr. Kamenetsky has held positions in the Main Botanical Garden and Institute of Botany in Kazakhstan and the J. Blaustein Institute for Desert Research, Ben-Gurion University of the Negev, Israel. Her main research areas are the mechanisms of the internal and environmental regulation of flowering and dormancy in geophytes, biodiversity, and the introduction of new ornamental plants and their cultivation in warm regions. She is the author of over 150 publications on geophytes.

REFERENCES

Arroyo, S. C., and D. F. Cutler. 1984. Evolutionary and taxonomic aspects of the internal morphology in Amaryllidaceae from South America and Southern Africa. *Kew Bull.* 39:467–498.

Avishai, M., G. Luria, and O. Fragman-Sapir. 2005. Perspectives in the domestication of native Israeli geophytes. *Herbertia* 58:47–74.

Badri, M. A., P. E. H. Minchin, and L. Lapointe. 2007. Effects of temperature on the growth of spring ephemerals: *Crocus vernus. Physiol. Plant.* 130:67–76.

Benschop, M., R. Kamenetsky, M. Le Nard, H. Okubo, and A. De Hertogh. 2010. The global flower bulb industry: Production, utilization, research. *Hort. Rev.* 36:1–115.

Botschantzeva, Z. P. 1982. *Tulips: Taxonomy, Morphology, Cytology, Phytogeography and Physiology.* Rotterdam, The Netherlands: A. A. Balkema. (Translated and edited by H. Q. Varenkamp).

Bridgen, M. P., E. Olate, and F. Schiappacasse. 2002. Flowering geophytes from Chile. *Acta Hort.* 570:75–80.

Bryan, J. E. 1989. *Bulbs.* Portland, OR: Timber Press.

Bryan, J. E. 1995. *Manual of Bulbs.* Portland, OR: Timber Press.

Bryan, J. E. 2002. *Bulbs* (Revised edition). Portland, OR: Timber Press.

Chase, M. W. 2004. Monocot relationship: An overview. *Am. J. Bot.* 91:1645–1655.

Cheremushkina, V. A. 1992. Evolution of life form of species in the subgenus *Rhizirideum* (Koch) Wendelbo, genus *Allium* L. In *The Genus Allium—Taxonomic Problems and Genetic Resources. Proceedings of an International Symposium, Gatersleben, Germany, 11–13 June 1991,* eds. P. Hanelt, K. Hammer, and H. Knupffer, 27–34. Gatersleben, Germany: Institut für Pflanzengenetik und Kulturpflanzenforschung.

Dafni, A., D. Cohen, and I. Noy-Meir. 1981a. Life-cycle variation in geophytes. *Ann. Missouri Bot. Gard.* 68:652–660.

Dafni, A., A. Shmida, and M. Avishai. 1981b. Leafless autumnal-flowering geophytes in the Mediterrenean region—Phytogeographical, ecological and evolutionary aspects. *Plant Syst. Evol.* 137:181–193.

De Hertogh, A. A., and M. Le Nard. 1993. Botanical aspects of flower bulbs. In *The Physiology of Flower Bulbs,* eds. A. De Hertogh, and M. Le Nard, 7–20. Amsterdam: Elsevier.

Desmet, P. G. 2007. Namaqualand—A brief overview of the physical and floristic environment. *J. Arid Environ.* 70:570–587.

Dole, J. M. 2003. Research approaches for determining cold requirements for forcing and flowering of geophytes. *HortScience* 38:341–346.

Doutt, R. L. 1994. *Cape Bulbs.* London: B. T. Batsford.

Duncan, G. D. 1998. Five new species of *Lachenalia* (Hyacinthaceae) from arid areas of Namibia and South Africa. *Bothalia* 28:131–139.

Du Plessis, N., and G. Duncan. 1989. *Bulbous Plants of Southern Africa—A Guide to their Cultivation and Propagation.* Cape Town: Tafelberg.

Ehlers, J. L., P. J. J. van Vuuren, and L. Morey. 2002. Flowering behaviour of four clones of *Veltheimia bracteata. Acta Hort.* 570:341–343.

Erwin, J. 2006. Factors affecting flowering in ornamental plants. In *Flower Breeding and Genetics. Issues, Challenges and Opportunities for the 21st Century,* ed. N. O. Anderson, 7–48. Dordrecht, The Netherlands: Springer.

Esler, K. J., P. W. Rundel, and P. Vorster. 1999. Biogeography of prostrate-leaved geophytes in semi-arid South Africa: Hypotheses on functionality. *Plant Ecol.* 142:105–120.

Fragman, O., and A. Shmida. 1997. Diversity and adaptation of wild geophytes along an aridity gradient in Israel. *Acta Hort.* 430:795–802.

Fritsch, R. M. 2001. Taxonomy of the genus *Allium* L.: Contribution from IPK Gatersleben. *Herbertia* 56:19–50.

Fukai, S., A. Hasegawa, M. Goi, and N. Yamasaki. 2002. Seed propagation of *Arisaema sikokianum* (Araceae). *Acta Hort.* 570:327–330.

Galil, J. 1981. Morpho-ecological studies on geophylic plants. Vegetative dispersal of *Gynandriris sisyrinchium* L. *Israel J. Bot.* 30:165–172.

Givnish, T. J. 1987. Comparative studies of leaf form: Assessing the relative roles of selective pressures and phylogenetic constraints. *New Phytol.* 106(Suppl. S1):131–160.

Goldblatt, P. 1986. *The Moraeas of Southern Africa. Annals of Kirstenbosch Botanical Gardens 14.* Cape Town: National Botanic Gardens in association with the Missouri Botanical Garden.

Goldblatt, P., and J. C. Manning. 2002. Plant diversity of the Cape Region of southern Africa. *Ann. Missouri Bot. Gard.* 89:281–302.

Goodwin, P. B. 1993. *Anigozanthos (Macropidia).* In *The Physiology of Flower Bulbs,* eds. A. De Hertogh, and M. Le Nard, 219–226. Amsterdam: Elsevier.

Halevy, A. H. 1990. Recent advances in control of flowering and growth habit of geophytes. *Acta Hort.* 266:35–42.

Halevy, A. H. 1999. New flower crops. In *Perspectives on New Crops and New Uses,* ed. J. Janick, 407–409. Alexandria, VA: ASHS Press.

Halevy, A. H. 2000. Introduction of native Israeli plants as new cut flowers. *Acta Hort.* 541:79–82.

Hall, T. 2007. *Iris narynensis. Curtis's Bot. Mag.* 24:34–41.

Hanelt, P., J. Schultze-Motel, R. Fritsch, J. Kruse, H. I. Maass, H. Ohle, and K. Pistrick. 1992. Infrageneric grouping of *Allium*—The Gatersleben approach. In *Proceedings of an International Symposium, Gatersleben, Germany 11–13 June 1991*, eds. P. Hanelt, K. Hammer, and H. Knupffer, 107–123. Gatersleben, Germany: Institut für Pflanzengenetik und Kulturpflanzenforschung.

Harder, L. D., J. D. Thomson, M. B. Cruzan, and R. S. Unnasch. 1985. Sexual reproduction and variation in floral morphology in an ephemeral vernal lily, *Erythronium americanum. Oecologia* 67:286–291.

Hartsema, A. M. 1961. Influence of temperatures on flower formation and flowering of bulbous and tuberous plants. In *Encyclopedia of Plant Physiology, vol. 16*, ed. W. Ruhland, 123–167. Berlin: Springer-Verlag.

Hilton-Taylor, C. 1987. Phytogeography and origins of the Karoo flora. In *The Karoo biome: A Preliminary Synthesis. Part 2: Vegetation and History. South African National Scientific Programme Report 142*, eds. R. M. Cowling, and P. W. Roux, 70–95. Pretoria: CSIR.

Holttum, R. E. 1955. Growth-habits of monocotyledons—Variations on a theme. *Phytomorphology* 5:399–413.

Kamenetsky, R. 1994. Life cycle, flower initiation, and propagation of the desert geophyte *Allium rothii. Int. J. Plant Sci.* 155:597–605.

Kamenetsky, R. 2005. Production of flower bulbs in regions with warm climates. *Acta Hort.* 673:59–66.

Kamenetsky, R., and R. Fritsch. 2002. Ornamental *Alliums*. In *Allium Crop Science: Recent Advances*, eds. H. D. Rabinowitch, and L. Currah, 459–492. Wallingford, UK: C.A.B International.

Kamenetsky, R., and H. D. Rabinowitch. 2006. The genus *Allium*: A developmental and horticultural analysis. *Hort. Rev.* 32:329–378.

Kapinos, D. B., and V. M. Dubrov. 1993. *Peonies in the Garden*. Tumen, Russia: Minion (In Russian).

Kariuki, W., and S. Kako. 1999. Growth and flowering of *Ornithogalum saundersiae* Bak. *Scientia Hortic.* 81:57–70.

Kawano, S. 1975. The productive and reproductive biology of flowering plants. II. The concept of life history strategy in plants. *J. Coll. Lib. Arts, Toyama Univ. Japan* 8:51–86.

Kawano, S. 1985. Life history characteristics of temperate woodland plants in Japan. In *The Population Structure of Vegetation*, ed. J. White, 515–549. Dordrecht, The Netherlands: Dr. W. Junk.

Kawano, S., M. Ohara, and F. H. Utech. 1992. Life history studies on the genus *Trillium* (Liliaceae) VI. Life history characteristics of three western North American species and their evolutionary-ecological implications. *Plant Species Biol.* 7:21–36.

Klimeš, L. 2003, Life-forms and clonality of vascular plants along an altitudinal gradient in E Ladakh (NW Himalayas). *Basic Appl. Ecol.* 4:317–328.

Klimeš, L., J. Klimešová, R. Hendriks, and J. van Groenendael. 1997. Clonal plant architecture: A comparative analysis of form and function. In *The Ecological and Evolution of Clonal Plants*, eds. H. de Kroon, and J. van Groenendael, 1–29. Leiden, The Netherlands: Backbuys Publ.

Körner, C. 1999. *Alpine Plant Life. Functional Plant Ecology of High Mountain Ecosystems*. Berlin: Springer.

Kruse, J. 1992. Growth form characters and their variation in *Allium* L. In *Proceedings of an International Symposium, Gatersleben, Germany 11–13 June 1991*, eds. P. Hanelt, K. Hammer, and H. Knupffer, 173–179. Gatersleben, Germany: Institut für Pflanzengenetik und Kulturpflanzenforschung.

Lapointe, L. 2001. How phenology influences physiology in deciduous forest spring ephemerals. *Physiol. Plant.* 113:151–157.

Lapointe, L., and S. Lerat. 2006. Annual growth of the spring ephemeral *Erythronium americanum* as a function of temperature and mycorrhizal status. *Can. J. Bot.* 84:39–48.

Le Nard, M., and A. A. De Hertogh. 1993. Bulb growth and development and flowering. In *The Physiology of Flower Bulbs*, eds. A. De Hertogh, and M. Le Nard, 29–43. Amsterdam: Elsevier.

McNeill, J. 1980. *Urginea* Staeinh. *Scilla*. In *Flora Europaea 5*, eds. T. G. Tutin, V. H. Heywood, N. A. Burges, D. H. Valentine, S. M. Walters and D. A. Webb, 40–43. Cambridge: Cambridge University Press.

Meerow, A. W., D. J. Lehmiller, and J. R. Clayton. 2003. Phylogeny and biogeography of *Crinum* L. (Amaryllidaceae) inferred from nuclear and limited plastid non-coding DNA sequences. *Bot. J. Linn. Soc.* 141:349–363.

Meerow, A. W., K. D. Preuss, and A. F. C. Tombolato. 2002. *Griffinia* (Amaryllidaceae), a critically endangered Brazilian geophyte with horticultural potential. *Acta Hort.* 570:57–64.

Meerow, A. W., M. Roh, and R. S. Lawson. 1992. Breeding of *Eucrosia* (Amaryllidaceae) for cutflower and pot plant production. *Acta Hort.* 325:555–560.

Minoletti, M. L., and R. E. J. Boerner. 1993. Seasonal photosynthesis, nitrogen and phosphorus dynamics, and resorption in the wintergreen fern *Polystichum acrostichoides* (Michx.) Schott. *Bull. Torrey Bot. Club* 120:397–404.

Muller, R. N. 1979. Biomass accumulation and reproduction in *Erythronium albidum*. *Bull. Torrey Bot. Club* 106:276–283.

Naqinezhad, A., F. Attar, A. Jalili, and K. Mehdigholi. 2010. Plant biodiversity of wetland habitats in dry steppes of Central Alborz Mts., N. Iran. *Aust. J. Basic Appl. Sci.* 4:321–333.

Nault, A., and D. Gagnon. 1988. Seasonal biomass and nutrient allocation patterns in wild leek (*Allium tricoccum* Ait.), a spring geophyte. *Bull. Torrey Bot. Club* 115:45–54.

Nault, A., and D. Gagnon. 1993. Ramet demography of *Allium tricoccum*, a spring ephemeral, perennial forest herb. *J. Ecol.* 81:101–119.

Niederwieser, J. G., R. Kleynhans, and F. L. Hancke. 2002. Development of a new flower bulb crop in South Africa. *Acta Hort.* 570:67–73.

Noroozi, J., H. Akhani, and S.-W. Breckle. 2008. Biodiversity and phytogeography of the alpine flora of Iran. *Biodivers. Conserv.* 17:493–521.

Noy-Porat, T., M. A. Flaishman, A. Eshel, D. Sandler-Ziv, and R. Kamenetsky. 2009. Florogenesis of the Mediterranean geophyte *Narcissus tazetta* and temperature requirements for flower initiation and differentiation. *Scientia Hortic.* 120:138–142.

Parsons, R. F., and S. D. Hopper. 2003. Monocotyledonous geophytes: Comparison of south-western Australia with other areas of Mediterranean climate. *Aust. J. Bot.* 51:129–133.

Peel, M. C., B. L. Finlayson, and T. A. McMahon. 2007. Updated world map of the Köppen–Geiger climate classification. *Hydrol. Earth Syst. Sci.* 11:1633–1644.

Perry, P. L., M. B. Bayer, and L. A. Wilbraham. 1979. Flora of the Karoo Botanic Garden. 2. Geophytes. *Veld Flora* 65:79–81.

Procheş, Ş., R. M. Cowling, and D. R. du Preez. 2005. Patterns of geophyte diversity and storage organ size in the winter-rainfall region of southern Africa. *Divers. Distrib.* 11:101–109.

Procheş, Ş., R. M. Cowling, P. Goldblatt, J. C. Manning, and D. A. Snijman. 2006. An overview of Cape geophytes. *Biol. J. Linn. Soc.* 87:27–43.

Raunkiær, C. 1934. *The Life Forms of Plants and Statistical Geography*. Oxford: Oxford University Press.

Rees, A. R. 1972. *The Growth of Bulbs. Applied Aspects of the Physiology of Ornamental Bulbous Crop Plants*. London: Academic Press.

Rees, A. R. 1989. Evolution of the geophytic habit and its physiological advantages. *Herbertia* 45:104–110.

Rees, A. R. 1992. *Ornamental Bulbs, Corms and Tubers*. Wallingford, Oxon, UK: C.A.B. International.

Roh, M. S., R. H. Lawson, K. C. Gross, and A. W. Meerow. 1992. Flower bud initiation and development of *Eucrosia* as influenced by bulb storage temperatures. *Acta Hort.* 325:105–111.

Rossa, B., and D. J. von Willert. 1999. Physiological characteristics of geophytes in semi-arid Namaqualand, South Africa. *Plant Ecol.* 142:121–132.

Rundel, P. W. 1996. Monocotyledonous geophytes in the California flora. *Madrono* 43:355–368.

Shaltout, K. H., M. G. Sheded, and A. I. Salem. 2010. Vegetation spatial heterogeneity in a hyper arid biosphere reserve area in north Africa. *Acta Bot. Croat.* 69:31–46.

Shmida, A. 1977. The tragacantic alpine vegetation of Mt. Hermon. PhD thesis, The Hebrew University of Jerusalem, Jerusalem, Israel.

Snijman, D. A. 1984. A revision of the genus *Haemanthus* (Amaryllidaceae). *J. S. Afr. Bot. Suppl.* 12:1–139.

Thomson, J. D. 2010. Flowering phenology, fruiting success and progressive deterioration of pollination in an early-flowering geophyte. *Philos. Trans. R. Soc. B.* 365:3187–3199.

Tissue, D. T., J. B. Skillman, E. P. McDonald, and B. R. Strain. 1995. Photosynthesis and carbon allocation in *Tipularia discolor* (Orchidaceae), a wintergreen understory herb. *Am. J. Bot.* 82:1249–1256.

Tojibaev, K. S. 2010. Flora of Southern-Western Tien-Shan (Uzbekistan territory). PhD dissertation, University of Andijan, Uzbekistan (In Russian).

Tombolato, A., and L. Matthes. 1998. Collection of *Hippeastrum* spp., *Alstroemeria* spp. and other Brazilian bulbous species. *Acta Hort.* 454:91–98.

Tombolato, A. F. C., R. B. Torres, and C. Azevedo. 1992. *Alstroemeria* collection in Brazil for a breeding program at Instituto Agronomica at Campinas—SP. *Acta Hort.* 325:873–877.

Uemura, S. 1994. Patterns of leaf phenology in forest understory. *Can. J. Bot.* 72:409–414.

Vidiella, P. E., J. J. Armesto, and J. R. Gutiérrez. 1999. Vegetation changes and sequential flowering after rain in the southern Atacama Desert. *J. Arid Environ.* 43:449–458.

Webb, D. A. 1980. *Narcissus* L., *Pancratium* L., *Gynandriris* Parl. In *Flora europaea 5*, eds. T. G. Tutin, V. H. Heywood, N. A. Burges, D. H. Valentine, S. M. Walters, and D. A. Webb, 78–84, 92. Cambridge: Cambridge University Press.

Werger, M. J. A., and H. Huber. 2006. Tuber size variation and organ preformation constrain growth responses of a spring geophyte. *Oecologia* 147:396–405.

Whigham, D. F. 2004. Ecology of woodland herbs in temperate deciduous forests. *Annu. Rev. Ecol. Evol. Syst.* 35:583–621.

Widmer, R. E., and R. E. Lyons. 1985. *Cyclamen persicum*. In *CRC Handbook of Flowering, Vol. II*, ed. A. H. Halevy, 382–390. Boca Baton, FL: CRC Press.

Yoshie, F. 2008. Dormancy of alpine and subalpine perennial forbs. *Ecol. Res.* 23:35–40.

Zohary, M. 1978. *The Plant World*. Tel Aviv: Am Oved (In Hebrew).

4 Botanical and Horticultural Aspects of Major Ornamental Geophytes

Hiroshi Okubo and Dariusz Sochacki

CONTENTS

I. INTRODUCTION

Although ornamental geophytes (flower bulbs) belong to more than 800 different genera, the industry is dominated by seven genera, *Tulipa, Lilium, Narcissus, Gladiolus, Hyacinthus, Crocus*, and *Iris* (Benschop et al. 2010). *Allium, Alstroemeria, Anemone, Freesia, Hippeastrum, Muscari, Ornithogalum, Ranunculus*, and *Zanthedeschia* are also prominent (Kamenetsky and Miller 2010; Chapter 1).

Comprehensive descriptions of the main crops (*Crocus*, Benschop 1993; *Gladiolus*, Cohat 1993; *Hyacinthus*, Nowak and Rudnicki 1993; *Iris*, De Munk and Schipper 1993; Easter lily, Miller 1993; other lilies, Beattie and White 1993; *Narcissus*, Hanks 1993; and *Tulipa*, Le Nard and De Hertogh 1993b) are available in the book *The Physiology of Flower Bulbs* (De Hertogh and Le Nard 1993a). The precise information on forcing culture techniques has been summarized by De Hertogh (1996). In addition, the roots of ornamental geophytes, their origin, morphology, and physiology are intensively reviewed by Kawa and De Hertogh (1992). The contribution of research to the production and utilization of major crops of ornamental geophytes was recently reviewed by Benschop et al. (2010).

In this chapter, the major botanical and horticultural traits of and recent research on the seven most popular genera are briefly described. The taxonomical status of geophytes is reviewed in Chapter 2.

Van Aatrijk (1995, 2000, 2005) provides detailed information and photographs on physiological disorders, pests, and diseases of ornamental geophytes. Thus, only the salient problems are summarized in this chapter.

II. CROCUSES

A. INTRODUCTORY REMARKS

The etymology of the name *Crocus* is derived from the Greek *kroke* (= saffron). Crocuses have been cultivated since ancient times in the Mediterranean region. Murals of crocus (saffron) gatherers painted in Minoan times were found on the island of Thera (Santorini), Greece (Margaris 2000). There is an illustration of *C. vernus* in a 1597 publication from Britain (Wilkins 1985). The first cultivars might have been developed by the early seventeenth century in The Netherlands. A crocus flower in a painting "Bouquet of Flowers" in the Louvre Museum by a Flemish-Dutch painter Ambrosius Bosschaert the Elder in about 1620 looks like a *C. chrysanthus* (Balkan and Asia Minor origin) cultivar. This plant resembles the modern cultivars 'Gypsy Girl', 'Saturn', or 'E. P. Bowles'.

There are autumn-, winter-, and spring-flowering *Crocus* species and cultivars, most of which are used as potted and garden plants. Horticulturally, the most important species are *C. vernus, C. tommasinianus, C. flavus*, and *C. chrysanthus* for spring flowering. Cultivars of *C. vernus* are also called Dutch crocuses. Many cultivars have been bred and are cultivated in The Netherlands. The saffron crocus (*C. sativus*) is one of the most popular autumn-flowering crocuses. It is cultivated for the dye, flavor, or medicine industries rather than used for ornamental purposes.

We often see misuses in the epithets of crocus and saffron. "Autumn crocus" is commonly applied to *Colchicum autumnale*, a plant of Colchicaceae. It has six stamens, whereas the *Crocus* species have three stamens. The *Colchicum* has another common name, "meadow saffron". The "Prairie crocus" is *Pulsatilla patens* or *P. ludoviciana*, a member of the family Ranunculaceae.

B. BOTANICAL CLASSIFICATION AND DISTRIBUTION

The genus *Crocus* belongs to the family Iridaceae and includes about 80–90 species (Mathew 1982; Van Scheepen 1991; Petersen et al. 2008). The species of the genus *Crocus* are found in the Mediterranean and Southwest Asia regions. The classification system of species, proposed by Mathew (1982), is widely accepted. The classification was based on morphology, chromosome and karyotype analyses, seed coat architecture, ecology, and distribution. According to this system, the genus *Crocus* is divided into two subgenera *Crocus* and *Crociris*, and the subgenus *Crocus* is further divided into two sections *Crocus* and *Nudiscapus* (Mathew 1982). Recently, the classification was revised, based on the chloroplast DNA sequences (Petersen et al. 2008; Mathew et al. 2009; Chapter 2).

The chromosome number varies between 2n = 6 and 64 among species (Mathew 1982), which may complicate interspecific hybridization. Autumn-flowering saffron (*C. sativus*) is triploid (2n = 3x = 24) and sterile (Karasawa 1933). It has been suggested by Frizzi et al. (2007) after using isozyme analysis that the autopolyploid origin of saffron is probably from *C. cartwrightianus*.

C.   Morphology and Annual Growth Cycle

The storage organ is a tunicated corm and is replaced every year. The plant is stemless. The leaves are grass-like and ensiform, generally with a silvery stripe along the leaf axis, and rise from the corm. The peduncle and ovary are located below the soil level. The presence or absence of prophyll, the style of branching, and how the tunic covers the corm are the major factors used for species classification. The flowers are cup-shaped and erect with a long and slender tube and six segments. The three stamens are attached in the throat of the perianth. The style is three-cleft, and the ovary is three-loculed. Flower colors vary from lilac, mauve to deep purple, white, yellow, pale-brown, and bicolors, but lack red. Petunidin, malvidin, and delphinidin are the anthocyanidins that contribute to the flower colors of the *Crocus* species (Nørbæk and Kondo 1998, 1999b).

Spring-flowering crocuses are generally synanthous. Flowering occurs when the leaves are green, followed by the senescence of the above-ground organs in late spring to early summer. Corms are annual, and new corms develop as replacements, either on top of the old ones or from their axillary buds. They are harvested in summer. The relationship between temperature and corm development in spring-flowering crocus has recently been reported. Since flowering of *C. vernus* occurs in late winter to early spring, growth at cooler temperatures has been attributed in this species to a longer leaf life and correlated with higher corm biomass (Badri et al. 2007). Lundmark et al. (2009) showed that the shift in carbon partitioning to cell wall material is responsible for the smaller final biomass of the corm at higher temperatures. They suggested that prolonged growth at lower temperatures results in delayed foliar senescence and the development of larger corms. Corms are "dormant" at harvest, and require high temperatures for breaking the dormancy. Flower initiation occurs during storage at 17–23°C (Benschop 1993). The minimum size of the corms for successful flower initiation is 5 cm, in circumference (Benschop 1993). The corms are planted in autumn. Forcing is possible by artificially compressing the natural annual cycle of temperature sequence of warm–cool–warm. Flower development in winter- to early spring-flowering species such as *C. sieberi*, *C. chrysanthus*, and *C. flavus* proceeds more slowly at lower temperatures after flower initiation. Therefore, high growing temperatures (>15°C) used in greenhouses for accelerating flowering may cause flower abortions. In Dutch crocuses (*C. vernus*), the forcing systems have been well established (see below).

Autumn-flowering crocuses are hysteranthous (Benschop 1993). Flowering is followed by a vegetative stage—leaf elongation throughout winter, the replacement of corms at the base of the shoots, and withering of leaves in spring. Corms are "dormant" at harvest in late spring to early summer. Flower initiation occurs during the summer and is similar to that of the spring-flowering crocus species. The optimal temperature for flower formation varies. For example, in *C. sativus* it ranges between 23°C and 27°C (Molina et al. 2005). The minimum corm size for flower initiation in *C. sativus* is 3–4 cm in circumference (Benschop 1993). The effects of temperature treatments, their duration and timing, as well as the effect of environmental conditions of the production areas on growth and flowering of *C. sativus* were well documented recently by Molina et al. (2004, 2005). The corms are planted in late summer to early autumn. Flower bud maturation proceeds in descending autumn temperatures. The optimum temperature for anthesis after planting is 15–20°C for the early-flowering *C. speciosus* and about 15°C for late-flowering *C. sativus* and *C. medius*.

D.   MAJOR SPECIES AND CULTIVARS

The major species and cultivars are listed in Table 4.1. In The Netherlands, more than 100 cultivars of *Crocus* are known (Van Scheepen 1991). Over 70 cultivars were cultivated in an area of 366 ha in the 2009/2010 season, among which 'Jeanne d'Arc' (57 ha), 'Flower Record' (42 ha), and 'King of the Striped' (30 ha) are the most prominent *C. vernus* cultivars (PT/BKD 2010a). *Crocus vernus* comprises about 55% of all crocus production, followed by *C. chrysanthus* (17%), *C. tommasinianus* (10%), and *C. flavus* (10%), represented by 'Romance' (22.9 ha), 'Ruby Giant' (21.5 ha), and 'Geel' (36.4 ha), respectively. All these cultivars are spring-flowering crocuses.

It is believed that *Crocus chrysanthus* 'Blue Perl' is a derivative of *C. biflorus* and that 'Cream Beauty' is a hybrid of *C. chrysanthus* with *C. biflorus* (Imanishi and Konoshima 1988). 'Stellaris' and 'Golden Yellow' have been shown to be hybrids originating from crosses between *C. flavus* and *C. angustifolius*. This was conformed with different combinations of chromosomes of the parent plants based on morphological traits and by genomic *in situ* hybridization (Ørgaard et al. 1995), as well as by flower pigment analysis (Nørbæk et al. 2002).

E.   PROPAGATION

Crocuses are asexually propagated by the developing of a new corm on top of the old one. In addition, new smaller corms are also formed by axillary buds. Multiplication by tissue culture is possible, but is not as common as that by seeds. Tissue culture of crocuses was reviewed by Kim and De Hertogh (1997).

TABLE 4.1
Major *Crocus* Cultivars

Series	Species	FS[a]	Major Cultivars
		Subgenus *Crocus* Section *Crocus*	
Verni	*tommasinianus*	S	Barr's Purple, Lilac Beauty, Roseus, Ruby Giant, Whitewell Purple
	vernus	S	Flower Record, Grand Maitre, Jeanne D'Arc, King of the Striped, Rickwick, Remembrance, Vanguard
Versicolores	*imperati*	S	De Jager
	versicolor	S	Picturatus
		Subgenus *Crocus* Section *Nudiscapus*	
Reticulati	*ancyrensis*	S	Golden Bunch
	sieberi	W	Firefly, Hubert Edelsten, Sublimis Tricolor, Violet Queen
Biflori	*biflorus*	S	Fairy, Miss Vain, Parkinsonii
	chrysanthus	S	Ard Schenk, Blue Perl[b], Cream Beauty[b], Dorothy, Fuscotinctus, Gypsi Girl, Romance, Zwanenburg Bronze
Orientales	*korolkowii*	S	Kiss of Spring
Flavi	*flavus*	S	Aureus, Geel
	olivieri	S	Balansae Zwanenburg
Speciosi	*pulchellus*	A	Albus, Zephyr
	speciosus	A	Cassiope, Conqueror, Oxonian
Laevigatae	*laevigatus*	A	Fontenayi

Source: Adapted from PT/BKD. 2010a. Bloembollen, voorjaarsbloeiers, beplante oppervlakten, 2006/'07 tot en 2009/'10.

[a]   Flowering season; A, autumn; W, winter; S, spring.

[b]   'Blue Perl' and 'Cream Beauty' are considered to be the derivatives of *C. biflorus* (see text).

F. Production of Spring-Flowering Crocuses

1. Agronomy

Crocuses can be grown in a wide range of soils, but the soils must be well drained. Sand and sandy loams with a pH range of 6–7 are preferable. When the corms are left in the soil in summer, the soil must be dry.

Although nitrogen usage varies according to the production region and country, the average application is around 100 kg/ha. High nitrogen levels tend to cause *Fusarium*. Potassium is applied in winter for new corm development. In The Netherlands, 700 kg of 7N-14P-28K fertilizer/ha are applied after planting in autumn, and 100 kg KNO_3 in February (Benschop 1993). Mulching and the deep planting of corms (7–10 cm) can avoid frost injury in outdoor production where the winter temperature is severe. Full sun to partial shade is required.

2. Corm Production

Corms are stored at 20–23°C until planting in autumn. They are planted in soil at a depth of 7–10 cm. The spacing is about 10–20 cm between rows and 10–20 cm within rows. Flowers are not removed. Since the ovary is located below the soil level, "topping"* does not prevent seed development. The corms are harvested after the foliage has senesced. The old, shrunken mother corms are removed by hand before grading and the newly formed corms are stored in well-ventilated and temperature-controlled rooms.

3. Flower Production

In general, spring-flowering crocuses can be forced by preplanting storage at low temperatures for 13–15 weeks. Dutch crocuses are very suitable for forcing. The corm size for forcing is 9/11 cm, in circumference (De Hertogh 1980). To have flowering by Christmas in the Northern Hemisphere, the corms are treated at 34°C for 1 week after lifting, 20°C for 2 weeks, 17°C until August 10, then precooled at 9°C for about 15 weeks until planting in early October (Benschop 1993). 'Flower Record', 'Jeanne d'Arc', and 'Remembrance' are well adapted for this forcing program. For early flowering at the end of January, the corms are stored at 17–20°C until the end of August or early September, followed by 9°C until planted in early October. For February flowering, the corms are stored at 17–20°C until planted in October. Subsequently, they are rooted at 9°C followed by 0–2°C for around 15 weeks, and then are grown at 13–15°C in greenhouses. The pots are marketed at the sprout stage and can be stored at 0–2°C, if necessary, before marketing. The pot life is about 2 weeks under favorable home conditions.

G. Physiological Disorders, Pests, and Diseases

Flower abortion during forcing is caused either by inappropriate cold storage (short storage period or high temperatures) or by high temperatures during the greenhouse stage.

Burkholderia gladioli ssp. *gladioli* (bacterial scab) and *Pectobacterium carotovorum* (bacterial soft rot) are the major bacterial pathogens. *Botrytis tulipae* (gray mold), *Botrytis gladiolorum*, *Fusarium oxysporum* f. sp. *gladioli* (dry rot), *Rhizoctonia solani* (root rot), *Penicillium crocicola* (blue mold), and *Uromyces croci* are the primary fungal pathogens. Viruses are *Iris mosaic virus* (IMV), *Arabis mosaic virus* (ArMV), *Bean yellow mosaic virus* (BYMV), and *Tobacco rattle virus* (TRV), and nematodes are *Aphelenchoides subtenuis*, *Ditylenchus destructor*, and *Pratylenchus coffeae*. Aphids are the main vectors of the viruses and rodents can feed on corms in storage and in the soil.

---

* Removal of the flower to shift photosynthetic sink. The terms "deheading" and "deflowering" are also commonly used.

III. DUTCH IRISES

A. INTRODUCTORY REMARKS

The name of the flower is derived from Iris, the goddess of the rainbow and a messenger of the gods in Greek mythology. Her role was to accompany souls to places of eternal peace, via the "road" marked by a rainbow—which is comprised of Iris's colors (Margaris 2000; Chapter 1). Iris plants have been known since ancient times. However, the cultivation history of the Dutch irises started at the end of the nineteenth century when the firm C. G. van Tubergen of Haarlem, The Netherlands crossed two varieties of *Iris xiphium* (var. *praecox* from Spain and var. *lusitanica* from France) with *I. tingitana* from North Africa (Hekstra and Broertjes 1968). 'Wedgwood' and its mutants 'Ideal' and 'White Wedgwood' are the hybrids of these species and varieties. 'Blue Magic' and 'Professor Blaauw' inherited the characteristics of *I. fontanesii*. A fertile spontaneous tetraploid of 'Wedgwood' was found in about 1952 (Eikelboom and Van Eijk 1990). It was crossed with cultivars of *I. xiphium* in order to produce several triploid cultivars, for example, 'Telstar'.

The majority of the Irises are the rhizomatous species best represented by the popular German iris (*Iris germanica*). Economically, however, the bulbous irises are the most important. Most of the commercial cultivars are the Dutch irises (*Iris × hollandica*) that were derived from crosses between the various bulbous species. In this chapter, only bulbous irises and in particular the Dutch irises are described.

B. BOTANICAL CLASSIFICATION AND DISTRIBUTION

Iris, the largest genus in the family Iridaceae, is comprised of more than 260 species and many infraspecific taxa (Wilson 2004; Chapter 2). The genus is separated into two main groups: rhizomatous and bulbous, and is currently divided into six subgenera, *Iris*, *Limniris*, *Xiphium*, *Nepalensis*, *Scorpiris*, and *Hermodactyloides* (Wilson 2004; Chapter 2), among which three, *Hermodactyloides*, *Scorpiris*, and *Xiphium* are bulbous. The major distribution areas of *Hermodactyloides* are in Central Asia, while those of *Scorpiris* are the Middle East to Central Asia and those of *Xiphium* are southern Europe and northern Africa (Mathew 1989).

Chromosome numbers are 2n = 34 in *I. xiphium* (distributed in northern Morocco and the Iberian Peninsula), 2n = 28 in *I. tingitana* (endemic to northwestern Morocco) (Hekstra and Broertjes 1968), 2n = 30 in Morroccan *I. filifolia*, 2n = 34 in Iberian *I. filifolia*, and 2n = 42 in *I. latifolia* (endemic to the Iberian Peninsula) (Martínez et al. 2010). 'Wedgwood' has 2n = 31 (Hekstra and Broertjes 1968), indicating its hybrid origin.

C. MORPHOLOGY AND ANNUAL GROWTH CYCLE

Iris bulbs consist of scales and leaf bases, and are covered by tunics. The scales do not form complete circles, but meet at their edges, the innermost one being a half-scale. The inflorescence usually has two flowers, which are enclosed within separate spathes, and open sequentially. A flower consists of three horizontal or partly erect perianth segments (falls) of the outer whorl and three smaller and erect perianth segments (standards) of the inner whorl. The ovary is inferior. The three anthers are inserted at the base between the outer perianth parts and the broad, petalloid, and bifid styles, which have two crests at the tip. The major anthocyanidin contributing to the blue coloration in Dutch iris flowers is delphinidin (Asen et al. 1970). The leaves are sword-shaped with parallel veining. The first leaf is radical and the others are on the emergent shoot. Dutch irises develop contractile roots.

In commercial production, the bulbs are planted in autumn, and flower in late spring. Daughter bulbs grow in early spring and continue their enlargement until early summer, when the leaves wither. Large bulbs produce flower stems and the daughter bulbs with a flat side like a tulip (flat bulbs). Small nonflowering bulbs produce a round bulb (round bulbs).

Freshly harvested bulbs are "dormant" for about 2 months and a temperature of >17°C is required for the release of dormancy. Flower initiation takes place after planting and is promoted by low temperatures with an optimum of 8–10°C for 6–9 weeks, depending on the cultivars (Imanishi 2005).

D. MAJOR SPECIES AND CULTIVARS

Two species, *I. xiphium* (vars. *praecox* and *lusitanica*) and *I. tingitana*, have contributed significantly to the development of Dutch iris cultivars (De Munk and Schipper 1993). *I. filifolia*, *I. latifolia*, and *I. fontanesii* were also employed in the hybridizations. *I. xiphium* produces small bulbs and has contributed to a wider range of flower colors (white, blue, yellow, and bronze), whereas *I. tingitana* produces large bulbs and has contributed to early flowering.

In The Netherlands, about 100 cultivars were cultivated on 280 ha in the 2009/2010 season, with 'Blue Magic' (87.4 ha) being the leading cultivar, followed by 'Telstar' (45.6 ha), 'Discovery' (12.7), 'Blue Diamond' (9.2 ha), 'Casablanca' (8.9 ha), 'Superstar' (6.4 ha), 'Danfordiae' (4.3 ha), and 'Professor Blaauw' (4.2 ha) (PT/BKD 2010a).

Depending on the critical bulb size for flowering, Dutch irises are divided into cultivars with large and small bulbs[*]. In The Netherlands, large-bulb (8/9 cm, in circumference) cultivars are the majority, while 'Purple Sensation', 'Sapphire Beauty', and 'White Excelsior' are the major small-bulb (5/6 cm) cultivars.

E. PROPAGATION

For commercial bulb production, Dutch irises are propagated only vegetatively. The propagation rate is low: only two to four new bulbs are produced annually by one mother bulb. *In vitro* propagation from inflorescence stems, leaves, and scales has been reviewed by Kim and De Hertogh (1997), but is not used commercially.

F. PRODUCTION

1. Agronomy

Dutch iris bulbs can be produced in different types of soils, but sandy and fertile soils are preferable. The soils must be well drained, with a pH ranging from 6 to 7. Cultural rotation of the fields with a minimum of 4-year intervals is recommended. Nutrient requirements per hectare are 150 kg nitrogen, 70–100 kg calcium, 20–30 kg magnesium, 25–40 kg phosphorus, and 100–200 kg potassium (De Munk and Schipper 1993). Aerial growth ceases at about <6°C, while high soil temperatures cause short stems. Poor light conditions may cause flower abortions.

2. Bulb Production

Marketable bulbs must be round and of flowering size. Intermediate-size bulbs are used for propagation, since smaller propagation material produces smaller bulbs. The minimal flowering size is about 7/8 cm for large-bulb cultivars and 4/5 cm for small-bulb cultivars. In The Netherlands, a high-temperature treatment at 30°C is recommended to prevent flower development within the bulb and to enhance bulb enlargement. This treatment is applied for 2 weeks during the storage period and prior to cooling at 9°C for the large-bulb cultivars (13–15°C for some cultivars such as 'Ideal')

---

[*] "Large" and "small" are the characteristics of cultivars, classified by the minimum critical size of bulbs for flowering.

or 5°C for small-bulb cultivars. The optimum growing temperatures in the field is 15–17°C with an allowable maximum and minimum range of 5°C and 25°C, respectively.

3. Flower Production

Flowers are available throughout the year by using a combination of accelerated, outdoor, and retarded flowering.

a. Accelerated Flowering

A period of warm storage of harvested bulbs is necessary for successful forced flowering. This requirement depends on the location of the bulb production. It is less important for the bulbs harvested in warmer-climate regions, for example, Israel and Japan. In The Netherlands, however, a treatment of 30°C for 1 week or 10 days is recommended to satisfy the requirement of the bulbs and to bring them into a state of physiological maturity (= maturation)[*] (De Munk and Schipper 1993). In order to promote reproductive development, bulbs are exposed to ethylene. The concentrations of ethylene and duration of treatments vary according to the cultivar. For example, ethylene application in a concentration of 10 μL/L for 3 h prior to cold treatment is effective for 'Blue Magic' (Yue and Imanishi 1988, 1990). In commercial forcing in Japan, exposure for 8–10 h at 10 μL/L is repeated three times (Imanishi 2005). This treatment facilitates the "maturation" or "preparation of flower initiation" of the bulbs between the stages of dormancy breaking by high temperatures and that of flower initiation at low temperatures.

The bulbs are transferred to dry storage at 9°C for 10 weeks for 'White Wedgwood', 'Blue Diamond', and 'Ideal', and 13 weeks for 'Blue Magic' and 'Professor Blaauw'. Subsequently, they are planted in greenhouses and grown at 13–18°C. Anthesis takes place after 7–9 weeks, depending on growth conditions.

The chemical control of plant height for potted irises has been reported. The immersion of bulbs in 20 mg/L of paclobutrazol reduced the plant heights of 'Casablanca' and 'Professor Blaauw' by 41–44% against the controls. Flowering, however, was delayed by 3–4 days (Francescangeli 2009).

b. Retarded Flowering

Respiration of iris bulbs is suppressed at 28–30°C, resulting in slow leaf development within the bulbs without losing their flowering ability. These bulbs are used for retarded flowering. Since dry matter content is reduced slowly during the long storage period at high temperatures, only large-sized bulbs are used for this purpose. Cold treatments and durations are similar to those for forcing.

G. Physiological Disorders, Pests, and Diseases

Blindness[†] is caused by the use of small bulbs or inappropriate temperatures for bulb maturation and cooling. Many factors can cause blasting[‡] and flower abortion[§]. For example, the use of small bulbs, diseased bulbs, poorly rooted bulbs, poor watering management, high planting density, or high growing temperatures. Development of long leaves and short floral scapes is usually caused by too long or too short a storage period at low temperatures after flower induction. Poor ventilation can also cause flower malformations.

Xanthomonas campestris pv. *tardicrescens* (bacterial blight), *Pectobacterium carotovorum* pv. *carotovorum* (bacterial soft rot), and *Pseudomonas marginalis* pv. *marginalis* are the main bacterial pathogens. Major fungal diseases are gray mold caused by *Botrytis cinerea*, leaf blight by

---

[*] Bulb maturation is the ability of the bulbs to respond to flower induction treatments.
[†] Blindness is the failure of a bulb to initiate any floral parts.
[‡] Blasting is the failure of a bulb to produce a marketable flower after the floral organs have been differentiated.
[§] Flower abortion occurs when flower buds wither and die at the latest stages of development, or wilt and turn brown prior to anthesis.

Mycosphaerella macrospora, basal rot or dry rot by *Fusarium oxysporum*, sheath blight by *Rhizoctonia solani*, blue mold by *Penicillium* spp., southern blight by *Sclerotium rolfsii*, and rust by *Puccinia iridis*.

Bean yellow mosaic virus (BYMV), *Iris mild mosaic virus* (IMMV), *Narcissus latent virus* (NLV), *Cucumber mosaic virus* (CMV), *Tomato spotted wilt virus* (TSWV), and *Iris severe mosaic virus* (ISMV) are the major virus diseases. *Aphelenchoides* spp., *Ditylenchus destructor*, *Ditylenchus dipsaci*, *Meloidogyne* spp., and *Pratylenchus penetrans* are the nematodes that can infest Dutch irises. Aphids during growth and mealy bugs during bulb storage are the major pests.

IV. GLADIOLI

A. INTRODUCTORY REMARKS

Gladiolus is commonly known as "sword lily" because of its leaf shape. Its etymology is from the Latin *gladius*, which means sword. The genus originated from South Africa. There are two cultivar types of gladioli: "spring flowering" and "summer flowering". The latter is more important and is widely used for landscaping and cut flower production. This chapter focuses on the summer-flowering gladioli. Although the production of gladiolus has been decreasing over the last 20 years worldwide (Benschop et al. 2010), the plant is still an important ornamental geophyte, particularly in warm-climate countries. For example, it is the second most important cut flower crop in India after roses (Export-Import Bank of India 2009).

According to Wilfret (1980), *Gladiolus* species were known more than 2000 years ago. Prior to 1730, the major garden species in England were *G. communis*, *G. segetum*, and *G. byzantinus*. Several South African species were carried to England, starting from 1737. The first important gladiolus hybrid was produced in 1823 at Colville's Nursery in Chelsea, England, where *G. tristis* var. *concolor* was pollinated by *G. cardinalis* to produce the Colvillei hybrids. These are spring-flowering types. Breeding of summer-flowering cultivars started with the introduction of *G. natalensis* (*psittacinus*)[*] from South Africa. It was crossed with *G. oppositiflorus* to produce *G. × gandavensis*. The history of *Gladiolus* hybridization and breeding is illustrated in Figure 4.1.

B. BOTANICAL CLASSIFICATION AND DISTRIBUTION

The genus *Gladiolus* is a member of the Iridaceae family, subfamily Ixioideae, and is classified into eight sections. The genus includes 255 species distributed in Africa, Madagascar, and Eurasia (Goldblatt and Manning 1998). Only 10 species originated from Mediterranean areas, western Asia, and Europe (Goldblatt and Manning 1998). They flower from autumn to spring. Examples are *G. byzantius*, *G. communis*, and *G. italicus*.

Many species distributed in the Cape region of South Africa, which is considered to be the center of diversification of the genus, are diploid (2n = 30). These are *G. alatus*, *G. oppositiflorus*, and *G. tristis*, whereas those in northern and central Africa have various ploidies, such as *G. dalenii* (2n = 60 and 90) and *G. papilio* (2n = 75) (Cohat 1993; Goldblatt et al. 1993). All Mediterranean species are polyploid. They are 2n = 90–180 in *G. communis* and 2n = 120 in *G. italicus*.

C. MORPHOLOGY AND ANNUAL GROWTH CYCLE

The storage organ is the corm, covered by several layers of brownish, fibrous tunics. Cormels, produced on the top end of the stolons, have the structure of corms, but possess a hard tunic. Usually, only one apical bud sprouts from each corm, but some cultivars with large corms develop two shoots. Generally, summer-flowering gladioli produce one inflorescence of 0.8–2 m height, whereas spring-flowering cultivars bear multiple shoots of 0.8–1 m height.

---

[*] *G. natalensis* is now included in *G. dalenii.*

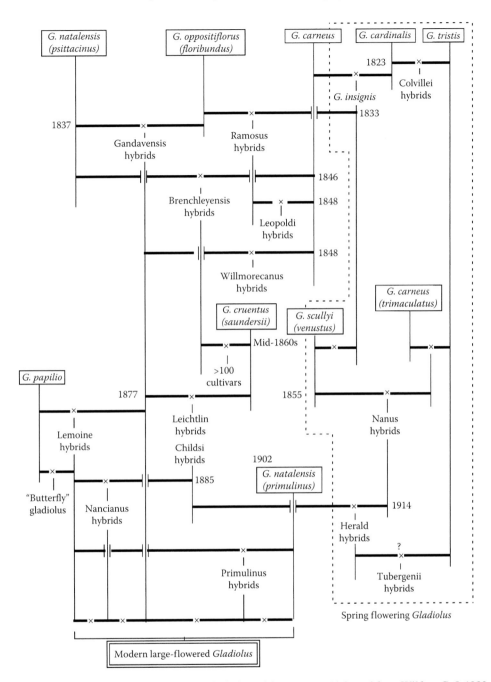

FIGURE 4.1 Development and breeding of *Gladiolus* cultivar groups. (Adapted from Wilfret, G. J. 1980. In *Introduction to Floriculture*, ed. R. A. Larson, 165–181. New York: Academic Press; In *The Physiology of Flower Bulbs*, Le Nard, M., and A. A. De Hertogh (eds). Plant breeding and genetics, 161–169, Copyright (1993a). Amsterdam: Elsevier.)

The stems are generally unbranched with up to 12 narrow, sword-shaped, longitudinally grooved leaves, enclosed in a sheath. The inflorescence is a spike with 30 or more florets. Individual florets are enclosed in two green spathe valves. The corolla is tubular, usually funnel-shaped, with three upper perianth segments, which are generally much larger than the three lower ones. Three stamens are inserted below the throat. The ovary is inferior.

Six major anthocyanidins are found in the genus. Thus, there is a wide range of flower colors in the modern gladiolus cultivars. The major anthocyanins of the purple flowers are malvidin glycosides together with petunidin glycoside as a minor component; red flowers are due to pelargonidin glycosides; and pink flowers contain various anthocyanins, pelargonidin, cyanidin, peonidin, petunidin, and malvidin glycosides (Takemura et al. 2008). However, anthocyanins were not detected in most yellow and white flowers. Kaempferol and quercetin were also found to contribute to the purple coloration in gladiolus (Takemura et al. 2005).

For commercial production, corms are harvested in autumn, when they are in deep "dormancy". The dormancy is gradually broken by low temperatures in winter.

There are no special temperature requirements for flower initiation and development. Flower differentiation of the first floret takes place when two foliage leaves appear after planting, while the top floret differentiates at the 6–7-leaf stage. Since florets develop acropetally, the first floret reaches the gynoecium stage when the top floret is being initiated. Therefore, florets at all the developmental stages are present in one inflorescence. The rate of flower development is generally parallel to that of visible leaf development. Pollen and embryo formation in the oldest florets are visible at the seven-leaf stage. When the flower stalk elongates rapidly to reach anthesis of the oldest florets, they have eight to nine foliage leaves. Short photoperiods and high light intensities accelerate flower development (Shillo and Halevy 1976a,b,c). Low light intensities, particularly during the four to seven foliage leaf stages, can cause either flower abortion or a lower number of florets at anthesis (Imanishi and Imae 1990).

A new corm is produced by the swelling of the base of the shoot at the 6–7-leaf stage, and at the same time, contractile roots are produced at the base of the new corm. The renewal corm grows rapidly after anthesis. Simultaneously, stolons develop from the buds situated between the mother and newly forming daughter corms to produce numerous cormels at their upper end.

D. MAJOR SPECIES AND CULTIVARS

Modern cultivars are considered to have been bred originally from only a relatively small number of species (Wilfret 1980; Benschop et al. 2010). They are *G. carneus*, *G. cruentus* (*saundersii*), *G. dalenii* (former *G. natalensis*, *G. psittacinus*), *G. oppositiflorus*, and *G. papilio* for summer-flowering cultivars and *G. cardinalis*, *G. carneus* (*trimaculatus*), *G. tristis*, and *G. scullyi* (*venustus*) for spring-flowering cultivars (Figure 4.1).

Over 10,000 cultivars have been recorded, although most are no longer available (Rees 1992). Cultivars that were produced in >10 ha plots in The Netherlands in 2009/2010 are 'Fidelico', 'Green Star', 'Jessica', 'Nova Lux', 'Peter Pears', 'Plum Tart', 'Priscilla', 'Traderhorn', and 'White Prosperity'.

Over the years, the complex and intense breeding efforts using interspecific hybridization have resulted in parentage confusion. Thus, cultivar classification is not defined according to the species of origin. Instead, the North American Gladiolus Council have classified them by assigned numbers, resulting in a three-digit designation according to flower size, color, and form (Table 4.2). The first parameter is the width of the fully expanded largest floret, the second is color code, and the third is the intensity of the color from a pale to a deep hue, which is indicated by even numbers, while odd numbers represent the presence of distinct markings. As an example, flowers of 'Pink Elegance' are large (4), pale pink (4) with a white throat (3); thus the code is 443. For the same reason, *Gladiolus × hybridus* is commonly used as the Latin name of summer-flowering gladioli.

E. PROPAGATION

Commercial propagation is done by cormels. After planting, one cormel produces one shoot, and subsequently one corm with numerous cormels. The division of corms with at least one axillary bud is possible for accelerating the propagation rate, which is also achieved by tissue culture (Ziv and

TABLE 4.2
Classification of *Gladiolus* Cultivars Used by the North American Gladiolus Council

Class	Designation	Diameter of the Lowest Floret without Spreading or Flattering of the Perianth Segments (cm)
100	Miniature	<6.4
200	Small or miniature	≥6.4 to <8.9
300	Decorative	≥8.9 to <11.4
400	Standard or large	≥11.4 to ≤14.0
500	Giant	>14.0

			Depth		
Color	Pale	Light	Medium	Deep	Other
White	00				
Green	02	04			
Yellow	10[a]	12	14	16	
Orange	20[b]	22	24	26	
Salmon	30	32	34	36[c]	
Pink	40	42	44	46	
Red		52[d]	54	56	58 black-red
Rose	60	62	64	66	68 black-rose
Lavender	70	72	74	76	78 purple
Violet	80	82	84	86	
Smokies		92	94	96	
Tan	90				98 brown

Source: Adapted from Fairchild, L. 1979. *How to Grow Glorious Gladiolus*. Edgewood, Maryland: The North American Gladiolus Council.

[a] Including cream.
[b] Including buff.
[c] Including orange-scarlet.
[d] Including red-scarlet.

Lilien-Kipnis 1990; Kim and De Hertogh 1997; Chapter 10). Seed propagation is not common, and is used only for breeding purposes.

F. PRODUCTION

1. Agronomy

Sandy loam soils with good drainage are essential for corm production. The optimum pH is 6.5 with the preferred range being between 6.0 and 7.5. A 3–5-year crop rotation or soil fumigation is required to avoid the carryover of soil-borne diseases. Adequate moisture of the soils is necessary from planting to leaf sprouting. An example of the nutrition in France is 122 kg/ha of N, 36 kg/ha of P_2O_5, 257 kg/ha of K_2O, 150 kg/ha of CaO, and 34 kg/ha of MgO (Cohat 1993). Ammonia from nitrogen and low pH may enhance *Fusarium*. Poor light conditions can cause flower abortions.

2. Corm Production

Depending on market preferences, gladiolus corm production might be combined with flower production. However, harvesting of the spikes for market and leaving only four leaves for photosynthesis resulted in the reduction of the corm yield by 30%, as compared with plants with intact foliage (Rees 1992).

This crop is not winter hardy and is planted in the colder regions of the Northern Hemisphere (e.g., The Netherlands) in spring, after overwinter storage. The research carried out by Grabowska (1978) in Poland showed that the highest total and marketable yields of gladiolus corms were obtained when planting was done in the second half of March. However, common protocols in The Netherlands and Poland indicate April as a time of gladioli planting due to the possibility of frost in spring. In warmer regions, such as Florida, USA or the Mediterranean countries, gladioli are often grown as a winter crop.

An effective chemical weed control program is essential for commercial corm production due to the long vegetative growth cycle of this crop, about 6–7 months. The topping of gladiolus increases the corm yield, but it is usually done after the opening of the first floret in order to allow rouging. The lifting date depends on the cultivar and climate conditions. Corms are generally lifted when the leaves are still green. Although later lifting results in higher yields, numerous disease problems can be encountered with late harvesting. After lifting, corms are initially dried at 20–30°C for a few days, then cleaned, and stored for 2–3 weeks at 15–23°C with good ventilation (Cohat 1993). From January onwards, the corms are stored at low temperatures, with the optimum at 5–8°C.

3. Flower Production

The majority of gladiolus flowers are produced in open fields. Corms of >2.5 cm diameter are used for flower production. They are planted continuously from spring to early summer for a continual harvest of floral spikes. Depending on the growing temperatures, it takes 60–120 days from planting to flowering.

In field production, a dormancy-breaking treatment is not essential, since corm dormancy is broken by low temperatures in winter. For forcing culture, a low temperature of <10°C is generally given after harvest, drying, and the removal of the old roots and corms. The duration of cold treatment is 4–8 weeks, depending on the cultivars, lifting dates, and location of the mother plant production. For flower forcing, it is suggested to keep the soil temperatures above 7–9°C until sprouting, and the air temperature at 25–30°C/15°C (day/night) until flowering (Cohat 1993). Retarded flowering is achieved by using corms stored at <5°C and then at 15–20°C for 1–2 weeks before planting.

G.   Physiological Disorders, Pests, and Diseases

Flower abortion is a major physiological disorder. Short photoperiods and low light intensities may cause blasting or a decrease in floret numbers in retarded flower production. Crooked stems are caused by high day temperatures.

The known fungal pathogens are *Botrytis gladiolorum* (gray mold), *Curvularia trifolii* f. sp. *gladioli* (brown spot), *Fusarium oxysporum* f. sp. *gladioli* (soft rot or brown rot), *Mycosphaerella macrospora* (leaf blight), *Penicillium gladioli* (corm rot), *Rhizoctonia solani* (collar rot, leaf-base rot), *Sclerotium rolfsii* (southern spot), *Stemphylium* sp. (leaf and stem spot), *Stromatinia gladioli* (dry rot), *Urocystis gladioli* (smut), and *Uromyces transversalis* (rust). Bacterial scab (neck rot) and bacterial blight are prevalent in *Burkholderia gladioli* ssp. *gladioli* and *Xanthomonas gummisudans*, respectively. Among them, *Fusarium* and *Stromatinia* are the most destructive diseases.

The gladiolus plants are highly susceptible to viral infections. These can be caused by *Bean yellow mosaic virus* (BYMV) and *Cucumber mosaic virus* (CMV), *Tomato ring spot virus* (TRSV), *Tobacco mosaic virus* (TMV), *Tobacco rattle virus* (TRV), *Tobacco ring spot virus* (TobRSV), *Arabis mosaic virus* (ArMV), and *Cycas necrotic stunt virus* (CNSV). Root knot is caused by several *Meloidogyne* species. *Aphelenchoides* sp. and *Ditylenchus dipsaci* (plant parasitic nematode) have also been reported. Aphids, thrips, wireworms (*Agriotes* spp.), corn earworm (*Heliothis zea*), and bulb mite (*Rhizoglyphus echinopus*) are the major pests.

V. HYACINTHS

A. INTRODUCTORY REMARKS

Hyacinths are very popular ornamental geophytes. They are used as garden plants and potted or cut flowers. They are widely used for interior decoration in winter and are adapted to hydroponic culture. The flowers have an intense but pleasant fragrance. In addition to the flowers, the foliage is also decorative, and numerous long unbranched white roots look beautiful in transparent glasses and pots (hydroponics).

According to Greek mythology, Hyakinthos was a beautiful youth loved by Apollo and Zephyr. When Hyakinthos and Apollo played throwing a discus, Hyakinthos was accidentally struck by it and died. It is believed that a jealous Zephyr blew Apollo's discus off course and, thus, it hit and killed Hyakinthos. Apollo made a flower from Hyakinthos's spilled blood, which became the hyacinth (see Chapter 1).

Hyacinthus orientalis has been cultivated in the Mediterranean region for centuries. Blue, white, and purple cultivars were imported into Western Europe in the middle of the sixteenth century (Doorenbos 1954). Double-flowered cultivars were described in 1612, with pink and reddish flowers appearing in 1709, and yellow cultivars in about 1760 (Doorenbos 1954). Breeding started in The Netherlands at the beginning of the eighteenth century, when hyacinths were the most fashionable flower bulb. In the early 1900s, Nicolaas Dames demonstrated that Dutch-grown bulbs could be forced into flower before Christmas by early lifting and special bulb treatments (Doorenbos 1954). Over 2000 cultivars were grown at the peak of its popularity. Cultivars adapted for forcing still include old cultivars such as 'L'Innocence' (bred in 1863), 'City of Haarlem' (1893), and 'Pink Pearl' (1922) (Van Scheepen 1991).

B. BOTANICAL CLASSIFICATION AND DISTRIBUTION

Hyacinthus orientalis (Hyacinthaceae), the only species of horticultural importance, is indigenous to Syria, Asia Minor, Greece, and Dalmatia (Nowak and Rudnicki 1993).

C. MORPHOLOGY AND ANNUAL GROWTH CYCLE

The perennial bulb is composed of swollen leaf bases and specialized scales that are replaced gradually over the years, with three generations of scales existing in one large bulb. It is covered with papery tunics, whose colors are often similar to those of the flowers. Four to six strap-shaped, thick and green leaves, 15–35 cm long and 1–3 cm wide, sprout from the basal whorl. The scape grows to 20–35 cm height. It bears a dense raceme with many fragrant flowers. Each floret is bell-shaped, 2–3.5 cm long, and six-lobed. There are six stamens and the ovary is superior.

In practice, the bulbs are lifted in early summer, when the apical meristem is vegetative. Flower initiation occurs within the bulb in summer and inflorescence development occurs at 20–28°C, with the optimal being 25°C. When the stage of the top floret development reaches anther differentiation, temperatures of 17–20°C accelerate inflorescence development. The minimum size of the bulb for flower initiation is >7 cm, in circumference, but >17/18 cm bulbs are preferable for forcing. After planting, they require 10–18 weeks of 9–13°C for subsequent growth and flowering. In bulbs planted outdoors in autumn, rooting occurs first and flowering takes place in the spring. Photoperiod has no effect on flower initiation and development nor on anthesis.

D. CULTIVARS

About 100 cultivars are currently in commercial production in The Netherlands. The major cultivars are 'Pink Pearl' (142.7 ha), 'Delft Blue' (112.6 ha), 'White Pearl' (109.2 ha), 'Fondant' (49.2 ha),

'Blue Star' (48.3 ha), 'Jan Bos' (46.8 ha), 'Blue Pearl' (43.4 ha), and 'Carnegie' (37.6 ha). In the 2009/2010 season in The Netherlands, bulbs were produced in an area of 869.3 ha (PT/BKD 2010a).

In contrast to other ornamental geophytes, all hyacinth cultivars have been bred from only one species, *H. orientalis*, and there are no interspecific hybrids. However, even with this level of genetic resources, progress has been made in breeding for flower colors, shape, and disease resistance. There are red, pink, orange, salmon, yellow, purple, blue, and white cultivars. The dominant antho-cyanidins are delphinidin in cultivars with blue flowers, and cyanidin or pelargonidin in cultivars with red or pink flowers (Hosokawa 1999). Flower colors can be roughly judged by the tunic colors before planting. Large morphological differences are probably reflected by the variability in chromosome numbers. 'Pink Perl', 'Lord Balfour', 'Gertrude', and 'Yellow Hammer' are diploids, having 16 chromosomes ($2n = 2x = 16$) (Doorenbos 1954). 'Bismarck', 'Jan Bos', and 'King of the Blues' are triploids ($2n = 3x = 24$), and 'Blue Giant' is tetraploid ($2n = 4x = 32$) (Darlington 1973). However, there are also a large number of heteroploid cultivars such as 'Rosalie' ($2n = 2x + 1 = 17$), 'City of Haarlem' ($2n = 3x - 1 = 23$), 'Ostara' ($2n = 3x + 2 = 25$), 'L'Innocence' ($2n = 3x + 3 = 27$), 'Carnegie' ($2n = 4x - 3 = 29$), 'Delft Blue' ($2n = 4x - 2 = 30$), and 'Blue Bird' ($2n = 4x - 1 = 31$) (Doorenbos 1954; Van Tuyl 1982).

E. PROPAGATION

Propagation is possible by dividing newly formed offsets (bulblets) from the mother bulbs. However, since this natural propagation rate is low, the techniques of "scooping" or "cross-cutting (scoring)" are used commercially (see Chapter 10). Scooping was developed around 1715 (Doorenbos 1954), and cross-cutting was first used in about 1935 (Rees 1992). While the number of new bulblets obtained by scooping (ca. 40–60) is higher than that obtained by cross-cutting (ca. 12–24), their size is smaller. It takes, therefore, about 3 years to obtain commercial-sized bulbs by scooping and only 2 years by cross-cutting. "Leaf cuttings" can also be used. The use of tissue culture for hyacinth propagation has been reviewed by Paek and Thorpe (1990), Bach (1992), and Kim and De Hertogh (1997) (see also Chapter 10). Seeds are not commercially used for propagation.

F. PRODUCTION

1. Agronomy

Light sandy loam soils enriched with organic matters with a soil pH between 6 and 7 are preferable. The soil must be well drained, but should retain moisture. Rotations of planting fields with intervals of at least 2–3 years are recommended to avoid disease carryover. In The Netherlands, 40 kg N, 120–160 kg P_2O_5, and 200–400 kg K_2O per hectare are applied before planting and an additional application of 80–100 kg N/ha is used in spring (Nowak and Rudnicki 1993).

Hyacinth requires a period of lower temperatures for subsequent growth and flowering, but during winter, mulching is necessary in very cold regions (climate zones of 3–4). High temperatures immediately applied after low temperatures can cause "spitting". Hyacinth grows best in full sun to partly shade.

2. Bulb Production

Most hyacinth bulbs are produced in The Netherlands. The bulbs are planted in late September to mid-November at a depth of 12 cm with a planting density of about 15–18 cm, depending on the bulb size. Flower stalks and/or florets are not removed during bulb production. Bulbs with a few newly formed offsets are lifted after foliage senescence. The bulbs are stored dry at 25.5°C in ethylene-free, well-ventilated conditions. In The Netherlands, to prevent "yellow disease" of the bulbs, which is caused by *Xanthomonas hyacinthi*, small-sized bulbs are stored at 30°C until 1 September, then at 38°C for 2 weeks, followed by 44°C for 3 days, and then again at 25.5°C (Nowak and Rudnicki 1993).

3. Flower Production

Hyacinths are primarily used as potted plants, but they can be used for cut flowers and landscaping. Forcing techniques have been developed in The Netherlands. By using the two types of bulbs, "prepared" (PR) and "regular" (RG) bulbs, for early and late forcing, respectively, and following the programming chart as shown in Figure 4.2, a constant production of flowering pots from 20 December to 20 April is possible. As sensitivity to low temperatures varies according to cultivar, cultivar selection is critical in forcing (Table 4.3).

Height control is important in potted hyacinth, and the use of chemicals has been reported. Ethephon spray of whole plants with 1000 or 2000 mg/L before floret coloration is used to prevent excessive stem elongation and stem topple (Shoub and De Hertogh 1975; De Hertogh 1996). Flurprimidol preplant bulb soaks were shown to be more effective in reducing plant height than uniconazole or paclobutrazol, while ethephon and flurprimidol foliar sprays were ineffective, and flurprimidol drenches had a minimal effect (Krug et al. 2005c; see also Chapter 11). Cultivars vary in their response to plant growth regulators (Krug et al. 2006).

Potted hyacinths are marketed at the stage of coloration of the lowermost florets. If stored at 18°C, the plant remains at this stage for up to 17 days in 'Ostara' and 10 days in 'Amsterdam' (De Hertogh 1996). They can also be stored at the green bud stage without any loss of ornamental value at 0–2°C for up to 4 weeks before marketing. For cut flowers, stems can be stored at 0–2°C upright with the bulbs attached.

G. PHYSIOLOGICAL DISORDERS, PESTS, AND DISEASES

Stem topple (a collapse of inflorescence stalk at flowering) is a well-known disorder in hyacinths. The process can be controlled by ethephon spray (Shoub and De Hertogh 1975). However, newly registered cultivars 'Baltic Sea', 'Deep Sea', and 'Woodbells' possess strong floral stalks and produce quality flower pots (Benschop et al. 2010). Spitting, straw-nails (abortion of the uppermost floret), green tops, crooked flower clusters, and top flowering are other physiological disorders; however, spitting is the main one. In general, all physiological disorders are caused by improper storage or forcing conditions.

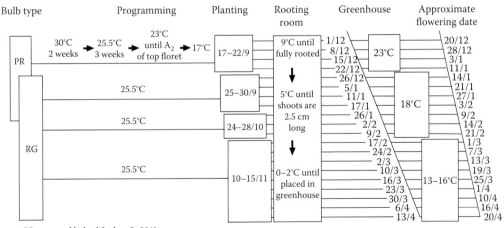

PR-prepared bubs, lifted on 5–20/6.
RG-regular bulbs, early cultivars, lifted on 25/6–5/7,
late cultivars, lifted on 5–15/7.

FIGURE 4.2 Programming chart for hyacinth forcing. (Adapted from De Hertogh, A. 1996. *Holland Bulb Forcer's Guide 5th ed.* Hillegom, The Netherlands: The International Flower Bulb Centre and The Dutch Bulb Exporters Association.)

TABLE 4.3
Major Hyacinth Cultivars Suitable for Particular Flowering Periods (Shown by ×) by Forcing

	Flowering Period in Forcing					
From	20 December	14 January	9 February	1 March	19 March	10 April
To	13 January	8 February	28 February	18 March	9 April	20 April
Cultivar (color)						
Jan Bos (red)	×	×	×	–	–	–
L'Innocence (white)	×	×	×	–	–	–
Anna Marie (pink)	×	×	×	×	–	–
Ostara (blue)	×	×	×	×	–	–
Delft Blue	×	×	×	×	×	–
Amsterdam (red)	×	×	×	×	×	×
Pink Pearl	×	×	×	×	×	×
White Pearl	×	×	×	×	×	×
Carnegie (white)	–	×	×	×	×	×
Blue Jacket	–	–	×	×	×	×
Lady Derby (pink)	–	–	×	×	×	–
Yellow Queen	–	–	×	×	×	–
Amethyst (violet)	–	–	–	–	×	×
Marconi (pink)	–	–	–	–	×	×
Queen of Pinks	–	–	–	–	×	×

Source: Adapted from De Hertogh, A. 1996. *Holland Bulb Forcer's Guide 5th ed.* Hillegom, The Netherlands: The International Flower Bulb Centre and The Dutch Bulb Exporters Association.

The bacteria *Xanthomonas hyacinthi* causes yellow disease, while *Pectobacterium carotovorum* causes soft-rot disease. The fungi *Penicillium* spp. and *Aspergillus niger* cause blue mold and black mold, respectively. Top rot is caused by *Rhizoctonia solani*, and fire is caused by *Botrytis hyacinthi*.

Hyacinth mosaic virus (HMV) and *Tobacco rattle virus* (TRV) affect hyacinths. Nematodes of *Ditylenchus dipsaci* and *Pratylenchus penetrans* attack the whole plant and roots, respectively. *Meloidogyne* sp. is known to cause root knot. Aphids, thrips, and mites can also affect hyacinths.

VI. LILIES

A. INTRODUCTORY REMARKS

The etymology of the name *Lilium* is from the Latin *li* (= white) + *lium* (= flower). Probably because of its beauty, the sign of purity (the pure white flowers of *L. candidum* and *L. longiflorum*), or charm, "lily" has also become a girl's given name, such as Lilian in English, or Susan, from the lily (Shoshan) in Hebrew. The Japanese names Yuri and Sayuri are from the Japanese "yuri", which means lily.

Lily has a long history in human life and culture in various regions of the world. In Europe, lily figures can be seen in Cretan vases and frescoes of the Middle Minoan IIIA-B period (ca. 1750–1675 B.C.). It is *L. candidum*, which has long been important in Christianity. The flowers often appeared in the paintings of the Annunciation by many painters, including Leonardo da Vinci (Galleria degli Uffizi, Firenze).

In China, a set of two Chinese letters pronounced "Pai Ho" that corresponds to *L. brownii* var. *colchesteri* first appeared in the book Shen Nong Ben Cao Jing (Shen Nong's Herbal), compiled by Tao Hongjing (A.D. 453–536) during the Southern and Northern dynasties. In this book, the lily is considered a natural medicine. The illustration of the lily probably first appeared in the book *Jiu Huang Ben Cao* by Zhu Di in 1406, which contains the descriptions and figures of more than 400 wild and cultivated plant species that can be consumed for diet or in case of hunger. To date, *L. brownii* var. *colchesteri* is commercially cultivated for its bulbs for herbal medicine and as an edible plant. Bulbs of *L. lancifolium* and *L. davidii* are also used for table dishes in China, as well as *L. leichtlinii* var. *maximowiczii* in Japan.

The first mention of lily in Japan is found in the oldest history (or mythology) book *Kojiki* (A.D. 712). The lily is considered to be *L. japonicum*. Later, *L. auratum* appeared in the history book "Nihonshoki" (A.D. 720). Eleven poems about lilies are included in the first poem anthology "Man-yo-shu" (A.D. 759). Various species have long been used in paintings and the literature and, later, in Ikebana (flower arrangements). The first scientific description of Japanese lilies is found in *Amoenitatum Exoticarum* published in 1712 by Engelbelt Kempfer. However, no commercial lily production existed in Japan until about 1882, when the export of bulbs began.

The introduction of Asian lilies into Europe started with *L. dauricum* from Siberia in 1743. Later, C. P. Thunberg described Japanese *L. longiflorum*, *L. japonicum*, *L. speciosum*, and *L. callosum* in *Flora Japonica* published in 1784. In 1830, P. F. B. von Siebold brought live bulbs of seven species to The Netherlands from Japan, but only *L. speciosum* flowered in 1832. Further introductions from Japan included *L. auratum*, *L. japonicum*, and *L. rubellum*. The Chinese species *L. lancifolium* and *L. brownii* were introduced into England in 1804, and *L. regale* in 1905. These Asian species were exotic to Europeans and were highly praised. John Lindley (1837) wrote on *L. speciosum*: "All the lilies previously seen in Europe, however beautiful they may be, are quite thrown into the shade by this glorious species ... Not only is it handsome beyond all we before knew in gardens, on account of the clear, deep rose-color of its flowers, which seem all rugged with rubies and garnets, and sparkling with crystal points ... for surely if there is anything not human, which is magnificent in beauty, it is this plant".

Lily breeding in Europe was initiated in the middle of the nineteenth century, but major breeding research occurred in Europe and the United States only after World War II. The history of recent cultivar development is described by Benschop et al. (2010) and Van Tuyl et al. (see Chapter 6 in this volume) and is also illustrated in Figure 4.3.

The current popularity and importance of lily flowers have resulted not only from breeding new cultivars with large diversity in plant architecture, flower shapes, colors, sizes, and fragrance, but also from the wordwide efforts to establish year-round production systems.

B. Botanical Classification and Distribution

The genus *Lilium* belongs to the family Liliaceae and comprises approximately 100 species native to North America, Europe, and Asia (Beattie and White 1993) between the latitudes of 63°N in Kamchatka, Russia (*L. tenuifolium* and *L. medeoloides*) and 11°N in southern India (*L. neilgherrense*) (Rees 1992). Among them, 39–47 (Liang 1980; Xia et al. 2005) species are found in China and 15 in Japan, seven of which are endemic (Shimizu 1987). About 25 species originate in North America and about 12 in Europe. The classification of the genus is described in Chapter 2.

According to the classification by the Royal Horticultural Society (2009), lily species and cultivars are classified into nine divisions (Table 4.4). In this chapter, the emphasis is on four horticulturally major groups, Asiatic hybrids (Division I), Longiflorum lilies (Division V), Oriental hybrids (Division VII), and Other hybrids (Division VIII).

The basic chromosome number is $x = 12$, and the wild *Lilium* species are mostly diploid ($2n = 2x = 24$). The exception is the tiger lily (*L. lancifolium*) with a ploidy complex species of diploid and triploid ($2n = 3x = 36$), with the triploid being the majority in distribution (Noda 1978; Kim et al. 2006).

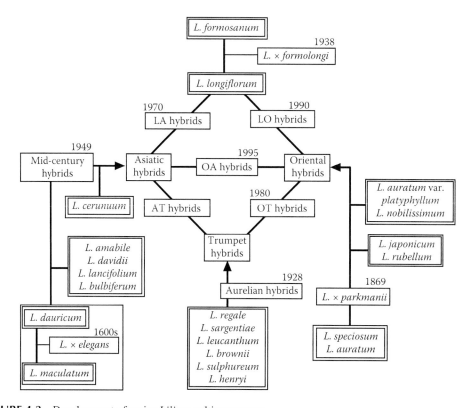

FIGURE 4.3 Development of major *Lilium* cultivar groups.

C. MORPHOLOGY AND ANNUAL GROWTH CYCLE

The storage organ of most of the species is the nontunicated bulb, but there are also stoloniferous (e.g., *L. canadense*), rhizomatous (e.g., *L. pardalinum*), and stoloniform (e.g., *L. duchartrei*) species. Bulblets can be formed not only on the underground stem, but also in the leaf or floral axis (aerial bulblets = bulbils). The leaf shape is lanceolate or linear. The leaf arrangement varies according to species and cultivar and can be scattered, whorled, spirally arranged, or alternate. Most species are deciduous, but a few species (e.g., *L. candidum*) have rosette leaves through the winter.

The individual flower has a long perianth tube comprising two fused whorls and six isomorphic perianth segments. It is stalked and subtended by a bract. The six anthers are borne on long filaments attached near the base of the perianth tube. The ovary is superior and trilocular. Flower shapes are trumpet (funnel) (e.g., *L. longiflorum*), bowl (e.g., *L. bulbiferum*), star (e.g., *L. concolor*), and Turk's cup (e.g., *L. martagon*). The orientation of the flowers varies and can be upward-facing, outward-facing (horizontal), or pendant.

Flowers of some *Lilium* species contain cyanidin-3-rutinoside as a major anthocyanin (Banba 1967), carotenoid pigments β-carotene, cryptoxanthin, zeaxanthin, capsanthin, capsorubin, and echinenone-like carotenoid (Banba 1968). Recently, cyanidin 3-O-β-rutinoside-7-O-β-glucoside and cyanidin 3-O-β-rutinoside were identified in Asiatic and Oriental hybrids (Nørbæk and Kondo 1999a). Carotenoids in Asiatic hybrid lilies are antheraxanthin, (9Z)-violaxanthin, *cis*-lutein, while violaxanthin is present in yellow flowers and capsanthin in red flowers (Yamagishi et al. 2010).

Most species possess two root systems, basal roots that develop at the base of the bulb and stem roots that are adventitiously initiated on the stem in the soil above the bulbs. There are exceptions like *L. candidum* and *L. concolor* which only produce basal roots. In addition, many species develop contractile roots which serve to pull the bulb down into the soil. In *L. longiflorum* 'Nellie White', the contraction occurs by the formation of epidermal wrinkles, starting at the base of the root and

TABLE 4.4
Horticultural Classification of Lilies

Division	Species and Origin of Interspecific Hybrids	Main Characters	Major Cultivars
I. Asiatic hybrids	amabile, bulbiferum, callosum, cernuum, concolor, dauricum, davidii, × hollandicum, lancifolium (syn. tigrinum), lankongense, leichtlinii, maculatum, pumilum, × scottiae, wardii, wilsonii	Flowers usually small to medium sized. Flower color often uniform, or with contrasting perianth segment tips and/or throat. Spots absent, or when present, well defined, and often rounded. Conspicuous brush marks sometimes present. Perianth segment margins smooth or slightly ruffled. Flowers with little or no scent.	Ariadne, Black Out, Brunello[a], Mont Blanc, Navona[a], Tresor[a], Vermeer
II. Martagon hybrids	× dalhansonii, hansonii, martagon, medeoloides, tsingtauense	Flowers small, often numerous, mostly down-facing, typically with rather thick, recurved perianth segments and often of Turk's cap form. Spots present, numerous, often on at least three-quarters of each perianth segment. Flowers with little or an unpleasant scent. Buds hairy. Bulbs mauve or orange-yellow. Early flowering.	Cadense, Claude Shride, Marhan
III. Euro-Caucasian hybrids	candidum, chalcedonicum, kesselringianum, monadelphum, pomponium, pyrenaicum, × testaceum	Flowers mostly small to medium sized, often bell-shaped to Turk's cap-shaped, down-facing. Flower color in rather pale muted shades. Spots absent to numerous. Perianth segment margins smooth, gently reflexed. Flowers often scented.	Moonlight Madonna, Zeus
IV. American hybrids	bolanderi, × burbankii, canadense, columbianum, grayi, humboldtii, kelleyanum, kelloggii, maritimum, michauxii, michiganense, occidentale, × pardaboldtii, pardalinum, parryi, parvum, philadelphicum, pikinense, superbum, vollmeri, washingtonianum, wigginsii	Flowers small to medium sized, mostly down-facing. Flower color strong yellow to orange or orange-red, with contrasting center and perianth segment tips. Spots very conspicuous, over at least half of each perianth segment, surrounded by a paler halo. Perianth segments rather narrow, margins smooth, usually gently to strongly reflexed. Flowers with little scent.	Lake Tulare, Shuksan

continued

TABLE 4.4   (continued)
Horticultural Classification of Lilies

Division	Species and Origin of Interspecific Hybrids	Main Characters	Major Cultivars
V. Longiflorum lilies	*formosanum, longiflorum, philippinense, wallichianum*	Flowers medium sized to large, trumpet shaped. Flower color typically uniform (white). Perianth segment margins smooth. Flowers usually scented.	Ace, Hinomoto, Lorina, Nellie White, Snow Queen, White American, White Heaven[a]
VI. Trumpet and Aurelian hybrids	*× aurelianense, brownii, × centigale, henryi, × imperiale, × kewense, leucanthum, regale, rosthornii, sargentiae, sulphureum, × sulphurgale* (excluding hybrids of *henryi* with all species listed in Division VII. Aurelian hybrids are derived from a combination of *henryi* and trumpet lilies)	Flowers medium sized to large, in all flower forms. Flower color white, cream, yellow to orange or pink, often with a contrasting star-shaped throat and/or strong bands of color outside. Trumpets usually scented, without markings; other types with spotting as small streaks in two bands at the base. Perianth segment margins smooth or twisted with irregular ruffling, tips reflexed.	African Queen, Golden Splendor, Pink Perfection
VII. Oriental hybrids	*auratum, japonicum, nobilissimum, × parkmanii, rubellum, speciosum* (excluding all hybrids of these with *henryi*)	Flowers medium sized to very large. Inner perianth segments very broad and overlapping at the base to give a "closed" center. Flower color mostly white to pink to purplish red, some golden yellow; ground color often white, with a contrasting central ray. Spots absent to numerous, sometimes over half of each perianth segment. Papillae and nectaries conspicuous. Flowers scented. Late-flowering.	Acapulco[a], Aktiva, Casa Blanca[a], Corvara, Crystal Blanca, Legend, Marlon, Medusa, Mero Star[a], Nova Zembla[a], Rialto[a], Santander[a], Siberia[a], Sorbonne[a], Star Gazer[a], Tiber[a]

VIII. Other hybrids	Hybrids not covered by any of the previous Divisions and all interdivisional hybrids, such as Asiatic/Trumpet (Asiapets or AT) hybrids, *longiflorum*/Asiatic (LA) hybrids, *longiflorum*/Oriental (LO) hybrids, Oriental/Asiatic (OA) hybrids and Oriental/Trumpet (Orienpets or OT) hybrids. Hybrids of *henryi* with *auratum*, *japonicum*, *nobilissimum*, × *parkmanii*, *rubellum* and *speciosum* (excluded from Divisions VI and VII)	AT: Oberhof LA: Advantage, Brindisi[a], Ceb Dazzle[a], Courier[a], Golden Tycoon, Litouwen[a], Menorca[a], Nashville, Original Love[a], Pavia[a], Salmon Classic LO: Chiara, Sky Treasure, Triumphator[a], Vendella, White Triumph Zanlotriumph OA: Elegant Crown, First Crown, Sunny Crown OT: Belladonna, Conca d'Or[a], Manissa, Robina[a], Yelloween[a]
IX. Species and cultivars of species	All species and their subspecies, varieties and forms, and cultivars selected therefrom (excluding those derived exclusively from *formosanum*, *longiflorum*, *philippinense* and *wallichianum* (Division V)	Chocolate Chips (*L. canadense* var. *editorum*), Crimson Beauty (*L. auratum* var. *rubrovittatum*), Splendens (*L. lancifolium*)

Source: Adapted from Royal Horticultural Society. 2009. *The International Lily Register and Checklist 2007. First Supplement.* London: The Royal Horticultural Society.

[a] See Table 6.1 in Chapter 6 for production acreage of these cultivars in The Netherlands.

advancing toward the root tip. The movement function occurs in shallowly planted bulbs and is stimulated by light (Jaffe and Leopold 2007).

Seed germination types in lilies are classified as epigeal (above-ground) and hypogeal (underground). In addition, germination can be immediate or delayed (Jefferson-Brown and Howland 1995).

During the annual development of *L. longiflorum* in warm-climate areas (e.g., in Japan), sprouting and rosette development occur in autumn, soon after planting in open fields, but the shoot does not elongate until spring. Leaf unfolding begins when the average temperatures increase above 5°C. Floral induction is completed by late winter, and flower initiation occurs in early spring. In cold areas (e.g., The Netherlands), bulbs are planted in spring and the shoot emerges and elongates when the temperature reaches 11–12°C. Flower initiation occurs simultaneously. Concurrent with flower formation and development, new inner scales are formed, and consequently bulb size increases. Following the scale formation, leaf primordia for the next season are initiated within the bulb.

Lilium longiflorum requires a period of "vernalization" for flower initiation and development in spring. The vernalization period can be partly replaced by a long photoperiod given at the shoot emergence stage. A long photoperiod can substitute for cold requirements for flowering, but not for bulb development (Weiler and Langhans 1972).

The annual developmental cycle in Asiatic, Oriental, and other hybrids is basically similar to that in *L. longiflorum* with differences in the timing of each developmental stage. The bulbs sprout in spring and flower in late spring to early summer (Asiatic) or summer (Oriental). In some earlier cultivars of Asiatic hybrids, flowering was initiated prior to bulb harvesting. However, in most of the newly released cultivars, flowering is initiated during active growth in spring. Oriental and other hybrid cultivars initiate flowering in spring also. A cold period is necessary for flower initiation in all types of hybrids.

D. Major Species and Cultivars

1. Species

Natural populations of *Lilium longiflorum* are found from Yakushima Island, Japan (30°30′N, 130°60′E) to Lanyu Island, Taiwan (22°00′N, 121°50′E) at the altitudes of <10–60 m a.s.l. (Shimizu 1987; Hiramatsu et al. 2001). Cultivars of *L. longiflorum* are self-incompatible, and this trait is one of the barriers in breeding. It was found recently that self-compatible individuals are dominant in the populations of *L. longiflorum* in both the north and south boundaries of the Ryukyu Archipelago of Japan and in Taiwan (Sakazono et al. 2006, 2009, 2012).

Lilium formosanum, a close relative to *L. longiflorum*, is naturally distributed in the mainland of Taiwan from sea level to 3500 m in altitude (Yang 2000), and is now widely found in nature in western Japan. Based on isozyme diversity, Hiramatsu et al. (2001) revealed that *L. formosanum* is a derivative species from *L. longiflorum* in the southern distribution area. The two species can be crossed to obtain viable F_1 seeds. Cultivars of an interspecific hybrid of *L. formosanum* × *L. longiflorum* (= *L. × formolongi*) have been incorporated into the year-round cultivation system of the Easter lily in Japan and it is marketed in the summer season (Imanishi 2005). Unlike other *Lilium* species, *L. formosanum* flowers within 12 months from seed germination and does not require special temperature control during its annual cycle (Hiramatsu et al. 2002). This trait can be used for breeding in order to establish the cultivars reproduced by seeds and to avoid possible virus transmission through conventional vegetative propagation (Saruwatari et al. 2008).

Lilium brownii, distributed in China, is classified into the section *Archelirion* (Comber 1949) or *Leucolirion* (Nishikawa et al. 2001). In fact, *L. brownii* var. *colchesteri* can be crossed with *L. formosanum* and with *L. × formolongi* to obtain F_1 progenies through sliced ovule culture (Hai et al. 2010). The species has been less used in breeding programs. Details of *L. brownii* and its varieties have recently been described by Okubo et al. (in press).

Asiatic hybrid cultivars were derived from parentage that includes *L. amabile*, *L. bulbiferum*, *L. callosum*, *L. cernuum*, *L. concolor*, *L. dauricum*, *L. davidii*, *L.* × *hollandicum*, *L. lancifolium*, *L. lankongense*, *L. leichtlinii*, *L. maculatum*, *L. pumilum*, *L.* × *scottiae*, *L. wardii*, and *L. wilsonii*, most of which originated from Japan, Korea, and China (Table 4.4, Figure 4.3). *Lilium* × *elegans* is considered to be a natural hybrid of *L. dauricum* (distributed widely in Sakhalin, Siberia, and northeast China) and *L. maculatum* (endemic to north Honshu, Japan), both of which are the parents of Asiatic hybrids. About 150 cultivars of *L.* × *elegans* were developed in Japan in the mid-seventeenth century during the Yedo period (1603–1867).

Lilium auratum, *L. japonicum*, *L. nobilissimum*, *L. rubellum*, and *L. speciosum* contributed to the establishment of Oriental hybrids (Table 4.4, Figure 4.3). All these species, except *L. speciosum*, are endemic to Japan. *Lilium auratum* is distributed in the central and northern parts of Honshu Island, *L. japonicum* in western Honshu, *L. nobilissimum* only in Kuchinoshima Island, Kagoshima Prefecture, and *L. rubellum* in northern Honshu. *Lilium speciosum* grows in Kyushu and Shikoku Islands in southern Japan as well as in Taiwan and China. *Lilium* × *parkmanii* was also used in the breeding of Oriental hybrids (Table 4.4). A hybrid between *L. speciosum* and *L. auratum* was bred in 1869 in the United States.

2. Cultivars

More than 7000 cultivars have been released to date (Leslie 1982) and over 100 new cultivars have been released annually over the last two decades. The major cultivars of each division are listed in Table 4.4. The development of breeding techniques (see Chapters 6 and 7) has made interdivisional crossings possible and new cultivars are being extensively bred. They are Asiapets or AT hybrids (Asiatic × Trumpet), LA hybrids (*longiflorum* × Asiatic, produced in 1970), LO hybrids (*longiflorum* × Oriental, 1990), OA hybrids (Oriental × Asiatic, 1995), and Orienpets or OT hybrids (Oriental × Trumpet, 1980). To emphasize the characteristics of one of the parents of inbred cultivars, they are often described as LAA, LLO, and so on. Ultimately, we can expect LAOT hybrids resulting from four-way crossings. Due to an increasing number of hybrid cultivars derived from the crosses of Trumpet and Aurelian hybrids (Division VI) and Oriental hybrids (Division VII), an independent division named Orienpet hybrids has been proposed by the North American Lily Society (McRae 1998).

Bulbs of more than 500 cultivars were produced in The Netherlands in 2010 (PT/BKD 2010b). The total growing area was 3180 ha, of which 1504.3 ha (47.3%) were Oriental hybrids, followed by LA (834.7 ha, 26.2%), Asiatic (409.5 ha, 12.9%), OT (307.3 ha, 9.7%), LO (46.3 ha, 1.5%), and Longiflorum (27.0 ha, 0.8%) hybrids. The most popular cultivars are 'Sorbonne' (Oriental), followed by 'Siberia' (Oriental), 'Robina' (OT), 'Rialto' (Oriental), and 'Navona' (Asiatic) (see Table 6.1 of Chapter 6 for the production acreage of the top 25 cultivars in 2009). Cultivars' acreage is changing dramatically year by year. 'Star Gazer' (Oriental hybrids) was produced on 296 ha in 2000 in The Netherlands, but it decreased to 33 ha in 2009. Asiatic hybrids are now being replaced by LA hybrids. In addition, there are significant production increases in OT, LA, and LO hybrids.

The major cultivars are diploid (2n = 2x = 24), triploid (2n = 3x = 36), and tetraploid (2n = 4x = 48) (see Chapter 6). For example, 'Sorbonne', 'Siberia', 'Rialto', and 'Tiber' (Oriental) are diploids, 'Robina', 'Navona', 'Brindisi' (LA), and 'Conca d'Or' (OT) are triploids, and 'Tresor' (Asiatic) and 'Brunello' (Asiatic) are tetraploids. Breeding of high-level polyploids, including octoploids, is one of the important goals in the lily breeding programs.

E. PROPAGATION

Lilies can be propagated by seeds, bulb division, cuttings, scaling, stem bulblets, and micropropagation. Most of the lily cultivars are propagated by scaling (see also Chapter 10), whereas the cultivars of *L.* × *formolongi* are commercially propagated by seeds. The advantage of propagation by bulblets is that flowering occurs one year earlier than in the plants propagated by scaling. It is, however, not used in commercial production, mainly due to the labor cost of bulblet collecting. Propagation by tissue culture in various species and hybrids has been reviewed by Kim and De Hertogh (1997). Currently, *Lilium* is the only bulbous genus that is micropropagated on a significant scale. Approximately 23×10^6 *Lilium* plants were produced by Dutch micropropagation labs alone in 1990 (George 1996). Although the current total production is lower than what it was 20 years ago, lilies are commonly propagated *in vitro* in Dutch and Polish labs.

F. PRODUCTION

1. Bulb Production

Lilies require fertile and well-drained soils with a pH range of 6.0–6.5. The large assortment of lily cultivars allows for the selection of suitable crops for different climatic zones. For example, it was shown that among three cultivars of Oriental lilies ('Siberia', 'Sorbonne', and 'Tiber') 'Tiber' can maintain a higher photosynthesis rate within a wide range of temperatures and, thus, is very suitable for the climate of the central areas of Yunnan Province (Chang et al. 2008). Covering the crop with shading nets in southern Italy facilitates the acclimation of many cultivars to high-intensity light even under high-temperature conditions (Sorrentino et al. 1997).

Lilies are grown in a 1-year cycle, but a 2-year cycle is recommended for the cultivation of bulblets originated from scaling. In general, the 2-year culture is not recommended as routine propagation protocols for lilies, since the number and weight of bulbs produced are lower in comparison with those obtained in two separate 1-year cycles (Sochacki and Mynett 1996).

A crop rotation of at least 5 years is required to reduce viral and disease contamination, pest and weed problems. Small bulbs (less than 10 cm), bulbils, and bulblets are used as commercial planting stocks. Bulbs are planted either in October–November or in March–April following overwinter storage at $-1°C$. One of the most important advantages of spring planting is the flexibility of the planting date and the avoidance of frost damage at leaf emergence.

Nutritional requirements for lilies vary by soil nutrient buffering capacity, seasonal temperatures, rainfall patterns, and stages of the plant's development. In the United States, for example, 140 kg N–280 kg P_2O_5–200 kg K_2O/ha/year is practiced for *L. longiflorum* (Miller 1993). In *L. longiflorum* 'Nellie White', the absence of N and P fertilization had a greater negative impact on plant growth and development than the absence of K during the first phase of vegetation (Niedziela et al. 2008).

Topping prior to the flowering stage improves the utilization of photosynthesis products for the growth and development of the bulbs.

A hot water treatment (HWT) of the bulbs at $41°C$ for 2 h in combination with 4 days of pre- and posttreatment storage at $20°C$ has been proposed to prevent nematode or mite infestation (Kok et al. 2008).

2. Flower Production

a. *Cut Flowers*

Bulbs of >10/12 cm (Longiflorum and Asiatic hybrids) and 12/14 cm (Oriental, LA, OT, LO, and OA hybrids) in size are used for cut flower production. Cut flower production usually takes place in greenhouses, but outdoor production is possible in warm-climate regions.

Cut flower production using "frozen bulbs" is applicable to all hybrid groups and this system has enabled year-round flower production. After harvest, bulbs are placed in moist peat, exposed to prefreezing acclimation treatment at $0–1°C$ for 4–6 weeks, and then frozen at $-2°C$ (Asiatic hybrids), $-1.5°C$ to $-2°C$ (LA), $-1°C$ to $-0.5°C$ (Oriental), or $-1.5°C$ (Longiflorum, OT, LO, and OA hybrids). The required temperatures must be kept uniform, since even small variations may cause shoot growth or frost damage of the bulbs. Overly long storage may decrease the growing capabilities and reduce flower quality. Bonnier et al. (1997) reported that the regeneration ability of the Asiatic hybrid 'Enchantment' decreased after 1 year of storage at $-2°C$, and was completely lost after

5 years. Prior to planting, the bulbs are thawed at 10–12°C. Planting is possible in any season. For recently developed long storage techniques, see also Chapter 11.

For flower production in greenhouses, temperatures of 10–12°C for a few weeks are employed, and then are raised to 14–15°C for Asiatic and LA hybrids, 15–17°C for Oriental, OT, and OA hybrids, and 14–16°C for Longiflorum and LO hybrids. Higher temperatures can result in shorter stems and fewer flower buds per stem, and can increase the risk of disease infection. Lower temperatures can result in flower deformation.

b. Potted Flowers

Generally, the desired stem length for the stems in potted lilies is 30–40 cm and this can often be obtained by the application of plant growth regulators. Research on the measures to reduce plant height has been extensively conducted (see also Chapter 11). In *L. longiflorum*, preplant bulb soaks of 'Nellie White' in 120 mg/L paclobutrazol for 15 min resulted in reducing stem length without affecting the number of days to flowering or the flower bud number (Currey and Lopez 2010). Height control by cool water was also demonstrated by Blom et al. (2004). In 'Star Gazer' (Oriental hybrids), drenches of flurprimidol at 0.5 mg/pot and preplant bulb soaks of uniconazol at 5 and 10 mg/L and flurprimidol at 25 mg/L resulted in plants 24, 30, 37, and 23% shorter than the untreated controls (Krug et al. 2005a). Ranwala et al. (2002) showed that LA hybrid lilies are more responsive to growth-retardant treatments than Oriental hybrids. Negative DIF (e.g., growing at 14–15°C/18–19°C (day/night)) is also effective in shortening stem length (Miller et al. 1993; Blom and Kerec 2003; Steininger and Pasian 2003). Genetically short cultivars, such as, 'Orange Pixie' (Asiatic), 'Souvenir' (Oriental), 'White American' (Longiflorum), 'Dream Catcher' (LA) or 'Eyellow' (OT), have been developed for potted plant production and do not require a chemical treatment.

G. Physiological Disorders, Pests, and Diseases

Leaf scorch and tip burn are major physiological disorders. A necrotic disorder on the upper leaves ("upper leaf necrosis") of many Oriental hybrids is a Ca deficiency disorder, and its severity was greatly increased by light reduction, as leaf transpiration was reduced (Chang and Miller 2003, 2005). A major bacterial pathogen is *Pectobacterium carotovorum* pv. *carotovorum* (bacterial soft rot). Fungal pathogens are *Aspergillus niger* (black mold), *Botrytis cinerea* (gray mold), *Botrytis elliptica* (botrytis blight), *Cercosporella inconspicua* (leaf spot), *Colletotrichum liliacearum* (black scale rot), *Cylindrocarpon destructans* (scale tip rot), *Fusarium oxysporum* f. sp. *lilii* (bulb, root and stem rot), *Helicobasidium mompa* (violet root rot), *Penicillium* spp. (blue mold, dry rot), *Phyllosticta liliicola* (leaf spot), *Phytophthora* spp. (blight), *Rhizoctonia solani* (foot rot), *Rhizopus* spp. (bulb rot), *Sclerotium cepivorum* (black rot), *Sclerotium rolfsii* (southern blight), *Septoria lilii* (leaf spot), and *Uromyces holwayi* (rust).

Lily symptomless virus (LSV), *Cucumber mosaic virus* (CMV), *Lily virus X* (LVX), and *Tulip breaking virus* (TBV) are the major viruses in *Lilium* species causing symptoms. Others are *Narcissus yellow stripe virus* (NYSV), *Tobacco rattle virus* (TRV), *Arabis mosaic virus* (ArMV), and *Tobacco ring spot virus* (TRSV). *Aphelenchoides* spp., *Colletotrichum liliacearum*, *Criconemoides* sp., *Helicotylenchus* sp., *Hirschmanniella imamuri*, *Meloidogyne hapla*, and *Pratylenchus penetrans* are nematodes that can infest lilies. Aphids and bulb mites can also affect lilies.

VII. NARCISSI

A. Introductory Remarks

Narcissus is one of the most important spring-flowering bulbs, widely cultivated in the world. Bulb production is dominated by the United Kingdom, The Netherlands, and the United States, and is ranked third after tulips and lilies in The Netherlands. "Narcissus" is used both as the name of a botanical genus (written in italic) and as a common English name (nonitalic), which includes all species. "Daffodil" is also generally used interchangeably with "Narcissus", but Rees (1985) has suggested that it should be strictly reserved for the yellow trumpet narcissi (*N. pseudonarcissus*). "Jonquil" and "Paperwhites" are used for *N. jonquilla* and *N. tazetta*, respectively.

The genus name *Narcissus* comes from Narcissus, a son of the river god Cephisus and a forest nymph in Greek Mythology (Margaris 2000) (see also Chapter 1).

Being indigenous in Europe, narcissus had been cultivated for hundreds of years (Doorenbos 1954). Until the end of the nineteenth century, the only daffodils grown on a large scale in The Netherlands were the double 'Van Sion', and varieties selected from *Narcissus tazetta*, which was imported in 1557. Breeding was started between 1835 and 1855 in the United Kingdom and at the end of the nineteenth century in The Netherlands (Wylie 1952; Doorenbos 1954).

An excellent book, *Narcissus and Daffodil: The Genus Narcissus,* has recently been published by Hanks (2002).

B.   Botanical Classification and Distribution

The genus *Narcissus* belongs to the family Amaryllidaceae and comprises about 60 species, classified into 10 sections (Hanks 1993; Kington 2008; Chapter 2). It originates primarily on the Iberian Peninsula and in Europe, with the northern limit being England, the southern Canary Islands, and Asia Minor and Syria probably representing the easterly extremes of distribution (Wylie 1952; Hanks 1993). In contrast to its name, the Chinese sacred lily (*N. tazetta* var. *chinensis*) originated from the Mediterranean region and was introduced to China and to Japan in ancient times. The basic chromosome number is 7 in all the species except for *N. tazetta* and the closely related *N. elegans* and *N. broussonetii*, in which it is 10 or the derived 11 (Wylie 1952). Recently, more exceptions, such as *N. papyraceus* 4x = 22, and *N. dubius* 6x = 50, have been reported (Zonneveld 2008).

C.   Morphology and Annual Development

The storage organ of the *Narcissus* is a perennial tunicated bulb consisting of scales and swollen leaf bases. A large flowering-sized bulb is composed of a number of branched bulbs (= bulb units). Each meristem produces one bulb unit, which is normally composed of two to four scales, two to three leaves (the innermost being semisheathing), and the inflorescence. The bud in the flower's axil forms a terminal unit, while another bulb unit forms in the axil of the next leaf (lateral unit). The terminal unit initiates flowering in the following season. Each bulb unit has a life span of about 4 years.

Narcissus pseudonarcissus has a solitary flower on a scape, while *N. tazetta* has inflorescence with several florets. Each flower comprises two whorls of three perianth segments and two whorls of three anthers each. The tricarpellate ovary is inferior. The sequence of floral initiation and development is described in Chapter 8. Two distinct flower organs, spathe and paracorolla, are characteristic of the *Narcissus* species. The spathe is initiated prior to the first whorl of perianth (P_1 stage), while paracorolla is the last flower organ to differentiate. A paracorolla (corona, cup, or trumpet), specific to Amaryllidaceae, is located between the perianth parts and the androecium. Flower colors range from white to yellow, pink, and orange.

In general, the *Narcissus* species can be physiologically divided into two groups; one that requires low temperatures for flowering and another that does not (Rees 1992; Theron and De Hertogh 2001). Most species belong to the first group, whereas *N. tazetta* cultivars belong to the latter.

Flower initiation of *N. pseudonarcissus* cultivars starts in late spring, before the bulbs are lifted, and is normally completed by mid-summer, during storage. The optimal temperature for flower initiation is 20°C, with a subsequent decrease to 13°C; however, flower initiation can also occur at temperatures between 9 and 28°C. High summer temperatures during storage delays flower development. Cultivars of *N. poeticus* start flower initiation before bulb lifting, at about the same time as the *N. pseudonarcissus* cultivars. Flower initiation of the *N. tazetta* cultivars occurs in summer, after bulb harvesting, at an optimum of 25°C. Due to low cold requirements, in temperate climates of the Northern Hemisphere cultivars of *N. tazetta* flower in December–February. *N. pseudonarcissus* cultivars flower in April, followed by the *N. poeticus* cultivars.

D. Major Species and Cultivars

Narcissus bulbocodium, N. cyclamineus, N. jonquilla, N. poeticus, N. pseudonarcissus, N. serotinus, and *N. tazetta* are the major species, but many other species have also contributed to the establishment of modern cultivars. Horticulturally, the cultivars are classified into 12 divisions, based on the ratio of corona (trumpet, cup) to perianth segment length, characteristics of the parent species, or other distinct features (Table 4.5). The "split corona" (Division 11) is the latest group and was created in 1969 (Kington 2008).

TABLE 4.5
Horticultural Classification of Major Cultivars of *Narcissus*

	Division	Characteristics	Major Cultivars[a]
1.	Trumpet	One flower to a stem; corona ("trumpet") as long as or longer than the perianth segments ("petals")	Dutch Master, Golden Harvest (1927), King Alfred (1889), Mount Hood, Standard Value
2.	Large-cupped	One flower to a stem; corona ("cup") more than one-third, but less than or equal to the length of the perianth segments ("petals")	Carlton (1927), Fortune (1923), Gigantic Star, Ice Follies, Salome
3.	Small-cupped	One flower to a stem; corona ("cup") not more than one-third the length of the perianth segments ("petals")	Altruist, Barrette Browning, Edna Earl, Verger (1930)
4.	Double	One or more flowers to a stem, with doubling of the perianth segments or the corona or both	Bridal Crown, Cheerfulness (1923), Dick Wilden, Replete, Sir Winston Churchill, Tahiti, Van Sion (1603)
5.	Triandrus	Characteristics of *N. triandrus* are clearly evident: usually two or more pendent flowers to a stem; perianth segments reflexed	Hawera, Petrel, Rijnveld's Early Sensation, Thalia, Tresamble
6.	Cyclamineus	Characteristics of *N. cyclamineus* are clearly evident: one flower to a stem; perianth segments significantly reflexed; flower at an acute angle to the stem, with a very short pedicel ("neck")	February Gold, Itzim, Jack Snipe, Jetfire, Peeping Tom
7.	Jonquilla and Apodanthus	Characteristics of sections *Jonquilla* or *Apodanthi* clearly evident: one to five (rarely eight) flowers to a stem; perianth segments spreading or reflexed; corona cup-shaped, funnel-shaped or flared, usually wider than long; flowers usually fragrant	Martinette, Pipit, Pueblo, Quail, Sun Disc, Sweetness
8.	Tazetta	Characteristics of section *Tazettae* clearly evident: usually 3–20 flowers to a stout stem; perianth segments spreading not reflexed; flowers usually fragrant	Cragford, Elvira, Galilee, Geranium (1930), Golden Dawn, Grand Soleil d'Or, Hoopoe, Laurens Kosta (1923), Minnow, Sheleg, Ziva

continued

TABLE 4.5   (continued)
Horticultural Classification of Major Cultivars of _Narcissus_

	Division	Characteristics	Major Cultivars[a]
9.	Poeticus	Characteristics of the *N. poeticus* group: usually one flower to a stem; perianth segments pure white; corona very short or disc-shaped, usually with a green and/or yellow center and a red rim, but sometimes of a single color; flowers usually fragrant	Actaea (1927), Angel Eyes, Dactyl, Green Pearl
10.	Bulbcodium	Characteristics of section *Bulbocodium* clearly evident: usually one flower to a stem; perianth segments insignificant compared with the dominant corona; anthers dorsifixed (i.e., attached more or less centrally to the filament); filament and style usually curved	Golden Bells, Kenellis, Spoirot
11.	Split corona	(a) Collar Daffodils: Split-corona daffodils with the corona segments opposite the perianth segments; the corona segments usually in two whorls of three segments each	Cassata, Orangery, Palmares, Parisienne, Tripartite
		(b) Papillon Daffodils: Split-corona daffodils with the corona segments alternate to the perianth segments; the corona segments usually in a single whorl of six segments	Broadway Star, Lemon Beauty, Marie-Jose, Papillon Blanc, Space Shuttle
12.	Others	Daffodil cultivars which do not fit the definition of any other division	Jessamy, Tête-à-Tête
13.	Species	Daffodils distinguished solely by botanical name	

Source: Adapted from Kington, S. 2008. *The Daffodil Register and Classified List*. Woking, UK: RHS Publ.

[a]   Cultivars with year in parentheses are examples of some old cultivars still commercially produced in The Netherlands.

More than 27,000 cultivar names of narcissi have accumulated in the RHS files (Kington 2008), but there is much synonymy. The real number of cultivars is estimated to be 18,000 (Hanks 1993). However, the number of commercially grown cultivars is far fewer. About 470 cultivars were cultivated on 1578 ha in the 2009/2010 season in The Netherlands. The major cultivars produced are 'Tête-à-Tête' (cyclamineus[*], 663.3 ha), 'Carlton' (large-cupped, 54.3 ha), 'Bridal Crown' (double, 50.6 ha), 'Dutch Master' (trumpet, 47.2 ha), 'Jetfire' (cyclamineus, 42.2 ha), 'Ice Follies' (large-cupped, 35.8 ha), 'Minnow' (tazetta, 35.4 ha), 'Dick Wilden' (double, 24.3 ha), 'Thalia' (triandrus, 17.7 ha), 'Standard Value' (trumpet, 15.1 ha), 'Tahiti' (double, 13.8 ha), 'Mount Hood' (trumpet, 13.5 ha), 'February Gold' (cyclamineus, 12.7 ha), 'Hawera' (triandrus, 12.5 ha), 'Replete' (double, 12.0 ha), 'Sir Winston Churchill' (double, 11.8 ha), and 'Quail' (jonquilla, 11.6 ha) (PT/BKD 2010a). It is notable that some old cultivars are still commercially produced in The Netherlands (Table 4.5). In Israel, about 25 million bulbs of *N. tazetta*, mainly 'Ziva' and 'Galilee', were exported in 2003 (Kamenetsky 2005).

---

[*]   'Tête-à-Tête' was moved to Division 12 (Others) (Kington 2008), while it is still classified as Cyclamineus in practice in The Netherlands.

E. PROPAGATION

Twin-scaling and chipping have been successfully used in the propagation of *Narcissus* (see Chapter 10 for details). *In vitro* multiplication of the plants has been reviewed by Seabrook (1990) and Kim and De Hertogh (1997), but is not common in commercial production. On the other hand, micro-propagation is used to establish commercial stocks, especially for new cultivars or pathogen-free plants (Sochacki 2011).

F. PRODUCTION

1. Bulb Production

Narcissi are grown in open fields for both bulb production and outdoor production of cut flowers. Depending on market preferences, flowers and bulbs can be produced in the same crop. This is possible because the harvest of narcissus flowers—even with leaves—is not as crucial for the photosynthetic area of plants and bulb yield as in other geophytes, for example in tulips (Grabowska and Mynett 1971; Hanks 1993).

Narcissus bulbs can be grown in regions with a mild maritime climate, where earlier and longer vegetation seasons favor the crop. However, higher rainfall and temperatures in these regions may cause disease, pest and weed problems, and complicate bulb lifting and drying (Hanks 2002). When grown in regions with continental climates, narcissi are sensitive to frost. In the United Kingdom, the soils seldom freeze deeply enough to damage the bulbs. In The Netherlands and in Poland, crops are covered with straw to protect them from frost (Hanks 2002).

In the United Kingdom, narcissi are grown in ridges in a 2-year cycle, whereas in The Netherlands and many other countries, they are produced in a 1-year cycle in beds or ridges. The first system saves on labor costs for lifting, processing, storage, and replanting. On the other hand, annual lifting allows improved bulb yield and better disease and weed control (Rasmussen 1976; Sochacki and Mynett 1996).

Soils for narcissus bulb production should be fertile and well drained. However, they must retain moisture, have an organic matter content higher than 3%, and a pH between 6.0 and 7.5. A crop rotation of at least 4 years is obligatory to maintain good soil structure and reduce disease, pest, and weed problems. A longer rotation must be implemented if nematodes are found in the field (Rees 1992).

The level of fertilization depends on the growing conditions and nutrient content of the soil. In the case of a low nutritional status, Rees (1992) recommends the following rates of fertilizers: N 125 kg, P_2O_5 150 kg, K_2O 300 kg, and Mg 60–90 kg per hectare. Slightly lower doses are given by Hanks (2002) in accordance with the recommendations of the British Ministry of Agriculture, Fisheries and Food (MAFF 1994): N 100 kg, P_2O_5 125 kg, K_2O 250 kg, and Mg 150 kg per hectare. An organic fertilizer may be used, but high rates should not be applied shortly before planting.

Prior to planting, the bulbs are subjected to hot water treatment (HWT) to control nematodes and other pests. The standard UK treatment is 3 h at 44.4°C. The proper time for utilizing an HWT is after complete flower differentiation and before root elongation (Hanks 2002).

2. Flower Production (Except for *N. tazetta* Cultivars)

Bulbs larger than 12 cm in size are generally used for forcing. For early forcing as cut flowers, bulbs are harvested in July and a high-temperature treatment of 34°C is applied for a week. They are then kept at 17–20°C until precooling in August at 9°C for 15–16 weeks. Forcing takes place in greenhouses at 13–15°C. For potted flowers, the bulbs are treated similarly to those for cut flowers but, instead of 9°C, the temperature is lowered to 5°C for 15 weeks, followed by forcing at 16–18°C in greenhouses. For mid- and late-forcing, bulbs are harvested in late July or August and stored at 17–20°C until planting. After planting, the bulbs are exposed to 9°C for 17–18 weeks for cut flowers or for 14–16 weeks for potted flowers. Temperature treatments can also be applied to dry bulbs prior to planting. This system is commonly practiced in Japan.

3. Flower Production of *N. tazetta* Cultivars

For October flowering in the Northern Hemisphere, the bulbs are lifted and dried in May. They are planted immediately after treatment at 30°C for 3 weeks, which facilitates forcing, and are then stored at 25°C for 12 weeks before planting. Ethylene treatment at a concentration of 10 mL/L also promotes early flowering, as well as flowering in small bulbs, which normally do not flower (Imanishi 1983). For retarded flowering, bulbs can be stored for long times at 5–10°C before the starting of flower initiation within the bulb. In Israel, storage at 25°C, and then 2 weeks at 9°C prior to planting for early and mid-forcing, or prolonged storage at 30°C for late forcing, is recommended (Cohen et al. 2011).

G. PHYSIOLOGICAL DISORDERS, PESTS, AND DISEASES

Flower abortion can be caused by the use of small bulbs, a very early application of cold treatment, placing the plants in the greenhouses too early, high growth temperatures, or water deficit during forcing. Spathe discoloration can be caused by placing the plants into greenhouses too early or by a lack of moisture during forcing.

The primary disease in *Narcissus* is basal rot caused by *Fusarium oxysporum* f. sp. *narcissi*. Other pathogens are *Rhizopus* spp. (soft rot), *Penicillium* sp. (blue mold), *Stagonospora curtisii* (leaf scorch), *Sclerotinia narcissicola* (smolder), *Ramularia vallisumbrosae* (white mold), *Sclerotinia polyblastis* (fire), *Sclerotinia bulborum* (black slime), *Cylindrocarpon destructans* (root rot), *Aecidium narcissi* (rust), *Rosellinia necatrix* (white root rot), *Trichoderma narcissi* (green mold), *Aspergillus niger* (black mold), and *Phoma* sp. (leaf tip blight). Neck rot is caused by a combination of *Stagonospora curtisii* and *Sclerotinia narcissicola*. Bacterial diseases are not very prevalent in *Narcissus*, but *Pseudomonas* (bacterial streak) and *Pectobacterium carotovorum* sp. *carotovorum* (bacterial soft rot) have been found.

The most serious and widespread viruses are *Narcissus yellow stripe virus* (NYSV) and *Narcissus degeneration virus* (NDV). *Arabis mosaic virus* (ArMV), *Broad bean wilt virus* (BBWV), *Cucumber mosaic virus* (CMV), *Narcissus symptomless virus* (NSV), *Narcissus late season yellow virus* (NLSYV), *Narcissus latent virus* (NLV), *Narcissus mild mottle virus* (NMMV), *Narcissus mosaic virus* (NMV), *Narcissus tip necrosis virus* (NTNV), *Narcissus white streak virus* (NWSV), *Tomato black ring virus* (TBRV), *Tomato ring spot virus* (TomRSV), and *Tobacco rattle virus* (TRV) also infect narcissi. Some of these viruses occur in narcissus in restricted geographical locations (such as BBWV reported only in Japan), whereas others are worldwide.

Aphelenchoides subtenuis, *Ditylenchus dipsaci*, *Platylenchus penetrans*, *Longidorus* spp., *Paratrichodorus* spp., *Trichodorus* spp., and *Xiphinema* spp. are known nematodes, while flies, mites, aphids, slugs, and moths are common pests. A hot water treatment at 43°C for 5 h (there are variations of temperature × warm storage of the bulbs at 30°C for 1 week) has been used to control stem nematode in *Narcissus* bulbs (Hanks 1993, 1995).

VIII. TULIPS

A. INTRODUCTORY REMARKS

Undoubtedly, tulip is the most popular spring-blooming ornamental geophyte because of its colorful and attractive cup-shaped flower. The etymology of the genus is the Turkish *tulipan* (= turban).

The history of tulip, from cultivation in Turkey and its introduction into Europe until the collapse of "Tulipomania", has been well documented (Lodewijk 1979; Pavord 1999; Dash 2001; see also Chapters 1 and 19). The introduction of tulip into Europe in the middle of the sixteenth century is generally ascribed to A. G. de Busbecq, a Belgian diplomat representing Ferdinand I of the Holy Roman Empire to the court of Suleyman the Magnificent of the Ottoman Empire. He saw the plants flowering and sent samples to Europe. The first flowering of tulip was recorded in 1559 in Bavaria

(Dash 2001). Tulips were rare in the early 1600s and then gained popularity, which led to the famous "Tulipomania". This was the great Dutch speculative stock market frenzy and sudden crash that took place in 1620–1637. Although the bulbs were in the field, they were traded as futures at extremely high prices. In the eighteenth century, tulips regained popularity in The Netherlands.

Darwin tulips (a former cultivar group, which is now included in Single late tulips, Table 4.6) appeared in 1889, and Mendel tulips (now included in Triumph tulips) in 1921. In 1904, the Dutch firm G. G. van Tubergen discovered *Tulipa fosteriana* in Bukhara (Central Asia). This species was crossed with *T. gesneriana*, which resulted in a broader cultivar range and the establishment of Darwin hybrid cultivars. The Central Asian *T. kaufmanniana*, *T. greigii*, and *T. tubergeniana* were also introduced by the firm for cross-breeding. Triumph tulips first appeared in 1923.

The first scientific studies of tulips were carried out in The Netherlands by Professor A. H. Blaauw (1882–1942) and his coworkers. They established the growth cycles of several ornamental geophytes (Lodewijk 1979). Scientific research on tulip genetics and breeding started in Europe in about 1960 (Benschop et al. 2010). New cultivars are released yearly and over 1500 have been registered in the last 10 years (Benschop et al. 2010). Most of these were bred in The Netherlands, but a few were also bred in France (INRA), the Czech Republic, Japan, Latvia, and Poland.

The Netherlands produces more than 4 billion tulip bulbs annually, and about 2.3 billion (53%) are exported to different countries for cut flower production (Buschman 2005). In 2002/2003, tulip bulbs were produced in Japan (300 ha), France (293 ha), Poland (200 ha), Germany (155 ha), and New Zealand (122 ha). However, production decreased in The Netherlands from 10,138 ha in 2006/2007 to 9856 ha in 2009/2010 (PT/BKD 2010a), as well as in Japan (to 261 ha in 2006, MAFF 2007).

B. Botanical Classification and Distribution

The genus *Tulipa* belongs to the family Liliaceae. According to different authors, the genus comprises from 40 to more than 100 species (Van Scheepen 1996; Zonneveld 2009; Chapter 2). The genus is distributed widely in Central Asia, extending from the region of Tien-Shan and Pamir-Alai to the north and northeast (Siberia, Mongolia, and China), south to Kashmir and India, and west to Afghanistan, Iran, the Caucasus, and Turkey (Hoog 1973).

Although the Latin name of most tulip cultivars is referred to as *Tulipa gesneriana*, the wild type of this species has not been found. It seems that tulips (later *T. gesneriana*) were already hybrids when introduced into Europe. The species name *gesneriana* is at present applied also to Darwin hybrid cultivars, which are known to be hybrids with *T. fosteriana*.

Most species of *Tulipa* are diploid ($2n = 2x = 24$) (Kroon and Van Eijk 1977; Zonneveld 2009). However, triploidy ($2n = 3x = 36$) is found in *T. clusiana*, *T. kaufmanniana*, tetraploidy ($2n = 4x = 48$) in *T. biflora*, *T. clusiana*, pentaploidy in *T. clusiana* ($2n = 5x = 60$) and hexaploidy in *T. polychroma* ($2n = 6x = 72$) (Kroon and Jongerius 1986; Zonneveld 2009).

C. Morphology and Annual Growth Cycle

The storage organ is an annual tunicated bulb consisting of two to six concentric fleshy scales. Normally, one flower develops on each scape, but some species (e.g., *T. praestans*) and cultivars (e.g., 'Orange Bouquet') are multiflowered. A single flower is comprised of two whorls of three perianth parts and two whorls of anthers. The six perianth segments often have darker markings near the base. The six distinct, basifixed stamens are shorter than the perianth segments, and each stigma of the flower has three distinct lobes. The ovary is superior. The color, size, and form of the flowers vary considerably among cultivars. It has been reported by Nieuwhof et al. (1990) that carotenoids and anthocyanidins are the most important pigments determining the color of tulip flowers; white flowers do not contain these pigments; yellow flowers contain only carotenoids; orange and red flowers contain carotenoids and cyanidin; pink and purple flowers contain cyanidin; and most flowers, except for purple, do not contain delphinidin.

TABLE 4.6
Horticultural Classification of Tulips and Major Cultivars

	Group, Class	Description	Major Cultivars
		I. Early Flowering—Middle March to Early April	
1.	Single early (SE)	Cup-shaped single flowers, 15–45 cm tall	Candy Prince, Christmas Dream, Christmas Marvel[a], Coquette, Sunny Prince, Flair
2.	Double early (DE)	Bowl-shaped double flowers, 30–40 cm tall	Abba[a,b], Monsella, Monte Carlo[a,b], Orange Princess, Verona, Viking[a,b]
		II. Mid-Season Flowering—Early to Middle April	
3.	Triumph (T)	Mainly the result of hybridization between single early- and late-flowering tulips. The former Mendel tulips are included in this group. Cup-shaped single flowers, 35–60 cm tall	Ben van Zanten[a,b], Couleur Cardinal[a], Debutante[a,b], Dynasty[a,b], Gander's Rhapsody, Ile de France[a,b], Kees Nelis, Leen van der Mark[a,b], Lustige Witwe[a], Negrita[a,b], Orange Cassini[a,b], Prinses Irene[a,b], Prominence[a], Purple Flag[a,b], Purple Prince[a,b], Silver Dollar[a], Strong Gold[a,b], White Dream[a], Yellow Flight[a,b], Yokohama[a,b]
4.	Darwin hybrids (DH)	Mainly the result of hybridization between Darwin tulips with *T. fosteriana* and the result of hybridization between other tulips and botanical tulips, which have the same habit and in which the wild plant is not evident. Ovoid-shaped single flowers, 50–70 cm tall	Ad Rem[a], Apeldoorn[a,b], Golden Apeldoorn, Golden Parade, Oxford, Parade, Van Eijk, World's Favourite
		III. Late Flowering—Late April to Early May	
5.	Single late (SL)	Includes older Darwin tulips and Cottage tulips. Owing to hybridization, the borderlines between the former classes are not visible. Cup or goblet-shaded flowers, 45–75 cm tall. Some cultivars produce multiflowering stems	Maureen, Menton, Pink Diamond, Queen of Night
6.	Lily-flowered (L)	Flowers with pointed reflected perianth segments	Ballerina, Claudia, Pretty Woman, Synaeda Orange
7.	Fringed (Fr)	Perianth segments are edged with crystal-shaped fringes	Arma, Davenport, Fabio
8.	Viridiflora (V)	Perianths are partly greenish	Doll's Minuet, Hollywood, Spring Green

9.	Rembrandt (R)[c]	Broken-color tulips, striped or marked brown, bronze, black, red, pine or purple on red, white or yellow background	
10.	Parrot (P)	Lacerate flowers, generally late-flowering	Blumex Favourite, Bright Parrot, Libretto Parrot, Rococo[a], Super Parrot, Topparrot
11.	Double late (DL)	Peony-flowered tulips with ca. 40 perianths	Angelique, Double Price, Freeman, Upstar, Wirosa, Yellow Pomponnette

IV. Species and Their Hybrids

12.	Kaufmanniana (K)	*T. kaufmanniana* and its hybrids. Very early flowering, sometimes with mottled foliage	Fashion, Giuseppe Verdi, Showwinner
13.	Fosteriana (F)	*T. fosteriana* and its hybrids. Large, early flowering, some cultivars with mottled or striped foliage	Candela, Orange Emperor, Purissima
14.	Greigii (G)	*T. greigii* and its hybrids. Always with mottled or striped foliage, flowering later than *kaufmanniana*	Plaisir, Red Riding Hood, Toronto
15.	Other species and their varieties and hybrids		

Source: Adapted from Van Scheepen, J. 1996. *Classified List and International Register of Tulip Names.* Hillegom, The Netherlands: Royal General Bulbgrowers' Association (KAVB); PT/BKD. 2010a. Bloembollen, voorjaarsbloeiers, beplante oppervlakten, 2006/'07 tot en 2009/'10.

[a] See Table 6.2 in Chapter 6 for production acreage of these cultivars.

[b] Seventeen leading cultivars grown in >100 ha in 2009/2010 in The Netherlands.

[c] Since the unusual marbled colorings in Rembrandt tulip cultivars were actually caused by viruses, the original Rembrandt tulips are no longer grown commercially. However, there are virus-free and genetically stable cultivars which show the stripes or streaks similar to Rembrandt tulips. Examples are 'Keizerskroon' (actually SE), 'Princess Irene' (SE), 'Flaming Parrot' (P), 'Ice Follies' (T), 'Mona Lisa' (L), and 'Sorbet' (SL). They are often sold as Rembrandt tulips.

The leaves are waxy-coated, somewhat fleshy, and linear to oblong in shape. They are usually light-to-medium green. Three to five leaves develop on a scape in a large bulb. In contrast, a small-sized bulb produces only one leaf. The roots are nonbranching and noncontractile and generally do not have root hairs (Kawa and De Hertogh 1992).

At planting, in autumn, one flowering-sized bulb contains three generations: (1) a mother bulb in which an apical bud has already developed leaves and flower, (2) daughter bulbs which replace the mother bulb in the next season, and (3) a meristem of each granddaughter bulb inside the daughter bulbs, which will develop during winter to early summer of the following year. The mother bulb produces roots after planting in autumn, but, during winter, the growth of the apical bud is slow and there are no above-ground shoots. When the temperature increases in spring, rapid scape elongation and flowering takes place. During this period, the mother bulb shrivels and is replaced by the daughter bulbs. Even if the new bulbs reach a minimal size of 6/9 cm, in circumference, which is large enough for flower initiation, the apical meristem is vegetative at harvest in early summer. Root primordia are initiated shortly after the bulb harvest and are complete by September. During summer storage, the apical bud develops 3–4 leaves and flower initiation and differentiation is completed. The initiation of the scape occurs concurrently with flower development, while the pollen and ovule develop after planting.

Flower initiation and development within the bulb occur at warm temperatures of 17–20°C or higher, while cold temperatures are required for the subsequent stem elongation and flowering in spring. Anthesis occurs in early-to-late spring, depending on the cultivars and on climate conditions. At the end of spring, the aerial organs senesce and daughter bulb growth is completed.

D. Major Species and Cultivars

Besides *T. gesneriana*, the most important species in the industry are *T. kaufmanniana*, *T. fosteriana*, and *T. greigii*, each having given rise to an independent cultivar group (Table 4.6). Commercial cultivars are divided into 15 groups (Van Scheepen 1996) based on the time of outdoor flowering, morphology of the perianth parts, and their origin (Table 4.6).

It is estimated that about 8000 cultivars have existed. In 2009/2011, over 1800 cultivars were grown on 9856 ha in The Netherlands, while 17 cultivars (Table 4.6) were grown on more than 100 ha each and occupied about 32% of the total acreage (3136 ha out of 9856 ha) (PT/BKD 2010a). About 800 cultivars were produced on >1 ha (International Flower Bulb Centre 2007b). Among the groups, Triumph (T) occupied 59% of the tulip bulb production in the 2009/2010 season, followed by Double early (DE) (9%), Darwin hybrid (DH) (8%), Single early (SE) (4%), Parrot (P) (4%), and Double late (DL) (4%) (PT/BKD 2010a). The production of other groups, Single late (SL), Lily-flowered (L), Fringed (Fr), Viridiflora (V), Rembrandt (R), and cultivars of *T. kaufmanniana* (K), *T. fosteriana* (F), and *T. greigii* (G), was <4%. Old cultivars, such as 'Keizerskroon' (SE, 1750), 'Couleur Cardinal' (T, 1815), and 'Brilliant Star' (SE, 1908), remain in commercial production.

The majority of tulip cultivars are diploid (2n = 2x = 24), but the Darwin hybrids (DH) are almost all triploid (2n = 3x = 36) and sterile (Le Nard and De Hertogh 1993b). Darwin hybrids were developed by an interspecific cross of Darwin tulips (now included in SL) and *T. fosteriana* (2n = 2x = 24) (Doorenbos 1954), whereas *T. fosteriana* 'Red Emperor' was primarily used as the male parent. It was concluded that 'Red Emperor' supplied the unreduced (2n) gametes (Kroon and Van Eijk 1977) to produce spontaneously occurring triploids of DH. Recently, it was suggested by Marasek et al. (2006) that *T. gesneriana* supplied the diploid gamete (2n). 'World's Favourite' is a triploid DH cultivar, which was bred by the cross of tetraploid *T. gesneriana* seedling ('Denbola' (T) × 'Lustige Witwe' (T)) × diploid *T. fosteriana* (see also Chapter 6). Besides DH cultivars, some SE (e.g., 'Keizerskroon'), T (e.g., 'Dreaming Maid'), P (e.g., 'Orange Favorite', 'Texas Gold'), and SL (e.g., 'Maureen', 'Renown') cultivars are triploid, and 'Mrs. John T. Scheepers' (SL) is tetraploid (Zeilinga and Schouten 1968).

E. PROPAGATION

Tulips can be propagated by seeds, offsets (daughter bulbs), or micropropagation. Seeds are not commercially used for propagation purposes, since they do not result in true-to-type progenies and the juvenile phase of seedlings is 5–6 years. Thus, seed propagation is used only in breeding. Offsets (daughter bulbs) are the most common material for vegetative propagation. Commercial bulb growers harvest the bulbs in late spring to early summer and grade them by circumference. Large bulbs of flowering size are prepared for sale, while small ones are used for replanting. *In vitro* propagation has been actively studied in Japan, Great Britain, France, Poland, and The Netherlands with some positive results (De Hertogh and Le Nard 1993b; Kim and De Hertogh 1997; Podwyszynska 2006; Podwyszynska and Rojek 2008; see also Chapter 10). *In vitro* propagation techniques are, however, not currently and routinely used in tulip breeding programs, nor in commercial bulb production.

F. PRODUCTION

1. Bulb Production

Tulips are grown in a 1-year cycle. They require well-drained light soils with a pH not lower than 6.5 and a low clay content (Rees 1992). Bulb production is based on planting small- and medium-sized bulbs (from 4/5 to 10/11 cm, in circumference). Small bulbs are sometimes grown on a 2-year cycle in order to reduce the costs of labor. Tulips are grown either in beds in sandy soils, or in ridges in heavier soils (Le Nard and De Hertogh 1993b). In the Northern Hemisphere, tulips are planted from October to early December. Late planting has a few advantages—it prolongs the period of warm dry storage, avoiding the risk of fungal pathogens, delays leaf emergence in the field, decreases the risk of frost damage, and the risk of *Fusarium* infection, which is favored by warm soils (Rees 1992). On the other hand, in areas with warm autumns and mild winters (e.g., southern France), planting delayed up to mid-December reduced daughter bulb production (Le Nard 2002).

Before planting, basic fertilizers are applied. The mineral nutrition of tulips depends on the type of soil, its acidity and mineral content, the tulip cultivar(s), and the size and weight of the planted bulbs. Nutrient requirements for bulb production are 140–150 kg of nitrogen, 40–50 kg of phosphorus, 140–150 kg of potassium, and 110–120 kg of calcium per hectare (Bakker 1991). When organic matter content is below 3%, particularly in sandy and salty soils, organic manure can be used before plowing (Le Nard and De Hertogh 1993b).

During the spring growth period, all non-true-to-type or diseased plants are removed and destroyed. This process is called roguing and is always performed during the flowering period. Roguing is labor intensive, but can be accomplished now by means of special herbicide applicators instead of physically digging up the plants. The mechanical removal of flowers ("topping") is beneficial for avoiding fungal diseases and improving bulb yield.

The bulb sizes for forcing are 11/12 and 12/up cm, in circumference, for the retail trade 10/11 cm and above, and those for replanting stocks are 7/8, 8/9, and 9/10 cm. During bulb storage, temperature, relative humidity, and ethylene concentration are controlled. Bulbs for replanting are usually kept at 20°C and an RH of 75%. Storage at higher temperatures decreases the effect of apical dominance and results in the production of a higher number of small daughter bulbs (Koster 1980).

2. Flower Production

Various methods of forcing are employed for early and late flower production. Forcing techniques, used in The Netherlands, are illustrated in Figure 4.4. Similar methods are used in other countries with variations according to different conditions and demands. Whatever the methods used, the procedure mimics the natural development cycle of the plant, that is, growth in warm–cold–warm sequence of temperatures, and the use of large bulbs that have received warm temperatures for flower development prior to planting. Bulbs of 12/up cm, in circumference, produce the largest

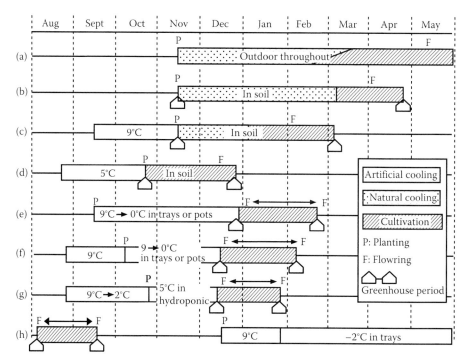

FIGURE 4.4 Programs of tulip flower production in The Netherlands. For details, see text on temperature programming during cold treatments from 9°C to 2°C or to 0°C. (a) Outdoor production. (b) Forcing of non-precooled bulbs in soil in a greenhouse. (c) Forcing of precooled (9°C) bulbs in soil in a greenhouse. (d) Forcing of precooled (5°C) bulbs in soil in a greenhouse. (e) Forcing non-precooled bulbs in trays or pots. (f) Forcing of precooled (9°C) bulbs in trays or pots. (g) Forcing of precooled bulbs in trays or pots on water (hydroponics). (h) Forcing of "Ice tulips" in trays. (After International Flower Bulb Centre. 2012. *The Forcing of Tulips.* http://www.bloembollencentrum.nl/ibc-jsp/binaries/pdf-bestanden/nl/pbt_e-lr.pdf.)

flowers and are suitable for the earliest flowering. Bulbs that are 11/12 cm, in circumference, are commonly used for later forcing. The critical developmental stage of flower initiation in bulbs for early forcing is "Stage G" (the formation of gynoecium), which must be achieved prior to exposure to low temperatures. Cultivar selection is another limiting factor for the control of flowering; some are suitable for early forcing and others for late forcing. Rees (1977) assessed the cold requirements of 116 cultivars, and they varied from 99 days to 171 days. Tulip bulbs are very susceptible to ethylene and must be stored in well-ventilated rooms and containers (see Chapter 11).

For outdoor cultivation, bulbs are planted in October to November and flowering occurs in April to May depending on cultivars and climates (Figure 4.4a). This is not suitable for regions where winter is not cold enough. For production in unheated greenhouses (Figure 4.4b), the non-precooled bulbs are planted in November, experience a natural cold period, and flower in early April. When dry bulbs are stored at 9°C from mid-September to November before planting in greenhouses, flowering is accelerated to early February (Figure 4.4c). For flowering before Christmas, bulbs are dry-stored (precooled) at 5°C or 2°C from August to October before being planted in greenhouses (Figure 4.4d). For flower production in trays or pots, non-precooled bulbs are planted in rooting rooms at 9°C from mid-September, and moved to greenhouses in late December. Flowers are harvested from December to February (Figure 4.4e). Alternatively, dry bulbs receive a 9°C precooling treatment before being transferred to rooting rooms (Figure 4.4f). Generally, the optimal cooling temperatures and periods for the forcing of 9°C (precooled) and dry bulbs in trays are at 9°C until 20 October, 7°C until 10 November, 5°C until 1 December, and then 2–0°C depending on shoot length. This sequence is also applicable in hydroponic culture (Figure 4.4g).

A remarkable change in tulip cultivation in the last two decades is the increase in hydroponic culture systems, started in 1990 in The Netherlands. Only 0.8% of tulips, produced for cut flowers in The Netherlands during the 1997/1998 season, were cultivated hydroponically. This percentage has increased to 75–80% in 2011 (International Flower Bulb Centre 2007a; Chapter 11).

The recommended greenhouse temperatures for early forcing is 17–18°C and a little lower for flower production from February to April. A pH of 6 and an EC of 1.5–1.8 mScm^2 are recommended for hydroponic production.

In potted tulip production, height control is important. Certain cultivars, e.g., 'Brilliant Star' (SE) and 'Flair' (SE) are used for potted plant production. To keep the flowers short, the rule that shorter cold periods produce shorter stems is applicable for certain cultivars, such as 'Abra' (T), 'Arma' (Fr), 'Kikomachi' (T), 'Monte Carlo' (DE), 'Prinses Irene' (T), and 'Seadov' (T). It is also possible to control tulips height by applying plant growth regulators. A 100 mL solution of 10 mg/L paclobutrazol to each 12 cm pot is applied routinely by producers of potted plants. Recently, the effects of chemical control of plant height in 'Prominence' (T) tulip were evaluated by Krug et al. (2005b). Substrate drenches of ancymidol, flurprimidol, and paclobutrazol resulted in adequate control, using concentrations of 0.5 and 1 mg per pot, respectively. Flurprimidol foliar sprays at <80 mg/L were ineffective in controlling height during greenhouse forcing, but during postharvest evaluation application of 80 mg/L resulted in plants 14% shorter than the untreated controls. Preplant bulb soaks of flurprimidol, paclobutrazol, and uniconazole at concentrations of 25, 50, and 10 mg/L, respectively, effectively controlled plant height.

Delayed flowering can be achieved by using the technology of "Ice tulips" (Figure 4.4h). The bulbs are planted in trays in November and rooted at 9°C for 2–4 weeks. Subsequently, they are frozen at −1.5 to −2°C until placed in the greenhouses. Flowering takes place under warm conditions; however, the quality of the flowers can be reduced. Bulbs produced in the Southern Hemisphere can replace the Ice tulip technique for off-season production of fresh flowers (see Chapter 11).

G.   Physiological Disorders, Pests, and Diseases

Flower abortion (flower blasting) and stem topple are critical disorders in the flower production of tulips. Flower abortion can be caused by improper temperature treatments at various stages of forcing and growth. Stem topple is reported to be caused by calcium deficiency. It is also caused by either very high or very low temperatures during forcing.

Botrytis cinerea (leaf tip necrosis), *Botrytis tulipae* (fire), *Fusarium oxysporum* f. sp. *tulipae* (bulb rot), and *Rhizoctonia tuliparum* (gray bulb rot) are the most common fungal pathogens. *Penicillium* spp. (blue mold), *Pythium ultimum* (bulb or root rot), *Phytophthora* spp. (blossom blight), *Sclerotium rolfsii* (blight), *Sclerotinia bulborum*, and *Sclerotiolum* sp. have also been observed. Bacterial brown rot and bacterial soft rot are caused by *Burkholderia gladioli* and *Pectobacterium carotovorum*, respectively.

The physiological disorders and diseases specific to hydroponics include leaf topple caused by calcium deficiency ('Leen van der Mark' (T), 'Purple Prince' (T), 'Monte Carlo' (DE), and 'Christmas Marvel' (SE)), bacterial slime growth ('Ben van Zanten' (T) and 'Debutante' (T)), and root rot caused by *Pythium* ('Leen van der Mark', 'Debutante', and 'Ile de France' (T)). Calcium deficiency in hydroponic tulips was investigated by Nelson and Niedziela (1998a, b) and Nelson et al. (2003).

Tulips are rather susceptible to viral infections. The main viruses are *Tulip breaking virus* (TBV), *Tulip virus X* (TVX), *Lily symptomless virus* (LSV), *Tobacco mosaic virus* (TMV), *Tomato bushy stunt virus* (TBSV), *Tobacco rattle virus* (TRV), *Arabis mosaic virus* (ArMV), *Tobacco ring spot virus* (TRSV), *Tomato black ring virus* (TBRV), *Cucumber mosaic virus* (CMV), and *Tobacco necrosis virus* (TNV). *Aphelenchoides fragariae*, *Ditylenchus destructor*, *Ditylenchus dipsaci*, and *Pratylenchus penetrans* are the nematodes that attack tulip plants. Aphids are the primary insects.

ACKNOWLEDGMENTS

The authors are grateful to Drs. Rina Kamenetsky and August A. De Hertogh for their critical review of this chapter.

CHAPTER CONTRIBUTORS

Hiroshi Okubo is a professor of horticultural science at Kyushu University, Fukuoka, Japan. He was born in Japan and received his BS, MS, and PhD in horticultural science from Kyushu University. His major expertise includes the physiology of growth and development in horticultural crops, particularly in ornamental geophytes. He also conducts research on tropical plant species. Dr. Okubo is the author of more than 200 professional papers, book chapters, technical publications, and popular press articles. He is the director of the University Farm of Kyushu University and a recipient of an award of the Japanese Society for Horticultural Science (1988).

Dariusz Sochacki is a senior researcher at the Research Institute of Horticulture, Skierniewice, Poland. He was born in Poland and received his MS from Warsaw University of Life Sciences and his PhD from the Research Institute of Pomology and Floriculture, Skierniewice. He has held research positions at the Research Institute of Pomology and Floriculture, (1992–2010) and the Research Institute of Horticulture, Skierniewice, Poland (2011–to date). Dr. Sochacki is also the secretary of the Polish Association of Flower Bulb Growers (1994–to date). His expertise includes propagation of flower bulbs, virus detection methods, and virus eradication *in vitro* and genetic resources. Dr. Sochacki is the author of 60 refereed publications and conference communications, two cultivars of narcissus, and one cultivar of tulip.

REFERENCES

Asen, S., R. N. Stewart, K. H. Norris, and D. R. Massie. 1970. A stable blue non-metallic co-pigment complex of delphinidin and *C*-glycosylflavones in Prof. Blaauw iris. *Phytochemistry* 9:619–627.

Bach, A. 1992. Micropropagation of hyacinths (*Hyacinthus orientalis* L.). In *Biotechnology in Agriculture and Forestry, Vol. 20, High-Tech and Micropropagation IV*, ed. Y. P. S. Bajaj, 144–159. Berlin: Springer-Verlag.

Badri, M. A., P. E. H. Minchin, and L. Lapointe. 2007. Effects of temperature on the growth of spring ephemerals: *Crocus vernus*. *Physiol. Plant.* 130:67–76.

Bakker, M. 1991. *Teelt Info.*, 442–449. Lisse, The Netherlands: CNB Info.

Banba, H. 1967. Pigments of lily flowers. I. Survey of anthocyanin. *J. Japan. Soc. Hort. Sci.* 36:433–437 (In Japanese with English summary).

Banba, H. 1968. Pigments of lily flowers. II. Survey of carotenoid. *J. Japan. Soc. Hort. Sci.* 37:368–378 (In Japanese with English summary).

Beattie, D. J., and J. W. White. 1993. *Lilium*—Hybrids and species. In *The Physiology of Flower Bulbs*, eds. A. De Hertogh, and M. Le Nard, 423–454. Amsterdam: Elsevier.

Benschop, M. 1993. *Crocus*. In *The Physiology of Flower Bulbs*, eds. A. De Hertogh, and M. Le Nard, 257–272. Amsterdam: Elsevier.

Benschop, M., R. Kamenetsky, M. Le Nard, H. Okubo, and A. De Hertogh. 2010. The global flower bulb industry: Production, utilization, research. *Hort. Rev.* 36:1–115.

Blom, T. J., and D. R. Kerec. 2003. Effects of far-red light/temperature DIF and far-red light/temperature pulse combinations on height of lily hybrids. *J. Hort. Sci. Biotech.* 78:278–282.

Blom, T., D. Kerec, W. Brown, and D. Kristie. 2004. Irrigation method and temperature of water affect height of potted Easter lilies. *HortScience* 39:71–74.

Bonnier, F. J. M., F. A. Hoekstra, C. H. R. de Vos, and J. M. van Tuyl. 1997. Viability loss and oxidative stress in lily bulbs during long-term cold storage. *Plant Sci.* 122:133–140.

Buschman, J. C. M. 2005. Globalisation—flower—flower bulbs—bulb flowers. *Acta Hort.* 673:27–33.

Chang, Y.-C., and W. B. Miller. 2003. Growth and calcium partitioning in *Lilium* 'Star Gazer' in relation to leaf calcium deficiency. *J. Am. Soc. Hort. Sci.* 128:788–796.

Chang, Y.-C., and W. B. Miller. 2005. The development of upper leaf necrosis in *Lilium* 'Star Gazer'. *J. Am. Soc. Hort. Sci.* 130:759–766.

Chang, W., S. Li, H. Hu, and Y. Fan. 2008. Photosynthetic characteristics of three varieties of *Lilium* "Oriental Hybrids" in the central areas of Yunnan Province, China. *Front. Biol. China* 3:453–458.

Cohat, J. 1993. *Gladiolus.* In *The Physiology of Flower Bulbs*, eds. A. De Hertogh, and M. Le Nard, 297–320. Amsterdam: Elsevier.

Cohen, D., D. S. Ziv, C. Fintea, A. Ion, A. Cohen, and R. Kamenetsky. 2011. Forcing new varieties of paperwhite narcissus for early flowering. *Acta Hort.* 886:427–434.

Comber, H. F. 1949. A new classification of the genus *Lilium*. *R.H.S. Lily Year Book* 13:86–105.

Currey, C. J., and R. G. Lopez. 2010. Paclobutrazol pre-plant bulb dips effectively control height of 'Nellie White' Easter lily. *HortTechnology* 20:357–360.

Darlington, C. D. 1973. *Chromosome Botany and the Origin of Cultivated Plants. 3rd ed.* London: George Allen & Unwin Ltd.

Dash, M. 2001. *Tulipomania: The Story of the World's Most Coveted Flower & the Extraordinary Passions It Aroused.* New York: Three Rivers Press.

De Hertogh, A. 1980. Bulbous plants. In *Introduction to Floriculture*, ed. R. A. Larson, 215–235. New York: Academic Press.

De Hertogh, A. 1996. *Holland Bulb Forcer's Guide 5th ed.* Hillegom, The Netherlands: The International Flower Bulb Centre and The Dutch Bulb Exporters Association.

De Hertogh, A., and M. Le Nard. 1993a. *The Physiology of Flower Bulbs.* Amsterdam: Elsevier.

De Hertogh, A. A., and M. Le Nard. 1993b. Production systems for flower bulbs. In *The Physiology of Flower Bulbs*, eds. A. De Hertogh, and M. Le Nard, 45–52. Amsterdam: Elsevier.

De Munk, W. J., and J. Schipper. 1993. *Iris*—bulbous and rhizomatous. In *The Physiology of Flower Bulbs*, eds. A. De Hertogh, and M. Le Nard, 349–379. Amsterdam: Elsevier.

Doorenbos, J. 1954. Notes on the history of bulb breeding in The Netherlands. *Euphytica* 3:1–11.

Eikelboom, W., and J. P. van Eijk. 1990. Prospects of interspecific hybridization in Dutch iris. *Acta Hort.* 266:353–356.

Export-Import Bank of India. 2009. *Floriculture—A Sector Study.* Mumbai, India: Export-Import Bank of India.

Fairchild, L. 1979. *How to Grow Glorious Gladiolus.* Edgewood, Maryland: The North American Gladiolus Council.

Francescangeli, N. 2009. Paclobutrazol and cytokinin to produce iris (*Iris × hollandica* Tub.) in pots. *Chilean J. Agric. Res.* 69:509–515.

Frizzi, G., M. Miranda, C. Pantani, and F. Tammaro. 2007. Allozyme differentiation in four species of the *Crocus cartwrightianus* group and in cultivated saffron (*Crocus sativus*). *Biochem. Syst. Ecol.* 35:859–868.

George, E. F. 1996. Micropropagation in practice. In *Plant Propagation by Tissue Culture, Part 2, in Practice*, ed. E. F. George, 834–1222. Edington, Wiltshire, UK: Exegetics Ltd.

Goldblatt, P., and J. Manning. 1998. *Gladiolus in Southern Africa.* Vleaberg, South Africa: Fernwood Press.

Goldblatt, P., M. Takei, and Z. A. Razzaq. 1993. Chromosome cytology in tropical African *Gladiolus* (Iridaceae). *Ann. Missouri Bot. Gard.* 80:461–470.

Grabowska, B. 1978. Effect of the time of planting of gladiolus cormels on the yield of corms. *Prace Instytutu Sadownictwa i Kwiaciarstwa, Ser. B* 3:15–22 (In Polish with English abstract).

Grabowska, B., and K. Mynett. 1971. The effect of photosynthetic area on tulip bulb yield. *Biuletyn Instytutu Hodowli i Aklimatyzacji Roslin* 3–4:37–38 (In Polish with English abstract).

Hai, N. T. L., M. Hiramatsu, J. H. Kim, and H. Okubo. 2010. Breeding new type lily cultivars by interspecific crosses between *Lilium brownii* var. *colchecteri* and its close relatives. *Abstract of 2010 IHC Lisboa*, 50.

Hanks, G. R. 1993. *Narcissus.* In *The Physiology of Flower Bulbs*, eds. A. De Hertogh, and M. Le Nard, 463–558. Amsterdam: Elsevier.

Hanks, G. R. 1995. Prevention of hot-water treatment damage in narcissus bulbs by pre-warming. *J. Hort. Sci. Biotech.* 70:343–355.

Hanks, G. R. 2002. *Narcissus and Daffodil: The Genus Narcissus.* Boca Raton, FL: CRC Press.

Hekstra, G. and C. Broertjes. 1968. Mutation breeding in bulbous iris. *Euphytica* 17:345–351.

Hiramatsu, M., K. Ii, H. Okubo, K. L. Huang, and C. W. Huang. 2001. Biogeography and origin of *Lilium longiflorum* and *L. formosanum* (Liliaceae) endemic to the Ryukyu Archipelago and Taiwan as determined by allozyme diversity. *Am. J. Bot.* 88:1230–1239.

Hiramatsu, M., H. Okubo, K. Yoshimura, K. L. Huang, and C. W. Huang. 2002. Biogeography and origin of *Lilium longiflorum* and *L. formosanum* II—Intra- and interspecific variation in stem leaf morphology, flowering rate and individual net production during the first year seedling growth. *Acta Hort.* 570:331–334.

Hoog, M. H. 1973. On the origin of *Tulipa*. In *Lilies and Other Liliaceae*, eds. E. Napier, and J. N. O. Platt, 47–64. London: Royal Horticultural Society.

Hosokawa, K. 1999. Variations among anthocyanins in the floral organs of seven cultivars of *Hyacinthus orientalis*. *J. Plant Physiol.* 155:285–287.

Imanishi, H. 1983. Effects of exposure of bulbs to smoke and ethylene on flowering of *Narcissus tazetta* cultivar 'Grand Soleil d'Or'. *Scientia Hortic.* 21:173–180.

Imanishi, H. 2005. *Flowering Control of Bulbous Plants*. Tokyo: Nousangyoson Bunkakyoukai (In Japanese).

Imanishi, H., and Y. Imae. 1990. Effects of low light intensity and low temperature given at different developmental stages on flowering of gladiolus. *Acta Hort.* 266:189–196.

Imanishi, H., and H. Konoshima. 1988. *Crocus* L. In *The Grand Dictionary of Horticulture Volume 2*, ed. Y. Tsukamoto, 187–191. Tokyo: Shogakukan (In Japanese).

International Flower Bulb Centre. 2007a. *Flower Bulb Power Education*. International Flower Bulb Centre, http://www.flowerbulbpower.com.

International Flower Bulb Centre. 2007b. *Tulip Picture Book*. Hillegom, The Netherlands: International Flower Bulb Centre.

International Flower Bulb Centre. 2012. *The Forcing of Tulips*. http://www.bloembollencentrum.nl/ibc-jsp/binaries/pdf-bestanden/nl/pbt_e-lr.pdf (accessed: May 2012).

Jaffe, M. J., and A. C. Leopold. 2007. Light activation of contractile roots of Easter lily. *J. Am. Soc. Hort. Sci.* 132:575–582.

Jefferson-Brown, M., and H. Howland. 1995. *The Gardener's Guide to Growing Lilies*. Portland, OR: Timber Press.

Kamenetsky, R. 2005. Production of flower bulbs in regions with warm climates. *Acta Hort.* 673:59–66.

Kamenetsky, R., and W. B. Miller. 2010. The global trade in ornamental geophytes. *Chronica Hort.* 50(4):27–30.

Karasawa, K. 1933. On the triploidy of *Crocus sativus*, L. and its high sterility. *Jap. J. Genet.* 9:6–8.

Kawa, L., and A. A. De Hertogh. 1992. Root physiology of ornamental flowering bulbs. *Hort. Rev.* 14:57–88.

Kim, J. H., H. Y. Kyung, Y. S. Choi, J. K. Lee, M. Hiramatsu, and H. Okubo. 2006. Geographic distribution and habitat differentiation in diploid and triploid *Lilium lancifolium* of South Korea. *J. Fac. Agr., Kyushu Univ.* 51:239–243.

Kim, K. W., and A. A. De Hertogh. 1997. Tissue culture of ornamental flowering bulbs (geophytes). *Hort. Rev.* 18:87–169.

Kington, S. 2008. *The Daffodil Register and Classified List*. Woking, UK: RHS Publ.

Kok, H., C. Conijn, and H. van Aanholt. 2008. An improved hot water treatment for controlling nematodes and mites in oriental lilies. *Book of Abstracts, 10th International Symposium on Flower Bulbs and Herbaceous Perennials*, 88.

Koster, J. 1980. Storage of tulip planting stock in relation to growth and yield. *Acta Hort.* 109:49–56.

Kroon, G. H., and M. C. Jongerius. 1986. Chromosome numbers of *Tulipa* species and the occurrence of hexaploidy. *Euphytica* 35:73–76.

Kroon, G. H., and J. P. van Eijk. 1977. Polyploidy in tulips (*Tulipa* L.). The occurrence of diploid gametes. *Euphytica* 26:63–66.

Krug, B. A., B. E. Whipker, and I. McCall. 2005a. Flurprimidol is effective at controlling height of 'Star Gazer' Oriental lily. *HortTechnology* 15:373–376.

Krug, B. A., B. E., Whipker, I. McCall, and J. M. Dole. 2005b. Comparison of flurprimidol to ancymidol, paclobutrazol, and uniconazole for tulip height control. *HortTechnology* 15:370–373.

Krug, B. A., B. E., Whipker, I. McCall, and J. M. Dole. 2005c. Comparison of flurprimidol to ethephon, paclobutrazol, and uniconazole for hyacinth height control. *HortTechnology* 15:872–874.

Krug, B. A., B. E. Whipker, and I. McCall. 2006. Hyacinth height control using preplant bulb soaks of flurprimidol. *HortTechnology* 16:370–375.

Le Nard, M. 2002. Effects of bulb planting date on growth of tulip 'Don Quichotte' under mild winter conditions. *Acta Hort.* 570:153–156.

Le Nard, M., and A. A. De Hertogh. 1993a. Plant breeding and genetics. In *The Physiology of Flower Bulbs*, eds. A. De Hertogh, and M. Le Nard, 161–169, Amsterdam: Elsevier.

Le Nard, M., and A. A. De Hertogh. 1993b. *Tulipa*. In *The Physiology of Flower Bulbs*, eds. A. De Hertogh, and M. Le Nard, 617–682. Amsterdam: Elsevier.

Leslie, A. C. 1982. *The International Lily Register. 3rd edition*. London: The Royal Horticultural Society.

Liang, S. Y. 1980. *Lilium* L. In *Flora Reipublicae Popularis Sinicae vol. 14*, eds. F.-T. Wang, and T. Tang, 116–157. Beijing: Science Press (In Chinese).

Lindley, J. 1837. *Lilium speciosum*. The crimson Japan lily. *Edwards's Botanical Register* 23:2000.

Lodewijk, T. 1979. *The Book of Tulips*. London: Cassel.

Lundmark, M., V. Hurry, and L. Lapointe. 2009. Low temperature maximizes growth of *Crocus vernus* (L.) Hill via changes in carbon partitioning and corm development. *J. Exp. Bot.* 60:2203–2213.

MAFF. 1994. *Fertiliser Recommendations for Agricultural and Horticultural Crops. Reference Book 209. 6th edition*. London: Her Majesty's Stationery Office.

MAFF. 2007. *Statistics of Agriculture, Forestry and Fisheries*. Ministry of Agriculture, Forestry and Fisheries, Japan.

Marasek, A., H. Mizuochi, and K. Okazaki. 2006. The origin of Darwin hybrid tulips analyzed by flow cytometry, karyotype analyses and genomic *in situ* hybridization. *Euphytica* 151:279–290.

Margaris, N. S. 2000. Flowers in Greek mythology. *Acta Hort.* 541:23–29.

Martínez, J., P. Vargas, M. Luceño, and Á. Cuadrado. 2010. Evolution of *Iris* subgenus *Xiphium* based on chromosome numbers, FISH of nrDNA (5S, 45S) and *trnL-trnF* sequence analysis. *Plant Syst. Evol.* 289:223–235.

Mathew, B. 1982. *The Crocus. A Revision of the Genus Crocus (Iridaceae)*. London: B. T. Batsford.

Mathew, B. 1989. *The Iris*. London: Batsford.

Mathew, B., G. Petersen, and O. Seberg. 2009. A reassessment of *Crocus* based on molecular analysis. *Plantsman New Series* 8:50–57.

McRae, E. A. 1998. *Lilies: A Guide for Growers and Collectors*, Portland, OR: Timber Press.

Miller, W. B. 1993. *Lilium longiflorum*. In *The Physiology of Flower Bulbs*, eds. A. De Hertogh, and M. Le Nard, 391–422. Amsterdam: Elsevier.

Miller, W. B., P. A. Hammer, and T. I. Kirk. 1993. Reversed greenhouse temperatures alter carbohydrate status in *Lilium longiflorum* Thunb. 'Nellie White'. *J. Am. Soc. Hort. Sci.* 118:736–740.

Molina, R. V., M. Valero, Y. Navarro, A. García-Luis, and J. L. Guardiola. 2004. The effect of time of corm lifting and duration of incubation at inductive temperature on flowering in the saffron plant (*Crocus sativus* L.). *Scientia Hortic.* 103:79–91.

Molina, R. V., M. Valero, Y. Navarro, J. L. Guardiola, and A. García-Luis. 2005. Temperature effects on flower formation in saffron (*Crocus sativus* L.). *Scientia Hortic.* 103:361–379.

Nelson, P. V., and C. E. Niedziela Jr. 1998a. Effects of calcium source and temperature regime on calcium deficiency during hydroponic forcing of tulip. *Scientia Hortic.* 73:137–150.

Nelson, P. V., and C. E. Niedziela Jr. 1998b. Effect of ancymidol in combination with temperature regime, calcium nitrate, and cultivar selection on calcium deficiency symptoms during hydroponic forcing of tulip. *Scientia Hortic.* 74:207–218.

Nelson, P. V., W. Kowalczyk, C. E. Niedziela Jr., N. C. Mingis, and W. H. Swallow. 2003. Effects of relative humidity, calcium supply, and forcing season on tulip calcium status during hydroponic forcing. *Scientia Hortic.* 98:409–422.

Niedziela, Jr., C. E., S. H. Kim, P. V. Nelson, and A. A. De Hertogh. 2008. Effects of N-P-K deficiency and temperature regime on the growth and development of *Lilium longiflorum* 'Nellie White' during bulb production under phytotron conditions. *Scientia Hortic.* 116:430–436.

Nieuwhof, M., L. W. D. van Raamsdonk, and J. P. van Eijk. 1990. Pigment composition of flowers of *Tulipa* species as a parameter for biosystematic research. *Biochem. Syst. Ecol.* 18:399–404.

Nishikawa T., K. Okazaki, K. Arakawa, and T. Nagamine. 2001. Phylogenetic analysis of section *Sinomartagon* in genus *Lilium* using sequences of the internal transcribed spacer region in nuclear ribosomal DNA. *Breed. Sci.* 51:39–46.

Noda, S. 1978. Chromosomes of diploid and triploid forms found in the natural populations of tiger lily in Tsushima. *Bot. Mag., Tokyo* 91:279–283.

Nørbæk, R., K. Brandt, J. K. Nielsen, M. Ørgaard, and N. Jacobsen. 2002. Flower pigment composition of *Crocus* species and cultivars used for a chemotaxonomic investigation. *Biochem. Syst. Ecol.* 30:763–791.

Nørbæk, R., and T. Kondo. 1998. Anthocyanins from flowers of *Crocus* (Iridaceae). *Phytochemistry* 47:861–864.

Nørbæk, R., and T. Kondo. 1999a. Anthocyanins from flowers of *Lilium* (Liliaceae). *Phytochemistry* 50:1181–1184.

Nørbæk, R., and T. Kondo. 1999b. Further anthocyanins from flowers of *Crocus antalyensis* (Iridaceae). *Phytochemistry* 50:325–328.

Nowak, J., and R. M. Rudnicki. 1993. *Hyacinthus*. In *The Physiology of Flower Bulbs*, eds. A. De Hertogh, and M. Le Nard, 335–347. Amsterdam: Elsevier.

Okubo, H., M. Hiramatsu, J. Masuda, and S. Sakazono. In press. New insight of *Lilium brownii* var. *colchesteri*. In *Floriculture and Ornamental Biotechnology*, ed. J. M. van Tuyl. Isleworth, UK: Global Science Books.

Ørgaard, M., N. Jacobsen, and J. S. Heslop-Harrison. 1995. The hybrid origin of two cultivars of *Crocus* (Iridaceae) analysed by molecular cytogenetics including genomic southern and *in situ* hybridization. *Ann. Bot.* 76:253–262.

PT/BKD. 2010a. Bloembollen, voorjaarsbloeiers, beplante oppervlakten, 2006/'07 tot en 2009/'10 (In Dutch).

PT/BKD. 2010b. Lelie opgeplante oppervlakten 2010 (In Dutch).

Paek, K. Y., and T. A. Thorpe. 1990. Hyacinth. In *Handbook of Plant Cell Culture, Vol. 5, Ornamental Species*, eds. P. V. Ammirato, D. R. Evans, W. R. Sharp, and Y. P. S. Bajaj, 479–508. New York: McGraw-Hill.

Pavord, A. 1999. *The Tulip*. London: Bloomsbury Publ. Plc.

Petersen, G., O. Seberg, S. Thorsøe, T. Jørgensen, and B. Mathew. 2008. A phylogeny of the genus *Crocus* (Iridaceae) based on sequence data from five plastid regions. *Taxon* 57:487–499.

Podwyszynska, M. 2006. Improvement of bulb formation in micropropagated tulips by treatment with NAA and paclobutrazol or ancymidol. *Acta Hort.* 752:679–684.

Podwyszynska, M., and A. Rojek. 2008. Effect of low temperature and GA treatments on *in vitro* shoot multiplication of tulip (*Tulipa gesneriana* L.). *Zesz. Probl. Post. Nauk Rol.* 524:107–113.

Rasmussen, E. 1976. Afstandsforsog og forsog med 1. og 2. ars kulturer af narcisser, kombineret med forskellig laeggetid og forskellig afstand. *Tidsskrift for Planteavl* 80:20–30 (In Danish).

Ranwala, A. P., G. Legnani, M. Reitmeier, B. B. Stewart, and W. B. Miller. 2002. Efficacy of plant growth retardants as preplant bulb dips for height control in LA and Oriental hybrid lilies. *HortTechnology* 12:426–431.

Rees, A. R. 1977. The cold requirement of tulip cultivars. *Scientia Hortic.* 7:383–389.

Rees, A. R. 1985. *Narcissus*. In *CRC Handbook of Flowering vol. 1*, ed. A. H. Halevy, 268–271. Boca Raton, FL: CRC Press.

Rees, A. R. 1992. *Ornamental Bulbs, Corms and Tubers*. Wallingford, Oxon, UK: C.A.B. International.

Royal Horticultural Society. 2009. *The International Lily Register and Checklist 2007. First Supplement*. London: The Royal Horticultural Society.

Sakazono, S., M. Hiramatsu, K. L. Huang, C. L., Huang, and H. Okubo. 2012. Phylogenetic relationship between degree of self-compatibility and floral traits in *Lilium longiflorum* Thunb. (Liliaceae). *J. Japan. Soc. Hort. Sci.* 81:80–90.

Sakazono, S., M. Hiramatsu, and H. Okubo. 2006. Geographic distribution of self-compatibility and -incompatibility in *Lilium longiflorum* Thunb. *Abstract of 27th IHC, Seoul*, 29.

Sakazono, S., M. Hiramatsu, and H. Okubo. 2009. Variation of pollen tube behavior and seed set in self-pollination of *Lilium longiflorum* Thunb. *J. Fac. Agr., Kyushu Univ.* 54:37–40.

Saruwatari, H., Y. Shuto-Nakano, K. Nakano, M. Hiramatsu, Y. Ozaki, and H. Okubo. 2008. Interspecific lily hybrids with the ability to flower precociously and to produce multiple flower stalks from *Lilium formosanum*. *J. Japan. Soc. Hort. Sci.* 77:312–317.

Seabrook, J. E. A. 1990. *Narcissus* (daffodil). In *Handbook of Plant Cell Culture, vol. 5, Ornamental Species*, eds. P. V. Ammirato, D. R. Evans, W. R. Sharp, and Y. P. S. Bajaj, 577–597. New York: McGraw-Hill.

Shillo, R., and A. H. Halevy. 1976a. The effect of various environmental factors on flowering of gladiolus. I. Light intensity. *Scientia Hortic.* 4:131–137.

Shillo, R., and A. H. Halevy. 1976b. The effect of various environmental factors on flowering of gladiolus. II. Length of the day. *Scientia Hortic.* 4:139–146.

Shillo, R., and A. H. Halevy. 1976c. The effect of various environmental factors on flowering of gladiolus. III. Temperature and moisture. *Scientia Hortic.* 4:147–155.

Shimizu, M. 1987. *The Lilies of Japan. Species and Hybrids*. Tokyo: Seibundo Shinkosha (In Japanese).

Shoub, J., and A. A. De Hertogh. 1975. Floral stalk topple: A disorder of *Hyacinthus orientalis* L. and its control. *HortScience* 10:26–28.

Sochacki, D. 2011. The use of ELISA in the micropropagation of virus-free *Narcissus*. *Acta Hort.* 886:253–258.

Sochacki, D., and K. Mynett. 1996. The effect of one-year and two-year crop of lilies, tulips and narcissus on bulb and flower yield. *Zeszyty Naukowe Inst. Sad. i Kwiac.* 3:141–152 (In Polish with English abstract).

Sorrentino, G., L. Cerio, and A. Alvino. 1997. Effect of shading and air temperature on leaf photosynthesis, fluorescence and growth in lily plants. *Scientia Hortic.* 69:259–273.

Steininger, J., and C. C. Pasian. 2003. Prediction of development of Asiatic lilies based on air temperature and thermal units. *HortScience* 38:1100–1103.

Takemura, T., Y. Takatsu, M. Kasumi, W. Marubashi, and T. Iwashina. 2005. Flavonoids and their distribution patterns in the flowers of *Gladiolus* cultivars. *Acta Hort.* 673:487–493.

Takemura, T., Y. Takatsu, M. Kasumi, W. Marubashi, and T. Iwashina. 2008. Anthocyanins of *Gladiolus* cultivars and their contribution to flower colors. *J. Japan. Soc. Hort. Sci.* 77:80–87.

Theron, K. I., and A. A. De Hertogh. 2001. Amaryllidaceae: Geophytic growth, development, and flowering. *Hort. Rev.* 25:1–70.

Van Aartrijk, J. 1995. *Ziekten en Afwijkingen Bij Bolgewassen. Deel II: Amaryllidaceae, Araceae, Begoniaciae, Cannaceae, Compositae, Iridaceae, Oxalidaceae, Ranunculaceae.* Lisse, The Netherlands: Tweede druk, Ministerie van Landbouw, Natuurbeheer en Visserij, het Laboratorium voor Bloembollenondeerzoek en het Informatie—en Kennis Centrum Landbouw (In Dutch).

Van Aartrijk, J. 2000. *Ziekten en Afwijkingen Bij Bolgewassen. Deel I: Liliaceae.* Lisse, The Netherlands: Derde druk, Ministerie van Landbouw, Natuurbeheer en Visserij, het Laboratorium voor Bloembollenondeerzoek en het Informatie—en Kennis Centrum Landbouw (In Dutch).

Van Scheepen, J. 1991. *International Checklist for Hyacinths and Miscellaneous Bulbs.* Hillegom, The Netherlands: Royal General Bulbgrowers' Association (KAVB).

Van Scheepen, J. 1996. *Classified List and International Register of Tulip Names.* Hillegom, The Netherlands: Royal General Bulbgrowers' Association (KAVB).

Van Tuyl, J. M. 1982. Breeding for resistance to yellow disease of hyacinths. II. Influence of flowering time, leaf characters, stomata and chromosome number on the degree of resistance. *Euphytica* 31:621–628.

Weiler, T. C., and R. W. Langhans. 1972. Growth and flowering responses of *Lilium longiflorum* Thunb. 'Ace' to different daylengths. *J. Am. Soc. Hort. Sci.* 97:176–177.

Wilfret, G. J. 1980. *Gladiolus.* In *Introduction to Floriculture*, ed. R. A. Larson, 165–181. New York: Academic Press.

Wilkins, H. F. 1985. *Crocus vernus, Crocus sativus.* In *CRC Handbook of Flowering vol. II*, ed. A. H. Halevy, 350–355. Boca Raton, FL: CRC Press.

Wilson, C. A. 2004. Phylogeny of *Iris* based on chloroplast *matK* gene and *trnK* intron sequence data. *Molec. Phylogen. Evol.* 33:402–412.

Wylie, A. P. 1952. The history of the garden narcissi. *Heredity* 6:137–156.

Xia, Y. P., H. J. Zheng, and C. H. Huang. 2005. Studies on the bulb development and its physiological mechanism in *Lilium* Oriental hybrids. *Acta Hort.* 673:91–98.

Yamagishi, M., S. Kishimoto, and M. Nakayama. 2010. Carotenoid composition and changes in expression of carotenoid biosynthetic genes in tepals of Asiatic hybrid lily. *Plant Breed.* 129:100–107.

Yang, S.-S. 2000. *Lilium.* In *Flora of Taiwan, Second Edition, Volume Five*, ed. T.-C. Huang, 49–52. Taipei, Taiwan: National Taiwan University.

Yue, D., and H. Imanishi. 1988. Effects of ethylene exposure on flowering for early forcing of Dutch iris 'Blue Magic'. *J. Japan. Soc. Hor. Sci.* 57:289–294 (In Japanese with English summary).

Yue, D., and H. Imanishi. 1990. Influence of storage temperature and its duration before or after ethylene exposure on the formation of flower buds in Dutch iris cultivar 'Blue Magic'. *Scientia Hortic.* 43:331–337.

Zeilinga, A. E., and H. P. Schouten. 1968. Polyploidy in garden tulips. I. A survery of *Tulipa* varieties for polyploids. *Euphytica* 17:252–264.

Ziv, M., and H. Lilien-Kipnis. 1990. *Gladiolus.* In *Handbook of Plant Cell Culture, Vol. 5, Ornamental Species*, eds. P. V. Ammirato, D. R. Evans, W. R. Sharp, and Y. P. S. Bajaj, 461–478. New York: McGraw-Hill.

Zonneveld, B. J. M. 2008. The systematic value of nuclear DNA content for all species of *Narcissus* L. (Amaryllidaceae). *Plant Syst. Evol.* 275:109–132.

Zonneveld, B. J. M. 2009. The systematic value of nuclear genome size for all species of *Tulipa* L. (Liliaceae). *Plant Syst. Evol.* 281:217–245.

5 Introduction and Development of New Ornamental Geophytes

Mark P. Bridgen

CONTENTS

I. INTRODUCTION

Plant introduction, breeding, and development are relatively recent activities in human history with their origins dating back to Gregor Mendel's work on inheritance in the 1900s, followed by the Green Revolution in the 1960s and 1970s, and genetic engineering and gene transformation most recently.

The introduction of new germplasm and the development of improved plants have contributed to an increased productivity and quality of those crops that are used for food, medicine, fiber, animal feed, and ornamental purposes. When plant breeding began thousands of years ago, it was a simple selection process based on straightforward visual observations. Today, plant improvements are the result of intensive research, state-of-the-art procedures, and often close cooperation among experts from many different disciplines. Universities, private enterprises, governments, and individuals are all involved in the detailed development and selection of valuable and beautiful plants.

Many opportunities exist for the development and introduction of new field crops, fruits, vegetables, and ornamental plants (Bridgen et al. 2009). The selection of an appropriate plant is the key decision that a plant breeder must eventually make. This selection begins with the genetic variation that is available and continues until the limits of the genetics are reached. Activities may include plant collection and evaluation, followed by improvement through genetics and breeding, and eventually product evaluation, propagation, technology transfer, and commercial production. The process to develop new plants takes significant time and resources before a new product can be successfully grown commercially.

This chapter will focus on the introduction and development of new plants, especially ornamental geophytes, including germplasm collection and maintenance, plant breeding, and technology transfer.

II. GERMPLASM COLLECTION AND MAINTENANCE

Plant collection and introduction is a valuable process because plant genetic resources are the basis for sustained plant improvement. Acquisition of superior plants by importing them from other places assists formal breeding programs and agricultural development. This procedure requires securing genetically desirable plants, properly growing and maintaining these plants, and using them effectively for genetic improvement. The genetic resources that are collected, commonly called the plant germplasm, are used to develop new plants; it is one of the most important natural resources in the world.

A. PLANT COLLECTION

As the population of the world increases and spreads to areas that have native vegetation, there is a concern that genetic diversity will be lost. The collection of plants from countries around the world has been an important method of germplasm enhancement. Collectors gather both wild and cultivated plants that have diverse genotypes and the potential to act as breeding material. Typically plant collections are made in areas that have special environments where, due to natural selection and evolution, unique genetics, or genotypes, and high levels of genetic diversity exist. The acquiring of new germplasm is not the only reason to collect native plants; species may be in danger of extinction or genetic erosion, botanical collections may need to add to their genetic diversity, plants may be required for biological studies or use in plant improvement, or more information about taxonomy, ecological adaptation, or geographical distribution is needed about the plant species (Guarino et al. 1995).

Beautiful hybrid tulips began as species that were collected in Turkey, Iran, Afghanistan, and Central Asia (Hoog 1973). More than 125 different species have been collected, but only a few are grown on a commercial scale (Le Nard and De Hertogh 1993). Most tulips that are grown today are modern hybrid cultivars that were bred from native species. Similarly, the successful breeding of winter-hardy garden *Alstroemeria* began from collecting native species from Chile and Brazil (Bridgen 1991). Cold-hardy cultivars of *Alstroemeria* were obtained by hybridizing with the Chilean species, *Alstroemeria aurea*, and the fragrant cultivar, 'Sweet Laura', was obtained by hybridizing with the Brazilian species, *Alstroemeria caryophyllacea*.

The objectives of a plant collection expedition are to collect the maximum genetic diversity between and within populations and to collect the most useful material. There are limits to the number of samples that can be collected and handled effectively so that the maximum amount of genetic diversity can be obtained. Therefore, decisions need to be made when developing a collecting strategy; plants to be targeted should be prioritized based on their potential.

Protecting biodiversity is in our self-interest. Collecting genetic resources should be based on populations, sampling strategies need to be determined, and practical feasibility and resources will determine decisions regarding the number of specimens to collect and the geographical area to be examined (Bridgen 2001). Due diligence is needed as preparations for the plant collection expedition evolve; some countries require permits before collection is allowed, obtaining the permission of land owners is often necessary, and background work with herbaria collections may be needed.

World laws and regulations within each county are also important to follow when collecting plants. In 1992, at the Earth Summit in Rio de Janeiro, world leaders agreed on a comprehensive strategy for sustainable development to meet the needs of plant breeders while leaving a viable world. One of the key agreements adopted at Rio was the Convention on Biological Diversity (CBD, http://www.cbd.int/convention/text/); this pact made laws and commitments by world governments for maintaining plants for future generations. The main goals of the Convention were the conservation of biological diversity, the sustainable use of its components, and the fair and equitable sharing of the benefits from the use of genetic resources. These regulations prevent the abuse by plant collectors as historically observed. For example, plant collectors who were looking for new tulips have visited locations in Turkey and Central Asia, where a limited number of plants are located, and have collected all the geophytes present. In some cases, entire species have been removed with the threat of extinction.

One of the recommendations of the international convention is to institute some form of benefit sharing between countries that collaborate with breeding programs. Currently, benefit sharing for flower bulbs is difficult because the genetic material of ornamental geophytes has been in the public domain for centuries. Therefore, the original custodians of the genetic material cannot claim ownership according to the international treaties. Benefit sharing in the form of intellectual property rights is only possible if genetic material is improved to the cultivar level through breeding and selection. For example, numerous potential geophyte crops from southern Africa are under development (see Chapter 18) and their commercialization should result in benefit sharing (Coetzee 2002). Therefore, international cooperation and joint research programs are necessary for breeding programs to be effective.

The requirements for benefit sharing between countries are still undocumented and confusing. Often, depending on the country, it is a casual agreement between cooperating individuals. For example, *Alstroemeria* breeding in the United States has been possible because of the native species that were collected in Chile. In return, the benefit sharing that the United States gives to Chile is training and educational programs, release from royalty payments on propagules of patented plants, collaborative research, and exchange programs for Chilean students.

B. Maintenance

Plants from collecting expeditions are very valuable; once they are acquired, they must be maintained and propagated in such a manner as to allow them to survive and reproduce. Plants can be propagated and multiplied sexually or asexually. Sexual reproduction requires the production of viable seeds from either self-pollination or cross pollination between species. Naturally self-pollinating species and their progenies are true breeding and do not show inbreeding depression or reduced vigor over time.

Some species require cross pollination and are either monoecious where the pollen and ovaries are produced by different flowers on the same plant, or dioecious where there are separate male and female flowers on different plants. Cross pollination may also result from self-incompatibility or morphological and developmental barriers. Whatever the reason, cross-pollinated species are often heterozygous with different gene characteristics on their chromosomes; this will lead to these plants not breeding true. If a cross-pollinating species is also self-pollinated, inbreeding depression may result in abnormalities forming or a lack of seeds produced. As a result, this kind of plant must be propagated asexually to maintain the genetic identity. Vegetative or asexual propagation of geophytes can be achieved by many techniques such as cuttings, chipping, scoring, divisions, and *in vitro* micropropagation.

Maintaining collections and making them available for breeding is an expensive and complicated task. At present, plant germplasm is stored in many different regions of the world, usually in national storage facilities, in vegetative forms, seeds, or cryopreservation of tissue cultures.

III. PLANT BREEDING

Plant breeding is practiced worldwide with the ultimate goal of changing the genetics of plants. Modern plant breeders aim to improve native plants in order to use them for commercial purposes. Breeding goals include disease resistance, increased length of flowering, stress tolerance, variety diversity, better postharvest life, and so on (Cadic and Widehem 2001; Bridgen et al. 2002; Snijder and Van Tuyl 2002; see also Chapters 6 and 7). For the ornamental geophytes, forcing ability and short production cycle are important breeding goals. An increasing consumers' concern for an improved quality of products and environmental protection are among the major trends in ornamental horticulture.

The strategy of plant breeding is to identify the valuable or new characteristics that are desired, search for the genetics that will supply these characteristics, combine the desired traits into an improved cultivar, assess the performance of the new breeding lines, and eventually introduce the superior, new cultivars. All these require adequate time to grow and evaluate plants, sufficient numbers of plants to cross, and selection pressures. Proper record keeping and plant labeling are the other important components of a successful breeding program. These activities maintain organization, but later when new plants are developed, they also help to identify and register the pedigree.

Plant breeders must understand the objectives of the breeding program and what is required to achieve them. Without clear objectives, it will be difficult to accomplish those goals or the desired end result. When breeding geophytes, there are some unique objectives that are desirable to obtain, depending on if the crops are to be grown as a potted plant, as a cut flower, or as a garden plant (De Hertogh 1990). In order to develop superior cultivars, breeders need to know and understand fundamental information about plants, including taxonomy, morphology, physiology, pathology, genetics, propagation, and entomology.

The term "traditional plant breeding" often indicates that genetic modifications or genetically modified organisms (GMO) are not produced as a result of the procedure. It does not mean, however, that modern or sophisticated procedures such as plant tissue culture and mutagenesis are not used. The term "recombinant DNA" (rDNA) may be more appropriate to make a distinction between traditional and biotechnological breeding because hybrid plants are indeed genetically modified. For the purpose of this chapter, "genetic engineering" will be used to identify biotechnological breeding via rDNA or GMO.

A. TRADITIONAL BREEDING

Traditional, sometimes called classical, plant breeding is time consuming and tedious but very rewarding. There are multiple plant breeding methods that can be used in a traditional program, but due to space limitations, this article will highlight only the main techniques.

For self-pollinated species, the most successful techniques include mass selection, pure line selection, and hybridization. Each of these methods has a foundation based on self-pollination or backcrossing to a homozygous parent which leads to homozygosity. Self-compatible species can be inbred with little or no genetic deterioration.

The diverse breeding behaviors of outcrossing or cross-pollinating species prevent the breeding methods from being as neatly categorized as with the self-pollinated species. Backcross breeding, hybridization of inbred lines, recurrent selection, and mass selection are acceptable methods to employ. Outcrossing groups of plants are heterozygous and this heterozygosity must be maintained during a breeding program or restored at the end of the program.

Plant breeders need to control the pollination procedure in order to prevent cross pollination and unwanted hybrids. This is usually accomplished by isolating groups of plants or by using a physical barrier, such as a bag, to prevent the spread of pollen. The control of pollination is also needed to make specific, desirable crosses. The goal is to place functional pollen from the desired male strain on receptive stigmas of the desired seed parent at the correct time. Protection is often provided against selfing by removing the anthers from the female parent before they mature.

Many of the ornamental geophytes have difficulties with interspecific hybridization such as recalcitrant embryo formation, pollination and fertilization barriers, and delayed embryogenesis. However, *in vitro* techniques such as embryo culture, somatic embryogenesis, and somaclonal variation have allowed successful hybrids to be developed (Bridgen et al. 1990; Chapters 6 and 7).

B. Biotechnological Breeding

Genetic engineering is a new type of genetic modification that can target specific genes, the part of the plant that will pass on genetic traits, and add these to the genome of a different plant without concern about whether these plants have the ability to cross hybridize. Genetic engineering can physically remove the DNA from one plant and transfer it into another. The requirement for two species to be able to cross fertilize is overcome and the problem of multiple traits being transferred is also avoided. In addition, perhaps the most exciting part of genetic engineering is that there is a possibility to transfer almost any trait from any living organism into a plant.

Genetic engineering requires certain steps to be accomplished before the transfer of genes can be achieved. The desired gene needs to be identified and the DNA that carried this trait needs to be extracted. As DNA is extracted, it needs to be cloned or copied, and replicated. Once the gene has been replicated, the gene is dissected by enzymes and reassembled in a desired form so that the new cell can replicate and function inside the host plant. The modified gene is now ready for transformation, also known as gene insertion. Plant tissue culture is often used for this process because it can be used to generate large numbers of undifferentiated plant cells, called callus, to which the new gene will be added. Various techniques can be used to insert these genes into the cells of the callus; these include *Agrobacterium* transformation, the gene gun, virus insertion, protoplast fusion, and electroporation. Once a plant has been transformed with new DNA, these transgenic plants can be grown and evaluated to determine that the gene has been inserted and is functioning. The amount of time for the propagation and evaluation process of genetically engineered plants is almost the same as of traditionally bred plants, taking a minimum of 5 years before the new transgenic hybrid is ready for release. Plants with rDNA are strictly regulated by governments to ensure safety for humans and the environment; this process will also affect the time for the release of new plants.

As exciting as genetic engineering is for the potential of plant breeding and development, it has its limitations due to the training and expense involved and also has legal limitations. At present, molecular techniques have not reached the stage where they can be used routinely by the average plant breeder, but they have successfully produced valuable, new plants.

IV. TECHNOLOGY TRANSFER

The release of new cultivars takes time and involves the propagation and distribution of the new plants to commercial enterprises. The procedure for release depends on the institution at which it was bred, the technique of propagation for the plant (sexually vs. asexually), protection of the variety, and laws that may be involved.

A. Assessment

Before a new plant is released, performance data should be obtained over several years to convince commercial investors that the product has value. The breeder should first evaluate the plant's potential in a manner that is adequate and reasonable for commercial release. It is desirable to conduct trials at several locations. This requires that the breeder coordinates the evaluations, determine how best to propagate and distribute the plants rapidly, and protect the new plant by using testing agreements. Ironically, the ultimate decision to release a new cultivar is usually not made by the breeder; it is made by the company or university that sponsored the research or by the business that is willing to invest in the plant.

The distribution of a new cultivar is an important consideration. The propagule, whether seed or vegetative, needs to be carefully and accurately reproduced. Public institutions are usually not involved in the direct sale of a new plant to the end user. They distribute stock plants to private companies that then propagate and market the plants. If the breeder works for a private breeding company, it will maintain control over the multiplication. There are also companies that specialize in the propagation and distribution of new plants and take responsibility of multiplication for commercial growers.

B. PLANT PROTECTION

Back in the seventeenth century, tulip bulbs were in high demand in The Netherlands and traded on the stock exchange at very high prices. This phenomenon, known as "Tulipmania", may have been the beginning of an interest in protecting breeders' rights. In 1961, the International Convention for the Protection of New Varieties of Plants (UPOV) was held in Geneva, Switzerland and developed an international system for the protection of new varieties of plants by an intellectual property right. This effective and internationally recognized system is an incentive to stimulate new breeding work and to encourage plant breeding. The system can also provide important benefits in an international context by removing barriers to trade in varieties, thereby increasing domestic and international market scope. Before plant protection was available, new cultivars were not protected and the breeders only received a one-time base payment or payments from licenses to propagators.

For plant breeders' rights to be granted, the new plant must meet four criteria under the rules established by UPOV. The new plant must be novel, distinct from other available varieties, it must display homogeneity, and the trait or traits that are unique to the new cultivar must be stable, so that the plant remains true to type after repeated cycles of propagation. If new plants are propagated asexually, they can be protected by plant patents. If the new plants are propagated sexually, by seeds, they can be protected by the Plant Variety Protection Act. Seed-propagated plants must be able to be reproduced for multiple generations without changes to their characteristics.

Once the new plants are protected, they can be distributed in an exclusive manner, where a single company manages the distribution, or in a nonexclusive manner where seeds or vegetative propagules are sold to several companies and each of them propagates and merchandises the plants. Each propagule that is produced for a protected variety has a royalty charge that is given back to the breeder or the breeder's company. This process encourages further breeding and assists with the expenses of the process.

C. PROMOTION OF NEW PLANTS

The names of new plants can be trademarked in addition to being given a cultivar name. The owner of the trademark cannot restrict the use, propagation, or distribution of the plant, but controls the use of the trademark name. This is used as a marketing strategy by some companies to avoid unattractive names that do not appeal to commercial attention. The trademark name can be more attractive, "catchy", or marketable and hopefully will encourage greater sales. Only the person or business that owns the trademark name can sell it under that name. For example, the name 'Globemaster' has been a well-known cultivar of *Allium giganteum*, while the other species and cultivars remain lesser known because of the lack of promotion (De Hertogh and Zimmer 1993).

Breeders will use other techniques to market new varieties. For example, Cornell University has been using alliterations when naming their new *Alstroemeria* cultivars: 'Mauve Majesty' and 'Tangerine Tango' were the two most recent introductions (Bridgen 2007). Another example of using marketing to make a rather unknown specialty plant a worldwide phenomenon is with the genus *Hippeastrum*. This plant became commercially known as Amaryllis, which is a more consumer-friendly name (Okubo 1993).

V. CONCLUSIONS

Working to introduce and develop new crops is exciting and rewarding. There are multiple approaches to reach the final objective of a novel and productive new plant. However, it must be remembered that science and art are both involved in this complicated field, and it takes more than plant breeding to eventually commercialize and market a new cultivar.

The process of new plant development occurs through the collection, selection, hybridization, propagation, and commercialization of new cultivars. Hybridization of genetically related plants became the building block for breeders because genes could be combined through cross pollination and introduced into new plants. In recent years, state-of-the-art *in vitro* and molecular techniques and genetic engineering have been able to hasten the process. All these procedures have led to higher yielding or more attractive plants that can be resistant to diseases, pests, or environmental conditions.

CHAPTER CONTRIBUTOR

Mark P. Bridgen is a professor of horticulture at Cornell University, Ithaca, New York. He was born in the United States and received BS from Pennsylvania State University, University Park, Pennsylvania, MS from Ohio State University, Columbus, Ohio, and PhD from V.P.I. University, Blacksburg, Virginia. He held faculty positions at the University of Connecticut and Cornell University, where he directed programs in ornamental plant breeding, primarily of Chilean geophytes. His major research areas are plant breeding, micropropagation, and plant tissue culture. Dr. Bridgen is the author of more than 350 refereed journal articles, book chapters, published abstracts, patents, and conference papers. He holds seven plant patents with the geophyte *Alstroemeria*. He is the recipient of the Herbert Medal from the International Bulb Society (2008), the New York Farmer's Medal from the New York Farmers Society (2008), and the Academic Excellence Award from the Perennial Plant Association (2006), is a fellow of the International Plant Propagators' Society, Eastern Region (2003), and a Fulbright fellow (1999), has been awarded the Excellence in Teaching Award from the College of Agriculture Alumni Association (1998), the Program of the Year Award from the University of Connecticut (1996), and the Booker/Wrobleski Faculty Advisor Award (1995).

REFERENCES

Bridgen, M. P. 1991. *Alstroemeria*—A promising pot plant. In *Ball Red Book, 15th edn.*, ed. V. Ball, 316–318. W. Chicago, IL: Geo. J. Ball Publ.

Bridgen, M. 2001. A nondestructive harvesting technique for the collection of native geophyte plant species. *Herbertia* 56:51–60.

Bridgen, M. P. 2007. *Alstroemeria* 'Mauve Majesty'. Patent No. PP18,183.

Bridgen, M. P., R. Craig, and R. Langhans. 1990. Biotechnological breeding techniques for *Alstroemeria*. *Proc. Int. Symp. Bulbous Cormous Plants. Herbertia* 45(1,2):93–96.

Bridgen, M., E. P. Kollman, and C. Lu. 2009. Interspecific hybridization of *Alstroemeria* for the development of new, ornamental plants. *Acta Hort.* 836:73–78.

Bridgen, M. P., E. Olate, and F. Schiappacasse. 2002. Flowering geophytes from Chile. *Acta Hort.* 570:75–80.

Cadic, A., and C. Widehem. 2001. Breeding goals for new ornamentals. *Acta Hort.* 552:75–86.

Coetzee, J. H. 2002. Benefit sharing from flower bulbs—Is it still possible? *Acta Hort.* 570:21–27.

De Hertogh, A. A. 1990. Basic criteria for selecting flower bulbs for North American markets. *N. C. Hort. Res. Series No. 85.*

De Hertogh, A. A., and K. Zimmer. 1993. *Allium*—Ornamental species. In *The Physiology of Flower Bulbs*, eds. A. De Hertogh, and M. Le Nard, 187–200. Amsterdam: Elsevier.

Guarino, L., V. Ramanatha Rao, and R. Reid. 1995. *Collecting Plant Genetic Diversity. Technical Guidelines*. Wallingford, UK: CAB International on behalf of IPGRI in association with FAO/IUCN/UNEP.

Hoog, M. H. 1973. On the origin of *Tulipa*. In *Lilies and other Liliaceae*, eds. E. Napier, and J. N. O. Platt, 47–64. London: Royal Horticultural Society.

Le Nard, M., and A. A. De Hertogh. 1993. *Tulipa*. In *The Physiology of Flower Bulbs*, eds. A. De Hertogh, and M. Le Nard, 617–682. Amsterdam: Elsevier.

Okubo, H. 1993. *Hippeastrum (Amaryllis)*. In *The Physiology of Flower Bulbs*, eds. A. De Hertogh, and M. Le Nard, 321–334. Amsterdam: Elsevier.

Snijder, R. C., and J. M. van Tuyl. 2002. Breeding for resistance in *Zantedeschia* spp. (Araceae) against soft rot caused by *Erwinia carotovora* ssp. *carotovora*. *Acta Hort.* 570:263–266.

6 Breeding and Genetics of Ornamental Geophytes

Jaap M. van Tuyl, Paul Arens, and
Agnieszka Marasek-Ciołakowska

CONTENTS

I. INTRODUCTION

Interspecific and intergeneric hybridization, often in combination with polyploidization, is the most important practice for the introduction of genetic variation in ornamental crops. Many of the cultivars of ornamental geophytes originated from complex species crosses that have given rise to a broad range of shapes and colors of plants and flowers (Benschop et al. 2010; *Alstroemeria*: Bridgen et al. 1989; *Gladiolus*: Ohri and Khoshoo 1983a,b; *Hippeastrum*: Meerow 2009; *Lilium*: Lim and

Van Tuyl 2006; *Narcissus*: Wylie 1952; Brandham 1986; *Tulipa*: Van Eijk et al. 1991; *Zantedeschia*: Snijder 2004).

This chapter focuses on recent advances in breeding and genetics of some of the most important ornamental geophytes (*Lilium*, *Tulipa*, *Narcissus*, *Hyacinthus*, *Alstroemeria*, and *Zantedeschia*). Many examples were chosen from *Lilium*, because lily is frequently used as a model plant for style manipulations and *in vitro* culture methods for breeding. The breeding histories of lily, tulip, narcissus, hyacinth, and the background of the main cultivars are briefly described. The major breeding goals, the role of interspecific hybridization and polyploidization and the use of mutation breeding in assortment development are also reviewed. Finally, recent developments in molecular breeding for the model crop, lily, are assessed.

II. INTERSPECIFIC HYBRIDIZATION

For almost all ornamental geophytes, interspecific hybridization is the most important source of new variations in the assortment (Van Tuyl et al. 2002b). In this respect, hyacinth is an exception, since only one species, *Hyacinthus orientalis*, has been used in breeding (Darlington et al. 1951). Most modern cultivars have originated from complex species crosses (Wylie 1952; Ohri and Khoshoo 1983b; Van Eijk et al. 1991; Van Tuyl 1997). Successful interspecific hybridization depends to a great extent on good knowledge of the relationships within the genera and a well-documented classification system. In general, closer relationships correspond to higher crossability. Therefore, for wider intersectional or intergeneric crosses, more advanced techniques are needed to overcome crossing barriers. Stebbins (1958) divided the crossing barriers into pre- and postfertilization barriers.

A. Prefertilization Barriers

The prefertilization barriers, caused in nature by temporal and spatial isolation of the parents, can be effectively overcome through pollen storage for the purpose of the intentional breeding. Stigmatal and stylar barriers were found in many studies and can be overcome by the application of pollination techniques as exampled in lily (Asano 1981; Van Tuyl et al. 1988; Van Tuyl and De Jeu 1997). Cut-style pollination is successfully applied in lily for intersectional hybridization (Asano and Myodo 1977a; Van Tuyl et al. 1991; Okazaki and Murakami 1992) and in *Fritillaria* (Wietsma et al. 1994) to make crosses between *Fritillaria imperialis* and *F. raddeana*. However, in tulip this technique did not allow the overcoming of prefertilization barriers (Van Creij et al. 1997a). The grafted style technique was successfully applied in lily as an improvement over the cut-style pollination technique (Van Tuyl et al. 1991).

B. Postfertilization Barriers

A breakthrough in overcoming the main postfertilization barrier is the development of embryo rescue techniques. These techniques are applied in many species and are adopted as routine methods in the breeding of *Lilium* (Myodo 1963; Asano and Myodo 1977b; Asano 1978, 1980a,b,c; Van Tuyl et al. 1991; Okazaki et al. 1992, 1994), *Tulipa* (Van Creij et al. 1997a,b, 1999, 2000; Custers et al. 1995), *Hippeastrum* (Bell 1972), *Nerine* (Van Tuyl et al. 1990), and *Alstroemeria* (Bridgen et al. 1989; Buitendijk et al. 1995; De Jeu and Jacobsen 1995). A range of techniques was developed to overcome problems related to embryo and/or endosperm degeneration in various types of crosses, for example, ovary, ovary-slice, ovule, embryo sac, and embryo culture. In lily all these techniques are used depending on the stage at which *in vitro* culture is applied. Thus, ovary culture can be applied before pollination has occurred, ovary-slice is employed within one week after pollination, ovule culture is used between 20 and 40 days after pollination, and embryo and embryo sac culture is used 40–70 days after pollination (Kanoh et al. 1988; Van Tuyl et al. 1990, 1991). In interspecific

Hippeastrum crosses, embryo culture was applied successfully (Bell 1972; Meerow 2009). Intergeneric crosses between *Nerine bowdenii* and *Amaryllis belladonna* were successful through ovary culture (Coertze and Louw 1990; Van Tuyl et al. 1990).

III. CYTOGENETICS

A. INTRODUCTION

Extensive cytological studies, focusing on chromosome number and chromosome morphology, have been carried out on *Lilium* (Stewart 1947; Noda 1991; Marasek and Orlikowska 2003; Marasek et al. 2005), *Tulipa* (Woods and Bamford 1937; Sayama et al. 1982; Asano 1982b; Kroon and Jongerius 1986; Van Raamsdonk and De Vries 1995), *Alstroemeria* (Tsuchiya et al. 1987; Hang and Tsuchiya 1988), *Narcissus* (Brandham and Kirton 1987), *Gladiolus* (Ohri and Khoshoo 1983a,b), *Crocus* (Brighton et al. 1973; Brighton 1976), and *Hyacinthus* (Darlington 1926, 1929, 1932; Darlington et al. 1951). Chromosome analysis has been improved by using techniques enabling longitudinal differentiation of the chromosomes, such as Giemsa staining (C-banding) revealing heterochromatic regions on the chromosomes. C-banding was used to study the chromosome structure of different species, for example, *Lilium* (Holm 1976; Smyth et al. 1989; Smyth 1991), *Tulipa* (Filion 1974; Blakey and Vosa 1982; Van Raamsdonk and De Vries 1995), and *Alstroemeria* (Buitendijk and Ramanna 1996). Fluorescence *in situ* hybridization (FISH) with the 5S and 45S rDNA sequences as probes has improved chromosome identification in the genus *Lilium* (Lim et al. 2001c; Siljak-Yakovlev et al. 2003; Marasek et al. 2004), *Tulipa* (Mizuochi et al. 2007; Marasek and Okazaki 2008; Figure 6.1), and *Alstroemeria* (Kamstra et al. 1999). Physical mapping of rRNA gene loci was used to elucidate interspecific relationships among wild *Lilium* species (Sultana et al. 2010). Molecular cytogenetic analysis, using genomic *in situ* hybridization (GISH), has been applied to assess the parentage and origin of hybrids in *Tulipa* (Marasek et al. 2006; Marasek and Okazaki 2007), *Lilium* (Barba-Gonzalez et al. 2005b,c; Zhou et al. 2008a,b; Xie et al. 2010a,b), *Narcissus* (Lifante et al. 2009), and *Crocus* (Ørgaard et al. 1995) as well as for analyzing the process of intergenomic recombination (Figure 6.2).

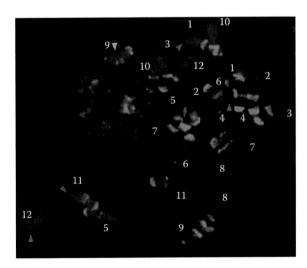

FIGURE 6.1 (See color insert.) Examples of the use of the FISH method for the identification of specific chromosomes. Double-target FISH of 5S rDNA (green fluorescence) and 45S rDNA (red fluorescence) probes on the chromosomes of *Tulipa gesneriana* cultivar 'Prominence'. Distributions of the 45S and 5S rDNA sites enabled individual chromosomes identification in a metaphase spread. (From Marasek-Ciolakowska, A., and M. Podwyszynska. 2008. *Floriculture Ornamental Biotechnol.* 2:65–72. Reproduced with the kind permission of Global Science Books.)

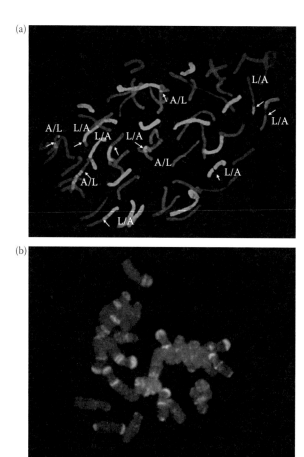

FIGURE 6.2 (**See color insert.**) Examples of the use of the GISH method for genome painting in interspecific hybrids. (a) Genome differentiation in pentaploid LA lily hybrid (2n = 5x = 60) showing 18 L and 42 A chromosomes. Recombination chromosomes are marked with arrows indicating two types of recombinant chromosomes, that is, L/A centromere of Longiflorum genome with Asiatic recombinant segments or vise versa (A/L). (Photo courtesy Nadeem Khan.) (b) Discrimination of chromosomes originating from *Tulipa gesneriana* (green fluorescence) and *T. fosteriana* (red fluorescence) in the genomes of 'Shirayukihime' (2n = 2x = 24). (Photo courtesy Marasek-Ciolakowska.)

B. POLYPLOIDIZATION

In tulip the commercial assortment mainly consists of diploid genotypes (2n = 2x = 24), as well as some triploid (2n = 3x = 36) and tetraploid (2n = 4x = 48) Darwin hybrid cultivars (Van Tuyl and Van Creij 2006). In contrast, a large majority of lily cultivars (>80%) is polyploid, including triploid (2n = 3x = 36), tetraploid (2n = 4x = 48) and a few aneuploid cultivars (Zhou 2007; Zhou et al. 2008c). Recently, allopolyploid lily cultivars, including hybrids between Longiflorum × Asiatic (LA), Longiflorum × Oriental (LO), Oriental × Asiatic (OA), Oriental × Trumpet (OT) hybrid groups, have become increasingly popular. Inter-ploidy crosses were used to obtain polyploid progenies both in *Tulipa* (Straathof and Eikelboom 1997; Okazaki and Nishimura 2000) and *Lilium* (Lim et al. 2003a; Zhou et al. 2008c). In tulip, crossing tetraploid *Tulipa gesneriana* seedling ('Denbola' × 'Lustige Witwe') with diploid *T. fosteriana* resulted in the strong growing triploid 'World's Favourite' (Straathof and Eikelboom 1997). The crossing of diploid Darwin hybrid tulips, producing both fertile n gametes and 2n gametes, with diploid *T. gesneriana* cultivars resulted in diploid, triploid, and tetraploid progenies (Marasek and Okazaki 2007). In lily, crosses of triploids with diploids and tetraploids produced aneuploid or near diploid and pentaploid progenies, respectively (Lim et al. 2003a; Barba-Gonzalez 2005), whereas triploid progenies were derived from 2x–5x crosses (Zhou et al. 2008c; Figure 6.3).

The production of tetraploids using laughing gas (nitrous oxide, N_2O) treatment was described in lily (Barba-Gonzalez et al. 2006a,b; Akutsu et al. 2007) and tulip (Zeilinga and Schouten 1968; Okazaki et al. 2005).

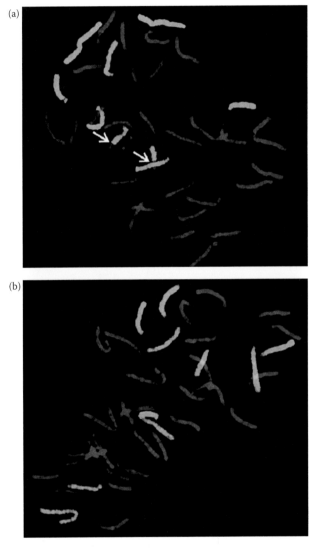

FIGURE 6.3   **(See color insert.)** GISH analysis in the triploid progenies of *Lilium* (2n = 3x = 36) derived from 2x–5x crosses of AA (12A) × ALALA (24L + 36A). (a) Somatic metaphase chromosome complements showing 27 chromosomes of A genome (red fluorescence) and 9 chromosomes of L genome (green fluorescence) including two L/A recombinant chromosomes. Recombinant break points are indicated by arrows. (b) Somatic metaphase chromosome complements showing 27 chromosomes of A (red fluorescence) and 9 chromosomes of L genome (green fluorescence) with no recombinant chromosomes. (From Zhou, S. et al. 2008c. In *Floriculture, Ornamental and Plant Biotechnology, Vol. V,* ed. J. A. Teixeira da Silva, 152–156. Isleworth, UK: Global Science Books. Reproduced with the kind permission of Global Science Books.)

C. Mitotic and Meiotic Polyploidization

In almost all cases, F_1 interspecific hybrids are highly sterile. The fertility of these F_1 hybrids can be restored by doubling the chromosomes, also called "mitotic doubling", using oryzalin or colchicine to produce amphidiploids (Emsweller and Brierley 1940; Asano 1982a; Van Tuyl 1989; Van Tuyl et al. 1989). In lily, fertility was restored in F_1 hybrids from crosses like *L. henryi* × *L. candidum*, *L. longiflorum* × Asiatic hybrids, *L. longiflorum* × *L. candidum*. Using these tetraploids, backcrossing was performed on Asiatic and Oriental hybrids (Van Tuyl 1997). Mitotic doubling to restore F_1-fertility has also been used in *Alstroemeria* for *A. aurea* × *A. caryophyllacea* progeny (Lu and Bridgen 1997) and in *Iris* for crosses *I. laevigata* × *I. ensata* (Yabuya 1985) and *I. hollandica* × *I. tingitana* (Van Eijk and Eikelboom 1990). Cohen and Yao (1996) applied *in vitro* chromosome doubling in nine *Zantedeschia* cultivars. Van Tuyl and Van Creij (2006) reported the production of tetraploid tulip cultivars using *in vitro* oryzalin and colchicine treatment of tulip stem explants excised from bulbs prior to elongation. The disadvantage of mitotic polyploidization is that it is difficult to accomplish intergenomic recombination in the allopolyploids due to autosyndetic chromosome pairing during meiosis (Karlov et al. 1999; Lim et al. 2001a; Ramanna and Jacobsen 2003; Figure 6.4). For instance, Xie et al. (2010a) analyzed a population comprising 26 lily neo-polyploids, obtained through crossing an allotriploid Longiflorum × Oriental hybrid (LLO) with allotetraploid Longiflorum × Trumpet hybrid (LLTT), both of which were derived from somatic chromosome doubling, in which there was no evidence of chromosomal rearrangements (Figure 6.4). Similarly, no recombination was observed in allotriploid Longiflorum × Rubellum (LLR) BC_1 progenies. The meiotic behavior of these hybrids showed preferential pairing as bivalent between L–L and univalents R (Lim et al. 2001b; Figure 6.5).

An alternative to the mitotic polypoidization is the use of unreduced gametes with somatic chromosome numbers (2n-gametes) referred to as meiotic polyploidization. The application of meiotic polyploidization in interspecific breeding programs proved to be very important for introgression of desired traits. The occurrence of 2n-gametes was reported in interspecific hybrids of *Alstroemeria* (Kamstra et al. 1999), interspecific hybrids of *Lilium* (Asano 1984; Van Tuyl et al. 1989; Lim et al. 2001a, 2004; Barba-Gonzalez et al. 2004), species of *Tulipa* (Kroon and Jongerius 1986), *Narcissus* (Brandham 1986), and *Gladiolus* (Ohri and Khoshoo 1983a,b). Spontaneously occurring triploids, resulting from unreduced gametes, have been recorded in *Lilium* (Noda 1986)

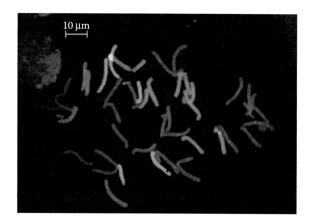

FIGURE 6.4   (See color insert.) Mitotic chromosome painting of LLO × LLTT (076928-21), an aneuploid (2n = 4x–5 = 43) derived from the crossing of allotriploid LLO and allotetraploid LLTT (both were derivatives of somatic chromosomes doubling), showing no chromosomal interchanges. GISH clearly identified the chromosomes of the three genomes: T = red (biotin labeled and detected with Cy-3), O = green (digoxigenin labeled and detected with the anti-digoxigenin FITC system), and L = blue (DAPI counterstaining). (Photo courtesy Songlin Xie.)

(a) 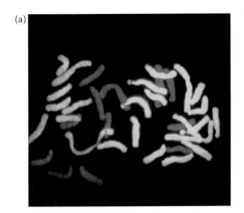 (b)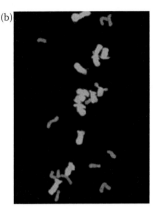

FIGURE 6.5 (**See color insert.**) GISH analysis in allotriploid Longiflorum × Rubellum (LLR) BC$_1$ progenies. (a) Mitotic chromosome constitution of LLR-lily plant, 24 L chromosomes (yellow fluorescence) and 12 R chromosomes (red fluorescence). Note the total absence of the recombinant segment in this plant. (b) The meiotic metaphase I stage in LLR hybrid showing 12 bivalents of L-L (yellow fluorescence) and 12 univalents of R (red fluorescence). The absence of trivalent formation in such plants confirms that there is no homoeologous recombination. (From Lim, K.-B. et al. 2001b. *Acta Hort.* 552:65–72. Reproduced with the kind permission of ISHS.)

and *Crocus* (Ørgaard et al. 1995). In breeding programs it is possible to select genotypes which produce a high frequency of 2n gametes. In *Lilium*, intersectional hybrids producing 2n pollen were extensively used to produce progeny (Lim et al. 2000b, 2001a,c, 2003a,b, 2008; Barba-Gonzalez et al. 2004, 2005a,b,c, 2006b; Zhou 2007; Zhou et al. 2008a). According to Kroon and Van Eijk (1977), triploid tulips are likely to have arisen as a result of the occurrence of diploid gametes in diploid cultivars. The origin of the triploid Darwin hybrid tulip 'Yellow Dover' developed from interspecific crosses of *T. gesneriana* and *T. fosteriana* cultivars was analyzed by flow cytometry, karyotype analyses, and GISH. All methods confirmed that *T. gesneriana* provided diploid gametes (Marasek et al. 2006; Figure 6.6). Similarly, tetraploid varieties of tulip such as

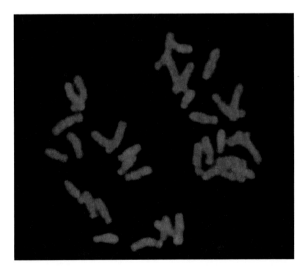

FIGURE 6.6 (**See color insert.**) Genome differentiation in triploid Darwin hybrid tulip 'Yellow Dover' (2n = 3x = 36) using *in situ* hybridization with total genomic DNA of *Tulipa gesneriana* 'Queen of Night' as a block DNA and *T. fosteriana* 'Red Emperor' as a probe (red fluorescence). The chromosomes were counterstained with DAPI (blue). (Photo courtesy Marasek-Ciolakowska.)

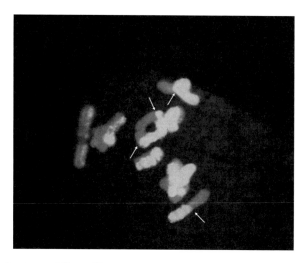

FIGURE 6.7 (**See color insert.**) Part of late metaphase in the pollen mother cells of the F_1 *Lilium longiflorum* × Asiatic hybrid (LA) showing univalents and half-bivalents. Recombination break-points are shown by arrows. Chromosomes are hybridized with total genomic DNA of *L. longiflorum* detected with anti-digoxigenin FITC (*yellow*) and counterstained with DAPI (Asiatic genome blue). (Photo courtesy Nadeem Khan.)

'Riant', 'Beauty of Canada', and 'Peerless Yellow' originated after crossing between diploid and tetraploid varieties where the former provided diploid egg cells, whereas in the case of the tetraploid hybrid 'Sunburst' both parents were diploids and the pollen and egg cells involved in this cross were 2n (Kroon and Van Eijk 1977). The most important advantage of meiotic polyploidization is that homoeologous recombination occurs during meiosis. The intergenomic recombination was shown, for example, in *Alstroemeria* (Kamstra et al. 1999) and *Lilium* (Karlov et al. 1999; Lim et al. 2001b; Khan 2009; Khan et al. 2010; Xie et al. 2010a; Figure 6.7). In *Lilium* and *Alstroemeria* extensive studies using GISH and FISH have been conducted to discover the mechanism leading to 2n gametes formation (Lim et al. 2001a; Ramanna et al. 2003; Barba-Gonzalez et al. 2004, 2006c; Zhou et al. 2008b). Generally, based on the particular meiotic stage at which nuclear restitution occurs, two main restitution mechanisms of 2n gametes formation known as First Division Restitution (FDR) and Second Division Restitution (SDR) have been recognized in monocots. In the case of FDR the homoeologous chromosomes failed to pair and the whole chromosome complement divided equationally before telophase I, followed by cytokinesis, leading to the formation of a dyad without further division. In the case of SDR a normal first meiotic division occurred, producing a dyad. Instead of the second division, the chromatids divided, but the nuclei restituted in each of the two cells of a dyad (Lim et al. 2004; Barba-Gonzalez et al. 2005b). A third mechanism called Indeterminate Meiotic Restitution (IMR) has been identified in LA lily hybrids (Lim et al. 2001a). In this case, during the first meiotic division some bivalents disjoined reductionally and, at the same time, some of the univalents divided, equationally leading to nuclear restitution (Figure 6.8).

Besides the afore-mentioned methods of polyploidization, induction of 2n gametes in a diploid by the application of laughing gas has also been used to overcome F_1 hybrid sterility in *Lilium* and *Tulipa* (Okazaki 2005; Okazaki et al. 2005; Barba-Gonzalez et al. 2006a,b).

D. Introgression

Introgression is one of the main goals of interspecific hybridization, to introduce desirable traits into the breeding pool, such as, for instance, disease resistance, a short forcing period or new flower

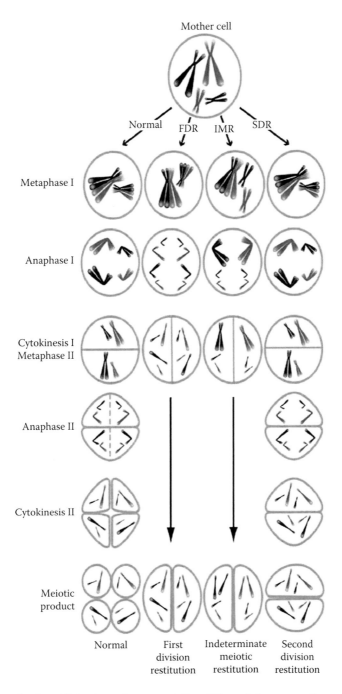

FIGURE 6.8 (**See color insert.**) A schematic representation of the meiotic process, the normal sequence, and three restitution mechanisms in microsporogenesis in species having the monocotyledonous successive type of meiotic division. FDR, first division restitution; IMR, indeterminate meiotic restitution; SDR, second division restitution. (From Barba-Gonzalez, R. et al. 2008. In *Floriculture, Ornamental and Plant Biotechnology*, *Vol. V*, ed. J. A. Teixeira da Silva, 138–145. Isleworth, UK: Global Science Books. Reproduced with the kind permission of Global Science Books.)

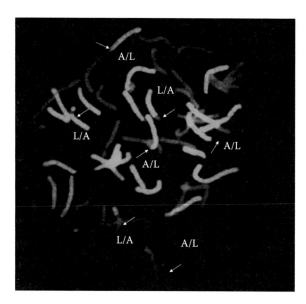

FIGURE 6.9 (**See color insert.**) Chromosomes complement of tetraploid (2n = 4x = 48) LA lily hybrid (064525-29) obtained from bilateral sexual polyploidization showing 25 L and 23 A chromosomes. Recombination chromosomes are marked with arrows indicating two types of recombinant chromosomes, that is, L/A centromere of Longiflorum genome with Asiatic recombinant segments or vice versa (A/L). (Photo courtesy Nadeem Khan.)

shapes and colors. It has played an important role in the improvement of the main ornamental geophytes, such as lily (Lim 2000; Lim et al. 2000a,b; Van Tuyl et al. 2002a; Beers et al. 2005; Barba-Gonzalez 2005; Zhou 2007), tulip (Van Eijk et al. 1979), and *Alstroemeria* (Kamstra et al. 1999, 2004). For introgression of the desired characters into the breeding pool, homoeologous recombination is essential for the creation of new trait combinations. Therefore, introgression breeding requires a technique that enables the monitoring of the presence of alien chromosomes to confirm hybrids and the recombination between genomes in F_1 hybrids and backcross progenies. The technique enabling the detection of alien chromosomes and chromosomal recombination is GISH. GISH has been successfully applied to estimate the extent and position of homoeologous recombination in distant hybrids of *Alstroemeria* (Kamstra et al. 1997, 1999, 2004; Ramanna et al. 2003), *Tulipa* (Marasek and Okazaki 2008; Marasek-Ciolakowska et al. 2009), and *Lilium* (Karlov et al. 1999; Lim 2000; Lim et al. 2003b; Barba-Gonzalez et al. 2006b; Khan et al. 2009a, 2010; Xie et al. 2010a; Figure 6.9). A chromosomal recombination map has been constructed for three genomes of lily using GISH (Khan et al. 2009b) that shows large disparities in the number of recombinations between chromosomes. Marasek-Ciolakowska et al. (2009) have shown that, in tulip, fertile n gametes with recombinant chromosomes seem to be ideal for introgression breeding at the diploid level. In this study all BC_1 plants resulted from the crossing of diploid *T. gesneriana* cultivar 'Yellow Flight' and fertile F_1 hybrids *T. gesneriana* × *T. fosteriana* (GF) were diploids (2n = 2x = 24) with recombinant chromosomes. Similarly, the presence of recombinant chromosomes was shown in diploid BC_1 progenies (2n = 2x = 24) of Darwin hybrid tulip 'Purissima' (Marasek and Okazaki 2008; Marasek-Ciolakowska et al. 2011; Figure 6.10).

E. HAPLOIDIZATION

The production of *in vitro* haploid plants using microspore culture is a technique used in order to produce homozygous plants. Haploid culture of tulip (*T. gesneriana*) through microspore embryogenesis without an intermediate callus phase was first published by Van den Bulk et al. (1994). Since

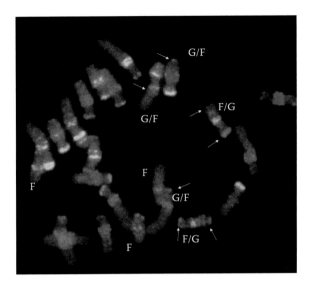

FIGURE 6.10 **(See color insert.)** Discrimination of chromosomes originating from *Tulipa gesneriana* (red fluorescence) and *T. fosteriana* (green fluorescence) in the genomes of BC_1 'Purissima' hybrids. Recombinant chromosomes are defined as F/G and G/F indicating a *T. fosteriana* centromere with *T. gesneriana* chromosome segment(s) and a *T. gesneriana* centromere with *T. fosteriana* chromosome segment(s), respectively. Recombinant break-points are indicated by arrows. (Photo courtesy Marasek-Ciolakowska.)

that time, embryogenesis in tulip microspores has been induced in several *Tulipa* species and cultivars (Van den Bulk and Van Tuyl 1997). Custers et al. (1997) reported that high-temperature pretreatments of bulbs containing fully developed inflorescence had a positive effect on embryo formation from microspores.

The formation of embryos from isolated microspores has not been described in the genus *Lilium* so far. There are few reports on anther culture in lily. Sharp et al. (1971) reported the formation of haploid callus from anther cultures of *L. longiflorum*; however, the haploidy was lost after four to six subcultures. Diploid and mixoploid plants with chromosome numbers ranging from 11 to 26 were observed by Qu et al. (1988) among plants derived from anther cultures of several *L. longiflorum* genotypes. Gu and Cheng (1982) reported the regeneration of haploid, diploid, and aneuploid plants from anther cultures of *L. davidii*, whereas Han et al. (1997) observed the regeneration of haploid, diploid, and mixoploid plants from the anther cultures of the Asiatic hybrid lily 'Connecticut King'.

IV. BREEDING

A. DEVELOPMENT OF CULTIVAR ASSORTMENT

1. Lily

Compared with tulip and narcissus, the breeding history of lily is rather short. Less than one century ago, the first crosses involving *Lilium maculatum, L. davidii, L. dauricum, L. bulbiferum,* and *L. tigrinum* within the section *Sinomartagon* were produced in Japan and in the United States. Various hybrid groups were developed, such as the Preston hybrids, the Mid-Century hybrids, the Patterson hybrids, the Harlequin hybrids, and so on (Rockwell et al. 1961; De Graaff 1970; Leslie 1982; McRae 1998). A breakthrough for the lily culture was the mid-century hybrid 'Enchantment' bred by Jan de Graaff in 1944. In 1977, 760 ha of this cultivar (74% of the total acreage of lily bulbs) were grown in The Netherlands. The first polyploid (triploid and tetraploid) Asiatic lilies were introduced in the eighties of the twentieth century. Nowadays, most of the Asiatic hybrids are polyploid (Table 6.1).

TABLE 6.1

The Most Widely Grown Lily Hybrids (According to Bulb Acreage) Grown in The Netherlands in 2000, 2005, and 2009, Their Hybrid Group, Ploidy Level, the Breeder, and Year of Introduction (Dutch Flower Bulb Inspection Service, BKD 2000, 2005, and 2009)

Cultivar	2000	2005	2009	Group	Ploidy	Breeder	Introduction
Sorbonne	136	201	192	O	2	Vletter Haan	1994
Siberia	154	173	185	O	2	Mak	1994
Robina		18	78	OT	3	Marklily	2004
Rialto		46	73	O	2	Vletter Haan	2002
Navona	93	89	70	A	3	Vletter Haan	1994
Tiber	56	80	66	O	2	Vletter Haan	1994
Brindisi		48	64	LA	3	Vletter Haan	2002
Conca d'Or		47	62	OT	3	Vletter Haan	2002
Tresor	14	51	57	A	4	Vletter Haan	1997
Litouwen			48	LA	3	Vletter Haan	2005
Pavia		14	47	LA	3	Vletter Haan	2003
Casa Blanca	113	61	46	O	2	Vletter Haan	1977
Yelloween		79	40	OT	3	World Breeding	2005
Acapulco	76	53	39	O	2	Vletter Haan	1990
Merostar	81	44	36	O	2	Imanse	1991
Brunello	63	51	36	A	4	Vletter Haan	1994
Triumphator		27	35	LO	3	Van Zanten	2003
Original Love			34	LA	3	Mak	2006
Dazzle	29	52	33	LA	3	Mak	1997
White Heaven	14	46	33	L	2	World Breeding	1998
Courier		45	31	LA	3	Vletter Haan	2002
Nova Zembla		33	34	O	2	Mak	2001
Santander			35	O	2	Vletter Haan	2007
Star Gazer	296	135	33	O	2	Woodriff	1977
Menorca		59	24	LA	3	Vletter Haan	2002

Note: Type of hybrids: A, Asiatic; O, Oriental; L, Longiflorum; LA, Longiflorum × Asiatic; OT, Oriental × Trumpet.

After World War II, Dutch companies became involved in most breeding activities. In the first few years (1970–1990) that lily culture gained significance, only Asiatic hybrids were grown. In the subsequent period, the Oriental hybrids, originating from crosses of mainly *L. auratum* and *L. speciosum* from the *Archelirion* section, were developed. These hybrids, with large, fragrant, mainly pink and white flowers, became the most important hybrid group during the 1990s. 'Star Gazer' was the first upfacing Oriental hybrid and for more than 25 years it was the most important cultivar in this group. From that time onwards, this group has been the most important group of lilies. Meanwhile, intersectional hybridization, due to advanced interspecific hybridization methods (see Section II), yielded its first commercial products, starting around 1986 with 'Salmon Classic'. The first group of intersectional hybrids, the LA hybrids, was developed by interspecific hybridization between hybrids of different sections. These LA hybrids are backcrosses from a 2n-producing or a chromosome-doubled F_1 LA hybrid with an Asiatic hybrid and are, therefore, in general, triploids (Table 6.1). Triploids possess superior growth vigor and plant habit that make them the most favorable ploidy level. Over the last 10 years, this group has become more important than the Asiatic hybrid group and has in large part replaced them in the industry. Similarly, other intersectional hybrids have been developed, such as the LO, LR (Longiflorum × Rubellum), OT, and OA hybrids (Barba-Gonzalez et al. 2004; Lim and Van Tuyl 2004, 2006). Like the LA hybrids have mainly replaced the Asiatic hybrids and the LO hybrids have replaced the Longiflorums, it can be expected that, in the future, the Oriental hybrids will be partially replaced by the OTs. Table 6.1 presents the most widely grown lily hybrids, by bulb acreage grown in The Netherlands in 2000, 2005, and 2009 as published by the Dutch Flower Bulb Inspection Service (BKD 1970–2010). It is clear that over the last 10–15 years dramatic changes in the assortment have taken place.

2. Tulip

The history of tulip breeding dates back to the twelfth or thirteenth century when it was grown as a garden plant in Turkey. But almost nothing is known about the origin of the tulip (Pavord 1999). *T. gesneriana*, a source of many cultivars, probably has an interspecific origin. In the sixteenth and seventeenth centuries, thousands of cultivars were grown in The Netherlands, including single, double, and parrot tulips. In the Hortus Bulborum in Limmen, The Netherlands, hundreds of old cultivars originating from the fifteenth to the nineteenth century are still preserved.

According to the classified list of the International Register of tulip names (Van Scheepen 1996), tulips are divided into 14 groups. The first nine groups belong to the species *T. gesneriana* (Single Early, Double Early, Triumph, Single Late, Lily-flowered, Fringed, Viridiflora, Parrot, Double Late), the cultivars of the Darwin group are hybrids between *T. gesneriana* and *T. fosteriana*, the Kaufmanniana group was derived from *T. kaufmanniana*, the Fosteriana group from *T. fosteriana*, and the Greigii group from *T. greigii*. The Species group contains all other cultivated species (for details, see Chapter 4).

Through the centuries, the breeding of tulip focused mainly on ornamental characteristics such as color, flower size, form, and so on. After their growing importance as cut and pot flower, forcing characteristics were given top priority. Nowadays, the use of tulip as a garden plant is no longer a primary goal for most breeders.

A breakthrough in tulip breeding was achieved by D. W. Lefeber, who produced the first crosses between *T. fosteriana* and *T. gesneriana* about 70 years ago. The resulting progeny was called Darwin hybrid tulips. Most of these hybrids are triploid, exhibiting large flowers and leaves with high virus resistance, but being mostly sterile. Well known is the cultivar 'Apeldoorn', which was a dominant cultivar in the seventies, eighties, and nineties of the twentieth century (Table 6.2). However, due to bad forcing traits this cultivar has become less important. Recently, another breakthrough was been achieved by the discovery of fertile Darwin hybrids and through backcrosses to *T. gesneriana* (Marasek et al. 2011). It is expected that the progeny of these backcrosses will eventually combine virus resistance with good forcing characteristics (Marasek et al. 2011).

3. Narcissus

The genus *Narcissus* is classified into 10 sections, of which the sections *Tazetta*, *Narcissus*, *Jonquilla*, *Bulbocodium*, and *Pseudonarcissus* have contributed most to the cultivated assortment (Fernandes 1968). The first breeding of narcissi was carried out in the seventeenth century (Wylie 1952), and all modern cultivars are complex interspecific and intersectional hybrids. No embryo rescue methods were employed in the breeding process, and spontaneous polyploidization occurred during the domestication of this crop. Brandham (1986), who investigated the history of the origin of polyploids in this genus, reports that before 1885 there were only a few diploid ($2n = 2x = 14$) or triploid ($2n = 3x = 21$) cultivars, and that the first tetraploid was introduced in 1887. From about the 1920s, there was an explosive increase in the number of polyploids. At present, triploid and tetraploid cultivars predominate in *Narcissus* production. In the Daffodil Register and Classified List of the Royal Horticultural Society (Kington 1998), more than 7000 cultivars are described. They are classified into 11 divisions, of which the trumpet ('Dutch Master'), the large-cupped ('Carlton'), the double, the split corona, the tazetta and the cyclamineus types are the most important. In the last division, the development of the triploid 'Tête-à-Tête', a hybrid between *N. tazetta* × *N. cyclamineus*, bred more than 60 years ago, is remarkable; its acreage grew in The Netherlands from less than 1 ha in 1981 to 744 ha in 2010, which is 45% of all the narcissi grown in The Netherlands. In 2009,

TABLE 6.2

The Most Widely Grown Tulip Hybrids, According to Bulb Acreage Grown in The Netherlands in 1982, 1990, 2000, and 2010 Including the Hybrid Group, the Breeder, and Year of Introduction (Dutch Flower Bulb Inspection Service, BKD 1982, 1990, 2000, and 2010)

Cultivar	1982	1990	2000	2010	Group	Breeder	Introduction
Leen van der Mark	4	95	372	441	TT	Kon Mark	1968
Strong Gold	0	0	49	410	TT	Vd Berg	1989
Ile de France	15	26	193	320	TT	Blom & Padding	1968
Yokohama	14	80	375	248	TT	Vd Berg	1961
Purple Prince	0	1	134	197	TT	Schoorl	1987
Ben van Zanten	1	6	74	182	TT	Van Zanten	1967
Yellow Flight	0	0	36	156	TT	Boots	1994
Debutante	0	7	63	148	TT	IVT	1984
Purple Flag	0	0	39	143	TT	Schoorl	1983
Apeldoorn (+sps)	1300	680	495	143	DHT	Lefeber	1951
Viking (sp MC)	0	8	117	142	DVT	Unknown	1984
Prinses Irene (sp CC)	49	60	206	134	TT	Unknown	1949
Abba (sp MC)	1	25	61	132	DVT	Unknown	1978
Dynasty	0	0	9	126	TT	Borst	1996
Orange Cassini	0	10	65	103	TT	Segers	1944
Monte Carlo (MC)	104	408	649	101	DVT	Nijssen	1955
Negrita	21	83	154	100	TT	Bik	1970
Rococo (sp CC)	0	1	186	99	PT	Unknown	1942
Silver Dollar	0	5	82	99	TT	IVT	1984
Ad rem	11	47	168	98	DHT	Kon Mark	1960
White Dream	33	172	190	94	TT	Vd Berg	1972
Christmas Marvel	179	272	154	23	EVT	Schoorl	1954
Lustige Witwe	165	230	54	21	TT	Vd Mey	1942
Couleur Cardinal (CC)	35	27	26	18	TT	Unknown	1845
Prominence	204	281	154	3	TT	Van Kooten	1943

Note: Type of hybrids: TT, Triumph Tulip; DHT, Darwin Hybrid Tulip; DVT, Double Early Tulip; PT, Parrot Tulip; EVT, Single Early Tulip; sp, sport; +sps, including all sports.

20 million pots of 'Tête-à-Tête' were sold at the Dutch auctions. 'Carlton' and 'Dutch Master' are the main narcissus sold in the Dutch market as cut flowers.

4. Hyacinth

The origin of the cultivated *Hyacinthus* dates back to at least five centuries ago. *Hyacinthus* was introduced into cultivation in western Europe from Turkey around 1560 (Doorenbos 1954). Only one species, *H. orientalis* (2n = 16), is involved in breeding (Doorenbos 1954). While wild *H. orientalis* is blue, a range of colors was developed during hybridization: pink, red, white, purple, violet, yellow, and almost black. Already in the seventeenth century, hundreds of double-flowered cultivars were known. In the breeding of hyacinths the role of polyploidy is remarkable. The ploidy level varies from diploid ('Pink Pearl'), to triploid ('Jan Bos', 'Anna Marie'), to tetraploid ('Aiolus', 'Atlas', 'Blue Giant'). A number of cultivars are aneuploid ('Rosalie' 2n = 17, 'Kronos' 2n = 22, 'City of Haarlem' 2n = 23, 'Ostara' 2n = 25, 'Blue Jacket' 2n = 27, 'Carnegie' 2n = 29, 'Delft Blue' 2n = 30, 'Vuurbaak' 2n = 30 (Van Scheepen 1991)).

Following the discovery of the effect of storage temperatures on early forcing in hyacinths by Nicolas Dames (Doorenbos 1954), the trait of forcing ability became most important in breeding. Although hyacinth is one of the most important bulbous crops for potted plant production during the Christmas period, over the last 15 years cut flowers of hyacinths have become equally important. Turnover of hyacinth cut flowers at the Dutch auctions reached US$ 63.8 million, as compared with US$ 58.5 million for potted hyacinths.

5. Other Ornamental Geophytes

Similar to *Narcissus*, the breeding of *Freesia* (Goemans 1979), *Gladiolus* (Ohri and Khoshoo 1983a,b), *Dahlia* and *Hippeastrum* (Meerow 2009) started with diploid species, but interspecific crosses and natural polyploidization resulted in large assortments of modern tetraploid cultivars. In *Zantedeschia* (Snijder et al. 2008), interspecific hybridization resulted in diploid assortment. Plastome–genome incompatibility plays a key role in *Zantedeschia* breeding barriers, thus resulting in hybrids with albino or variegated leaves, caused by biparental plastid inheritance and hampering breeding progress (Yao et al. 1995). In *Nerine*, the triploid (*Nerine* × *Amaryllis*) × *Nerine*, obtained after *in vitro* ovary culture followed by chromosome doubling of F_1 hybrids and backcrossing 20 years ago, resulted in the development of new cultivars, which have since been introduced in the market (Van Tuyl et al. 1992).

B. CURRENT BREEDING OBJECTIVES

Commercial breeders of flower bulbs focus on a number of characteristics. The main ornamental traits that are often judged at the seedling stage (when only single flowering bulbs are available) are flower color, size, shape, and fragrance, in combination with the plant size, number of flowers, postharvest quality, and foliage attractiveness. At the clonal stage (when the selected seedlings produce larger clonal populations), further selection focuses on physiological and agronomic traits, that is, year-round flowering ability, flower longevity, vigorous growth, bulb production, resistance to pests and diseases, and tolerance to abiotic stress factors (salt, high/low temperature, light, and climate conditions).

1. Postharvest Quality and Flower Longevity

Postharvest quality and flower longevity of tulips were studied by Van Eijk et al. (1977) and Van Eijk and Eikelboom (1976, 1986). They found large differences in keeping quality in a wide range of species and cultivars. Several key characteristics were investigated, for example, the number of days between the onset of anthesis and perianth drop, 50% discoloration of perianth, and loss of turgor in the leaves and stem. However, none of these parameters could be used as exclusive criteria for postharvest quality. Van der Meulen-Muisers et al. (1996, 1997, 1998, 1999, 2001) studied the segregation of flower longevity in lily and tulip and found a large variation within populations tested at the individual plant level. They used standardized conditions in a climate-controlled room at 17°C (for lily) and 14°C (for tulip). While the individual flower longevity of the parental genotypes ranged from 4 to 8 days in lily and from 8 to 16 days for tulip, the range for the progenies in lily was from 2 to 11 days and in tulip from 6 to 22 days. High correlations were found between offspring and parents, which makes breeding for these characters very promising.

2. Low-Light and Low-Temperature Conditions

Breeding for low-light conditions in Asiatic lilies was carried out by Van Tuyl et al. (1985). During Dutch winter conditions, additional lighting was required for Asiatic hybrid lilies in order to prevent flower bud abscission and flower bud blasting. Growth room experiments showed considerable differences among eight lily cultivars in their response to low light. Among these cultivars 'Connecticut King' and 'Enchantment' appeared to be the most sensitive to low light conditions, while 'Uncle Sam' and 'Scout' were the least sensitive. For a number of traits, for example, flower bud abortion, leaf scorch, forcing time and plant length, significant General Combining Abilities were found, indicating a wide variation in the characteristics affected by light conditions. This variation could certainly be used for future breeding for low light tolerance.

Breeding of *L. longiflorum* for Dutch climate conditions was examined by Van Tuyl (1985, 1988). Before 1982, the market for Longiflorum bulbs was dependent on the import of bulbs from Japan (the origin of native *L. longiflorum*). One of the problems in growing *L. longiflorum* under the relatively cool conditions in The Netherlands was the lack of vigor and the premature sprouting of the daughter bulbs (summer sprouting). Therefore, a scale propagation method, adapted to the conditions in The Netherlands, was developed (Van Tuyl 1983). Using diallel crossing populations, bulb growth, summer sprouting, and year-round forcing characteristics were studied (Van Tuyl 1988). The material was released to the industry, resulting in an increase in Longiflorum bulb acreage from 5 ha in 1980 to around a stable 200 ha in 1992. Over the last 5 years, the acreage decreased to around 50 ha due to the replacement of Longiflorum cultivars by LA hybrids.

3. Disease Resistance

Breeding for disease resistance has been carried out for a number of fungal and bacterial diseases, for example, *Fusarium* (in lily, tulip, narcissus, gladiolus, and *Nerine*), *Botrytis* (lily, tulip), *Stromatinia* (gladiolus), *Xanthomonas* (hyacinth), and *Pectobacterium* (*Erwinia*) (*Zantedeschia*).

a. Fusarium

Fusarium oxysporum f. sp. *lilii* causes bulb and scale rot in lily. Resistance to *Fusarium* in the lily assortment was studied, and a standardized screening method for determining partial resistances in lily clones was developed (Straathof et al. 1993, 1996; Straathof 1994; Straathof and Van Tuyl 1994). Sources of resistance were found, especially in species and Asiatic hybrids belonging to the *Sinomartagon* section. A clear selection response was found in a seedling test (Straathof and Löffler 1994), and segregation of *Fusarium* resistance was analyzed in an interspecific cross between *L. longiflorum* and *L. dauricum* (Löffler et al. 1996).

In tulip, *F. oxysporum* f. sp. *tulipae* causes dry rot in bulbs, and a test for the screening and evaluation of genotypic differences in *Fusarium* resistance has been developed (Van Eijk et al. 1978; Van Eijk and Eikelboom 1983). It was shown that resistance is based on additive gene action that can also be transmitted to the offspring (Van Eijk et al. 1979). The most resistant parental cultivars showed high General Combining Ability values, which means that they transmit a high degree of resistance to their progenies. Preselection for *Fusarium* resistance at the juvenile stage appeared to be effective (Van Eijk et al. 1978).

Testing for resistance to *F. oxysporum* f. sp. *narcissi* in *Narcissus* was investigated by Tompsett (1986). 'Golden Harvest' and 'Carlton' are very susceptible to this pathogen, while *N. tazetta* 'Soleil d'Or' showed a high level of resistance. In screening parents and progenies, Bowes et al. (1992) showed that similar to tulip, an additive gene model fits in *Narcissus*.

In *Gladiolus*, several clonal tests for the determination of *F. oxysporum* resistance have been described and used (McClellan and Pryor 1957; Palmer and Pryor 1958; Wilfret and Woltz 1973; Jones and Jenkins 1975; Chandra et al. 1985; Löffler et al. 1997). High levels of resistance were found in cultivars and species, while absolute resistance was found in *G. dalenii*. In a seedling test, the resistance level of the parents was associated with the General Combining Ability (Straathof et al. 1997). *Fusarium sacchari* var. *elongatum* causes bulb rot in *N. bowdenii*. A screening method was developed by Van Tuyl et al. (1986a), and a large variation in the degree of resistance was demonstrated.

b. Botrytis

Botrytis elliptica and *Botrytis tulipae* cause the gray mould, which destroys the leaves in lily and tulip, respectively. Testing methods have been developed in lily by Doss et al. (1984, 1986), Beers et al. (2005), and Balode (2009) and in tulip by Straathof et al. (2002). In both lily and tulip, a large variation in resistance was found among species and cultivars.

c. Stromatinia

Stromatinia gladioli causes dry-rot disease in gladiolus. A test method for the selection of resistance to dry-rot was developed by Van Eijk et al. (1990). *Gladiolus italicus* showed a high degree of resistance, while commercial cultivars showed much lower levels of resistance. A seedling test, carried out by sowing seeds in infested soil, showed a low correlation ($r = 0.49$) with corms planted in infested soil. This was probably due to the developmental differences between germinating seedlings and corms.

d. Xanthomonas

The bacterial disease *Xanthomonas hyacinthii* causes yellow disease in hyacinth. A disease test, using the spraying of a bacterial suspension on leaves, was developed by Van Eijk et al. (1976). The degree of leaf infection appeared to be a good selection criterion. Diallel analysis for resistance to *X. hyacinthii* demonstrated a good General Combining Ability of the cultivar 'King of the Blues' (Van Tuyl and Toxopeus 1980; Van Tuyl 1982; Van Tuyl et al. 1986b) in comparison with 13 other cultivars. It should be noted that diploid hyacinth cultivars tend to be more resistant to the pathogen than tetraploid (Van Tuyl and Toxopeus 1980).

e. Pectobacterium (Erwinia) carotovorum

Infestation with the bacterium that causes soft rot *Pectobacterium carotovorum* ssp. *carotovorum* (formerly *Erwinia carotovora* ssp. *carotovora*) has devastating effects on *Zantedeschia* plants. Different tests to determine resistance in *Zantedeschia* spp. to soft rot were evaluated (Snijder and Van Tuyl 2002). A new non-destructive method was used on seedlings and involved the immersion of leaf disks in a bacterial suspension. Genetic variation was demonstrated among *Zantedeschia* accessions, from highly susceptible (*Z. pentlandii* and some accessions of *Z. albomaculata* ssp. *albomaculata*), to partially resistant (*Z. rehmannii* and other accessions of *Z. albomaculata* ssp. *albomaculata*), and almost completely resistant (some accessions of *Z. aethiopica*) (Snijder et al. 2004a,b). Breeding for partial resistance controlled by multiple genes appeared to be possible, but complicated by the plastome–genome incompatibility caused by biparental plastid inheritance in interspecific hybrids in *Zantedeschia* (Snijder et al. 2008).

f. Tulip Breaking Virus and Lily Mottle Virus

From the experience of bulb growers, it is known that resistance to *Tulip breaking virus* (TBV) can be found in *T. fosteriana*. The resistance level to TBV was determined in tulip species and cultivars using either virus transmission by aphids or mechanical inoculation (Romanow et al. 1991; Eikelboom et al. 1992). *T. fosteriana* 'Princeps' showed a high resistance, while all accessions of *T. gesneriana* appeared to be very susceptible, as compared with *T. fosteriana* in general. The seedling populations of diallel crosses between different *T. gesneriana* and *T. fosteriana* were screened for TBV resistance and resistant individuals were detected (Straathof et al. 1997). Introgression of *T. fosteriana* in *T. gesneriana* was monitored using GISH by Marasek-Ciolakowska et al. (2009), and an opportunity for introgression of TBV resistance into *T. gesneriana* breeding material was demonstrated. Currently, molecular markers based on ESTs are being developed (Shahin et al. unpublished). In lily, resistance to *Lily mottle virus* (LMoV) was found in certain cultivars of the Asiatic hybrid group (e.g., 'Connecticut King'). LMoV resistance is considered a monogenic trait and segregates in a population of 'Connecticut King' × 'Orlito' in a 1:1 ratio (Van Heusden et al. 2002; Shahin et al. 2009, 2011).

C. Mutation Breeding

1. Sports: Natural Mutants

Sports, the result of spontaneous mutations, are known to occur frequently in many ornamental crops, such as rose, chrysanthemum, tulip, hyacinth, iris, and so on. A good example is tulip. The heterozygosity of tulip is large and, therefore, it is not surprising that by spontaneous mutations many sports are selected. It is known that these sports are often periclinal chimeras in which only the cells of the L$_1$ outer layer mutate and the mutation is not passed on to the progeny. These periclinal mutants are often induced in mutation breeding programmes (Langton 1980). Examples of cultivars which produced large numbers of sports are the cultivar 'Murillo' (Van Scheepen 1996) and Darwin hybrids 'Apeldoorn' and 'Pink Impression'. Sports of 'Apeldoorn' that are still in production are: 'Apeldoorn's Elite', 'Blushing Apeldoorn', 'Golden Apeldoorn' and 'Striped Apeldoorn'. Sports of 'Pink Impression' are 'Apricot Impression', 'Design Impression', 'Red Impression', and 'Salmon Impression'. In hyacinth, from 'Pink Pearl', the leading cultivar, a number of sports are grown, for example, 'Blue Pearl', 'Early Pink Pearl', 'Early White Pearl', 'Scarlet Pearl', 'Violet Pearl', and 'White Pearl'.

2. Radiation-Induced Mutations

Radiation-induced mutation breeding was very popular from the 1960s to the 1980s (Broertjes and Ballego 1967; Broertjes and Alkema 1970). The Mutation Breeding Newsletter (http://www-pub.iaea.org/mtcd/publications/), a journal of the International Atomic Energy Agency in Vienna, includes lists of hundreds of released sports of radiation-induced cultivars of ornamentals crops. For crops such as *Chrysanthemum* and *Alstroemeria* (Broertjes 1966) mutation breeding is a routine method for developing a range of colors, after the value of a cultivar is determined. When an adventitious bud technique is available, for example, using *in vitro* techniques (Broertjes and Lock 1985) or using scales in lily (Van Tuyl 1983), a mutation is usually a single-cell event, which results in solid mutants, but not in chimeras. Most of the bulbous geophytes are irradiated for the induction of mutations. However, this method produced only a few cultivars.

3. Somaclonal Variation

Somaclonal variation is a type of mutation occurring in cell tissue cultures (Larkin and Scowcroft 1981). Van Harmelen et al. (1997) studied somaclonal variation in callus of lily, which had been cultured for a period of 3 years. The callus was obtained by incubating the scales slices of *L. longiflorum* 'Gelria' on a modified MS medium with the addition of NAA and BAP. Regeneration was induced on a hormone-free medium and, via shoots, lily bulbs were obtained *in vitro*. Several mutants were found, for example, plants with partially colored leaves (albinism) and plants with shorter or wider or curled leaves. Furthermore, one dwarf mutant and six male sterile individuals were detected (Van Harmelen et al. 1997).

Similarly, somaclonal variation has been observed in long-term micro-propagated tulip in the presence of thidiazuron (Podwyszynska 2005). The occurrence of somaclonal changes and especially variegation increases during the period of maintenance of the progeny lines *in vitro*. Marked changes, such as abnormal flowers, and malformed stamens and anthers, were observed after 4, 5, or 6 years. Somaclonal variation in tulip has also been analyzed by DNA markers (e.g., random amplified polymorphic DNA, ISSR (inter simple sequence repeats), Podwyszyñska et al. 2006). FISH with 5S and 45S repetitive DNA in off-type tulip plants and their original cultivar 'Prominence' exhibited variations in the number of loci and in the size of hybridization signals. Thus FISH karyotyping with cloned 5S and 45S rDNAs was shown to be useful for the analysis of genome restructuring in long-term micropropagated tulips (Marasek-Ciolakowska and Podwyszynska 2008).

D. Molecular Breeding

In contrast with developments in vegetable crops, where genetic mapping and marker-assisted breeding (MAB) have become an elementary skill in the breeding routine, only a few studies have been reported on genetic mapping in the main bulbous crops.

In lilies, the construction of a molecular map and the mapping of either ornamental traits such as flower color, flower pigmentation, and flower longevity (Yamagishi 1995; Abe et al. 2002) or disease resistance traits (Van Heusden et al. 2002) or both (Shahin et al. 2011) have been reported. Not surprisingly, the first attempt at marker conversion into simple PCR markers aiming at utilizing linked markers for MAB has also been described for lily (Shahin et al. 2009). Recent advances in sequencing and SNP (single nucleotide polymorphism) detection have led to a sharp decrease in the cost of mapping projects and, thus, facilitated data processing. It is likely that in the near future molecular tools will become a frequently used breeding tool for ornamental geophytes (Van Heusden and Arens 2010).

V. CONCLUDING REMARKS

The breeding advances of the past 50 years have been most effective in creating new variations in the shapes, colors, flower and plant sizes, as well as growth characteristics of ornamental geophytes. Many breeding methods have enlarged the breeder's "toolbox", enabling interspecific hybridization and ploidy level manipulations and the creation of new varieties. But these efforts have been made mainly in the major crops, for example, lily and tulips, where long-term breeding programs within an integrated approach exist.

In many crops, the potential for creating new variations by including distantly related species or genera into the cultivated breeding pool through interspecific hybridization is evident, but still awaits application (e.g., *Allium*, *Crocus*, *Iris*, *Lilium*, *Muscari*, and *Narcissus*). With the transition of agricultural practices toward more sustainable production, breeding for resistance to pathogens and tolerance to biotic and abiotic stresses will likely become more important and will result in higher demands on selection methods in the breeding process. In order to optimize selection methods, it must be emphasized that, for these more complex traits, molecular selection methods will have to find their way into the breeding process and be integrated with the established conventional selection procedures for other traits.

CHAPTER CONTRIBUTORS

Jaap M. van Tuyl is a senior researcher, Ornamental Plant Breeding, Wageningen University and Research Center, Wageningen, The Netherlands. He grew up in The Netherlands and received his BS, MS, and PhD from Wageningen University. He held research positions at the Institute for Horticultural Plant Breeding (IVT), CPO (Centre for Plant Breeding Research), and CPRO (Centre for Plant Breeding and Reproduction Research) (Wageningen). Dr. Van Tuyl has experience in the commercial ornamental industry as a flower breeder and a specialist in interspecific hybridization. His areas of expertise include hybridization, polyploidization, *in vitro* pollination and embryo rescue techniques, development of molecular marker techniques in lily and tulip, resistance breeding, flower longevity, and genetic resources of bulbous plants. He carries out ornamental breeding projects in cooperation with many Dutch and international companies. He is the author of more than 200 refereed journal articles, books or chapters, and conference papers and the recipient of the Wilson Award of The North American Lily Society (2008) and of the John Dix Penning Award of The Dutch Bulb Growers Association (2009).

Paul Arens is a researcher in ornamental plant breeding, Wageningen University and Research Center, Wageningen, The Netherlands. He graduated from the University for Higher Professional Education in Wageningen and obtained his PhD from the State University of Groningen. His area of expertise includes genetic diversity, the development of molecular markers, cultivar identification and EDV (essentially derived cultivars), genetic mapping, and resistance breeding. He carries out ornamental breeding projects in cooperation with Dutch and international companies. Dr. Arens is the author of more than 60 refereed journal articles, books or chapters, and conference papers.

Agnieszka Marasek-Ciołakowska is a researcher, Research Institute of Horticulture, Skierniewice, Poland. She grew up in Poland and received her MS from the University of Łódz and her PhD from the Research Institute of Pomology and Floriculture, Skierniewice, Poland. She held research positions at the Research Institute of Horticulture, Skierniewice, Poland (1994–the present) and post-doctorate positions at the University of Wales Aberystwyth, Great Britain (2004); Niigata University, Niigata, Japan (2005–2007); Ornamental Plant Breeding, Wageningen University and Research Center, Wageningen, The Netherlands (2007–2011). Her areas of expertise include molecular cytogenetics and plant genome analysis on the chromosomal level in lily and tulip. Dr. Marasek-Ciołakowska has published more than 50 refereed journal articles, books or chapters, and conference papers. She is the recipient of the Stefan Barbacki Award from the Polish Academy of Science (2004).

REFERENCES

Abe, H., M. Nakano, A. Nakatsuka, M. Nakayama, M. Koshioka, and M. Yamagishi. 2002. Genetic analysis of floral anthocyanin pigmentation traits in Asiatic hybrid lily using molecular linkage maps. *Theor. Appl. Genet.* 105:1175–1182.

Akutsu, M., S. Kitamura, R. Toda, I. Miyajima, and K. Okazaki. 2007. Production of 2*n* pollen of Asiatic hybrid lilies by nitrous oxide treatment. *Euphytica* 155:143–152.

Asano, Y. 1978. Studies on crosses between distantly related species of lilies. III. New hybrids obtained through embryo culture. *J. Japan. Soc. Hort. Sci.* 47:401–414.

Asano, Y. 1980a. Studies on the crosses between distantly related species of lilies. IV. The culture of immature hybrid embryos 0.3–0.4 mm long. *J. Japan. Soc. Hort. Sci.* 49:114–118.

Asano, Y. 1980b. Studies on crosses between distantly related species of lilies. V. Characteristics of newly obtained hybrids through embryo culture. *J. Japan. Soc. Hort. Sci.* 49:241–250.

Asano, Y. 1980c. Studies on crosses between distantly related species of lilies. VI. Pollen-tube growth in inter-specific crosses on *Lilium longiflorum* (I). *J. Japan. Soc. Hort. Sci.* 49:392–396.

Asano, Y. 1981. Pollen-tube growth in interspecific crosses of *Lilium longiflorum* Thunb. (II). *J. Japan. Soc. Hort. Sci.* 50:350–354.

Asano, Y. 1982a. Overcoming interspecific hybrid sterility in *Lilium. J. Japan. Soc. Hort. Sci.* 51:75–81.

Asano, Y. 1982b. Chromosome association and pollen fertility in some interspecific hybrids of *Lilium. Euphytica* 31:121–128.

Asano, Y. 1984. Fertility of a hybrid between distantly related species in *Lilium. Cytologia* 49:447–456.

Asano, Y., and H. Myodo. 1977a. Studies on crosses between distantly related species of lilies. I. For the intra-stylar pollination technique. *J. Japan. Soc. Hort. Sci.* 46:59–65.

Asano, Y., and H. Myodo. 1977b. Studies on crosses between distantly related species of lilies. II. The culture of immature hybrid embryos. *J. Japan. Soc. Hort. Sci.* 46:267–273.

Balode, A. 2009. Breeding for resistance against *Botrytis* in lily. *Acta Hort.* 836:143–148.

Barba-Gonzalez, R. 2005. The use of 2*n* gametes for introgression breeding in Oriental × Asiatic lilies. PhD thesis, Wageningen University, Wageningen, The Netherlands.

Barba-Gonzalez, R., K.-B. Lim, M. S. Ramanna, and J. M. van Tuyl. 2005a. Use of 2*n* gametes for inducting intergenomic recombination in lily hybrids. *Acta Hort.* 673:161–166.

Barba-Gonzalez, R., K.-B. Lim, M. S. Ramanna, R. G. F. Visser, and J. M. van Tuyl. 2005b. Occurrence of 2*n* gametes in the F₁ hybrids of Oriental × Asiatic lilies (*Lilium*): Relevance to intergenomic recombination and backcrossing. *Euphytica* 143:67–73.

Barba-Gonzalez, R., K.-B. Lim, S. Zhou, M. S. Ramanna and J. M. van Tuyl. 2008. Interspecific hybridization in lily: The use of 2*n* gametes in interspecific lily hybrids. In *Floriculture, Ornamental and Plant Biotechnology, Vol. V*, ed. J. A. Teixeira da Silva, 138–145. Isleworth, UK: Global Science Books.

Barba-Gonzalez, R., A. C. Lokker, K.-B. Lim, M. S. Ramanna, and J. M. van Tuyl. 2004. Use of 2*n* gametes for the production of sexual polyploids from sterile Oriental × Asiatic hybrids of lilies (*Lilium*). *Theor. Appl. Genet.* 109:1125–1132.

Barba-Gonzalez, R., C. T. Miller, M. S. Ramanna, R. and J. M. van Tuyl. 2006a. Induction of 2*n* gametes in for overcoming F₁-sterility in lily and tulip. *Acta Hort.* 714:99–106.

Barba-Gonzalez, R., C. T. Miller, M. S. Ramanna, R. and J. M. van Tuyl. 2006b. Nitrous oxide (N₂O) induces 2*n* gametes in sterlies F₁ hybrids between Oriental × Asiatic lily (*Lilium*) hybrids and leads to interge-nomic recombination. *Euphytica* 148:303–309.

Barba-Gonzalez, R., M. S. Ramanna, R. G. F. Visser, and J. M. van Tuyl. 2005c. Intergenomic recombination in F₁ lily hybrids (*Lilium*) and its significance for genetic variation in the BC₁ progenies as revealed by GISH and FISH. *Genome* 48:884–894.

Barba-Gonzalez, R., A. A. van Silfhout, R. G. F. Visser, M. S. Ramanna, and J. M. van Tuyl. 2006c. Progenies of allotriploids of Oriental × Asiatic lilies (*Lilium*) examined by GISH analysis. *Euphytica* 151:243–250.

Beers, C. M., R. Barba-Gonzalez, A. A. van Silfhout, M. S. Ramanna, and J. M. van Tuyl. 2005. Mitotic and meiotic polyploidization in lily hybrids for transferring *Botrytis* resistance. *Acta Hort.* 673:449–452.

Bell, W. D. 1972. Culture of immature *Amaryllis* embryos. *Plant Life* 28:72–76.

Benschop, M., R. Kamenetsky, M. Le Nard, H. Okubo, and A. De Hertogh. 2010. The global flower bulb industry: Production, utilization, research. *Hort. Rev.* 36:1–115.

BKD. 1970–2010. Beplante oppervlakten Bloembollen: lelie, tulp, narcis, hyacinth (Annually published statistics by The Dutch Bloembollenkeuringsdienst).

Blakey, D. H., and C. G. Vosa. 1982. Heterochromatin and chromosome variation in cultivated species of *Tulipa* subg. *Leiostemones* (Liliaceae). *Plant Syst. Evol.* 139:163–178.

Bowes, S. A., R. N. Edmondson, C. A. Linfield, and F. A. Langton. 1992. Screening immature bulbs of daffodil (*Narcissus* L.) crosses for resistance to basal rot disease caused by *Fusarium oxysporum* f. sp. *narcissi*. *Euphytica* 63:199–206.

Brandham, P. E. 1986. Evolution of polyploidy in cultivated *Narcissus* subgenus *Narcissus*. *Genetica* 68:161–167.

Brandham, P. E., and P. R. Kirton 1987. The chromosomes of species, hybrids and cultivars of *Narcissus* L. (Amaryllidaceae). *Kew Bull.* 42:65–102.

Brighton, C. A. 1976. Cytological problems in the genus *Crocus* (Iridaceae): I. *Crocus vernus* Aggregate. *Kew Bull.* 31:33–46.

Brighton, C. A., B. Mathew, and C. J. Marchant. 1973. Chromosome counts in the genus *Crocus* (Iridaceae). *Kew Bull.* 28:451–464.

Bridgen, M. P., R. Langhans, and R. Graig. 1989. Biotechnological breeding techniques for *Alstroemeria*. *Herbertia* 45:93–96.

Broertjes, C. 1966. Mutation breeding of chrysanthemums. *Euphytica* 15:156–162.

Broertjes, C., and H. Y. Alkema. 1970. Mutation breeding in flower bulbs. *Acta Hort.* 23:407–412.

Broertjes, C., and J. M. Ballego. 1967. Mutation breeding of *Dahlia variabilis*. *Euphytica* 16:171–176.

Broertjes, C., and C. A. M. Lock. 1985. Radiation-induced low-temperature tolerant solid mutants of *Chrysanthemum morifolium* Ram. *Euphytica* 34:97–103.

Buitendijk, J. H., N. Pinsonneaux, A. C. van Donk, M. S. Ramanna, and A. A. M. van Lammeren. 1995. Embryo rescue by half-ovule culture for the production of interspecific hybrids in *Alstroemeria*. *Scientia Hortic.* 64:65–75.

Buitendijk, J. H., and M. S. Ramanna. 1996. Giemsa C-banded karyotypes of eight species of *Alstroemeria* L. and some of their hybrids. *Ann. Bot.* 78:449–457.

Chandra, K. J., S. S. Negi, S. B. S. Raghava, and T. V. R. S. Sharma. 1985. Evaluation of gladiolus cultivars and hybrids for resistance to *Fusarium oxysporum* f. sp. *gladioli*. *Indian J. Hort.* 42:304–305.

Coertze, A. F., and E. Louw 1990. The breeding of interspecies and intergenera hybrids in the Amaryllidaceae. *Acta Hort.* 266:349–352.

Cohen, D., and J.-L. Yao. 1996. *In vitro* chromosome doubling of nine *Zantedeschia* cultivars. *Plant Cell, Tiss. Org. Cult.* 47:43–49.

Collard, B. C. Y., and D. J. Mackill. 2008. Marker-assisted selection: An approach for precision plant breeding in the twenty-first century. *Philos. Trans. Royal Soc. B.* 363:557–572.

Custers, J. B. M., W. Eikelboom, J. H. W. Bergervoet, and J. P. van Eijk. 1995. Embryo-rescue in the genus *Tulipa* L.; successful direct transfer of *T. kaufmanniana* Regel germplasm into *T. gesneriana* L. *Euphytica* 82:253–261.

Custers, J. B. M., E. Ennik, and W. Eikelboom. 1997. Embryogenesis from isolated microspores of tulip; towards developing F₁ hybrid varieties. *Acta Hort.* 430:259–266.

Darlington, C. D. 1926. Chromosome studies in the Scilleae. *J. Genet.* 16:237–251.

Darlington, C. D. 1929. Meiosis in polyploids II. Aneuploid hyacinths. *J. Genet.* 21:17–27, 52–56.

Darlington, C. D. 1932. The origin and behavior of chiasmata: VI. *Hyacinthus amethystinus*. *Biol. Bull.* 63:368–371.

Darlington, C. D., J. B. Hair, and R. Hurcombe. 1951. The history of the garden hyacinths. *Heredity* 5:233–252.

De Graaff, J. 1970 Looking backwards. *Lily Yearbook North Amer. Lily Soc.* 23:7–20.

De Jeu, M. J., and E. Jacobsen. 1995. Early postfertilization ovule culture in *Alstroemeria* L. and barriers to interspecific hybridization. *Euphytica* 86:15–23.

Doorenbos, J. 1954. Notes on the history of bulb breeding in The Netherlands. *Euphytica* 3:1–11.

Doss, R. P., G. A. Chastagner, and K. L. Riley. 1984. Techniques for inoculum production and inoculation of lily leaves with *Botrytis elliptica*. *Plant Dis.* 68:854–856.

Doss, R. P., G. A. Chastagner, and K. L. Riley. 1986. Screening ornamental lilies for resistance to *Botrytis elliptica*. *Scientia Hortic.* 30:237–246.

Eikelboom, W., J. P. van Eijk, D. Peters, and J. M. van Tuyl. 1992. Resistance to *Tulip breaking virus* (TBV) in tulip. *Acta Hort.* 325:631–636.

Emsweller, S. L., and P. Brierley. 1940. Colchicine-induced tetraploidy in *Lilium*. *J. Hered.* 31:223–230.

Fernandes, A. 1968. Improvements in the classification of the genus *Narcissus* L. *Plant Life* 24:51–57.

Filion, W. G. 1974. Differential Giemsa staining in plants. I. Banding patterns in three cultivars of *Tulipa*. *Chromosoma* 49:51–60.

Goemans, R. A., 1979. The history of the modern freesia. *Symposium Proceedings "Research in monocots of horticultural importance"*, 161–170.

Gu, Z.-P., and K.-C. Cheng. 1982. Studies on induction of pollen plantlets from the anther cultures of lily. *Acta Bot. Sin.* 24:28–32.

Han, D.-S., Y. Niimi, and M. Nakano. 1997. Regeneration of haploid plants from anther culture of the Asiatic hybryd lily 'Connecticut King'. *Plant Cell, Tiss. Org. Cult.* 47:153–158.

Hang, A., and T. Tsuchiya. 1988. Chromosome studies in the genus *Alstroemeria*. II. Chromosome constitutions of eleven additional cultivars. *Plant Breed.* 100:273–279.

Holm, P. B. 1976. The C and Q banding patterns of chromosomes of *L. longiflorum* (Thumb). *Carlsberg Res. Commun.* 41:217–224.

Jones, R. K., and J. M. Jenkins, Jr. 1975. Evaluation of resistance in *Gladiolus* sp. to *Fusarium oxysporum* f. sp. *gladioli*. *Phytopathology* 65:481–484.

Kamstra, S. A., J. H. de Jong, E. Jacobsen, M. S. Ramanna, and A. G. J. Kuipers. 2004. Meiotic behaviour of individual chromosomes in allotriploid *Alstroemeria* hybrids. *Heredity* 93:15–21.

Kamstra, S. A., A. G. J. Kuipers, M. J. de Jeu, M. S. Ramanna, and E. Jacobsen. 1997. Physical localisation of repetitive DNA sequences in *Alstroemeria*: Karyotyping of two species with species specific and ribosomal DNA. *Genome* 40:652–658.

Kamstra, S. A., A. G. J. Kuipers, M. J. de Jeu, M. S. Ramanna, and E. Jacobsen. 1999. The extent and position of homoeologous recombination in a distant hybrid of *Alstroemeria*: A molecular cytogenetic assessment of first generation backcross progenies. *Chromosoma* 108:52–63.

Kanoh, K., M. Hayashi, Y. Serizawa, and T. Konishi. 1988. Production of interspecific hybrids between *Lilium longiflorum* and *L. × elegance* by ovary slice culture. *Japan. J. Breed.* 38:278–282.

Karlov, G. I., L. I. Khrustaleva, K. B. Lim, and J. M. van Tuyl. 1999. Homoeologous recombination in 2*n*-gamete producing interspecific hybrids of *Lilium* (Liliaceae) studied by genomic *in situ* hybridization (GISH). *Genome* 42:681–686.

Khan, M. N. 2009. A molecular cytogenetic study of intergenomic recombination and introgression of chromosomal segments in lilies (*Lilium*). PhD thesis, Wageningen University, Wageningen, The Netherlands.

Khan, N., R. Barba-Gonzalez, M. S. Ramanna, P. Arens, R. G. F. Visser, and J. M. van Tuyl. 2010. Relevance of unilateral and bilateral sexual polyploidization in relation to intergenomic recombination and introgression in *Lilium* species hybrids. *Euphytica* 171:157–173.

Khan, N., R. Barba-Gonzalez, M. S. Ramanna, R. G. F. Visser, and J. M. van Tuyl. 2009a. Construction of chromosomal recombination maps of three genomes of lilies (*Lilium*) based on GISH analysis. *Genome* 52:238–251.

Khan, N., S. Zhou, M. S. Ramanna, P. Arens, J. Herrera, R. G. F. Visser, and J. M. van Tuyl. 2009b. Potential for analytic breeding in allopolyploids: An illustration from Longiflorum × Asiatic hybrid lilies (*Lilium*). *Euphytica* 166:399–409.

Kington, S. 1998. *The International Daffodil Register and Classified List 1998*, London: The Royal Horticultural Society.

Kroon, G. H., and M. C. Jongerius. 1986. Chromosome numbers of *Tulipa* species and the occurrence of hexaploidy. *Euphytica* 35:73–76.

Kroon, G. H., and J. P. van Eijk. 1977. Polyploidy in tulips (*Tulipa* L.). The occurrence of diploid gametes. *Euphytica* 26:63–66.

Langton, F. A. 1980. Chimerical structure and carotenoid inheritance in *Chrysanthemum morifolium* (Ramat.). *Euphytica* 29:807–812.

Larkin, P. J., and W. R. Scowcroft. 1981. Somaclonal variation—A novel source of variability from cell culture. *Theor. Appl. Genet.* 60:197–214.

Leslie, A. C. 1982. *The International Lily Register.* 3rd edn. London: The Royal Horticultural Society.

Lifante, Z. D., C. A. Camacho, J. Viruel, and A. C. Caballero. 2009. The allopolyploid origin of *Narcissus obsoletus* (Alliaceae): Identification of parental genomes by karyotype characterization and genomic *in situ* hybridization. *Bot. J. Linn. Soc.* 159:477–498.

Lim, K.-B. 2000. Introgression breeding through interspecific polyploidisation in lily: A molecular cytogenetic study. PhD thesis, Wageningen University, Wageningen, The Netherlands.

Lim, K.-B., R. Barba-Gonzalez, S. Zhou, M. S. Ramanna, and J. M. van Tuyl. 2008. Interspecific hybridization in lily (*Lilium*): Taxonomic and commercial aspects of using species hybrids in breeding. In *Floriculture, Ornamental and Plant Biotechnology, Vol. V,* ed. J. A. Teixeira da Silva, 146–151. Isleworth, UK: Global Science Books.

Lim, K. B., J. D. Chung, B. C. E. van Kronenburg, M. S. Ramanna, J. H. de Jong, and J. M. van Tuyl. 2000a. Introgression of *Lilium rubellum* Baker chromosomes into *L. longiflorum* Thunb.: A genome painting study of the F$_1$ hybrid, BC$_1$ and BC$_2$ progenies. *Chromosome Res.* 8:119–125.

Lim, K.-B., M. S. Ramanna, J. H. de Jong, E. Jacobsen, and J. M. van Tuyl. 2001a. Indeterminate meiotic restitution (IMR): A novel type of meiotic nuclear restitution mechanism detected in interspecific lily hybrids by GISH. *Theor. Appl. Genet.* 103:219–230.

Lim, K.-B., M. S. Ramanna, E. Jacobsen, and J. M. van Tuyl. 2003a. Evaluation of BC$_2$ progenies derived from 3*x*-2*x* and 3*x*-4*x* crosses of *Lilium* hybrids: A GISH analysis. *Theor. Appl. Genet.* 106:568–574.

Lim, K.-B., M. S. Ramanna, and J. M. van Tuyl. 2001b. Comparison of homoelogous recombination frequency between mitotic and meiotic polyploidization in BC$_1$ progeny of interspecific lily hybrids. *Acta Hort.* 552:65–72.

Lim, K.-B., M. S. Ramanna, and J. M. van Tuyl. 2003b. Homoelogous recombination in interspecific hybrids of *Lilium. Kor. J. Breed.* 35:8–12.

Lim, K.-B., T. M. Shen, R. Barba-Gonzalez, M. S. Ramanna, and J. M. van Tuyl. 2004. Occurrence of SDR 2N-gametes in *Lilium* hybrids. *Breed. Sci.* 54:13–18.

Lim, K.-B., and J. M. van Tuyl. 2004. A pink longiflorum lily cultivar, 'Elegant Lady' suitable for cut flower forcing. *Kor. J. Breed.* 36:123–124.

Lim, K.-B., and J. M. van Tuyl. 2006. Lily, *Lilium* hybrids. In *Flower Breeding and Genetics: Issues, Challenges and Opportunities for the 21st Century,* ed. N. O. Anderson, 517–537. Dordrecht: Springer.

Lim, K.-B., J. M. van Tuyl, L. I. Khrustaleva, G. I. Karlov, and J. H. de Jong. 2000b. Introgression of interspecific hybrids of lily using genomic *in situ* hybridization (GISH). *Acta Hort.* 508:105–111.

Lim, K.-B., J. Wennekes, J. H. de Jong, E. Jacobsen, and J. M. van Tuyl. 2001c. Karyotype analysis of *Lilium longiflorum* and *Lilium rubellum* by chromosome banding and fluorescence *in situ* hybridisation. *Genome* 44:911–918.

Löffler, H. J. M., H. Meijer, Th. P. Straathof, and J. M. van Tuyl. 1996. Segregation of *Fusarium* resistance in an interspecific cross between *Lilium longiflorum* and *Lilium dauricum. Acta Hort.* 414:203–208.

Löffler, H. J. M., Th. P. Straathof, P. C. L. van Rijbroek, and E. J. A. Roebroeck. 1997. *Fusarium* resistance in *Gladiolus:* The development of a screening assay. *J. Phytopath.* 145:465–468.

Lu, C., and M. P. Bridgen. 1997. Chromosome doubling and fertility study of *Alstroemeria aurea* × *A. caryophyllacea. Euphytica* 94:75–81.

Marasek, A., R. Hasterok, K. Wiejacha, and T. Orlikowska. 2004. Determination by GISH and FISH of hybrid status in *Lilium. Hereditas* 140:1–7.

Marasek, A., H. Mizuochi, and K. Okazaki. 2006. The origin of Darwin hybrid tulips analyzed by flow cytometry, karyotype analyses and genomic *in situ* hybridization. *Euphytica* 151:279–290.

Marasek, A., and K. Okazaki. 2007. GISH analysis of hybrids produced by interspecific hybridization between *Tulipa gesneriana* and *T. fosteriana. Acta Hort.* 743:133–137.

Marasek, A., and K. Okazaki. 2008. Analysis of introgression of the *Tulipa fosteriana* genome into *Tulipa gesneriana* using GISH and FISH. *Euphytica* 160:217–230.

Marasek, A., and T. Orlikowska. 2003. Karyology of nine lily genotypes. *Acta Biologica Cracoviensia, Ser. Botanica* 45:165–174.

Marasek, A., E. Sliwinska, and T. Orlikowska. 2005. Cytogenetic analysis of eight lily genotypes. *Caryologia* 58:359–366.

Marasek-Ciolakowska, A., and M. Podwyszynska. 2008. Somaclonal variation in long-term micropropagated tulips (*Tulipa gesneriana* L.) determined by FISH analysis. *Floriculture Ornamental Biotechnol.* 2:65–72.

Marasek-Ciolakowska, A., M. S. Ramanna, and J. M. van Tuyl. 2009. Introgression breeding in genus *Tulipa* analysed by GISH. *Acta Hort.* 836:105–110.

Marasek-Ciolakowska, A., M. S. Ramanna, and J. M. van Tuyl. 2011. Introgression of chromosome segments of *Tulipa fosteriana* into *T. gesneriana* detected through GISH and its implications for breeding virus resistant tulips. *Acta Hort.* 886:175–182.

McClellan, W. D., and R. L. Pryor. 1957. Susceptibility of gladiolus varieties to *Fusarium*, *Botrytis*, and *Curvularia*. *Plant Dis. Rep.* 41:47–50.

McRae, E. A. 1998. *Lilies: A Guide for Growers and Collectors*. Portland, OR: Timber Press.

Meerow, A. W. 2009. Tilting at windmills: 20 years of *Hippeastrum* breeding. *Israel J. Plant Sci.* 57:303–313.

Mizuochi, H., A. Marasek, and K. Okazaki. 2007. Molecular cloning of *Tulipa fosteriana* rDNA and subsequent FISH analysis yields cytogenetic organization of 5S rDNA and 45S rDNA in *T. gesneriana* and *T. fosteriana*. *Euphytica* 155:235–248.

Myodo, H. 1963. Experimental studies on the sterility of some *Lilium* species. *J. Fac. Agr. Hokkaido Univ.* 52:70–122.

Noda, S. 1986. Cytogenetic behavior, chromosomal differences, and geographic distribution in *L. lancifolium* (Liliaceae). *Plant Species Biol. (Kyoto)* 1:69–78.

Noda, S. 1991. Chromosomal variation and evolution in the genus *Lilium*. In *Chromosome Engineering in Plants: Genetics, Breeding, Evolution. Part B*, eds. T. Tsuchiya, and P. K. Gupta, 507–524. Amsterdam: Elsevier.

Ohri, D., and T. N. Khoshoo. 1983a. Cytogenetics of garden gladiolus, III. Hybridization. *Z. Pflanzenzüchtg.* 91:46–60.

Ohri, D., and T. N. Khoshoo. 1983b. Cytogenetics of garden gladiolus, IV. Origin and evolution of ornamental taxa. *Proc. Indian Natl. Sci. Acad. Part B: Biol. Sci.* 49(3):279–294.

Okazaki, K. 2005. New aspects of tulip breeding: Embryo culture and polyploid. *Acta Hort.* 673:127–140.

Okazaki, K., Y. Asano, and K. Oosawa. 1994. Interspecific hybrids between *Lilium* "Oriental" hybrid and *L.* "Asiatic" hybrid produced by embryo culture with revised media. *Breed. Sci.* 44:59–64.

Okazaki, K., K. Kurimoto, I. Miyajima, A. Enami, H. Mizuochi, Y. Matsumoto, and H. Ohya. 2005. Induction of 2n pollen in tulips by arresting the meiotic process with nitrous oxide gas. *Euphytica* 143:101–114.

Okazaki, K., and K. Murakami. 1992. Effects of flowering time (in forcing culture), stigma excision, and high temperature on overcoming of self-incompatibility in tulip. *J. Japan. Soc. Hort. Sci.* 61:405–411.

Okazaki, K., and M. Nishimura. 2000. Ploidy of progenies crossed between diploid, triploid and tetraploid in tulip. *Acta Hort.* 522:127–134.

Okazaki, K., Y. Umada, O. Urashima, J. Kawada, M. Kunishige, and K. Murakami. 1992. Interspecific hybrids of *Lilium longiflorum* and *L.* × *formolongi* with *L. rubellum* and *L. japonicum* through embryo culture. *J. Japan. Soc. Hort. Sci.* 60:997–1002.

Ørgaard, M., N. Jacobsen, and I. S. Heslop-Harrison. 1995. The hybrid origin of two cultivars of *Crocus* (Iridaceae) analysed by molecular cytogenetics including genomic southern and *in situ* hybridization. *Ann. Bot.* 76:253–262.

Palmer, J. G., and R. L. Pryor. 1958. Evaluation of 160 varieties of *Gladiolus* for resistance to *Fusarium* yellows. *Plant Dis. Rep.* 42:1405–1407.

Pavord, A. 1999. *The Tulip*. London: Bloomsbury.

Podwyszynska, M. 2005. Somaclonal variation in micropropagated tulips based on phenotype observation. *J. Fruit Ornam. Plant Res.* 13:109–122.

Podwyszyñska, M., K. Niedoba, M. Korbin, and A. Marasek. 2006. Somaclonal variation in micropropagated tulips determined by phenotype and DNA markers. *Acta Hort.* 714:211–219.

Qu, Y., M. C. Mok, D. W. S. Mok, and J. R. Stang. 1988. Phenotypic and cytological variation among plants derived from anther cultures of *Lilium longiflorum*. *In Vitro Cell. Dev. Biol.—Plant* 24:471–476.

Ramanna, M. S., and E. Jacobsen. 2003. Relevance of sexual polyploidization for crop improvement—A review. *Euphytica* 133:3–18.

Ramanna, M. S., A. G. J. Kuipers, and E. Jacobsen. 2003. Occurrence of numerically unreduced (2n) gametes in *Alstroemeria* interspecific hybrids and their significance for sexual polyploidisation. *Euphytica* 133:95–106.

Rockwell, F. F., E. C. Grayson, and J. de Graff. 1961. *The Complete Book of Lilies*. New York: Doubleday and Co., Inc.

Romanow, L. R., J. P. van Eijk, W. Eikelboom, A. R. van Schadewijk, and D. Peters. 1991. Determining levels of resistance to *Tulip breaking virus* (TBV) in tulip (*Tulipa* L.) cultivars. *Euphytica* 51:273–280.

Sayama, H., T. Moue, and Y. Nishimura. 1982. Cytological study in *Tulipa gesneriana* and *T. fosteriana*. *Japan. J. Breed.* 32:26–34.

Shahin, A., P. Arens, A. W. van Heusden, G. van der Linden, M. van Kaauwen, N. Khan, H. J. Schouten, W. E. van de Weg, R. G. F. Visser, and J. M. van Tuyl. 2011. Genetic mapping in *Lilium*: Mapping of major genes and quantitative trait loci for several ornamental traits and disease resistances. *Plant Breed.* 130:372–382.

Shahin, A., P. Arens, A. W. van Heusden, and J. M. van Tuyl. 2009. Conversion of molecular markers linked to *Fusarium* and virus resistance in Asiatic lily hybrids. *Acta Hort.* 836:131–136.

Sharp, W. R., R. S. Raskin, and H. E. Sommer. 1971. Haploidy in *Lilium*. *Phytomorphology* 21:334–337.

Siljak-Yakovlev, S., S. Peccenini, E. Muratović, V. Zoldoš, O. Robin, and J. Vallès. 2003. Chromosomal differentiation and genome size in three European mountain *Lilium* species. *Plant Syst. Evol.* 236:165–173.

Smyth, D. R. 1991. Chromosome bands in *Lilium*. *The Lily Yearbook North Am. Lily Soc.* 44:103–105.

Smyth, D. R., K. Kongsuwan, and S. Wisudharomn. 1989. A survey of C-band patterns in chromosomes of *Lilium* (Liliaceae). *Plant Syst. Evol.* 163:53–69.

Snijder, R. C. 2004. Genetics of *Erwinia* resistance in *Zantedeschia*. Impact of plastome-genome incompatibility. PhD thesis, Wageningen University, Wageningen, The Netherlands.

Snijder, R. C., F. S. Brown, and J. M. van Tuyl. 2008. The role of plastome–genome incompatibility and biparental plastid inheritance in interspecific hybridization in the genus *Zantedeschia* (Araceae). *Floricult. Ornamental Biotechnol.* 1:150–157.

Snijder, R. C., H.-R. Cho, M. M. W. B. Hendriks, P. Lindhout, and J. M. van Tuyl. 2004a. Genetic variation in *Zantedeschia* spp. (Araceae) for resistance to soft rot caused by *Erwinia carotovora* subsp. *carotovora*. *Euphytica* 135:119–128.

Snijder, R. C., P. Lindhout, and J. M. van Tuyl. 2004b. Genetic control of resistance to soft rot caused by *Erwinia carotovora* subsp. *carotovora* in *Zantedeschia* spp. (Araceae), section *Aestivae*. *Euphytica* 136:319–325.

Snijder, R. C., and J. M. van Tuyl. 2002. Evaluation of tests to determine resistance of *Zantedeschia* spp. (Araceae) to soft rot caused by *Erwinia carotovora* subsp. *carotovora*. *Eur. J. Plant Pathol.* 108:565–571.

Stebbins, G. L. 1958. The inviability, weakness, and sterility of interspecific hybrids. *Adv. Genet.* 9:147–215.

Stewart, R. N. 1947. The morphology of somatic chromosomes in *Lilium*. *Am. J. Bot.* 34:9–26.

Straathof, Th. P. 1994. Studies on the *Fusarium*–lily interaction: A breeding approach. PhD thesis, Wageningen University, Wageningen, The Netherlands.

Straathof, Th. P., and W. Eikelboom. 1997. Tulip breeding at PRI. *Daffodil and Tulip Yearbook* 8:27–33.

Straathof, Th. P., J. Jansen, and H. J. M. Löffler. 1993. Determination of resistance to *Fusarium oxysporum* in *Lilium*. *Phytopathology* 83:568–572.

Straathof, Th. P., J. Jansen, E. J. A. Roebroeck, and H. J. M. Löffler. 1997. *Fusarium* resistance in *Gladiolus*: Selection in seedling populations. *Plant Breed.* 116:283–286.

Straathof, Th. P., and H. J. M. Löffler. 1994. Screening for *Fusarium* resistance in seedling populations of Asiatic hybrid lily. *Euphytica* 78:43–51.

Straathof, Th. P., J. J. Mes, W. Eikelboom, and J. M. van Tuyl. 2002. A greenhouse screening assay for *Botrytis tulipae* resistance in tulips. *Acta Hort.* 570:415–421.

Straathof, Th. P., and J. M. van Tuyl. 1994. Genetic variation in resistance to *Fusarium oxysporum* f. sp. *lilii* in the genus *Lilium*. *Ann. Appl. Biol.* 125:61–72.

Straathof, Th. P., J. M. van Tuyl, B. Dekker, M. J. M. van Winden, and J. M. Sandbrink. 1996. Genetic analysis of inheritance of partial resistance to *Fusarium oxysporum* in Asiatic hybrids of lily using RAPD markers. *Acta Hort.* 414:209–218.

Sultana, S., S.-H. Lee, J.-W. Bang, and H.-W. Choi. 2010. Physical mapping of rRNA gene loci and inter-specific relationships in wild *Lilium* distributed in Korea. *J. Plant Biol.* 53:433–443.

Tompsett, A. A. 1986. Narcissus varietal susceptibility to *Fusarium oxysporum* (basal rot). *Acta Hort.* 177:77–83.

Tsuchiya, T., A. Hang, W. E. Healy Jr., and H. Hughes. 1987. Chromosomes studies in the genus *Alstroemeria*. I. Chromosome numbers in 10 cultivars. *Bot. Gaz.* 148:519–524.

Van Creij, M. G. M., D. M. F. J. Kerckhoffs, and J. M. van Tuyl. 1997a. Interspecific crosses in the genus *Tulipa* L.: Identification of pre-fertilization barriers. *Sex. Plant Reprod.* 10:116–123.

Van Creij, M. G. M., D. M. F. J. Kerckhoffs, and J. M. van Tuyl. 1999. The effect of ovule age on ovary—Slice culture and ovule culture in intraspecific and interspecific crosses with *Tulipa gesneriana* L. *Euphytica* 108:21–28.

Van Creij, M. G. M., D. M. F. J. Kerckhoffs, S. M. de Bruijn, D. Vreugdenhil, and J. M. van Tuyl. 2000. Ovary-slice culture and ovule culture in intraspecific and interspecific crosses with *Tulipa gesneriana*: Influence of culture date. *Plant Cell, Tiss. Org. Cult.* 60:61–67.

Van Creij, M. G. M., J. L. van Went, and D. M. F. J. Kerckhoffs. 1997b. The progamic phase, embryo and endosperm development in an intraspecific *Tulipa gesneriana* L. cross and in the incongruent interspecific cross *T. gesneriana* × *T. agenensis* DC. *Sex. Plant Reprod.* 10:241–249.

Van den Bulk, R. W., H. P. J. de Vries-van Hulten, J. B. M. Custers, and J. J. M. Dons. 1994. Induction of embryogenesis in isolated microspores of tulip. *Plant Sci.* 104:101–111.

Van den Bulk, R. W., and J. M. van Tuyl. 1997. *In vitro* induction of haploid plants from the gametophytes of lily and tulip. In *In Vitro Haploid Production in Higher Plants, vol. 5. Oil, Ornamental and Miscellaneous Plants*, eds. S. M. Jain, S. K. Sapory and R. E. Veilleux, 73–88. Dordrecht: Kluwer Academic Publ.

Van der Meulen-Muisers, J. J. M., J. C. van Oeveren, J. Jansen, and J. M. van Tuyl. 1999. Genetic analysis of postharvest flower longevity of Asiatic hybrid lilies. *Euphytica* 107:149–157.

Van der Meulen-Muisers, J. J., J. C. van Oeveren, J. M. Sandbrink, and J. M. van Tuyl. 1996. Molecular markers as a tool for breeding for flower longevity in Asiatic hybrid lilies. *Acta Hort.* 420:68–71.

Van der Meulen-Muisers, J. J. M., J. C. van Oeveren, L. H. W. van der Plas, and J. M. van Tuyl. 2001. Postharvest flower development in Asiatic hybrid lilies as related to tepal carbohydrate status. *Postharvest Biol. Tech.* 21:201–211.

Van der Meulen-Muisers, J. J. M., J. C. van Oeveren, and J. M. van Tuyl. 1997. Breeding as a tool for improving postharvest quality characters of lily and tulip flowers. *Acta Hort.* 430:569–575.

Van der Meulen-Muisers, J. J. M., J. C. van Oeveren, and J. M. van Tuyl. 1998 Genotypic variation in post harvest flower longevity of Asiatic hybrid lilies. *J. Am. Soc. Hort. Sci.* 123:283–287.

Van Eijk, J. P., B. H. H. Bergman, and W. Eikelboom. 1978. Breeding for resistance to *Fusarium oxysporum* f. sp. *tulipae* in tulip (*Tulipa* L.). 1. Development of a screening test for selection. *Euphytica* 27:441–446.

Van Eijk, J. P., and W. Eikelboom. 1976. Possibilities of selection for keeping quality in tulip breeding. *Euphytica* 25:353–359.

Van Eijk, J. P., and W. Eikelboom. 1983. Breeding for resistance to *Fusarium oxysporum* f. sp. *tulipae* in tulip (*Tulipa* L.). 3. Genotypic evaluation of cultivars and effectiveness of pre-selection. *Euphytica* 32:505–510.

Van Eijk, J. P., and W. Eikelboom. 1986. Aspects of breeding for keeping quality in *Tulipa*. *Acta Hort.* 181:237–243.

Van Eijk, J. P., and W. Eikelboom. 1990. Evaluation of breeding research on resistance to *Fusarium oxysporum* in tulip. *Acta Hort.* 266:357–364.

Van Eijk, J. P., W. Eikelboom, and L. D. Sparnaaij. 1977. Possibilities of selection for keeping quality in tulip breeding. *Euphytica* 26:825–828.

Van Eijk, J. P., F. Garretsen, and W. Eikelboom. 1979. Breeding for resistance to *Fusarium oxysporum* f. sp. *tulipae* in tulip (*Tulipa* L.). 2. Phenotypic and genotypic evaluation of cultivars. *Euphytica* 28:67–71.

Van Eijk, J. P., A. C. van der Giessen, and S. J. Toxopeus. 1976. Perspectives of breeding for resistance to "yellow disease", *Xanthomonas hyacinthii* (Wakker) Dowson, in hyacinths (*Hyacinthus* L.). *Euphytica* 25:131–138.

Van Eijk, J. P., L. W. D. van Raamsdonk, W. Eikelboom, and R. J. Bino. 1991. Interspecific crosses between *Tulipa gesneriana* cultivars and wild *Tulipa* species—A survey. *Sex. Plant Reprod.* 4:1–5.

Van Eijk, J. P., A. van Zaayen, and W. Eikelboom. 1990. Developing a test method for selection on resistance to dry-rot disease (*Stromatinia gladioli*) in gladiolus. *Acta Hort.* 266:365–374.

Van Harmelen, M. J., H. J. M. Löffler, and J. M. van Tuyl. 1997. Somaclonal variation in lily after *in vitro* cultivation. *Acta Hort.* 430:347–350.

Van Heusden, A. W., and P. Arens. 2010. Marker development in ornamental plants. *Acta Hort.* 855:137–141.

Van Heusden, A. W., M. C. Jongerius, J. M. van Tuyl, Th. P. Straathof, and J. J. Mes. 2002. Molecular assisted breeding for disease resistance in lily. *Acta Hort.* 572:131–138.

Van Raamsdonk, L. W. D., and T. de Vries. 1995. Species relationships and taxonomy in *Tulipa* subg. *Tulipa* (Liliaceae). *Plant Syst. Evol.* 195:13–44.

Van Scheepen, J. 1991. *International Checklist for Hyacinths and Miscellaneous Bulbs*. Hillegom, The Netherlands: Royal General Bulbgrowers' Association (KAVB).

Van Scheepen, J. 1996. *Classified List and International Register of Tulip Names*. Hillegom, The Netherlands: Royal General Bulbgrowers' Association (KAVB).

Van Tuyl, J. M. 1982. Breeding for resistance to yellow disease of hyacinths. II. Influence of flowering time, leaf characters, stomata and chromosome number on the degree of resistance. *Euphytica* 31:621–628.

Van Tuyl, J. M. 1983. Effect of temperature treatments on the scale propagation of *Lilium longiflorum* 'White Europe' and *Lilium* × 'Enchantment'. *HortScience* 18:754–756.

Van Tuyl, J. M. 1985. Effect of temperature on bulb growth capacity and sensitivity to summer sprouting in *Lilium longiflorum* Thunb. *Scientia Hortic.* 25:177–187.

Van Tuyl, J. M. 1988. Dutch-grown *Lilium longiflorum* a reality. *The Lily Yearbook North Am. Lily Soc.* 41:33–37.

Van Tuyl, J. M. 1989. Research on mitotic and meiotic polyploidization in lily breeding. *Herbertia* 45:97–103.

Van Tuyl, J. M. 1997. Interspecific hybridization of flower bulbs: A review. *Acta Hort.* 430:465–476.

Van Tuyl, J. M., R. J. Bino, and J. B. M. Custers. 1990. Application of *in vitro* pollination, ovary culture, ovule culture and embryo rescue techniques in breeding of *Lilium, Tulipa* and *Nerine.* In *Proceedings of the Eucarpia Symposium, Integration of in vitro techniques in ornamental plant breeding*, ed. J. de Jong, 86–97. Wageningen, The Netherlands: Wageningen University

Van Tuyl, J. M., M. Y. Chung, J. D. Chung, and K. B. Lim. 2002a. Introgression with *Lilium* hybrids: Introgression studies with the GISH method on *L. longiflorum* × Asiatic, *L. longiflorum* × *L. rubellum* and *L. auratum* × *L. henryi. The Lily Yearbook North Am. Lily Soc.* 55:17–22, 70–72.

Van Tuyl, J. M., and M. J. de Jeu. 1997. Methods for overcoming interspecific crossing barriers. In *Pollen Biotechnology for Crop Production and Improvement*, eds. V. K. Shivanna, and K. R. Sawhney, 273–293. Cambridge: Cambridge University Press.

Van Tuyl, J. M., J. N. de Vries, R. J. Bino, and T. A. M. Kwakkenbos. 1989. Identification of 2n-pollen producing interspecific hybrids of *Lilium* using flow cytometry. *Cytologia* 54:737–745.

Van Tuyl, J. M., K. B. Lim, and M. S. Ramanna. 2002b. Interspecific hybridization and introgression. In *Breeding for Ornamentals: Classical and Molecular Approaches*, ed. A. Vainstein, 85–103. Dordrecht/Boston: Kluwer Academic Publ.

Van Tuyl, J. M., C. A. M. Lock, and A. C. van der Giessen. 1986a. Selection for *Fusarium* resistance in *Nerine bowdenii. Acta Hort.* 177:597–600.

Van Tuyl, J. M., B. Meijer, and M. P. van Diën. 1992. The use of oryzalin as an alternative for colchicine in in-vitro chromosome doubling of *Lilium* and *Nerine. Acta Hort.* 325:625–630.

Van Tuyl, J. M., T. P. Straathof, R. J. Bino, and A. A. M. Kwakkenbos. 1988. Effect of three pollination methods on embryo development and seedset in intra- and interspecific crosses between seven *Lilium* species. *Sex. Plant Reprod.* 1:119–123.

Van Tuyl, J. M., and S. J. Toxopeus. 1980. Breeding for resistance to yellow disease of hyacinths. I. Investigations on F_1's from diallel crosses. *Euphytica* 29:555–560.

Van Tuyl, J. M., and M. G. M. van Creij. 2006. Tulip *Tulipa gesneriana* and *T.* hybrids. In *Flower Breeding and Genetics: Issues, Challenges and Opportunities for the 21st Century*, ed. N. O. Anderson, 623–641. Dordrecht: Springer.

Van Tuyl, J. M., M. P. van Diën, M. G. M. van Creij, T. C. M. van Kleinwee, J. Franken, and R. J. Bino. 1991. Application of *in vitro* pollination, ovary culture, ovule culture and embryo rescue for overcoming incongruity barriers in interspecific *Lilium* crosses. *Plant Sci.* 74:115–126.

Van Tuyl, J. M., J. P. van Eijk, and T. A. M. Kwakkenbos. 1986b. Breeding for resistance to yellow disease of hyacinths. *Acta Hort.* 177:585–589.

Van Tuyl, J. M., J. E. van Groenestijn, and S. J. Toxopeus. 1985. Low light intensity and flower bud abortion in Asiatic hybrid lilies. I. Genetic variation among cultivars and progenies of a diallel cross. *Euphytica* 34:83–92.

Wietsma, W. A., K. Y. de Jong, and J. M. van Tuyl. 1994. Overcoming prefertilization barriers in interspecific crosses of *Fritillaria imperialis* and *F. raddeana. Plant Cell Incompatibility Newslett.* 26:89–92.

Wilfret, G. J., and S. S. Woltz. 1973. Susceptibility of corms of gladiolus cultivars to *Fusarium oxysporum* f. sp. *gladioli* Snyd. & Hans. at different temperatures. *Proc. Florida State Hortic. Soc.* 86:376–378.

Woods, M. W., and R. Bamford. 1937. Chromosome morphology and number in *Tulipa. Am. J. Bot.* 24:175–184.

Wylie, A. P. 1952. The history of the garden narcissi. *Heredity* 6:137–156.

Xie, S., N. Khan, M. S. Ramanna, L. Niu, A. Marasek-Ciolakowska, P. Arens, and J. M. van Tuyl. 2010a. An assessment of chromosomal rearrangements in neopolyploids of *Lilium* hybrids. *Genome* 53:439–446.

Xie, S., M. S. Ramanna, and J. M. van Tuyl. 2010b. Simultaneous identification of three different genomes in *Lilium* hybrids through multicolour GISH. *Acta Hort.* 855:299–303.

Yabuya, T. 1985. Amphidiploids between *Iris laevigata* Fisch. and *I. ensata* Thunb. induced through *in vitro* culture of embryos treated with colchicine. *Japan. J. Breed.* 35:136–144.

Yamagishi, M. 1995. Detection of section-specific random amplified polymorphic DNA (RAPD) markers in *Lilium. Theor. Appl. Genet.* 91:830–835.

Yao, J.-L., D. Cohen, and R. E. Rowland. 1995. Interspecific albino and variegated hybrids in the genus *Zantedeschia. Plant Sci.* 109:199–206.

Zeilinga, A. E., and H. P. Schouten. 1968. Polyploidy in garden tulips. II. The production of tetraploid. *Euphytica* 17:303–310.

Zhou, S. 2007. Intergenomic recombination and introgression breeding in Longiflorum × Asiatic lilies (*Lilium*). PhD thesis, Wageningen University, Wageningen, The Netherlands.

Zhou, S., K. B. Lim, R. Barba-Gonzalez, M. S. Ramanna, and J. M. van Tuyl. 2008c. Interspecific hybridization in lily (*Lilium*): Interploidy crosses involving interspecific F_1 hybrids and their progenies. In *Floriculture, Ornamental and Plant Biotechnology, Vol. V*, ed. J. A. Teixeira da Silva, 152–156. Isleworth, UK: Global Science Books.

Zhou, S., M. S. Ramanna, R. G. F. Visser, and J. M. van Tuyl. 2008a. Genome composition of triploid lily cultivars derived from sexual polyploidization of Longiflorum × Asiatic hybrids (*Lilium*). *Euphytica* 160:207–215.

Zhou, S., M. S. Ramanna, R. G. F. Visser, and J. M. van Tuyl. 2008b. Analysis of the meiosis in the F_1 hybrids of Longiflorum × Asiatic (LA) of lilies (*Lilium*) using genomic *in situ* hybridization. *J. Genet. Genomics* 35:687–695.

7 Biotechnology for the Modification of Horticultural Traits in Geophytes

Kathryn K. Kamo, Frans A. Krens, and Meira Ziv

CONTENTS

I. INTRODUCTION

Plant biotechnology is defined as techniques that use live organisms (bacteria, viruses, fungi, yeast, plant, and animal cells) to make or modify a product, to improve plants, or to engineer plants for specific uses. Modern plant biotechnology holds considerable promise to meet the challenges in agricultural and horticultural production. It encompasses genetic engineering, enzyme and protein engineering, plant tissue culture technology, biosensors for biological monitoring, bioprocessing, and fermentation technology. Biotechnology is an interdisciplinary field of research encompassing biochemistry, molecular biology, microbiology, genetics, immunology, pathology, developmental biology, and so on.

The contribution of plant biotechnology to geophyte production includes clonal propagation (see also Chapter 10), virus elimination, breeding, and crop improvement through somaclonal variation, haploid production, embryo rescue, and *in vitro* fertilization. *In vitro* methods enable to perform breeding and screening programs in a shorter period of time and in a more economical and efficient way. The use of new propagation and breeding techniques has already contributed significantly to the flower bulb industry. Propagation systems through enhanced axillary bud development, organogenesis, and somatic embryogenesis have been reported for even the most recalcitrant species. Liquid culture in bioreactors was found to be an efficient large-scale propagation system for *Lilium*, *Gladiolus*, *Nerine*, *Brodiaea*, *Narcissus*, *Ornithogalum*, *Cyclamen*, and other geophytes.

Conventional breeding techniques using wild species and superior cultivars as the gene source have been used successfully to improve many quality traits in a wide range of ornamental plant species. However, these methods are slow for the needs of the flower bulb industry and limited by the traits that can be introduced by conventional breeding. Genetic engineering offers an opportunity for the improvement of horticultural traits in any plant species, including geophytes. Transformation systems using either *Agrobacterium tumefaciens* or the gene gun have been developed for several ornamental geophytes, including *Lilium*, *Gladiolus*, *Zantedeschia*, *Muscari*, *Hyacinthus*, *Narcissus*, *Ornithogalum*, *Iris*, and *Alstroemeria*. Most transformation studies involving geophytes have used both a reporter gene and a selectable marker gene for the isolation of transgenic plants. However, only a few geophyte species have a transformation efficiency that is high enough to effectively test many genes. In recent years, genetic engineering of *Lilium* spp. with ornamental traits of potential interest for the consumer has been employed (see also Chapter 6). Several studies were also dedicated to the possible use of biotechnology for improving their biotic resistance to viruses, bacteria, and fungi in geophytes.

Knowledge of the elements that constitute the genome and determine the genome organization, and isolation of genes of interest are essential for the improvement of species or cultivars by genetic modification. During the last two decades, this field of knowledge in genomics has advanced tremendously, and the number of plant species with completely sequenced genomes is rapidly increasing. For whole-genome sequencing, preferably a species is chosen that is diploid, homozygous, and has a small-sized genome. However, the list of fully sequenced plant species does not yet contain a representative of ornamental geophytes, since most geophyte species are characterized by a huge genome size along with many large repetitive regions within and among the chromosomes. These large repetitive regions will hamper the establishment of contigs after sequencing and the assembly of the contigs into putative chromosomes. At present, whole-genome sequencing in geophytes is not yet a feasible option because in most species the required high-density maps, as well as full BAC libraries, are lacking. Twenty-three genes and ten promoters have been isolated from geophyte species so far. Transformation systems available for geophytes have enabled the researcher to test for expression of a promoter or gene of interest.

This chapter reviews three major directions: (1) recent achievements in the modification of horticultural traits using *in vitro* technologies, (2) current studies on the transformation of plants for virus, fungal, and bacterial resistance with emphasis on the pathogens that affect geophytes, and (3) an overview of genomics, including expressed sequence tags (ESTs), and its application to alternative strategies for transgenesis.

II. *IN VITRO* BIOTECHNOLOGIES: FROM VIRUS ELIMINATION TO GENETICALLY MODIFIED GEOPHYTES

A. INTRODUCTORY REMARKS

Regeneration of isolated organs, tissues, or cells under *in vitro* conditions is a unique capacity of higher plants. This ability is based on two major traits, plasticity and totipotency, that are the foundation of plant cell culture and regeneration. It is well known that most of the processes involved in plant growth and development are regulated by environmental signals. Plasticity allows plants to divert their growth and adapt to their surroundings, including growth conditions and environmental cues. In plant tissue culture, plasticity allows the initiation of cell division from different types of plant cells, regeneration of new organs, or alteration of the developmental pathways in response to particular stimuli. In response to specific stimuli, plant cells are able to express the total genetic potential of the parent plant—"totipotency" (Hussey 1975; Ziv 1999). In cell culture, totipotency and organ regeneration depend on media manipulations to direct dedifferentiation, redifferentiation, and development of the isolated plant cells. The exposed cells undergo dedifferentiation from mature cells to meristematic cells. Following the formation of meristematic centers and a polar gradient, the cells redifferentiate and develop into new cells, tissues, organs, and entire shoots or somatic embryos directly on the explant, depending on the stimulus and tissue of origin. Dedifferentiating cells can also develop to callus tissue and initiate meristematic centers at some point in time. Following the establishment of new hormonal and nutritional gradients within the callus tissue, internal signals together with external stimuli induce organogenesis and/or somatic embryogenesis depending on the species. Shoots produced can be induced to develop roots, while somatic embryos having shoot and root meristems develop into a complete plant under specific culture conditions (Ziv 1999; Ziv and Chen 2007).

Recently, various new *in vitro* techniques were developed to overcome hand manipulations and reduce the production costs of propagation of elite and selected ornamental species. These include automation, large-scale plant production in bioreactors, the use of growth retardants, acclimatization *in vitro*, and, specifically in geophytes, the induction of storage organ formation for efficient transplanting and establishment *ex vitro* (Ziv 1995a,b, 2005, 2008; Chen and Ziv 2005; De Klerk and Ter Brugge 2010). All these morphogenetic and regeneration pathways *in vitro* are applied for the effective and rapid propagation and production of genetically modified (GM) geophytes.

The general aspects and recent advances in micropropagation of geophytes are reviewed in detail in Chapter 10. In this section, the most important achievements of *in vitro* technologies and their contribution to breeding and improvement of horticultural traits in geophytes are discussed.

B. VIRUS ELIMINATION

In vitro techniques are commonly used for the elimination of viruses and other pathogens from vegetatively propagated crops. Virus-free plants can be obtained by culturing isolated meristem tips 0.2–0.5 mm in size. This procedure has been improved for many species by combining it with thermotherapy and chemotherapy. Virus elimination has been achieved in *Lilium, Iris, Tulipa* (Asjes 1990), *Lilium* (Allen et al. 1980; Takahashi et al. 1992), *Gladiolus* (Aminuddin and Singh 1985; Logan and Zettler 1985), *Iris* (Anderson et al. 1990), and several clones of *Narcissus* by adventitious shoot regeneration (Sochacki and Orlikowska 2005). For detailed reviews, see Kim and De Hertogh (1997) and Lawson (1990).

Antiviral compounds such as virazole, thiouracil, and acyclovir have been used during attempts to eliminate viruses in cultured meristem tips of *Narcissus* (Phillips 1990). Virazole was effective in eliminating *Lily symptomless virus* (LSV) and *Tulip breaking virus* (TBV) from lilies (Blom-Barnhoorn and Van Aartrijk 1985). However, virazole was not effective in meristem culture of *Hyacinthus* (Blom-Barnhoorn et al. 1986). Virus elimination was achieved in *Canna indica* by combining heat treatment with meristem culture (Alfasi 2009). Once a number of virus-tested plants are available, the great potential of micropropagation can be exploited to produce commercial quantities of virus-free stocks.

C. IN VITRO POLLINATION, FERTILIZATION, AND EMBRYO RESCUE

Interspecific fertilization is often interrupted by incomplete embryo development and endosperm or nucellus degeneration, which results in sexual incompatibility. *In vitro* pollination, fertilization, and embryo rescue have been used to overcome pre- and postfertilization barriers in interspecific crosses in geophytes to facilitate the production of hybrid plants (Van Tuyl et al. 1986, 1991; Van Tuyl and De Jeu 1997). Embryo rescue involves the isolation and *in vitro* culture of immature embryos following hybridization. When the isolation of small immature embryos is complicated, the culture of halved ovules can provide initial support for the propagation of developing embryos. In *Alstroemeria*, postfertilization barriers in interspecific hybrids were overcome by culturing the embryos in halved ovules prior to endosperm degeneration (Buitendijk et al. 1995). Later, ovule culture and embryo rescue were used for the production of interspecific hybrids of *Alstroemeria* (Ishikawa et al. 1997; Burchi et al. 2000). Development of triploid daylily (*Hemerocallis*) germplasm by embryo rescue was reported by Li et al. (2009). The effect of medium composition on ovary-slice culture and ovule culture in interspecific *Tulipa gesneriana* crosses was an important factor in increasing the germination percentage of embryos rescued *in vitro* (Van Creij et al. 2000). Embryo rescue was also employed in tulips to produce polyploidy in plants using nitric oxide as an inducing agent (Okazaki 2005). Ovule culture has been reported for the production of *Lilium* hybrids (Van Tuyl et al. 1990, 1991; Yoon 1991) and for *Zephyranthes* (Sachar and Kapoor 1959). In order to overcome sterility of the interspecific *Lilium* hybrids, tetraploid plants were obtained by chromosome doubling with a low concentration of oryzalin, which recovered 40% of the interspecific hybrids (Rhee and Kim 2008).

D. HAPLOID PLANTS

Isolated anthers (the somatic tissue that surrounds and contains the pollen) or microspores are used as explants to produce haploid plants in culture. Haploid tissue cultures can also be initiated *in vitro* from the female gametophyte, the ovaries, or unfertilized ovules (gynogenesis) (Shivanna and Sawhney 1997; Chen et al. 2011).

Callus and embryos can be produced from the anthers or pollen grains on solid or liquid medium. Both methods, anther and microspore culture, have advantages and disadvantages. An advantage of anther culture is that the somatic tissue supports the microspores; however, some of the embryos produced from anther culture may originate from the somatic tissue rather than the haploid microspore cells. If isolated pollen grains are cultured, the production of haploid tissue is guaranteed, but the efficiency of plant regeneration is low and the process time-consuming.

Regeneration from microspore explants can be obtained either by direct embryogenesis or via a callus stage with subsequent embryo development. In microspore culture, the physiological status, juvenile or mature, of the donor plant is of critical importance, as is the timing of microspore isolation. Pretreatments, such as a cold treatment, are often found to increase pollen response (Sato et al. 2002).

Plants regenerated from haploid cultures may not always be haploid as a result of chromosome doubling during the culture period. Chromosome doubling may be an advantage, since in many cases dihaploid plants are the desired regenerated product yielding homozygous diploid plants. These can be used effectively in breeding programs for trait improvement.

Haploid plants were obtained in *Lilium* as early as 1972 (Sharp et al. 1972) and in *Ranunculus* in 1965 (Konar and Nataraja 1965). In *Gladiolus* (Bajaj et al. 1983) and in *Freesia* (Bajaj and Pierik 1974), callus was believed to be haploid when obtained from anthers cultured *in vitro*. Isolated microspores from young flower buds of tulip developed into haploid embryos, depending on the variety (Custers et al. 1997). Unfertilized ovules were used as a source for haploid plants in Oriental lilies (Prakash and Giles 1986). Haploid plants were regenerated from anthers of the Asiatic *Lilium* hybrid 'Connecticut King' (Han et al. 1997, 1999) and *Lilium longiflorum* 'Georgia' (Jang et al. 2002). Both haploid and doubled haploid plants were induced from anther-derived callus of *Lilium formosanum*. The doubled haploids' flowering plants were produced by colchicine-treated haploid callus, but their pollen was sterile (Han and Niimi 2005). In *Lilium longiflorum*, variation was observed in anther-derived plants (Qu et al. 1988). In Chinese narcissus (*Narcissus tazetta* var. *chinensis*), a high percentage of anthers at the early- to mid-uninucleate microspore stages were responsive to balanced growth regulators in the medium, and the callus produced regenerated plants with bulblets (Chen et al. 2005). In summary, anther or microspores in culture are a well-proven system for the production of haploid and dihaploid plants.

E.　Callus Culture and Somaclonal Variation

The isolation and culture of explants *in vitro* frequently induce cell dedifferentiation, unorganized cell proliferation, and eventually the formation of callus. Callus formation is species dependent, and is generally favored by a high auxin-to-cytokinin ratio. Continuous callus culture over several subculture generations has been reported to cause genetic variation and the loss of regeneration potential. Callus cultures have been used for plant regeneration, production of cell suspensions, and isolation of somaclonal variants in geophytes (Kamo et al. 1990; Kim and De Hertogh 1997; Tribulato et al. 1997a). In *Lilium longiflorum* 'Gelria' limited somaclonal variation was observed, and variant regenerants were selected as a source for disease resistance (Van Harmelen et al. 1997). However, most reports indicate that in general the callus culture of geophytes is genetically stable. This may be due to the fact that in many cases the callus is highly organized, forming meristematic centers (Ziv et al. 1994), rather than loose, friable unorganized callus. Nevertheless, in *Hemerocallis* (Chen and Goeden-Kallemeyn 1979), *Freesia* (Stimart and Ascher 1982), and some irises (Laublin et al. 1991), genetic variation in the cell ploidy of regenerated plants was observed.

F.　Variation through Polyploidization and Mutagenesis

In vitro techniques are a useful and efficient system to induce variation in horticultural traits through mutagenesis and polyploidization. Polyploidy and mutagenesis can be used in an attempt to restore quality traits lost during breeding programs and to introduce variation in domesticated species. Polyploid induction in tissue culture was achieved using oryzalin, colchicine, nitrous oxide, and various plant extracts which double, triple, or produce polyploids with various, odd chromosome numbers (aneuploids). In geophytes, flowering and flower quality were investigated in doubled haploids and polyploid plants (Van Tuyl and De Jeu 1997). Oryzalin was used to induce polyploidy in the Asiatic lily (*Lilium* 'Pollyanna') and was more efficient than other mutagens tested (Chandanie et al. 2011). Tetraploid interspecific *Lilium* hybrids treated with oryzalin restored their pollen fertility (Rhee and Kim 2008). *Gladiolus* × *grandiflorus* cormels were irradiated with gamma rays which affected callus formation, somatic embryogenesis, and flower color (Kasumi et al. 2001).

G.　Protoplast Culture and Somatic Hybrids

Protoplasts are wall-free cells commonly isolated from either leaf mesophyll cells or cell suspensions. *In vitro* protoplast culture and fusion has been used for transformation and for obtaining somatic hybrids in many plant species. However, in geophytes, successful protoplast culture and formation of cell colonies is so far limited to a small number of species (Kim and De Hertogh 1997).

Plants were regenerated from protoplasts isolated from *Lilium* (Famelaer et al. 1997; Horita et al. 2002, 2003; Komai et al. 2006), black iris (Shibli and Ajlouni 2000), and *Muscari neglectum* (Karamian and Ranjbar 2011). Successful protoplast fusion has been reported only for *Lilium* (Maeda et al. 1979; Tanaka et al. 1987; Ueda et al. 1990; Mii et al. 1994; Horita et al. 2003; Anderson et al. 2011).

H. TRANSFORMATION AND GENETIC MANIPULATION

Genetic engineering methods have recently been used in several crops to introduce genes conferring pathogen resistance as an effective biocontrol method. However, as in other ornamental plants, in geophytes, the genes are yet to be identified and isolated, and since many of the important floricultural characteristics are under multigenic control, it is difficult to introduce and integrate them in the target organism. The subsequent regeneration *in vitro*, specifically from the altered cells, is often the most difficult step in plant transformation studies.

The first transformation studies for geophytes using *Agrobacterium tumefaciens* aimed to demonstrate that *A. tumefaciens* is capable of infecting geophytes, most of which are monocots thought to be nonhosts and thereby making them more difficult to infect than dicots. Plants were not regenerated from the transformed tissue in these initial studies that used *Chlorophytum* and *Narcissus* (Hooykaas-van Slogteren et al. 1984), *Gladiolus* (Graves and Goldman 1987), *Tulipa* (Wilmink et al. 1992), an Asiatic *Lilium* hybrid 'Harmony' (Langeveld et al. 1995), *Lilium longiflorum* 'Snow Queen' (Tribulato et al. 1997b), and *Cyclamen* (Aida et al. 1999). In later studies, transgenic plants of several geophyte species were obtained using *A. tumefaciens* to deliver a reporter gene and a selectable marker gene. The *uidA* reporter gene that codes for GUS expression was used to transform *Alstroemeria* (Akutsu et al. 2004; Kim et al. 2007) and the *Lilium* Oriental hybrid 'Acapulco' (Hoshi et al. 2004). More recent studies have demonstrated a high efficiency of transformation by using basal plates of the *Lilium* Oriental hybrid 'Star Gazer' (Núñez de Cáceres et al. 2011) and the *Lilium longiflorum* 'Tiepao' (Liu et al. 2011) and bulb scales of *Lilium longiflorum* × *L. formosanum* (Li et al. 2008) rather than callus as used in previous studies. Other factors that improved the transformation of *Lilium* × *formolongi* and *Lilium* 'Acapulco' included using a buffer to control the pH of the cocultivation medium and omitting specific macronutrients from the cocultivation and inoculation media (Ogaki et al. 2008; Azadi et al. 2010a).

The biolistic particle delivery system (gene gun) has been used for the transformation of *Gladiolus*, *Ornithogalum*, and *Alstroemeria* to regenerate plants with the *uidA* reporter gene and the *phosphinothricin acetyltransferase (ppt)* selectable marker gene (Kamo et al. 1995; De Villiers et al. 2000; Lin et al. 2000). Biolistic-mediated transformation was used on callus for the introduction of both the *uidA* reporter gene and either the *nptII* or *ppt* selectable marker gene for *Lilium longiflorum* 'Snow Queen' (Watad et al. 1998; Cohen et al. 2004), *Lilium speciosum* 'Rubrum' (Langeveld et al. 1997), and *Lilium longiflorum* (Irifune et al. 2003). Transgenic plants were successfully regenerated from the callus. Production of transgenic lilies via pollen-mediated transformation has been reported by Van der Leede-Plegt et al. (1997).

Only a few studies have applied the transformation technology to geophytes using genes of interest other than just reporter genes (see next sections). Virus resistance has been targeted for *Ornithogalum* and *Gladiolus* (Lipsky et al. 2002; Cohen et al. 2005; Kamo et al. 2005, 2010; Van Emmenes et al. 2008), aphid resistance for Oriental *Lilium* cultivars (Krens et al. 2009), fungus resistance for *Hyacinthus* (Popowich et al. 2007), bacteria resistance for *Zantedeschia* (Yip et al. 2007), changes in morphology for *Lilium longiflorum* 'Snow Queen' (Mercuri et al. 2003), metabolic engineering of carotene for *Narcissus tazzeta* (Lu et al. 2007), and carotenoids for *Lilium* × *formolongi* (Azadi et al. 2010b). Transient transformation has allowed the characterization of promoter elements in geophytes (Wilmink et al. 1995; Joung and Kamo 2006), and levels of tissue-specific expression for various promoters can be learned from transgenic plants (Kamo and Blowers 1999; Kamo et al. 2000).

III. GENETIC ENGINEERING FOR DISEASE RESISTANCE

A. RESISTANCE TO VIRUS

1. Strategies for Virus Resistance

The strategy of parasite-derived resistance was first proposed by Sanford and Johnston (1985). Their approach suggested that key gene products of the parasite's genome when present in the host either in excess or in a dysfunctional form should disrupt the life of the parasite. Based on this proposal, various viral gene sequences have been used in genetic engineering of plants for virus resistance as reviewed by Prins et al. (2008). Coat protein-mediated resistance uses the translatable coat protein gene isolated from a virus to engineer plants. It is thought that coat protein-mediated protection operates similarly to cross-protection, although the mechanism(s) for coat protein-mediation protection is not clearly resolved. Cross-protection occurs when a plant infected with one viral strain is protected from infection by another related virus. Protection relies on the expression of the protecting virus that is thought to interfere with uncoating of the challenge virus, and this may be the reason why virus resistance often correlates with high levels of coat protein expression. Another form of resistance—replicase-mediated resistance—uses either a mutated or an intact region of the replicase gene that is involved in viral replication to transform the plant (Anderson et al. 1992). Transformation with replicase frequently conferred resistance to virus more effectively than coat protein-mediated resistance; however, the replicase-mediated resistance was specific for the type of virus that the replicase gene had been isolated from. The mechanism of replicase-mediated protection remains unclear. A broad spectrum of virus resistance occurred in plants transformed with a mutated movement protein gene isolated from the virus. Viral movement proteins function to allow viruses to pass through cell walls. It has been suggested that the expression of a movement protein transgene within a plant cell competes with the infecting virus for cellular factors that are required by the infecting virus for its movement and subsequent spread throughout the plant (Prins et al. 2008).

The discovery of RNA silencing or posttranscriptional gene silencing (PGTS) by cosuppression or antisense approaches, followed by RNA interference (RNAi) in which double-stranded RNA (dsRNA) molecules are formed, opened up new possibilities for engineering virus resistance. Research on RNA silencing has led to the application of microRNAs (miRNA) and artificial microRNAs (amiRNA) that target a short, approximately 22-nucleotide-long sequence of the viral pathogen. This results in degradation of the invading virus for achieving virus resistance as reviewed by Lu et al. (2008), Ossowski et al. (2008), Frizzi and Huang (2010), and Zhang et al. (2011).

Virus resistance has been engineered in plants using genes isolated from nonviral sources such as antibody genes (Prins et al. 2008). Antibodies have been made to target specific regions of the virus, and expression of the antibody gene or a portion of the antibody (single-chain variable fragment; scFv) in plants has made the plant resistant to virus.

More recently, studies have demonstrated virus resistance using plant defense genes such as the *Rx* gene that encodes a member of the nucleotide-binding site leucine-rich repeat (NBS-LRR). Resistance using *Rx* is specific to potexviruses, and confers resistance prior to the plant's hypersensitive response (Baures et al. 2008). Transcription factor genes involved in defense signaling have also been tested in transgenic plants for their ability to confer resistance to viruses (Ren et al. 2010; Shi et al. 2010). Unfortunately, the defense pathways and their genes have not been well characterized in ornamental geophytes, but it is likely that many of the genes identified in well-characterized systems such as *Arabidopsis* will be conserved in the geophytes, and much of what has been learned with model plants can be applied to geophytes in the future.

2. Genes for Virus Resistance

Virus resistance was achieved in 1986 when Powell-Abel et al. (1986) used the coat protein gene of *Tobacco mosaic virus* (TMV) for coat protein-mediated resistance of TMV in tobacco plants. Since then, this approach has been used to engineer virus resistance in many plant species. Many initial studies used the translatable viral coat protein gene, and later studies used untranslatable coat protein genes as in sugarcane to confer resistance to the *Sugarcane yellow leaf virus* (ScYLV) (Ingelbrecht et al. 1999; Zhu et al. 2011). A broader spectrum of resistance to several viral groups was achieved by using a dysfunctional movement protein gene to disrupt local and systemic movement of the invading virus throughout the plant (Prins et al. 2008). In 1998, the highest percentage of plants resistant, 90%, was achieved in tobacco and rice using inverted repeats of sequences isolated from the viral genome (Waterhouse et al. 1998). It was thought that the viral sequence of the inverted repeat resulted in sequence-specific RNA degradation of the infecting virus. Chen et al. (2004) demonstrated that it was possible to obtain a high-frequency, 50–100%, resistance to *Cucumber mosaic virus* (CMV) in *Nicotiana benthamiana* using an inverted repeat construct of either the CMV coat protein or RNA 2 sequence isolated from *Lilium*.

Plant defense genes have also resulted in virus resistance. The coat proteins of three potexviruses, *Narcissus mosaic virus*, *White clover mosaic virus*, and *Cymbidium mosaic virus*, were shown to be recognized by *Rx* (Baures et al. 2008). A small, 90 amino acid region was identified in the coat protein of *Narcissus mosaic virus* as an elicitor in the recognition response by *Rx*. At least 12 genes of the NBS-LRR subfamily of resistance genes isolated from *Arabidopsis thaliana* have been found to confer resistance to viruses. *RCY1*, another NBS-LRR gene isolated from *Arabidopsis*, was used to transform *Arabidopsis*, and high levels of *RCY1* expression resulted in a high level of resistance to CMV (Sekine et al. 2008). Ribosome-inactivating proteins (RIPs) are implicated in plant defense, and transgenic tobacco plants expressing type 1 and type 2 RIPs isolated from *Iris* showed local, but not systemic, protection from TMV (Vandenbussche et al. 2004). The mitogen-activated protein kinase gene, *GhMPK*, isolated from *Gossypium hirsutum*, and *WRKY*, a transcription factor involved in defense signaling, have been shown to confer virus resistance in *Nicotiana* spp. (Ren et al. 2010; Shi et al. 2010).

The antibody approach in conferring resistance in plants is now feasible, as was demonstrated in *Nicotiana* spp. giving resistance to *Tomato spotted wilt virus* (TSWV) infection (Prins et al. 2005; Zhang et al. 2008). Other studies confirmed the effectiveness of using antibodies against the *Citrus tristeza virus* and *Potato virus Y* that infect Mexican lime and potato plants, respectively (Gargouri-Bouzid et al. 2006; Cervera et al. 2010).

3.   Transformation of Geophytes for Virus Resistance

Viral infections present one of the major problems in the commercial production of ornamental geophytes, but there has been relatively little research done on generating transgenic geophytes for virus resistance. *Ornithogalum* has been transformed with either the *Ornithogalum mosaic virus* (OMV) coat protein or the OMV replicase gene using biolistics and later *Agrobacterium*-mediated transformation (Cohen et al. 2005; Van Emmenes et al. 2008). A field trial is currently under progress in South Africa to determine if these transgenic *Ornithogalum* plants are resistant to OMV (Dr. Lynne van Emmenes 2010, pers. comm.).

Gladiolus has been transformed with several antiviral genes using biolistics. Transgenic *Gladiolus* plants, containing the *Bean yellow mosaic virus* (BYMV) coat protein gene in either sense or antisense orientation, were shown to have short-term, one-month resistance to the virus. However, long-term resistance throughout a summer outdoor cultivation was not achieved (Kamo et al. 2005). Previously, this same BYMV coat protein gene had been used to transform *N. benthamiana*, and there was a delay in viral symptoms (Hammond and Kamo 1995a). The antisense construct of the BYMV coat protein was found to be more effective than the sense orientation in conferring resistance in *N. benthamiana* (Hammond and Kamo 1995b). One out of 10 transgenic *N. benthamiana* plants transformed with the antisense BYMV coat protein was found to be highly resistant to BYMV as the virus was absent in the leaves of the challenged plant one month following inoculation.

It is unclear as to why this same antisense coat protein gene did not produce long-term resistance in *Gladiolus*, especially as the BYMV coat gene had been cloned from a *Gladiolus* isolate, and expression of the transgene continued during the following summer when plants were grown outdoors for natural virus challenge. This study showed that results will differ between plant species as has been noted by others. One possibility for the difference in virus resistance between *Gladiolus* and *N. benthamiana* is that the *N. benthamiana* plants used for challenge were homozygous T2 plants, whereas the *Gladiolus* plants were heterozygous T0 plants. On the other hand, there were multiple copies of the transgene in *Gladiolus* resulting from gene gun bombardment.

The transgenic *Gladiolus* plants with the BYMV coat protein gene had been challenged using controlled aphid inoculation followed by immunoelectron microscopy for the detection of virus particles. Both methods are very laborious. An easier, more consistent method to inoculate *Gladiolus* was developed using BioRad's hand-held gene gun to shoot purified virus particles into the meristem of cormels growing *in vitro* and then checking for the presence of virus using an ELISA assay (Aebig et al. 2005). One disadvantage of this technique is that cormels require 6–9 months to develop.

Most of the flower bulb crops are susceptible to CMV, which is not surprising, as CMV is known to infect more than 1000 plant species (Chen et al. 2001; Chen 2003). Although CMV has a broad host range, the CMV sequences of *Lilium* isolates from various locations throughout the world are similar (Chen et al. 2001). The CMV isolate HL from an edible *Lilium* in Japan was used to identify the RNA sequences necessary for infection (Yamaguchi et al. 2005). It was determined that the 5′-untranslated region and the region between nucleotides 1546 and 2604 of RNA1 that code for an integral component of the viral replicase were needed for systemic infection.

Both *Lilium* and *Gladiolus* have been transformed for CMV resistance. *L. longiflorum* has been transformed with a replicase gene from a tomato isolate of CMV, and the lilies were not resistant to CMV (Lipsky et al. 2002). Transgenic *Gladiolus* plants were transformed with the CMV coat protein subgroup I gene, the CMV coat protein subgroup II gene, both the CMV subgroup I and II genes, the defective CMV replicase gene, or a combination of both the CMV replicase and CMV coat protein subgroup II genes (Kamo et al. 2010). Transgenic plants were challenged using the hand-held gene gun. Three out of 21 independently transformed plants containing the CMV subgroup II coat protein were resistant to CMV subgroup II, and 3 out of 19 plants with the replicase gene were found to be resistant to the CMV subgroup I. None of the 18 plants with the CMV coat protein subgroup 1 was resistant. Replicase-mediated resistance is reported to be strain specific. Resistance to CMV was apparently possible in *Gladiolus* using the tomato isolate of the CMV replicase gene, probably because CMV isolated from *Gladiolus* has a very high homology (96%) to the tomato isolate. Levels of transgene expression as determined by real-time PCR showed that a high level of expression was important for coat protein-mediated resistance in *Gladiolus* but not for replicase-induced resistance.

A panel of monoclonal antibodies was made into a CMV isolate from *Cymbidium* and is now marketed by Agdia (Hsu et al. 2000). The variable regions of heavy- and light-chain genes were cloned from mouse hybridoma cell lines that produced these monoclonal antibodies to either CMV subgroup I or II (Aebig et al. 2006). An scFv fragment from CMV subgroup II was placed under control of the duplicated CaMV 35S promoter, and an scFv from CMV subgroup I was placed under either the duplicated CaMV 35S promoter or the sugarcane ubiquitin promoter, *Ubi9*. Ninety transgenic *Gladiolus* plants were developed with an scFv to either CMV subgroup I or II, and these lines are currently under analysis for resistance and levels of transgene expression (Kamo et al. unpublished).

These studies on virus resistance in geophytes are only the beginning of what needs to be accomplished for successful virus resistance. Typically, a geophyte species is infected with more than one virus, and the symptoms of infection such as streaking of flower petals and stunted plant growth make the flowers and bulbs unmarketable. An approach that will result in resistance to multiple viruses is important for healthy crops of ornamental geophytes. One possible approach is to engineer for aphid resistance as aphids are a major vector for viruses. Krens et al. (2009) have transformed Oriental *Lilium* hybrids with both a proteinase inhibitor and a gene coding for a monoterpenoid repellent in an effort to provide resistance against aphids. Future studies will determine the effectiveness of this approach.

4. Risk Assessment of Virus-Resistant Plants

Transgenic plants with antiviral genes have been around for at least 15 years, making it possible to conduct both laboratory and field studies to assess the risks associated with growing transgenic plants with antiviral genes. Some concerns about the release of transgenic plants with virus resistance genes are heteroencapsidation (encapsidation of one virus' genome by the coat protein of another), possible recombination (exchange of genetic material between two RNA molecules or between a resident virus and an invading virus during replication), transgene movement by pollen flow, and effect of virus-resistant transgenic crops on nontarget organisms. These concerns have been addressed by Fuchs and Gonsalves (2007) who concluded that extensive safety assessment studies have shown that there is "limited, if any, evidence for heteroencapsidation, recombination, and impact on nontarget organisms beyond background events". Fuchs and Gonsalves recommended that gene flow from virus-resistant transgenic crops to wild relatives remain a priority for risk assessment in the future. They noted that one unpredicted benefit of growing transgenic, virus-resistant papaya in Hawaii was that it enhanced the production of nontransgenic papaya, apparently because the virus titer where nontransgenic papayas were growing decreased when transgenic papayas were in their vicinity.

Potential recombination between two viruses has been addressed in a study by Chung et al. (2007). Hybrid viruses intentionally created using *Potato virus X* and *Tobacco rattle virus* were used to inoculate plants and were found to be unstable. Inserted sequences in the hybrid virus were lost 2–4 weeks after their inoculation into either *N. tabacum* or *N. benthamiana*. A study conducted to assess the potential toxicity and allergenicity in transgenic tomatoes containing the replicase gene from *Tomato leaf curl virus* concluded that genetically engineered tomatoes showed no toxicity in mice, and allergenicity of the transgenic tomatoes was comparable to that of nontransformed ones (Singh et al. 2009). It is anticipated that it will be faster obtaining regulatory approval for ornamental crops as compared with food crops since the ornamental plants are not meant for consumption as edible crops.

B. Resistance to Fungi

1. Strategies for Fungal Resistance

There have been many laboratory and several greenhouse studies on plant species that were transformed for fungal resistance as reviewed by Hammond et al. (2006). A large group of genes that have been tested in transgenic plants are the pathogenesis-related (PR) proteins that include the chitinases, plant glucanases, and thaumatins. PR proteins accumulate following infection by a pathogen and may contribute to immunity within the plant. Plant chitinases are PR proteins found to lyse the hyphal tips of fungi. The chitinases are digestive enzymes that break down glycosidic bonds in chitin, one of the main components of fungal cell walls. Consequently, many studies have employed chitinase genes from both plant and fungal sources to engineer resistance to fungi. Glucanases are PR proteins that hydrolyze β-1,3-glucan, another component of fungal cell walls. Thaumatin is a class V PR protein implicated in plant defense.

Other genes found to be involved in plant defense include stilbene synthase and the nitrogen permease reactivator (*NPR1*) gene. *NPR1* has been shown to play a critical role in activating the systemic acquired resistance (SAR) pathway. An invading pathogen will cause *NPR1* to translocate to the nucleus where it interacts with a family of DNA-binding proteins that result in the expression of the *PR1* (PR) gene (Makandar et al. 2006). Stilbene synthase is found in only a few plant species and is used in genetic engineering to stimulate the production of resveratrol, a phytoalexin involved in plant defense. The WRKY transcription factor functions as a major regulatory protein because it binds to the W-box of several SAR gene promoters (Maleck et al. 2000).

Antimicrobial peptides or defensins are small 45–54 amino acid, cysteine-rich peptides that have been shown to effectively confer resistance to a variety of fungal pathogens, typically by first binding to and then causing permeabilization of the fungus' membrane (Thevissen et al. 2000). Thionins are another group of low-molecular-weight polypeptides known for their toxicity to bacteria, fungi, and both plant and animal cells (Broekaert et al. 1995).

The amount of protection appears to correlate with the level of transgene expression at the protein and RNA level in several studies (Shin et al. 2008; Chen et al. 2009; Portieles et al. 2010). The ribosome inactivating proteins (RIPs) are inhibitors of protein synthesis that can affect either prokaryotic and eukaryotic ribosomes or primarily eukaryotic ribosomes depending on the specific RIP. Tobacco plants transformed with *Curcin 2*, a RIP isolated from *Jatropha curcas* leaves, showed reduced symptoms following challenge with *Rhizoctonia solani* that correlated with the level of *curcin 2* expression (Huang et al. 2008).

2. Genes for Fungal Resistance

Fungi can be a devastating problem for the geophytes that can be rapidly destroyed when infected (see also Chapter 13). The main fungal pathogen that infects many geophytes is *Fusarium oxysporum*. *Botrytis cinerea* infects *Lilium, Tulipa, Crocus, Narcissus*, and *Iris*, while *Botrytis elliptica* and *Botrytis tulipae* infect *Lilium* and *Tulipa*, respectively. *Sphaerotheca pannosa* (powdery mildew) infects *Crocus, Freesia*, and *Dahlia. Rhizoctonia solani* infects *Muscari* and *Alstroemeria. Pythium* is a pathogen of *Alstroemeria* and *Zantedeschia. Phytophthora* infects *Alstroemeria* and *Muscari*. Resistance to the same fungal pathogens was already addressed in plant species other than geophytes, using various transgenes. In the following section, some examples are provided.

There are several reports on successful genetic engineering to *Fusarium oxysporum* using plant-derived chitinases and glucanases, the *NPR1* gene, antimicrobial peptides, and an antibody that recognized the cell walls of *F. oxysporum*. Cotransformation of tomato with a tobacco class I chitinase and a tobacco class I glucanase genes had synergistic effects against *F. oxysporum* (Jongedijk et al. 1995). More recently, tomatoes, transformed with a wheat endochitinase gene, showed resistance to *Fusarium* (Girhepuje and Shinde 2011). Cotton plants expressing *NPR1* had significant resistance to several fungal pathogens, including *F. oxysporum* and *Rhizoctonia solani* (Parkhi et al. 2010). *Arabidopsis* transformed with a thionin showed resistance to *F. oxysporum* (Epple et al. 1997). Defensin isolated from Wasabi (*Wasabia japonica*) conferred resistance to *F. oxysporum* in transgenic melon 'Egusi' (Ntui et al. 2010). Peschen et al. (2004) were the first to demonstrate the possibility of using the antibody approach for fungal resistance. They generated an antibody to antigens on the cell wall of *F. oxysporum* and were able to obtain a high level of protection to *F. oxysporum* in transgenic *Arabidopsis* plants when the antibody gene was combined with an antifungal protein from either *Raphanus sativus* or *Aspergillus giganteus*, or a wheat class I chitinase.

In recent years, a variety of transgenes have been tested and shown to confer resistance to *Botrytis*. Some of these genes such as chitinases (Lorito et al. 1998; Gentile et al. 2007; Distefano et al. 2008), glucanases (Flors et al. 2007), *NPR1* (Wally et al. 2009), and antimicrobial peptides (Bi et al. 1999; Khan et al. 2006) were the same as those used to confer resistance to *Fusarium*. Resistance to *Botrytis cinerea* increased when either a chitinase or glucanase transgene was combined with a second antifungal gene. For example, the combination of *chi-2*, a barley chitinase, and a wheat lipid transfer protein in carrot plants resulted in a 95% reduction in disease to *B. cinerea* whereas the disease reduction was only 40–50% when a single gene was used for transformation (Jayaraj and Punja 2007). Resistance to *B. cinerea* was more effective when two genes, the rice chitinase and alfalfa defensin *alfAFP*, were expressed in transgenic tomato rather than when there was only a single transgene present (Chen et al. 2009).

Other approaches for resistance have also been found to result in resistance to *Botrytis*. A recently identified family of proteins, the cerato-platanin family, has been recognized as an elicitor in inducing a disease resistance response in plants. *MgSM1*, a cerato-platanin family protein isolated from the fungus *Magnaporthe grisea*, conferred resistance in *Arabidopsis* to *Botrytis cinerea* (Yang et al. 2009). *ATAF1* is one of the first transcription factors identified in *Arabidopsis* as a negative regulator, and transgenic *Arabidopsis* with a repressor of *ATAF1* showed enhanced resistance to *B. cinerea* (Wang et al. 2009). These studies indicate the complexity involved with a plant's response to fungal infection. *Botrytis* infection was reduced by 47–69% in transgenic tobacco plants using a *Vitis vinifera* polygalacturonase-inhibiting protein, VvPGIP1, which inhibits the polygalacturonases secreted by the invading fungi.

The cell wall plays an active role in protecting the plant from invading pathogens. Pectin is a major component of the primary cell walls of plants, and pectin methylesterase (PME) is responsible for demethylesterification of pectin when secreted in a methylesterified form. This makes PME an important factor in the defense response of the cell wall to pathogen-infecting enzymes such as endopolygalacturonases that soften the cell wall. Transgenic *Arabidopsis* with the genes *AtPMEI-1* and *AtPMEI-2* showed resistance to *Botrytis cinerea* (Lionetti et al. 2007). Because the cell walls of monocots have a lower pectin content than those of dicots, *AtPMEI-1* and *AtPMEI-2* may not function as effectively in ornamental geophytes as in *Arabidopsis*.

Transgenic tobacco overexpressing the *WRKY* gene isolated from grapevine, *VvWRKY1*, showed resistance to both *Pythium* and powdery mildew (Marchive et al. 2007). A combination of two protease inhibitors, sporamin, which is a major storage protein from sweet potato, and *CeCPI*, a phytocystatin from taro (*Colocasia esculenta*), were used to transform tobacco, and the plants were resistant to *Pythium* (Senthilkumar et al. 2010). An antimicrobial peptide isolated from onion, *Ace-AMP1*, was used to transform roses, and six out of seven rose lines were more resistant to powdery mildew than control plants when tested in the greenhouse (Li et al. 2003).

Several studies, including a field study, have shown that resistance to *Rhizoctonia* can be achieved. Chitinases, glucanases, and antimicrobial proteins have been shown to confer resistance to this pathogen. Tobacco plants expressing three barley transgenes, a class II chitinase, a β-1,3-glucanase, and a type I RIP, had much more protection from *Rhizoctonia solani* than tobacco plants with any one of these genes (Jach et al. 1995). Transgenic rice plants expressing *Dm-AMP1*, a defensin isolated from *Dahlia merckii* seeds, suppressed the growth of *Rhizoctonia solani* (Jha and Chattoo 2010). The first field study to demonstrate resistance to a fungal pathogen, *Rhizoctonia solani*, in the field was conducted by Howie et al. (1994) using transgenic tobacco plants expressing *chiA*, a chitinase gene, isolated from *Serratia marcescens*.

Several genes have been shown to confer resistance to *Phytophthora*. Transgenic tobacco plants expressing either of the class IV PR genes, *Wheatwin1* or *Wheatwin2*, showed resistance to *Phytophthora nicotianae* (Fiocchetti et al. 2008). Resistance to *Phytophthora cactorum* was demonstrated in strawberry and tomato plants transformed with a *rolC* gene (Bettini et al. 2003; Landi et al. 2009). Programmed cell death plays a role in a plant's immunity and resistance to pathogens. Apoptosis is a form of programmed cell death, and the Bax protein is a positive regulator of cell death. Tobacco plants transformed with a mouse-derived pro-apoptotic *bax* gene were more resistant to *Phytophthora parasitica* than nontransformed plants (Suomeng et al. 2008). In a field study, *Nicotiana tabacum* and *Solanum tuberosum* plants transformed with an antimicrobial peptide, *NmDefO₂*, from *Nicotiana megalosiphon*, showed enhanced resistance to *Phytophothora infestans*, the agent of potato late blight disease (Portieles et al. 2010).

3. Transformation of Geophytes for Fungal Resistance

There has been little work done on the transformation of flower bulb crops for fungal resistance. Hyacinth cultivars 'Chine Pink' and 'Edisson' were transformed with a thaumatin II gene using *Agrobacterium*-mediated transformation (Popowich et al. 2007). Five transgenic lines of hyacinth were selected using 200 mg/L kanamycin. Thaumatin II was expressed at 0.06–0.28% of the total soluble protein, and this was visualized on Western blots. Despite the relatively high levels of thaumatin II expression, there were no significant differences observed in disease symptoms when transgenic plants were challenged with *Fusarium culmorum* and *Botrytis cinerea* other than that the fungal disease developed more slowly in the transgenic lines than in the control.

On the basis of the significant results achieved in various plant species and crops, we propose a list of promising genes for genetically engineering ornamental geophytes for fungus resistance (Table 7.1). It should be noted that several groups of the discussed genes are not included in this list. For example,

TABLE 7.1

Potential Antifungal Genes to Consider for Genetic Engineering of Geophytes

Antifungal Genes	Fungal Pathogen	Crop of Demonstrated Resistance	Reference
Wheat *chi194*	*Fusarium*	Tomato	Girhepuje and Shinde (2011)
Tulip chitinase-1	ND	ND	Yamagami et al. (2000)
Gladiolus chitinase	ND	ND	Yamagami et al. (1997)
Arabidopsis thionin	*Fusarium*	*Arabidopsis*	Epple et al. (1997)
Tobacco *chi1* with tobacco class I glucanase	*Fusarium*	Tomato	Jongedijk et al. (1995)
Antibody to *Fusarium* with wheat class I chitinase or other antifungal protein	*Fusarium*	*Arabidopsis*	Peschen et al. (2004)
Wasabi defensin	*Fusarium, Botrytis*	Melon, potato	Ntui et al. (2010); Khan et al. (2006)
Arabidopsis NPR1	*Fusarium*	Cotton	Parkhi et al. (2010)
Dahlia merckii Dm-AMP1	*Botrytis, Rhizoctonia*	Geranium, rice	Bi et al. (1999); Jha and Chattoo (2010)
Frog skin *MsrA2* and temporin A	*Fusarium, Botrytis, Pythium*	Tobacco	Yevtushenko and Misra (2007)
Grapevine *VvPGIP1*	*Botrytis*	Tobacco	Joubert et al. (2006)
Magnaporthe (fungus) *MgSM1*	*Botrytis*	*Arabidopsis*	Yang et al. (2009)
Tomato *Cel1, Cel2*	*Botrytis*	Tomato	Flors et al. (2007)
Arabidopsis ATAF1 repressor	*Botrytis*	*Arabidopsis*	Wang et al. (2009)
Barley *chi2* with wheat *ltp*	*Botrytis*	Carrot	Jayaraj and Punja (2007)
Rice *chi* with alfalfa *alfAFP*	*Botrytis*	Tomato	Chen et al. (2009)
Trichoderma (fungus) *chi42*	*Botrytis*	Lemon	Gentile et al. (2007); Distefano et al. (2008)
Trichoderma (fungus) *ThEn-42*	*Botrytis, Rhizoctonia*	Tobacco, potato	Lorito et al. (1998)
Tobacco *NmDefO2*	*Phytophthora*	Tobacco, potato	Portieles et al. (2010)
Raphanus Rs-AFP2	*Rhizoctonia*	Rice	Jha and Chattoo (2010)
Agrobacterium rolC	*Phytophthora*	Strawberry	Landi et al. (2009)
Mouse *bax*	*Phytophthora*	Tobacco	Suomeng et al. (2008)
Sweet potato sporamin with taro *CeCPI*	*Pythium*	Tobacco	Senthilkumar et al. (2010)
Grapevine *VvWRKY1*	*Pythium*, powdery mildew	Tobacco	Marchive et al. (2007)
Onion Ace-*AMP1*	Powdery mildew	Roses	Li et al. (2003)
Jatropha curcas curcin2	*Rhizoctonia*	Tobacco	Huang et al. (2008)
Barley *chi2*, barley β-1,3-glucanase, barley type I RIP	*Rhizoctonia*	Tobacco	Jach et al. (1995)
Bacteriophage T4 lysozyme	*Rhizoctonia*	Tall fescue	Dong et al. (2008)
Serratia (bacteria) *chiA*	*Rhizoctonia*	Tobacco	Howie et al. (1994)

Note: ND, no data.

genes not included are RIPs because transgenic tobacco plants with the type-2 *Iris* RIP had phenotypic abnormalities (Vandenbussche et al. 2004). Transcription factors and defense pathway genes were also not included because they have not been characterized at all in geophytes.

Chitinase genes have been cloned from *Tulipa*, *Crocus sativus*, and *Gladiolus* (Yamagami et al. 1997, 2000; Castillo et al. 2004). A class I chitinase identified in *C. sativus* was shown to be expressed in roots and corms with significantly lower levels of expression in petals, stigmas, stamens, and callus (Castillo et al. 2004). Although chitinase genes have potential antifungal activity, the antifungal activity of the chitinase genes isolated from geophytes has not been proven.

Following to laboratory research, it is important to demonstrate resistance in the field for several seasons, and preferably in several field locations, as studies have shown that fungus resistance in the greenhouse does not necessarily carry over to the field (Anand et al. 2003). Only six field studies using transgenic plants engineered for fungus resistance demonstrated resistance in the field (Howie et al. 1994; Grison et al. 1996; Gao et al. 2000; Schlaich et al. 2006; Mackintosh et al. 2007; Portieles et al. 2010). Genetic engineering for effective, stable resistance to fungi is difficult to achieve although there has been much research on this topic. Not surprisingly, there are no transgenic plants with antifungal genes that have been commercialized to date.

C. Resistance to Bacteria

1. Strategies for Bacterial Resistance

Some pathogenic bacteria produce an enzyme that inactivates their own toxin, and isolation of this enzyme provides a means of engineering plants expressing resistance to the bacteria. The three sources of genes used in transgenic studies of plant resistance to bacterial pathogens include (1) antimicrobial peptides or proteins, (2) enzymes that detoxify or desensitize plants to bacterial toxins, and (3) plant defense genes (Table 7.2) (Hammond et al. 2006). Antimicrobial peptides such as cecropins,

TABLE 7.2
Genes to Consider for Genetic Engineering of Geophytes for Bacterial Resistance

Genes (Source)	Bacterial Pathogen	Crop of Demonstrated Resistance	Reference
Antimicrobial Proteins			
Cecropins and attacins (giant silk moth)	*Ralstonia*	Tobacco	Jaynes et al. (1987)
Lysozyme (T4 bacteriophage)	*Erwinia carotovora*	Potato	Düring et al. (1993)
Tachyplesin1 (horseshoe crab)	*E. carotovora*	Potato	Allefs et al. (1996)
Lactoferrin (human and bovine sources)	*Ralstonia*	Tobacco	Mitra and Zhang (1994)
Type I thionins (wheat and barley)	*Pseudomonas syringae*	Tobacco	Carmona et al. (1993); Florack et al. (1994)
msr, a synthetic cecropin A (bee venom)	*E. carotovora*	Potato	Osusky et al. (2000)
CaAMP (*Capsicum annuum*)	*P. syringae*	*Arabidopsis*	Lee et al. (2008)
Detoxifying/Desensitizing Enzymes			
Tabtoxin-detoxifying enzyme (*Pseudomonas syringae*)	*P. syringae*	Tobacco	Anzai et al. (1989)
Phaseolotoxin-insensitive enzyme (*P. syringae* pv. *phaseolicola*)	*P. syringae*	Bean	De la Fuente-Martínez et al. (1992)
albD (*Xanthomonas albilineans*)	*Xanthomonas*	Sugarcane	Zhang et al. (1999)
Defense Genes—R Genes			
Pto and *Prf* (tomato)	*P. syringae*	Tomato	Ronald et al. (1992); Rommens et al. (1995); Oldroyd and Staskawicz (1998)
Glucose oxidase (*Aspergillus niger*)	*E. carotovora*	Potato	Wu et al. (1995)
CALTPII (*Capsicum annuum*)	*P. syringae*	Tobacco	Jung et al. (2003)

attacins, tachyplesin, and *CaAMP* act by permeabilizing the inner and outer membranes of bacteria while the enzymatic activity of lysozyme hydrolyzes the peptidoglycan cell wall of bacteria.

2.   Genes for Bacterial Resistance

Several antimicrobial genes have been tested for their effectiveness against *Pectobacterium caroto-vorum* (formerly *Erwinia carotovora*), the bacterial pathogen of both *Ornithogalum* and *Zantedeschia*. Transgenic potatoes expressing the T4 bacteriophage lysozyme, tachyplesin, or synthetic cecropin A genes showed more resistance than nontransformed plants (Düring et al. 1993; Allefs et al. 1996; Osusky et al. 2000). *CaAMP*, isolated from *Capsicum annuum*, is an antimicrobial gene that showed broad-spectrum resistance in transgenic *Arabidopsis* plants to *Pseudomonas syringae*, *Hyaloperonospora parasitica*, *Fusarium oxysporum*, and *Alternaria brassicicola*, demonstrating the usefulness of using antimicrobial genes for engineering both bacterial and fungal resistance (Lee et al. 2008).

Complete resistance to specific bacteria species has been achieved using enzymes that detoxify the toxins produced by bacteria. Tobacco plants engineered to express an enzyme that detoxified tabtoxin were resistant to *Pseudomonas syringae* pv. *tabaci*, and transgenic bean plants were made resistant to *Pseudomonas syringae* pv. *phaseolicola* by expressing an enzyme that confers insensitivity to phaseolotoxin (Anzai et al. 1989). The albicidin-detoxifying gene *albD* was used to transform sugarcane plants that became resistant to *Xanthomonas albilineans* (Zhang et al. 1999).

The defense gene, glucose oxidase, was used to transform potatoes, and strong resistance to *Pectobacterium carotovorum* was demonstrated (Wu et al. 1995). Activation of plant defense genes requires a lipid signal, and tobacco plants transformed with *CALTPII* were resistant to *Phytophthora nicotinae* and *Pseudomonas syringae* (Sarowar et al. 2009).

3.   Transgenic Resistance to Bacteria in Geophytes

Zantedeschia are plagued by soft rot caused by *Pectobacterium carotovorum*. Yip et al. (2007) were the first to transform *Zantedeschia elliottiana* using *Agrobacterium*-mediated transformation of the shoot's basal meristem. Transgenic *Zantedeschia* expressed a plant ferredoxin-like protein (PFLP) isolated from sweet peppers that was shown to result in resistance to *Pectobacterium*. Nontransformed leaves of calla lily were completely rotten 7 days after inoculation while transgenic plant leaves had a hypersensitive reaction localized at the site of infection. *Ornithogalum* is also affected by *P. carotovorum*, and transgenic *Ornithogalum* plants containing the antimicrobial *tachyplesin1* gene showed varying levels of resistance to a highly virulent isolate of *P. carotovorum* from calla lily (Cohen et al. 2011).

Geophytes may contribute as a source of antimicrobial genes in the future. Potent antibacterial activity has been demonstrated *in vitro* for tuliposides, the major secondary metabolites found in species of the Liliaceae and Alstroemeriaceae. A water extract of tulip anthers showed antimicrobial activity for *E. coli*, *Salmonella enteritidis*, several *Pseudomonas* species, and *Bacillus subtilis*, and the active compound was identified as 6-tuliposide B (Shoji et al. 2005). It would be useful to test these antimicrobial genes for antifungal activity because other antimicrobials, such as the defensin gene *Cm-AMP-1* isolated from *Dahlia merckii* seeds, were shown to confer resistance to *Botrytis* and *Rhizoctonia* (Bi et al. 1999; Jha and Chattoo 2010).

In conclusion, there are quite a few genes identified for engineering virus, fungal, and bacterial resistance in geophytes. It appears that virus and bacterial resistance can be achieved in geophytes, but fungal resistance remains difficult to achieve in all plant species.

IV.   GENOMICS

A.   Introductory Remarks

Phenotypical, i.e., morphological, physiological, and biochemical characteristics are determined by the expression of the approximately 30,000 genes that are generally present within living organisms.

The remainder of the DNA can be noncoding and structural or consist of retrotransposons or code for microRNAs. Such sequences can be very repetitive. Genomic research is focused on the expressed parts of the genome, in particular the genes, through the generation of ESTs or through next-generation sequencing of cDNA samples prepared from isolated RNA. Genome walking can be used to identify or acquire full coding regions and regulatory sequences such as promoters and terminators. In this chapter, we will summarize what is known at present about the genome organization, EST libraries, promoters, and genes in geophytes. Most research is on gene function analysis, and the best-studied genes are engaged in flower development, metabolism, and pathogen resistance. Genes and promoters isolated from geophytes were predominantly characterized ectopically in other, mostly dicotyledonous model crops. Only a limited number have been subsequently applied to the genetic modification of ornamental geophytes.

B.   GENOME ORGANIZATION

The genome size of geophytes can vary considerably among families, genera, and species, but generally they are very large. In Table 7.3, the genome sizes of some reference species are presented of which the smaller ones have been fully sequenced. They can be compared to the average sizes of some representative species belonging to the class of ornamental geophytes. The numbers are given in megabasepairs for the haploid genome. The champion for plants is *Paris japonica*, a member of the Melanthiaceae family, order Liliales. With its 150,000 Mbp, it is approximately 1000 times larger than the model crop *Arabidopsis thaliana* with a genome size of 125 Mbp.

Large variation among species within a genus can occur as demonstrated by the genus *Ornithogalum. Ornithogalum maculatum* has a genome size of 4000 Mbp, while *Ornithogalum narbonense* reaches 32,781 Mbp (Zonneveld et al. 2005). This applies to other genera presented in Table 7.3 (data not shown). The sources of such large differences in genome size among organisms have been discussed extensively in the literature (for reviews, see King 2002; Carels 2005; Vitte and Panaud 2005; Grover et al. 2008). A few factors engaged in genome organization leading to an increase in genome size are sequence duplications, genome duplication, repetitive DNA, retrotransposons, and

---

TABLE 7.3
Genome Sizes of Geophyte Species and Reference Crops

Geophyte Species	Genome Size in Mbp	Reference Species	Genome Size in Mbp
Gladiolus	1100	*Arabidopsis*[a]	125
Freesia	1300	Peach[a]	230
Zantedeschia	2200	Rice[a]	430
Muscari	4200	Grape[a]	475
Crocus	10,000	Poplar[a]	520
Iris	12,000	Apple[a]	750
Hippeastrum	14,000	Maize	2500
Tulipa	24,000	Barley	5300
Ornithogalum	4000–32,000[b]	Wheat	16,000
Alstroemeria	30,000		
Scilla	36,000		
Hyacinthus	40,000		
Lilium	40,000		
Fritillaria	130,000		
Paris japonica	150,000		

[a] These crops have been sequenced.
[b] This genus was chosen to demonstrate the differences that can occur among species within a genus.

microRNAs. Genes can become duplicated in evolution, leading to gene families with pseudogenes and paralogs. Chromosome segments or regions can be duplicated as was studied in rice linkage blocks (Devos and Gale 2000), and ultimately the entire genome can be duplicated, leading to polyploidy. In many geophyte species, diploids, triploids, and tetraploids occur in the range of cultivars or hybrids, either bred or obtained spontaneously. Within genes, the introns, sequences that are not translated into protein, can also show variation in size and number, for example, by the presence or absence of (retro)transposons. The major part of the so-called nongenic DNA in large genomes is constituted by repetitive sequences interspersing genes or regions with gene clusters. These repetitive sequences, for example, transposons, tandem repeats, simple sequence repeats (SSRs), and long terminal repeats (LTRs) can be species specific. In *Crocus vernus*, repetitive sequences were studied by Frello and Heslop-Harrison (2000) and Frello et al. (2004). Because of their ability to move around and to make copies of themselves, (retro)transposons are considered to be primarily responsible for variation and increase in genome size (Vitte and Panaud 2005; Zou et al. 2009). Evidence for the existence of retrotransposons was found in *Alstroemeria* (Kuipers et al. 1998) and *Lilium* (Leeton and Smyth 1993), where a non-LTR retrotransposon called del2 accounted for 4% of the genome being present in 250,000 copies.

MicroRNAs (miRNAs) have recently been identified as approximately 21–24-nucleotides-long RNA molecules that play a significant role in regulating gene expression (Fahlgren et al. 2010; Frazier et al. 2010). Their influence extends to processes involved in development, differentiation, stress response, and signal transduction, and their presence is not limited to plants. miRNAs can be highly conserved as well as species specific, and hundreds of small RNA molecules have been cloned and identified in plants, some of which can be grouped into miRNA families (Jones-Rhoades et al. 2006; Li and Mao 2007). The DNA that encodes these small RNAs is believed to be generated by duplications of gene parts or chromosome segments and can account for part of the genome's functions and of its size. In tobacco, 259 conserved miRNAs have been identified belonging to 65 families (Frazier et al. 2010). In bulbous ornamentals, no reports have been published on the occurrence of miRNAs and we do not know as yet the extent to which they contribute to the large genome size of these crops.

C. ESTs

Since whole-genome sequencing and contig assembly using BAC[*] clones are not feasible due to the highly repetitive nature of the geophyte genome, genomics has focused on the expressed part of the genomes. For this, RNA is isolated and converted into cDNA by reverse transcriptase. cDNA libraries can be used in the large-scale sequencing of stretches of 100–500 bps, resulting in ESTs. The sequence information from ESTs can be used to identify gene functions from homology-based searches in databases or they can be used to develop molecular markers based on them, usually EST-SSRs (simple sequence repeats). When the RNA is isolated from particular plant organs or from specific stages in development, the gene functions to be identified or the markers to be developed can also be tissue or developmental stage specific. From *Lilium*, anther- and tepal-specific ESTs have been isolated from 2000 cDNA clones and their putative function was deduced from homology searches using BLAST programs (Yu et al. 2003). A similar approach was followed by Okada et al. (2006) for generative cells of *Lilium longiflorum*, checking 886 ESTs. In *Iris*, 6530 ESTs were produced from leaf and root cDNA to develop EST-SSR molecular markers (Tang et al. 2009). The markers were used to investigate genetic diversity among ecotypes and cultivars. *Crocus sativus* (saffron) ESTs have been produced for gene function analysis (6603 ESTs from stigmas; D'Agostino et al. 2007) as well as for marker development (6803 ESTs; Fluch et al. 2010).

---

[*] BAC, a bacterial artificial chromosome, is a DNA construct based on a functional fertility plasmid (F-plasmid), used for transforming and cloning in bacteria, usually *E. coli*. BACs are often used to sequence the genome of organisms.

D. PROMOTERS

Promoters are the 5′-upstream DNA sequences which determine the level, time, and place of expression of the genes they are linked to. CAAT and TATA boxes exist, but other domains or elements involved in stimulating or hampering the transcription of the coding region into messenger RNA are also present. Such domains can serve as docking sites for transcription factors that, once bound, can influence the expression of even multiple genes simultaneously. The study of promoters is aimed at elucidating their ability to stimulate expression and at identifying domains interacting with transcription factors. The primary goal is to understand specific processes; a secondary goal could be to combine a well-performing promoter exhibiting a specific expression profile with a particular gene in order to change or redirect the gene's original performance by genetic modification, thus complying with the desires of the molecular plant breeder for the trait at hand.

Table 7.4 provides a survey of promoters isolated from geophytes and their characteristics. Most promoters studied are engaged in flower or flower organ development, in male sterility, and in

TABLE 7.4
Geophyte Promoter Studies

Plant Species	Gene Promoter	Characteristics	Research Goal	Reference
Lilium longiflorum	Meiotin-1	Anther specific, flower development	Tissue specificity, domain identification	Hasenkampf et al. (2000)
Lilium longiflorum	LIM (lily induced at meiosis)	Anther specific, flower development	Transposition induction in rice	Morita et al. (2003)
Lilium longiflorum	Histone gcH3	Sperm cell specific, flower development	Tissue specificity, domain identification in lily[a]	Okada et al. (2005)
Hyacinthus orientalis	α-Amylase	Shoot specific, metabolism	Cold induction, domain identification	Sato et al. (2006)
Lilium	Chalcone synthase	Flower specific, flavonoid biosynthesis	Tissue specificity, in *Arabidopsis*	Chang et al. (2009)
Crocus sativus	Floral homeotic genes	Flower development	Technique development, domain identification	Tsaftaris et al. (2010)
Crocus sativus	Lycopene-β-cyclase	Flower specific, carotenoid biosynthesis	Paralogue divergence	Ahrazem et al. (2010a)
Crocus sativus	Carotenoid cleavage dioxygenase	Flower specific, carotenoid metabolism	Tissue specificity, domain identification, in *Arabidopsis*	Ahrazem et al. (2010b)
Gladiolus 'Jenny Lee'	Ubiquitin	Protein catabolism	High expression levels in monocots	Joung and Kamo (2006)
Gladiolus × *grandiflorus*	Expansin	Organ expansion	Development stage specificity, in *Arabidopsis* and gladiolus[a]	Azeez et al. (2010)

[a] Examples of functional analysis in the donor geophyte species itself (partly).

carbohydrate, flavonoid, or carotenoid synthesis or metabolism. Specific functions or expression profiles are usually tested ectopically in nongeophyte species, for example, rice or *Arabidopsis*.

E. Genes

Genes code for proteins or enzymes that eventually lead to specific phenotypes at the morphological, physiological, and biochemical levels. When breeders want to improve traits in a crop, it is a prerequisite to know the genetic background of the trait, its inheritance, monogenic or polygenic nature, linkage between traits, availability of associated markers, characteristics of the genes themselves, and allelic variations. In general, to understand a trait, one needs to understand the genes involved and their respective roles, and the process of studying genes begins by unraveling their DNA sequence. A major topic within genomics is to identify and isolate genes and to annotate their functions. Despite the fact that no geophyte has been completely sequenced, individual genes have been isolated and characterized from bulbous ornamental crop species. Isolation was feasible using degenerate primers in PCR reactions based on sequence homology with genes from other crops.

Table 7.5 provides a survey of genes isolated from geophyte species. The main goal was gene function analysis, and the primary targets of investigation were reproductive development, flower color modifications, and metabolism. A few examples are aimed at pathogen resistance or oxidative stress. Many articles can be found in the literature on *Galanthus nivalis* agglutinin (GNA). The first paper describing the isolation and cloning of the gene was published in 1991 (Van Damme et al. 1991), and a few years later the negative effect on population size and growth of sap-sucking insects of transgenic plants carrying the *GNA* gene was reported (Hilder et al. 1995). Lectins showing substantial homology to the GNA from other geophytes, such as *Narcissus* and *Hippeastrum*, have recently been characterized for their mannose-binding capacity (Liu et al. 2009). A state-of-the-art report on the anti-insect potential of plant lectins and more specifically of GNA is given by Michiels et al. (2010). The *GNA* gene is an example of a gene isolated from a geophyte species that has been widely applied to other species after genetic modification in order to introduce insect resistance.

Other genes with a potentially large application range are the carotenoid biosynthesis genes phytoene synthase (*PSY*) and lycopene-β-cyclase (β-*LCY*) isolated from *Narcissus pseudonarcissus* and introduced into rice in order to enhance the nutritional value of that crop by increasing provitamin A levels (Burkhardt et al. 1997; Beyer et al. 2002). The so-called Golden Rice developed in this way was meant to be an example of consumer-oriented genetic modification, alleviating vitamin A (retinol) deficiency in the diets of people living in the third world. For this, holders of patents on elements within the required protocol to produce the Golden Rice waived their rights on licenses and royalties (Potrykus 2003, 2005).

Due to the ornamental importance of geophytes, it will not come as a surprise that most of the genes studied and isolated are involved in reproductive traits, flower organ development, or pollen development. A relatively low number of genes responsible for color, one of the most important traits in flowers, have been the target for research and these from only a very limited number of species. Emphasis has been on flavonoid biosynthesis and more specifically on anthocyanin production and stabilization. The dihydroflavonol 4-reductase (*DFR*) gene from *Iris* was used ectopically in rose to alter flower color in that crop (Katsumoto et al. 2007). Flower development received the most attention with the MADS box genes being the most prominent class of genes that were studied. The primary goal was to understand the mechanisms behind flower development in bulbous monocots, referring to the general homeotic flower development ABC model (Coen and Meyerowitz 1991) or to modified or extended ABCDE models (Van Tunen et al. 1993; Theißen 2001; Benedito et al. 2004a). Potentially, all genes could be used to alter flower color or morphology (e.g., into doubled-flowers) in geophytes, provided a sufficiently effective transformation protocol has been established for the target crop or cultivar. The use of genes or promoters from the geophyte species itself or of DNA sequences based on known geophyte gene sequences, for example, in RNAi

TABLE 7.5

Gene Characterization in Geophyte Species

Plant Species	Gene (Class)	Characteristics	Research Goal	Reference
Galanthus nivalis	Snowdrop lectin	Endosperm storage protein	Resistance against insects	Van Damme et al. (1991); Hilder et al. (1995)
Tulipa bakeri	*Chitinase-1*	Defense	Resistance against fungi, cloning	Yamagami et al. (2000)
Iris × hollandica	Ribosome-inactivating protein	Defense, RNA *N*-glycosidase	Resistance against viruses	Desmyter et al. (2003); Vandenbussche et al. (2004)
Crocus vernus	Cv agglutinin	Endosperm storage protein	Domain identification in carbohydrate binding	Van Damme et al. (2000)
Narcissus pseudonarcissus	Phytoene synthase, *PSY*	β-Carotene synthesis	Nutritional value, rice	Burkhardt et al. (1997)
Narcissus pseudonarcissus	Lycopene-β-cyclase, *LCY*	β-Carotene synthesis	Nutritional value, rice, together with *PSY*	Beyer et al. (2002)
Crocus sativus	Zeaxanthin 7,8-cleavage dioxygenase	Carotenoid biosynthesis	Nutritional value, tomato	Qiu et al. (2007)
Crocus sativus	*LCY*	β-Carotene synthesis	Genome organization, tissue-specific expression	Ahrazem et al. (2010a)
Narcissus tazetta	*PSY* antisense	β-Carotene synthesis	Flower color modification[a]	Lu et al. (2007)
Muscari armeniacum	Flavonoid 3′,5′-hydroxylase, F3′5′H	Flavonoid biosynthesis	Flower color modification, expression studies	Mori et al. (2005)
Iris × hollandica	Dihydroflavonol 4-reductase, *DFR*	Flavonoid biosynthesis	Flower color modification, ectopically	Katsumoto et al. (2007)
Iris × hollandica	Anthocyanin acyltransferase	Anthocyanin acetylation	Flower color modification	Yoshihara et al. (2006)
Iris × hollandica	Anthocyanin 3-*O*- and 5-*O*-glucosyltransferase	Anthocyanin glycosylation	Flower color modification	Imayama et al. (2004); Yoshihara et al. (2005)
Tulipa gesneriana	Vacuolar iron transporter, *VIT*	Fe translocation	Flower color modification	Momonoi et al. (2009)
Tulipa gesneriana	Ferritin	Fe storage	Flower color modification	Shoji et al. (2010)
Hyacinthus orientalis	α-Amylase	Shoot specific, metabolism	Cold induction, domain identification	Sato et al. (2006)
Lily Oriental hybrid	Cu/Zn superoxide dismutase, SOD	Reactive oxygen species	Oxidative stress, ethylene biosynthesis, potato	Kim et al. (2008)

Species	Gene	Process	Application	References
Lilium spp., for example, *longiflorum*, ×*formolongi*	Homeotic MADS box genes	Flower development	Mechanism study, gene function, ectopically	Tzeng and Yang (2001); Tzeng et al. (2002, 2003); Benedito et al. (2004b); Zhang et al. (2004); Chen et al. (2008); Akita et al. (2008); Thiruvengadam and Yang (2009); Hsu et al. (2010)
Narcissus tazetta	Homeotic MADS box genes	Flower development	Mechanism study, gene function, ectopically	Wang et al. (2006); Gao et al. (2008, 2009)
Crocus	Homeotic MADS box genes	Flower development	Mechanism study, gene function	Tsaftaris et al. (2004, 2005, 2006); Kalivas et al. (2007)
Lilium longiflorum	Anther-specific genes	Flower organ development	Meiosis, cell cycle, sterility	Wang and Poovaiah (1999); Morohashi et al. (2003); Yang et al. (2008); Tzeng et al. (2009a,b)
Alstroemeria aurea	Anther specific gene	Flower organ development	Male gametogenesis	Igawa et al. (2009)
Gladiolus × *grandiflorus*	Expansin	Organ expansion	Development stage specificity, in *Arabidopsis* and gladiolus[a]	Azeez et al. (2010)

[a] Examples of functional analysis in the donor geophyte species itself (partly).

approaches, opens the way to alternative genetic modification methods, such as cisgenesis or intra-genesis (Schouten et al. 2006; Rommens et al. 2007).

F. ALTERNATIVE METHODS IN GENETIC MODIFICATION

Although genetically modified (GM) crops are grown on a large area worldwide, the number of cul-tivated GM species is still rather low and limited to fodder or processed crops (James 2010). The only GM crop that is directly eaten by consumers is transgenic papaya that is marketed mainly in the United States. In Europe, the reluctance to accept GM crops presents a barrier to large-scale cultiva-tion or applications of multiple crops or cultivars. Some of the concerns of consumers dealing with possible toxicity or allergenicity issues could be considered irrelevant in ornamental crops, since ornamentals are not generally eaten. In fact, of the eight GM crops authorized for marketing (but not for cultivation) within the EU, two represent ornamentals—the carnation cultivars with altered flower color. Of the other five crops, still pending authorization, three are carnations. All GM carnations were selected for in the gene transfer protocol by herbicide (chlorsulfuron) resistance based on the acetolactate synthase gene ($SuRB$) from tobacco (see the website of the European Commission, Joint Research Centre; http://gmoinfo.jrc.ec.europa.eu/gnc_browse.aspx). No antibiotic resistance genes were used, nor are they present in the final products, which is another of the possible objections that consumers might have. In several studies, for example, those of Lusk and Sullivan (2002), consumers have expressed their preference for the use in GM foods or products of genes coming from the species itself rather than from other plants, let alone other organisms such as bacteria or viruses. This limits the use of other selection systems, as most of the genes involved in alternative selection systems are derived from bacteria, for example, the phosphomannose isomerase system (Joersbo et al. 1998) or d-amino acid metabolism genes (Erikson et al. 2004). Avoiding the use of any selection system what-soever requires a very efficient transformation and regeneration protocol (De Vetten et al. 2003; Krens et al. 2004), which is lacking in geophytes. Cotransformation of two T-DNAs, one with the gene-of-interest and one with the selectable marker gene(s), followed by subsequent segregation after sexual crossings, is not feasible in crops with long generation times or vegetatively propagated crops. Many geophytes fall in those categories. An alternative is offered by systems based on marker exci-sion by recombination to generate marker-free GM plants (Schaart et al. 2004; Krens et al. 2004; for a review, see Schaart et al. 2011). Here, selectable marker genes are first introduced and used to select for GM plants; then, when they are no longer required, the selection genes are removed after inducing recombinase activity. Schaart et al. (2004) presented a system allowing selection for successful exci-sion events using a dual positive/negative selection system.

With such a system, marker-free GM plants can be produced carrying only the genes of interest as functional units, transcribed into gene products. In case the genes of interest are derived from the recipient plant species itself or from sexually compatible relatives and have their original configura-tion, that is, promoter, introns, exons, terminator, and sense orientation, this particular form of GM is called cisgenesis (Schouten et al. 2006). When new combinations of promoters and coding regions or cDNAs, or antisense or RNAi sequences are used, but all from the species itself, this is called intragenesis (Rommens et al. 2007). From the Eurobarometer 2010 (Gaskell et al. 2010), which presents the outcome of questionnaires on the views of Europeans on technological developments, it was clear that 55% of the people interviewed perceived cisgenic apples as positive, 22% more than those looking favorably on transgenic apples.

Cisgenesis or intragenesis in geophytes requires the availability of genes, promoters and other regulatory elements, and gene sequence information from the bulbous ornamental crop species themselves and the availability of transformation protocols. At present, modification of the flower morphology in lily and narcissus through cis- or intragenesis is feasible. Figure 7.1 shows some spontaneous geophyte flower morphology mutants, of which one even made it to a cultivar, the lily Oriental hybrid 'Miss Lucy'. Double-flowered cultivars such as this could be "easily" produced by genetic modification. Flower color modification might also be close to realization because it should

FIGURE 7.1 **(See color inert.)** Double-flowered mutants. Left: *Lilium* Oriental hybrid, 'Miss Lucy'; right: spontaneous mutant in *Hippeastrum*.

be possible to isolate the desired genes involved in the flavonoid biosynthetic pathway and to use them directly, in combination with other promoters, or use the sequence information to knock them out by means of RNAi approaches. On the other hand, specific enzymes from the pathway might differ in substrate specificity from one species to another, necessitating the knockout of resident genes and the introduction of a homolog from another species (Katsumoto et al. 2007). This by definition would not be considered as cis- or intragenic.

In conclusion, what can be said about the perspectives for bringing GM ornamental geophytes to the European market? A precedent for an ornamental crop exists. GM carnations are authorized for import and processing in Europe, but not for cultivation, and those carnations are not marker-free but carry a selectable marker gene. However, in this authorized case, the selectable marker gene was not an antibiotic resistance gene, but a herbicide resistance gene and not derived from bacteria but from a plant, *Nicotiana tabacum*. Marker-free or cisgenic is not an absolute requirement for GM crops, but it could help reduce the time or costs involved in getting authorization for market introduction or even for cultivation, at least within Europe. A clear trait beneficial or appealing to the consumer, such as color, flower morphology, or environmentally friendly cultivation will also help acceptance by consumers. They seem to prefer GMs with only genes of interest and no extra genes, so there might be a future for marker-free systems. Because there are so many ornamental crops and different cultivars, and small breeders around the world, it all boils down to the costs associated with the process of bringing ornamentals to the market and the revenues that can be expected from the improved cultivars. General public acceptance is a prerequisite. The creation of catchy and attractive GM crops, which will serve as examples and impress the consumers, will probably pave the way for the next generation of ornamental products.

V. CONCLUSIONS AND FUTURE PROSPECTS

It is generally accepted that transgenic plants may contribute to solving some problems of modern agriculture, as they can contribute to sustainable production of crops, pathogen resistance, and improved biological and production traits.

Currently, there are many difficulties when attempting to bring GM products to the marketplace, particularly in the European Union. The general public needs to be better educated as to exactly what a GM is and its advantages and disadvantages. In response to public concern about introducing "foreign genes", scientists are now using cisgenesis, intragenesis, and selected excision of unnecessary genes into plants. This response to the public concern about GMs is critical because biotech research and development is costly. As funding for research becomes increasingly limited, there is an increasing demand to justify the research funded, and ultimately the industry and public will determine whether or not funding is to be allocated to plant biotech research.

History has shown that with horticultural crops, plants engineered for disease resistance are not likely to be considered for commercialization unless the disease is of such economic importance that a transgenic plant is one of the few options remaining for solving the crisis. This occurred when the papaya industry in Hawaii was in a rapid decline due to *Papaya ringspot virus*, and transgenic papayas with antiviral genes were successfully developed and introduced in Hawaii to save the papaya industry. Another recent example is the development of virus-resistant transgenic plums that were developed by the U.S. Department of Agriculture in response to the *Plum pox virus* outbreak in Europe. These transgenic plums have been approved and can be introduced into the marketplace anytime, but there are no plans for their commercialization unless *Plum pox virus* spreads to the United States and becomes a threat to the U.S. plum industry.

Modern biotechnological approaches, utilized in various plant species, have also been applied to geophytes, as summarized in this chapter. The main problem in developing transgenic geophytes is the lack of high-frequency transformation systems and the inability to efficiently integrate a gene into a specific region of the genome. Ornamental geophytes have some of the largest genomes, making it difficult to study and therefore to conduct this type of research. Also, new cultivars of flowers are continually being bred and introduced, and this makes it difficult to apply genetic engineering and have an impact on a single genus of bulb crop. These problems can undoubtedly be solved, but it is questionable as to whether adequate funding will be given for geophytes since in comparison with large edible and forage crops, they are niche crops with relatively small production value. Nonetheless, geophytes have unique characteristics and genes in the plant kingdom. Biotechnological research on geophytes broadens our understanding of plant development, genomics, evolution, physiology, and biochemistry.

Although transgenic flower bulb crops will probably not help solve global problems of the future, the flower bulb crops together with other ornamentals contribute significantly to enhancing our well-being around the world and in many cultures. In addition, they are an important source of revenue for many small growers and some countries in particular.

CHAPTER CONTRIBUTORS

Kathryn K. Kamo is a research plant physiologist at the U.S. Department of Agriculture. She was born and brought up in the United States, received her MS and PhD from Indiana University, and has held a postdoctoral position at Purdue University. Her area of expertise includes transformation technology and virus resistance of ornamental crops. Dr. Kamo is the author of numerous publications on *Gladiolus*, *Lilium*, *Rosa hybrida*, *Zea mays*, and *Papaver somniferum*. She is a USDA Biotech Patent Committee member, *Plant Cell Reports* editor, and a member of the *Plant Science* editorial board.

Frans A. Krens is a senior scientist at Wageningen University and Research Centre (Wageningen UR) Plant Breeding. He was brought up in the western part of The Netherlands, close to the North Sea, and the bulb cultivation areas. His BSc, MSc, and PhD are from Leiden University, The Netherlands. He held a postdoctoral position at the University of Leiden, and research positions in Wageningen UR. Since 2005, he is a research leader of the Ornamentals, Tissue Culture and Gene Transfer Group of Wageningen UR Plant Breeding. His expertise includes the transformation of flower bulbs, gene function analysis, disease, and (insect) resistance. Dr. Krens has published more than 100 refereed journal articles, book chapters, and conference papers.

Meira Ziv is a professor emeritus of the R.H. Smith Institute of Plant Sciences and Genetics, The Hebrew University of Jerusalem, Israel. She was born in Israel and received her BS from the University of California, Davis, her MS from the University of Minnesota, USA, and her PhD from The Hebrew University of Jerusalem, Israel. She has held academic positions at the University of Queensland, Sydney University, Australia; The Rupin Institute, Israel; and the Faculty of Agriculture, The Hebrew University of Jerusalem. Her area of expertise includes the developmental physiology of geophyte storage organs *in vivo* and *in vitro*, propagation and bulb production in tissue culture of geophytes in bioreactors, and *in vitro* acclimatization. Dr. Ziv has published more than 150 refereed journal articles and symposia papers, chapters, and books. She was the editor of *Plant Tissue Culture and Biotechnology*, and the recipient of the Smithsonian Institute, the Florizen, and the Israel Flower Growers awards.

REFERENCES

Aebig, J. A., H. H. Albert, B. L. Zhu, J. S. Hun, and H. T. Hsu. 2006. Cloning and construction of single-chain variable fragments (scFv) to *Cucumber mosaic virus* and production of transgenic plants. *Acta Hort.* 722:129–135.

Aebig, J. A., K. Kamo, and H.-T. Hsu. 2005. Biolistic inoculation of gladiolus with cucumber mosaic cucumovirus. *J. Virol. Methods* 123:89–94.

Ahrazem, O., A. Rubio-Moraga, R. Castillo López, and L. Gómez-Gómez. 2010a. The expression of a chromoplast-specific lycopene beta cyclase gene is involved in the high production of saffron's apocarotenoid precursors. *J. Exp. Bot.* 61:105–119.

Ahrazem, O., A. Trapero, M.D. Gómez, A. Rubio-Moraga, and L. Gómez-Gómez. 2010b. Genomic analysis and gene structure of the plant carotenoid dioxygenase 4 family: A deeper study in *Crocus sativus* and its allies. *Genomics* 96:239–250.

Aida, R., Y. Hirose, S. Kishimoto, and M. Shibata. 1999. *Agrobacterium tumefaciens*-mediated transformation of *Cyclamen persicum* Mill. *Plant Sci.* 148:1–7.

Akita, Y., Y. Horikawa, and A. Kanno. 2008. Comparative analysis of floral MADS-box genes between wild-type and a putative homeotic mutant in lily. *J. Hort. Sci. Biotech.* 83:453–461.

Akutsu, M., T. Ishizaki, and H. Sato. 2004. Transformation of the monocotyledonous *Alstroemeria* by *Agrobacterium tumefaciens. Plant Cell Rep.* 22:561–568.

Alfasi, R. 2009. Virus elimination and bud regeneration from meristem culture in *Canna* sp. *in vitro*. MS thesis, The Hebrew University of Jerusalem, Jerusalem, Israel.

Allefs, S. J. H. M., E. R. De Jong, D. E. A. Florack, C. Hoogendoorn, and W. J. Stiekema. 1996. *Erwinia* soft rot resistance of potato cultivars expressing antimicrobial peptide tachyplesin I. *Mol. Breed.* 2:97–105.

Allen, T. C., O. Ballantyne, J. Goodell, W. C. Anderson, and W. Lin. 1980. Recent advances in research on lily symptomless virus. *Acta Hort.* 109:479–486.

Aminuddin, and B. P. Singh. 1985. *In vitro* gladiolus propagation for virus elimination through tissue cultures. *Indian Phytopathol.* 38:375–377.

Anand, A., T. Zhou, T. H. N. Trick, B. S. Gill, W. W. Bockus, and S. Muthukrishnan. 2003. Greenhouse and field testing of transgenic wheat plants stably expressing genes for thaumatin-like protein, chitinase and glucanase against *Fusarium graminearum. J. Exp. Bot.* 54:1101–1111.

Anderson, J. M., P. Palukaitis, and M. Zaitlin. 1992. A defective replicase gene induces resistance to cucumber mosaic virus in transgenic tobacco plants. *Proc. Natl. Acad. Sci. USA* 89:8759–8763.

Anderson, N. O., A. Plattes, E. Opitz, and A. Younis. 2011. Transgressive segregant, interspecific hybrids between *Lilium × formolongi* and *L. martagon* with unique morphology. *Acta Hort.* 900:181–187.

Anderson, W. C., P. N. Miller, K. A. Mielke, and T. Allen. 1990. *In vitro* bulblet propagation of virus-free Dutch iris. *Acta Hort.* 266:77–81.

Anzai, H., K. Yoneyama, and I. Yamaguchi. 1989. Transgenic tobacco resistance to a bacterial disease by the detoxification of a pathogenic toxin. *Mol. Gen. Genet.* 219:492–494.

Asjes, C. J. 1990. Production for virus freedom of some principal bulbous crops in The Netherlands. *Acta Hort.* 266:517–529.

Azadi, P., D. P. Chin, K. Kuroda, R. S. Khan, and M. Mii. 2010a. Macro elements in inoculation and co-cultivation medium strongly affect the efficiency of *Agrobacterium*-mediated transformation in *Lilium. Plant Cell, Tiss. Org. Cult.* 101:201–209.

Azadi, P., N. V. Otang, D. P. Chin, I. Nakamura, M. Fujisawa, H. Harada, N. Misawa, and M. Mii. 2010b. Metabolic engineering of *Lilium × formolongi* using multiple genes of the carotenoid biosynthesis pathway. *Plant Biotechnol. Rep.* 4:269–280.

Azeez, A., A. P. Sane, S. K. Tripathi, D. Bhatnagar, and P. Nath. 2010. The gladiolus *GgEXPA1* is a GA-responsive alpha-expansin gene expressed ubiquitously during expansion of all floral tissues and leaves but repressed during organ senescence. *Postharvest Biol. Tech.* 58:48–56.

Bajaj, Y. P. S., and R. L. M. Pierik. 1974. Vegetative propagation of *Freesia* through callus cultures. *Nether. J. Agr. Sci.* 22:153–159.

Bajaj, Y. P. S., M. M. S. Sidhu, and A. P. S. Gill. 1983. Some factors affecting the *in vitro* propagation of gladiolus. *Scientia Hortic.* 18:269–275.

Baures, I., T. Candresse, A. Leveau, A. Bendahmane, and B. Sturbois. 2008. The *Rx* gene confers resistance to a range of Potexviruses in transgenic *Nicotiana* plants. *Mol. Plant-Microbe Int.* 21:1154–1164.

Benedito, V. A., G. C. Angenent, J. M. van Tuyl, and F. A. Krens. 2004a. *Lilium longiflorum* and molecular floral development; the ABCDE model. *Acta Hort.* 651:83–89.

Benedito, V. A., P. B. Visser, J. M. van Tuyl, G. C. Angenent, S. C. de Vries, and F. A. Krens. 2004b. Ectopic expression of *LLAG1*, an *AGAMOUS* homologue from lily (*Lilium longiflorum* Thunb.) causes floral homeotic modifications in *Arabidopsis. J. Exp. Bot.* 55:1391–1399.

Bettini, P., S. Michelotti, D. Bindi, R. Giannini, M. Capuana, and M. Buiatti. 2003. Pleiotropic effect of the insertion of the *Agrobacterium rhizogenes rolD* gene in tomato (*Lycopersicon esculentum* Mill.). *Theor. Appl. Genet.* 107:831–836.

Beyer, P., S. Al-Babili, X. D. Ye, P. Lucca, P. Schaub, R. Welsch, and I. Potrykus. 2002. Golden Rice: Introducing the beta-carotene biosynthesis pathway into rice endosperm by genetic engineering to defeat vitamin A deficiency. *J. Nutr.* 132:506S–510S.

Bi, Y.-M., B. P. A. Cammue, P. H. Goodwin, S. KrishnaRaj, and P. K. Saxena. 1999. Resistance to *Botrytis cinerea* in scented geranium transformed with a gene encoding the antimicrobial protein *Ace*-AMP1. *Plant Cell Rep.* 18:835–840.

Blom-Barnhoorn, G. J., and J. van Aartrijk. 1985. The regeneration of plants free of LSV and TBV from infected *Lilium* bulb-scale explants in the presence of virazole. *Acta Hort.* 164:163–168.

Blom-Barnhoorn, G. J., J. van Aartrijk, and P. C. G. van der Linde. 1986. Effect of virazole on the production of hyacinth plants free from hyacinth mosaic virus (HMV) by meristem culture. *Acta Hort.* 177:571–574.

Broekaert, W. F., F. R. G. Terras, B. P. A. Cammue, and R. W. Osborn. 1995. Plant defensins: Novel antimicrobial peptides as components of the host defense system. *Plant Physiol.* 108:1353–1358.

Buitendijk, J. H., N. Pinsonneaux, A. C. van Donk, M. S. Ramanna, and A. A. M. van Lammeren. 1995. Embryo rescue by half-ovule culture for the production of interspecific hybrids in *Alstroemeria. Scientia Hortic.* 64:65–75.

Burchi, G., A. Mercuri, C. Bianchini, R. Bregliano, and T. Schiva. 2000. New interspecific hybrids of *Alstroemeria* obtained through *in vitro* embryo-rescue. *Acta Hort.* 508:233–235.

Burkhardt, P. K., P. Beyer, J. Wünn, A. Klöti, G. A. Armstrong, M. Schledz, J. von Lintig, and I. Potrykus. 1997. Transgenic rice (*Oryza sativa*) endosperm expressing daffodil (*Narcissus pseudonarcissus*) phytoene synthase accumulates phytoene, a key intermediate of provitamin A biosynthesis. *Plant J.* 11:1071–1078.

Carels, N. 2005. The genome organization of angiosperms. In *Recent Research Developments in Plant Science, vol. 3*, ed. S. G. Pandalai, 129–194. Trivandrum, Kerala, India: Research Signpost.

Carmona, M. J., A. Molina, J. A. Fernández, J. J. López-Fando, and F. García-Olmedo. 1993. Expression of the α-thionin gene from barley in tobacco confers enhanced resistance to bacterial pathogens. *Plant J.* 3:457–462.

Castillo, R., L. Gómez-Gómez, and J.-A. Fernández. 2004. Characterization of a class I chitinase from saffron (*Crocus sativus* L.). *Acta Hort.* 650:165–171.

Cervera, M., O. Esteban, M. Gil, M. T. Gorris, M. C. Martínez, L. Peña, and M. Cambra. 2010. Transgenic expression in citrus of single-chain antibody fragments specific to *Citrus tristeza virus* confers virus resistance. *Transgenic Res.* 19:1001–1015.

Chandanie, M. A., S. K. Singh, S. S. Sindhu, A. Singh, S. M. S. Tomar, and K. V. Prasad. 2011. Efficacy of oryzalin as a potent chemical for *in vitro* induction of polyploids in asiatic lily (*Lilium hybrida* L.) var. *polyanna. Indian J. Gen. Plant Breed.* 71:262–268.

Chang, X.-l., Y.-l. Liu, Y.-j. Wang, W.-r. Xu, and Z.-q. Zhang. 2009. Construction and function analysis of the expression vector of the flower-specific promoter PchsA from lily. *J. Northwest A&F Univ.—Natural Science Edition* 37:135–140.

Chen, Y.-K. 2003. Occurrence of Cucumber Mosaic Virus in ornamental plants and perspectives of transgenic control. PhD dissertation, Wageningen University, Wageningen, The Netherlands.

Chen, J.-F., L. Cui, A. A. Malik, and K. G. Mbira. 2011. *In vitro* haploid and dihaploid production via unfertilized ovule culture. *Plant Cell, Tiss. Org. Cult.* 104:311–319.

Chen, Y. K., A. F. L. M. Derks, S. Langeveld, R. Goldbach, and M. Prins. 2001. High sequence conservation among cucumber mosaic virus isolates from lily. *Arch. Virol.* 146:1631–1636.

Chen, C. H., and Y. C. Goeden-Kallemeyn. 1979. *In vitro* induction of tetraploid plants from colchicine-treated diploid daylily callus. *Euphytica* 28:705–709.

Chen, M.-K., I.-C. Lin, and C.-H. Yang. 2008. Functional analysis of three lily (*Lilium longiflorum*) *APETALA1*-like MADS box genes in regulating floral transition and formation. *Plant Cell Physiol.* 49:704–717.

Chen, S. C., A. R. Liu, F. H. Wang, and G. J. Ahammed. 2009. Combined overexpression of chitinase and defensin genes in transgenic tomato enhances resistance to *Botrytis cinerea. Afr. J. Biotechnol.* 8:5182–5188.

Chen, Y.-K., D. Lohuis, R. Goldbach, and M. Prins. 2004. High frequency induction of RNA-mediated resistance against *Cucumber mosaic virus* using inverted repeat constructs. *Mol. Breed.* 14:215–226.

Chen, L. J., X. Y. Zhu, L. Gu, and J. Wu. 2005. Efficient callus induction and plant regeneration from anther of Chinese narcissus (*Narcissus tazetta* L. var. *chinensis* Roem). *Plant Cell Rep.* 24:401–407.

Chen, J., and M. Ziv. 2005. The effects of storage conditions on starch metabolism and regeneration potential of twin-scale and inflorescence stem explants *of Narcissus tazetta. In Vitro Cell. Dev. Biol.—Plant* 41:816–821.

Chung, B.-N., T. Canto, and P. Palukaitis. 2007. Stability of recombinant plant viruses containing genes of unrelated plant viruses. *J. Gen. Virol.* 88:1347–1355.

Coen, E. S., and E. M. Meyerowitz. 1991. The war of the whorls: Genetic interactions controlling flower development. *Nature* 353:31–37.

Cohen, A., A. Lipsky, T. Arazi, A. Ion, R. Stav, D. Sandler-Ziv, C. Fintea, R. Barg, Y. Salts, S. Shabtai, V. Gaba, and A. Gera. 2004. Efficient genetic transformation of *Lilium longiflorum* and *Ornithogalum dubium* by particle acceleration followed by prolonged selection in liquid medium. *Acta Hort.* 651:131–138.

Cohen, A., A. Lipsky, T. Arazi, A. Ion, R. Stav, D. Sandler-Ziv, C. Fintea, V. Gaba, and A. Gera. 2005. Particle bombardment-mediated transformation of *Ornithogalum dubium* for Ornithogalum mosaic virus resistance. *Acta Hort.* 673:183–190.

Cohen, A., A. Lipsky, T. Arazi, A. Ion, R. Stav, D. Sandler-Ziv, C. Fintea, I. Yedidia, N. Gollop, and S. Manulis. 2011. Molecular breeding of *Ornithogalum* for *Erwinia* resistance. *Acta Hort.* 886:49–58.

Custers, J. B. M., E. Ennik, W. Eikelboom, J. J. M. Dons, and M. M. van Lookeren Campagne. 1997. Embryogenesis from isolated microspores of tulip; towards developing F_1 hybrid varieties. *Acta Hort.* 430:259–266.

D'Agostino, N., D. Pizzichini, M. L. Chiusano, and G. Giuliano. 2007. An EST database from saffron stigmas. *BMC Plant Biol.* 7:53(1–8).

De Klerk, G.-J., and J. ter Brugge. 2010. Micropropagation of *Alstroemeria* in liquid medium using slow release medium components. *Propagation of Ornamental Plants* 10:246–252.

De la Fuente-Martínez, J. M., G. Mosqueda-Cano, A. Alvarez-Morales, and L. Herrera-Estrella. 1992. Expression of a bacterial phaeseolotoxin-resistant ornithyl transcarbamylase in transgenic tobacco confers resistance to *Pseudomonas syringae* pv. *phaseolicola. Nat. Biotechnol.* 10:905–909.

Desmyter, S., F. Vandenbussche, Q. Hao, P. Proost, W. J. Peumans, and E. J. M. van Damme. 2003. Type-1 ribosome-inactivating protein from iris bulbs: A useful agronomic tool to engineer virus resistance? *Plant Mol. Biol.* 51:567–576.

De Vetten, N., A.-M. Wolters, K. Raemakers, I. van der Meer, R. ter Stege, E. Heeres, P. Heeres, and R. Visser. 2003. A transformation method for obtaining marker-free plants of a cross-pollinating and vegetatively propagated crop. *Nat. Biotechnol.* 21:439–442.

De Villiers, S. M., K. Kamo, J. A. Thomson, C. H. Bornman, and D. K. Berger. 2000. Biolistic transformation of chincherinchee (*Ornithogalum*) and regeneration of transgenic plants. *Physiol. Plant.* 109:450–455.

Devos, K. M., and M. D. Gale. 2000. Genome relationships: The grass model in current research. *Plant Cell* 12:637–646.

Distefano, G., S. La Malfa, A. Vitale, M. Lorito, Z. Deng, and A. Gentile. 2008. Defence-related gene expression in transgenic lemon plants producing an antimicrobial *Trichoderma harzianum* endochitinase during fungal infection. *Transgenic Res.* 17:873–879.

Dong, S., H. D. Shew, L. P. Tredway, J. Lu, E. Sivamani, E. S. Miller, and R. Qu. 2008. Expression of the bacteriophage T4 lysozyme gene in tall fescue confers resistance to gray leaf spot and brown patch diseases. *Transgenic Res.* 17:47–57.

Düring, K., P. Porsch, M. Fladung, and H. Lörz. 1993. Transgenic potato plants resistant to the phytopathogenic bacterium *Erwinia carotovora. Plant J.* 3:587–598.

Epple, P., K. Apel, and H. Bohlmann. 1997. Overexpression of an endogenous thionin enhances resistance of *Arabidopsis* against *Fusarium oxysporum. Plant Cell* 9:509–520.

Erikson, O., M. Hertzberg, and T. Näsholm. 2004. A conditional marker gene allowing both positive and negative selection in plants. *Nat. Biotechnol.* 22:455–458.

Fahlgren, N., S. Jogdeo, K. D. Kasschau, C. M. Sullivan, E. J. Chapman, S. Laubinger, L. M. Smith, M. Dasenko, S. A. Givan, D. Weigel, and J. C. Carrington. 2010. MicroRNA gene evolution in *Arabidopsis lyrata* and *Arabidopsis thaliana. Plant Cell* 22:1074–1089.

Famelaer, I, M. Bordas, E. Baliu, E. Ennik, H. Meijer, J. M. van Tuyl, and J. Creemers-Molenaar. 1997. The use of morphogenic suspension cultures for the development of a protoplast regeneration system in lily. *Acta Hort.* 430:339–345.

Fiocchetti, F., R. D'Amore, M. De Palma, L. Bertini, C. Caruso, C. Caporale, A. Testa, G. Cristinzio, F. Saccardo, and M. Tucci. 2008. Constitutive over-expression of two wheat pathogenesis-related genes enhances resistance of tobacco plants to *Phytophthora nicotianae*. *Plant Cell, Tiss. Org. Cult.* 92:73–84.

Florack, D. E. A., W. G. Dirkse, B. Visser, F. Heidekamp, and W. J. Stiekema. 1994. Expression of biologically active hordothionins in tobacco. Effects of pre- and pro-sequences at the amino and carboxyl termini of the hordothionin precursor on mature protein expression and sorting. *Plant Mol. Biol.* 24:83–96.

Flors, V., M. de la O. Leyva, b. Vicedo, I. Finiti, M. D. Real, P. García-Agustín, A. B. Bennett, and C. González-Bosch. 2007. Absence of the endo-β-1,4-glucanases Cel1 and Cel2 reduces susceptibility to *Botrytis cinerea* in tomato. *Plant J.* 52:1027–1040.

Fluch, S., K. Hohl, M. Stierschneider, D. Kopecky, and B. Kaar. 2010. *Crocus sativus* L.—Molecular evidence on its clonal origin. *Acta Hort.* 850:41–46.

Frazier, T. P., F. Xie, A. Freistaedter, C. E. Burklew, and B. Zhang. 2010. Identification and characterization of microRNAs and their target genes in tobacco (*Nicotiana tabacum*). *Planta* 232:1289–1308.

Frello, S., and J. S. Heslop-Harrison. 2000. Repetitive DNA sequences in *Crocus vernus* Hill (Iridaceae): The genomic organization and distribution of dispersed elements in the genus *Crocus* and its allies. *Genome* 43:902–909.

Frello, S., M. Ørgaard, N. Jacobsen, and J. S. Heslop-Harrison. 2004. The genomic organization and evolutionary distribution of a tandemly repeated DNA sequence family in the genus *Crocus* (Iridaceae). *Hereditas* 141:81–88.

Frizzi, A., and S. Huang. 2010. Tapping RNA silencing pathways for plant biotechnology. *Plant Biotechnol. J.* 8:655–677.

Fuchs, M., and D. Gonsalves. 2007. Safety of virus-resistant transgenic plants two decades after their introduction: Lessons from realistic field risk assessment studies. *Annu. Rev. Phytopathol.* 45:173–202.

Gao, A.-G., S. M. Hakimi, C. A. Mittanck, Y. Wu, B. M. Woerner, D. M. Stark, D. M. Shah, J. Liang, and C. M. T. Rommens. 2000. Fungal pathogen protection in potato by expression of a plant defensin peptide. *Nat. Biotechnol.* 18:1307–1310.

Gao, Z.-m., D.-f. Chen, X.-p. Li, C.-j. Cai, and Z.-h. Peng. 2008. Cloning and sequence analysis of a flowering-related MADS-box gene in *Narcissus tazette* var. *chinensis* Roem. *Acta Hort. Sin.* 35:295–300.

Gao, Z. M., D. F. Chen, and Z. H. Peng. 2009. Cloning and expression analysis of *NtMADS3* gene from *Narcissus tazetta* var. *chinensis*. *J. Beijing For. Univ.* 31(2):96–101.

Gargouri-Bouzid, R., L. Jaoua, S. Rouis, M. N. Saidi, D. Bouaziz, and R. Ellouz. 2006. PVY-resistant transgenic potato plants expressing an anti-*Nla* protein scFv antibody. *Mol. Biotechnol.* 33:133–140.

Gaskell, G., S. Stares, A. Allensdottir, N. Allum, P. Castro, Y. Esmer, C. Fischler, J. Jackson, N. Kronberger, J. Hampel, N. Mejlgaard, A. Quintanilha, A. Rammer, G. Revuelta, P. Stoneman, H. Torgersen, and W. Wagner. 2010. *Europeans and Biotechnology in 2010. Winds of Change?* Brussels: European Commission.

Gentile, A., Z. Deng, S. La Malfa, G. Distefano, F. Domina, A. Vitale, G. Polizzi, M. Lorito, and E. Tribulato. 2007. Enhanced resistance to *Phoma tracheiphila* and *Botrytis cinerea* in transgenic lemon plants expressing a *Trichoderma harzianum* chitinase gene. *Plant Breed.* 126:146–151.

Girhepuje, P. V., and G. B. Shinde. 2011. Transgenic tomato plants expressing a wheat endochitinase gene demonstrate enhanced resistance to *Fusarium oxysporum* f. sp. *lycopersici*. *Plant Cell, Tiss. Org. Cult.* 105:243–251.

Graves, A. C. F., and S. L. Goldman. 1987. *Agrobacterium tumefaciens*-mediated transformation of the monocot genus *Gladiolus*: Detection of expression of T-DNA-encoded genes. *J. Bacteriol.* 169:1745–1746.

Grison, R., B. Grezes-Besset, M. Schneider, N. Lucante, L. Olsen, J.-J. Leguay, and A. Toppan. 1996. Field tolerance to fungal pathogens of *Brassica napus* constitutively expressing a chimeric chitinase gene. *Nat. Biotechnol.* 14:643–646.

Grover, C. E., J. S. Hawkins, and J. F. Wendel. 2008. Phylogenetic insights into the pace and pattern of plant genome size evolution. (Genome dynamics vol. 4). In *Plant Genomes*, ed. J. N. Volff, 57–68. Basel: S. Karger AG.

Hammond, J., H.-T. Hsu, Q. Huang, R. Jordan, K. Kamo, and M. Pooler. 2006. Transgenic approaches to disease resistance in ornamental crops. *J. Crop Improvement* 17:155–210.

Hammond, J., and K. K. Kamo. 1995a. Resistance to bean yellow mosaic virus (BYMV) and other potyviruses in transgenic plants expressing BYMV antisense RNA, coat protein, or chimeric coat proteins. In *Biotechnology and Plant Protection: Viral Pathogenesis and Disease Resistance*, eds. D. D. Bills, and S.-D. Kung, 369–389. Singapore: World Scientific.

Hammond, J., and K. K. Kamo. 1995b. Effective resistance to potyvirus infection conferred by expression of antisense RNA in transgenic plants. *Mol. Plant-Microbe Int.* 8:674–682.

Han, D.-S., and Y. Niimi. 2005. Production of haploid and double haploid plants from anther-derived callus of *Lilium formosanum. Acta Hort.* 673:389–393.

Han, D.-S., Y. Niimi, and M. Nakano. 1997. Regeneration of haploid plants from anther cultures of the Asiatic hybrid lily 'Connecticut King'. *Plant Cell, Tiss. Org. Cult.* 47:153–158.

Han, D.-S., Y. Niimi, and M. Nakano. 1999. Production of doubled haploid plants through colchicine treatment of anther-derived haploid calli in the Asiatic hybrid lily 'Connecticut King'. *J. Japan. Soc. Hort. Sci.* 68:979–983.

Hasenkampf, C. A., A. A. Taylor, N. U. Siddiqui, and C. D. Riggs. 2000. *Meiotin-1* gene expression in normal anthers and in anthers exhibiting prematurely condensed chromosomes. *Genome* 43:604–612.

Hilder, V. A., K. S. Powell, A. M. R. Gatehouse, J. A. Gatehouse, L. N. Gatehouse, Y. Shi, W. D. O. Hamilton, A. Merryweather, C. A. Newell, J. C. Timans, W. J. Peumans, E. van Damme, and D. Boulter. 1995. Expression of snowdrop lectin in transgenic tobacco plants results in added protection against aphids. *Transgenic Res.* 4:18–25.

Hooykaas-van Slogteren, G. M. S., P. J. J. Hooykaas, and R. A. Schilperoort. 1984. Expression of Ti plasmid genes in monocotyledonous plants infected with *Agrobacterium tumefaciens. Nature* 311:763–764.

Horita, M., H. Morohashi, and F. Komai. 2002. Regeneration of flowering plants from difficile lily protoplasts by means of a nurse culture. *Planta* 215:880–884.

Horita, M., H. Morohashi, and F. Komai. 2003. Production of fertile somatic hybrid plants between Oriental hybrid lily and *Lilium × formolongi. Planta* 217:597–601.

Hoshi, Y., M. Kondo, S. Mori, Y. Adachi, M. Nakano, and H. Kobayashi. 2004. Production of transgenic lily plants by *Agrobacterium*-mediated transformation. *Plant Cell Rep.* 22:359–364.

Howie, W., L. Joe, E. Newbigin, T. Suslow, and P. Dunsmuir. 1994. Transgenic tobacco plants which express the *chiA* gene from *Serratia marcescens* have enhanced tolerance to *Rhizoctonia solani. Transgenic Res.* 3:90–98.

Hsu, H. T., L. Barzuna, Y. H. Hsu, W. Bliss, and K. L. Perry. 2000. Identification and subgrouping of *Cucumber mosaic virus* with mouse monoclonal antibodies. *Phytopathology* 90:615–620.

Hsu, H.-F., W.-P. Hsieh, M.-K. Chen, Y.-Y. Chang, and C.-H. Yang. 2010. C/D class MADS box genes from two monocots, orchid (*Oncidium* Gower Ramsey) and lily (*Lilium longiflorum*), exhibit different effects on floral transition and formation in *Arabidopsis thaliana. Plant Cell Physiol.* 51:1029–1045.

Huang, M.-X., P. Hou, Q. Wei, Y. Xu, and F. Chen. 2008. A ribosome-inactivating protein (curcin 2) induced from *Jatropha curcas* can reduce viral and fungal infection in transgenic tobacco. *Plant Growth Regul.* 54:115–123.

Hussey, G. 1975. Totipotency in tissue explants and callus of some members of the Liliaceae, Iridaceae, and Amaryllidaceae. *J. Exp. Bot.* 26:253–262.

Igawa, T., Y. Hoshino, and Y. Yanagawa. 2009. Isolation and characterization of the plant glsA promoter from *Alstroemeria. Plant Biol.* 11:878–885.

Imayama, T., N. Yoshihara, M. Fukuchi-Mizutani, Y. Tanaka, I. Ino, and T. Yabuya. 2004. Isolation and characterization of a cDNA clone of UDP-glucose: Anthocyanin 5-*O*-glucosyltransferase in *Iris hollandica. Plant Sci.* 167:1243–1248.

Ingelbrecht, I. L., J. E. Irvine, and T. E. Mirkov. 1999. Posttranscriptional gene silencing in transgenic sugarcane. Disssection of homology-dependent virus resistance in a monocot that has a complex polyploid genome. *Plant Physiol.* 119:1187–1197.

Irifune, K., Y. Morimoto, and M. Uchihama. 2003. Production of herbicide resistant transgenic lily plants by particle bombardment. *J. Japan. Soc. Hort. Sci.* 72:511–516.

Ishikawa, T., T. Takayama, H. Ishizawa, K. Ishikawa, and M. Mii. 1997. Production of interspecific hybrids between *Alstroemeria ligtu* L. hybrid and *A. pelegrina* L. var. *rosea* by ovule culture. *Breed. Sci.* 47:15–20.

Jach, G., B. Görnhardt, J. Mundy, J. Logemann, E. Pinsdorf, R. Leah, J. Schell, and C. Maas. 1995. Enhanced quantitative resistance against fungal disease by combinatorial expression of different barley antifungal proteins in transgenic tobacco. *Plant J.* 8:97–109.

James, C. 2010. Global status of commercialized Biotech/GM crops: 2009. The first fourteen years, 1996–2009. *ISAAA Brief* 41–2009. http://www.isaaa.org/resources/publications/briefs/41/.

Jang, Y. S., C. W. Kim, Y. B. Kim, D. Y. Hyun, I. H. Choi, B. C. Jeong, J. H. Park, and S. H. Park. 2002. Regeneration of haploid plants through anther culture in *Lilium longiflorum* Thunb. 'Georgia'. *J. Kor. Soc. Hort. Sci.* 43:231–234.

Jayaraj, J., and Z. K. Punja. 2007. Combined expression of chitinase and lipid transfer protein genes in transgenic carrot plants enhances resistance to foliar fungal pathogens. *Plant Cell Rep.* 26:1539–1546.

Jaynes, J. M., K. G. Xanthopoulos, L. Destéfano-Beltrán, and J. H. Dodds. 1987. Increasing bacterial disease resistance in plants utilizing antibacterial genes from insects. *BioEssays* 6:263–270.

Jha, S., and B. B. Chattoo. 2010. Expression of a plant defensin in rice confers resistance to fungal phytopathogens. *Transgenic Res.* 19:373–384.

Joersbo, M., I. Donaldson, J. Kreiberg, S. G. Petersen, J. Brunstedt, and F. T. Okkels. 1998. Analysis of mannose selection used for transformation of sugar beet. *Mol. Breed.* 4:111–117.

Jones-Rhoades, M. W., D. P. Bartel, and B. Bartel. 2006. MicroRNAs and their regulatory roles in plants. *Annu. Rev. Plant Biol.* 57:19–53.

Jongedijk, E., H. Tigelaar, J. S. C. van Roekel, S. A. Bres-Vloemans, I. Dekker, P. J. M. van den Elzen, B. J. C. Cornelissen, and L. S. Melchers. 1995. Synergistic activity of chitinases and β-1,3-glucanases enhances fungal resistance in transgenic tomato plants. *Euphytica* 85:173–180.

Joubert, D. A., A. R. Slaughter, G. Kemp, J. V. W. Becker, G. H. Krooshof, C. Bergmann, J. Benen, I. S. Pretorius, and M. A. Vivier. 2006. The grapevine polygalacturonase-inhibiting protein (VvPGIP1) reduces *Botrytis cinerea* susceptibility in transgenic tobacco and differentially inhibits fungal polygalacturonases. *Transgenic Res.* 15:687–702.

Joung, Y. H., and K. Kamo. 2006. Expression of a polyubiquitin promoter isolated from *Gladiolus*. *Plant Cell Rep.* 25:1081–1088.

Jung, H. W., W. Kim, and B. K. Hwang. 2003. Three pathogen-inducible genes encoding lipid transfer protein from pepper are differentially activated by pathogens, abiotic, and environmental stresses. *Plant, Cell and Environment* 26:915–928.

Kalivas, A., K. Pasentsis, A. N. Polidoros, and A. S. Tsaftaris. 2007. Heterotopic expression of B-class floral homeotic genes *PISTILLATA/GLOBOSA* supports a modified model for crocus (*Crocus sativus* L.) flower formation. *DNA Sequence* 18:120–130.

Kamo, K., and A. Blowers. 1999. Tissue specificity and expression level of *gusA* under *rolD*, mannopine synthase and translation elongation factor 1 subunit α promoters in transgenic *Gladiolus* plants. *Plant Cell Rep.* 18:809–815.

Kamo, K., A. Blowers, and D. McElroy. 2000. Effect of the cauliflower mosaic virus 35S, actin, and ubiquitin promoters on *UidA* expression from a *Bar-uidA* fusion gene in transgenic *Gladiolus* plants. *In Vitro Cell. Dev. Biol.—Plant* 36:13–20.

Kamo, K., A. Blowers, F. Smith, J. van Eck, and R. Lawson. 1995. Stable transformation of *Gladiolus* using suspension cells and callus. *J. Am. Soc. Hort. Sci.* 120:347–352.

Kamo, K., J. Chen, and R. Lawson. 1990. The establishment of cell suspension cultures of *Gladiolus* that regenerate plants. *In Vitro Cell. Dev. Biol.—Plant.* 26:425–430.

Kamo, K., A. Gera, J. Cohen, J. Hammond, A. Blowers, F. Smith, and J. van Eck. 2005. Transgenic *Gladiolus* plants transformed with the bean yellow mosaic virus coat-protein gene in either sense or antisense orientation. *Plant Cell Rep.* 23:654–663.

Kamo, K., R. Jordan, M. A. Guaragna, H.-t. Hsu, and P. Ueng. 2010. Resistance to *Cucumber mosaic virus* in *Gladiolus* plants transformed with either a defective replicase or coat protein subgroup II gene from *Cucumber mosaic virus*. *Plant Cell Rep.* 29:695–704.

Karamian, R., and M. Ranjbar. 2011. Somatic embryogenesis and plantlet regeneration from protoplast culture of *Muscari neglectum* Guss. *Afr. J. Biotech.*10:4602–4607.

Kasumi, M., Y. Takatsu, T. Manabe, and M. Hayashi. 2001. The effects of irradiating gladiolus (*Gladiolus × grandiflora* hort.) cormels with gamma rays on callus formation, somatic embryogenesis and flower color variations in the regenerated plants. *J. Japan. Soc. Hort. Sci.* 70:126–128 (In Japanese with English summary).

Katsumoto, Y., M. Fukuchi-Mizutani, Y. Fukui, F. Brugliera, T. A. Holton, M. Karan, N. Nakamura, K. Yonekura-Sakakibara, J. Togami, A. Pigeaire, G.-Q. Tao, N. S. Nehra, C.-Y. Lu, B. K. Dyson, S. Tsuda, T. Ashikari, T. Kusumi, J. G. Mason, and Y. Tanaka. 2007. Engineering of the rose flavonoid biosynthetic pathway successfully generated blue-hued flowers accumulating delphinidin. *Plant Cell Physiol.* 48:1589–1600.

Khan, R. S., M. Nishihara, S. Yamamura, I. Nakamura, and M. Mii. 2006. Transgenic potatoes expressing wasabi defensin peptide confer partial resistance to gray mold (*Botrytis cinerea*). *Plant Biotechnol.* 23:179–183.

Kim, K. W., and A. A. De Hertogh. 1997. Tissue culture of ornamental flowering bulbs (geophytes). *Hort. Rev.* 18:87–169.

Kim, Y.-S., H.-S. Kim, Y.-H. Lee, M.-S. Kim, H.-W. Oh, K.-W. Hahn, H. J., and J.-H. Jeon. 2008. Elevated H_2O_2 production via overexpression of a chloroplastic Cu/ZnSOD gene of lily (*Lilium* oriental hybrid 'Marco Polo') triggers ethylene synthesis in transgenic potato. *Plant Cell Rep.* 27:973–983.

Kim, J. B., C. J. J. M. Raemakers, E. Jacobsen, and R. G. F. Visser. 2007. Efficient production of transgenic *Alstroemeria* plants by using *Agrobacterium tumefaciens*. *Ann. Appl. Biol.* 151:401–412.

King, G. J. 2002. Through a genome, darkly: Comparative analysis of plant chromosomal DNA. *Plant Mol. Biol.* 48:5–20.

Komai, F., H. Morohashi, and M. Horita. 2006. Application of nurse culture for plant regeneration from protoplasts of *Lilium japonicum* Thunb. *In Vitro Cell. Dev. Biol.—Plant* 42:252–255.

Konar, R. N., and K. Nataraja. 1965. Production of embryoids from the anthers of *Ranunculus sceleratus* L. *Phytomorphology* 15:245–248.

Krens, F. A., T. R. Menzel, C. Liu, D. C. T. Dees, and B. C. E. van Kronenburg. 2009. Oriental lily hybrids engineered to resist aphid attack. *Acta Hort.* 836:253–257.

Krens, F. A., K. T. B. Pelgrom, J. G. Schaart, A. P. M. den Nijs, and G. J. A. Rouwendal. 2004. Clean vector technology for marker-free transgenic ornamentals. *Acta Hort.* 651:101–105.

Kuipers, A. G. J., J. S. (Pat) Heslop-Harrison, and E. Jacobsen. 1998. Characterisation and physical localisation of Ty1-*copia*-like retrotransposons in four *Alstroemeria* species. *Genome* 41:357–367.

Landi, L., F. Capocasa, E. Costantini, and B. Mezzetti. 2009. *ROLC* strawberry plant adaptability, productivity, and tolerance to soil-borne disease and mycorrhizal interactions. *Transgenic Res.* 18:933–942.

Langeveld, S. A., M. M. Gerrits, A. F. L. M. Derks, P. M. Boonekamp, and J. F. Bol. 1995. Transformation of lily by *Agrobacterium*. *Euphytica* 85:97–100.

Langeveld, S. A., S. Marinova, M. M. Gerrits, A. F. L. M. Derks, and P. M. Boonekamp. 1997. Genetic transformation of lily. *Acta Hort.* 430:290.

Laublin, G., H. S. Saini, and M. Cappadocia. 1991. *In vitro* plant regeneration via somatic embryogenesis from root culture of some rhizomatous irises. *Plant Cell, Tiss. Org. Cult.* 27:15–21.

Lawson, R. H. 1990. Production and maintenance of virus-free bulbs. *Acta Hort.* 266:25–34.

Lee, S. C., I. S. Hwang, H. W. Choi, and B. K. Hwang. 2008. Involvement of the pepper antimicrobial protein *CaAMP1* gene in broad spectrum disease resistance. *Plant Physiol.* 148:1004–1020.

Leeton, P. R. J., and D. R. Smyth. 1993. An abundant LINE-like element amplified in the genome of *Lilium speciosum*. *Mol. Gen. Genet.* 237:97–104.

Li, X., K. Gasic, B. Cammue, W. Brockaert, and S. S. Korban. 2003. Transgenic rose lines harboring an antimicrobial protein gene, *Ace-AMP1*, demonstrate enhanced resistance to powdery mildew (*Sphaerotheca pannosa*). *Planta* 218:226–232.

Li, Q.-H., B. Hong, Z. Tong, C. Ma, A.-N. Guan, J.-J. Yu, and J.-P. Gao. 2008. Establishment of regeneration system and transformation of *Zm401* gene in *Lilium longiflorum* × *L. formosanum*. *Chinese J. Agricul. Biotech.* 5:113–119.

Li, A., and L. Mao. 2007. Evolution of plant microRNA gene families. *Cell Res.* 17:212–218.

Li, Z., L. Pinkham, N. F. Campbell, A. C. Espinosa, and R. Conev. 2009. Development of triploid daylily (*Hemerocallis*) germplasm by embryo rescue. *Euphytica* 169:313–318.

Lin, H.-S., C. van der Toorn, K. J. J. M. Raemakers, R. G. F. Visser, M. J. de Jeu, and E. Jacobsen. 2000. Genetic transformation of *Alstroemeria* using particle bombardment. *Mol. Breed.* 6:369–377.

Lionetti, V., A. Raiola, L. Camardella, A. Giovane, N. Obel, M. Pauly, F. Favaron, F. Cervone, and D. Bellincampi. 2007. Overexpression of pectin methylesterase inhibitors in *Arabidopsis* restricts fungal infection by *Botrytis cinerea*. *Plant Physiol.* 143:1871–1880.

Lipsky, A., A. Cohen, V. Gaba, K. Kamo, A. Gera, and A. Watad. 2002. Transformation of *Lilium longiflorum* plants for cucumber mosaic virus resistance by particle bombardment. *Acta Hort.* 568:209–214.

Liu, W. F., C. Hou, J. Gao, and X. Shen. 2009. Cloning and sequence analysis of the lectin ortholog NTA from Chinese narcissus (*Narcissus tazetta* var. *chinensis* Roem) and structure prediction the protein. *Genomics Appl. Biol.* 28:216–222.

Liu, J., J. Zhang, B. Xu, C. Jia, J. Zhang, G. Tan, and Z. Jin. 2011. Regeneration and production of transgenic *Lilium longiflorum* via *Agrobacterium tumefaciens*. *In Vitro Cell. Dev. Biol.—Plant.* 47:348–356.

Logan, A. E., and F. W. Zettler. 1985. Rapid *in vitro* propagation of virus-indexed gladioli. *Acta Hort.* 164:169–180.

Lorito, M., S. L. Woo, I. Gercia-Fernandez, G. Colucci, G. E. Harman, J. A. Pintor-Toro, E. Filippone, S. Muccifora, C. B. Lawrence, A. Zoina, S. Tuzun, and F. Scala. 1998. Genes from mycoparasitic fungi as a source for improving plant resistance to fungal pathogens. *Proc. Natl. Acad. Sci. USA* 95:7860–7865.

Lu, Y.-d., Q.-h. Gan, X.-y. Chi, and S. Qin. 2008. Roles of microRNA in plant defense and virus offense interaction. *Plant Cell Rep.* 27:1571–1579.

Lu, G., Q. Zou, D. Guo, X. Zhuang, X. Yu, X. Xiang, and J. Cao. 2007. *Agrobacterium tumefaciens*-mediated transformation of *Narcissus tazetta* var. *chinensis*. *Plant Cell Rep.* 26:1585–1593.

Lusk, J. L., and P. Sullivan. 2002. Consumer acceptance of genetically modified foods. *Food Technol.* 56:32–37.

Mackintosh, C. A., J. Lewis, L. E. Radmer, S. Shin, S. J. Heinen, L. A. Smith, M. N. Wyckoff, R. Dill-Macky, C. K. Evans, S. Kravchenko, G. D. Baldridge, R. J. Zeyen, and G. J. Muehlbauer. 2007. Overexpression of defense response genes in transgenic wheat enhances resistance to *Fusarium* head blight. *Plant Cell Rep.* 26:479–488.

Maeda, M., M. Yoshioka, and M. Ito. 1979. Studies on the behavior of meiotic protoplasts. IV. Protoplasts isolated from microsporocytes of Liliaceous plants. *Bot. Mag., Tokyo* 92:111–121.

Makandar, R., J. S. Essig, M. A. Schapaugh, H. N. Trick, and J. Shah. 2006. Genetically engineered resistance to *Fusarium* head blight in wheat by expression of *Arabidopsis NPR1*. *Mol. Plant-Microbe Int.* 19:123–129.

Maleck, K., A. Levine, T. Eulgem, A. Morgan, J. Schmid, K. A. Lawton, J. L. Dangl, and R. A. Dietrich. 2000. The transcriptome of *Arabidopsis thaliana* during systemic acquired resistance. *Nat. Genet.* 26:403–410.

Marchive, C., R. Mzid, L. Deluc, F. Barrieu, J. Pirrello, A. Gauthier, M.-F. Corio-Costet, F. Regad, B. Cailleteau, S. Hamdi, and V. Lauvergeat. 2007. Isolation and characterization of a *Vitis vinifera* transcription factor, VvWRKY1, and its effect on responses to fungal pathogens in transgenic tobacco plants. *J. Exp. Bot.* 58:1999–2010.

Mercuri, A., L. De Benedetti, S. Bruna, R. Bregliano, C. Bianchini, G. Foglia, and T. Schiva. 2003. *Agrobacterium*-mediated transformation with *rol* genes of *Lilium longiflorum* Thunb. *Acta Hort.* 612:129–136.

Michiels, K., E. J. M. van Damme, and G. Smagghe. 2010. Plant-insect interactions: What can we learn from plant lectins? *Arch. Insect Biochem Physiol.* 73:193–212.

Mii, M., Y. Yuzawa, H. Suetomi, T. Motegi, and T. Godo. 1994. Fertile plant regeneration from protoplasts of a seed-propagated cultivar of *Lilium × formolongi* by utilizing meristematic nodular cell clumps. *Plant Sci.* 100:221–226.

Mitra, A., and Z. Zhang. 1994. Expression of a human lactoferrin cDNA in tobacco cells produces antibacterial protein(s). *Plant Physiol.* 106:977–981.

Momonoi, K., K. Yoshida, S. Mano, H. Takahashi, C. Nakamori, K. Shoji, A. Nitta, and M. Nishimura. 2009. A vacuolar iron transporter in tulip, TgVit1, is responsible for blue coloration in petal cells through iron accumulation. *Plant J.* 59:437–447.

Mori, S., M. Nakano, M. Kondo, Y. Hoshi, and H. Kobayashi. 2005. Isolation and characterization of a cytochrome P450 gene from *Muscari armeniacum*. *Acta Hort.* 673:429–435.

Morita, R., Y. Hattori, S. Yokoi, H. Takase, K. Minami, K. Hiratsuka, and K. Toriyama. 2003. Assessment of utility of meiosis-associated promoters of lily for induction of germinal *Ds* transposition in transgenic rice. *Plant Cell Physiol.* 44:637–642.

Morohashi, K., M. Minami, H. Takase, Y. Hotta, and K. Hiratsuka. 2003. Isolation and characterization of a novel GRAS gene that regulates meiosis-associated gene expression. *J. Biol. Chem.* 278:20865–20873.

Ntui, V. O., G. Thirukkumaran, P. Azadi, R. S. Khan, I. Nakamura, and M. Mii. 2010. Stable integration and expression of wasabi defensin gene in 'Egusi' melon (*Colocynthis citrullus* L.) confers resistance to *Fusarium* wilt and *Alternaria* leaf spot. *Plant Cell Rep.* 29:943–954.

Núñez de Cáceres, F. F., M. R. Davey, and Z. A. Wilson. 2011. A rapid and efficient *Agrobacterium*-mediated transformation protocol for *Lilium*. *Acta Hort.* 900:161–167.

Ogaki, M., Y. Furuichi, K. Kuroda, D. P. Chin, Y. Ogawa, and M. Mii. 2008. Importance of co-cultivation medium pH for successful *Agrobacterium*-mediated transformation of *Lilium × formolongi*. *Plant Cell Rep.* 27:699–705.

Okada, T., P. L. Bhalla, and M. B. Singh. 2005. Transcriptional activity of male gamete-specific histone *gcH3* promoter in sperm cells of *Lilium longiflorum*. *Plant Cell Physiol.* 46:797–802.

Okada, T., P. L. Bhalla, and M. B. Singh. 2006. Expressed sequence tag analysis of *Lilium longiflorum* generative cells. *Plant Cell Physiol.* 47:698–705.

Okazaki, K. 2005. New aspects of tulip breeding: Embryo culture and polyploidy. *Acta Hort.* 673:127–140.

Oldroyd, G. E. D., and B. J. Staskawicz. 1998. Genetically engineered broad-spectrum disease resistance in tomato. *Proc. Natl. Acad. Sci. USA* 95:10300–10305.

Ossowski, S., R. Schwab, and D. Weigel. 2008. Gene silencing in plants using artificial microRNAs and other small RNAs. *Plant J.* 53:674–690.

Osusky, M., G. Zhou, L. Osuska, R. E. Hancock, W. W. Kay, and S. Misra. 2000. Transgenic plants expressing cationic peptide chimeras exhibit broad-spectrum resistance to phytopathogens. *Nat. Biotechnol.* 18:1162–1166.

Parkhi, V., V. Kumar, L. M. Campbell, A. A. Bell, J. Shah, and K. S. Rathore. 2010. Resistance against various fungal pathogens and reniform nematodes in transgenic cotton plants expressing Arabidopsis *NPR1*. *Transgenic Res.* 19:959–975.

Peschen, D., H.-P. Li, R. Fischer, F. Kreuzaler, and Y.-C. Liao. 2004. Fusion proteins comprising a *Fusarium*-specific antibody linked to antifungal peptides protect plants against a fungal pathogen. *Nat. Biotechnol.* 22:732–738.

Phillips, S. 1990. The efficacy of four antiviral compounds in the elimination of narcissus viruses during meristem tip culture. *Acta Hort.* 266:531–538.

Popowich, E. A., A. P. Firsov, T. Y. Mitiouchkina, V. L. Filipenya, S. V. Dolgov, and V. N. Reshetnikov. 2007. *Agrobacterium*-mediated transformation of *Hyacinthus orientalis* with thaumatin II gene to control fungal diseases. *Plant Cell, Tiss. Org. Cult.* 90:237–244.

Portieles, R., C. Ayra, E. Gonzalez, A. Gallo, R. Rodriguez, O. Chacón, Y. López, M. Rodriguez, J. Castillo, M. Pujol, G. Enriquez, C. Borroto, L. Trujillo, B. P. H. J. Thomma, and O. Borrás-Hidalgo. 2010. *NmDef02*, a novel antimicrobial gene isolated from *Nicotiana megalosiphon* confers high-level pathogen resistance under greenhouse and field conditions. *Plant Biotechnol. J.* 8:678–690.

Potrykus, I. 2003. Nutritionally enhanced rice to combat malnutrition disorders of the poor. *Nutr. Rev.* 61:S101–S104.

Potrykus, I. 2005. Is GMO over-regulation costing lives? (The World Life Sciences Forum 2005: Volume 2) In *Agriculture and Nutrition: Analyses and Recommendations*, ed. The World Life Sciences Forum, 61–74. Weinheim: Wiley-VCH Verlag.

Powell-Abel, P., R. S. Nelson, B. De, N. Hoffmann, S. G. Rogers, R. T. Fraley, and R. N. Beachy. 1986. Delay of disease development in transgenic plants that express the tobacco mosaic virus coat protein gene. *Science* 232:738–743.

Prakash, J., and K. L. Giles. 1986. Production of doubled haploids in oriental lilies. In *Genetic Manipulation in Plant Breeding*, eds. W. Horn, C. J. Jensen, W. Odenbach, and O. Schieder, 335–337. Berlin: Walter de Gruyter & Co.

Prins, M., M. Laimer, E. Noris, J. Schubert, M. Wassenegger, and M. Tepfer. 2008. Strategies for antiviral resistance in transgenic plants. *Mol. Plant Pathol.* 9:73–83.

Prins, M., D. Lohuis, A. Schots, and R. Goldbach. 2005. Phage display-selected single-chain antibodies confer high levels of resistance against *Tomato spotted wilt virus*. *J. Gen. Virol.* 86:2107–2113.

Qiu, D., G. Diretto, R. Tavarza, and G. Giuliano. 2007. Improved protocol for *Agrobacterium* mediated transformation of tomato and production of transgenic plants containing carotenoid biosynthetic gene *CsZCD*. *Scientia Hortic.* 112:172–175.

Qu, Y., M. C. Mok, D. W. S. Mok, and J. R. Stang. 1988. Phenotypic and cytological variation among plants derived from anther cultures of *Lilium longiflorum*. *In Vitro Cell. Dev. Biol.—Plant* 24:471–476.

Ren, X.-J., W.-D. Huang, W.-Z. Li, and D.-Q. Yu. 2010. Tobacco transcription factor WRKY4 is a modulator of leaf development and disease resistance. *Biol. Plant.* 54:684–690.

Rhee, H. K., and K. S. Kim. 2008. Interspecific hybridization and polyploidization in lily breeding. *Acta Hort.* 766:441–447.

Rommens, C. M., M. A. Haring, K. Swords, H. V. Davies, and W. R. Belknap. 2007. The intragenic approach as a new extension to traditional plant breeding. *Trends in Plant Sci.* 12:397–403.

Rommens, C. M. T., J. M. Salmeron, G. E. D. Oldroyd, and B. J. Staskawicz. 1995. Intergeneric transfer and functional expression of the tomato disease resistance gene *Pto*. *Plant Cell* 7:1537–1544.

Ronald, P. C., J. M. Salmeron, F. M. Carland, and B. J. Staskawicz. 1992. The cloned avirulence gene *avrPto* induces disease resistance in tomato cultivars containing the *Pto* resistance gene. *J. Bacteriol.* 174:1604–1611.

Sachar, R. C., and M. Kapoor. 1959. *In vitro* culture of ovules of *Zephyranthes*. *Phytomorphology* 9:147–156.

Sanford, J. C., and S. A. Johnston. 1985. The concept of parasite-derived resistance—Deriving resistance genes from the parasite's own genome. *J. Theor. Biol.* 113:395–405.

Sarowar, S., Y. J. Kim, K. D. Kim, B. K. Hwang, S. H. Ok, and J. S. Shin. 2009. Overexpression of lipid transfer protein (LTP) genes enhances resistance to plant pathogens and LTP functions in long-distance systemic signaling tobacco. *Plant Cell Rep.* 28:419–427.

Sato, S., N, Katoh, S. Iwawi, and M. Hagimori. 2002. Effects of low temperature pretreatment of buds or inflorescences on isolated microspore culture in *Brassica rapa* (syn. *B. campestris*). *Breed. Sci.* 52:23–26.

Sato, A., H. Okubo, and K. Saitou. 2006. Increase in the expression of an alpha-amylase gene and sugar accumulation induced during cold period reflects shoot elongation in hyacinth bulbs. *J. Am. Soc. Hort. Sci.* 131:185–191.

Schaart, J. G., F. A. Krens, K. T. B. Pelgrom, O. Mendes, and G. J. A. Rouwendal. 2004. Effective production of marker-free transgenic strawberry plants using inducible site-specific recombination and a bifunctional selectable marker gene. *Plant Biotechnol. J.* 2:233–240.

Schaart, J. G., F. A. Krens, A.-M. A. Wolters, and R. G. F. Visser. 2011. Transformation methods for obtaining marker-free genetically modified plants. In *Plant Transformation Technologies*, eds. C. N. Stewart, A. Touraev, V. Citovsky, and T. Tzfira, 229–242. Oxford: Wiley-Blackwell Publ.

Schlaich, T., B. M. Urbaniak, N. Malgras, E. Ehler, C. Birrer, L. Meier, and C. Sautter. 2006. Increased field resistance to *Tilletia caries* provided by a specific antifungal virus gene in genetically engineered wheat. *Plant Biotechnol. J.* 4:63–75.

Schouten, H. J., F. A. Krens, and E. Jacobsen. 2006. Cisgenic plants are similar to traditionally bred plants. *EMBO Rep.* 7:750–753.

Sekine, K.-T., S. Kawakami, S. Hase, M. Kubota, Y. Ichinose, J. Shah, H.-G. Kang, D. F. Klessig, and H. Takahashi. 2008. High level expression of a virus resistance gene, *RCY1*, confers extreme resistance to *Cucumber mosaic virus* in *Arabidopsis thaliana*. *Mol. Plant-Microbe Int.* 21:1398–1407.

Senthilkumar, R., C.-P. Cheng, and K.-W. Yeh. 2010. Genetically pyramiding protease-inhibitor genes for dual broad-spectrum resistance against insect and phytopathogens in transgenic tobacco. *Plant Biotechnol. J.* 8:65–75.

Sharp, W. R., R. S. Raskin, and N. E. Sommer. 1972. Haploidy in *Lilium. Phytomorphology* 21:334–337.

Shi, J., H.-L. An, L. Zhang, Z. Gao, and X.-Q. Guo. 2010. *GhMPK7*, a novel multiple stress-responsive cotton group C MAPK gene, has a role in broad spectrum disease resistance and plant development. *Plant Mol. Biol.* 74:1–17.

Shibli, R. A., and M. M. Ajlouni. 2000. Somatic embryogenesis in the endemic black iris. *Plant Cell, Tiss. Org. Cult.* 61:15–21.

Shin, S., C. A. Mackintosh, J. Lewis, S. J. Heinen, L. Radmer, R. Dill-Macky, G. D. Baldridge, R. J. Zeyen, and G. J. Muehlbauer. 2008. Transgenic wheat expressing a barley class II chitinase gene has enhanced resistance against *Fusarium graminearum. J. Exp. Bot.* 59:2371–2378.

Shivanna, K. R., and V. K. Sawhney. 1997. *Pollen Biotechnology for Crop Production and Improvement*. Cambridge, UK: Cambridge University Press.

Shoji, K., K. Momonoi, and T. Tsuji. 2010. Alternative expression of vacuolar iron transporter and ferritin genes leads to blue/purple coloration of flowers in tulip cv. 'Murasakizuisho'. *Plant Cell Physiol.* 51:215–224.

Shoji, K., M. Ubukata, K. Momonoi, T. Tsuji, and T. Morimatsu. 2005. Anther-specific production of antimicrobial tuliposide B in tulips. *J. Japan. Soc. Hort. Sci.* 74:469–475.

Singh, A. K., S. Praveen, B. P. Singh, A. Varma, and N. Arora. 2009. Safety assessment of leaf curl virus resistant tomato developed using viral derived sequences. *Transgenic Res.* 18:877–887.

Sochacki, D., and T. Orlikowska. 2005. The obtaining of narcissus plants free from potyviruses via adventitious shoot regeneration *in vitro* from infected plants. *Scientia Hortic.* 103:219–225.

Stimart, D. P., and P. D. Ascher. 1982. Plantlet regeneration and stability from callus cultures of *Freesia × hybrida* Bailey cultivar 'Royal'. *Scientia Hortic.* 17:153–157.

Suomeng, D., Z. Zhengguang, Z. Xiaobo, and W. Yuanchao. 2008. Mammalian pro-apoptotic *bax* gene enhances tobacco resistance to pathogens. *Plant Cell Rep.* 27:1559–1569.

Takahashi, S., K. Matsubara, H. Yamagata, and T. Morimoto. 1992. Micropropagation of virus free bulblets of *Liliun longiflorum* by tank culture—1. Development of liquid culture method and large scale propagation. *Acta Hort.* 319:83–88.

Tanaka, I., C. Kitazume, and M. Ito. 1987. The isolation and culture of lily pollen protoplasts. *Plant Sci.* 50:205–211.

Tang, S. X., R. A. Okashah, M.-M. Cordonnier-Pratt, L. H. Pratt, V. E. Johnson, C. A. Taylor, M. L. Arnold, and S. J. Knapp. 2009. EST and EST-SSR marker resources for *Iris. BMC Plant Biol.* 9:72(1–11).

Theißen, G. 2001. Genetics of identity. *Nature* 414:491.

Thevissen, K., B. P. A. Cammue, K. Lemaire, J. Winderickx, R. C. Dickson, R. L. Lester, K. K. A. Ferket, F. van Even, A. H. A. Parret, and W. F. Broekaert. 2000. A gene encoding a sphingolipid biosynthesis enzyme determines the sensitivity of *Saccharomyces cerevisiae* to an antifungal plant defensin from dahlia (*Dahlia merckii*). *Proc. Natl. Acad. Sci. USA* 97:9531–9536.

Thiruvengadam, M., and C.-H. Yang. 2009. Ectopic expression of two MADS box genes from orchid (*Oncidium* Gower Ramsey) and lily (*Lilium longiflorum*) alters flower transition and formation in *Eustoma grandiflorum. Plant Cell Rep.* 28:1463–1473.

Tribulato, A., P. C. Remotti, and H. J. M. Löffler. 1997a. Lily regenerative callus and cell cultures for transformation. *Acta Hort.* 430:299–306.

Tribulato, A., P. C. Remotti, H. J. M. Löffler, and J. M. van Tuyl. 1997b. Somatic embryogenesis and plant regeneration in *Lilium longiflorum* Thunb. *Plant Cell Rep.* 17:113–118.

Tsaftaris, A., K. Pasentzis, and A. Argiriou. 2010. Rolling circle amplification of genomic templates for inverse PCR (RCA-GIP): A method for 5'- and 3'-genome walking without anchoring. *Biotechnol. Lett.* 32:157–161.

Tsaftaris, A. S., K. Pasentsis, I. Iliopoulos, and A. N. Polidoros. 2004. Isolation of three homologous *AP1*-like MADS-box genes in crocus (*Crocus sativus* L.) and characterization of their expression. *Plant Sci.* 166:1235–1243.

Tsaftaris, A. S., K. Pasentsis, and A. N. Polidoros. 2005. Isolation of a differentially spliced C-type flower specific *AG*-like MADS-box gene from *Crocus sativus* and characterization of its expression. *Biol. Plant.* 49:499–504.

Tsaftaris, A. S., A. N. Polidoros, K. Pasentsis, and A. Kalivas. 2006. Tepal formation and expression pattern of B-class paleo*AP3*-like MADS-box genes in crocus (*Crocus sativus* L.). *Plant Sci.* 170:238–246.

Tzeng, J.-D., S.-W. Hsu, M.-C. Chung, F.-L. Yeh, C.-Y. Yang, M.-C. Liu, Y.-F. Hsu, and C.-S. Wang. 2009a. Expression and regulation of two novel anther-specific genes in *Lilium longiflorum. J. Plant Physiol.* 166:417–427.

Tzeng, T.-Y., H.-Y. Chen, and C.-H. Yang. 2002. Ectopic expression of carpel-specific MADS box genes from lily and lisianthus causes similar homeotic conversion of sepal and petal in *Arabidopsis. Plant Physiol.* 130:1827–1836.

Tzeng, T.-Y., C.-C. Hsiao, P.-J. Chi, and C.-H. Yang. 2003. Two lily *SEPALLATA*-like genes cause different effects on floral formation and floral transition in *Arabidopsis. Plant Physiol.* 133:1091–1101.

Tzeng, T.-Y., L.-R. Kong, C.-H. Chen, C.-C. Shaw, and C.-H. Yang. 2009b. Overexpression of the lily p70[s6k] gene in *Arabidopsis* affects elongation of flower organs and indicates TOR-dependent regulation of AP3, PI and SUP translation. *Plant Cell Physiol.* 50:1695–1709.

Tzeng, T. Y., and C. H. Yang. 2001. A MADS box gene from lily (*Lilium longiflorum*) is sufficient to generate dominant negative mutation by interacting with PISTILLATA (PI) in *Arabidopsis thaliana. Plant Cell Physiol.* 42:1156–1168.

Ueda, K., Y. Miyamoto, and I. Tanaka. 1990. Fusion studies of pollen protoplasts and generative cell protoplasts in *Lilium longiflorum. Plant Sci.* 72:259–266.

Van Creij, M. G. M., D. M. F. J. Kerckhoffs, S. M. de Bruijn, D. Vreugdenhil, and J. M. van Tuyl. 2000. The effect of medium composition on ovary-slice culture and ovule culture in intraspecific *Tulipa gesneriana* crosses. *Plant Cell, Tiss. Org. Cult.* 60:61–67.

Van Damme, E. J. M., C. H. Astoul, A. Barre, P. Rougé, and W. J. Peumans. 2000. Cloning and characterization of a monocot mannose-binding lectin from *Crocus vernus* (family Iridaceae). *Eur. J. Biochem.* 267:5067–5077.

Van Damme, E. J. M., H. Kaku, F. Perini, I. J. Goldstein, B. Peeters, F. Yagi, B. Decock, and W. J. Peumans. 1991. Biosynthesis, primary structure and molecular cloning of snowdrop (*Galanthus nivalis* L.) lectin. *Eur. J. Biochem.* 202:23–30.

Vandenbussche, F., W. J. Peumans, S. Desmyter, P. Proost, M. Ciani, and E. J. M. van Damme. 2004. The type-1 and type-2 ribosome-inactivating proteins from *Iris* confer transgenic tobacco plants local but not systemic protection against viruses. *Planta.* 220:211–221.

Van der Leede-Plegt, L. M., B. C. E. Kronenburg-v.d. Ven, J. Franken, J. M. van Tuyl, A. J. van Tunen, and H. J. M. Dons. 1997. Transgenic lilies *via* pollen-mediated transformation. *Acta Hort.* 430:529–530.

Van Emmenes, L., A. Veale, A. Cohen, and T. Arazi. 2008. *Agrobacterium*-mediated transformation of the bulbous flower *Ornithogalum. Acta Hort.* 766:477–484.

Van Harmelen, M. J., H. J. M. Löffler, and J. M. van Tuyl. 1997. Somaclonal variation in lily by *in vitro* cultivation. *Acta Hort.* 430:347–350.

Van Tunen, A. J., M. Busscher, J. Franken, and G. C. Angenent. 1993. Molecular flower breeding; the ABC of floral organogenesis. *Agro Food Ind. Hi Tech.* 4:15–17.

Van Tuyl, J. M., and M. J. de Jeu. 1997. Methods for overcoming interspecific genetic barriers. In *Pollen Biotechnology for Crop Production and Improvement*, eds. K. R. Shivanna, and V. K. Sawhney, 273–292. Cambridge, UK: Cambridge University Press.

Van Tuyl, J. M., J. Franken, R. C. Jongerius, C. A. M. Lock, and T. A. M. Kwakkenbos. 1986. Interspecific hybridization in *Lilium. Acta Hort.* 177:591–595.

Van Tuyl, J. M., K. van de Sande, K. van Diën, D. Straathof, and H. M. C. van Holsteijn. 1990. Overcoming interspecific crossing barriers in *Lilium* by ovary and embryo culture. *Acta Hort.* 266:317–322.

Van Tuyl, J. M., M. P. van Diën, M. G. M. van Creij, T. C. M. van Kleinwee, J. Franken, and R. J. Bino. 1991. Application of *in vitro* pollination, ovary culture, ovule culture and embryo rescue for overcoming incongruity barriers in interspecific *Lilium* crosses. *Plant Sci.* 74:115–126.

Vitte, C., and O. Panaud. 2005. LTR retrotransposons and flowering plant genome size: Emergence of the increase/decrease model. *Cytogenet. Genome Res.* 110:91–107.

Wally, O., J. Jayaraj, and Z. K. Punja. 2009. Broad-spectrum disease resistance to necrotrophic and biotrophic pathogens in transgenic carrots (*Daucus carota* L.) expressing an *Arabidopsis NPR1* gene. *Planta* 231:131–134.

Wang, X., B. M. V. S. Basnayake, H. Zhang, G. Li, W. Li, N. Virk, T. Mengiste, and F. Song. 2009. The *Arabidopsis* ATAF1, a NAC transcription factor is a negative regulator of defense responses against necrotrophic fungal and bacterial pathogens. *Mol. Plant-Microbe Int.* 22:1227–1238.

Wang, Z.-k., J. Gao, L.-b. Li, and Z.-h. Peng. 2006. Isolation and characterization of the *AGAMOUS* homologous gene *NTAG* in Chinese narcissus (*Narcissus tazetta* var. *chinensis* Roem). *For. Stud. China* 8:21–26.

Wang, W. Y., and B. W. Poovaiah. 1999. Interaction of plant chimeric calcium/calmodulin-dependent protein kinase with a homolog of eukaryotic elongation factor-1α. *J. Biol. Chem.* 274:12001–12008.

Watad, A. A., D.-J. Yun, T. Matusomoto, X. Niu, Y. Wu, A. K. Kononowicz, R. A. Bressan, and P. M. Hasegawa. 1998. Microprojectile bombardment-mediated transformation of *Lilium longiflorum*. *Plant Cell Rep.* 17:262–267.

Waterhouse, P. M., M. W. Graham, and M.-B. Wang. 1998. Virus resistance and gene silencing in plants can be induced by simultaneous expression of sense and antisense RNA. *Proc. Natl. Acad. Sci. USA* 95:13959–13964.

Wilmink, A., B. C. E. van de Ven, and J. J. M. Dons. 1992. Expression of the GUS-gene in the monocot tulip after introduction by particle bombardment and *Agrobacrerium*. *Plant Cell Rep.* 11:76–80.

Wilmink, A., B. C. E. van de Ven, and J. J. M. Dons. 1995. Activity of constitutive promoters in various species from the Liliaceae. *Plant Mol. Biol.* 28:949–955.

Wu, G., B. J. Shortt, E. B. Lawrence, E. B. Levine, K. C. Fitzsimmons, and D. M. Shah. 1995. Disease resistance conferred by expression of a gene encoding H_2O_2-generating glucose oxidase in transgenic potato plants. *Plant Cell* 7:1357–1368.

Yamagami, T., Y. Mine, Y. Aso, and M. Ishiguro. 1997. Purification and characterization of two chitinase isoforms from the bulbs of gladiolus (*Gladiolus gandavensis*). *Biosci. Biotech. Biochem.* 61:2140–2142.

Yamagami, T., K. Tsutsumi, and M. Ishiguro. 2000. Cloning, sequencing, and expression of the tulip bulb chitinase-1 cDNA. *Biosci. Biotech. Biochem.* 64:1394–1401.

Yamaguchi, N., Y. Seshimo, and C. Masuta. 2005. Mapping the sequence domain for systemic infection in edible lily on the viral genome of *Cucumber mosaic virus*. *J. Gen. Plant Pathol.* 71:373–376.

Yang, C.-Y., C.-H. Wu, G.-Y. Jauh, J.-C. Huang, C.-C. Lin, and C.-S. Wang. 2008. The LLA23 protein translocates into nuclei shortly before desiccation in developing pollen grains and regulates gene expression in *Arabidopsis*. *Protoplasma* 233:241–254.

Yang, Y., H. Zhang, G. Li, W. Li, X. Wang, and F. Song. 2009. Ectopic expression of MgSM1, a cerato-platanin family protein from *Magnaporthe grisea*, confers broad-spectrum disease resistance in *Arabidopsis*. *Plant Biotechnol. J.* 7:763–777.

Yevtushenko, D. P., and S. Misra. 2007. Comparison of pathogen-induced expression and efficacy of two amphibian antimicrobial peptides, MsrA2 and temporin A, for engineering wide-spectrum disease resistance in tobacco. *Plant Biotechnol. J.* 5:720–734.

Yip, M.-K., H.-E. Huang, M.-J. Ger, S.-H. Chiu, Y.-C. Tsai, C.-I. Lin, and T.-Y. Feng. 2007. Production of soft rot resistant calla lily by expressing a ferredoxin-like protein gene (*pflp*) in transgenic plants. *Plant Cell Rep.* 26:449–457.

Yoon, E. S. 1991. Ovary slice culture as a technique for hybridization between incompatible species of *Lilium*. I. Survival rate and germination rate of hybrid embryos. *Korean J. Plant Tiss. Cult.* 18:185–193.

Yoshihara, N., T. Imayama, M. Fukuchi-Mizutani, H. Okuhara, Y. Tanaka, I. Ino, and T. Yabuya. 2005. cDNA cloning and characterization of UDP-glucose: Anthocyanidin 3-*O*-glucosyltransferase in *Iris hollandica*. *Plant Sci.* 169:496–501.

Yoshihara, N., T. Imayama, Y. Matsuo, M. Fukuchi-Mizutani, Y. Tanaka, I. Ino, and T. Yabuya. 2006. Characterization of cDNA clones encoding anthocyanin 3-*p*-coumaroyltransferase from *Iris hollandica*. *Plant Sci.* 171:632–639.

Yu, H.-J., E.-J. Suh, B. W. Yae, B.-H. Han, and I.-G. Mok. 2003. Analysis of organ-specific and -preferentially expressed genes from anther and tepal of lily. *Acta Hort.* 625:71–77.

Zhang, X., H. Li, J. Zhang, C. Zhang, P. Gong, K. Ziaf, F. Xiao, and Z. Ye. 2011. Expression of artificial microRNAs in tomato confers efficient and stable virus resistance in a cell-autonomous manner. *Transgenic Res.* 20:569–581.

Zhang, Y., Q. Liu, Q. Ouyang, and W. Cai. 2004. Cloning of flower development associated MADS box gene fragments in *Lilium. Acta Hort. Sin.* 31:332–336.

Zhang, L., J. Xu, and R. G. Birch. 1999. Engineered detoxification confers resistance against a pathogenic bacterium. *Nat. Biotechnol.* 17:1021–1024.

Zhang, M.-Y., S. Zimmermann, R. Fischer, and S. Schillberg. 2008. Generation and evaluation of movement protein-specific single-chain antibodies for delaying symptoms of *Tomato spotted wilt virus* infection in tobacco. *Plant Pathol.* 57:854–860.

Zhu, Y. J., H. McCafferty, G. Osterman, S. Lim, R. Agbayani, A. Lehrer, S. Schenck, and E. Komor. 2011. Genetic transformation with untranslatable coat protein gene of *sugarcane yellow leaf virus* reduces virus titers in sugarcane. *Transgenic Res.* 20:503–512.

Ziv, M. 1995a. The control of bioreactor environment for plant propagation in liquid culture. *Acta Hort.* 393:25–38.

Ziv, M. 1995b. *In vitro* acclimatization. In *Automation and Environmental Control in Plant Tissue Culture*, eds. J. Aitken-Christie, T. Kozai, and M. A. L. Smith, 493–516. Dordrecht, The Netherlands: Kluwer Academic Publ.

Ziv, M. 1999. Developmental and structural patterns of *in vitro* plants. In *Morphogenesis in Tissue Culture*, eds. S. Woong-Young, and S. S. Bhojwani, 235–253. Dordrecht, The Netherlands: Kluwer Academic Publ.

Ziv, M. 2005. Simple bioreactors for mass propagation of plants. *Plant Cell, Tiss. Org. Cult.* 81:277–285.

Ziv, M. 2008. Paclobutrazol and xanthan gum involvement in proliferation and stress response of *Ornithogalum dubium* Houtt. bud clusters cultured in bioreactors. *Propagation of Ornamental Plants* 8:28–32.

Ziv, M., and J. Chen. 2007. The anatomy and morphology of tissue cultured plants. In *Plant Propagation by Tissue Culture, the Background, Vol 1, 3rd ed.*, eds. E. F. George, M. A. Hall, and G.-J. de Klerk, 465–478. Dordrecht, The Netherlands: Springer.

Ziv, M., S. Kahany, and H. Lilien-Kipnis. 1994. Scaled-up proliferation and regeneration of *Nerine* in liquid cultures part I. The induction and maintenance of proliferating meristematic clusters by paclobutrazol in bioreactors. *Plant Cell, Tiss. Org. Cult.* 39:109–115.

Zonneveld, B. J. M., I. J. Leitch, and M. D. Bennett. 2005. First nuclear DNA amounts in more than 300 angiosperms. *Ann. Bot.* 96:229–244.

Zou, J., H. Gong, T.-J. Yang, and J. Meng. 2009. Retrotransposons—A major driving force in plant genome evolution and a useful tool for genome analysis. *J. Crop Sci. Biotech.* 12:1–8.

8 Florogenesis

Rina Kamenetsky, Michele Zaccai, and Moshe A. Flaishman

CONTENTS

I. INTRODUCTION

Flowering is one of the most fascinating, yet complicated, processes in nature, ensuring seed production and species persistence. The process involves a variety of physiological, biochemical, and molecular mechanisms regulating the proper timing and correct development of the reproductive organs. Over the last few decades, the genetic and molecular mechanisms of flowering have been studied by analyzing genetic variation in model plants such as *Arabidopsis thaliana* and *Antirrhinum majus* (Boss et al. 2004; Sung and Amasino 2004; Amasino 2010; Irish 2010; Rijpkema et al. 2010; Wellmer and Riechmann 2010). These studies have led to the identification of components within individual signaling pathways that affect flowering and their positioning within molecular hierarchies. Furthermore, distinct signaling pathways have been shown to converge on activation of the same flowering-time genes (Mouradov et al. 2002; Wellmer and Riechmann 2010). A genetic survey of flowering mutants defined the inductive photoperiod, extended exposure to cold, and gibberellins (GAs) as major factors promoting flowering in *Arabidopsis* (Koornneef et al. 1991). Further studies identified the effects of light quality, ambient temperature, stress, and other phytohormones on flowering-time regulation (Domagalska et al. 2010).

In ornamental geophytes, the morphological and physiological aspects of florogenesis and the effects of environmental factors on flowering have been studied for major crops, thus providing a large scientific background for the diverse techniques of flowering manipulation (Rees 1972, 1992; De Hertogh and Le Nard 1993a; Theron and De Hertogh 2001) and for the significant increase in flower bulb production worldwide (Benschop et al. 2010). Florogenesis of geophytes is affected by the interfacing of environmental factors and the genetic background of the individual plant, leading to the plant's transition from vegetative to reproductive state, flower formation, and anthesis. However, only limited information on the genetic control of floral transition in the meristem and subsequent formation of the inflorescence or individual flowers and flower organs is available in herbaceous perennial plants in general and in geophytes in particular (Townsend et al. 2006; Albani and Coupland 2010; Kamenetsky 2011).

Although homologues of flowering genes from model plants have recently been isolated from geophytes such as *Narcissus*, *Tulipa*, and *Lilium*, the molecular regulation of flower development in these plants differs from that in model plants (Kanno et al. 2007; Rotem et al. 2007, 2011; Noy-Porat et al. 2010). One of the reasons why only limited molecular information is available on geophytes might be the size of the main genera's genomes. For example, the DNA content of the unreplicated haploid genome of *Tulipa* spp. is about 200 times larger than that of *Arabidopsis thaliana* (Arumuganathan and Earle 1991). Another reason is the lack of efficient transformation systems in geophytes and the consequent inability to perform functional analyses of the isolated genes.

In this chapter, recent research reports that have significantly impacted on our understanding of the detailed mechanisms of florogenesis in ornamental geophytes are reviewed. We also discuss the internal and external factors involved in geophyte florogenesis, with special emphasis on the prospects for future investigations of the biochemical and molecular mechanisms of flowering. Further elucidation of the genetic control of florogenesis in geophytes is essential for understanding their developmental biology, as well as for its agronomic and economic implications.

II. ECOLOGICAL BACKGROUND OF FLOROGENESIS PATTERNS IN ORNAMENTAL GEOPHYTES

The growth cycle of plant species consists of several important stages: vegetative growth, florogenesis, fruiting, and seed production. During this cycle, diverse growth phenomena react to environmental changes such that the developmental programs can be completed and unfavorable conditions survived. The evolution of geophytes under various climatic conditions has led to their adaptation to periods of low and high temperatures and/or drought, and to significant changes in annual cycles of growth, flowering, and dormancy* (Figure 8.1). The significant variation among species with regard to florogenesis during their annual life cycle falls into several distinct groups.

In the geophyte species of tropical and subtropical origin, meristematic activity and visible external growth continue year round, and the apical meristem produces vegetative and reproductive organs autonomously. Several coinciding growth cycles occur in a single plant. Although the growth rate of these species might change during their annual cycle, they have no specific photoperiod or temperature requirements for flowering (Figure 8.1a; Hartsema 1961; Rees 1972; Du Plessis and Duncan 1989; Okubo 2000). Well-known examples of nondeciduous geophytes are *Clivia*, *Hippeastrum*, *Crinum*, and *Zephyranthes*.

Species from temperate zones undergo external growth pauses as a reaction to short days and low temperatures (winter dormancy). These plants form new leaves throughout the year, and the low winter temperatures only slow down their development (Figure 8.1b; Kamenetsky and

---

* The state of dormancy is defined as a complex and dynamic physiological, morphological, and biochemical state of the plant during which there are no apparent external morphological changes. This state is also referred to, in the literature, as growth arrest, rest, quiescence, or the intrabulb developmental period (Rees 1972; Halevy 1990; Le Nard and De Hertogh 1993; Kamenetsky 1996, 2009; Okubo 2000).

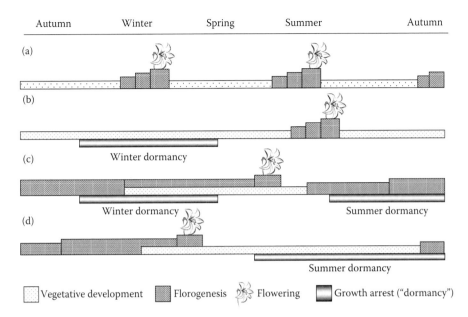

FIGURE 8.1 Patterns of flowering and dormancy in the annual life cycle of geophytes. (a) Subtopic type, (b) Temperate zone type, (c) Irano-Turanian, and (d) Mediterranean type. (Adapted from Flaishman, M., and R. Kamenetsky. 2006. In *Floriculture, Ornamental and Plant Biotechnology, Vol. I,* ed. J. A. Texiera de Silva, 33–43. Isleworth, UK: Global Science Books.)

Rabinowitch 2006). Flowering is promoted by photoperiod and occurs in the summer. Examples of this type are found mostly in the temperate perennial herbaceous genera (e.g., *Allium*, *Lilium*, and *Hosta*).

Geophytes with a sharp thermoperiodic cycle, originating mainly from the Irano-Turanian region, cease underground growth at relatively high temperatures (summer), but for further flowering these species require a period of low temperatures (winter). Meristems remain active during the dormancy period, and are able to produce vegetative and reproductive organs (Figure 8.1c). These species flower in the spring and their above-ground development takes approximately 3–5 months. *Tulipa* or *Paeonia* are good examples of this type.

Species from the Mediterranean and semiarid areas (such as *Cyclamen*, *Pancratium*, and *Bellevalia*) undergo a long summer dormancy period. When the outside summer temperatures are high, the vegetative meristems remain in a state of "stagnation" with no visible activity. Plants are released from dormancy when the temperatures decrease, then sprout and develop leaves and inflorescences during the mild winter. No cold induction is required for floral development or stalk elongation (Figure 8.1d; Widmer and Lyons 1985; Kamenetsky and Fritsch 2002).

Spring ephemerals of the temperate zones flower early in the spring before the surrounding trees sprout leaves (Lapointe 2001). After flowering, the leaves remain green for another month. Once the tree canopy fills out and reduces the available light, the plants enter summer dormancy. During the summer, the trees use most of the available moisture, thus drying out the surrounding soil. *Anemone nemorosa* (wood anemone) and *Trillium* are examples of this type.

Desert ephemerals have a short growth season corresponding to the timing and amount of precipitation. Flowering of these species is "opportunistic", and in some cases, prolonged dormancy in the absence of water, lasting 2–5 years, can occur (Rossa and von Willert 1999). Some examples of prolonged dormancy of perennial geophytes can be found in the Atacama desert of Chile (Vidiella et al. 1999) and the Namaqualand desert in South Africa (Rossa and von Willert 1999; Desmet 2007). Plant species with underground storage organs (bulbs in *Leucocoryne* and *Rhodophiala*,

Bulbinella, Ornithogalum; storage roots in *Chloraea*) can survive long periods of drought, and have a short period of vegetative growth and flowering.

A detailed description of geophyte morphology, biodiversity, and ecological adaptation can also be found in Chapter 3 of this volume.

III. MORPHOLOGICAL, PHYSIOLOGICAL, AND MOLECULAR ASPECTS OF FLORAL TRANSITION

The research of Anton Blaauw and his coworkers in the 1920s to 1950s in The Netherlands (Hartsema 1961) provided the scientific foundation for understanding bulb morphology, the physiology of flower initiation and development, and the temperature treatments required for flowering. The principles of flowering regulation in ornamental geophytes and the subsequent utilization of this knowledge for commercial forcing have been reviewed by Rees (1972, 1992), De Hertogh (1974), and Theron and De Hertogh (2001). De Hertogh and Le Nard (1993b) and their colleagues provide a comprehensive treatise on the physiology of flower bulbs.

In most geophytes, florogenesis can be divided into several consecutive steps: induction, initiation, differentiation (organogenesis), maturation and growth of floral organs, anthesis[*], and senescence (Figure 8.2; Halevy 1990; Bernier et al. 1993; Le Nard and De Hertogh 1993; Flaishman and Kamenetsky 2006). In many species, the differentiation stage involves two steps: (1) the formation of the inflorescence composed of individual flowers and (2) differentiation of the individual flowers and flower organs, including differentiation of sporogenous tissues in the anthers, meiosis, and pollen and embryo sac development.

When propagated from seeds, all geophytes need to reach a certain physiological age and critical mass before flowering can be initiated. In general, a plant passes through three distinct periods: juvenile, adult vegetative (competent), and adult reproductive (determined). The duration of the juvenile stage (physiological age required for flowering) ranges from a few months (e.g., *Ornithogalum dubium* and *Ranunculus asiaticus*) to several years (*Tulipa, Allium,* and *Narcissus*). The ability to flower also depends on the amount of reserves in the bulb. Thus, the minimum bulb circumference needed for flowering varies from 3–5 cm (*Triteleia, Freesia, Allium neapolitanum*) to 12–14 cm (*Tulipa, Narcissus*) and sometimes even 20–22 cm (*Allium giganteum*) (Figure 8.3; Le Nard and De Hertogh 1993; Kamenetsky and Fritsch 2002).

The critical difference between the juvenile and adult stages is the meristem's ability to produce reproductive organs, which is only observed in adult plants. Flowering is possible if a meristem is competent to perceive inductive signals. In addition, flowering competence may also depend on the size of the apical meristem (Halevy 1990; Le Nard and De Hertogh 1993). In many species, the shift

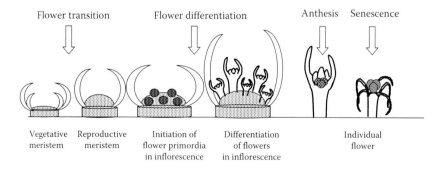

FIGURE 8.2 Schematic representation of the consecutive stages of flower development in geophytes.

---

[*] Anthesis (efflorescence) is the opening of the flower bud, or flowering.

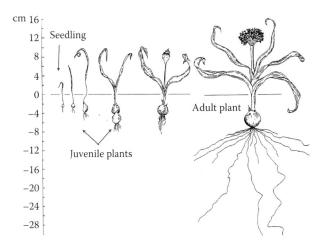

FIGURE 8.3 Development of *Allium rothii* from seedling to generative plant. First flowering occurs in the fifth year of development. (Adapted from Kamenetsky, R. 1994. *Int. J. Plant Sci.* 155:597–605.)

from juvenile to adult phase can be identified by the number of leaves that have unfurled since germination or sprouting. For this reason, identification of the competence to flower based on leaf number is commonly used as a general phase-change indicator in commercial floriculture production (Dole and Wilkins 1999).

Interactions between the plant's genetic background and environmental signals generate a series of molecular and biochemical processes leading to transition of the shoot apical meristem (SAM) from vegetative to reproductive development. At the vegetative stage, the SAM is usually flat and initiates leaf primordia from the periphery toward the center (Figure 8.4a). During flower initiation, the SAM ceases leaf production and shifts to reproductive development (Figure 8.4b). In many species, the transition of the SAM from vegetative to reproductive state is accompanied by spathe[*] formation. Juvenile plants exhibit a monopodial growth habit, becoming sympodial only after the formation of the first generative meristem.

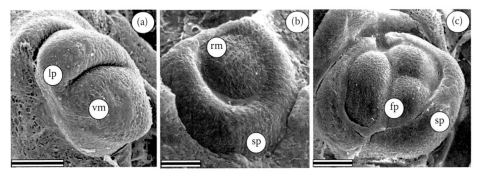

FIGURE 8.4 Scanning electron micrographs of apical meristems in *Allium triquetrum*, bar = 100 μm. (a) Initiation of leaf primordia (lp) in the vegetative meristem (vm); older leaf primordia removed. (b) Initiation of the reproductive meristem. The spathe (sp) differentiates around the reproductive meristem (rm), which is swollen and hemispheric. (c) Differentiation of the centers of development in an inflorescence meristem; first flower primordia (fp) and spathe (sp) are visible.

---

[*] A spathe is a large metamorphosed bract or a pair of bracts enclosing the inflorescence or flower.

Classical studies by Blaauw and coworkers on bulb morphogenesis demonstrated that in various species, flower initiation takes place at different periods of the year (Hartsema 1961). Five main types of florogenesis were identified in that research and later confirmed and elaborated upon by in-depth studies of flower physiology in various species and varieties (Halevy 1990; De Hertogh and Le Nard 1993b and references therein). The main florogenetic types of geophytes are as follows:

1. Flower initiation is usually autonomous and occurs concurrently with continual leaf formation (*Hippeastrum*, *Zephyranthes*, and *Crinum*). Flower buds at different stages of differentiation can be present in the same bulb. Optimal growth temperatures for these species are 20–28°C. The highest-quality flowers are produced at 22/18°C (day/night) under a long photoperiod (Okubo 1993; Vishnevetsky et al. 1997; De Hertogh and Gallitano 2000).
2. Flower initiation occurs a year in advance, and takes place immediately after anthesis of the parent plant, when temperatures are still relatively low (e.g., 3–10°C, *Galanthus*, Langeslag 1989). Further differentiation requires relatively high temperatures (20–25°C, *Leucojum*, Mori et al. 1991a), while lower temperatures, 13–15°C, are more favorable for scape* elongation and anthesis (*Leucojum*, Mori et al. 1991b; *Convallaria*, Le Nard and Verron 1993).
3. Flowers initiate after bulb maturation and harvest, during the summer storage period. The transition from vegetative to reproductive phase occurs without cold induction, and warm temperatures are required for flower differentiation (*Tulipa*, Hartsema 1961; *Ornithogalum dubium*, Roh and Joung 2004; *Narcissus tazetta*, Noy-Porat et al. 2009; *Paeonia*, Kamenetsky et al. 2003). In some species, a prolonged cold period at 4–9°C is required for dormancy release (*Tulipa*, *Hyacinthus*, *Crocus*, Rees 1992). Optimal temperatures for the elongation of flower stalks after planting and anthesis are 15–20°C (Le Nard and De Hertogh 1993).
4. Flowers initiate during winter storage or growth (*Lilium*, *Galtonia*, *Allium cepa*). Low temperatures promote flower initiation. Flower differentiation and flowering respond positively to a wide range of temperatures (13–27°C, *Lilium longiflorum*, Miller 1993), and a long photoperiod supports flower development.
5. Flowers form after planting in the spring (*Gladiolus*, *Freesia*). Flower initiation occurs in growing plants following the formation of several green leaves. Mild growth temperatures and a long photoperiod are usually essential for floral initiation and scape elongation (Halevy 1985).

In addition, a major distinction can be made between the geophytes in which flower initiation takes place within the bulb during the "dormancy" period prior to growth ("Tulip type", Figure 8.5a) and those in which flowers are initiated during active growth following the development of several leaves ("Lily type", Figure 8.5b). Environmental requirements and temperature control vary between these two types: in the first case, the formation of the reproductive meristem requires relatively high temperatures, while in the second, vernalization† and a long photoperiod are beneficial for florogenesis and further flowering.

It is interesting to note that *Lilium longiflorum* is of subtropical origin and, under natural condition, does not require low temperatures for flower initiation. For example, it grows wild and flowers along the coast of Ishigaki Island, Japan (24°20′N, 124°09′E), where the minimum temperature in January (the coldest month) is 18.3°C. Neither low temperatures in winter nor long photoperiods at sprouting in spring are available under these environmental conditions (H. Okubo 2011, pers. comm.). One can argue that the ability to initiate flowering in response to low temperatures was developed during a long selection process in a temperate climate. However, lily bulbs collected in

---

* Scapes are flowering stems, usually leafless, arising from the underground organ.
† Vernalization is the acquisition of a plant's ability to flower by extended exposure to low temperatures.

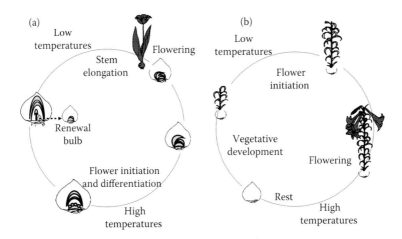

FIGURE 8.5 Schematic representation of the two types of geophytes, with distinct patterns of flower initiation and differentiation during an annual cycle. (a) "Tulip type": Floral initiation occurs within the bulb during the summer and requires relatively high temperature. Following flower differentiation, low temperatures are essential for further elongation of floral stem and anthesis. (b) "Lily type": Floral initiation occurs outside the bulb, following leaf development and stem elongation. Low temperatures (vernalization) are essential for floral initiation and subsequent flowering.

subtropical habitats respond positively to low temperatures and can be stored frozen for long periods of time, even though they never experience such conditions in nature. Therefore, we assume that the ability to require or survive low temperatures is embedded in the *L. longiflorum* genome.

Sensing and integration of external cues by the plant are necessary for optimal timing of flowering during the year, for synchronized flowering within a population, and to ensure successful seed development (Erwin 2006). In general, a number of primary external environmental cues affect flowering and allow plants to synchronize it with the seasons: temperature, photoperiod, irradiance, and stress (e.g., fire or water level). In addition, the lack of stress or supraoptimal levels of nutrients can also affect progress toward flowering. Therefore, horticultural manipulations aimed at directing flowering to specific periods usually make use of these cues (Erwin 2006).

A. Temperature

Temperature is the most potent factor affecting florogenesis in flower bulbs (Hartsema 1961; De Hertogh and Le Nard 1993b). It affects the timing and rate of flower initiation and differentiation, which are critical for commercial forcing in order to meet the market demand.

1. Ambient and High Temperature

In many geophytes, cold exposure is not required for floral initiation; on the contrary, warmer temperatures are essential for this process (Figure 8.5a). The first experimental evidence for this was provided by Blaauw (1923, 1924a,b), who concluded that flower formation in hyacinth requires a minimum of four leaves and relatively high temperatures. Luyten et al. (1932) were able to advance intrabulb flower formation in hyacinth with an optimal temperature of 25°C. Even higher temperatures (30–35°C) were found to hasten flowering in both *Tulipa* and *Hyacinthus* (Rees 1992). In saffron (*Crocus sativus*), flower formation occurs after the winter, when temperatures rise. The optimal temperature for flower initiation is 23°C, and no flower will form in corms stored below 9°C (Molina et al. 2005). In *Ornithogalum dubium*, flower initiation was obtained at 22°C (Roh and Joung 2004). Floral initiation and reproductive development in *Narcissus tazetta* were supported by high temperatures with an optimum of 25°C, but both sub- and supraoptimal temperatures (20°C and 30°C, respectively) reduced differentiation, while lower temperatures (12°C) inhibited florogenesis completely (Noy-Porat et al. 2009). In the Dutch iris, bulb storage at 30°C induced a higher number of flower buds than lower temperatures, and no flowers formed under 10°C (Yue and Imanishi 1990). From these examples, it is clear that temperatures well above 20°C are needed for intrabulb flower initiation in this group of species.

2. Vernalization

In general, vernalization has been defined as the acquisition or acceleration of the ability to flower by a chilling treatment (Chouard 1960). In many plant species originating from moderate-climate regions, prolonged exposure to temperatures between 0°C and 16°C for 6–12 weeks induces and synchronizes flowering. Presumably, the vernalization requirement ensures that a plant will not flower until the end of winter in temperate climates. In some cases, perennial species have a vernalization requirement followed by a requirement for a long photoperiod. There are two distinct phases within the period ranging from completion of the vernalization requirement (cold exposure) to flower initiation. The first phase, starting immediately after vernalization, is characterized by the capacity for reversal of flower induction, a phenomenon termed as "devernalization" (Lang 1965). The main factor in devernalization is exposure to warm temperatures (25–30°C), but low irradiance or short photoperiod (or their combination) can also cause devernalization (Anderson 2006). In the second phase, floral induction remains stable and cannot be reversed (Erwin 2006).

Conceptually, vernalization differs from dormancy release, which can also be achieved by cold exposure (Horvath 2009). Dormancy refers to a state of growth arrest during the annual developmental cycle, and dormancy release is needed to initiate regrowth (see also Chapter 9 in this volume). Vernalization refers to the promotion of flowering following cold exposure. For example, lilies which are constantly grown at warm temperatures are not dormant, but will remain in the vegetative stage, producing a large number of leaves (M. Zaccai, unpublished data). Exposure of such plants to cold temperature will enable their flowering. However, it has been hypothesized that in perennial plants, both processes—vernalization and endodormancy release—have common mechanisms of action (Horvath 2009).

Many ornamental geophytes require vernalization to flower (Figure 8.5b; De Hertogh and Le Nard 1993b; Dole 2003; Streck and Schuh 2005). Vernalization can be an obligatory factor for flower induction, as in some lily species (Miller 1993; Zlesak and Anderson 2009), or a nonobligatory factor when low temperature has only an accelerating effect on flowering, such as in *Ornithogalum thyrsoides* (Roh and Hong 2007).

The effect of vernalization has probably been studied most extensively in lilies, due to their commercial impact and obligatory requirement of cold exposure for effective flowering. Typically, the lily plant is receptive to vernalization at various physiological stages, from stored bulb to shoot elongation in growing plants (Roh and Willkins 1977b; Miller 1993). Cold storage of lily at 2–10°C accelerates flowering and reduces the height and the leaf and flower number in a quantitative manner, up to a saturation point of 6 weeks (Roh and Willkins 1977a,b,c; Miller 1993; Dole and Willkins 1994, 1999; Holcomb and Berghage 2001). After planting, temperatures higher than 21°C can cause devernalization of cold-treated bulbs, thereby delaying flowering (Miller 1993). Similarly, in *Eucharis grandiflora*, a short exposure to 15°C was required to induce flowering, while continuous growth at a constant temperature of 20°C led to the reversal of the apical meristem from reproductive to vegetative state (Doi et al. 2000).

B. Photoperiod

Photoperiodism occurs in both animals and plants and is defined as the ability to detect day length (or, more correctly, night length). Photoperiodic flowering responses can be divided into several groups: short day (SD), long day (LD), day-neutral, intermediate day, and ambiphotoperiodic (Thomas and Vince-Prue 1997). Within the SD and LD groups, there are plants that exhibit an obligate (qualitative) or facultative (quantitative) response (Lang 1965; Erwin 2006). Species with a facultative response will flower under any day-length condition, but flowering will be enhanced by the appropriate photoperiod. In contrast, species with an obligate or qualitative response will not flower unless they are exposed to the appropriate photoperiod.

In geophytes, photoperiodism can interact with temperature requirements and regulate growth processes, flowering, the formation of underground storage organs, and the onset of dormancy (De Hertogh and Le Nard 1993b, and references therein). For example, LD conditions enhance flower initiation and flowering in lilies. Although photoperiod cannot fully substitute for vernalization, it acts in an additive manner with vernalization in hastening flower initiation (Miller 1993; Dole and Wilkins 1994). SD conditions support scape elongation and flowering in the desert hysteranthous geophyte *Colchicum tunicatum* (Gutterman and Boeken 1988), although in general, short photoperiod induces dormancy and bulbing in most geophytes (De Hertogh and Le Nard 1993b; Masuda et al. 2006). Cyclamen is an example of a day-neutral plant (Erwin 2006). Other spring ephemerals and species from regions with thermoperiodic cycles, for example, *Tulipa* or *Narcissus*, are also not photoperiodic, as their growth cycle is relatively short, and flowering is controlled mainly by temperature (Le Nard and De Hertogh 1993; Anderson 2006).

C. Irradiance and Light Quality

Irradiance during vernalization is critical in some species, with higher light intensities tending to promote flowering. Supplementary lighting accelerates flowering in lily (Treder 2003), while low light intensity delays anthesis in *Gladiolus* (Cohat 1993). Light quality can also affect the plant's response to vernalization, for example, in *Lilium longiflorum*, a low red:far-red light ratio is antagonistic to vernalization in the context of flower initiation: plants exposed to a high red:far-red ratio were completely induced to flowering by a 4-week vernalization treatment, while vernalization for 8 weeks under light with a low red:far-red ratio was not inductive for flowering (Erwin 2006).

Light quality and spectrum have been shown to affect plant development and flowering in a number of ornamental cut-flower crops (Shahak 2008; Ovadia et al. 2009). Initial results with *Ornithogalum dubium* showed that plants growing under red-colored nets flower 3 weeks earlier than those growing under a black net (M. Oren-Shamir, pers. comm.).

D. Hormonal Regulation of Flower Induction

In general, several main classes of plant hormones have been characterized: gibberellic acids (GAs), auxins, cytokinins, ethylene, and abscisic acid (ABA). More recently, additional compounds, for example, jasmonates, brassinosteroids, and strigolactones, have been categorized as phytohormones as well. All these hormones have been reported to play various roles in the flowering process of different plant species (Davis 2009; Domagalska et al. 2010; Waldie et al. 2010). GA, in particular, has been linked to floral induction, and its exogenous application was proven to promote flowering in a biennial species under noninductive conditions (Lang 1957). However, GA has never been considered a universal flowering hormone and its role in the process of floral transition is still under investigation (Hisamatsu and King 2008; Mutasa-Göttgens and Hedden 2009; Domagalska et al. 2010).

In geophytes, the most accurate link between endogenous level of GA and floral transition has probably been provided by the work of Naor et al. (2008), who found that the GA level undergoes a strong, temporary increase in elongating buds of colored *Zantedeschia* during storage. This increase was observed about 15 days prior to the appearance of inflorescences, suggesting that GA is potentially involved in floral induction. In the view of the authors, the increase in GA prior to flower initiation was sharp and transient, and not observed during the whole period of elongation. Exogenous application of GA_3 increased the number of flowering stems per plant in *Zantedeschia*, while uniconazole, a GA inhibitor, led to a reduction in the rate of flowering plants (Naor et al. 2009). GA_3

application has been shown to increase the rate of flowering in *Zantedeschia* (Brooking and Cohen 2002) and to hasten flowering in gladiolus (Cohat 1993).

Additional studies on the endogenous presence of phytohormones, mainly ABA, in geophytes have been performed in the context of dormancy release during cold treatments of the bulbs (Xu et al. 2006; Kondrat'eva et al. 2009). However, as these investigations are not primarily related to floral induction, they are not further covered here.

The effects of phytohormones on flowering have mostly been studied via their exogenous application: treatments with growth regulators by bulb dip, foliar drench, foliar spray, or soil drench are routinely used to control growth and flowering in bulbous species (Le Nard and De Hertogh 1993).

The "florigen" theory, first proposed in the early twentieth century, states that the stimulus for flowering is a unique and universal substance, a "flowering hormone", produced in the leaves and acting in the SAM (Chaïlakhyan 1937, 1975; Machackova and Krekule 2002). Although at the time, and for many years afterward, this substance was not found, the idea of a florigen was never totally discarded. Various theories, involving multiple stimuli, have since been proposed, mainly in the context of photoperiod (Machackova and Krekule 2002). Recently, the concept of universal flowering stimulus has made a triumphant comeback, as the product of the gene *FLOWERING LOCUS T* (*FT*) has been defined as an element of the floral stimulus, leading to the switch from vegetative to reproductive stage at the meristem in various species (Corbesier and Coupland 2006; Kobayashi and Weigel 2007, see also Section III.F below).

Plant hormones belonging to the GA, auxin, and cytokinin groups have been shown to play an important role in the onset of flowering of several geophytes *in vitro* (reviewed by Ziv and Naor 2006). Although these studies provide valuable information on florogenesis in geophytes, the effects of the hormones might as well be different on flowering *in vivo*.

E. Autonomous Flower Initiation

In some ornamental geophytes, flower initiation occurs autonomously, that is, independent of environmental factors, as has been suggested for *Hippeastrum* (Okubo 1993) and *Anemone coronaria* (Ben-Hod et al. 1988). In certain tropical bulbs (Rees 1966) and in *Nerine* (Vishnevetsky et al. 1997), flower buds are produced at intervals with leaf production. Since several flower stalks are initiated in the same bulb, *Hippeastrum*, *Cyrtanthus*, and other Amarillidaceae species with similar patterns of florogenesis can simultaneously produce more than one flower stalk, which is a desirable trait for forcing (Okubo 1993; Slabbert 1997; Theron and De Hertogh 2001).

F. Molecular Regulation of Floral Transition in Geophytes

To date, most of our knowledge on the genetic and molecular mechanisms underlying flower initiation and formation in plants has been obtained from model species, mainly *Arabidopsis thaliana*. These studies have resulted in the molecular elucidation of environmental and internal pathways that induce or repress flowering (Figure 8.6a). The pathways are mediated by "integrator" genes which, in turn, upregulate floral meristem identity genes. This process enables the switch from vegetative to reproductive phase in the meristem (Mouradov et al. 2002). Floral inductive pathways are integrated into a flowering network, which contains several steps, representing a site of signal integration (Tremblay and Colasanti 2006; Tsuji et al. 2011). In *Arabidopsis*, these pathways involve an array of transcription factors [for example, *CONSTANS (CO)*, *FLOWERING LOCUS C (FLC)*, *SUPPRESSOR OF OVEREXPRESSION OF CONSTANS1 (SOC1)*, *LEAFY (LFY)*, *APETALA1 (AP1)*], regulators of chromatin structure [*VERNALISATION2 (VRN2)*], the putative kinase inhibitor *FT*, as well as many other genes (Simpson and Dean 2002; Song et al. 2010).

Day length, the circadian clock, and light intensity affect floral induction in *Arabidopsis* via the so-called photoperiodic flowering pathway (Figure 8.6a). The *CO* pathway, regulating the photoperiodic response, is quite conserved among species, although major modifications, responsible for the

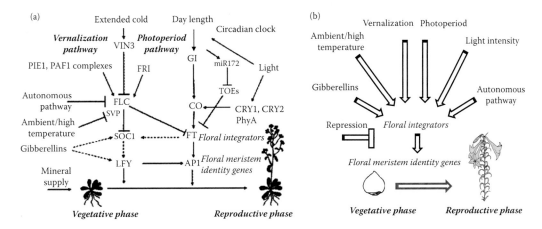

FIGURE 8.6 Schematic representation of the environmental and endogenous cues for the floral transition in model plants (a) and geophytes (b). (a) In *Arabidopsis*, numerous environmental and internal pathways are mediated by "integrator" genes which, in turn, upregulate floral meristem identity genes. This process enables the switch from vegetative to reproductive phase in the meristem. (Adapted from Flowering Time Control: From Natural Variation to Crop Improvement. 2012. http://www.flowercrop.uni-kiel.de, April 2012.) (b) In the geophytes, flower induction occurs autonomously or is induced by environmental signals. Homologues of the floral integrator and meristem identity genes from model plants have been isolated in several species (see text).

shifts from LD to SD flowering in plants such as *Oryza* (rice) and *Pharbitis* (Yano et al. 2000; Liu et al. 2001; Hayama et al. 2003; Hayama and Coupland 2004), have been reported.

CO regulates the expression of the floral integrator genes *FT* and *SOC1*, leading to floral initiation (Figure 8.6a). The *Arabidopsis FT* gene and its homologues[*] have proven to be powerful regulators of flowering, involved in most of the flowering pathways (Boss et al. 2004; Lee et al. 2006; Moyroud et al. 2009). Recent publications suggest that *FT* might be a general systemic regulator of plant growth (Shalit et al. 2009). It has also been shown in model plants that *FT* is activated by different environmental signals, including photoperiod, light quality, vernalization, and ambient temperature (Cerdán and Chory 2003; Henderson and Dean 2004; Balasubramanian et al. 2006; King et al. 2008; Turck et al. 2008). However, *FT*'s involvement in floral induction in *Arabidopsis* and tomato has mainly been studied in the context of the photoperiodic pathway (reviewed by Kobayashi and Weigel 2007; Turck et al. 2008). The light signal was shown to be perceived in the leaves, where a cascade of genes are induced leading to the formation of *FT* mRNA. FT protein is then created in the leaves and moves to the SAM, thus acting as a "florigen" (Corbesier et al. 2007; Jaeger and Wigge 2007; Lin et al. 2007; Tamaki et al. 2007; Turck et al. 2008). A bZIP transcription factor, FD, preferentially expressed in the shoot apex, is required for FT to promote flowering (Abe et al. 2005). Gene activation by the FT/FD complex is considered the earliest event in the floral transition to occur in the meristem itself (Turck et al. 2008).

FT homologues have been shown to play critical roles in the induction of flowering in other plant species, including LD, SD, and day-neutral plants (Ballerini and Kramer 2011). In the perennial herbaceous eudicot *Aquilegia formosa*, which is emerging as a new model for the investigation of plant ecology and evolution, the *FT* homologue *AqFT* is expressed before the transition to flowering under both LD and SD conditions. Although vernalization is critical to flowering in *Aquilegia*, low temperature is not strictly required for the transcriptional activation of *AqFT*.

The "integrator" genes, such as *FT* and *SOC1*, function together to upregulate floral meristem identity genes such as *LFY* and *AP1* (Figure 8.6a; Krizek and Fletcher 2005; Benlloch et al. 2007;

---

[*] A homologous gene is related to another gene by descent from a common ancestral DNA sequence. The term "homologue" may also apply to the relationship between genes separated by the event of speciation (orthologue) or to the relationship between genes separated by the event of genetic duplication (paralogue).

Kaufmann et al. 2010). These genes are mainly expressed in the SAM, the site at which the floral transition occurs. The activity of *LFY* and *API* is antagonized by *TFL1*, a gene required for the maintenance of SAM indeterminacy (Parcy 2005; Bernier and Perilleux 2005). At a later stage, the meristem identity genes activate the floral organ identity genes in discrete meristematic regions of the inflorescence, with the consequent differentiation of flower primordia. Collaborative action of the entire regulatory network guarantees the proper timing and correct spatial control of floral meristem specification, maintenance, and development (Vijayraghavan et al. 2005).

Significant progress has been made in understanding the molecular basis of vernalization in model plants (Figure 8.6a; Distelfeld et al. 2009; Amasino 2010). Basically, *FLC*, a MADS-box gene encoding a potent repressor of flowering, is active in meristems of *Arabidopsis* in the autumn. During the winter, vernalization allows meristem competence to flower by repressing *FLC* expression. Once it has been repressed by vernalization, *FLC* remains turned off for the rest of the plant's annual cycle, even after the return of warm conditions; in other words, the repression is epigenetic in the sense that it is mitotically stable in the absence of the inducing signal (cold exposure). It is interesting to note that clear orthologues of the *FLC* locus, which is critical for flowering in *Arabidopsis*, have not been identified outside the core eudicots (Reeves et al. 2007). A mechanism similar to that in *Arabidopsis* was found in cereals requiring vernalization: a flowering repressor [*VERNALIZATION 2* (*VRN 2*)] is turned off by the cold and *VRN1*, which encodes a MADS-box protein, promotes flowering (Yan et al. 2004).

The genetic pathways derived from studies of model plants define the basic mechanisms that are common to most plants. However, assignment of the main pathways does not necessarily mean that the same hierarchy exists in all plant species. Since the environmental stimuli perceived by plants vary according to their geographical habitats, distinct inductive mechanisms are expected to have evolved independently in different species (Tremblay and Colasanti 2006).

The in-depth knowledge accumulated on physiological mechanisms regulating flower initiation represents a valuable background for biochemical and molecular studies in ornamental geophytes. Critical analysis of the information on physiological mechanisms in geophytes has revealed several pathways of flower induction (Figure 8.6b). In general, these pathways resemble those found in model plants, but only little information is available on their molecular regulation. Most of the molecular studies reported to date have been based on the search for homologues of main floral regulators from *Arabidopsis*.

Recently, homologues of the flowering integrator gene *FT* have been isolated in two geophytes: onion and narcissus. In onion (*Allium cepa*) three *FT*-like (*AcFTL*) genes were isolated and characterized (Taylor 2009; Taylor et al. 2010). Their expression profiles showed no diurnal or circadian pattern, suggesting that they are not orthologues[*] of the *Arabidopsis FT*. Preliminary evidence suggests that there is a large family of *FT*-like genes in onion, as has been reported for other monocots, such as rice (Izawa et al. 2002).

The *FT* homologue *NtFT* was isolated from *Narcissus tazetta*, and its expression was monitored during the plant's annual cycle under different temperature conditions (Noy-Porat 2009). In addition, quantitative real-time PCR and RNA *in situ* hybridization were employed to analyze spatiotemporal expression patterns of *NtFT* during early stages of florogenesis. It was shown that *NtFT* expression correlates with the timing of floral induction under ambient or high temperatures. *NtFT* was not expressed in mature foliage leaves, but was detected in leaf primordia and the apical meristem inside the bulb. The gene's expression was relatively high in bulbs subjected to high or ambient temperatures during summer storage, whereas it was not detected in bulbs stored at 12°C. These findings suggest that *NtFT* might be a target of temperature control rather than photoperiodic regulation.

---

[*] Orthologues are genes in different species that evolved from a common ancestral gene by speciation. Normally, orthologues retain the same function in the course of evolution. Identification of orthologues is critical for the reliable prediction of gene function in newly sequenced genomes.

A study of *LFY* homologues in angiosperms revealed that, while the sequence of these genes is rather conserved, they play different roles among species, including regulation of plant and inflorescence architecture and leaf shape, in addition to their roles at the various stages of flower development (Moyroud et al. 2009). In *Titanotrichum oldhamii*, which produces flowers and bulbils sequentially, RT-PCR studies showed strong expression of the *LFY/FLO* homologue *GFLO* during normal flower development, in both the apical meristems of the inflorescences and the flower buds (Wang et al. 2004). However, when plants were exposed to SD conditions or in the autumn, bulbil formation began in the inflorescence instead of the flowers and the level of *GFLO* transcript was markedly decreased (Wang et al. 2004). In *Aquilegia formosa*, the expression patterns of the flowering genes *AqLFY* and *AqAGL24.2* suggest differential control of meristem progression from inflorescence to floral identity (Ballerini and Kramer 2011).

LFY homologues have been isolated from several flower bulbs (Rotem et al. 2007, 2011; Wang et al. 2008; Noy-Porat et al. 2010). In *Narcissus tazetta*, the expression pattern of the *LFY* homologue *NLF* was investigated during flower development at various temperatures. Under ambient summer conditions in Israel, *NLF* was detected in the apical meristem within the bulb throughout its development, from the vegetative state to fully differentiated flowers. A dramatic increase in its expression was observed twice: during meristem transition from vegetative to reproductive development and during differentiation of flower primordia in the inflorescence. *NLF* was also found in leaf primordia, but not in mature leaves, suggesting a role for this gene in young leaf development. Temperature conditions during florogenesis affected *NLF* expression in the apical meristem. Summer storage of *Narcissus* bulbs at a constant 30°C resulted in florogenesis and an expression pattern of *NLF* similar to that observed under ambient temperatures. At 12°C, meristems remained morphologically vegetative but relatively high expression of *NLF* was observed in these vegetative meristems. These results suggest that *NLF* plays a major role in floral development of *N. tazetta*, but that temperature does not affect *NLF* expression directly (Noy-Porat et al. 2010).

A search for genes involved in the control of flowering in garlic (*Allium sativum*) resulted in the identification of the *LFY* homologue *gaLFY*. Further comparative analyses of gene expression revealed two *gaLFY* transcripts differing in 64 nucleotides (Rotem et al. 2007). Spatio-temporal accumulation of *gaLFY* was strongly associated with reproductive organs, significantly increased during florogenesis and gametogenesis and down-regulated in the vegetative meristems and topsets of the inflorescence (Figure 8.7). The two alternative transcripts of the gene showed different expression patterns: a high level of the long *gaLFY* transcript coincided with floral transition, while further upregulation of this gene in the reproductive organs was associated mainly with the short *gaLFY* transcript. It was concluded that *gaLFY* is a multifunctional gene, possibly involved in different stages of the sexual reproduction of garlic (Rotem et al. 2011).

The *LFY* homologue in lily, *LiLFY*, was found to be expressed in the SAM, young flower buds, mature leaves, and mature floral organs (Wang et al. 2008). In *Hyacinthus*, the *LFY* homologue *HLY* was cloned, but no information was published apart from its sequence (GenBank accession no. AY520841). In the *Ophrys* orchids, *OrcLFY* was found to be a new and useful marker for reconstructing molecular phylogenies at low taxonomic levels (Schlüter et al. 2007). Taken together, the data accumulated on *LFY* homologues in geophytes suggest that, like in *Arabidopsis* and other dicots, these genes play a central role in the transition from vegetative to reproductive development, and can probably be used as markers for flower transition in geophytes. In addition, *LFY* homologues might be involved in various stages of vegetative and reproductive development.

In addition to the search for specific genes based on homology among species, sequencing of cDNA stretches to generate expressed sequence tags (ESTs) has proven to be a powerful, economical, and rapid approach to identifying genes that are preferentially expressed in certain tissues or at specific phases of development. The availability of a significant EST database offers the possibility of studying gene expression in different tissues and organs and can serve as a source for the isolation of homologous genes in target species and commercial crops. Recently, a saffron (*Crocus sativus*) gene collection, based on EST sequences, was constructed (http://www.saffrongenes.org/), to be

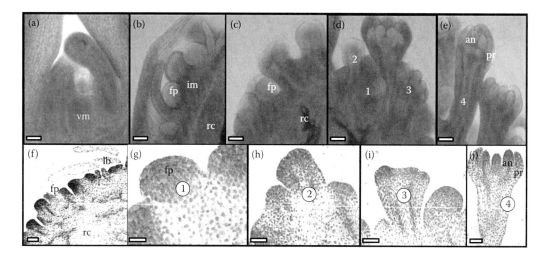

FIGURE 8.7 (**See color insert.**) *gaLFY* expression in the vegetative meristem and during early florogenesis stages in garlic. (a–e) FISH, as observed by confocal microscopy. (f–j) DIG-labeled *in situ* hybridization, as observed by light microscopy. Bars: (a–e) 200 μm, (f, h–j) 100 μm, (g) 50 μm. (a) Vegetative meristem (vm) in plants at the fourth foliage leaf stage of development, in November. (b) Meristem transition and inflorescence initiation. Note the high level of fluorescence signal in the receptacle (rc), inflorescence meristem (im) and flower primordia (fp). (c) Differentiation of flower primordia (fp). Note the weaker signal in the receptacle (rc). (d) Early developmental stages of flower buds (stages 1–3): *gaLFY* expression is evident in the center of the reproductive meristem at stage 1, in the apical and lateral parts of the primordium at stage 2, and in flower organ primordia at stage 3. (e) Early developmental stages of flower buds (stage 4). Perianth lobes (pr) and anthers (an) are visible. (f) Initiation of flower primordia in the inflorescence, comparable to b and c. Note the absence of *gaLFY* expression in the leafy bracts (lb). (g) Reproductive meristem prior to differentiation of flower organs, comparable to stage 1 in d. *gaLFY* expression is visible in the cytoplasm (blue staining). Note the small meristematic cells with large nuclei. (h) Initial differentiation of individual flowers. The apical and lateral parts of the primordium are stained blue, comparable with stage 2 in d. (i) Flower with perianth and anther primordia, comparable with stage 3 in d. (j) Individual flowers: perianth lobes (pr) and anthers (an) are differentiated, comparable with stage 4 in e. (From Rotem, N. et al. 2011. *Planta* 233:1063–1072. With permission.)

used for target transcript analyses during *C. sativus* stigma development (Moraga et al. 2009). In addition, a total of 21,595 ESTs collected from a cDNA library were used to construct GarlicESTdb, an integrated database and mining tool for large-scale garlic (*Allium sativum*) sequencing (Kim et al. 2009).

The research on *Arabidopsis* and cereals shows that although the general mechanism of vernalization inducing the down-regulation of flower inhibitors is conserved among distant species, the sequences of the main regulating genes vary in different species (Distelfeld et al. 2009; Amasino 2010). In this case, the search for homologous genes in lily and other species with vernalization requirements appears to be a flawed approach. Therefore, a more general strategy, based on differentially expressed genes, was taken to identify molecular factors involved in the vernalization response in *Lilium longiflorum* (Zaccai et al. 2010). Subtraction libraries were constructed from cDNA isolated from meristems of vernalized and nonvernalized lily bulbs. It was shown that the overall RNA transcription (ratio between mRNA and total RNA), as well as homology to known genes, was much higher in vernalized than in nonvernalized meristems. Accordingly, ESTs from vernalized bulbs were involved in a much larger number of molecular and metabolic mechanisms than those from nonvernalized lily bulbs (Zaccai et al. 2010). Clones from the subtraction libraries showed homology to genes involved in, among others, seed dormancy and chromatin modification in *Arabidopsis* (Ko et al. 2010), which are also known to play a key role in the vernalization response. In the same study (Zaccai et al. 2010), a homologue of *SOC1* from *Arabidopsis*, whose expression is quantitatively upregulated by vernalization (Sheldon et al. 2006), was identified in *L. longiflorum*. Its expression was strongly upregulated in the bulb during cold exposure, suggesting that this gene might be involved in the vernalization pathway in lily.

Despite the importance of understanding the molecular mechanisms of vernalization in ornamental geophytes, these processes remain largely unknown. An elucidation of genetic control of the flowering response would not only enable future production of flower bulbs with reduced cold requirements, but would also provide molecular markers of optimum vernalization treatments for the improvement of flowering time and flower quality.

IV. FLORAL DIFFERENTIATION AND DEVELOPMENT

The ornamental value of numerous popular species is based on their multiflowered inflorescences, which sometimes carry 200–500 flowers (e.g., species of *Allium*, *Eremurus*, and *Scilla*). However, some ornamental bulbs have only a few large flowers per inflorescence (e.g., *Lilium*, *Narcissus*, *Hippeastrum*, and *Amaryllis belladonna*). Inflorescences are usually terminal and represent spike, raceme, corymb, panicle, umbel, or cyme types. During inflorescence formation, the SAM goes through several significant steps/phases, from transition to the reproductive stage, to differentiation of the flowers and flower-bearing bracts (Table 8.1, Figure 8.8).

Morphological variability of the individual flowers of geophytes is also remarkable. Most flower bulbs belong to the botanical order Liliales, which produces flowers with six perianth lobes (tepals) arranged in outer and inner whorls, three or six stamens, and a tricarpellary pistil located in the center of the flower (Table 8.2, Figure 8.9). Ovaries of differentiated flowers often include the nectaries, consisting of secretory cells, on the outer ovary walls (Fritsch 1992). In Amarillidaceae and Liliaceae, the perianth is formed first (Table 8.2, Figure 8.9; Waterschoot 1927; Blaauw 1931; Le Nard and De Hertogh 1993; Kamenetsky and Rabinowitch 2002). In Dutch iris, the first stage after initiation is the formation of the first whorl of stamens (A1) (Cremer et al. 1974).

The environmental conditions required for flower differentiation (e.g., photoperiod, a minimum irradiance, or a certain range of temperatures) vary in many species from that for flower induction.

TABLE 8.1

Stages during Inflorescence Development in Bulbous Species of Liliaceae and Amaryllidaceae and Their Abbreviations

Stage Symbol	Inflorescence Developmental Stages
I	Vegetative apical meristem—leaf-forming stage
II	Transition to reproductive stage—doming of apex prior to flower initiation (prefloral stage)
Sp	Spathe initiation
Pr	Appearance of primordia of first flowers
Br	Appearance of initials of flower-bearing bracts or specialized leaves
Bo	Secondary bracts

Source: Adapted from Blaauw, A. H. 1931. *Koninklijke Akademie van Wetenschappen Amsterdam (Section 2)* 29:1–90; Beyer, J. J. 1942. *Mededeelingen van de Landbouwhoogeschool, Wageningen* 46:1–17; *The Physiology of Flower Bulbs*, Le Nard, M., and A. A. De Hertogh (eds). Blub growth and development and flowering, 29–43, Copyright (1993), Amsterdam: Elsevier.

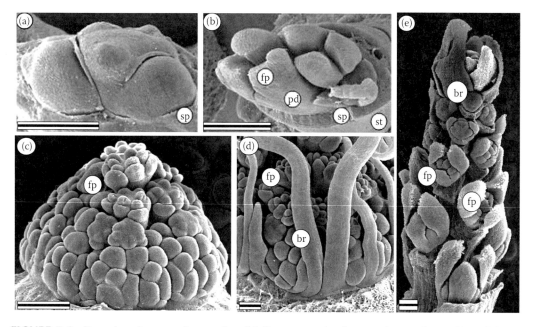

FIGURE 8.8 Scanning electron micrographs of inflorescence development in geophytes. Bar = 0.1 mm. (a) Initial stages of inflorescence differentiation in *Narcissus tazetta*. The spathe (sp) is removed. (b) A number of differentiating flower primordia (fp) are visible in the inflorescence of *Narcissus tazetta*. Pedicels (pd) and stem (st) are elongated; the spathe (sp) is removed. (c) Differentiation of flower primordia (fp) and individual flowers of *Allium scabriscapum*. (d) Differentiation of flower primordia (fp) and bracts (br) in the inflorescence of *Allium sativum*. (e) Differentiated inflorescence of *Ornithogalum dubium*.

A. EFFECT OF ENVIRONMENTAL CONDITIONS ON FLOWER DIFFERENTIATION

1. Temperature

In ornamental geophytes, temperature is a major factor affecting inflorescence and flower differentiation and production of floral organs. Optimal temperature is usually species specific, and interacts with photoperiod and irradiance level. Temperature might affect the later stages of

TABLE 8.2
Stages during Flower Development in Bulbous Species of Liliaceae and Amaryllidaceae and Their Abbreviations

Stage Symbol	Developmental Stages of Individual Flower
P1	Formation of first whorl of perianth (tepals)
P2	Formation of second whorl of perianth (tepals)
A1	Formation of first whorl of androecia (stamens)
A2	Formation of second whorl of androecia (stamens)
G	Formation of trilobed gynoecium (pistil)
G+	Style is distinct and lobes are discernable
PC	Paracorolla (e.g., trumpet of *Narcissus*)

Source: Adapted from Blaauw, A. H. 1931. *Koninklijke Akademie van Wetenschappen Amsterdam (Section 2)* 29:1–90; Beyer, J. J. 1942. *Mededeelingen van de Landbouwhoogeschool, Wageningen* 46:1–17; *The Physiology of Flower Bulbs*, Le Nard, M., and A. A. De Hertogh (eds). Blub growth and development and flowering, 29–43, Copyright (1993), Amsterdam: Elsevier.

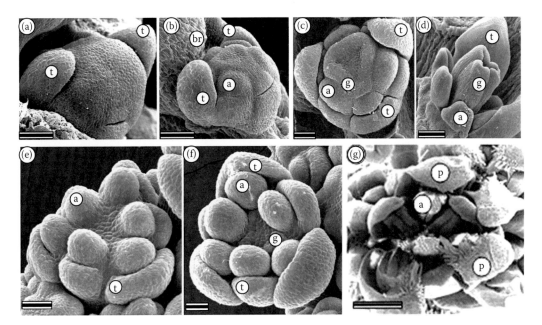

FIGURE 8.9 Scanning electron micrographs of the development of individual flowers. Bar = 0.1 mm. (a–d) Consequent stages of flower development in *Ornithogalum dubium*. Tepals (t), anthers (a) and gynoecium (g) are visible. (e–f) Differentiation of individual flowers of *Allium scabriscapum*. (g) Differentiated flower of *Narcissus tazetta*. Perianth (p) elongates and encloses anthers and gynoecium.

flower initiation, differentiation of floral organs, or flower bud abortion (De Hertogh and Le Nard 1993a).

High-temperature inhibition of flowering due to reduced overall carbon assimilation, increased respiration, or reduced ability of floral organs to recruit photoassimilates has been noted for many plant species (Erwin 2006). In addition to reducing the flower number, high temperature might induce floral sterility, resulting in reduced seed set, fruit size, and yield. Such a reduction in yield is a common problem in commercial horticulture production worldwide.

As a rule, flower differentiation in geophytes is enhanced by warm temperatures, as opposed to vernalization, which is beneficial for flower induction. For example, a temperature of 30°C promotes flower differentiation in iris (Yue and Imanishi 1990) and *Eucharis grandiflora* (Doi et al. 2000). In other plants, a larger number of florets per inflorescence were obtained at 25°C than at 30°C (*Ornithogalum dubium*, Roh and Joung 2004). The precise number of floral organs formed by a tulip flower is influenced by the temperature during bulb storage (De Hertogh 1974).

In comparison with tulips and hyacinths, the daffodil has two additional stages of flower formation (Tables 8.1 and 8.2). In 'King Alfred' daffodils, floral differentiation occurs during the spring and is almost complete at the time of bulb harvest (Huisman and Hartsema 1933). If the bulbs are lifted when the apical meristem has only reached stage I (initiation of scales and leaves; Table 8.1), flowers will not differentiate. Thus, if stage Sp (Table 8.1) has been reached prior to lifting of the bulbs, flower formation and/or flower development can be precisely controlled by postharvest storage temperatures (Gerritsen and Van der Kloot 1936). Hartsema and Blaauw (1935) found that storage of daffodils at 25–31°C for 16 weeks delays flowering, while temperatures above 31°C or 1–5°C induce flower abortion.

In *Lilium longiflorum*, the effect of temperature on inflorescence development was carefully monitored, and a specific range of 16–18°C was found optimal for flower number in the inflorescence (De Hertogh 1974). In *Gladiolus*, in which the inflorescence is formed after sprouting and leaf elongation, warmer temperatures (above 20°C) were found to positively affect organ differentiation (Cohat 1993).

2. Light

In addition to its major influence on flower initiation, photoperiod can also affect flower differentiation. A plant in which flowering has already been induced can "revert" to the vegetative stage if transferred to inappropriate photoperiodic conditions (Erwin 2006). LD conditions are beneficial for flower differentiation in *Gladiolus* (Cohat 1993), while short photoperiod during inflorescence development reduces floret numbers in this crop (Shillo and Halevy 1976b).

The effect of photoperiod on flower differentiation has also been investigated in the context of flower abortion. For example, LD conditions can induce floral abortion in irises (Elphinstone et al. 1986). In *Lilium candidum*, originating from short-day geographical zone, inflorescence formation occurs after stem elongation, and high rates of flower abortion were observed when plants were exposed to LD conditions after planting (Weingartern-Kenan 2005). However, the floral stage at which abortion occurs in *L. candidum* and the mechanism leading to this phenomenon have not been characterized.

Besides photoperiod, light intensity can also play a role in flower differentiation and anthesis. In Dutch iris, Hartsema and Luyten (1953, 1955) found higher flowering percentages when supplementary lighting was begun 40–50 days prior to the predicted date of flowering. Light intensity was correlated with inflorescence formation in *Gladiolus*: low intensities reduced the number of florets per inflorescence and induced floret shrinkage up to abortion of the whole inflorescence, while additional lighting promoted flower quality (Shillo and Halevy 1976a; Cohat 1993). Similar results have been obtained in lilies (Treder 2003; Duarte et al. 2004).

B. MOLECULAR REGULATION OF FLOWER DIFFERENTIATION

Following the shift to the reproductive stage and floral initiation, many genes, in several tiers, are switched on or off, both spatially and temporally, to ensure the process of flower differentiation (Zik and Irish 2003). In recent years, much progress has been made in understanding the genetic control of inflorescence and flower development in higher plants.

In higher eudicots, including the most popular model species *Arabidopsis* and *Antirrhinum*, the floral organs are arranged in four concentric whorls carrying, from the outside in, sepals, petals, stamens, and carpels, respectively (Figure 8.10a). Functional analyses of single genetic mutants of *Arabidopsis* and *Antirrhinum* led to the proposal of a model for flower differentiation, in which five classes (A, B, C, D, and E) of floral homeotic genes are involved in regulating this process (Coen and Meyerowitz 1991; Theißen et al. 2000; Theißen 2001; Litt and Kramer 2010). In this model, the class A genes [e.g., *APETALA1* (*AP1*) and *AP2* of *Arabidopsis*] control sepal formation; class A, B [*AP3*, *PISTILLATA* (*PI*)], and E [*SEPALLATA* (*SEP1/2/3*)] genes together regulate petal formation; class B, C [*AGAMOUS* (*AG*)], and E genes control stamen formation; class C and E genes regulate carpel formation; and class D genes [*FLORAL BINDING PROTEIN7* (*FBP7*) and *FBP11*] are involved in ovule development. Most of these genes belong to the family of MADS-box genes encoding transcription factors (Theißen et al. 2000).

Subsequent studies, based on an analysis of orthologues of the ABC MADS-box genes, revealed that this model is fairly well conserved in angiosperms (Ma and de Pamphilis 2000; Theißen and Melzer 2007). However, various modifications of the specific components of the model may have occurred in different lineages (Van Tunen et al. 1993; Kramer et al. 2003; Kanno et al. 2007; Kramer 2009). One recent significant conclusion is that the class A genes' function, as originally defined in *Arabidopsis*, does not represent a deeply conserved genetic program, unlike many aspects of the B and C programs (Litt and Kramer 2010).

In contrast to flowers of higher eudicots, the perianths of many Liliaceae have two whorls of almost identical petaloid organs, called tepals (Figure 8.10b). Most monocot flowers have three

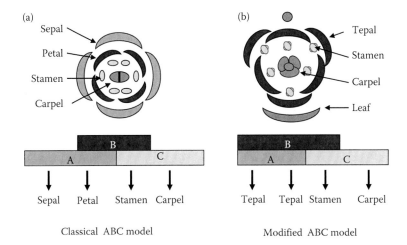

FIGURE 8.10 **(See color insert.)** Flower structures of the model species *Arabidopsis* (a) compared with typical representatives of ornamental geophyte *Tulipa* (b). Note the presence of two separate whorls of sepals and petals in the model dicots, while in tulip the perianth lobes are represented by two whorls of tepals. Classical ABC model suggests that in higher eudicots (e.g., *Arabidopsis thaliana*, Coen and Meyerowitz 1991), expression of class A genes specifies sepal formation in whorl 1, the combination of class A and B genes specifies the formation of petals in whorl 2, class B and C genes specify stamen formation in whorl 3, and expression of a class C gene alone determines the formation of carpels in whorl 4. According to the modified ABC model in *Tulipa* (Van Tunen et al. 1993), expression of class A genes specifies tepal formation in whorls 1 and 2; class B genes are expressed in whorls 1, 2 and 3; class B and C genes specify stamen formation in whorl 3, and expression of a class C gene determines whorl 4.

outer tepals, three inner tepals, three or six stamens, and three carpels (Dahlgren et al. 1985). To explain the morphology of wild-type and some mutant flowers of tulip (*Tulipa gesneriana*), where the first and second whorls of tepals are identical, Van Tunen et al. (1993) proposed a modified ABC model in which expression of class B genes is extended to the first floral whorl (Figure 8.10b). This modified model was subsequently supported by developmental genetics studies. To test the modified ABC model, putative class B genes from tulip were cloned and characterized (Kanno et al. 2003). The homologues of one *GLOBOSA* and two *DEFICIENS* genes were identified, named *TGGLO*, *TGDEFA*, and *TGDEFB*, respectively. Northern hybridization analysis showed that all these genes are expressed in whorls 1, 2, and 3 (outer and inner tepals and stamens), thus corroborating the modified ABC model (Kanno et al. 2003).

Further studies of a number of nongrass monocots, such as tulip, *Agapanthus*, *Muscari*, *Tricyrtis*, *Phalaenopsis*, *Crocus*, and *Dendrobium*, supported the proposed modification of the ABC model (Kanno et al. 2007). However, the data and model do not apply to some species, such as *Asparagus*, where the expression of class B genes is restricted to whorls 2 and 3, despite the presence of almost identical petaloid perianths in whorls 1 and 2 (Park et al. 2003, 2004). At the same time, in *Crocus sativus* and *Muscari armeniacum*, the class B genes are expressed in whorls 1, 2, and 3, which fits the modified ABC model, but are also expressed in whorl 4, deviating from the model (Kanno et al. 2007). Therefore, the modified ABC model proposed for tulip (Van Tunen et al. 1993) may not be universal for all Liliaceae species.

Characterization and functional analysis of several ABCDE genes were reported for the *Lilium* species. Thus, isolation and expression analysis of the class B gene *LLGLO1* from *Lilium longiflorum* showed that it is strongly expressed in the outer tepals, inner tepals, and stamens (whorls 1, 2, and 3, respectively) (Wu et al. 2010). Further, in transformed *Arabidopsis* plants overexpressing *LLGLO1*, homeotic conversion of sepals into petaloid sepals was induced in the first whorl. It was concluded that *LLGLO1* is one of the members of the *GLO/PI* subfamily, which is involved in the regulation of flower development in *L. longiflorum*.

Class B, C, and D MADS-box genes were isolated from *Lilium × formolongi*, and their expression patterns were compared between *L. × formolongi* and the double-flowered lily 'Aphrodite', in which stamens are completely changed into petaloid organs. The expression of the *AG*-like gene (class C) from 'Aphrodite' (*LaphAG1*) was observed only in whorl 4, but not in whorl 3 as in *L. × formolongi* (*LFAG1*), while genes of other classes (B and D) were expressed similarly in both genotypes (Akita et al. 2008). These results suggest that in double-flowered lilies, the transformation of stamens into petaloid organs is regulated by the class C genes.

Other types of double-flowered cultivars, which are valuable for horticulturists, are found among the *Lilium* species. Thus, in the double-flowered lily 'Elodie', stamens are homeotically converted into petaloid organs in whorl 3, and the strength of petaloidy varies from weak to strong (Akita et al. 2011). A class C *AG*-like gene was isolated from 'Elodie' (*LelAG1*), and its expression was detected in whorls 3 and 4. In flowers with weak petaloidy, *LelAG1* was expressed strongly in both whorls 3 and 4, whereas in flowers with strong petaloidy, its expression in whorl 3 was significantly decreased. In the intermediate phenotype, *LelAG1* expression was reduced by 60% in whorl 3. It was concluded that in 'Elodie', the expression of *AG*-like genes correlates with the degree of petaloidy of the stamens (Akita et al. 2011). Prior to this, a class C gene (*LLAG1*) had been isolated from *L. longiflorum* and functional studies showed that *LLAG1* and *AG* have similar functions (Benedito et al. 2004a,b).

Following the identification and characterization of E-function-related genes from monocots such as rice and maize, two *AGL2*-like MADS-box genes, *LMADS3* and *LMADS4*, were characterized from *L. longiflorum*, with extensive homology of *LMADS3* to the *Arabidopsis SEP3* (Tzeng et al. 2003). Both *LMADS3* and *LMADS4* mRNAs were detected in the inflorescence meristem and in floral buds at different developmental stages, as well as in all four whorls of the flower organs. *LMADS4* was also expressed in vegetative leaves and in the inflorescence stem, where *LMADS3* expression was absent. Ectopic expression of *LMADS3*, but not *LMADS4*, affected floral formation and floral transition in heterologous transgenic *Arabidopsis* plants: overexpression of *LMADS3* significantly promoted flowering by indirectly activating the flowering-time genes *FT* and *SOC1* (Benedito et al. 2004b). These studies provided useful information on the relationships between class C and D MADS-box genes and class E MADS-box genes in lily flower development, and thus served to establish the ABCDE flowering model in *L. longiflorum* (Figure 8.11; Benedito et al. 2004a,b).

Genes regulating floral organ formation have also been isolated from other geophytes, such as the *AG* orthologues *HAG1* from *Hyacinthus* (Li et al. 2002) and *NTAG1* from *Narcissus* (Deng et al. 2011). Functional analysis of the *Narcissus NTAG1* using ectopic expression tests in *Arabidopsis* showed novel phenotypes of the *NTAG1*-transgenic plants. These plants demonstrated early flowering, loss of inflorescence indeterminacy, and an increase in branch number, which are important

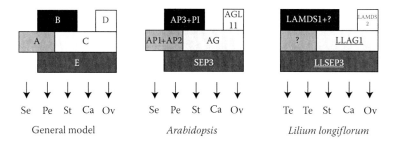

FIGURE 8.11 The ABCDE model for flower development in lily. The general ABCDE model states that five genetic functions act in an overlapping fashion to trigger the development of floral organs. Genes of each function were found in the model species *Arabidopsis*, and some of the ABCDE functions have already been characterized in *Lilium longiflorum*. Se, sepals; Pe, petals; St, stamens; Ca, carpel; Ov, ovules; Te, tepals. (From Benedito, V. A. et al. 2004a. *Acta Hort.* 651:83–89. With permission.)

characters in ornamental plants (Deng et al. 2011). A comparison of the deduced amino acid sequence of NTAG1 with the sequence of other MADS-box proteins showed 91.3% identity with HAG1 from *Hyacinthus orientalis*. Sequence analysis and alignment also showed significant similarity with other *AG* homologues. Gao et al. (2008) isolated an additional MADS-box protein from *Narcissus*, NTMADS1, which shows high homology to the MADS-box proteins from different species, especially *Asparagus virgatus* (Yun et al. 2004a,b). Phylogenetic tree analysis indicated that NTMADS1 also belongs to the AG subfamily (Gao et al. 2008). Thus, the *AG* homologues appear to be structurally conserved between dicots and monocots.

The flower of crocus (*Crocus sativus*) is bisexual and sterile, since crocus is a triploid species, and the gynoecium consists of a single compound pistil with three carpels, a single three-branched style, and an inferior ovary. The MADS-box transcription factors controlling flower development in cultivated crocus and a number of crocus flower mutants have been cloned and characterized (Tsaftaris et al. 2004; Kalivas et al. 2007, 2010). Three homologous *AP1*-like genes were found to be expressed in different amounts in leaves, and in mature tepals, stamen, and carpels (Tsaftaris et al. 2004). Expression of the crocus *AP3*-like class B MADS-box gene *CsatAP3* and the transcripts of *PI*-like *CsatPIc* were not restricted to the second and third whorls, but were also detected in the tepals of the first whorl and stigmata of the fourth whorl of the mature crocus flower (Kalivas et al. 2007). In addition, the *C. sativus AG* homologue *CsAG1*, a putative class C MADS-box gene, was isolated and characterized (Tsaftaris et al. 2005).

The presented data support the suggestion that, in ornamental geophytes, flower differentiation is controlled differently than in model plants (Kanno et al. 2007). However, to date, the molecular data available on flower organ development in geophytes have been based mainly on mRNA expression patterns, reflecting gene expression only at the transcriptional level, which is not sufficient to assess gene function. Such data need to be complemented by protein expression and localization data (Urbanus et al. 2009). Ultimately, analysis of gene function could be achieved pending the availability of suitable mutants in which rescue experiments could be performed. Alternatively, phenotypic analysis of transgenic plants, in which relevant genes are silenced, could be carried out, provided an efficient transformation system exists for the plant of interest.

C. Flower Reproductive Organs, Micro- and Megasporogenesis

Flowers contain reproductive structures, such as stamens and carpels, which allow fertilization and further seed production. Within the anther, male meiosis produces microspores, which develop into pollen grains, relying on both sporophytic and gametophytic gene functions. The mature pollen is released when the anther dehisces, allowing pollination. The diploid cells of the female flower organs (ovaries) undergo meiotic division and form megaspores.

In many plant species, unsuitable environmental conditions during differentiation of the reproductive organs disrupt micro- and megasporogenesis. For instance, in tomato, wheat, cowpea, rice, and *Arabidopsis*, high temperatures effected developmental abnormalities in pollen production and germination, as well as in the fertilization process, resulting in poor seed production (Saini et al. 1983; Currah 1990; Peet et al. 1998; Kim et al. 2001; Matsui and Omasa 2002). Unsuitable photoperiod during flower differentiation has been reported to decrease microspore viability in rice and maize (Moss and Heslop-Harrison 1968; Jiang et al. 2007). It is also known that in flowering geophytes, high growth temperatures and dryness can cause anther deterioration, abnormalities in meiosis, pollen production and germination, and complications in the fertilization process, followed by decreased seed production (E. Shemesh, pers. comm.).

The identification and characterization of the MADS-box organ identity genes, as well as of floral genes involved in the growth and patterning of flower organs, have helped elucidate the differentiation processes resulting in the unique reproductive tissues (Liu et al. 2009). Production of commercially important hybrids is continually being attempted in numerous ornamental geophytes, and clarification of the basic reproductive mechanism is therefore important for the future breeding of these crops.

The anthers are comprised of several tissue types, including the epidermis and endothecium, and the unique tapetum cells that surround the microsporocyte and are required for pollen development (Goldberg et al. 1993). A large number of genes expressed exclusively or predominantly in the anthers during microsporogenesis have been identified in model and nonmodel plants (Zik and Irish 2003; Wijeratne et al. 2007; Borg et al. 2009). In *Arabidopsis*, many of the characterized stamen-differentiation genes are required for tapetum development and/or microsporogenesis (Feng and Dickinson 2007). A number of these genes are also required for female reproductive development, indicating a certain similarity in these processes (Irish 2010). In recent years, new high-throughput technologies have enabled the analysis of male gametophyte gene expression on a global scale. Approximately 12,000 active genes have been shown to be expressed in the early phases of microsporogenesis in *Arabidopsis*, while this number progressively declines to over 7000 in mature pollen (Honys and Twell 2004).

In *Lilium longiflorum*, 22 individual cDNA clones induced by GA were identified in anthers at the microspore stage (Hsu et al. 2007). Sequence analysis revealed five novel anther-specific/predominant genes. The transcripts of anther-specific genes were differentially detected in the microspore development phase, while some of these genes accumulated as early as meiosis and showed a strong localization to the tapetum. Two stage-specific genes, *LLA-67* and *LLA-115*, from developing anthers of *L. longiflorum*, were anther specific and differentially expressed at the stage of microspore development (Tzeng et al. 2009). The two genes were negatively regulated by ethylene, and cross-talk between GA and ethylene was involved in their regulation.

The differentiation of vegetative and generative cells is a critical event during pollen development. *LlglsA* (*Lilium longiflorum glsA*), a reproductive factor expressed preferentially in the generative cell, was identified as the orthologue of *Arabidopsis GlsA*. *LlglsA* is localized in pollen and is considered to be involved in the development of the generative cell (Mori et al. 2003). Similarly, the *glsA* orthologue from *Alstroemeria* was also isolated. *AaglsA* (*Alstroemeria aurea glsA*) expression was progressively upregulated following pollen mitosis I (PMI). The expression level of the transcripts was significant at the beginning of generative-cell elongation (Igawa et al. 2009).

In several geophytes, the expression of *LFY* homologues has been observed in at least three phases of floral development: during meristem transition from vegetative to reproductive development, flower differentiation, and gametogenesis (Noy-Porat et al. 2010; Rotem et al. 2011). The expression of an *LFY* homologue in *Narcissus tazetta*, *NLF*, was observed at later stages of flower differentiation in the developing anthers (Noy-Porat et al. 2010). Similarly, temporal accumulation of *gaLFY* is strongly associated with reproductive organs of garlic (*Allium sativum*), significantly increasing during florogenesis and gametogenesis (Rotem et al. 2011). As already mentioned, it has been suggested that *LFY* and its homologues play numerous roles in plant development, including flower differentiation and gametogenesis (Rotem et al. 2011).

The gynoecium is one of the plant's most complex organs. It plays a multifunctional role in housing the female gametophyte and acting as a conduit for pollen tube growth (Ferrándiz et al. 1999; Balanzá et al. 2006; Irish 2010). In *Arabidopsis*, mutant screens and reverse-genetics approaches have identified a number of genes that function in specifying tissue types and growth during gynoecium development. Several MADS-box genes, including *AG*, *SHATTERPROOF1* and 2 (*SHP1, 2*) and *SEEDSTICK* (*STK*), and the *YABBY* domain gene *CRABS CLAW* (*CRC*), function specifically in the gynoecium (Irish 2010).

Despite the importance of the female organs to reproductive development, limited work has been done toward identifying gynoecium-specific genes in geophytes. In *Asparagus*, an *AG*-like gene, designated *AVAG1*, has been reported to be involved in the formation of both the stamens and carpels (Yun et al. 2004a). Phylogenetic analysis showed that the *AVAG1* gene is closely related to *HAG1* from *Hyacinthus* and *PeMADS1* from *Phalaenopsis*. *AVAG1* transcripts were detected in the floral meristem and the developing reproductive organs. Early in flower development, expression of *AVAG1* was restricted to the stamens and carpels, while its expression in the stamen decreased at later stages of flower development, and remained strong in the ovule.

Genetic control of the reproductive structures, such as stamens and carpels, in geophytes has been only poorly investigated. Although breeding of commercially important hybrids has great commercial value, limited data are available on the regulation of stamen and carpel function. Given that numerous key genes involved in regulating micro- and megasporogenesis have been identified in model and nonmodel plants, investigations in the next decade are expected to improve our understanding of this process in geophytes, thus contributing to the development of breeding techniques and the production of new ornamental hybrids.

D. FLORAL ABERRATIONS

Abnormalities or modifications in floral development may lead to malformations in inflorescence and flower structures, and to a reduction in the ornamental value of the plant. In some cases, however, mutations or deviations in florogenesis may give rise to new and unusual flower forms and, when inheritable, may encourage the breeding of new varieties. In many bulbous species, floral aberrations or malformations occur as a direct consequence of adverse conditions during floral initiation and differentiation (Le Nard and De Hertogh 1993; Kamenetsky and Rabinowitch 2002; Flaishman and Kamenetsky 2006). Indeed, high temperatures at the time of differentiation increased floral malformations in *Allium aflatunense* (Kamenetsky and Rabinowitch 2002). In *Eremurus*, the physiological disorder termed as "interrupted floral development" (IFD) is characterized by the drying of flowers in the upper part of the raceme or by partial flower abortion (Kamenetsky and Rabinowitch 1999). IFD is observed only in plants harvested at the stage of fully differentiated inflorescence and is caused by high air and soil temperatures prior to harvest. In *Iris*, anomalous flowers with either extra floral organs or their absence are produced when flower formation within the bulb is delayed by temperatures below 0°C. Omission of floral organs is sometimes caused by prolonged exposure to high storage temperatures (De Munk 1989). Anoxic conditions during bulb storage can also result in abnormal flowers in iris (De Munk and Schipper 1993). Adverse cultural conditions or hormonal balance during *in vitro* propagation can induce somaclonal variations, resulting in homeotic mutations or flower malformations (Figure 8.12).

Many *Allium* species (*A. caeruleum*, *A. carinatum*, *A. proliferum*, *A. scorodoprasum*, *A. vineale*) develop topsets (bulbils) that intermingle with flowers in the inflorescence (Gustafsson 1946/1947, cited by Etoh 1985). In garlic, vegetative meristems differentiate into flower primordia in the inflorescence, and the strong competition by the developing topsets causes garlic flowers to wither and die (Kamenetsky and Rabinowitch 2001). Environmental conditions, particularly temperature and photoperiod, strongly affect the posttransitional stages of garlic florogenesis, and normal flowering cannot be achieved if any of the developmental stages of florogenesis are retarded or proceed in the wrong direction (Figure 8.13; Kamenetsky et al. 2004).

Double-flower pattern occurs in numerous geophyte species, such as tulips, narcissus, and hyacinths, naturally or as a result of classical breeding, and is of great interest to flower breeders. This phenomenon is linked to mutations in the *AG* orthologue or to abnormal interactions with transregulatory elements (Figure 8.14; Roeder and Yanofsky 2001; Franks et al. 2002; Kanno et al. 2007). However, the double-flower pattern is rare and not completely understood. Knowledge of the mechanisms involved in this process will enable the creation of new varieties of ornamental bulbous crops with modified flower phenotype.

In different floral phenotypes of *Lilium longiflorum*, homeotic mutations of stamens into tepals and carpel into a new flower have been reported (Benedito et al. 2004a,b). Functional studies of *LLAG1*, the *AG* homologue from lily (Benedito et al. 2004b), in *Arabidopsis* suggested that homeotic changes of floral organs were caused by the constitutive overexpression of *LLAG1*. The observed modifications were entirely in accordance with reports on *AG* overexpression in *Arabidopsis* (Mizukami and Ma 1992) and hyacinth (Li et al. 2002). These results provided additional evidence for the capability of *in vivo* cross-interaction of proteins belonging to the ABC model from different species, even among those which are distantly related, such as *Arabidopsis* and lily. However, there

FIGURE 8.12 **(See color insert.)** Flower aberrations in *Ornithogalum dubium*, resulting from homeotic mutations and flower malformations in the plants propagated *in vitro*. (a) Normal inflorescence, six perianth lobes and six stamens are visible in the flower. (b) Individual flower with eight perianth lobes and eight stamens; three stamens are malformed and pollen is not formed. (c–d) Numerous malformations of the inflorescence. (Photo courtesy: R. Kamenetsky 2005).

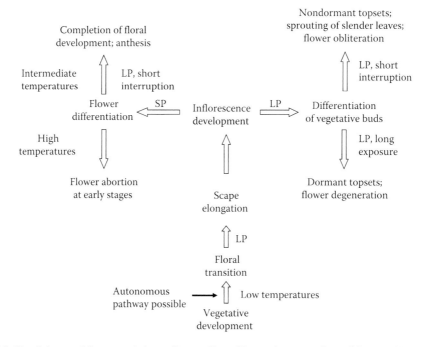

FIGURE 8.13 Scheme of florogenesis in garlic, as affected by environmental conditions. With the exception of the apex floral transition (which may occur autonomously), environmental conditions during both storage and growth affect the progress and direction of further development into either the reproductive or vegetative pathway. (From Kamenetsky, R. et al. 2004. *J. Am. Soc. Hort. Sci.* 129:144–151. With permission.)

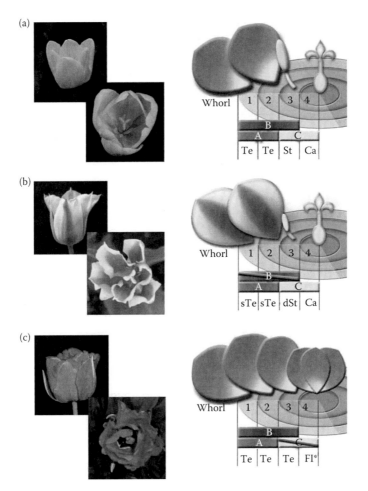

FIGURE 8.14 **(See color insert.)** Various tulip flowers (left) and the models explaining these flower morphologies (right). (a) The wild-type tulip flower has two whorls of petaloid tepals, stamens, and carpels. (b) The viridiflora tulip flower has two whorls of partially greenish tepals (sepaloid), degenerated stamens, and carpels. This floral morphology is explained by reduced expression of class B gene(s). (c) The double-flowered tulip has petaloid tepals in whorl 3 in addition to the outer two whorls and a new flower in the innermost whorl. This floral morphology is explained by reduced expression of the class C gene in whorls 3 and 4. Te, tepals; St, stamens; Ca, carpels; sTe, sepaloid tepals; dSt, degenerated stamens; Fl*, new flower. (From Kanno, A. et al. 2007. *TSW Dev. Embryol.* 2:17–28. With permission.)

are indications that MADS-box protein dimerization may occur in a different way in monocot species (Winter et al. 2002; Kanno et al. 2007).

Double-flower varieties of the oriental hybrids of *Lilium* spp. might result from *AG* loss of function, loss of flower determinacy, or having tepals instead of stamens and modified tepals instead of carpels. Indeed, as already mentioned, *AG-like* expression in the double-flowered lily 'Elodie' (*LelAG1*) appears to be correlated with the extent of stamen petaloidy (Akita et al. 2011). Clearly, this modification involves loss of C function in the ABCDE model (Roeder and Yanofsky 2001), especially due to its coupled abnormalities (homeotic mutation in C and loss of floral determinacy).

V. CONCLUSIONS AND FUTURE RESEARCH

In this chapter, we describe the current status of research in the morphological, physiological, and molecular mechanisms of flowering in ornamental geophytes. For centuries, people have enjoyed and celebrated the ravishing beauty of geophyte flowers. Extensive development of the bulb industry and globalization of the flower trade have led to the generation of a considerable amount of research data, aimed at producing high-quality flowers with the desired timing. It is only recently, however, that molecular studies, based on the substantial data on plant physiology and flower manipulation, have been launched to elucidate molecular aspects of flowering regulation in geophytes.

Recent studies have demonstrated that the control of flower development in ornamental geophytes differs from that in model plants such as *Arabidopsis* or *Antirrhinum* (Flaishman and Kamenetsky 2006). At the same time, the homologues of *FT* and *LFY*, the key conserved genes in floral initiation, have been found in several geophytes, and their spatio-temporal expression has been studied in *Narcissus* and garlic (Rotem et al. 2007, 2011; Noy-Porat et al. 2010). A further understanding of the mechanisms controlling the transition of apical meristem from the vegetative to reproductive stage could be used to monitor the effect of different growing conditions on this process. In addition, biochemical and molecular markers (e.g., the expression patterns of cold-regulated genes during vernalization) need to be developed for an accurate determination of the effect of low temperatures on optimal flower time and quality. Such markers might be culled from the ESTs isolated from vernalized bulbs, after validation of their regulation by the cold.

Several genes involved in the differentiation of individual flowers in *Lilium longiflorum*, *Tulipa*, and other geophytes have been shown to be homologous to the genes of the ABCDE model proposed for *Arabidopsis* and other model species (Benedito et al. 2004a,b; Kanno et al. 2007; Wu et al. 2010). A modified ABC model was proposed for monocot flowers and exemplified by tulip (Van Tunen et al. 1993). However, the modified model proposed for tulip may not apply to all Liliaceae species or to other geophytes (Kanno et al. 2007; Litt and Kramer 2010). In the future, molecular characterization of genes involved in flower morphology could lead to novel floral architectures in ornamental geophytes via classical breeding techniques or genetic manipulation using transformation systems.

In general, in the absence of a common nomenclature for plant genes (Price and Reardon 2001), various research groups and authors have adopted different names for the genes they isolate. Although this problem is not specific to the geophytes' genes, scientific discussions should be undertaken on this matter in order to develop a common genetic language for this group of plants.

Currently, there is no model geophyte species. To advance the investigation of geophyte biology, an alternative model plant, agreed upon by the scientific community in the field, should be chosen for genetic research. To date, most of the genetic studies on florogenesis have been performed on commercially important crops, such as lilies, narcissi, or tulips. However, in addition to being economically important, a model species has to meet several criteria, for example, a short juvenile period, easy pollination and seed germination, well-known morphological and physiological traits, an established transformation system, and rapid regeneration ability (Kamenetsky 2011). Basic developmental differences among ornamental bulbs also complicate the choice of a "model bulb". Possible candidates for such a model might be *Ornithogalum dubium*, an ornamental bulb from South Africa, or *Anemone coronaria*, a tuberous species from the Mediterranean region. *Gladiolus* was also proposed as a possible candidate for the model geophytes species. Such candidate might be a genetically controllable species with large phenotypic and genotypic variations in biological traits. For example, recently obtained populations of garlic seedlings (Shemesh et al. 2008) have revealed a large variation in vegetative and reproductive characters, including response to environmental conditions, bolting ability, flower differentiation, and seed production.

To date, two basic strategies have been used to isolate genes involved in florogenesis. The first is based on sequence homology between species. Since some flowering genes are conserved among distant species, this approach has yielded good results, for example, for genes involved in the initial stages of florogenesis or in differentiation of the floral organs. The second approach is broader and involves the determination of a large number of genes, using cDNA libraries and isolation of the genes expressed in specific tissue locations or physiological states (Kim et al. 2009). In the near future, expression studies of a large number of genes, using, for example, microarray technology and transcriptome profiling, are expected to play an important role in the isolation of candidate genes involved in the flowering of geophytes. Today, a growing number of nonmodel plants' genomes are being fully sequenced. Sunflowers, *Boechera* species and *Arabidopsis* relatives, wild relatives of edible crops (such as rice, tomato, and barley), and fruit trees (such as apple, peach, citrus, and grape) are on track for whole-genome sequencing. Population genomics, transcriptome profiling, proteomics, and other techniques will contribute to the identification of flowering genes in these nonmodel plants, marking the beginning of the research transition from model to nonmodel plants and horticultural crops (Song and Mitchell-Olds 2011).

Once flowering genes have been isolated from geophytes, functional analyses using transformation systems in geophytes are essential. The genetic transformation of commercial bulb crops, such as lily and tulip, is very difficult (Kanno et al. 2007). Therefore, it is important to isolate and characterize genes of interest in geophytes with established transformation systems, such as *Ornithogalum*, *Agapanthus*, *Muscari*, and *Gladiolus* (Flaishman and Kamenetsky 2006; see also Chapter 7). The discovery of the functions of certain genes will be made using phenotypic studies, and gene interaction, loss of function, gain of function, tracking, and expression analyses. One of the current challenges in geophyte research lies in determining how the flowering genes function at the cellular level and in the whole plant, and to understand the molecular regulation of flowering diversity under various environmental conditions.

Since timing and quality of flowering are the ultimate goals of the bulb industry, studies dealing with internal and environmental regulation of florogenesis are of particular interest. Great technological progress applied to geophytes, together with the rapid advances in our understanding of flower biology in model plants, are providing a solid background for rapid and effective studies of geophyte florogenesis, with valuable implications for the bulb industry.

ACKNOWLEDGMENTS

The authors wish to thank F. Krens (Plant Research International, Wagenengen, The Netherlands) and H. Okubo (Kyushu University, Fukuoka, Japan) for their critical revision of this chapter.

CHAPTER CONTRIBUTORS

Rina Kamenetsky is a senior researcher at the Agricultural Research Organization (ARO, The Volcani Center) and a professor at the Hebrew University at Jerusalem in Israel. She was born in Almaty, Kazakhstan, and earned her BSc and MSc from Kazakh State University and her PhD from the National Academy of Science of Kazakhstan. Dr. Kamenetsky has held positions in the Main Botanical Garden and Institute of Botany in Kazakhstan and the J. Blaustein Institute for Desert Research, Ben-Gurion University of the Negev, Israel. Her main research areas are the mechanisms of internal and environmental regulation of flowering and dormancy in geophytes, biodiversity, and the introduction of new ornamental plants and their cultivation in warm regions. She is the author of over 150 publications on geophytes.

Michele Zaccai is a senior lecturer at Ben Gurion University, Israel. She was born in Belgium and received her BS and MS from the Faculty of Agriculture, Hebrew University of Jerusalem and her PhD from The Weizmann Institute, Israel. Following postdoctoral work at Caltech, CA, USA, she joined Ben Gurion University, Israel, where she is leading a research group focusing on ornamentals. Her area of expertise includes physiological and molecular regulation of flowering and vernalization in *Lilium* species, as well as the introduction and characterization of new ornamental geophytes. Dr. Zaccai is the author of more than 50 articles, chapters, and conference papers and abstracts.

Moshe A. Flaishman is a senior scientist at the ARO, The Volcani Center, Israel. He was born in Israel and received his BS and MS from the Faculty of Agriculture at The Hebrew University of

Jerusalem and his PhD from Tel Aviv University, Israel. His expertise includes physiological and molecular regulation of flower development and dormancy. Dr. Flaishman is the author of more than 90 refereed journal articles, book chapters, and conference papers.

REFERENCES

Abe, M., Y. Kobayashi, S. Yamamoto, Y. Daimon, A. Yamaguchi, Y. Ikeda, H. Ichinoki, M. Notaguchi, K. Goto, and T. Araki. 2005. FD, a bZIP protein mediating signals from the floral pathway integrator FT at the shoot apex. *Science* 309:1052–1056.

Akita, Y., Y. Horikawa, and A. Kanno. 2008. Comparative analysis of floral MADS-box genes between wild-type and a putative homeotic mutant in lily. *J. Hort. Sci. Biotech.* 83:453–461.

Akita, Y., M. Nakada, and A. Kanno. 2011. Effect of the expression level of an *AGAMOUS*-like gene on the petaloidy of stamens in the double-flowered lily, 'Elodie'. *Scientia Hortic.* 128:48–53.

Albani, M. C. and G. Coupland. 2010. Comparative analysis of flowering in annual and perennial plants. *Curr. Top. Dev. Biol.* 91:323–348.

Amasino, R. 2010. Seasonal and developmental timing of flowering. *Plant J.* 61:1001–1013.

Anderson, N. O. 2006. *Flower Breeding and Genetics. Issues, Challenges and Opportunities for the 21st Century.* Dordrecht, The Netherlands: Springer.

Arumuganathan, K., and E. D. Earle. 1991. Nuclear DNA content of some important plant species. *Plant Mol. Biol. Rep.* 9:208–218.

Balanzá, V., M. Navarrete, M. Trigueros, and C. Ferrándiz. 2006. Patterning the female side of *Arabidopsis*: The importance of hormones. *J. Exp. Bot.* 57:3457–3469.

Balasubramanian, S., S. Sureshkumar, J. Lempe, and D. Weigel. 2006. Potent induction of *Arabidopsis thaliana* flowering by elevated growth temperature. *PLoS Genetics* 2:980–989.

Ballerini, E. S., and E. M. Kramer. 2011. Environmental and molecular analysis of the floral transition in the lower eudicot *Aquilegia formosa*. *EvoDevo* 2:1–20.

Benedito, V. A., G. C. Angenent, J. M. van Tuyl, and F. A. Krens. 2004a. *Lilium longiflorum* and molecular floral development: The ABCDE model. *Acta Hort.* 651:83–89.

Benedito, V. A., P. B. Visser, J. M. van Tuyl, G. C. Angenent, S. C. de Vries, and F. A. Krens. 2004b. Ectopic expression of *LLAG1*, an *AGAMOUS* homologue from lily (*Lilium longiflorum* Thunb.) causes floral homeotic modifications in *Arabidopsis*. *J. Exp. Bot.* 55:1391–1399.

Ben-Hod, G., J. Kigel, and B. Steinitz. 1988. Dormancy and flowering in *Anemone coronaria* L. as affected by photoperiod and temperature. *Ann. Bot.* 61:623–633.

Benlloch, R., A. Berbel, A. Serrano-Mislata, and F. Madueño. 2007. Floral initiation and inflorescence architecture: A comparative view. *Ann. Bot.* 100:659–676.

Benschop, M., R. Kamenetsky, M. Le Nard, H. Okubo, and A. De Hertogh. 2010. The global flower bulb industry: Production, utilization, research. *Hort. Rev.* 36:1–115.

Bernier, G., A. Havelange, C. Houssa, A. Petitjean, and P. Lejeune. 1993. Physiological signals that induce flowering. *Plant Cell* 5:1147–1155.

Bernier, G., and C. Périlleux. 2005. A physiological overview of the genetics of flowering time control. *Plant Biotechnol. J.* 3:3–16.

Beyer, J. J. 1942. De terminologie van de bloemaanleg der bloembolgewassen. *Mededeelingen van de Landbouwhoogeschool, Wageningen* 46:1–17.

Blaauw, A. H. 1923. De periodieke dikte-toename van den bol der hyacinthen. *Mededeelingen van de Landbouwhoogeschool, Wageningen* 27(2):1–103.

Blaauw, A. H. 1924a. The results of the temperature during flower formation for the whole hyacinth (first part). *Proceedings Section Science, Koninklijke Nederlandsche Akademie van Wetenschappen, Amsterdam, tweede sectie* 23:1–66.

Blaauw, A. H. 1924b. The results of the temperature during flower formation for the whole hyacinth (second part). *Proceedings Section Science, Koninklijke Nederlandsche Akademie van Wetenschappen, Amsterdam* 27:781–799.

Blaauw, A. H. 1931. Orgaanvorming en periodiciteit van *Hippeastrum hybridum*. Tweede gedeelte. De periodiciteit van *Hippeastrum*. *Koninklijke Akademie van Wetenschappen Amsterdam (Section 2)* 29:1–90 (In Dutch with English summary).

Borg, M., L. Brownfield, and D. Twell. 2009. Male gametophyte development: A molecular perspective. *J. Exp. Bot.* 60:1465–1478.

Boss, P. K., R. M. Bastow, J. S. Mylne, and C. Dean. 2004. Multiple pathways in the decision to flower: Enabling, promoting, and resetting. *Plant Cell.* 16:S18–S31.

Brooking, I. R. and D. Cohen. 2002. Gibberellin-induced flowering in small tubers of *Zantedeschia* 'Black Magic'. *Scientia Hortic.* 95:63–73.

Cerdán, P. D., and J. Chory. 2003. Regulation of flowering time by light quality. *Nature* 423:881–885.

Chaïlakhyan, M. K. 1937. *Gormonal'naya Teoriya Razvitiya Rastenii [Hormonal Theory of Plant Development].* Moskva-Leningrad: Akademia Nauk SSSR (In Russian).

Chaïlakhyan, M. Kh. 1975. Substances of plant flowering. *Biologia Plantarum* 17:1–11.

Chouard, P. 1960. Vernalization and its relation to dormancy. *Annu. Rev. Plant Physiol.* 11:191–238.

Coen, E. S., and E. M. Meyerowitz. 1991. The war of the whorls: Genetic interactions controlling flower development. *Nature* 353:31–37.

Cohat, J. 1993. *Gladiolus.* In *The Physiology of Flower Bulbs*, eds. A. De Hertogh, and M. Le Nard, 297–320. Amsterdam: Elsevier.

Corbesier, L., and G. Coupland. 2006. The quest for florigen: A review of recent progress. *J. Exp. Bot.* 57:3395–3403.

Corbesier, L., C. Vincent, S. Jang, F. Fornara, Q. Fan, I. Searle, A. Giakountis, S. Farrona, L. Gissot, C. Turnbull, and G. Coupland. 2007. FT protein movement contributes to long-distance signaling in floral induction of *Arabidopsis. Science* 316:1030–1033.

Cremer, C., J. J. Beijer, and W. J. de Munk. 1974. Developmental stages of flower formation in tulips, narcissi, irises, hyacinths, and lilies. *Mededeelingen Landbouwhoogeschool, Wageningen* 74:1–16.

Currah, L. 1990. Pollination biology. In *Onion and Allied Crops, Vol. I. Botany, Physiology, and Genetics*, eds. H. D. Rabinowitch, and J. L. Brewster, 135–149. Boca Raton, FL: CRC Press.

Dahlgren, R. M. T., H. T. Clifford, and P. F. Yeo. 1985. *The Families of the Monocotyledons: Structure, Evolution, and Taxonomy.* Berlin: Springer-Verlag.

Davis, S. J. 2009. Integrating hormones into the floral-transition pathway of *Arabidopsis thaliana. Plant Cell Environ.* 32:1201–1210.

De Hertogh, A. 1974. Principles for forcing tulips, hyacinths, daffodils, Easter lilies and Dutch irises. *Scientia Hortic.* 2:313–355.

De Hertogh, A. A., and L. Gallitano. 2000. Influence of photoperiod and day/night temperatures on flowering of Amaryllis (*Hippeastrum*) cv. Apple Blossom. *Acta Hort.* 515:129–134.

De Hertogh, A. A., and M. Le Nard. 1993a. Physiological and biochemical aspects of flower bulbs. In *The Physiology of Flower Bulbs*, eds. A. De Hertogh, and M. Le Nard, 53–70. Amsterdam: Elsevier.

De Hertogh, A., and M. Le Nard. 1993b. *The Physiology of Flower Bulbs.* Amsterdam: Elsevier.

De Munk, W. J. 1989. Thermomorphogenesis in bulbous plants. *Herbertia* 45:50–55.

De Munk, W. J., and J. Schipper. 1993. *Iris*—Bulbous and rhizomatous. In *The Physiology of Flower Bulbs*, eds. A. De Hertogh, and M. Le Nard, 349–379. Amsterdam: Elsevier.

Deng, X., L. Xiong, Y. Wang, Y. Sun, and X. Li. 2011. Ectopic expression of an *AGAMOUS* homolog *NTAG1* from Chinese narcissus accelerated earlier flowering and senescence in *Arabidopsis. Mol. Plant Breed.* 2:14–21.

Desmet, P. G. 2007. Namaqualand—A brief overview of the physical and floristic environment. *J. Arid Environments* 70:570–587.

Distelfeld, A., C. Li, and J. Dubcovsky. 2009. Regulation of flowering in temperate cereals. *Curr. Opin. Plant Biol.* 12:178–184.

Doi, M., N. Kawamura, T. Sugimoto, and H. Imanishi. 2000. Controlling the flowering of *Eucharis grandiflora* Planchon with ambient and regulated soil temperature. *Scientia Hortic.* 86:151–160.

Dole, J. M. 2003. Research approaches for determining cold requirements for forcing and flowering of geophytes. *HortScience* 38:341–346.

Dole, J. M., and H. F. Wilkins. 1994. Interaction of bulb vernalization and shoot photoperiod on 'Nellie White' Easter lily. *HortScience* 29:143–145.

Dole, J. M., and H. F. Wilkins. 1999. *Lilium*, Easter. In *Floriculture: Principles and Species*, eds. J. M. Dole, and H. F. Wilkins, 400–416. Upper Saddle River, NJ: Prentice Hall.

Domagalska, M. A., E. Sarnowska, F. Nagy, and S. J. Davis. 2010. Genetic analysis of interactions among gibberellin, abscisic acid, and brassinosteroids in the control of flowering time in *Arabidopsis thaliana. PLoS ONE* 5(11):1–8.

Duarte, A. R., A. B. Mendoza, L. B. Herrera, A. R. López, and R. K. Maiti. 2004. Effect of light intensity on flower bud abortions in lily (*Lilium* spp.). *Crop Res. (Hisar)* 28:68–75.

Du Plessis, N., and G. Duncan. 1989. *Bulbous plants of southern Africa—A guide to their cultivation and propagation*, Cape Town: Tafelberg.

Elphinstone, E. D., A. R. Rees, and J. G. Atherton. 1986. The effect of photoperiod and temperature on the development of Dutch iris flowers and daughter bulbs. *Acta Hort.* 177:613–618.

Erwin, J. 2006. Factors affecting flowering in ornamental plants. In *Flower breeding and genetics. Issues, challenges and opportunities for the 21st century*, ed. N. O. Anderson, 7–48. Dordrecht, The Netherlands: Springer.

Etoh, T. 1985. Studies on the sterility in garlic, *Allium sativum* L. *Mem. Fac. Agr., Kagoshima Univ.* 21:77–132.

Feng, X., and H. G. Dickinson. 2007. Packaging the male germline in plants. *Trends Genet.* 23:503–510.

Ferrándiz, C., S. Pelaz, and M. F. Yanofsky. 1999. Control of carpel and fruit development in *Arabidopsis. Annu. Rev. Biochem.* 68:321–354.

Flaishman, M., and R. Kamenetsky. 2006. Florogenesis in flower bulbs: Classical and molecular approaches. In *Floriculture, Ornamental and Plant Biotechnology, Vol. I*, ed. J. A. Texiera de Silva, 33–43. Isleworth, UK: Global Science Books.

Flowering Time Control: From Natural Variation to Crop Improvement. 2012. http://www.flowercrop.uni-kiel.de, April 2012.

Franks, R. G., C. Wang, J. Z. Levin, and Z. Liu. 2002. *SEUSS*, a member of a novel family of plant regulatory proteins, represses floral homeotic gene expression with *LEUNIG. Development* 129:253–263.

Fritsch, R. 1992. Septal nectaries in the genus *Allium*—Shape, position and excretory canals. In *The Genus Allium—Taxonomic Problems and Genetic Resources. Proceedings of an International Symposium Held at Gatersleben, Germany 11–13 June 1991*, eds. P. Hanelt, K. Hammer, and H. Knupffer, 77–85. Gatersleben, Germany: Institut für Pflanzengenetik und Kulturpflanzenforschung.

Gao, Z.-m., D.-f. Chen, X.-p. Li, C.-j. Cai, and Z.-h. Peng. 2008. Cloning and sequence analysis of a flowering-related MADS-box gene in *Narcissus tazetta* var. *chinensis* Roem. *Acta Hort. Sin.* 35:295–300.

Gerritsen, J. D., and W. G. van der Kloot. 1936. Verschillen in het bloemvormende vermogen van narcis en hyacinth. *Proceedings Section Science, Koninklijke Nederlandsche Akademie van Wetenschappen, Amsterdam* 39:404–413.

Goldberg, R. B., T. P. Beals, and P. M. Sanders. 1993. Anther development: Basic principles and practical applications. *Plant Cell* 5:1217–1229.

Gutterman, Y., and B. Boeken. 1988. Flowering affected by daylength and temperature in the leafless flowering desert geophyte *Colchicum tunicatum*, its annual life cycle and vegetative propagation. *Bot. Gaz.* 149:382–390.

Halevy, A. H. 1985. *Gladiolus*. In *CRC Handbook of Flowering Vol. III*, ed. A. H. Halevy, 63–70. Boca Raton, FL: CRC Press.

Halevy, A. H. 1990. Recent advances in control of flowering and growth habit of geophytes. *Acta Hort.* 266:35–42.

Hartsema, A. M. 1961. Influence of temperatures on flower formation and flowering of bulbous and tuberous plants. In *Encyclopedia of Plant Physiology. Vol. 16*, ed. W. Ruhland, 123–167. Berlin: Springer-Verlag.

Hartsema, A. M., and A. H. Blaauw. 1935. Verschuiving der periodiciteit door hooge temperaturen. Aanpassing en export voor het Zuidelijke Halfrond II. *Proceedings Section Science, Koninklijke Nederlandsche Akademie van Wetenschappen, Amsterdam* 38:722–734.

Hartsema, A. M., and I. Luyten. 1953. Snelle bloei van Hollandse irissen var. Imperator. IV. Invloed van temperatuur en licht. *Proceedings Section Science, Koninklijke Nederlandsche Akademie van Wetenschappen, Amsterdam, Series C* 56:81–105.

Hartsema, A. M., and I. Luyten. 1955. Early flowering of Dutch irises 'Imperator'. V. Light intensity and daylength. *Acta Bot. Neerl.* 4:370–375.

Hayama, R. and G. Coupland. 2004. The molecular basis of diversity in the photoperiodic flowering responses of Arabidopsis and rice. *Plant Physiol.* 135:677–684.

Hayama, R., S. Yokoi, S. Tamaki, M. Yano, and K. Shimamoto. 2003. Adaptation of photoperiodic control pathways produces short-day flowering in rice. *Nature* 422:719–722.

Henderson, I. R., and C. Dean. 2004. Control of *Arabidopsis* flowering: The chill before the bloom. *Development* 131:3829–3838.

Hisamatsu, T., and R. W. King. 2008. The nature of floral signals in *Arabidopsis*. II. Roles for *FLOWERING LOCUS T* (*FT*) and gibberellin. *J. Exp. Bot.* 59:3821–3829.

Holcomb, E. J. and R. Berghage. 2001. Photoperiod, chilling, and light quality during daylight extension affect growth and flowering of tissue-cultured Easter lily plants. *HortScience* 36:53–55.

Honys, D., and D. Twell. 2004. Transcriptome analysis of haploid male gametophyte development in Arabidopsis. *Genome Biol.* 5(11):R85.1–13.

Horvath, D. 2009. Common mechanisms regulate flowering and dormancy. *Plant Sci.* 177:523–531.

Hsu, Y.-F., C.-S. Wang, and R. Raja. 2007. Gene expression pattern at desiccation in the anther of *Lilium longiflorum. Planta* 226:311–322.

Huisman, E., and A. M. Hartsema. 1933. De periodieke ontwikkeling van *Narcissus pseudonarcissus* L. *Mededeelingen van de Landbouwhoogeschool, Wageningen* 37:1–55.

Igawa, T., Y. Hoshino, and Y. Yanagawa. 2009. Isolation and characterization of the plant *glsA* promoter from *Alstroemeria*. *Plant Biol.* 11:878–885.

Irish, V. F. 2010. The flowering of *Arabidopsis* flower development. *Plant J.* 61:1014–1028.

Izawa, T., T. Oikawa, N. Sugiyama, T. Tanisaka, M. Yano, and K. Shimamoto. 2002. Phytochrome mediates the external light signal to repress *FT* orthologs in photoperiodic flowering of rice. *Genes Dev.* 16:2006–2020.

Jaeger, K. E., and P. A. Wigge. 2007. FT protein acts as a long-range signal in *Arabidopsis. Curr. Biol.* 17:1050–1054.

Jiang, S.-Y., M. Cai, and S. Ramachandran. 2007. *ORYZA SATIVA MYOSIN XI B* controls pollen development by photoperiod-sensitive protein localizations. *Dev. Biol.* 304:579–592.

Kalivas, A., K. Pasentsis, A. Argiriou, and A. S. Tsaftaris. 2010. Isolation, characterization, and expression analysis of an NAP-like cDNA from crocus (*Crocus sativus* L.). *Plant Mol. Biol. Rep.* 28:654–663.

Kalivas, A., K. Pasentis, A. N. Polidoros, and A. S. Tsaftaris. 2007. Heterotopic expression of B-class floral homeotic genes *PISTILLATA/GLOBOSA* supports a modified model for crocus (*Crocus sativus* L.) flower formation. *DNA Sequence* 18:120–130.

Kamenetsky, R. 1994. Life cycle, flower initiation, and propagation of the desert geophyte *Allium rothii. Int. J. Plant Sci.* 155:597–605.

Kamenetsky, R. 1996. Life cycle and morphological features of *Allium* L. species in connection with geographical distribution. *Bocconea* 5:251–257.

Kamenetsky, R. 2009. Patterns of dormancy and florogenesis in herbaceous perennial plants: The environmental and internal regulation. *Crop Sci.* 49:2400–2404.

Kamenetsky, R. 2011. Florogenesis in geophytes: Classical and molecular approaches. *Acta Hort.* 886:113–118.

Kamenetsky, R., A. Barzilay, A. Erez, and A. H. Halevy. 2003. Temperature requirements for floral development of herbaceous peony cv. 'Sarah Bernhardt'. *Scientia Hortic.* 97:309–320.

Kamenetsky, R., and R. M. Fritsch. 2002. Ornamental *Alliums*. In *Allium Crop Science: Recent Advances*, eds. H. D. Rabinowitch, and L. Currah, 459–492. Wallingford, UK: CAB International.

Kamenetsky, R., I. London Shafir, H. Zemah, A. Barzilay, and H. D. Rabinowitch. 2004. Environmental control of garlic growth and florogenesis. *J. Am. Soc. Hort. Sci.* 129:144–151.

Kamenetsky, R., and E. Rabinowitch. 1999. Flowering response of *Eremurus* to post-harvest temperatures. *Scientia Hortic.* 79:75–86.

Kamenetsky, R., and H. D. Rabinowitch. 2001. Floral development in bolting garlic. *Sex. Plant Reprod.* 13:235–241.

Kamenetsky, R., and H. D. Rabinowitch. 2002. Florogenesis. In *Allium Crop Science: Recent Advances*, eds. H. D. Rabinowitch, and L. Currah, 31–58. Wallingford, UK: CAB International.

Kamenetsky, R., and H. D. Rabinowitch. 2006. The genus *Allium*: A developmental and horticultural analysis. *Hort. Rev.* 32:329–378.

Kanno, A., M. Nakada, Y. Akita, and M. Hirai. 2007. Class B gene expression and the modified ABC model in nongrass monocots. *TSW Dev. Embryol.* 2:17–28.

Kanno, A., H. Saeki, T. Kameya, H. Saedler, and G. Theißen. 2003. Heterotopic expression of class B floral homeotic genes supports a modified ABC model for tulip (*Tulipa gesneriana*). *Plant Mol. Biol.* 52:831–841.

Kaufmann, K., F. Wellmer, J. M. Muiño, T. Ferrier, S. E. Wuest, V. Kumar, A. Serrano-Mislata, F. Madueno, P. Krajewski, E. M. Meyerowitz, G. C. Angenent, and J. L. Riechmann. 2010. Orchestration of floral initiation by APETALA1. *Science* 328:85–89.

Kim, S. Y., C. B. Hong, and I. Lee. 2001. Heat shock stress causes stage-specific male sterility in *Arabidopsis thaliana. J. Plant Res.* 114:301–307.

Kim, D.-W., T.-S. Jung, S.-H. Nam, H.-R. Kwon, A. Kim, S.-H. Chae, S.-H. Choi, D.-W. Kim, R. N. Kim, and H.-S. Park. 2009. GarlicESTdb: An online database and mining tool for garlic EST sequences. *BMC Plant Biol.* 9:61(1–6).

King, R. W., T. Hisamatsu, E. E. Goldschmidt, and C. Blundell. 2008. The nature of floral signals in *Arabidopsis*. I. Photosynthesis and a far-red photoresponse independently regulate flowering by increasing expression of *FLOWERING LOCUS T* (*FT*). *J. Exp. Bot.* 59:3811–3820.

Ko, J.-H., I. Mitina, Y. Tamada, Y. Hyun, Y. Choi, R. M. Amasino, B. Noh, and Y.-S. Noh. 2010. Growth habit determination by the balance of histone methylation activities in *Arabidopsis. EMBO J.* 29:3208–3215.

Kobayashi, Y., and D. Weigel. 2007. Move on up, it's time for change—mobile signals controlling photoperiod-dependent flowering. *Genes Dev.* 21:2371–2384.

Kondrat'eva, V. V., M. V. Semenova, T. V. Voronkova, and N. N. Danilina. 2009. Changes in the carbohydrate and hormonal status in *Tulipa bifloriformis* bulbs forced into bloom in a greenhouse and in the open ground. *Russian J. Plant Physiol.* 56:428–435.

Koornneef, M., C. J. Hanhart, and J. H. van der Veen. 1991. A genetic and physiological analysis of late flowering mutants in *Arabidopsis thaliana*. *Mol. Gen. Genet.* 229:57–66.

Kramer, E. M. 2009. *Aquilegia*: A new model for plant development, ecology, and evolution. *Ann. Rev. Plant Biol.* 60:261–277.

Kramer, E. M., V. S. di Stilio, and P. M. Schlüter. 2003. Complex patterns of gene duplication in the *APETALA3* and *PISTILLATA* lineages of the Ranunculaceae. *Int. J. Plant Sci.* 164:1–11.

Krizek, B. A., and J. C. Fletcher. 2005. Molecular mechanisms of flower development: An armchair guide. *Nat. Rev. Genet.* 6:688–698.

Lang, A. 1957. The effect of gibberellin upon flower formation. *Proc. Nat. Acad. Sci. USA* 43:709–717.

Lang, A. 1965. Physiology of flower initiation. In *Encyclopedia of PlantPhysiology*, ed. W. Ruhland, 1371–1536. Berlin: Springer-Verlag.

Langeslag, J. J. 1989. *Teelt en Gebruiksmogelijkheden van Bijgoedgewassen. Tweede Uitgave*. Lisse, The Netherlands: Ministrie van Landbouw, Natuurbeheer en Visserij en Consulentschap Algemene Dienst Bloembollenteelt.

Lapointe, L. 2001. How phenology influences physiology in deciduous forest spring ephemerals. *Physiol. Plant.* 113:151–157.

Le Nard, M., and A. A. De Hertogh. 1993. Bulb growth and development and flowering. In *The Physiology of Flower Bulbs*, eds. A. De Hertogh, and M. Le Nard, 29–43. Amsterdam: Elsevier.

Le Nard, M., and P. Verron. 1993. *Convallaria*. In *The Physiology of Flower Bulbs*, eds. A. De Hertogh, and M. Le Nard, 249–256. Amsterdam: Elsevier.

Lee, J. H., S. M. Hong, S. J. Yoo, O. K. Park, J. S. Lee, and J. H. Ahn. 2006. Integration of floral inductive signals by flowering locus T and suppressor of overexpression of *Constans 1*. *Physiol. Plant.* 126:475–483.

Li, Q. Z., X. G. Li, S. N. Bai, W. L. Lu, and X. S. Zhang. 2002. Isolation of *HAG1* and its regulation by plant hormones during *in vitro* floral organogenesis in *Hyacinthus orientalis* L. *Planta* 215:533–540.

Lin, M.-K., H. Belanger, Y.-J. Lee, E. Varkonyi-Gasic, K.-I. Taoka, B. Xoconostle-Cázares, K. Gendler, R. A. Jorgensen, B. Phinney, T. J. Lough, and W. J. Lucas. 2007. FLOWERING LOCUS T protein may act as the long-distance florigenic signal in the Cucurbits. *Plant Cell* 19:1488–1506.

Litt, A., and E. M. Kramer. 2010. The ABC model and the diversification of floral organ identity. *Seminars in Cell & Developmental Biology* 21:129–137.

Liu, C., Z. Thong, and H. Yu. 2009. Coming into bloom: The specification of floral meristems. *Development* 136:3379–3391.

Liu, J., J. Yu, L. McIntosh, H. Kende, and J. A. D. Zeevaart. 2001. Isolation of a *CONSTANS* ortholog from *Pharbitis nil* and its role in flowering. *Plant Physiol.* 125:1821–1830.

Luyten, I., M. C. Versluys, and A. H. Blaauw. 1932. De optimale temperatuur van bloemaanleg tot bloei voor *Hyacinthus orientalis*. *Proceedings Section Science, Koninklijke Nederlandsche Akademie van Wetenschappen, Amsterdam, Natuurkunde, tweede sectie* 29:1–64.

Ma, H., and C. dePamphilis. 2000. The ABCs of floral evolution. *Cell* 101:5–8.

Machackova, I., and J. Krekule. 2002. Sixty-five years of searching for the signals that trigger flowering. *Russian J. Plant Physiol.* 49:451–459.

Masuda, J., T. Urakawa, Y. Ozaki, and H. Okubo. 2006. Short photoperiod induces dormancy in lotus (*Nelumbo nucifera*). *Ann. Bot.* 97:39–45.

Matsui, T., and K. Omasa. 2002. Rice (*Oryza sativa* L.) cultivars tolerant to high temperature at flowering: Anther characteristics. *Ann. Bot.* 89:683–687.

Miller, W. B. 1993. *Lilium longiflorum*. In *The Physiology of Flower Bulbs*, eds. A. De Hertogh, and M. Le Nard, 391–422. Amsterdam: Elsevier.

Mizukami, Y., and H. Ma. 1992. Ectopic expression of the floral homeotic gene *AGAMOUS* in transgenic Arabidopsis plants alters floral organ identity. *Cell* 71:119–131.

Molina, R. V., M. Valero, Y. Navarro, J. L. Guardiola, and A. García-Luis. 2005. Temperature effects on flower formation in saffron (*Crocus sativus* L.). *Scientia Hortic.* 103:361–379.

Moraga, Á. R., J. L. Rambla, O. Ahrazem, A. Granell, and L. Gómez-Gómez. 2009. Metabolite and target transcript analyses during *Crocus sativus* stigma development. *Phytochemistry* 70:1009–1016.

Mori, G., H. Kawabata, H. Imanishi, and Y. Sakanishi. 1991a. Growth and flowering of *Leucojum aestivum* L. and *L. autumnale* L. grown outdoors. *J. Japan. Soc. Hort. Sci.* 59:815–821.

Mori, G., H. Kawabata, H. Imanishi, and Y. Sakanishi. 1991b. Effects of temperature on flower initiation and development in *Leucojum aestivum* L. and *L. autumnale* L. *J. Japan. Soc. Hort. Sci.* 59:833–838.

Mori, T., H. Kuroiwa, T. Higashiyama, and T. Kuroiwa. 2003. Identification of higher plant GlsA, a putative morphogenesis factor of gametic cells. *Biochem. Biophysic. Res. Commun.* 306:564–569.

Moss, G. I., and J. Heslop-Harrison. 1968. Photoperiod and pollen sterility in maize. *Ann. Bot.* 32:833–846.

Mouradov, A., F. Cremer, and G. Coupland. 2002. Control of flowering time: Interacting pathways as a basis for diversity. *Plant Cell* 14:S111–S130.

Moyroud, E., G. Tichtinsky, and F. Parcy. 2009. The *LEAFY* floral regulators in angiosperms: Conserved proteins with diverse roles. *J. Plant Biol.* 52:177–185.

Mutasa-Göttgens, E., and P. Hedden. 2009. Gibberellin as a factor in floral regulatory networks. *J. Exp. Bot.* 60:1979–1989.

Naor, V., J. Kigel, Y. Ben-Tal, and M. Ziv. 2008. Variation in endogenous gibberellins, abscisic acid, and carbohydrate content during the growth cycle of colored *Zantedeschia* spp., a tuberous geophyte. *J. Plant Growth Regul.* 27:211–220.

Naor, V., J. Kigel, and M. Ziv. 2009. The relationships between gibberellin and organ size in colored *Zantedeschia* cv. 'Calla Gold'. *Israel J. Plant Sci.* 57:369–375.

Noy-Porat, T. 2009. Temperature effect on florogenesis in *Narcissus tazetta*: Morphological, physiological and biochemical aspects. Tel Aviv University, Tel-Aviv, Israel.

Noy-Porat, T., M. A. Flaishman, A. Eshel, D. Sandler-Ziv, and R. Kamenetsky. 2009. Florogenesis of the Mediterranean geophyte *Narcissus tazetta* and temperature requirements for flower initiation and differentiation. *Scientia Hortic.* 120:138–142.

Noy-Porat, T., R. Kamenetsky, A. Eshel, and M. Flaishman. 2010. Temporal and spatial expression patterns of the *LEAFY* homologue *NLF* during florogenesis in *Narcissus tazetta*. *Plant Sci.* 178:105–113.

Okubo, H. 1993. *Hippeastrum* (Amaryllis). In *The Physiology of Flower Bulbs*, eds. A. De Hertogh, and M. Le Nard, 321–334. Amsterdam: Elsevier.

Okubo, H. 2000. Growth cycle and dormancy in plants. In *Dormancy in Plants: From Whole Plant Behaviour to Cellular Control*, eds. J.-D. Viémont, and J. Crabbé, 1–22. Wallingford, Oxon, UK: C.A.B International.

Ovadia, R., I. Dori, A. Nissim-Levi, Y. Shahak, and M. Oren-Shamir. 2009. Coloured shade-nets influence stem length, time to flower, flower number and inflorescence diameter in four ornamental cut-flower crops. *J. Hort. Sci. Biotech.* 84:161–166.

Parcy, F. 2005. Flowering: A time for integration. *Int. J. Dev. Biol.* 49:585–593.

Park, J.-H., Y. Ishikawa, T. Ochiai, A. Kanno, and T. Kameya. 2004. Two *GLOBOSA*-like genes are expressed in second and third whorls of homochlamydeous flowers in *Asparagus officinalis* L. *Plant Cell Physiol.* 45:325–332.

Park, J.-H., Y. Ishikawa, R. Yoshida, A. Kanno, and T. Kameya. 2003. Expression of *AODEF*, a B-functional MADS-box gene, in stamens and inner tepals of the dioecious species *Asparagus officinalis* L. *Plant Mol. Biol.* 51:867–875.

Peet, M. M., S. Sato, and R. G. Gardner. 1998 Comparing heat stress effects on male-fertile and male-sterile tomatoes. *Plant, Cell and Environment* 21:225–231.

Price C. A., and Reardon E. M. 2001. Mendel, a database of nomenclature for sequenced plant genes. *Nucleic Acids Res.* 29:118–119.

Rees, A. R. 1966. The physiology of ornamental bulbous plants. *Bot. Rev.* 32:1–23.

Rees, A. R. 1972. *The Growth of Bulbs. Applied Aspects of the Physiology of Ornamental Bulbous Crop Plants*. London and New York: Academic Press.

Rees, A. R. 1992. *Ornamental Bulbs, Corms and Tubers*. Wallingford, Oxon, UK: C.A.B International.

Reeves, P. A., Y. H. He, R. J. Schmitz, R. M. Amasino, L. W. Panella, and C. M. Richards. 2007. Evolutionary conservation of the *FLOWERING LOCUS C*-mediated vernalization response: Evidence from the sugar beet (*Beta vulgaris*). *Genetics* 176:295–307.

Rijpkema, A. S., M. Vandenbussche, R. Koes, K. Heijmans, and T. Gerats. 2010. Variations on a theme: Changes in the floral ABCs in angiosperms. *Seminars in Cell & Developmental Biology* 21:100–107.

Roeder, A. H. K., and M. F. Yanofsky. 2001. Unraveling the mystery of double flowers. *Dev. Cell* 1:4–6.

Roh, M. S., and D.-K. Hong. 2007. Inflorescence development and flowering of *Ornithogalum thyrsoides* hybrid as affected by temperature manipulation during bulb storage. *Scientia Hortic.* 113:60–69.

Roh, M. S., and Y. H. Joung. 2004. Inflorescence development in an *Ornithogalum dubium* hybrid as influenced by bulb temperature treatments. *J. Hort. Sci. Biotech.* 79:576–581.

Roh, S. M., and H. F. Wilkins. 1977a. The effects of bulb vernalization and shoot photoperiod treatments on growth and flowering in *Lilium longiflorum* Thunb. cv. Nellie White. *J. Am. Soc. Hort. Sci.* 102:229–235.

Roh, S. M., and H. F. Wilkins. 1977b. Temperature and photoperiod effect on flower numbers in *Lilium longiflorum* Thunb. *J. Am. Soc. Hort. Sci.* 102:235–242.

Roh, S. M., and H. F. Wilkins. 1977c. Comparison of continuous and alternating bulb temperature treatments on growth and flowering in Lilium longiflorum Thunb. *J. Am. Soc. Hort. Sci.* 102:242–247.

Rossa, B., and D. J. von Willert. 1999. Physiological characteristics of geophytes in semi-arid Namaqualand, South Africa. *Plant Ecol.* 142:121–132.

Rotem, N., R. David-Schwartz, Y. Peretz, I. Sela, H. D. Rabinowitch, M. Flaishman, and R. Kamenetsky. 2011. Flower development in garlic: The ups and downs of *gaLFY* expression. *Planta* 233:1063–1072.

Rotem, N., E. Shemesh, Y. Peretz, F. Akad, O. Edelbaum, H. D. Rabinowitch, I. Sela, and R. Kamenetsky. 2007. Reproductive development and phenotypic differences in garlic are associated with expression and splicing of *LEAFY* homologue *gaLFY*. *J. Exp. Bot.* 58:1133–1141.

Saini, H. S., M. Sedgley, and D. Aspinall. 1983. Effect of heat stress during floral development on pollen tube growth and ovary anatomy in wheat (*Triticum aestivum* L.). *Aust. J. Plant Physiol.* 10:137–144.

Schlüter, P. M., G. Kohl, T. F. Stuessy, and H. F. Paulus. 2007. A screen of low-copy nuclear genes reveals the *LFY* gene as phylogenetically informative in closely related species of orchids (*Ophrys*). *Taxon* 56:493–504.

Shahak, Y. 2008. Photo-selective netting for improved performance of horticultural crops. A review of ornamental and vegetable studies carried out in Israel. *Acta Hort.* 770:161–168.

Shalit, A., A. Rozman, A. Goldshmidt, J. P. Alvarez, J. L. Bowman, E. Eshed, and E. Lifschitz. 2009. The flowering hormone florigen functions as a general systemic regulator of growth and termination. *Proc. Nat. Acad. Sci. USA* 106:8392–8397.

Sheldon, C. C., E. J. Finnegan, E. S. Dennis, and W. J. Peacock. 2006. Quantitative effects of vernalization on *FLC* and *SOC1* expression. *Plant J.* 45:871–883.

Shemesh, E., O. Scholten, H. D. Rabinowitch, and R. Kamenetsky. 2008. Unlocking variability: Inherent variation and developmental traits of garlic plants originated from sexual reproduction. *Planta* 227:1013–1024.

Shillo, R., and A. H. Halevy. 1976a. The effect of various environmental factors on flowering of gladiolus. I. Light intensity. *Scientia Hortic.* 4:131–137.

Shillo, R., and A. H. Halevy. 1976b. The effect of various environmental factors on flowering of gladiolus. II. Length of the day. *Scientia Hortic.* 4:139–146.

Simpson, G. G., and C. Dean. 2002. *Arabidopsis*, the Rosetta Stone of flowering time? *Science* 296:285–289.

Slabbert, M. M. 1997. Inflorescence initiation and development in *Cyrtanthus elatus* (Jacq. Traub). *Scientia Hortic.* 69:61–71.

Song, Y. H., S. Ito, and T. Imaizumi. 2010. Similarities in the circadian clock and photoperiodism in plants. *Curr. Opin. Plant Biol.* 13:594–603.

Song, B.-H., and T. Mitchell-Olds. 2011. Evolutionary and ecological genomics of non-model plants. *J. System. Evol.* 49:17–24.

Streck, N. A., and M. Schuh. 2005. Simulating the vernalization response of the "Snow Queen" lily (*Lilium longiflorum* Thunb.). *Sci. Agric. (Piracicaba, Braz.)* 62:117–121.

Sung, S., and R. M. Amasino. 2004. Vernalization and epigenetics: How plants remember winter. *Curr. Opin. Plant Biol.* 7:4–10.

Tamaki, S., S. Matsuo, H. L. Wong, S. Yokoi, and K. Shimamoto. 2007. Hd3a protein is a mobile flowering signal in rice. *Science* 316:1033–1036.

Taylor, A. 2009. Functional genomics of photoperiodic bulb initiation in onion (*Allium cepa*). PhD thesis. University of Warwick, Warwick, UK.

Taylor, A., A. J. Massiah, and B. Thomas. 2010. Conservation of *Arabidopsis thaliana* photoperiodic flowering time genes in onion (*Allium cepa* L.). *Plant Cell Physiol.* 51:1638–1647.

Theißen, G. 2001. Development of floral organ identity: Stories from the MADS house. *Curr. Opin. Plant Biol.* 4: 75–85.

Theißen, G., and R. Melzer. 2007. Molecular mechanisms underlying origin and diversification of the angiosperm flower. *Ann. Bot.* 100:603–619.

Theißen, G., A. Becker, A. di Rosa, A. Kanno, J. T. Kim, T. Munster, K. U. Winter, and H. Saedler. 2000. A short history of MADS-box genes in plants. *Plant Mol. Biol.* 42:115–149.

Theron, K. I., and A. A. De Hertogh. 2001. Amaryllidaceae: Geophytic growth, development, and flowering. *Hort. Rev.* 25:1–70.

Thomas, B., and D. Vince-Prue. 1997. *Photoperiodism in Plants (2nd ed.)*. San Diego: Academic Press.

Townsend, T., M. Albani, M. Wilkinson, G. Coupland, and N. Battey. 2006. The diversity and significance of flowering in perennials. In *Annual Plant Reviews, vol. 20. Flowering and Its Manipulation*, ed. C. Ainsworth, 181–201. Oxford: Blackwell Publ.

Treder, J. 2003. Effects of supplementary lighting on flowering, plant quality and nutrient requirements of lily 'Laura Lee' during winter forcing. *Scientia Hortic.* 98:37–47.

Tremblay, R., and J. Colasanti. 2006. Floral Induction. In *Annual Plant Reviews Vol. 20: Flowering and Its Manipulation*, ed. C. Ainsworth, 28–48. Oxford: Blackwell Publ.

Tsaftaris, A. S., K. Pasentsis, I. Iliopoulos, and A. N. Polidoros. 2004. Isolation of three homologous *AP1*-like MADS-box genes in crocus (*Crocus sativus* L.) and characterization of their expression. *Plant Sci.* 166:1235–1243.

Tsaftaris, A. S., K. Pasentsis, and A. N. Polidoros. 2005. Isolation of a differentially spliced C-type flower specific *AG*-like MADS-box gene from *Crocus sativus* and characterization of its expression. *Biol. Plant.* 49:499–504.

Tsuji, H., K.-I. Taoka, and K. Shimamoto. 2011. Regulation of flowering in rice: Two florigen genes, a complex gene network, and natural variation. *Curr. Opin. Plant Biol.* 14:45–52.

Turck, F., F. Fornara, and G. Coupland. 2008. Regulation and identity of florigen: FLOWERING LOCUS T moves center stage. *Ann. Rev. Plant Biol.* 59:573–594.

Tzeng, J.-D., S.-W. Hsu, M.-C. Chung, F.-L. Yeh, C.-Y. Yang, M.-C. Liu, Y.-F. Hsu, and C.-S. Wang. 2009. Expression and regulation of two novel anther-specific genes in *Lilium longiflorum*. *J. Plant Physiol.* 166:417–427.

Tzeng, T.-Y., C.-C. Hsiao, P.-J. Chi, and C.-H. Yang. 2003. Two lily *SEPALLATA*-like genes cause different effects on floral formation and floral transition in *Arabidopsis*. *Plant Physiol.* 133:1091–1101.

Urbanus, S. L., S. de Folter, A.V. Shchennikova, K. Kaufmann, R. GH Immink, and G. C. Angenent. 2009. In planta localisation patterns of MADS domain proteins during floral development in *Arabidopsis thaliana*. *BMC Plant Biology* 9:5(1–16).

Van Tunen, A. J., W. Eikelboom, and G. C. Angenent. 1993. Floral organogenesis in *Tulipa*. *Flowering Newsletter* 16:33–37.

Vidiella, P. E., J. J. Armesto, and J. R. Gutiérrez. 1999. Vegetation changes and sequential flowering after rain in the southern Atacama Desert. *J. Arid Environments* 43:449–458.

Vijayraghavan, U., K. Prasad, and E. Meyerowitz. 2005. Specification and maintenance of the floral meristem: Interactions between positively-acting promoters of flowering and negative regulators. *Curr. Sci.* 89:1835–1843.

Vishnevetsky, J., N. Azizbekova, M. Ziv, and H. Lilien-Kipnis. 1997. Development of the bulb and inflorescence in outdoor grown *Nerine sarniensis*. *Acta Hort.* 430:147–153.

Waldie, T., A. Hayward, and C. A. Beveridge. 2010. Axillary bud outgrowth in herbaceous shoots: How do strigolactones fit into the picture? *Plant Mol. Biol.* 73:27–36.

Wang, C.-N., M. Möller, and Q. C. B. Cronk. 2004. Altered expression of *GFLO*, the Gesneriaceae homologue of *FLORICAULA/LEAFY*, is associated with the transition to bulbil formation in *Titanotrichum oldhamii*. *Dev. Genes Evol.* 214:122–127.

Wang, A.-j., J.-f. Tang, X.-y. Zhao, and L.-h. Zhu. 2008. Isolation of *LiLFY1* and its expression in lily (*Lilium longiflorum* Thunb.). *Agric. Sci. China* 7:1077–1083.

Waterschoot, H. F. 1927. Results of the temperature during flower formation in *Hyacinthus*, 'L'Innocence' and 'La Victoire'. *Proceedings Section Science, Koninklijke Nederlandsche Akademie van Wetenschappen, Amsterdam* 31:31–49.

Weingartern-Kenan, E. 2005. Characterization of flowering in Madonna lily (*Lilium candicum*). Ms thesis. Ben Gurion University of the Negev, Israel.

Wellmer, F., and J. L. Riechmann. 2010. Gene networks controlling the initiation of flower development. *Trends Genet.* 26:519–527.

Widmer, R. E., and R. E. Lyons. 1985. *Cyclamen persicum*. In *CRC Handbook of Flowering Vol. II*, ed. A. H. Halevy, 382–390. Boca Raton, FL: CRC Press.

Wijeratne, A. J., W. Zhang, Y. Sun, W. Liu, R. Albert, Z. Zheng, D. G. Oppenheimer, D. Zhao, and H. Ma. 2007. Differential gene expression in Arabidopsis wild-type and mutant anthers: Insights into anther cell differentiation and regulatory networks. *Plant J.* 52:14–29.

Winter, K.-U., C. Weiser, K. Kaufmann, A. Bohne, C. Kirchner, A. Kanno, H. Saedler, and G. Theißen. 2002. Evolution of class B floral homeotic proteins: Obligate heterodimerization originated from homodimerization. *Mol. Biol. Evol.* 19:587–596.

Wu, X., J. Shi, M. Xi, Z. Luo, and X. Hu. 2010. A B functional gene cloned from lily encodes an ortholog of *Arabidopsis PISTILLATA* (*PI*). *Plant Mol. Biol. Rep.* 28:684–691.

Xu, R.-Y., Y. Niimi, and D.-S. Han. 2006. Changes in endogenous abscisic acid and soluble sugars levels during dormancy-release in bulbs of *Lilium rubellum*. *Scientia Hortic.* 111:68–72.

Yan, L., A. Loukoianov, A. Blechl, G. Tranquilli, W. Ramakrishna, P. SanMiguel, J.L. Bennetzen, and J. Dubcovsky. 2004. The wheat *VRN2* gene is a flowering repressor down-regulated by vernalization. *Science* 303:1640–1644.

Yano, M., Y. Katayose, M. Ashikari, U. Yamanouchi, L. Monna, T. Fuse, T. Baba, K. Yamamoto, Y. Umehara, Y. Nagamura, and T. Sasaki. 2000. *Hd1*, a major photoperiod sensitivity quantitative trait locus in rice, is closely related to the Arabidopsis Flowering Time gene *CONSTANS*. *Plant Cell* 12:2473–2484.

Yue, D., and H. Imanishi. 1990. Influence of storage temperature and its duration before or after ethylene exposure on the formation of flower buds in Dutch iris cultivar 'Blue Magic'. *Scientia Hortic.* 43:331–337.

Yun, P.-Y., T. Ito, S.-Y. Kim, A. Kanno, and T. Kameya. 2004a. The *AVAG1* gene is involved in development of reproductive organs in the ornamental asparagus, *Asparagus virgatus*. *Sex. Plant Reprod.* 17:1–8.

Yun, P.-Y., S.-Y. Kim, T. Ochiai, T. Fukuda, T. Ito, A. Kanno, and T. Kameya. 2004b. *AVAG2* is a putative D-class gene from an ornamental asparagus. *Sex. Plant Reprod.* 17:107–116.

Zaccai M., M. Lugassi-Ben Hamo, and I. Mazor, 2010. Studying the molecular regulation of vernalization in lily. *Second International Symposium on Genus Lilium (Symplitaly2010)*. Pescia, Italy, August 30–September 3, 2010.

Zik, M., and V. F. Irish. 2003. Flower development: Initiation, differentiation, and diversification. *Annu. Rev. Cell Dev. Biol.* 19:119–140.

Ziv, M., and V. Naor. 2006. Flowering of geophytes *in vitro. Propagation Ornamental Plants* 6:3–16.

Zlesak, D. C., and N. O. Anderson. 2009. Inheritance of non-obligate vernalization requirement for flowering in *Lilium formosanum* Wallace. *Israel J. Plant Sci.* 57:315–327.

9 Dormancy

Hiroshi Okubo

CONTENTS

I. INTRODUCTION

Dormancy occurs in seeds and vegetative propagules of many flowering plants (Lang 1996). Biologically, dormancy serves as a mechanism for ensuring plant survival during adverse environmental conditions, such as high and low temperatures and drought. Horticulturally, the occurrence of dormancy restricts the flowering season. On the other hand, a dormant period is advantageous for commercial production of many crops since it allows the commercial handling, storage, and transportation of the dormant organs. Thus, the understanding and control of dormancy in plants is critical for effectively managing the crop production, shipping, and usage of crops.

Since plant dormancy is a complicated and versatile mechanism, varying greatly in different species, one of the major problems of dormancy research is the lack of universal terminology and definition of this process. Lang et al. (1987) listed 54 terms, from the most frequent "dormancy", "rest", and "quiescence" to "true dormancy", "main rest" (Vegis 1964), "innate dormancy" (Sussman and Douthit 1973), "endogenous dormancy", "internal dormancy" (Taylorson and Hendricks 1977), "imposed dormancy" (Wareing and Saunders 1971; Rees 1981), "winter rest" (Fuchigami et al. 1977), and "correlative inhibition" (Hillman 1984). The definition also varies from a relatively simple designation, such as "a state in which growth is temporarily suspended" (Wareing and Phillips 1981), to a more detailed one, such as "an endogenously controlled but environmentally imposed temporary suspension of growth, accompanied by reduced metabolic activity, and relatively independent of ambient environmental conditions" (Amen 1968). At present, the definition "Dormancy is a temporary suspension of visible growth of any plant structure containing a meristem", proposed by Lang et al. (1987), has been widely accepted.

In geophytes, traditional studies on dormancy-related mechanisms have mainly focused on hormonal changes, along with environmental factors and carbohydrate metabolism. Studies on dormancy induction and release by environmental stimuli and the hormonal control of these processes have yielded a significant amount of information. However, the experimental hypotheses and conditions, as well as the species under study, varied greatly. The molecular nature and cellular basis of the processes of regulating dormancy are largely unknown. Modern approaches to molecular analyses are, without doubt, the most promising measures for future dormancy studies. However, the main reason for the limited molecular studies in geophytes is their huge genome sizes as well as the unavailability of a "model plant" for conducting such studies (see also Chapters 7 and 8).

This chapter reviews the concept and terminology of dormancy in ornamental geophytes and the mechanisms of its induction and termination. Also, universal terminology and definition of dormancy phenomena are proposed, which are applicable not only to geophytes but also to other higher plants.

II. CONCEPT OF DORMANCY

A. Evolutionary Aspects in Angiosperms

The common understanding of the history of plant evolution is that angiosperms evolved from woody gymnospermous ancestors during the Early Cretaceous period. The main location of this evolution is considered to be paleo-tropical rainforests of low latitudes, where temperatures and precipitation were high and stable throughout the year (Hickey and Doyle 1977; Crane and Lidgard 1989; Lupia et al. 1999).

The growth behavior of the paleoangiosperms can be assumed from the growth patterns of typical trees in modern tropic rainforests. McClure (1966) observed flowering and fruiting of many plant species at Ulu Gombak near Kuala Lumpur in a tropical rainforest of the Malay Peninsula (6°N). As an example, the fruiting of *Ficus sumatrana* was studied for 6 years during which time no yearly periodicity of growth was observed (Figure 9.1). This fruiting cycle was inherent in the physiology of the tree and could not be related to external stimuli. Therefore, in a typical tropical rainforest climate, plants continue their enduring growth without any seasonal quiescence.

During the Late Tertiary, the climatic zones were established and the flora was dominated by woody angiosperms that became segregated into climatic types. The plants were forced to develop novel adaptation systems for survival in the changing climates, with cool winters and hot and dry summers. Some woody species changed from an evergreen habit to a deciduous habit by developing winter adaptation. Others evolved into herbaceous plants with a shorter life cycle and smaller size. The herbaceous species developed the ability to survive unfavorable seasons by developing seeds, leaf rosettes, horizontal stems (aboveground runners or underground rhizomes), or special underground storage organs (bulbs, corms, tubers). In fact, geophytes are rarely found in tropical rainforests since the formation of storage organs itself is the adaptation to unfavorable environments.

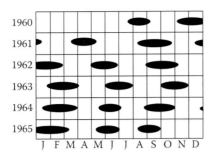

FIGURE 9.1 Observations over 6 years (1960–1965) of fruiting in *Ficus sumatrana* in the tropical rain forest of Ulu Gombak (3°N, 102°E), near Kuala Lumpur, Malaysia. (From McClure, H. E. 1966. *Malayan Forester* 29:182–203. Not copyrighted.)

B. RECENT RESEARCH IN PLANT DORMANCY

In seeds, endogenous and exogenous dormancy types are distinguished (Nikolaeva 1969). Endogenous dormancy is due to physiological (e.g., endogenous inhibitors) and/or morphological (e.g., immature embryo) characteristics of the seed embryo, while exogenous dormancy is caused by physical, chemical, or mechanical characteristics of the seed coat or fruit (e.g., hardseed coat). Baskin and Baskin (2004) have classified the physiological, morphological, morphophysiological, physical, and combinational dormancy in seeds. In addition, primary and secondary states of dormancy, which refer to the period of time when dormancy develops, are recognized in seeds (Foley 2001).

Seed dormancy is controlled by a balance between the hormones abscisic acid (ABA) and gibberellic acid (gibberellin, GA). Temperature, light, oxygen, and other environmental factors can affect this balance (Foley 2001). The concept was reinforced by recent molecular studies using model plants. In seeds of *Arabidopsis thaliana*, ABA regulation of dormancy was confirmed by expression patterns of key genes in ABA synthesis/catabolism, while dormancy loss was observed in the presence of fluridone (Cadman et al. 2006). It was shown that an ABA–GA hormone balance is dependent on the synthetic and catabolic pathways of both hormones. Many genes, expressed in the dormant state, are stress related. This indicates that stress and dormancy responses and ABA balance overlap significantly at the transcriptome level. Mechanisms of seed dormancy on molecular, physiological, and ecophysiological levels were reviewed by Finch-Savage and Leubner-Metzger (2006).

Lang et al. (1987) proposed the definition of dormancy in plants as a temporary suspension of visible growth of any plant structure containing a meristem, and classified three main types of dormancy: endodormancy, ecodormancy, and paradormancy. Endodormancy (innate dormancy) is growth suppression induced by an endogenous signal. Ecodormancy is considered to be a reduced growth response to an external stimulus, such as drought or cold, while removal of the stimulus results in an immediate resumption of growth. Paradormancy (correlative inhibition) is a reduced growth response induced by an inner signal, but not by external factors, and is transported to a target tissue within the individual plant. Removal of the signal results in a resumption of growth. This type of dormancy is usually exemplified by apical dominance, the phenomenon whereby the main central meristem of the plant is dominant over lateral meristems or buds.

Paradormancy, which occurs only in buds, has not been observed in seeds. The term "ecodormancy" also cannot be applied to seeds, since seed dormancy is measured only under favorable conditions, i.e., an adequate water supply, a suitable temperature, and a normal composition of the atmosphere (Junttila 1988). On the other hand, "ecodormant" buds are practically nondormant since they are capable of growth when external conditions permit (Dennis 1996). It is, therefore, apparent that the terms "ecodormancy" and "paradormancy" are not universal, and that they refer to different states of the plant development cycle. To avoid confusion, these states should be described by terms that do not include "dormancy". For example, the classical term "quiescence" can be used for reduced growth caused by external signals, while the removal of these factors results in the resumption of growth. In this chapter, only the state of endodormancy connotes the state of dormancy of the meristems.

Recent advances in bud dormancy mechanisms of woody and herbaceous perennials have been reviewed by Anderson et al. (2001, 2010) and Van der Schoot and Rinne (2011). Based on information gathered from annual and perennial plants, a dormancy model was proposed for the multiple pathways of dormancy regulation and flowering (Anderson et al. 2010). In *Populus*, apical meristems cease growth in response to short photoperiods. At this stage, leaf primordia differentiate into scales and endodormancy is induced (Anderson et al. 2010). It was found that FLOWERING LOCUS T (FT), TERMINAL FLOWER 1 (TFL1), and CONSTANS (CO) proteins, which are associated with flowering in *Arabidopsis*, are also associated with regulating growth cessation and endodormancy in *Populus*. Prior to the onset of dormancy, a short photoperiod induces downregulation of *FT* and a cessation of stem elongation, while a long photoperiod induces *FT* upregulation and inflorescence stem elongation in *A. thaliana* (Böhlenius et al. 2006; Ruonala et al. 2008). It was suggested that PHYTOCHROME A (PHYA), the protein involved in the entrainment of the circadian clock (Rodríguez-Falcón et al. 2006), plays a role in seasonal growth cessation and endodormancy induction through the regulation of FT and CENL1 (ortholog of TFL1) (Olsen et al. 1997; Ruonala et al. 2008; Anderson et al. 2010), while PHYTOCHROME B (PHYB) is involved in the perception of short photoperiods and the induction of tuberization in potato (Rodríguez-Falcón et al. 2006).

Using the *evergrowing* (*evg*) peach (*Prunus persica*), a mutant that fails to cease growth and enter dormancy under dormancy-inducing conditions, Bielenberg et al. (2008) identified six genes. They are a cluster of MIKC-type MADS-box transcription factors, similar to the genes that regulate meristem growth in vegetative tissues of *Ipomoea batatas* and *Solanum tuberosum* MADS-box. It was suggested that these genes might be involved in the regulation of growth cessation and terminal bud formation in peach in response to dormancy-inducing conditions, and they have been named dormancy-associated MADS-box (DAM) genes, which are closely related to *SHORT VEGETATIVE PHASE* (*SVP*) gene. The MADS-box genes *SUPPRESSOR OF CONSTANS1* (*SOC1*) and *FRUITFULL* (Melzer et al. 2008) determine the growth habit in *A. thaliana*. *PICKLE*, a gene that encodes a repressor of the cell cycle gene *AINTEGUMENTA* (*ANT*) in *A. thaliana* (Mizukami and Fischer 2000), and *GA-insensitive*, encoding a DELLA protein (Ruttink et al. 2007), are thought to function in suppressing shoot growth in the bud scales. It is possible that these genes are involved in the various steps of meristem transition into a dormant state.

The roles of plant hormones in the induction and release of bud dormancy in relation to gene expression have been reported in various species. For example, the interaction of circadian clock and hormone signaling in dormancy was reported in *Arabidopsis* plants (Covington and Harmer 2007; Robertson et al. 2008). An association of ethylene with the initiation of dormancy was found in woody and herbaceous perennials (Ruttink et al. 2007; Horvath et al. 2008). It was shown that ethylene may directly induce the key ABA biosynthetic gene *9-cis-epoxycarotenoid dioxygenase* (*NCED*) (Rodrigo and Alquezar 2006).

In general, comparative transcriptomics indicate that the numerous genes involved in multiple physiological, developmental, and biochemical responses show conserved patterns of expression during endodormancy transitions and during floral transitions in perennial plants (Anderson et al. 2010).

C. Dormancy in Geophytes

The concept of dormancy in geophytes is controversial. Kamerbeek et al. (1972) divided the major bulb species into three groups based on their dormancy traits. The species of the first group (e.g., *Gladiolus*) have a relatively long dormancy period, during which the differentiation of new organs, as well as their elongation, is directly interrupted. Those of the second group (e.g., *Tulipa*) differentiate flower buds inside the bulbs after bulb formation, but the rapid flower stem elongation is suppressed, while those of the third group (e.g., *Hippeastrum*) do not exhibit dormancy unless it is environmentally imposed. Rees (1981) suggested that only periods of meristem inactivity could be regarded as true dormancy and that no real dormancy occurs in tulip bulbs, since growth and the differentiation of the shoot continue slowly during the cold period. Kamenetsky (1994) refers to this period as "Intrabulb Development". Miller (1993) also proposed that bulbs have no dormant state, for example, *Lilium longiflorum* in which the meristem is continually initiating organs throughout the year.

Le Nard (1983) has suggested that the term "dormancy" should be applied to the entire bulb and not specifically to the apical bud or meristem. Therefore, bulb dormancy should be defined as "a complex and dynamic physiological, morphological and biochemical state during which there are no apparent external morphological changes or growth" (Le Nard 1983).

Later, Le Nard and De Hertogh (1993) suggested that in equatorial and subtropical areas, with relatively uniform environmental conditions, many genera show no marked rest periods and continuously produce foliage leaves (evergreen). Well-known examples of these nondeciduous bulbs are *Hippeastrum* and *Clivia*. Quiescence (ecodormancy) can be observed in these species, when grown under conditions with marked climatic changes. Geophyte species from climatic areas with seasonal changes have developed dormancy mechanisms in order to survive adverse climatic conditions of abnormally low or high temperatures and/or drought. Species of this group vary in the depth and longevity of their dormancy as well as their periodicity.

Okubo (2000) suggested that ancestral evergreen plants did not form underground storage organs, and that these morphological structures were developed during evolution as a result of the plant's adaptation to climate change. During the annual life cycle of these species, bulb formation precedes the dormancy phase and, therefore, the induction of dormancy and the induction of the storage organ (bulb, corm, tuber, etc.) formation are the same phenomenon. Similarly, Le Nard and De Hertogh (1993) note that the period of dormancy corresponds to the period of bulb enlargement (bulbing). In potato (*Solanum tuberosum*), dormancy cannot be separated from tuber initiation and enlargement (Fernie and Willmiter 2001).

III. INDUCTION AND DEVELOPMENT OF BULB FORMATION AND DORMANCY IN GEOPHYTES

A. TEMPERATURES

Low temperatures induce the formation of lateral bulbs in tulips (Le Nard and Cohat 1968; Aoba 1976), garlic (*Allium sativum*) (Aoba 1971), bulbous *Oxalis* (Aoba 1972), *Ornithogalum* (Aoba 1974a), and Dutch iris (Aoba 1974b). Bulb formation occurs during the spring growth period of the mother plant and terminates with the onset of dormancy at the beginning of summer. Bulb formation can be controlled by bulb storage and growth temperatures in tulip and Dutch iris. In tulip, when the bulbs are planted at 20°C in the absence of any previous low-temperature treatment, they do not exhibit any bulbing, and the lateral buds produce leaves (Le Nard and Cohat 1968). In Dutch iris, after bulb storage at 20°C, the vegetative apex produces a new bulb, when the mother plant is grown at 15°C, whereas continued growth at 25°C leads to the development of 10 or more leaves without bulbing (Okubo and Uemoto 1981). Nondormant new lateral shoots elongate instead of forming a bulblet (Figure 9.2). It was also shown that under bulb-inducing conditions, endogenous levels of ABA in the plants increase (Okubo and Uemoto 1981). These results clearly indicate that low temperatures induce bulbing in these plants.

Growing temperatures and seasonal changes also affect *Ornithogalum thyrsoides* and *Nerine sarniensis*—two deciduous bulbs that are native to South Africa. In temperate climates, these plants flower seasonally, the bulbs growing and enlarging occur at the end of the growing season, and then the plants become dormant. When grown in the tropics, these species remain evergreen. Thus, they grow and flower constantly and do not produce bulbs and off-set bulblets. The main factor separating the phenotypes seems to be temperature rather than the photoperiod, but this fact has not yet been clarified (Halevy 1990; Vishnevetsky et al. 1997).

FIGURE 9.2   Growth of 3–4 g Dutch iris bulbs at 25°C after storage at 20°C. The main shoot continued developing 10 or more leaves without forming a new bulb over a 12-month period, and two lateral shoots emerging. Normally, this weight class bulb develops 3–4 leaves and then the apex initiates a new round bulb; the lateral meristems form bulblets, from which the shoots emerge in the next season, after the release of dormancy. (From Okubo, H. 2000. In *Dormancy in Plants: From Whole Plant Behaviour to Cellular Control*, eds. J.-D. Viémont, and J. Crabbé, 1–22. Wallingford, Oxon, UK: CABI Publ. With permission from the publisher.)

Little information is available on temperature control of tuber and rhizome or corm induction in ornamental geophytes. Aoba (1970) reported that corm formation in the seedlings of *Freesia* and *Babiana* was accelerated by long exposure to temperatures of 5°C. It is possible that most tuberous, rhizomatous, and cormous species have photoperiod-sensitive or autonomous pathways for storage organ formation.

De Klerk and Gerrits (1996) provide three reasons affecting our understanding of the dormancy mechanism in whole plants. They are: (1) complex interactions may occur between the dormant organ and the other organs of the plant, (2) it is difficult to administer compounds to the organ when it is either attached or internal to the plant and a quantitative application is almost impossible, and (3) whole plants often require a large growing area and the effects of environmental factors, e.g., temperature or day length, can be studied only in an expensive facility like a phytotron. As an alternative, the employment of *in vitro* systems can simplify the experiments and data analyses and, therefore, they are often applied in dormancy research.

Bulb induction of *Hyacinthus orientalis* and *Lilium speciosum in vitro* is controlled by temperature. In experiments with the hyacinth 'Carnegie', explants taken from noncooled adventitious shoots regenerated new shoots at 23°C, whereas those cooled for 8 weeks at 4°C produced shoots and bulblets after transfer to 23°C (Bach 1992). It was demonstrated that ABA promoted the bulb formation of the explants, while fluridone suppressed the bulb formation in hyacinth and *L. speciosum* 'Rubrum' (De Klerk et al. 1992; Kim et al. 1994; Ii et al. 2002; Maehara et al. 2005). The degree of bulblet dormancy is also affected by culture temperatures. The bulblets generated *in vitro* at 15°C on scale explants of *L. speciosum* 'Rubrum' showed a higher emergence rate than those generated at 20°C and 25°C (Aguettaz et al. 1990).

Although *in vitro* systems can simplify the experiments and data analyses, a precise understanding of the entire mechanism cannot be gained without considering the effects of other tissues or organs. Unusual factors, such as unnatural nutrients and ambient conditions, and plant hormones that are added to the culture medium, often in extremely high concentrations, can distort the results or cause after-effects. Thus, *in vitro*-induced corms of summer-flowering *Gladiolus* 'Eurovision' (Ziv 1989)

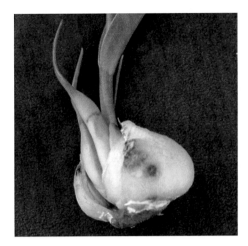

FIGURE 9.3   Lateral bud growth of a noncooled tulip bulb treated with gibberellin A₃ and TIBA. Root primordia were excised before the treatment. The lateral buds normally develop into new bulbs and sprout in the next season. (Photo by the author.)

and bulbs of *Lilium longiflorum* (Nhut 1998) and *Crinum variabile* (Fennell et al. 2001) did not experience dormancy, while under normal conditions these species do go through a state of dormancy. Although the simplified system can be useful in dormancy research, one must take into consideration the advantages and disadvantages of *in vitro* experiments.

Gibberellin and ABA are associated with temperature regulation in dormancy induction *in vitro* (De Klerk et al. 1992; Kim et al. 1994; Xu et al. 1998; Ii et al. 2002; Maehara et al. 2005). GAs also inhibit dormancy induction in whole tulip bulbs. A high concentration of GA₃ with 0.1% 2,3,5-triiodobenzoic acid (TIBA) in lanolin, applied at planting to noncooled tulip bulbs accelerated shoot growth of the lateral buds instead of bulbing (Figure 9.3; H. Okubo, unpublished).

B.   Photoperiod*

The fundamental roles of photoperiod, light intensity, and the circadian clock in the regulation of the life cycle of living organisms have been shown in numerous examples and studied in many plant species. Photoperiod was found to be a key factor in flowering and dormancy of the perennial plants (Horvath 2009; Anderson et al. 2010; Van der Schoot and Rinne 2011). In geophytes, short and long photoperiods provide the essential environmental signals for annual development, although geophytes species vary significantly in their reaction to this factor.

1.   Short Photoperiods

Short photoperiods affect the induction of tuberous roots in *Dahlia* (Zimmerman and Hitchcock 1929; Aoba et al. 1960; Moser and Hess 1968) and *Ranunculus asiaticus* (R. Kamenetsky, pers. comm.), and the tuberization of *Begonia evansiana* (Esashi and Nagao 1958) and *Helianthus tuberosus* (Hamner and Long 1939). In *Dahlia*, GA₃ inhibited the enlargement of the roots, whereas ABA and ethylene enhanced this process (Biran et al. 1972).

Psophocarpus tetragonolobus (winged bean, Leguminosae), a popular vegetable in southeast Asian countries, is a typical short-day (SD) plant for flowering (Herath and Ormrod 1979) and tuberous root formation (Wong 1981). Selections have been conducted in Kyushu University, Japan, to develop photo-insensitive cultivars for flowering that will allow young pod production under long-day (LD) conditions in temperate regions in summer. The original SD cultivars for flowering

---

* "Photoperiod" and "daylength" are usually used as synonyms.

also required short photoperiods for tuberous root formation, whereas the newly developed cultivars flowered and formed tuberous roots in long day (Okubo et al. 1992). The loss of photosensitivity for flowering was automatically linked to the loss of that for the induction of tuberous root formation, which suggests common mechanisms for flowering and dormancy induction in the winged bean.

Due to its economic importance, the potato is the most studied geophyte species, in terms of tuberization and dormancy mechanisms. It has been reported that short photoperiods perceived by the leaves induced underground stolons, which developed into tubers with axillary buds (Rodríguez-Falcón et al. 2006), which are endodormant (Sonnewald 2001). There is evidence that seasonal regulation of floral and tuberizing signaling pathways might also overlap at the molecular level (Rodríguez-Falcón et al. 2006) and that endodormancy maintenance may involve chromatin remodeling (Law and Suttle 2004).

Tuberization in potato is controlled by phytochrome B-mediated signal transduction. Phytochrome B and GA-mediated photoperiodic perception occur in the leaf. Subsequently, the RNA acts as a systemic signal in the long-distance signaling pathway and initiates tuberization in the subapical region of the underground stolon. There is evidence that the flowering and tuberization signals may be similar (Sarkar 2008).

Research on flowering mechanisms has led to the identification of FLOWERING LOCUS T (FT) as the main component of the florigen or mobile flowering promoting signal produced in the leaves (Turck et al. 2008; Zeevaart 2008). Recently, a similar mobile signal or tuberigen (a potato homolog of FT) has been reported to induce tuber formation in potato (Abelenda et al. 2011). Flowering regulators, such as CONSTANS and miR172, might also play a role in tuberization, although their pathways remain unclear.

An aquatic rhizomatous species of lotus *Nelumbo nucifera* was used as a model plant for dormancy research. Long photoperiods and high temperatures accelerated rhizome branching and elongation and leaf development, whereas SD was the main environmental factor affecting rhizome enlargement and induction of dormancy (Masuda et al. 2003, 2006; Figure 9.4).

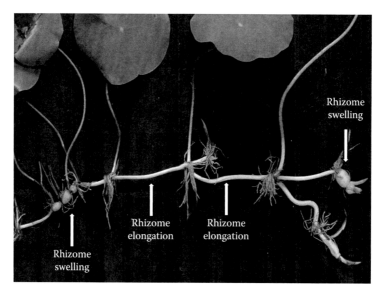

FIGURE 9.4 Photoperiod control of the reversibility in the swelling of the rhizome of *Nelumbo nucifera*. The seedling (left end) was initially grown in LD (16 h) on June 21, 2010 for 2 weeks, in SD (8 h) for the following 2 weeks, again in LD for the next 3 weeks, and again in SD for the following 2 weeks. The 5th internode of the main rhizome elongated, the next three (6th to 8th) internodes enlarged, the 9th to 12th internodes elongated again, and then the 13th enlarged (right end). The 1st to 4th internodes are always compact and neither elongate nor enlarge in any condition. (Produced and photographed by, and courtesy of Dr. J. Masuda.)

The critical photoperiod for rhizome transition from active growth to storage material accumulation is between 12 and 13 h. In plants grown under an 8-h photoperiod, a short night interruption with red (R) light resulted in the reverse of dormancy onset and subsequent rhizome elongation. A night break interruption with far-red (FR) light did not affect rhizome swelling. The effect of R light night break was reversed by the subsequent FR light (Table 9.1). These results indicate the key role of phytochrome in the photoperiodic response of rhizome swelling in the lotus plant (Masuda et al. 2007), and that phyB or phytochrome(s) with similar characteristics to phyB might be involved in the photoperiodic regulation in the lotus plant, as was proven in potato (Rodríguez-Falcón et al. 2006).

Under SD-inducing conditions, GA$_3$ inhibited rhizome enlargement. In contrast, ABA or GA biosynthesis inhibitors, uniconazol and paclobutrazol, enhanced rhizome enlargement under noninductive LD (Masuda et al. 2012). The responses to chemical applications were dose dependent. These data clearly indicated that GA and ABA, mediated by phytochrome, could control rhizome enlargement. In addition, the expression levels of *NCED*, involved in the biosynthesis of ABA, increased under SD, but not under LD (J. Masuda et al. unpublished data). Ethylene had no effect on rhizome enlargement either in SD or in LD (Masuda et al. 2010), although there are reports indicating the role of ethylene in inducing dormancy in potato (Suttle 1998b), *Populus* (Ruttink et al. 2007), and *Euphorbia esula* (Horvath et al. 2008). Figure 9.5 shows the proposed model of rhizome enlargement controlled by phytochrome-mediated plant hormones in *Nelumbo nucifera* (Masuda 2007).

2. Long Photoperiods

Among bulbous geophytes, some *Allium* species require long photoperiods for bulbing. For instance, a long photoperiod is the major environmental factor affecting bulb formation and development in onion (*Allium cepa*) (Brewster 1990; Taylor 2009). Treatments with ethephon caused bulb initiation in *A. cepa* under noninducing SD conditions (Levy and Kedar 1970). Lercari (1983) confirmed the ethephon effect on onion bulbs initiation in SD, but also found that silverthiosulfate (STS, an anti-ethylene complex) and aminoethoxyvinylglycine (AVG, an inhibitor of ethylene biosynthesis) did not affect photoperiod-induced bulb formation. This

---

TABLE 9.1

Maximum Rhizome Enlargement Index in Main Rhizomes of Lotus (*Nelumbo nucifera*) Responding to Sequential Red (R) and Far-Red (FR) Night Break Treatments in the Middle of the Dark Period of a Short Photoperiod

Night Break	Maximum Enlargement Index[a]
Control (8 h)	0.56
FR	0.46
R	0.04
R/FR	0.45
R/FR/R	0.18
R/FR/R/FR	0.28

Source: Adapted from Masuda, J. et al. 2007. *Planta* 226:909–915, with permission from the publisher.

Note: Plants were grown under a constant 8 h photoperiod with FR, R, and sequential R/FR, R/FR/R, and R/FR/R/FR irradiations in the middle of the dark period after 2 weeks of growth. Each irradiation was provided for 5 min (sequences up to 20 min).

[a] Maximum internode diameter/internode length.

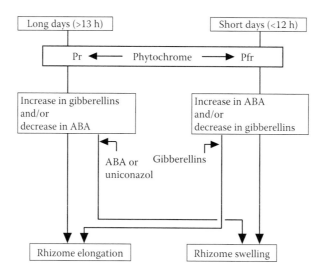

FIGURE 9.5 The proposed model of rhizome enlargement controlled by phytochrome-mediated plant hormones in *Nelumbo nucifera*. (Redrawn from Masuda, J. 2007. Studies on rhizome enlargement in lotus (*Nelumbo nucifera*). PhD thesis, Kyushu Univeristy, Fukuoka, Japan. Not copyrighted.)

suggests that ethephon-induced bulb formation might not be related to the endogenous control of bulbing.

Allium × *wakegi*, an F_1 hybrid of Japanese bunching onion (*A. fistulosum*) × shallot (*A. cepa*), develops bulbs in response to long photoperiods (Okubo et al. 1981). During this process, endogenous ABA contents in the shoots and scales increased. In comparison, ABA content in plants grown under SD did not change. Sprouting of the formed bulbs was inhibited by exogenous ABA, while an application of fluridone reduced the levels of endogenous ABA and accelerated bulb sprouting (Yamazaki et al. 1999a).

Similarly to tuberous and rhizomatous geophytes, phytochrome-mediated control has been suggested for the bulbing process. Onion plants (*A. cepa* 'Kaizuka-wase') in which the critical day length for bulb formation is between 12 and 12.5 h continued developing leaves without bulb formation in 8 h sunlight followed by 8 h R light, whereas those receiving an 8 h FR light formed bulbs (Terabun 1965). Mondal et al. (1986) showed that, in onion, the lower the R:FR ratio used, the higher scale initiation and bulbing ratio was obtained. An FR-intercepting acrylic film inhibited an increase in the bulbing index of *Allium* × *wakegi* in field experiments (Yamazaki et al. 1998).

Recently, three *FT*-like genes were found in onion (Taylor 2009), and the expression of onion homologs of the *Arabidopsis* flowering genes *GIGANTEA* (*GI*), *FLAVIN-BINDING, KELCH REPEAT F-BOX (FKF1)*, and *ZEITLUPE (ZTL)* was examined under SD and LD conditions (Taylor et al. 2010). The authors hypothesized that the photoperiodic control of flowering and bulbing in onion might share molecular pathways and have similar molecular controlling mechanisms.

C. OTHER FACTORS INDUCING DORMANCY

In addition to temperature and light, there are many other factors that can stimulate the formation of storage organs and the induction of dormancy. Physical conditions and plant growth regulators have also been shown to affect this process.

The formation of tuberous roots in sweet potato (*Ipomoea batatas*) is completely inhibited when the root system is immersed in water owing to O_2 deprivation by poor gas exchange and hypoxia

(Eguchi and Yoshida 2004, 2007). When the basal part of the root was exposed to air and the distal part was immersed in a nutrient solution, normal tuberous roots were produced (Eguchi and Yoshida 2004).

High levels of endogenous sucrose are necessary for tuberous root formation. In the presence of high sucrose concentrations, cytokinins can trigger tuberous root formation, for example, the sweet potato. Cytokinin (trans-zeatin riboside [t-ZR]) induced tuberous root formation in sweet potato cuttings in a hydroponic culture when the roots contained approximately 9 µg/mg FW sucrose, but did not induce morphogenesis when the sucrose level in the root was lowered to approximately 2 µg/mg FW (Eguchi and Yoshida 2008).

Several reports proposed that jasmonates (jasmonic acid, methyl jasmonate, and their derivatives) might take part in dormancy induction and tuberization of geophytes (potato, Pelacho and Mingo-Castel 1991; onion, Nojiri et al. 1992; garlic, Ravnikar et al. 1993). These data, however, are controversial. In potato, *in vitro* tuber formation, in combination with jasmonic acid, occurred only in the lateral buds but not at the top end of the stolons. Jackson and Willmitzer (1994) concluded that jasmonic acid does not induce tuberization of potato. Sarkar et al. (2006) also showed that jasmonates did not have a role in the number of tubers induced but promoted tuber growth. In onions and garlic, an analysis of endogenous jasmonic acid was performed after bulbing was initiated (Nojiri et al. 1992; Ravnikar et al. 1993).

A role for methyl jasmonate in corm formation was recently suggested for *Gladiolus* × *hybridus* (He et al. 2008). However, in this study, the control plants, without methyl jasmonate treatment, also formed corms, and the difference between the effects of the chemicals and the control was only quantitative. Thus, it can be concluded that jasmonates might be involved in the onset of dormancy, but that they are not direct inducers of the storage organ formation and dormancy in geophytes.

D. INHERITANCE OF DORMANCY INDUCTION

The induction of dormancy and bulb formation can be studied using segregating populations of the same species, including bulbing and nonbulbing phenotypes and/or fertile interspecific hybrids. This approach is challenging since interspecific F_1 hybrids are usually sterile and cannot be crossed to obtain the progenies of F_2 or BC_1 generations. However, if these crosses are possible, the hybrids can be used for dormancy studies.

The inheritance of bulb formation in F_1 and BC_1 progenies of chives (*Allium schoenoprasum*) × *A. schoenoprasum* var. *foliosum* was investigated (Xiao et al. 2010) using two parents that vary in bulbing ability. Chives have weak dormancy, constantly produce leaves, and do not bulb (Poulsen 1990) even under continuous 24 h light (R. Kamenetsky, pers. comm.). In contrast, *A. schoenoprasum* var. *foliosum*, an edible minor crop in Japan, ceases its growth and forms bulbs with leaf withering in summer. In F_1 progenies, the morphology of the basal part of the leaf sheaths resembled that of chives, and bulb formation was not observed. However, in the BC_1 progenies, bulb-forming and non-bulb-forming phenotypes were found. Three peaks of the segregation patterns were observed in maximum thickness (MT) of scale leaves (Figure 9.6). In the first group of genotypes, MT values of 1.0–1.5 mm were close to those of chives (0.7 mm) and the F_1 (1.1 mm). The group with thick scales (>5 mm) was similar to the second parent var. *foliosum* (6.4 mm), while the intermediate group included values of 3.0–3.5 mm. This segregation indicates the existence of at least two genes that control scale thickness and bulb formation. The fertile progenies of all the groups are available for molecular research on bulb formation. The results also suggest that the parameter "bulbing ratio", often used in the study of bulb formation, might be not reliable when studying the processes of bulbing and dormancy.

Other geophyte species, for example, fertile garlic genotypes (Kamenetsky et al. 2004; Shemesh et al. 2008) or species of *Oxalis*, *Iris*, or *Begonia*, can be used in the future for interspecific crosses and evaluation of bulbing and dormancy inheritance.

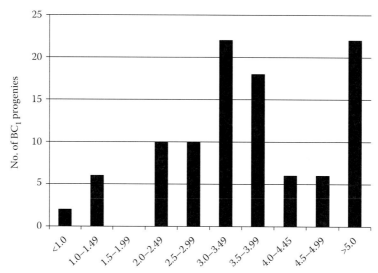

FIGURE 9.6 Frequency of distribution of the maximum thickness of scale leaves in BC_1 progenies of (chives × *A. schoenoprasum* var. *foliosum*) × *A. schoenoprasum* var. *foliosum*. The values in chives, *A. schoenoprasum* var. *foliosum*, and their F_1 were, respectively, 0.7, 6.4, and 1.1 mm. (From Xiao, J. et al. 2010. *J. Japan. Soc. Hort. Sci.* 79:282–286. With permission from the publisher.)

IV. RELEASE FROM DORMANCY

A. ENVIRONMENTAL FACTORS IN DORMANCY RELEASE

Similarly to the induction of dormancy, temperature is the main factor in release from dormancy in many geophyte species. For example, in Dutch iris, newly developed bulbs are dormant at harvest, and do not sprout for about 2 months. High temperatures cause bulb release from dormancy. Thus, bulbs planted in autumn emerge and develop leaves before winter. In tulip bulbs, the dormancy period is long and newly developed bulbs do not sprout until spring. During this period, they require warm temperatures for leaf and flower initiation inside the bulbs in the summer and low temperatures for subsequent rapid emergence and anthesis in the spring. The factor controlling dormancy release in the tulip is, therefore, low temperature. Similarly, in gladioli, newly formed corms are dormant in summer and the low temperatures of winter cause dormancy release. Although various techniques for dormancy release during bulb forcing have been developed, little is known about the biological mechanisms of this phenomenon. In general, the successful control of dormancy release can be achieved by the artificial application of temperature sequence in the natural annual cycle of each species. This is one of the basic methods of bulb forcing.

The application of forcing temperature, timing, and duration varies according to species and cultivars. In addition, it is also affected by the production country and the agrotechnique used, climatic conditions before and after bulb harvest, marketing programs, and so on. A detailed description of dormancy release by temperature treatments is available, for the various genera and species, in *The Physiology of Flower Bulbs* (De Hertogh and Le Nard 1993) and *Holland Bulb Forcer's Guide* (De Hertogh 1996). Some newer reports on dormancy release by temperature are also available, for example, for *Zephyra elegans* (Kim et al. 1996; Yañez et al. 2005).

B.   HORMONAL CONTROL OF DORMANCY RELEASE

1.   Gibberellins

Many reports suggest that GAs are involved in dormancy release in geophytes. For example, exogenously applied GA induced shoot growth and flowering of noncooled tulip bulbs (Saniewski et al. 1999). Similar results with exogenous GA were reported for micropropagated *Lilium japonicum* bulblets (Nishimura and Atsumi 2001).

Based on the results obtained by exogenous applications, endogenous content of GAs has been studied in dormant storage organs. Research on the role of endogenous GAs in dormancy release in geophytes has led to two opposing hypotheses. The first one denies the GA function in bulb dormancy release. Rebers et al. (1996) analyzed endogenous GAs in the shoots and basal plates of tulip 'Apeldoorn' stored at 5°C or 17°C for 12 weeks, and did not find a direct correlation between cold-stimulated growth and a change in the endogenous GA status. They also indicated that the free GA content in shoots or basal plates at the end of bulb storage cannot be used as a marker for the success-ful cold treatment of tulip bulbs. Similarly, Yamazaki et al. (2002) suggested that the state of bulb dormancy is independent of GA concentrations in the basal leaf sheaths or bulbs of *Allium × wakegi*.

The second hypothesis suggests that GAs play a major role in breaking dormancy. Takayama et al. (1993) identified 12 GAs in the bulbs of *Lilium* 'Connecticut King' during postharvest cold storage. They concluded that GA_4 might have a key role in dormancy release. Unfortunately, the authors did not compare the changes in these GAs in these bulbs without cold treatment.

2.   Auxins

There is little evidence suggesting the role of auxins in dormancy release of geophytes. Sukhova et al. (1993) found no changes in free indoleacetic acid (IAA) levels during the storage of potato tubers and only slight declines associated with sprouting. Rietveld et al. (2000) proposed that an increase in sensitivity to auxin, which is the main factor in stem elongation after planting, is the essential developmental change in the lowermost internode (foot) of tulip during cold treatment. Other studies showed that auxin transport inhibitors TIBA and naphthylphthalamic acid (NPA) caused early sprouting and flowering of tulip bulbs without cold treatment, particularly when applied together with GA_3 (Geng et al. 2005a,b). These facts imply a role of auxin in early sprouting during bulb storage, but this role is unclear.

3.   Cytokinins

Benzyl adenine (BA) is known to break the dormancy of freesia corms (Nakamura et al. 1974; Masuda and Asahira 1978; Uyemura and Imanishi 1987), but BA-induced sprouts subsequently exhibited poor growth. That is the reason why BA is not commercially used. Cytokinin treatment broke the dormancy rapidly in the tubers of the potato 'Majestic', which has a long dormancy period (Hemberg 1970). Eight endogenous cytokinins were identified in the apical bud tissues of the potato 'Russet Burbank' tubers (Suttle 1998a). In tubers incubated at 20°C, the total content of cytokinins increased sevenfold, and the loss of dormancy was preceded by significant increases in the endoge-nous levels of zeatin (Z), zeatin riboside (ZR), isopentenyl adenosine-5′-monophosphate (IPMP), and isopentenyl adenine-9-glucoside (IP-9-G). Continuous storage at 3°C (a growth-inhibiting tempera-ture) is associated with an increase in ZR, IP-9-G, and isopentenyl adenine + isopentenyl adenosine. Exogenous injection of Z, ZR, and IPMP resulted in the rapid and complete termination of tuber dormancy. It was concluded that cytokinins are endogenous dormancy-terminating agents in potato tuber buds (Suttle 1998a).

The question of whether GAs and/or auxins are more effective in the breaking of dormancy in bulbs while cytokinins are active in corms and tubers is unsolved. In tulips, the growth of the cen-tral bud after the completion of the cold requirement relies on cell elongation because the cell divi-sions in the central bud are complete (Gilford and Rees 1973). This can be accelerated by GAs, whereas cytokinins are not essential. Therefore, the fact that GAs are more effective than cytokinins may be explained by the bud developmental stage prior to dormancy release. On the other hand, in tubers, the termination of dormancy precedes cell division in the bud meristems (Suttle 2000). This might be the reason why cytokinins play an important role in dormancy release. However, additional evidence is needed.

4.   ABA and Ethylene

Endogenous ABA, measured by various methods, in *Lilium rubellum* (Xu et al. 2006) and *Allium × wakegi* bulbs (Yamazaki et al. 1995b), and potato tubers (Suttle 1995) decreased as the storage period increased. However, in two cultivars of *A. × wakegi*, one with deep, and the other with weak bulb dormancy, the initial ABA content was comparable and decreased equally in both cultivars during storage. This fact suggests that not only the amount but also the sensitivity to ABA are associated with dormancy. A cultivar with weak dormancy might have a lower sensitivity to ABA than a cultivar with a deeper dormancy (Yamazaki et al. 1999b).

Microtubers of the potato 'Russet Burbank' generated from single-node explants were dormant for >12 weeks and contained 10–50 times higher free ABA content than those developed with fluridone and showed an early release from dormancy (Suttle and Hultstrand 1994). Later, Suttle (1995) examined endogenous ABA changes in intact potato tubers of the same cultivar. In tubers stored at 20°C, dormancy was broken and sprout elongation began after 35–50 days of storage, whereas the same process was observed only after 50–80 days when tubers were stored at 3°C and then transferred to 20°C. Endogenous levels of free ABA were highest in tubers stored at 3°C, intermediate in tubers transferred from 3°C to 20°C, 7 days prior to analysis, and lowest in tubers stored at 20°C. Free ABA levels declined with increasing lengths of storage, regardless of temperature, and the onset of sprouting was not associated with changes in the free ABA levels. It was, therefore, concluded that a decrease in endogenous ABA is not required for the release of potato tuber from dormancy and sprout growth (Suttle 1995). However, Destefano-Beltrán et al. (2006) suggested that catabolism of ABA during dormancy release in potato may, in part, be correlated to decreased levels of a key ABA biosynthetic gene *NCED*.

Dormancy release by ethylene is known and is used commercially in freesia corms (Imanishi 1993) and narcissi bulbs (Hanks 1993). However, the mechanisms of ethylene's effect on dormancy release have not been clarified. Exposure of dormant corms of *Triteleia laxa* to ethylene has resulted in early sprouting (Han et al. 1990), whereas a lesser effect of ethylene has been reported on dormancy breaking in gladiolus corms (Imanishi 1996). It is possible that ethylene is effective only in some species and interacts with various mechanisms of dormancy release. For example, dormancy of freesia and *Triteleia* corms is released by high temperatures in summer, while in gladiolus corms, low temperatures in winter are required for dormancy release. Endogenous ethylene levels in freesia corms increased during storage at high temperatures in summer and were followed by a decrease in endogenous ABA in conjunction with dormancy release (Uyemura and Imanishi 1983, 1987). From these results, Imanishi (1996) concluded that the ethylene effect on dormancy release is a substitute for high temperatures, but not for low temperatures. However, Keren-Paz et al. (1989) reported that the storage of dormant corms of *Liatris spicata* 'Callilepsis' at 3°C for about 9 weeks resulted in a complete break of dormancy, accompanied by an increase in the rate of ethylene production, mainly by the buds. An application of ethephon also accelerated dormancy release and increased sprouting (Keren-Paz et al. 1989). Ethylene might be involved in the dormancy release by both high and low temperatures.

Interactions between ethylene and ABA were reported in the internodal growth of deepwater rice (Hoffmann-Benning and Kende 1992), growth and seed germination in *Arabidopsis thaliana* (Ghassemian et al. 2000; Cheng et al. 2002), and root elongation in maize (Spollen et al. 2000). Exogenously applied ABA did not affect the endogenous ethylene levels, whereas the application of ethylene reduced the endogenous levels of ABA in the heterophyllous leaf formation of *Ludwigia arcuata* (Kuwabara et al. 2003). Similar mechanisms might also be applicable for dormancy release of freesia corms and narcissus bulbs, since their dormancy is released by a decrease in endogenous

ABA, which is affected by ethylene treatment. No such evidence, however, has been found in dormancy regulation in geophytes.

C. CELL CYCLE DURING RELEASE OF DORMANCY

The assumption that the cell cycle is arrested during dormancy is based on the absence of cell division at certain phases of dormancy (Campbell 2006). In eukaryotes, the cell cycle consists of four phases: Gap 1 (G_1), Synthesis (S), Gap 2 (G_2), and Mitosis (M). DNA duplication occurs during the S phase, while the M phase consists of cell division. G_1 and G_2 are the periods of cell growth and energy accumulating for the following phases. The arrest of the cell cycle can occur at two meiotic points during the transition from G_1 to S and from G_2 to M. The cell cycle is controlled by the activity of cyclin-dependent kinase (CDK) complexes (Dewitte and Murray 2003), and cyclins (CYCs) are the primary regulators of the activity of CDKs (Wang et al. 2004).

In dormant potato buds, 77% and 13% of the nuclei remained in phases G_1 and G_2, respectively, whereas in the actively growing sprouts of potato, only 27% and 40% were found in phases G_1 and G_2, suggesting that potato dormancy is associated with a G_1 arrest (Campbell et al. 1996; Campbell 2006). It is known that, in underground adventitious buds of *Euphorbia esula*, GA induces S phase-specific gene expression, but does not affect M phase-specific genes (Horvath et al. 2002). In *Arabidopsis thaliana*, exogenously applied cytokinin induces cell division through induction of CycD3, a D-type cyclin, at the G_1–S cell cycle phase transition (Riou-Khamlichi et al. 1999). These findings suggest that plant hormones could activate and/or inactivate the cell cycle.

An analysis of the molecular mechanisms regulating the cell cycle could provide one of the approaches for understanding the pathways of geophyte dormancy. This approach might be applicable for some species, for example, *Gladiolus*, in which the meristem enters an arrested period for a few weeks or months. However, in other species, for example, *Lilium longiflorum*, a meristem is continuously initiating new scale, leaf, or flower primordia throughout the year (Blaney and Roberts 1966). This indicates that the cell cycle is never arrested. In addition, this approach is not applicable to the stage of dormancy induction of geophytes in which cell division occurs to initiate storage organs.

D. OTHER FACTORS ASSOCIATED WITH DORMANCY RELEASE

Breaking the dormancy of geophytes by various chemical agents has been reported in relation to endogenous plant hormone levels.

The treatment of dormant potato tubers with bromoethane resulted in rapid loss of dormancy with increases in histone H3 and H4, multiacetylation and small increases in RNA synthesis (Law and Suttle 2004). Yoshida et al. (1986) found that the dipping of Easter lily (*Lilium longiflorum*) bulbs in running water can break their dormancy, and suggested that some unknown inhibitors of sprouting may be dissolved in the water. Amaki and Amaki (2005) followed up these experiments, applying cool water dipping, airtight packing, and packing with dry ice to the dormant bulbs of the Easter lily 'Hinomoto'. These treatments resulted in 100%, 100%, and 92% sprouting, respectively, whereas no sprouting was observed in the control bulbs. One of the reasons for the dormancy release could be the lowering of O_2 concentration and anaerobic conditions of the bulb storage at low temperatures.

Changes in carbohydrates associated with dormancy release and sprouting of various geophytes were intensively studied. In tulip, alpha-amylase activity increased gradually during storage at low temperatures and slightly increased and then significantly increased during shoot elongation (Moe and Wickstrøm 1973; Lambrechts et al. 1994; Komiyama et al. 1997). In Easter lily 'Nellie White', a substantial decrease in starch and large accumulations of sucrose, mannose, fructose, and oligosaccharides were observed during storage at $-1°C$, whereas storage at $4.5°C$ resulted in lesser accumulations in these carbohydrates (Miller and Langhans 1990). In noncooled bulbs, the alpha-amylase activity decreased slightly or remained at its initial level (Miller and Langhans 1990). Similarly, long cold storage induced the net breakdown of starch and the accumulation of sucrose and glucose in bulblets of *Lilium* regenerated *in vitro*. Alpha- and beta-amylase exhibited higher activities in both cultivars after cold treatment (Shin et al. 2002).

In hyacinth, it was found that the process from an increase in alpha-amylase in bulb scales to the translocation and accumulation of sugars in the shoots reached completion during cold storage (Sato et al. 2006). Alpha-amylase activity increased during the storage period at 5°C, but remained stable throughout storage at 25°C for 12 weeks and the subsequent growth at 25°C. Carbohydrate analyses showed that low storage temperatures caused a decrease in starch content, but not in sucrose accumulation in the scales. These results suggest that sugars produced by the breakdown of the starch in the scales are transported from the scales to the shoot during the cold period. This mechanism allows for fast flowering in hyacinth when temperatures increase. The alpha-amylase gene *HoAmy1A* was isolated from the hyacinth bulbs, and was shown to exhibit increased expression in the scale during the cold storage period, but lower expression in the scale during shoot elongation. The promoter region of *HoAmy1A* contained a CArG element, which is associated with the response to low temperature (Sato et al. 2006).

Changes in water status during dormancy release was studied using magnetic resonance imaging (MRI) in *Allium* × *wakegi* (Yamazaki et al. 1995a), tulip (Okubo et al. 1997; Van der Toorn et al. 2000; Kamenetsky et al. 2000, 2003), and *Allium aflatunense* (Zemah et al. 1999). Kamenetsky et al. (2003) suggested that the degradation of storage polysaccharides to low-molecular-weight sugar molecules during release from dormancy might be accompanied by the local release of water molecules tightly bound to the polysaccharide granules in the bulk water, or by an influx of free water molecules due to increased osmotic potential caused by the higher sugar concentration.

E. Markers for Termination of Dormancy

Reliable markers for endodormancy termination in geophytes are one of the most important issues in horticultural practice. In addition to plant hormones, possible biochemical markers for dormancy release are chalcones (Gorin et al. 1990; Franssen and Kersten 1992), polyamines (Kollöffel et al. 1992), amino acids (Le Nard and Fiala 1990; Tonecki and Gorin 1990), and the 4-methyleneglutamine:asparagine ratio (Lambrechts et al. 1992). The development of nondestructive and time-saving methods, based on computerized imaging, has also been investigated. As an example, MRI was proposed to serve not only as the parameter dormancy status but also as a practical and critical sensor for monitoring bulb health and maturity (Okubo et al. 1997; Van der Toorn et al. 2000; Kamenetsky et al. 2003).

Light- or sound-reflecting sensors, similar to a nondestructive monitoring system, have been developed for fruits (Abbott et al. 1997). Similar molecular markers are reliable and useful in bulb production and forcing.

V. DORMANCY MODEL

Based on the eco-morphological approach in geophyte evolution and recent research on dormancy induction and release in various plant species, a new model for dormancy in bulbous geophytes (e.g., tulip) has been proposed. This model can also serve for future approaches to the general mechanisms of dormancy and growth in angiosperms.

In autumn (Figure 9.7a), one flowering-sized tulip bulb contains three generations: (1) a mother bulb, in which an apical bud has developed leaves and flower(s), (2) lateral buds, which will become the daughter bulbs and replace the mother bulb in the next season, and (3) primordia of a few lateral buds inside the daughter bulbs, which will develop during winter to early summer of the following year. The growth of an apical bud (differentiated leaves and a flower bud[s]) of the mother bulb in autumn is slow and no above-ground sprouting occurs. After receiving low temperatures in winter, flowering takes place in spring, during which time the mother bulb shrives and is replaced by the daughter bulbs (Figure 9.7b). The apical meristem of the daughter bulb (now it becomes the mother

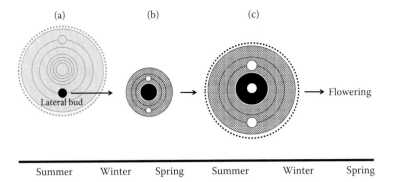

FIGURE 9.7 Schematic illustration of the development of a tulip bulb.

bulb) is still vegetative in early summer. During the summer, the apical bud initiates 3–4 leaves and then progresses to flower initiation and differentiation (Figure 9.7c). A flowering-sized bulb cannot sprout during the period from its harvest (late spring) to early spring of the following year even under adequate environmental conditions, although major morphological and physiological changes are occurring inside the bulb. Therefore, the bulb structure is "nesting" at least three generations of the buds, each at different states of dormancy and active growth. It is apparent that low winter temperatures play a dual role: dormancy induction in the lateral bud and dormancy release in the apical bud (Le Nard and Cohat 1968). This factor should be taken into consideration in studies of gene expression induced by low temperatures.

Figure 9.8 presents the comparison of the time-sequential dormancy process of tulip with that of a woody perennial *Populus* (Anderson et al. 2010). In *Populus*, growth cessation, bud set[*], and dormancy

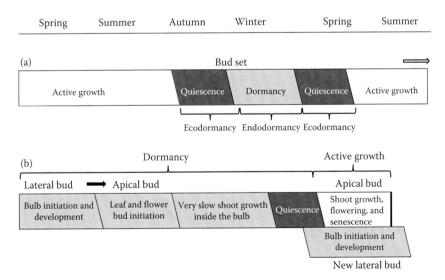

FIGURE 9.8 Comparison of dormancy models in woody perennials and geophytes. (a) *Populus* (woody perennial). Bud set occurs during transition from quiescence to dormancy (eco- and endodormancy as per Lang et al. 1987). (Adapted from Anderson, J. V. et al. 2010. In *Dormancy and Resistance in Harsh Environments, Topics in Current Genetics 21*, eds. E. Lubzens, J. Cerda, and M. Clark, 69–90. Berlin: Springer-Verlag). (b) *Tulipa* (geophyte). A lateral bud of the current year becomes the apical bud of the following year. Dormancy includes the period when organ differentiation begins and proceeds inside the bulb, but no shoot elongation are visible outside of the bulb.

---

[*] Developmental process of scale intiation, maturation, and the formation of an enclosed embryonic shoot.

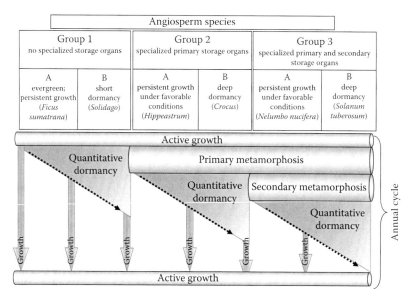

FIGURE 9.9 Proposed dormancy model in angiosperm species.

of the meristem occur sequentially in response to the environment, mostly photoperiod and low temperatures (Figure 9.8a). In the period from growth cessation to bud set, the meristem is ecodormant. Later, the meristem is arrested and is unable to resume growth, even under growth-promoting conditions. This state is defined as endodormancy, that is, when the development of the meristem cannot be reversed (Van der Schoot and Rinne 2011). Bud formation precedes dormancy, while scale initiation is the first event of bud formation and is followed by subsequent maturation of scales and the formation of a dormant bud. This process requires sufficient reserves, stored in the overwintering organs to facilitate cold acclimation (Van der Schoot and Rinne 2011).

The morphological and physiological homology between the bud set of woody plants and the bulb initiation and development in bulbous species is evident (Figure 9.8b). We can consider that the meristem has already entered endodormancy after the lateral bud had passed the "point-of-no-return" and continued bulb development.

Generalization of the proposed dormancy concept for the angiosperm species classifies the species into three groups (Figure 9.9). The plants of Group 1 develop no special storage organs. They are evergreen and grow continuously, such as, for example, the trees of tropical rain forests (A) or enter dormancy without forming any special organs (B) (e.g., epicotyl dormancy of *Quercus* seedlings (Wareing and Phillips 1981) and vegetative dormancy of *Solidago* (Shefferson 2009)). Geophytes belong to Group 2 or 3, which form special organs (e.g., bulbs, corms, and rhizomes) (primary metamorphosis). Species of subgroup 2A form special organs but possess weak dormancy and can grow under favorable conditions (e.g., *Hippeastrum*), while species of the subgroup 2B have deep dormancy and require a cold period for the breaking of dormancy (e.g., *Crocus*). The species of Group 3 initially form specialized organs (e.g., rhizomes (stolon) of potato plants—primary metamorphosis) similar to those of Group 2 and then, in addition, form storage organs (e.g., potato tubers—secondary metamorphosis) in response to environmental conditions (e.g., photoperiod). For example, an enlarged lotus rhizome can sprout immediately after planting (3A). In contrast, potato tubers require a specific period of low temperature for dormancy breaking (3B).

VI. CONCLUSIONS AND FUTURE RESEARCH

In general, plant dormancy is a complex and sophisticated process, highly regulated by external and internal signals. Geophyte species are extremely variable and, at this stage, it is impossible to propose a complete and universal model mechanism of geophyte dormancy. Based on classical and modern research on the morphological, physiological, and biochemical aspects of geophytes dormancy, it can be concluded that

1. The geophyte species remain in a state of dormancy as long as their storage organs do not exhibit any external growth.
2. Dormancy initiation in geophytes corresponds to the initiation and formation of a storage organ.
3. Only the so-called endodormancy (Lang et al. 1987) can be defined as in a "state of dormancy" in geophytes. To avoid confusion in terminology, the term "quiescence" is proposed for the so-called ecodormancy and the term "apical dominance" instead of "paradormancy" (in the sense of Lang et al. 1987).

Recent studies suggest that similar to the seeds and vegetative structures of other species, dormancy in geophytes is regulated by plant hormones. ABA and GAs are the primary hormones that control plant dormancy induction, development, and release (Koornneef et al. 2002; Finkelstein et al. 2008; Holdsworth et al. 2008).

An interaction between dormancy and florogenesis was suggested more than 50 years ago by Chouard (1960). Recent studies on the molecular level suggest that there are common pathways for vernalization, flower initiation, and dormancy induction. These may be controlled by common genes regulated by temperature or photoperiod (Horvath 2009; Anderson et al. 2010; and references therein). However, flowering and dormancy play very different roles in plant development. During florogenesis and flowering, the meristem progresses toward senescence, particularly in monocarpic species, whereas dormancy occurs at various stages of growth, even in seeds and at the seedling stage, and its role is to ensure survival and future plant growth. In the case of the common genes that control both flowering and dormancy induction, additional genetic mechanisms might differentiate between these two processes. As an example, in a large bulb of Dutch iris, low temperatures induce, concurrently, flowering of the apical bud and bulb formation (dormancy) of the lateral buds. Therefore, a system of control and directing of different pathways in the same plant must exist. This regulator(s) will direct meristem development from leaf development to either flowering or scale formation and subsequent dormancy.

In some geophytes, flowering and dormancy are opposite processes. In *Ranunculus asiaticus*, SD induces flowering while LD affects the formation of the tuberous roots (R. Kamenetsky, pers. comm.). In tulips, flower initiation and development within the bulb occur at warm temperatures of 17–20°C or higher (Hartsema 1961), whereas low temperatures induce the formation of the lateral bulbs (Le Nard and Cohat 1968). Therefore, in these species, pathways of dormancy and flowering are regulated by different environmental signals, and different genetic pathways might be involved.

Another important aspect of dormancy regulation is the differential response of various plant organs to environmental signals. For example, *Muscari armeniacum* does not require chilling for leaf elongations, but low temperatures are essential for inflorescence scape growth and flowering (Rudnicki and Novak 1993). When it is grown in a mild winter climate (Fukuoka, Japan; 33°35′N, 130°24′E), the leaves continue growing from planting in autumn and reach >60 cm at anthesis, whereas the flower stalk length is 15–25 cm (Figure 9.10). In the monocarpic shoots of *Hippeastrum*, leaves and flowers differentiate simultaneously, but the leaves elongate much earlier than the flower stem, and the latter appears only with the leaves of the next-generation bud. Another example is the hysteranthous and synanthous species of the genus *Crocus* (Benschop 1993). In both types, flower initiation occurs in summer and the initiation order of the leaves and flowers are similar. However, a time interval takes place between leaf elongation and anthesis in hysteranthous species, when compared to synanthous species. These examples suggest that different organs might vary in their dormancy requirements and that the dormancy process can be regulated not only by external signals but also by internal signals.

During the last few decades, significant progress has been made in dormancy studies, from morphological and developmental aspects to the molecular approach. It is evident that the application

FIGURE 9.10 *Muscari armeniacum* plants at anthesis with extremely elongated leaves grown outdoors in Fukuoka, Japan (33°35′N, 130°24′E). (Photo by the author.)

of molecular techniques is indispensable for clarifying the precise mechanism(s) of dormancy induction, development, maintenance, and release. Although model geophytes species have not been selected, the establishment of such a model system is crucial for further research on bulb dormancy (Benschop et al. 2010; Kamenetsky 2011).

Future research on the basic mechanisms of dormancy induction and release should result in the implementation of this knowledge in production techniques.

ACKNOWLEDGMENTS

The author is grateful to Drs. A. De Hertogh, M. Le Nard, and R. Kamenetsky for their critical review of this chapter.

CHAPTER CONTRIBUTOR

Hiroshi Okubo is a professor of horticultural science at Kyushu University, Fukuoka, Japan. He was born in Japan and received his BS, MS, and PhD in horticultural science from Kyushu University. His major expertise includes the physiology of growth and development in horticultural crops, in particular in ornamental geophytes. He also conducts research on tropical plant species. Dr. Okubo is the author of more than 200 professional papers, book chapters, technical publications, and popular press articles. He is the director of the University Farm of Kyushu University and a recipient of an award from the Japanese Society for Horticultural Science (1988).

REFERENCES

Abbott, J. A., R. Lu, B. L. Upchurch, and R. L. Stroshine. 1997. Technologies for nondestructive quality evaluation of fruits and vegetables. *Hort. Rev.* 20:1–120.

Abelenda, J. A., C. Navarro, and S. Prat. 2011. From the model to the crop: Genes controlling tuber formation in potato. *Curr. Opin. Biotechnol.* 22:287–292.

Aguettaz, P., A. Paffen, I. Delvallée, P. van der Linde, and G.-J. de Klerk. 1990. The development of dormancy in bulblets of *Lilium speciosum* generated *in vitro*. 1. The effects of culture conditions. *Plant Cell, Tiss. Org. Cult.* 22:167–172.

Amaki, W., and W. Amaki. 2005. Dormancy release of Easter lily bulb by low O_2 concentration treatments. *Acta Hort.* 673:591–594.

Amen, R. D. 1968. A model of seed dormancy. *Bot. Rev.* 34:1–31.

Anderson, J. V., W. S. Chao, and D. P. Horvath. 2001. A current review on the regulation of dormancy in vegetative buds. *Weed Sci.* 49:581–589.

Anderson, J. V., D. P. Horvath, W. S. Chao, and M. E. Foley. 2010. Bud dormancy in perennial plants: A mechanism for survival. In *Dormancy and Resistance in Harsh Environments, Topics in Current Genetics 21*, eds. E. Lubzens, J. Cerda, and M. Clark, 69–90. Berlin: Springer-Verlag.

Aoba, T. 1970. Effect of low temperature on the bulb or corm formation in some ornamental plants. *J. Japan. Soc. Hort. Sci.* 39:369–374 (In Japanese with English summary).

Aoba, T. 1971. Studies on the bulb formation in garlic plants (II). On the effect of low temperature. *J. Yamagata Agric. Foresty Soc.* 28:35–40 (In Japanese with English summary).

Aoba, T. 1972. Effect of temperature on bulb- and tuber-formation in bulbous and tuberous plants II. On bulb formation in bulbous oxalis. *J. Japan. Soc. Hort. Sci.* 41:393–397 (In Japanese with English summary).

Aoba, T. 1974a. Effects of temperature on bulb- and tuber-formation in bulbous and tuberous crops. (V) Effect of chilling treatment on bulb formation in *Ornithogalum arabicum* L., *O. thyrsoides* JARQ, and *O. umbellatum* L. Bull. *Yamagata Univ., Agr. Sci.* 7:229–237 (In Japanese with English summary).

Aoba, T. 1974b. Effect of temperature on bulb- and tuber-formation in bulbous and tuberous crops VI. On the bulb formation in bulbous iris. *J. Japan. Soc. Hort. Sci.* 43:273–280 (In Japanese with English summary).

Aoba, T. 1976. Effects of temperature on bulb- and tuber-formation in bulbous and tuberous crops IX. On the bulb formation in tulip. *Bull. Yamagata Univ., Agr. Sci.* 7:387–399 (In Japanese with English summary).

Aoba, T., S. Watanabe, and C. Saito. 1960. Studies on tuberous root formation in dahlia I. Periods of tuberous root formation in dahlia. *J. Japan. Soc. Hort. Sci.* 29:247–252 (In Japanese with English summary).

Bach, A. 1992. Micropropagation of hyacinths (*Hyacinthus orientalis* L.). In *Biotechnology in Agriculture and Forestry, Vol. 20, High-Tech and Micropropagation IV*, ed. Y. P. S. Bajaj, 144–159. Berlin: Springer-Verlag.

Baskin, J. M., and C. C. Baskin. 2004. A classification system for seed dormancy. *Seed Science Research* 14:1–16.

Benschop, M. 1993. *Crocus*. In *The Physiology of Flower Bulbs*, eds. A. De Hertogh, and M. Le Nard, 257–272. Amsterdam: Elsevier.

Benschop, M., R. Kamenetsky, M. Le Nard, H. Okubo, and A. De Hertogh. 2010. The global flower bulb industry: Production, utilization, research. *Hort. Rev.* 36:1–115.

Bielenberg, D. G., Y. Wang, Z. Li, T. Zhebentyayeva, S. Fan, G. L. Reighard, R. Scorza, and A. G. Abbott. 2008. Sequencing and annotation of the evergrowing locus in peach [*Prunus persica* (L.) Batsch] reveals a cluster of six MADS-box transcription factors as candidate genes for regulation of terminal bud formation. *Tree Genet. Genomes* 4:495–507.

Biran, I., I. Gur, and A. H. Halevy. 1972. The relationship between exogenous growth inhibitors and endogenous levels of ethylene, and tuberization of dahlias. *Physiol. Plant.* 27:226–230.

Blaney, L. T., and A. N. Roberts. 1966. Growth and development of the Easter lily bulb, *Lilium longiflorum* Thunb. 'Croft'. *Proc. Am. Soc. Hort. Sci.* 89:643–650.

Böhlenius, H., T. Huang, L. Charbonnel-Campaa, A. M. Brunner, S. Jansson, S. H. Strauss, and O. Nilsson. 2006. CO/FT regulatory module controls timing of flowering and seasonal growth cessation in trees. *Science* 312:1040–1043.

Brewster, J. L. 1990. Physiology of crop growth and bulbing. In *Onion and Allied Crops. I*, eds. J. L. Brewster, and H. D. Rabinowitch, 53–88. Boca Raton, FL: CRC Press.

Cadman, C. S. C., P. E. Toorop, H. W. M. Hilhorst, and W. E. Finch-Savage. 2006. Gene expression profiles of *Arabidopsis* Cvi seeds during dormancy cycling indicate a common underlying dormancy control mechanism. *Plant J.* 46:805–822.

Campbell, M. A. 2006. Dormancy and the cell cycle. In *Genetic Engineering, Vol. 27*, ed. J. K. Setlow, 21–33. Berlin: Springer.

Campbell, M. A., J. C. Suttle, and T. W. Sell. 1996. Changes in cell cycle status and expression of p34[cdc2] kinase during potato tuber meristem dormancy. *Physiol. Plant.* 98:743–752.

Cheng, W.-H., A. Endo, L. Zhou, J. Penney, H.-C. Chen, A. Arroyo, P. Leon, E. Nambara, T. Asami, M. Seo, T. Koshiba, and J. Sheen. 2002. A unique short-chain dehydrogenase/reductase in *Arabidopsis* glucose signaling and abscisic acid biosynthesis and functions. *Plant Cell* 14:2723–2743.

Chouard, P. 1960. Vernalization and its relations to dormancy. *Annu. Rev. Plant Physiol.* 11:191–238.

Covington, M. F., and S. L. Harmer. 2007. The circadian clock regulates auxin signaling and responses in *Arabidopsis. PLoS Biol.* 5(8):e222.

Crane, P. R., and S. Lidgard. 1989. Angiosperm diversification and paleolatitudinal gradients in Cretaceous floristic diversity. *Science* 246:675–678.

De Hertogh, A. 1996. *Holland Bulb Forcer's Guide. 5th ed.* Hillegom, The Netherlands: Int. Flower Bulb Centre.

De Hertogh, A., and M. Le Nard. 1993. *The Physiology of Flower Bulbs.* Amsterdam: Elsevier.

De Klerk, G.-J. M., and M. M. Gerrits. 1996. Development of dormancy in tissue-cultured lily bulblets and apple shoots. In *Plant Dormancy: Physiology, Biochemistry and Molecular Biology*, ed. G. A. Lang, 115–131. Wallingford, Oxon, UK: CAB International.

De Klerk, G.-J., K. S. Kim, M. van Schadewijk, and M. Gerrits. 1992. Growth of bulblets of *Lilium speciosum in vitro* and in soil. *Acta Hort.* 325:513–520.

Dennis, Jr., F. G. 1996. A physiological comparison of seed and bud dormancy. In *Plant Dormancy: Physiology, Biochemistry and Molecular Biology*, ed. G. A. Lang, 47–56. Wallingford, UK: CAB International.

Destefano-Beltrán, L., D. Knauber, L. Huckle, and J. C. Suttle. 2006. Effects of postharvest storage and dormancy status on ABA content, metabolism, and expression of genes involved in ABA biosynthesis and metabolism in potato tuber tissues. *Plant Mol. Biol.* 61:687–697.

Dewitte, W., and J. A. H. Murray. 2003. The plant cell cycle. *Annu. Rev. Plant Biol.* 54:235–264.

Eguchi, T., and S. Yoshida. 2004. A cultivation method to ensure tuberous root formation in sweetpotatoes (*Ipomoea batatas* (l.) Lam.). *Environ. Cont. Biol.* 42:259–266.

Eguchi, T., and S. Yoshida. 2007. Effects of gas exchange inhibition and hypoxia on tuberous root morphogenesis in sweetpotato (*Ipomoea batatas* (l.) Lam.). *Environ. Cont. Biol.* 45:103–111.

Eguchi, T., and S. Yoshida. 2008. Effects of application of sucrose and cytokinin to roots on the formation of tuberous roots in sweetpotato (*Ipomoea batatas* (L.) Lam.). *Plant Root* 2:7–13.

Esashi, Y., and M. Nagao. 1958. Studies on the formation and sprouting of aerial tubers in *Begonia evansiana* Andr. I. Photoperiodic conditions for tuberization. *Science Reports of the Tohoku University Fourth Series (Biology)* 24:81–88.

Fennell, C. W., N. R. Crouch, and J. van Staden. 2001. Micropropagation of the river lily, *Crinum vaiablile* (Amaryllidaceae). *S. Afr. J. Bot.* 67:74–77.

Fernie, A. R., and L. Willmiter. 2001. Molecular and biochemical triggers of potato tuber development. *Plant Physiol.* 127:1459–1465.

Finch-Savage, W. E., and G. Leubner-Metzger. 2006. Seed dormancy and the control of germination. *New Phytol.* 171:501–523.

Finkelstein, R., W. Reeves, T. Ariizumi, and C. Steber. 2008. Molecular aspects of seed dormancy. *Annu. Rev. Plant Biol.* 59:387–415.

Foley, M. E. 2001. Seed dormancy: An update on terminology, physiological genetics, and quantitative trait loci regulating germinability. *Weed Sci.* 49:305–317.

Franssen, J. M., and C. H. Kersten. 1992. Chalcones: A possible parameter to test the cold duration of tulip (*Tulipa gesneriana* cv. Apeldoorn) bulbs. *Acta Hort.* 325:259–266.

Fuchigami, L. H., M. Hotze, and C. J. Weiser. 1977. The relationship of vegetative maturity to rest development and spring bud break. *J. Am. Soc. Hort. Sci.* 102:450–452.

Geng, X. M., K. Ii-Nagasuga, H. Okubo, and M. Saniewski. 2005a. Effects of TIBA on growth and flowering of non pre-cooled tulip bulbs. *Acta Hort.* 673:207–215.

Geng, X. M., H. Okubo, and M. Saniewski. 2005b. Cultivar and seasonal differences in the response of non pre-cooled tulip bulbs to gibberellin, TIBA and root excision. *J. Fac. Agr., Kyushu Univ.* 50:503–509.

Ghassemian, M., E. Nambara, S. Cutler, H. Kawaide, Y. Kamiya, and P. McCourt. 2000. Regulation of abscisic acid signaling by the ethylene response pathway in *Arabidopsis. Plant Cell* 12:1117–1126.

Gilford, J. McD., and A. R. Rees. 1973. Growth of the tulip shoot. *Scientia Hortic.* 1:143–156.

Gorin, N., R. Sütfield, J. Tonecki, J. H. Franssen, and N. Haanappel. 1990. Histochemical test for presence or absence of chalcones in anthers from bulbs of tulip cv. Apeldoorn precooled at 5°C or kept at 17°C. *Acta Hort.* 266:221–227.

Halevy, A. H. 1990. Recent advances in control of flowering and growth habit of geophytes. *Acta Hort.* 266:35–42.

Hamner, K. C., and E. M. Long. 1939. Localization of photoperiodic perception in *Helianthus tuberosus. Bot. Gaz.* 101:81–90.

Han, S. S., A. H. Halevy, R. M. Sachs, and M. S. Reid. 1990. Enhancement of growth and flowering of *Triteleia laxa* by ethylene. *J. Am. Soc. Hort. Sci.* 115:482–486.

Hanks, G. R. 1993. *Narcissus*. In *The Physiology of Flower Bulbs*, eds. A. De Hertogh, and M. Le Nard, 463–558. Amsterdam: Elsevier.

Hartsema, A. M. 1961. Influence of temperatures on flower formation and flowering of bulbous and tuberous plants. In *Encyclopedia of Plant Physiology. Vol. 16*, ed. W. Ruhland, 123–167. Berlin: Springer-Verlag.

He, X.-L., L.-W. Shi, Z.-H. Yuan, Z. Xu, Z.-Q. Zhang, and M.-F. Yi. 2008. Effects of lipoxygenase on the corm formation and enlargement in *Gladiolus hybridus*. *Scientia Hortic*. 118:60–69.

Hemberg, T. 1970. The action of some cytokinins on the rest-period and the content of acid growth-inhibiting substances in potato. *Physiol. Plant*. 23:850–858.

Herath, H. M. W., and O. P. Ormrod. 1979. Effects of temperature and photoperiod on winged beans (*Psophocarpus tetragonolobus* (L.) D.C.). *Ann. Bot*. 43:729–736.

Hickey, L. J., and J. A. Doyle. 1977. Early Cretaceous fossil evidence for angiosperm evolution. *Bot. Rev*. 43:3–104.

Hillman, J. R. 1984. Apical dominance. In *Advanced Plant Physiology*, ed. M. B. Wilkins, 127–148. London: Pitman.

Hoffmann-Benning, S., and H. Kende. 1992. On the role of abscisic acid and gibberellin in the regulation of growth in rice. *Plant Physiol*. 99:1156–1161.

Holdsworth, M. J., L. Bentsink, and W. J. J. Soppe. 2008. Molecular networks regulating *Arabidopsis* seed maturation, after-ripening, dormancy and germination. *New Phytol*. 179:33–54.

Horvath, D. 2009. Common mechanisms regulate flowering and dormancy. *Plant Sci*. 177:523–531.

Horvath, D. P., W. S. Chao, and J. V. Anderson. 2002. Molecular analysis of signals controlling dormancy and growth in underground adventitious buds of leafy spurge. *Plant Physiol*. 128:1439–1446.

Horvath, D. P., W. S. Chao, J. C. Suttle, J. Thimmapuram, and J. V. Anderson. 2008. Transcriptome analysis identifies novel responses and potential regulatory genes involved in seasonal dormancy transitions of leafy spurge (*Euphorbia esula* L.). *BMC Genomics* 9:536–552.

Ii, K., H. Okubo, and T. Matsumoto. 2002. Control of bulb dormancy in hyacinth—a molecular biological approach. *Acta Hort*. 570:241–246.

Imanishi, H. 1993. *Freesia*. In *The Physiology of Flower Bulbs*, eds. A. De Hertogh, and M. Le Nard, 285–296. Amsterdam: Elsevier.

Imanishi, H. 1996. Effects of ethylene on dormancy breaking and flowering in flower bulbs. *Chemical Regulation of Plants* 31:161–170 (In Japanese).

Jackson, S. D., and L. Willmitzer. 1994. Jasmonic acid spraying does not induce tuberisation in short-day-requiring potato species kept in non-inducing conditions. *Planta* 194:155–159.

Junttila, O. 1988. To be or not to be dormant: Some comments on the new dormancy nomenclature. *HortScience* 23:805–806.

Kamenetsky, R. 1994. Life cycle, flower initiation, and propagation of the desert geophyte *Allium rothii*. *Int. J. Plant Sci*. 155:597–605.

Kamenetsky, R. 2011. Florogenesis in geophytes: Classical and molecular approaches. *Acta Hort*. 886:113–118.

Kamenetsky, R., I. L. Shafir, H. Zemah, A. Barzilay, and H. D. Rabinowitch. 2004. Environmental control of garlic growth and florogenesis. *J. Am. Soc. Hort. Sci*. 129:144–151.

Kamenetsky, R., H. Zemah, A. P. Ranwala, F. Vergeldt, N. L. Ranwala, W. B. Miller, H. van As, and P. Bendel. 2003. Water status and carbohydrate pools in tulip bulbs during dormancy release. *New Phytol*. 158:109–118.

Kamenetsky, R., H. Zemah, A. van der Toorn, H. van As, and P. Bendel. 2000. Morphological structure and water status in tulip bulbs during their transition from dormancy to active growth: Visualization by NMR imaging. In *Dormancy in Plants*, eds. J. D. Viémont, and J. Crabbé, 121–138. Wallingford, Oxon, UK: CAB International.

Kamerbeek, G. A., J. C. M. Beijersbergen, and P. K. Schenk. 1972. Dormancy in bulbs and corms. *Proceedings of the 18th International Horticultural Congress*, Tel Aviv, 5:233–239.

Keren-Paz, V., A. Borochov, and S. Mayak. 1989. The involvement of ethylene in liatris corm dormancy. *Plant Growth Regul*. 8:11–20.

Kim, H., K. Ohkawa, and K. Sakaguchi. 1996. Effects of storage temperature and duration on flower bud development, emergence and flowering of *Zephyra elegans* D. Don. *Scientia Hortic*. 67:55–63.

Kim, K. S., E. Davelaar, and G.-J. de Klerk. 1994. Abscisic-acid controls dormancy development and bulb formation in lily plantlets regenerated *in-vitro*. *Physiol. Plant*. 90:59–64.

Kollöffel, C., J. Geuns, and H. Lambrechts. 1992. Changes in free polyamine contents in tulip bulbs cv. Apeldoorn during dry storage. *Acta Hort*. 325:247–252.

Komiyama, S., T. Yamazaki, E. Hori, Y. Shida, A. Murayama, T. Ikarashi, and T. Ohyama. 1997. Degradation of storage starch in tulip bulb scales induced by cold temperature. *Japan. J. Soil Sci. Plant Nutr*. 68:23–29 (In Japanese with English abstract).

Koornneef, M., L. Bentsink, and H. Hilhorst. 2002. Seed dormancy and germination. *Curr. Opin. Plant Biol.* 5:33–36.

Kuwabara, A., K. Ikegami, T. Koshiba, and T. Nagata. 2003. Effects of ethylene and abscisic acid upon hetero-phylly in *Ludwigia arcuata* (Onagraceae). *Planta* 217:880–887.

Lambrechts, H., J. M. Franssen, and C. Kollöffel. 1992. The 4-methylene- glutamine: Asparagines ratio in the shoot of tulip bulbs cv. Apeldoorn as a criterion for dry storage duration at 5°C. *Scientia Hortic.* 52:105–112.

Lambrechts, H., F. Rook, and C. Kollöffel. 1994. Carbohydrate status of tulip bulbs during cold-induced flower stalk elongation and flowering. *Plant Physiol.* 104:515–520.

Lang, G. A. 1996. *Plant Dormancy—Physiology, Biochemistry and Molecular Biology.* Wallingford, Great Britain: CAB International.

Lang, G. A., J. D. Early, G. C. Martin, and R. L. Darnell. 1987. Endo-, para-, and ecodormancy: Physiological terminology and classification for dormancy research. *HortScience* 22:371–377.

Law, R. D., and J. C. Suttle. 2004. Changes in histone H3 and H4 multi-acetylation during natural and forced dormancy break in potato tubers. *Physiol. Plant.* 120:642–649.

Le Nard, M. 1983. Physiology and storage of bulbs: Concepts and nature of dormancy in bulbs. In *Post-Harvest Physiology and Crop Preservation. N.A.T.O. Advanced Study Institute Series; Series A Life Science 46*, ed. M. Lieberman, 191–230. New York: Plenum Publ.

Le Nard, M., and J. Cohat. 1968. Influence des températures de conservation des bulbes sur l'élongation, la floraison et la bulbification de la Tulipa (*Tulipa gesneriania* L.). *Ann. de l'Amélioration des Plantes* 18:181–215.

Le Nard, M., and A. A. De Hertogh. 1993. Bulb growth and development and flowering. In *The Physiology of Flower Bulbs*, eds. A. De Hertogh, and M. Le Nard, 29–43. Amsterdam: Elsevier.

Le Nard, M., and V. Fiala. 1990. Post-harvest variation of free arginine in basal plate tissues of tulip bulbs; relation to bulb physiological evolution. *Acta Hort.* 266:293–298.

Lercari, B. 1983. The role of ethylene in photoperiodic control of bulbing in *Allium cepa. Physiol. Plant.* 59:647–650.

Levy, D., and N. Kedar. 1970. Effect of ethrel on growth and bulb initiation in onion. *HortScience* 5:80–82.

Lupia, R., S. Lidgard, and P. R. Crane. 1999. Comparing palynological abundance and diversity: Implications for biotic replacement during the Cretaceous angiosperm radiation. *Paleobiology* 25:305–340.

Maehara, K., K. Ii-Nagasuga, and H. Okubo. 2005. Search for the genes regulating bulb formation in *Lilium speciosum* Thunb. *Acta Hort.* 673:583–589.

Masuda, J. 2007. Studies on rhizome enlargement in lotus (*Nelumbo nucifera*). PhD thesis, Kyushu Univeristy, Fukuoka, Japan.

Masuda, J., Y. Ozaki, S. Matsuda, and H. Okubo. 2003. Effect of long day treatment on the growth of rhizome-enlarged lotus. *J. Japan. Soc. Hort. Sci.* 72(suppl. 1):253 (In Japanese).

Masuda, J., Y. Ozaki, I. Miyajima, and H. Okubo. 2010. Ethylene is not involved in rhizome transition to stor-age organ in lotus (*Nelumbo nucifera*). *J. Fac. Agr., Kyushu Univ.* 55:231–232.

Masuda, J., Y. Ozaki, and H. Okubo. 2007. Rhizome transition to storage organ is under phytochrome control in lotus (*Nelumbo nucifera*). *Planta* 226:909–915.

Masuda, J., Y. Ozaki, and H. Okubo. 2012. Gibberellin and abscisic acid regulate rhizome transition to storage organ in lotus (*Nelumbo nucifera*). *J. Japan. Soc. Hort. Sci.* 81:67–71.

Masuda, J., T. Urakawa, Y. Ozaki, and H. Okubo. 2006. Short photoperiod induces dormancy in lotus (*Nelumbo nucifera*). *Ann. Bot.* 97:39–45.

Masuda, M., and T. Asahira. 1978. Changes in endogenous cytokinin-like substances and growth inhibitors in freesia corms during high-temperature treatment. *Scientia Hortic.* 8:371–382.

McClure, H. E. 1966. Flowering, fruiting and animals in the canopy of a tropical rain forest. *Malayan Forester* 29:182–203.

Melzer, S., F. Lens, J. Gennen, S. Vanneste, A. Rohde, and T. Beeckman. 2008. Flowering-time genes modulate meristem determinacy and growth form in *Arabidopsis thaliana. Nat. Genet.* 40:1489–1492.

Miller, W. B. 1993. *Lilium longiflorum.* In *The Physiology of Flower Bulbs*, eds. A. De Hertogh, and M. Le Nard, 391–422. Amsterdam: Elsevier.

Miller, W. B., and R. W. Langhans. 1990. Low temperature alters carbohydrate metabolism in Easter lily bulbs. *HortScience* 25:463–465.

Mizukami, Y., and R. L. Fischer. 2000. Plant organ size control: AINTEGUMENTA regulates growth and cell numbers during organogenesis. *Proc. Natl. Acad. Sci. USA* 97:942–947.

Moe, R., and A. Wickstrøm. 1973. The effect of storage temperature on shoot growth, flowering, and carbohy-drate metabolism in tulip bulbs. *Physiol. Plant.* 28:81–87.

Mondal, M. F., J. L. Brewster, G. E. L. Morris, and H. A. Butler. 1986. Bulb development in onion (*Allium cepa* L.) II. Influence of red:far-red spectral ratio and of photon flux density. *Ann. Bot.* 58:197–206.

Moser, B. C., and C. E. Hess. 1968. The physiology of tuberous root development in dahlia. *Proc. Am. Soc. Hort. Sci.* 93:593–603.

Nakamura, S., S. Yoshida, and K. Akiyama. 1974. Studies on the dormancy of bulbs and corms. II. Effect of calcium cyanamide and benzyl adenine on the corms and cormels of freesia and gladiolus. *Bull. Fac. Agric. Yamaguchi Univ.* 25:857–866 (In Japanese with English summary).

Nhut, D. T. 1998. Micropropagation of lily (*Lilium longiflorum*) via *in vitro* stem node and pseudo-bulblet culture. *Plant Cell Rep.* 17:913–916.

Nikolaeva, M. G. 1969. *Physiology of Deep Dormancy in Seeds.* Leningrad: Izdatel'stvo Nauka (In Russian, translated by Z. Shapiro).

Nishimura, H., and S. Atsumi. 2001. Breaking dormancy in micropropagated bulblets of *Lilium japonicum* using gibberellic acid$_4$. *J. Japan. Soc. Hort. Sci.* 70:522–524 (In Japanese with English summary).

Nojiri, H., H. Yamane, H. Seto, I. Yamaguchi, N. Murofushi, T. Yoshihara, and H. Shibaoka. 1992. Qualitative and quantitative analysis of endogenous jasmonic acid in bulbing and non-bulbing onion plants. *Plant Cell Physiol.* 33:1225–1231.

Okubo, H. 2000. Growth cycle and dormancy in plants. In *Dormancy in Plants: From Whole Plant Behaviour to Cellular Control,* eds. J.-D. Viémont, and J. Crabbé, 1–22. Wallingford, Oxon, UK: CABI Publ.

Okubo, H., S. Adaniya, K. Takahashi, and K. Fujieda. 1981. Studies on the bulb formation of *Allium wakegi* Araki. *J. Japan. Soc. Hort. Sci.* 50:37–43 (In Japanese with English summary).

Okubo, H., M. Iwaya-Inoue, K. Motooka, N. Ishada, H. Kano, and M. Koizumi. 1997. Monitoring the cold requirement in tulip bulbs by ^{1}H-NMR imaging. *Acta Hort.* 430:411–417.

Okubo, H., T. Masunaga, H. Yamashita, and S. Uemoto. 1992. Effects of photoperiod and temperature on tuberous root formation in winged bean (*Psophocarpus tetragonolobus*). *Scientia Hortic.* 49:1–8.

Okubo, H., and S. Uemoto. 1981. Changes in the endogenous growth regulators in bulbous iris in bulb-forming and nonbulb-forming aspects. *Plant Cell Physiol.* 22:297–301.

Olsen, J. E., O. Junttila, J. Nilsen, M. E. Eriksson, I. Martinussen, O. Olsson, G. Sandberg, and T. Moritz. 1997. Ectopic expression of oat phytochrome A in hybrid aspen changes critical daylength for growth and prevents cold acclimatization. *Plant J.* 12:1339–1350.

Pelacho, A. M., and A. M. Mingo-Castel. 1991. Jasmonic acid induces tuberization of potato stolons cultured *in vitro. Plant Physiol.* 97:1253–1255.

Poulsen, N. 1990. Chives, *Allium schoenoprasum* L. In *Onion and Allied Crops. III,* eds. J. L. Brewster, and H. D. Rabinowitch, 231–250. Boca Raton, FL: CRC Press.

Ravnikar, M., J. Zel, I. Plaper, and A. Spacapan. 1993. Jasmonic acid stimulates shoot and bulb formation of garlic *in vitro. J. Plant Growth Regul.* 12:73–77.

Rebers, M., E. Vermeer, E. Knegt, and L. H. W. van der Plas. 1996. Gibberellin levels are not a suitable indicator for properly cold-treated tulip bulbs. *HortScience* 31:837–838.

Rees, A. R. 1981. Concepts of dormancy as illustrated by the tulip and other bulbs. *Ann. Appl. Biol.* 98:544–548.

Rietveld, P. L., C. Wilkinson, H. M. Franssen, P. A. Balk, L. H. W. van der Plas, P. J. Weisbeek, and A. D. de Boer. 2000. Low temperature sensing in tulip (*Tulipa gesneriana* L.) mediated through an increased response to auxin. *J. Exp. Bot.* 51:587–594.

Riou-Khamlichi, C., R. Huntley, A. Jacqmard, and J. A. H. Murray. 1999. Cytokinin activation of *Arabidopsis* cell division through a D-type cyclin. *Science* 283:1541–1544.

Robertson, F. C., A. W. Skeffington, M. J. Gardner, and A. A. R. Webb. 2008. Interactions between circadian and hormonal signaling in plants. *Plant Mol. Biol.* 69:419–427.

Rodrigo, M. J., and B. Alquezar. 2006. Cloning and characterization of two 9-*cis*-epoxycarotenoid dioxygenase genes, differentially regulated during fruit maturation and under stress conditions, from orange (*Citrus sinensis* L. Osbeck). *J. Exp. Bot.* 57:633–643.

Rodríguez-Falcón, M., J. Bou, and S. Prat. 2006. Seasonal control of tuberization in potato: Conserved elements with the flowering response. *Annu. Rev. Plant Biol.* 57:151–180.

Rudnicki, R. M., and J. Nowak. 1993. *Muscari.* In *The Physiology of Flower Bulbs,* eds. A. De Hertogh, and M. Le Nard, 455–462. Amsterdam: Elsevier.

Ruonala, R., P. L. H. Rinne, J. Kangasjärvi, and C. van der Schoot. 2008. *CENL1* expression in the rib meristem affects stem elongation and the transition to dormancy in *Populus. Plant Cell* 20:59–74.

Ruttink, T., M. Arend, K. Morreel, V. Storme, S. Rombauts, J. Fromm, R. P. Bhalerao, W. Boerjan, and A. Rohde. 2007. A molecular timetable for apical bud formation and dormancy induction in poplar. *Plant Cell* 19:2370–2390.

Saniewski, M., L. Kawa-Miszczak, E. Wegrzynowicz-Lesiak, and H. Okubo. 1999. Gibberellin induces shoot growth and flowering in nonprecooled derooted bulbs of tulip (*Tulipa gesneriana* L.). *J. Fac. Agr., Kyushu Univ.* 43:411–418.

Sarkar, D. 2008. The signal transduction pathways controlling in planta tuberization in potato: An emerging synthesis. *Plant Cell Rep.* 27:1–8.

Sarkar, D., S. K. Pandey, and S. Sharma. 2006. Cytokinins antagonize the jasmonates action on the regulation of potato (*Solanum tuberosum*) tuber formation *in vitro*. *Plant Cell, Tiss. Organ Cult.* 87:285–295.

Sato, A., H. Okubo, and K. Saitou. 2006. Increase in the expression of an alpha-amylase gene and sugar accumulation induced during cold period reflects shoot elongation in hyacinth bulbs. *J. Am. Soc. Hort. Sci.* 131:185–191.

Shefferson, R. P. 2009. The evolutionary ecology of vegetative dormancy in mature herbaceous perennial plants. *J. Ecol.* 97:1000–1009.

Shemesh, E., O. Scholten, H. D. Rabinowitch, and R. Kamenetsky. 2008. Unlocking variability: Inherent variation and developmental traits of garlic plants originated from sexual reproduction. *Planta* 227:1013–1024.

Shin, K. S., D. Chakrabarty, and K. Y. Paek. 2002. Sprouting rate, change of carbohydrate contents and related enzymes during cold treatment of lily bulblets regenerated *in vitro*. *Scientia Hortic.* 96:195–204.

Sonnewald, U. 2001. Control of potato tuber sprouting. *Trends Plant Sci.* 6:333–335.

Spollen, W. G., M. E. LeNoble, T. D. Samuels, N. Bernstein, and R. E. Sharp. 2000. Abscisic acid accumulation maintains maize primary root elongation at low water potentials by restricting ethylene production. *Plant Physiol.* 122:967–976.

Sukhova, L. S., L. S. Machakova, J. Edeer, N. D. Bibik, and N. P. Korableva. 1993. Changes in levels of free IAA and cytokinins in potato tubers during dormancy and sprouting. *Biologia Plantarum* 35:387–391.

Sussman, A. S., and H. A. Douthit. 1973. Dormancy in microbial spores. *Annu. Rev. Plant Physiol.* 24:311–352.

Suttle, J. C. 1995. Postharvest changes in endogenous ABA levels and ABA metabolism in relation to dormancy in potato tubers. *Physiol. Plant.* 95:233–240.

Suttle, J. C. 1998a. Postharvest changes in endogenous cytokinins and cytokinin efficacy in potato tubers in relation to bud endodormancy. *Physiol. Plant.* 103:59–69.

Suttle, J. C. 1998b. Involvement of ethylene in potato microtuber dormancy. *Plant Physiol.* 118:843–848.

Suttle, J. C. 2000. The role of endogenous hormones in potato tuber dormancy. In *Dormancy in Plants: From Whole Plant Behaviour to Cellular Control*, eds. J.-D. Viémont, and J. Crabbé, 211–226. Wallingford, Oxon, UK: CABI Publ.

Suttle, J. C., and J. F. Hultstrand. 1994. Role of endogenous abscisic acid in potato microtuber dormancy. *Plant Physiol.* 105:891–896.

Takayama, T., T. Toyomasu, H. Yamane, N. Murofushi, and H. Yajima. 1993. Identification of gibberellins and abscisic acid in bulbs of *Lilium elegans* Thunb. and their quantitative changes during cold treatment and the subsequent cultivation. *J. Japan. Soc. Hort. Sci.* 62:189–196.

Taylor, A. 2009. Functional genomics of photoperiodic bulb initiation in onion (*Allium cepa*). PhD thesis. University of Warwick, Warwick, UK.

Taylor, A., A. J. Massiah, and B. Thomas. 2010. Conservation of *Arabidopsis thaliana* photoperiodic flowering time genes in onion (*Allium cepa* L.) *Plant Cell Physiol.* 51:1638–1647.

Taylorson, R. B., and S. B. Hendriks. 1977. Dormancy in seeds. *Annu. Rev. Plant Physiol.* 28:331–354.

Terabun, M. 1965. Studies on the bulb formation in onion plants. I. Effects of light quality on the bulb formation and the growth. *J. Japan. Soc. Hort. Sci.* 34:196–204.

Tonecki, J., and N. Gorin. 1990. Further studies on the use of free amino acids in anthers from tulip bulbs cultivar Apeldoorn as indicators about cold treatment at 5°C. *Scientia Hortic.* 42:133–140.

Turck, F., F. Fornara, and G. Coupland. 2008. Regulation and identity of florigen: FLOWERING LOCUS T moves center stage. *Annu. Rev. Plant Biol.* 59:573–594.

Uyemura, S., and H. Imanishi. 1983. Effects of gaseous compounds in smoke on dormancy release in freesia corms. *Scientia Hortic.* 20:91–99.

Uyemura, S., and H. Imanishi. 1987. Changes in abscisic acid levels during dormancy release in freesia corms. *Plant Growth Regul.* 5:97–103.

Van der Schoot, C., and P. L. H. Rinne. 2011. Dormancy cycling at the shoot apical meristem: Transitioning between self-organization and self-arrest. *Plant Sci.* 180:120–131.

Van der Toorn, A., H. Zemah, H. van As, P. Bendel, and R. Kamenetsky. 2000. Developmental changes and water status in tulip bulbs during storage visualization by NMR imaging. *J. Exp. Bot.* 51:1277–1287.

Vegis, A. 1964. Dormancy in higher plants. *Annu. Rev. Plant Physiol.* 15:185–224.

Vishnevetsky, J., H. Lilien-Kipnis, N. Azizbekova, and M. Ziv. 1997. Bulb and inflorescence development in *Nerine sarniensis*. *Israel J. Plant Sci.* 45:13–18.

Wang, G., H. Kong, Y. Sun, X. Zhang, W. Zhang, N. Altman, C. dePamphilis, and H. Ma. 2004. Genome-wide analysis of the cyclin family in Arabidopsis and comparative phylogenetic analysis of plant cyclin-like proteins. *Plant Physiol.* 135:1084–1099.

Wareing, P. F., and I. D. J. Phillips. 1981. *Growth and Differentiation in Plants 3rd ed.* Oxford, UK: Pergamon Press.

Wareing, P. F., and P. F. Saunders. 1971. Hormones and dormancy. *Annu. Rev. Plant Physiol.* 22:261–288.

Wong, K. C. 1981. Environmental factors affecting the growth, flowering and tuberization in winged bean (*Psophocarpus tetragonolobus* (L.) D.C.). Paper presented at the 2nd International Seminar of Winged Bean, Colombo.

Xiao, J., K. Ureshino, M. Hosoya, H. Okubo, and A. Suzuki. 2010. Inheritance of bulb formation in *Allium schoenoprasum* L. *J. Japan. Soc. Hort. Sci.* 79:282–286.

Xu, R.-Y., Y. Niimi, and D.-S. Han. 2006. Changes in endogenous abscisic acid and soluble sugars levels during dormancy-release in bulbs of *Lilium rubellum*. *Scientia Hortic.* 111:68–72.

Xu, X., A. A. van Lammeren, E. Vermeer, and D. Vreugdenhil. 1998. The role of gibberellin, abscisic acid, and sucrose in the regulation of potato tuber formation *in vitro*. *Plant Physiol.* 117:575–584.

Yamazaki, H., N. Ishida, N. Katsura, H. Kano, T. Nishijima, and M. Koshioka. 1995a. Changes in carbohydrate composition and water status during bulb development of *Allium wakegi* Araki. *Bull. Natl. Res. Inst. Veg., Ornam. Plants & Tea, Japan A* 10:1–11.

Yamazaki, H., T. Nishijima, and M. Koshioka. 1995b. Changes in abscisic acid content and water status in bulbs of *Allium wakegi* Araki throughout the year. *J. Japan. Soc. Hort. Sci.* 64:589–598.

Yamazaki, H., T. Nishijima, M. Koshioka, and H. Miura. 2002. Gibberellins do not act against abscisic acid in the regulation of bulb dormancy of *Allium wakegi* Araki. *Plant Growth Regul.* 36:223–229.

Yamazaki, H., T. Nishijima, Y. Yamato, M. Hamano, M. Koshioka, and H. Miura. 1999b. Involvement of abscisic acid in bulb dormancy of *Allium wakegi* Araki. II. A comparison between dormant and nondormant cultivars. *Plant Growth Regul.* 29:195–200.

Yamazaki, H., T. Nishijima, Y. Yamato, M. Koshioka, and H. Miura. 1999a. Involvement of abscisic acid (ABA) in bulb dormancy of *Allium wakegi* Araki I. Endogenous levels of ABA in relation to bulb dormancy and effects of exogenous ABA and fluridone. *Plant Growth Regul.* 29:189–194.

Yamazaki, H., R. Oi, T. Nishijima, and H. Miura. 1998. Inhibition of bulb development in *Allium wakegi* under a high red/far-red photon ration. *J. Japan. Soc. Hort. Sci.* 67:337–340.

Yañez, P., H. Ohno, and K. Ohkawa. 2005. Temperature effects on corm dormancy and growth of *Zephyra elegans* D. Don. *Scientia Hortic.* 105:127–138.

Yoshida, H., Y. Sakai, and T. Matsukawa. 1986. Effects of running water dipping on the breaking dormancy in *Lilium longiflorum* Thunb. *Kyushu Nougyou Kenkyu* 48:260 (In Japanese).

Zeevaart, J. A. 2008. Leaf-produced floral signals. *Curr. Opin. Plant Biol.* 11:541–547.

Zemah, H., P. Bendel, H. D. Rabinowitch, and R. Kamenetsky. 1999. Visualization of morphological structure and water status during storage of *Allium aflatunense* bulbs by NMR imaging. *Plant Sci.* 147:65–73.

Zimmerman, P. W., and A. E. Hitchcock. 1929. Root formation and flowering of dahlia cuttings when subjected to different day lengths. *Bot. Gaz.* 87:l–l3.

Ziv, M. 1989. Enhanced shoot and cormlet proliferation in liquid cultured gladiolus buds by growth retardants. *Plant Cell, Tiss. Org. Cult.* 17:101–110.

10 Propagation of Ornamental Geophytes
Physiology and Management Systems

Anna Bach and Dariusz Sochacki

CONTENTS

I. INTRODUCTION

The use of ornamental geophytes in commercial floriculture and gardening requires constant and accelerated reproduction. Although most geophyte species propagate vegetatively, they often have a low multiplication rate due to a restricted number of axillary meristems. Therefore, introduction of newly bred cultivars or pathogen-free propagation material requires a long period of time for the creation of a commercial stock.

In natural populations of geophytes, seed propagation dominates over vegetative reproduction. Natural rates of annual vegetative multiplication vary from slow in narcissus ($\times 1.6$ of daughter bulbs per parent plant) to medium in tulip and iris ($\times 5.0$) and to a large number of bulblets in the lily or cormels produced by gladioli (Rees 1992).

In horticultural practice, growers use vegetative propagation to build up large stocks of geophyte cultivars, using both natural and artificial methods to produce a large number of units (Buschman 2005). Propagation techniques vary from simple methods, for example, separating and growing individual propagules, to more complicated systems of micropropagation under sterile *in vitro* conditions, which allow for the production of 100,000 plants from one bulb in 6 months.

General aspects of propagation physiology in geophytes were reviewed by De Hertogh and Le Nard (1993a), Mansfield (1994), Davies and Santamaria (2000), and Borkowska (2003). In addition, propagation methods and systems were reviewed for major crops (Rees 1992; Le Nard and De Hertogh 1993a; Kim and De Hertogh 1997; Langens-Gerrits and De Klerk 1999; Ellis 2001; Kim et al. 2001; Hanks 2002; Takayama and Akita 2005; Ziv 2005; Nhut et al. 2006; Ascough et al. 2008; De Klerk 2009).

In this chapter, salient advances in multiplication techniques and optimal environmental and culture conditions for the propagation of ornamental geophytes are reviewed.

II. SEED REPRODUCTION

Geophyte propagation from seed is not widely used commercially, except for some individual species or genera. On the other hand, sexual reproduction is essential for the breeders, who create new cultivars of species, normally propagated vegetatively. There are two main disadvantages of raising bulbs from seed. First, most genera (e.g., *Tulipa* or *Narcissus*) have an extended period of juvenility of up to 5–7 years before they start flowering. The second disadvantage is that a sexual type of propagation is strictly associated with genetic variability. This trait is widely used in breeding and selection, but disrupts commercial production of true-to-type cultivars.

Apart from being required for breeding, seed propagation is used only where there is a possibility of obtaining a large number of seeds, where seedlings have a short juvenile period (e.g., *Hyacinthoides*, *Galanthus nivalis*, *Muscari*, and *Crocus*) and where there is no requirement to avoid the greater variability introduced into the stock as compared with vegetative propagation (Rees 1992). An additional advantage of seed propagation is that in almost all cases, bulbous, cormous, or tuberous plants produced from seed are virus-free in comparison with vegetatively propagated material. Ornamental plants of the following genera are commercially propagated by seed: *Allium*, *Anemone*, *Begonia*, *Chionodoxa*, *Cyclamen*, *Eranthis*, *Fritillaria*, *Liatris*, *Puschkinia*, *Scilla*, *Sparaxis*, *Ranunculus*, and *Tigridia*.

For some species, for example, hybrids of tuberous *Begonia*, seedlings are produced under greenhouse conditions and then transplanted to the field (De Hertogh and Le Nard 1993a). Others—such as *Allium*, *Chionodoxa*, or *Fritillaria*—require moist chilling (stratification) to germinate. Harvested seeds of *Chionodoxa* are stored in boxes lined with paper in layers about 2 cm thick at a temperature of 17°C until sowing. In Holland, seeds are sown in sterilized soil in September, covered with 1–3 cm of soil and protected with straw during the winter. To obtain bulbs, a 2-year growing cycle is required (Le Nard and De Hertogh 1993b). Commercial fields of bulbous ornamental *Allium* species (e.g., *A. aflatunense*, *A. cristophii*, and *A. karataviense*) are sown in the autumn, and germination occurs in the spring after seed exposure to low winter temperatures. The juvenile period lasts 2–3 or 3–5 years in species with small or large bulbs, respectively (Kamenetsky and Fritsch 2002). *Ranunculus asiaticus* and *Anemone coronaria* produce flowering-size propagules after 1 year of growing from seeds in Mediterranean areas, South Africa, California, and Japan (Kamenetsky 2005).

III. NATURAL VEGETATIVE PROPAGATION

Natural vegetative propagation includes the initiation of axillary meristems and the further growth and development of propagules, for example, daughter bulbs, offsets[*], bulblets[†], bulbils[‡], cormels, or branched rhizomes.

---

[*] Offsets—new small bulbs that form naturally in the interior of the parent bulb.
[†] Bulblet—a small new bulb born in the underground leaf axils or from a bulb scale.
[‡] Bulbil—aerial bulblet produced on the upper portion of the stem or in an inflorescence.

In annual bulbs or corms, such as tulips, crocuses, and gladioli, the mother storage organ is replaced annually by one or more daughter progeny. In the case of hyacinths, muscari, or hippeastrum, mother bulbs persist and flower for several years, and offsets are formed on the basal plate. Many species of the genus *Lilium*, for example, *L. henryi*, *L. longiflorum*, *L. regale*, *L. speciosum*, and their hybrids, form bulblets at leaf nodes of the underground part of the stem. Other lily species such as *L. bulbiferum*, *L. tigrinum*, and *L. sulphureum* form aerial bulbils in the above-ground leaf axils of the stem. Lily of the valley (*Convallaria*) and rhizomatous iris (*Iris*) produce new buds on the underground rhizomes (De Hertogh and Le Nard 1993a). Stolons, produced by some *Allium* and *Oxalis* species, as well as by *Scilla adlami*, *Erythronium*, and *Achimenes,* serve both for vegetative propagation and for population expansion.

Natural rates of multiplication vary between genera and species. In many cases, differences between cultivars of the same species are also noticeable. Sometimes, as in *Zephyra elegans*, only one daughter corm replaces the mother corm (Kim et al. 1997). In *Nerine bowdenii*, 1–4 lateral buds develop into bulblets during each growing cycle (Van Brenk and Benschop 1993). Tulip or iris bulbs form more daughter bulbs, often one or more in the axil of each scale. Under proper cultivation conditions, one saffron corm produces, on an average, four corms of medium sizes (Ahmad et al. 2011). Gladiolus has an exceptionally high natural multiplication rate. A hundred or more cormels can be produced by a single large gladiolus plant (Figure 10.1a). In the case of *Lilium*, the multipli-

FIGURE 10.1 **(See color insert.)** Vegetative propagation of ornamental geophytes. (a) *Gladiolus* corm after harvest; numerous cormels are produced; (b) bulblets obtained on lily scale; (c) scooping process in hyacinth bulb; (d) plantlets obtained on cross-cutted (scored) hyacinth's bulb.

cation rate, based on the natural splitting of the bulbs into two parts, is intensified by the formation of stem and scale bulblets and bulbils. *Allium dictyoprasum* produces up to 40 small bulbs on stolons, while *A. aschersonianum* does not produce any daughter bulbs and, in nature, propagates only by seeds (Kamenetsky and Rabinowitch 2006).

The rate of natural vegetative multiplication, however, is generally low in geophytes. Thus, horticultural techniques are employed for increased propagation efficiency for commercial production.

IV. ARTIFICIAL VEGETATIVE PROPAGATION

A. TRADITIONAL PROPAGATION METHODS

1. Scaling

This old technique is based on the ability of an individual detached bulb scale to produce a new plant. Scaling is commercially used for lilies but can also be employed for hyacinths (*Hyacinthus orientalis*), some *Fritillaria*, *Muscari*, and *Scilla* (Figure 10.1b; Rees 1992).

Scales are broken away from the bulbs and, following dipping in a fungicide, are planted in a sterilized rooting medium, or in a thin polythene bag filled with moist vermiculite (Rees 1992) or a 50/50 mixture of coarse sand and peat moss (Ellis 2001). Moist perlite or a 50/50 mixture of perlite and peat moss can also be used. Incubation temperatures and durations are typically 24°C for 8 weeks, followed by 17°C for 3 weeks, depending on the species or the cultivar (Rees 1992). In some cases, precooling is necessary to break dormancy. This also promotes leaf emergence. Before planting, the scales with one or more new bulblets, formed at its base, are separated from the medium, dusted with a systemic fungicide, and grown outdoors in beds for two growing seasons in order to obtain the first batch of commercially sized bulbs (Rees 1992; Bryan 2002).

According to Bryan (2002), different groups of lily cultivars require different periods of incubation to produce bulblets. Asiatics and Oriental hybrids require 6–8 weeks at 16°C, followed by 6 weeks of precooling. Trumpet hybrids are incubated for 12 weeks at 16°C and then kept for a short time at 10°C just before planting. Rees (1992) reports that incubation temperatures are typically 24°C for 8 weeks, followed by 17°C for 3 weeks. Experiments done by Park (1996) showed that the optimum scaling temperature was 20°C for the cultivar 'Connecticut King' (Asiatic hybrid group) and 25°C for 'Stargazer' (Oriental hybrid group). Asiatic lily 'Polyanna' requires a short period (3–6 weeks) at 22°C, followed by a long cold treatment (8 or more weeks). The best treatment for Longiflorum hybrid 'Snow Queen' was 10 weeks at 22°C, followed by 4 weeks at 0°C (Grassotti 1997).

Many reports show that the temperatures during storage of the initial (donor) bulbs and during scale incubation determine the number, weight, and size of the bulbs obtained from bulblets during subsequent cultivation. High storage temperatures of the donor bulbs of *Lilium longiflorum* 'White American' promote leaf emergence from the new scale bulblets, while low temperatures delay leaf emergence of the obtained scale bulblets (Matsuo and Van Tuyl 1984). The time of scaling can also affect the leaf emergence of scale bulblets. In Easter lily (*Lilium longiflorum*), the scaling done in March resulted in 10% of sprouting bulblets after 90 days. In comparison, scaling in September resulted in leaf emergence after 22 days, and over 80% of the bulblets produced leaves (Okubo et al. 1988).

The type of plant developing from the scale bulblet is of great importance. Epigeal plants form stems and flowers, have larger leaf surfaces, better resistance to diseases, and a longer growing period. This type is preferable to hypogeal types, which form rosette foliage and are not commercially demanded (Matsuo et al. 1982).

2. Cross-Cutting and Scooping

A true bulb consists of a basal plate—a compressed stem flattened into a disk, surrounded by modified leaves and scales. New buds develop in the axils of the scales attached to the basal plate (Bryan

2002). The techniques of cross-cutting and scooping are used to destroy apical dominance within the bulb and to induce bulblet formation on the wounded surface of the base plate (Figure 10.1c and d).

Scooping has been known since 1715 (Doorenbos 1954), and was developed commercially for hyacinths in the nineteenth century (Rees 1992). The base of the bulb is wounded by cutting out a deep cone-shaped piece from the bottom of the bulb, using a curved or spoon-shaped knife (Bryan 2002). The basal plate must be removed without any damage to the base of the scales, which retain the capacity for regenerating new meristems (Rees 1992).

The second method, cross-cutting or scoring, first used for *Hippeastrum* in about 1935, involves v-shaped cuts into the basal plate with a cross-pattern or the making of 3–4 incisions across the diameter deep enough to ensure that the main growing point is completely damaged (Rees 1992; Ellis 2001).

After treatment with a fungicide, scooped or scored bulbs are stored upside down in dry sand or open trays for a week in a warm (21–25°C), dry, well-ventilated place to encourage the formation of callus tissue. After that, the treated bulbs are kept at a high level of relative humidity (about 90%). Numerous bulblets form at the base of the bulb scales or along the cuts after 3–4 months (Ellis 2001). Depending on the cultivar and the size of the mother bulb, a scooped bulb produces 40–60 bulblets, while a cross-cut bulb produces 12–24 bulblets. The latter are larger and are saleable a year earlier (Rees 1992; Nowak and Rudnicki 1993). The bulbs with their bulblets attached are planted in late autumn. In the spring, they produce a lot of grassy foliage and a few flowers (Bryan 2002). After three growing cycles, about 70% of the bulblets of hyacinth attain the size of 12 cm or larger.

These techniques are also used for other species, including *Scilla sibirica* and *Fritillaria imperialis* (Rees 1992; Le Nard and De Hertogh 1993b). The bulbs of *Nerine sarniensis* can be multiplied successfully by means of the cross-cutting technique, as shown by the results obtained in Japan (Mori et al. 1997). This report states that the formation of bulblets was not affected by the period of cross-cutting during the annual cycle, and larger mother bulbs produced larger bulblets following incubation for about 6 months.

3. Chipping and Twin-Scaling

The chipping technique, sometimes referred to as bulb cutting, is a useful way of propagating true bulbs that are slow to produce offsets. The mother bulbs can be cut into 4, 8, or 16 equal vertical sections using a sharp knife, each portion including a piece of the basal plate. These sections (chips) are disinfected by soaking in a liquid fungicide and kept in polythene bags filled with moist vermiculite or media similar to those used in scaling. Mid-summer is the best time for the majority of bulbs to be propagated by chipping. The treated chips are incubated in a dark, warm (20°C) place for 2–3 months, the actual time varying according to the species. After this period, bulblets are formed around the basal plate of each chip (Ellis 2001; Bryan 2002). The chips of narcissus with attached bulblets are grown in a frost-free polythene tunnel for 2 years (Rees 1992).

This method is commercially used for *Narcissus*, snowdrops (*Galanthus*), and Dutch irises (*Iris × hollandica*). Many experiments on the chipping technique for narcissus were conducted in The Netherlands and the United Kingdom, and both the process and the controlling factors have already been well described (Flint 1982; Vreeburg 1986; Langton and Hanks 1993). Some research with local cultivars of narcissus was also conducted in Poland (Sochacki and Orlikowska 2000). The equipment developed to facilitate the multiplication of large batches of narcissus bulbs includes either static or spinning blades able to cut a trimmed bulb into 16 chips with one movement, a procedure that is highly efficient (several hundreds cut of bulbs per hour). This method also involves simpler handling methods, including direct planting and planting in nets. Current recommendations for narcissus chipping are to cut the bulbs into eight segments because bigger chips produce bigger bulblets (Rees 1992).

Many other species can be propagated by the chipping technique, which can be very useful, especially in the case of new crops or new cultivars, or in the case of difficulties with multiplication by other means. The method has been examined for *Chionodoxa luciliae* 'Pink Giant', *Muscari azureum*, *Scilla mischtschenkoana*, *Eucomis*, and *Veltheimia* (Van Leeuwen and Van der Weijden

1997). Interesting innovations in the propagation of hippeastrum by various bulb cutting techniques were introduced in experiments conducted in Israel (Sandler-Ziv et al. 1997). *Allium aschersonianum*, which normally does not propagate vegetatively, produced axillary meristems and bulblets after chipping and following incubation at 20–22°C for 5 weeks (Gilad et al. 2001).

Twin-scaling is an old technique, known since 1935, when "stem cuttage" of *Hippeastrum* was described in *The American Gardener's Book of Bulbs* (Everett 1954). This technique was used at the British Glasshouse Crops Research Institute in the early 1960s for propagating virus-free narcissi, and was improved by Alkema (1975) at the Bulb Research Centre in Lisse, The Netherlands (Rees 1992). While twin-scaling is a simple technique, it is time consuming and the propagules are small and delicate (Hanks 2002). Twin-scaling is developed well for *Narcissus*. Chips are cut further with a sterilized scalpel so that only two scales are left intact with a fragment of the basal plate. A large bulb of *Narcissus* (12–14 cm in circumference) can produce 20–45 bulblets. However, too high a number of twin-scales cut results in smaller bulblets (Hanks and Jones 1986). Incubation and further growing are carried out under conditions similar to those for chips. In twin-scaling of hippeastrum, the thickness and length of the outer scale affected the rate of bulblet formation and leaf development, while those of the inner scale did not (Huang et al. 1990).

Both chipping and twin-scaling techniques require high levels of sanitation. All working surfaces must be clean and all instruments must be sterilized periodically while working. It is also necessary to protect twin-scales from drying out and from fungal attack during incubation. Twin-scales should be treated by dipping in a high concentration of fungicidal solution or dusting with fungicidal powder (Rees 1992). The bulblets obtained need 3–4 years to reach flowering size.

4. Cuttings

Some ornamental geophytes have a capacity for regenerating plantlets from leaf cuttings. Whole mature leaves, removed from the growing plant, or leaf segments are placed in a sterile rooting medium. It is essential to perform the cutting at a high level of relative humidity in a warm and shaded place. Cutting across the prominent veins can help induce plantlet formation on the wounds. Small plantlets form within a few weeks, and then—after rooting—they can be transplanted (Rees 1992). The method has been used successfully for *Begonia × tuberhybrida* (Shimada et al. 2005), *Haemanthus* (De Hertogh and Le Nard 1993a), *Hyacinthus orientalis* (Krause 1980), *Lachenalia* (Niederwieser and Ndou 2002), *Lilium longiflorum* (Roh 1982), *Ornithogalum* (Blomerus and Schreuder 2002), and many others.

Two popular genera of tuberous plants—dahlia (*Dahlia × cultorum*) and begonia (*Begonia × tuberhybrida*)—are propagated commercially by stem cuttings. The tuberous roots of dahlias are planted in February and March and are grown in a greenhouse at 13–15°C for 2 weeks and then at 20–25°C. The cuttings, with a piece of the crown, are taken when the stems reach a length of 7.5–15 cm, are dipped in a rooting hormone, and rooted in moist sandy soil (De Hertogh and Le Nard 1993b). As soon as good root growth is observed, the cuttings are replanted in a more nutritious planting mixture and grown at a lower humidity level. After several weeks, cuttings can be grown outdoors (Bryan 2002). Tuberous begonias do not produce offsets, but can be cut into a few sections with a sharp clean knife, each section containing at least one bud. The cut surface should be dusted with a fungicide to prevent infection. The cut sections can be grown like regular tubers. The main disadvantage of this method is that the scar tissue does not produce roots, so the root system is smaller than that of a full-sized, intact tuber. Despite this drawback, cutting is a common method of propagation (Bryan 2002).

Stem cuttings of lilies, made shortly after flowering, form bulblets at the axils of the leaves and then produce roots and small shoots (Hartmann et al. 2002). In gladiolus, corms can be cut for rapid propagation of new planting material (McKay at al. 1981; Memon et al. 2009). An effective system for the propagation of colored calla lilies (*Zantedeschia*) by cutting mother tubers into 10, 15, 20, and 25 segments was also proposed (Sandler-Ziv et al. 2011). In this case, the cutting protocol is combined with the growth of propagation units in a heated medium.

Division and rhizome cutting is the most common procedure for propagating rhizomatous species, such as *Convallaria majalis*, rhizomatous *Iris*, and *Alstroemeria*.

B. PROPAGATION *IN VITRO*

The term "*in vitro* culture" embraces numerous procedures involved in the maintenance and development of plant explants under experimental or industrial sterile conditions. Explants can be prepared from organ or tissue fragments, single cells or protoplasts which, after being isolated from a donor plant, are placed in artificial media under precisely controlled light (wavelength, intensity, and photoperiod), and temperature and humidity conditions. Plant regeneration from somatic cells is based on the phenomenon of totipotency, which relies on the plant cells' ability to resume mitotic divisions and differentiation into other cell types and, as a consequence, to regenerate specialized organs and the whole organism.

The goals of this technology vary from tissue propagation and differentiation in various organs or somatic embryos, for the mass multiplication of the most desirable cultivars and the conservation of endangered species. Plant "gene banks", created increasingly in recent years, today preserve not only seeds but also plant tissues and organs, by the use of the cryopreservation technique (Wang and Perl 2006; Baghdadi et al. 2010; Keller and Senula 2010; Jevremovic et al. 2011).

In vitro techniques have been used for the production of ornamental geophytes since the late 1950s. In combination with other propagation techniques, for example, scaling (lily), scoring or scooping (hyacinth), and chipping (narcissus), tissue culture provides a valuable means of building up commercial stocks of ornamental geophytes with desired characteristics, especially new genotypes or pathogen-free plants. In addition, the development of a successful system of *in vitro* regeneration of ornamental geophytes from genetically modified cells can be an important step in plant transformation methods (Anderson 2007; see also Chapter 7).

1. Physiological Conditions

The factors affecting the regeneration process of ornamental geophytes *in vitro* can be divided into the physiological status of the tissue/explant and the growing conditions of the culture. In addition, genetic variation significantly affects the rate of propagation of various species and cultivars of geophytes.

a. *Plant Tissues*

The main factors related to plant tissues are: the physiological state of the donor plant and its hormonal balance, the origin and size of explants, their polarity, and pretreatments. Many sources of tissue explants are used to generate the initial step of ornamental bulbous plants culture, including shoot tips (apical and axillary buds), bulb scales, leaves, stems, and various parts of the inflorescence (Chung et al. 1981; Kim and De Hertogh 1997; Ziv and Lilien-Kipnis 2000; Blomerus and Schreuder 2002; Bacchetta et al. 2003). Shoot tips are usually isolated from the apical and axillary buds of tubers or corms (Cohen 1981; Bach 1992b; Chang et al. 2003; Ascough et al. 2009), while in bulbous plants (tulip, lily, and hyacinth) this type of explant is only used occasionally.

Organogenesis, that is, the formation of shoots or storage organs on the explants, depends on the position of the dissected tissue in the plant or the position of the explant in the medium (Saniewski 1977; Leshem et al. 1982; Park 1996; Langens-Gerrits et al. 1997; Malik and Bach 2010). It was shown that although the root formation in *Hyacinthus* culture was induced by an auxin in the medium, the formation of bulblets in the same explants depended on the orientation of the segments and took place only at the morphologically basal end of the explants and in the presence of a cytokinin and auxin (Saniewski 1977).

The temperatures and storage periods of the parent plant are the main factors of bulblet induction *in vitro*. The successful result, that is, faster regeneration of plantlets and larger bulblets, was obtained after prolonged storage of the donor hyacinth or tulip bulbs at 5–9°C (Paek et al. 1983;

Chung et al. 2006; Ascough et al. 2008). The low-temperature treatments of the donor bulbs decreased the content of abscisic acid (ABA) and enhanced the level of the endogeneous gibberellins and cytokinins in the bulb tissue (Franssens and Voskens 1992, Saniewski and Kawa-Miszczak 1992). The base of the flower stem, isolated from tulip bulbs and chilled for 12 weeks, proved to be the best explants for embryogenic callus formation (Ptak and Bach 2007) and for tulip shoot regeneration (Alderson and Taeb 1990). It has also been shown that flowering is the best period for starting a culture from the explants taken from leaves or a flower stalk of hyacinth (Chung et al. 1984).

Endogenous contamination of donor storage organs by microorganisms is often a crucial culture problem. The application of hot water treatment (43–50°C for 1–4 h) helps to avoid the reduction of growth in the tissue culture. Such a treatment of bulbs (narcissi and lilies) prior to disinfection with sodium hypochlorite resulted in a reduction in microbial contamination (Hol and Van der Linde 1992; Langens-Gerrits et al. 1998).

b.   Culture Media

Culture media contain components essential for life and normal development of plants: minerals (macro- and microelements), carbohydrates as a carbon source, vitamins, phytohormones, and other growth regulators. A basic mineral composition of a medium, suited to the requirements of herbaceous plants and geophytes, was proposed by Murashige and Skoog (1962) and is known as the universal MS medium. Among the carbohydrates, sucrose is the most frequently used, whereas glucose, fructose, mannitol, or sorbitol is used less often, and their concentrations range from 1% to 5%. Sucrose can regulate the induction of bulb dormancy, the process of tuberization (filling of storage organs), the regeneration of adventitious roots and the maturation of somatic embryos. The optimum concentration of sucrose in culture medium was found to be 3.0–5.0% and it depends on the stage of plantlet growth and development *in vitro*. The number of proliferating shoots of hyacinth, narcissus, and tulip was highest at a low sucrose concentration or on the media containing glucose, fructose, and their combination, while a higher sucrose concentration was required for bulblets formation (Bach et al. 1992; Bach and Swiderski 2000; Kumar et al. 2005; Staikidou et al. 2005; Chung et al. 2006; Staikidou et al. 2006; Cheesman et al. 2010). A 6–9% sucrose concentration is necessary for the growth and maturation of somatic embryos and during the tuberization process in tulip (Custers et al. 1992; Bach and Ptak 2001).

Plant growth regulators play the role of trigger compounds in the media for increasing the potential number of regenerated shoots, roots, and bulbs. Growth regulators include plant hormones and their synthetic analogs, and are classified into six groups: cytokinins, auxins, gibberellins, growth retardants, ABA, and ethylene (Davies 1987; George 2008). Cytokinins are derivatives of purines or urea and stimulate cell divisions. The best known of them include: benzylaminopurine (BAP), kinetin (KIN), zeatin (ZEA), isopentenyladenine (2iP), and tidiazuron (TDZ) (Davies 1987). Auxins regulate both growth and cell divisions. The natural auxin, indoleacetic acid (IAA), is used rarely in cultures of ornamental geophytes due to its susceptibility to light-induced decomposition. Thus, its analogs, naphthaleneacetic acid (NAA) and dichlorophenoxyacetic acid (2,4-D), are more useful (Ongaro and Leyser 2008; Shimizu-Sato et al. 2009; Ferguson and Beveridge 2009). The use of exogenous auxins and cytokinins *in vitro* yields very diverse results. Different proportions of auxins and cytokinins can either stimulate or inhibit the development process in geophytes, depending on the explant type, its origin, and the type of phytohormones (Figure 10.2). Ishimori et al. (2007, 2009) observed that BAP stimulates the phase transitions from juvenile to vegetative adult in bulblets of lily.

A growth retardant, paclobutrazol, added to a cytokinin (especially TDZ)-containing medium, significantly enhanced the response of tulip explants, resulting in a higher number of leaf-like shoot structures. The most intensive shoot formation was noted with a TDZ concentration of 0.5–2 mg/L and paclobutrazol at 0.05–0.01 mg/L (Podwyszynska and Marasek 2003). Methyl jasmonic acid (MeJA) and gibberellin reduced dormancy development in regenerated bulblets of lilies (Langens-Gerrits et al. 2003; Jasik and De Klerk 2006). The role of ABA in culture of geophytes *in vitro*

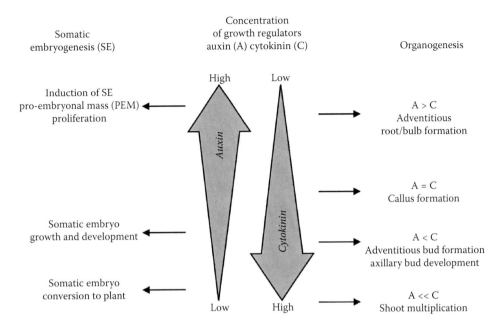

FIGURE 10.2 Effect of auxins and cytokinins on the regeneration processes in *in vitro* cultures of geophytes.

depends on the seasonal variation of ABA content in the parent bulbs. ABA controls dormancy onset both in lily bulb and in the somatic embryo of tulip (Okubo 1992; Kim et al. 1994; Gabryszewska 2001; Maslanka and Bach 2007).

In addition to the chemical composition of the medium, its physical properties, for example, the state of solidification, acidity, or the volume of culture flasks, which affects the intensity of the gaseous exchange between explants and environment, are important for the development of explants. Optimal medium pH for a majority of plant species ranges from 5.5 to 6.0 (Kim and De Hertogh 1997). In practice, solid, semiliquid, and liquid media are used. Most often, agar (0.6–0.8%) and less often gelrite (0.3–0.4%) are applied as solidifiers. The chemical and physical properties of agar can control the availability of water and ions in the medium, resulting in different regeneration pathways of plant explants. The growth and multiplication rates of *Ranunculus asiaticus* shoots were different in media solidified with three different commercial brands of agar (Beruto and Curir 2006).

c. Light

Light is an important factor in the morphogenesis of ornamental geophytes (Leshem et al. 1982; Bach and Pawlowska 2006; Hoshino and Cuello 2006). Both the quality and the intensity of light can influence the growth and development of geophytes through biosynthesis, transport, and metabolism of endo- and exogenous growth regulators. Blue light stimulated the rapid growth and elongation of adventitious buds and shoots from existing meristems in hyacinth cultures. On the other hand, red light and darkness inhibited the growth of adventitious buds from callus tissue, but promoted embryogenic callus initiation and proliferation in *Cyclamen, Freesia, Hyacinthus, Galanthus*, and *Lilium* (Hvoslef-Eide and Munster 1999; Bach and Krol 2001). The number of regenerated bulbs in tulip and lily was highest when explants were irradiated with red or yellow light, or cultured in darkness (Rice et al. 1983; Pelkonen and Kauppi 1999). It has been suggested that light quality might affect the cytokinin metabolism (Hunter and Burritt 2003).

Moreover, the type of light might interact with the balance of the different carbohydrates in the medium, and thus affect the development and growth of the adventitious shoots in geophytes during cultivation. Red light and darkness negatively affect the formation of hyacinth shoots cultured on media supplemented with sucrose, while under the same light conditions, the induction of adventitious shoots is promoted on medium with glucose (Bach et al. 1992). In general, the use of glucose as the carbohydrate source and cultivation under blue or white light are beneficial for the production of adventitious shoots *in vitro*. In contrast, mature, properly developed bulbs of hyacinth are produced on the medium supplemented with sucrose and under red light or darkness conditions. It was concluded that red and far-red light induced bulb dormancy, while blue light supported shoot multiplication and inhibited the maturation of adventitious bulbs, and their outer bulb scales formed leaf-shaped structures (Bach and Swiderski 2000). This effect might be associated with the ability of blue light to decrease the amount of ABA in plant tissues and to prevent the induction of dormancy (Gude and Dijkema 1992; Reynolds and Crawford 1997; Gabryszewska 2001).

The quality of light affects the induction and development of the embryogenic tissues of geophytes *in vitro* (Hoshino and Cuello 2006). Blue and white light, as well as UV irradiation, inhibited the initiation of embryogenic callus in *Tulipa* and *Galanthus* cultures, but stimulated the development of globular and torpedo embryos in hyacinth culture. Red light improved the maturation of the torpedo stage somatic embryos in cultures of *Tulipa*, *Hyacinthus*, and *Galanthus* (Bach and Pawlowska 2006).

A correlation between the quantity of pigments in regenerated tissues and the varying light treatment was noticed. Blue light enhanced the total amount of anthocyanins, while both blue light and UV irradiation promoted the production of chlorophyll a and b in *Hyacinthus* (Bach and Swiderski 2000).

d. Temperature

Temperature is one of the major factors in regeneration, bulbing, and dormancy of geophytes in tissue culture (Van Aartrijk and Blom-Barnhoorn 1984; Okubo 1992; Langens-Gerrits et al. 2003; Chung et al. 2006; Ascough et al. 2008). Low-temperature treatment and high sucrose concentration in a medium are prerequisites for the induction of *in vitro* bulbing of shoots in tulip (Nishiuchi 1980; Bach and Ptak 2005). Similarly, in hyacinth culture, bulblet initiation requires a period of 6–8 weeks at low temperature (4°C), while the swelling of bulblets is possible only after their cultivation at a relatively higher temperature (23°C) (Bach 1992a). This is in agreement with the argument of Le Nard and Cohat (1968) that tulip bulbing consists of two main processes: bulb induction and bulb filling, while the filling rate increases proportionally to temperature increases. The swelling of the bulblet at higher temperatures following cold treatment may also be associated with a low respiration rate and optimum thermal conditions for starch synthesis (Rees 1992).

Regenerated *in vitro* bulblets require the breaking of dormancy for further development. A similar cold treatment as used for the initiation of bulblets is used to break dormancy in regenerated bulblets of lily, hyacinth, narcissus, and tulip (Van Aartrijk and Van der Linde 1986; Gerrits et al. 1992; Paek and Murthy 2002; Langens-Gerrits et al. 2003). On the other hand, dormancy of iris bulblets is broken at a high temperature of 30°C (Van der Linde and Schipper 1992).

2. Developmental Processes

The development of geophyte explants *in vitro*—morphogenesis—can occur via a direct route by the propagation of the already existing axillary (lateral) bud meristems, the formation of adventitious meristems of bulbs, shoots, or roots (organogenesis), or by the development of somatic embryos (somatic embryogenesis). The above-mentioned processes, except for axillary bud development, can also take place via an indirect route with the intermediate proliferation of callus tissue.

The initiation and course of these processes depend on the types and proportions of growth regulators present in the medium, on the type of explant and its physiological status, and on culture conditions (Figure 10.3; Tribulato et al. 1997; Pelkonen and Kauppi 1999; Selles et al. 1999).

Axillary (lateral) bud development (proliferation) occurs when apical meristems, shoot tips (shoot apex), or axillary (lateral) buds are used as primary explants. Shoot tips can also be isolated from zygotic embryos. Vegetative propagation by axillary buds guarantees high genetic stability.

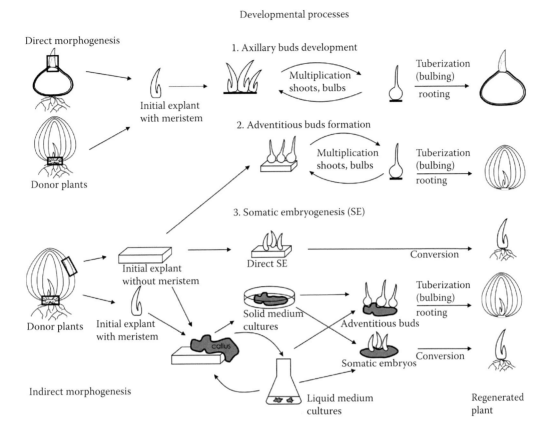

FIGURE 10.3 Scheme of the developmental processes in geophytes *in vitro*.

This type of propagation was reported for *Freesia, Gladiolus, Ranunculus, Zantedeschia*, and *Gloriosa* (Cohen 1981; Hussey 1982; Van Aartrijk and Van der Linde 1986; Beruto and Curir 2006; Roy and Sayeed Hassan 2008; Ascough et al. 2009). After being placed on a hormone-free medium, buds usually do not branch and develop a single shoot plantlet. This is the result of the activity of an endogenous auxin produced in the meristems (Shimizu-Sato et al. 2009). When cytokinins counteract the action of the auxin (apical dominance), the development of the existing axillary meristems, as well as the formation of new shoots, is possible. BAP and KIN are the most commonly used cytokinins. However, in certain plants axillary bud formation is difficult (e.g., tulip), therefore 2iP, ZEA, and TDZ are applied in order to stimulate the growth and development of the axillary buds (Farrow and Emery 2008). The number of buds increases with increasing cytokinin concentration; however, too high a concentration causes short shoots and lowers the ability to produce bulbs or roots.

Adventitious buds of shoots or roots are formed directly as new meristems from cells of primary and secondary explants or from the callus cells (the indirect pathway of morphogenesis). They can develop on different plant organs, usually on explants isolated from bulb scales or flower stalks, leaves, and roots. The ability of explants to regenerate adventitious shoots *in vitro* is affected by the ratio between auxin and cytokinin, both in the culture medium and within the explants (De Klerk 2009). Adventitious bud formation is used for propagation when axillary buds are not easily formed, for example, in the case of ornamental bulbous plants of the Liliaceae, Iridaceae, and Amaryllidaceae families (Chung et al. 1984; Kim and De Hertogh 1997; Niederwieser and Ndou 2002; Podwyszynska and Marasek 2003; Kumar et al. 2005; Naik and Nayak 2005; Minas 2007; Ascough et al. 2008, 2009). In cultures of lily, hyacinth, or tulip, the formation of adventitious plantlets consists of at least two stages of *in vitro* regeneration. The first stage is the induction of buds and shoots, that is, unipolar structures; the second stage is the regeneration of adventitious bulbs or roots. In different species, the second stage of regeneration can occur either *in vitro* or during greenhouse hardening (Bach 1992a; Langens-Gerrits et al. 1997; Langens-Gerrits and De Klerk 1999; Zaidi et al. 2000; Niederwieser and Ndou 2002). Adventitious root formation under *in vitro* conditions is influenced by auxins, ethylene, phenols, and the effects of mechanical tissue damage as the key factors. The rooting of axillary shoots usually occurs on media supplemented with auxins (IBA, IAA, NAA) or is spontaneous with the stimulation provided by an endogenous auxin. Rooting is also stimulated by growth retardants, floroglucinol, vitamin D, potassium humate, amino acids, chrysanthemic acid, and ethylene inhibitors (Gabryszewska 2001; Ilczuk et al. 2005; Podwyszynska 2006).

Morphogenesis via an indirect route can also occur with cell dedifferentiation on various explants and the intermediate proliferation of callus tissue. Cell dedifferentiation and callus formation occur in response to wounding, and are defined as the loss of the specialized features of differentiated cells and the reverting to the meristematic stage. Auxins are usually necessary for dedifferentiation and the promotion of callus formation, while cytokinins stimulate the redifferentiation of shoots or somatic embryos in callus cultures (Tribulato et al. 1997; Pelkonen and Kauppi 1999; Selles et al. 1999; Mori et al. 2005; Kedra and Bach 2005).

Somatic embryogenesis is a morphogenetic pathway for vegetative propagation *in vitro* that consists of the development of embryos from somatic cells (Von Arnold et al. 2002). Under specific conditions, most living cells are capable of division and becoming embryogenic, that is, are able to form somatic embryos. The factors determining somatic embryogenesis depend on explant competence, medium composition, and culture conditions. The best explants can be isolated from the young leaves, bulb scales, hypocotyls and seedling cotyledons, or undeveloped flower buds (Koster 1993; Sage and Hammatt 2002; Ruffoni and Savona 2006; Ptak 2010). Somatic embryos can also develop on adventitious root fragments and on shoot sections (Gude and Dijkema 1997; Ho et al. 2006). Somatic embryogenesis induced in young tissues, for example, in immature zygotic embryos, is usually direct (without the callus stage), while that induced in older explants involves callus tissue formation (Figure 10.4).

The triggering of somatic embryo development in a single cell or cell group is called embryogenic induction commitment with the formation of proembryonal masses (PEMs). There are many common features between somatic and zygotic embryo formation, and the individual embryonic developmental stages (globular, heart, and torpedo shape stages) in many plants are similar to that of the ontogeny of the species to which they belong. Somatic embryos are bipolar, that is, they possess identifiable shoots and root meristems. Somatic embryogenesis can lead to the development of a complete plant. This process is called conversion of the embryos to plants or plantlets (emblings) with one or two cotyledons (in mono- and dicotyledons, respectively), epicotyl, and embryonic root. This process has been described for tulip (Gude and Dijkema 1997; Ptak and Bach 2007), gladiolus (Stefaniak 1994), lily (Tribulato et al. 1997; Nhut et al. 2002), freesia (Bach 1992b), cyclamen (Winkelmann 2010), nerine (Lilien-Kipnis et al. 1994), and narcissus (Sage and Hammatt 2002; Sage 2005).

Somatic embryogenesis can arise on solid (solidified with agar) or liquid media. In practice, somatic embryogenesis is often induced on solid media, and embryonic tissue is later transferred to a liquid medium in order to accelerate tissue growth and increase the proliferation index (Takayama and Misawa 1982; Lilien-Kipnis et al. 1994; Kim et al. 2001; Takayama and Akita 2005; Ziv 2005; Malik 2008; Winkelmann 2010).

The developmental stages of somatic embryos are affected by the presence and proportions of auxins and cytokinins in the medium. For a majority of geophytes, the induction is elicited by a specific auxin, for example, 2,4-D, Picloram, less often by NAA, IAA, or others (Stefaniak 1994; Ptak and Bach 2007). The effect of 2,4-D can also be imitated by lowering the medium pH to about 4.0 (Mutafschiev et al. 1987; Krikorian and Smith 1992). In addition, the process of somatic embryogenesis is stimulated by increasing the osmotic potential of the media or by their supplementation with salicylic acid or antibiotics (Famelaer et al. 1996). Auxins induce the dedifferentiation process in primary explants and stimulate the embryogenic potential in the cells. These cells then regain embryogenetic ability and undergo rapid divisions, forming meristematic aggregates. Consecutive

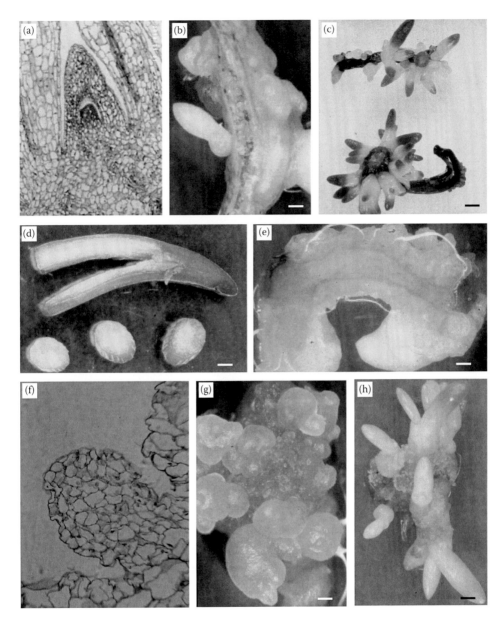

FIGURE 10.4   (See color insert.) Morphogenesis of *Hyacinthus orientalis* in *in vitro* cultures. (a) Longitudinal section of meristematic tissue in adventitious shoot tip (bar = 1 mm); (b), (c) direct formation of adventitious buds and shoots on leaf explants (bars = 2 mm, b; 5 mm, c); (d) initial leaf explants (bar = 5 mm); (e) development of callus tissue; (f) longitudinal section of somatic embryo in globular phase (bar = 0.1 mm); (g) somatic embryos in globular phase (bar = 1 mm); (h) mature somatic embryos (bar = 2 mm). ((b)–(e) Adapted from Bach, A., and B. Pawlowska. 2009. In *Plant Biotechnology*. ed. S. Malepszy, 21–40. Warsaw: Wydawnictwo Naukowe PWN (In Polish).)

developmental stages of somatic embryos usually occur when the auxin is removed from the medium or its concentration level is lowered. Embryo conversion and the development of plants can occur on media free of growth regulators or enriched in cytokinins (*Lilium*, Wang et al. 2003; *Narcissus*, Sage et al. 2000). Embryo maturation is facilitated by medium supplementation with ABA, which simultaneously limits the number of embryos (e.g., *Tulipa*, Maslanka and Bach 2007).

Somatic embryogenesis can be induced under both light and dark regimes, while irradiation of the explants from some species (*Hyacinthus*, *Tulipa*) with red light was beneficial for the development of embryogenic callus. On the other hand, the conversion of embryos into plantlets requires maintaining the cultures in light (Hoshino and Cuello 2006).

3. Management of Mass Propagation Technologies (Culture Systems)

The choice of media and physical culture conditions are decisive in determining the direction and rate of the explant's developmental response. These parameters can be modified according to the desired effect of the *in vitro* culture. It is reflected by the large number of detailed reports and procedures developed for different geophyte species and culture types.

a. Micropropagation

The system for vegetative (clonal) plant propagation in *in vitro* cultures is called micropropagation. It utilizes the ability of geophytes to regenerate either by organogenesis or by somatic embryogenesis, and includes several stages (Table 10.1). In contrast to other plant species in which the final stage of micropropagation involves the rooting of shoots, micropropagation in many geophytes culminates in the formation of storage organs—tubers, bulbs, or corms (Van Aartrijk and Van der Linde 1986; Langens-Gerrits and De Klerk 1999; Slabbert and Niederwieser 1999; Nhut et al. 2002, 2006; Ascough et al. 2008; Roy and Sayeed Hassan 2008).

The first stage involves the preparation of the parent plants for explant excision (Table 10.1). In order to limit contamination, the plants, and especially their storage organs used for the explants excision, have to be healthy and undamaged. To limit lethal infections, *Narcissus* bulbs are subjected to hot water treatment for 1 h at 54°C, while *Lilium* bulbs require treatment at temperatures of 43°C (Langens-Gerrits and De Klerk 1999).

TABLE 10.1
Major Stages of Micropropagation of Ornamental Geophytes

I. Preparation of donor plants	
Organogenesis	**Somatic Embryogenesis**
II. Initiation	
a. Development of axillary buds from the initial explants containing meristem: apical bud, axillary bud, seeds	Induction of embryogenicity (preferentially juvenile explants)
b. Formation of adventitious buds/bulbs from the initial explants not containing meristem: leaf blade, cotyledon, bulb scale, flower stalk	Induction of embryogenesis
III. Culture proliferation	
Cultures of axillary shoots or adventitious shoots and bulbs	Proliferation of pro-embryonal mass (PEM)
	Formation of somatic embryos
IV. Maturation	
Rooting of shoots or induction and formation of bulbs	Maturation of somatic embryos, conversion to plants
	Induction and formation of adventitious bulbs
V. Dormancy release, acclimatization and hardening	

Stage II is the culture initiation, which can progress either via direct organogenesis or via induction of the somatic embyogenesis. In *Freesia*, *Gladiolus*, and *Zantedeschia*, axillary shoots develop from apical meristems (Van Aartrijk and Van der Linde 1986; Chang et al. 2003). In *Hyacinthus* and *Lilium*, adventitious shoots or bulbs develop on the explants, isolated from the bulb scales or fragments of leaves (Bach 1992a; Kumar et al. 2005). In tulip, adventitious shoots regenerate on the initial explants, e.g., fragments of inflorescence stalks (Rice et al. 1983; Le Nard et al. 1987), ovary (Bach and Ptak 2005), or directly on fragments of the pedicels (Kuijpers and Langens-Gerrits 1997). The regeneration medium at the initiation stage should be supplemented with an auxin (NAA, IAA, or IBA), a cytokinin (BAP, TDZ, or ZEA), and optionally, a growth retardant, for example, paclobutrazol in tulip cultures (Podwyszynska and Marasek 2003). In the case of somatic embryogenesis, starting from various explants, callus is induced and propagated in a medium containing 2,4-D or Picloram. Transfer to a hormone-free medium or a medium containing a low concentration of 2iP or BAP resulted in the differentiation of the somatic embryos, which can convert into plantlets (emblings) on the same medium or on a medium containing BAP and NAA (see also Section IV.B.2).

At stage III, shoots, and storage organs are cyclically propagated on a media of a similar composition as the initial medium (Table 10.1). Cyclical proliferation enables the propagation of *Freesia* and *Gladiolus* in axillary shoot cultures and the propagation of *Tulipa* in adventitious shoot cultures (Podwyszynska and Marasek 2003; Ascough et al. 2009). In a narcissus culture, the proliferation of a PEM of cells and the formation of numerous somatic embryos occur at this stage (Sage 2005).

Stage IV of micropropagation via direct organogenesis involves the rooting of shoots or the induction and formation of bulbs. Following separation into individual shoots, the storage organs or adventitious roots are formed (Ascough et al. 2008). The most important factors influencing this process include the level of carbohydrates in the medium and proper culture temperature. The key inducers of tulip and hyacinth bulb development *in vitro* are the increase in the level of sucrose in the medium to 6–7%, at least an 8-week-long low-temperature (5°C) treatment, and the elimination of cytokinins from the medium (Podwyszynska 2006). Bulb induction in a majority of geophytes requires dark conditions (Ascough et al. 2008). The treatment of shoots with growth retardants and auxin (NAA) before cooling and the addition of activated charcoal into the medium have a beneficial effect on bulb formation (Takayama and Misawa 1980). It was also demonstrated that jasmonates (e.g., jasmonic acid (JA), and its derivatives, including jasmonic acid methyl ester (MeJA)) play a significant role in affecting the efficacy of the tuberization process (Podwyszynska 2006). In tulip, the number and weight of the obtained bulbs were enhanced by the application of jasmonates in tissue culture 8 weeks after cooling. *Tulipa* bulb formation occurs under light or in the dark and lasts 8–12 weeks (Podwyszynska 2006). Santos and Salema (2000) observed the promotion of bulb formation by jasmonic acid in narcissus shoot cultures. At this stage, the somatic embryos mature and convert to plantlets and the induction and formation of bulblets occur (Sage 2005).

Bulbous, tuberous, and rhizomatous plants obtained by micropropagation are usually characterized by dormancy (Okubo 1992, 2000). At the proliferation stage, dormancy is not desirable, since it can lower the propagation index. However, the acclimation of plantlets and their further growth *ex vitro* require induction and then the release from dormancy. Under *in vitro* conditions, dormancy can be controlled by growth regulators and environmental factors. In *Lilium* bulbs, *in vitro* bulb dormancy was enhanced by a higher concentration of sucrose in the medium, high temperatures, or exogenous ABA (Gerrits et al. 1992; Langens-Gerrits et al. 2003).

The final stage of micropropagation includes the hardening and acclimation of the plantlets, and their transition to the *ex vitro* stage (Pospisilova et al. 1999). This process also corresponds to the breaking of the dormancy stage (Table 10.1). In *Tulipa*, *Hyacinthus*, *Lilium*, and *Lachenalia*, dormancy in bulbs obtained *in vitro* can be broken by the rooting of the storage organs in the dark at 9°C for about 3 months (Bach 1992a; Slabbert and Niederwieser 1999; Langens-Gerrits et al. 2003; Ascough et al. 2008). *In vitro* treatment with gibberellin for 24 h broke the dormant state in bulbs of *Lilium speciosum* (Gerrits et al. 1992).

Following the micropropagation process, *in vitro* plantlets are often hyperhydric and possess malformed, abnormal shoots and nonfunctioning roots. Their leaves might lack cuticular layers and contain nonfunctioning stomata and deformed mesophyll (Ziv and Chen 2007). These *in vitro* malformations are affected by high humidity, high concentrations of calcium and nitrogen, the concentration of agar and its rigidity, and the medium's osmotic potential (Ziv 1991, 1992, 2010). Abnormal morphogenesis in liquid cultures has been alleviated by the use of growth retardants, such as paclobutrazol, uniconazole, or ancymidol (Ziv 1992; Chen and Ziv 2001), xanthan gum (Ziv and Lilien-Kipnis 1997), and silicon (Ziv 2010), to reduce shoot elongation and enhance the development of the meristematic or bud clusters (Ziv 1990, 1992). These clusters can be separated and inoculated mechanically, either to hardening media in which corms or bulbs are produced or to bioreactors for further proliferation and biomass production.

Tissue culture conditions can positively affect plant development *ex vitro*. Ziv and Naor (2006) have shown that *in vitro* conditions can shorten the juvenile period in geophytes. Thus, lily bulblets, regenerated on medium with high sucrose or law salt concentrations, produced elongated adult stems and, therefore, demonstrated more rapid establishment after being planted in soil, in comparison with juvenile plants (Langens-Gerrits et al. 2003). However, *ex vitro* establishment can be the most challenging stage of micropropagation, as was shown for gladiolus shoot culture. Plant appearance *ex vitro* was improved after the application of a growth retardant, Triadimefon, to the potting media (Sheena and Sheela 2010). Untreated gladiolus plants showed a survival rate of 47.5%, while plantlets treated with triazole (at a concentration of 4 mg/L) showed a 55.8% rate of survival.

b.   Geophyte Propagation in Liquid Media

Numerous systems for the vegetative propagation of geophytes in liquid and biphasic (solid and liquid) media have been developed. The use of traditional methods in *in vitro* plant propagation on solid media depends on intensive and expensive labor, which accounts for 70–80% of the total costs under commercial conditions. In contrast, the use of the liquid media raises the bud and shoot propagation index, synchronizes the development of somatic embryos, and enables the automation of the propagation procedures, thereby reducing production costs (Aitken-Christie et al. 1995; Levin et al. 1997; Ziv 2005).

Propagation in liquid media employs cultures in bioreactors and temporary immersion systems. A bioreactor comprises a tank for culture, an aeration and mixing system, and a set of measuring devices as standard equipment (Takayama and Akita 2005; Ziv 2005). Currently, three main techniques of liquid culture (batch, fed batch, and continuous) are used in geophytes production on an industrial scale. The batch method consists of a cyclic exchange of the medium associated with biomass reduction. The fed batch (periodical) technique involves replenishment of nutrients during the reproductive cycle. In the continuous method, the medium is constantly replaced (Preil 2005). Recently, an efficient system, providing a slow release of nutrients and cytokinins to the depleting liquid medium, was found to enhance growth in *Alstroemeria* (De Klerk and Ter Brugge 2011).

Large-scale liquid cultures in bioreactors can be used for micropropagation via organogenesis (Ziv 1992, 2005; Paek et al. 2005; Preil 2005; Takayama and Akita 2005) or somatic embryogenesis (Lilien-Kipnis et al. 1994; Nhut et al. 2006; Malik 2008). *Lilium, Ixia, Hippeastrum, Gladiolus*, and *Narcissus* can be propagated in bioreactors using protocols developed on a laboratory scale: callus embryogenic or meristematic culture, adventitious shoot or bulbs culture, or a culture of embryos and protoplasts (Kim et al. 2001; Lian et al. 2003; Ruffoni et al. 2005; Takayama and Akita 2005; Malik 2008).

This technology is most effective for geophytes, when the plantlets or embryos are induced to form storage organs, either at the liquid stage or after subculture (Ziv 1990; Ilan et al. 1995; Takayama and Akita 2005). Bulbs, corms, or tubers can be easily transplanted *ex vitro*, shipped, or kept in storage, as this has been reported for *Lilium* (Takayama et al. 1991; Takahashi et al. 1992), *Crocus sativus* (Plessner and Ziv 1998), *Gladiolus* (Ziv 1990), *Nerine* (Lilien-Kipnis et al. 1994), and *Ornithogalum* (Ziv and Lilien-Kipnis 1997). Although bioreactors can provide solutions for scaling-up cultures, the problem of malformed tissues and organs in liquid cultures often hinders *ex vitro* transplanting.

In the temporary immersion bioreactor systems (RITA®, Vitropic, France), the tissues and organs of geophytes are cyclically immersed in the medium. This technique protects the culture from oxidative stress, frequently occurring in cultures that are constantly submerged in the medium (Berthouly and Etienne 2005). Oxidative stress can cause physiological and anatomical disorders, while excessive hydration can lower the survival rate of regenerates during plant acclimatization to *in vivo* conditions. The temporary immersion technique proved to be efficient for the commercial micropropagation of *Narcissus*, *Cyclamen*, and *Hippeastrum* via organogenesis or the proliferation of the embryogenic tissue (Ilczuk et al. 2005; Sage 2005; Anbari et al. 2007; Malik 2008).

c. Cryopreservation

Traditional preservation of geophyte germplasm in field gene banks is expensive and requires large cultivation areas. Therefore, cryopreservation of plant material, in an ultra-low temperature of liquid nitrogen (−196°C), in which most of the cellular divisions and metabolic processes are arrested, is an efficient and appropriate method for the long-term storage of germplasm. Genetic stability is one of the most important criteria for the successful maintenance of clonal genotypes. Hence, preservation of the cultures of adventitious meristems and cell suspensions is preferable over that of callus cultures. The classical technique, based on freezing-induced dehydration, has been proven successfully for *Lilium*, widely propagated *in vitro* through axillary or adventitious meristems (Bouman and De Klerk 1990). The freezing of embryogenic cell suspension cultures of *Cyclamen persicum* has resulted in a 75% rate of regrowth of the cells after cryostorage (Winkelmann et al. 2004). Over the last 15 years, new techniques of cryopreservation have been established: encapsulation–dehydration, encapsulation–vitrification, vitrification, pregrowth, pregrowth–desiccation, and droplet (reviewed by Wang and Perl 2006). These novel techniques are based on a physical process—vitrification, in which a concentrated cryoprotective solution in the cells is cooled to extremely low temperatures and solidifies into a metastable glass state without crystallization. The vitrification procedure has been successfully applied in shoot apex cultures of *Lilium japonicum* (Matsumoto et al. 1995) and *Iris pumila* (Jevremovic et al. 2011).

4. Quality of Regenerants

It is expected that plant material obtained by micropropagation will be free of pathogens and remain in an active physiological state, thus guaranteeing complete hardening and adaptation for further development. The quality of plants obtained from *in vitro* cultures has been evaluated for morphological anatomical, biochemical, and genetic traits (Van der Linde 2000; De Klerk 2007).

The physiological status of the plants is evaluated using microprobes, which determine the status of water in individual cells, characterization of the cellular solutions, and the properties of the cell wall. Enzymatic activity, related to the cell wall, or photosynthesis is also used as a physiological marker of the quality of the plants (Davies and Santamaria 2000; Borkowska 2003).

Podwyszynska (2005) has examined somaclonal variation in micropropagated tulip plants. Her results, based on phenotype observations, showed that the occurrence of somaclonal changes and variegation increases with the time that the progeny lines are maintained *in vitro*. Marked changes were observed after 4–6 years of *in vitro* cyclic multiplication of tulip adventitious shoots. By limiting this period to 3 years, the risk of somaclonal variation can be reduced.

In addition to traditional techniques, a morphological assessment can be performed in a noninvasive way, using a video camera. This technique is used, for instance, to examine the stomatal apparatus, and thus, indirectly, to assess transpiration and photosynthesis (Mansfield 1994). Comparative studies of the seedling material obtained by traditional breeding methods and by an *in vitro* technique demonstrated differences in the structure and functioning of the stomatal apparatus between these two groups (Davies and Santamaria 2000). Lily and tulip plants from sterile cultures showed a different distribution of starch and chloroplasts, while the stomata closed more slowly, requiring higher humidity during the beginning of the acclimation process. The humidity was then gradually reduced from 95% to 55% (Langens-Gerrits et al. 1997; Minas 2007).

Hyperhydricity of regenerants can cause a 60% loss of plant production, but the etiology of this disorder is complex. A majority of the factors limiting hyperhydricity (lower temperature, cytokinin level, humidity or ammonium ion concentration in the medium, bottom cooling, and replacing sucrose with fructose) can also adversely affect plant propagation *in vitro*. This phenomenon is especially important in liquid culture and in bioreactors that are used for mass propagation of geophytes (Ziv 1991; Olmos 2006). De Klerk et al. (2010) suggested that hyperhydricity is associated with apoplast waterlogging, which causes hypoxia and leads to oxidative stress. Other biological indicators of stress in geophyte plants propagated *in vitro* include the level of phenolic acids and the amino acid proline, and the accumulation of storage proteins and fats (De Klerk 2007; Bach et al. 2010).

The quality of the regenerates also depends on their genetic stability and can be assessed by the determination of ploidy based on the DNA content in the cells, using a flow cytometry, or estimated by other molecular techniques, e.g., AFLP, RFLP, RAPD, SSR, ISSR[*] (Pawlowska et al. 2007; De Klerk and Smulders 2011).

V. CONCLUSIONS

In the search for commercially efficient methods of geophytes propagation, many new methods have been developed at over the last 20 years. A close look at the publications on vegetative propagation and the production of ornamental geophytes by traditional or modern methods shows that most research concentrates mainly on defining the optimal conditions for the sequence of phase change process in plants and the development of culture environment control.

Modern propagation *in vitro* requires precisely controlled environmental factors: light, temperature, state of the medium, osmotic regulation, and plant hormones balance. Moreover, the development of cryopreservation techniques seems to be a very promising method for the long-term storage of the germplasm of geophytes. Research conducted on the reversible processes of rejuvenation or maturation of tissues (juvenile–mature or dormant–nondormant states) will improve the multiplication rates of known crops and allow the establishment of new efficient ways of propagation of new genera.

We conclude that the modern research, based on recent achievements in biochemistry, physiology, and molecular biology of plant tissue and environment stress conditions, will result in further developments in geophyte propagation and bulb production.

ACKNOWLEDGMENTS

Thanks are due to Dr. Meira Ziv (The Hebrew University of Jerusalem, Israel) for the information she shared with us, and to Dr. Bozena Szewczyk-Taranek for the graphic design of the figures.

CHAPTER CONTRIBUTORS

Anna Bach is a professor of horticulture at the University of Agriculture in Krakow, Poland. She was born in Krakow, and earned her MS and PhD from the University of Agriculture in Krakow. She is the head of the Department of Ornamental Plants (since 1994) and the section president of Ornamental Plants in the Committee of Horticultural Science, Polish Academy of Sciences (since 2011). She has been the vice president of the Polish Society of Horticultural Sciences (2001–2011).

---

[*] RFLP, restriction fragment length polymorphism; AFLP, amplified fragment length polymorphism; RAPD, random amplified polymorphic DNA; SSR, simple sequence repeat; ISSR, inter-simple sequence repeat.

Her expertise relates to flower bulb cultures and developmental processes *in vitro*, and geophyte physiology and quality. Dr. Bach is the author of more than 90 refereed journal articles and chapters in books and 60 conference papers.

Dariusz Sochacki is a senior researcher of the Research Institute of Horticulture, Skierniewice, Poland. He was born in Poland and received his MS from Warsaw University of Life Sciences and his PhD from the Research Institute of Pomology and Floriculture in Skierniewice. He has held research positions at the Research Institute of Pomology and Floriculture (1992–2010) and the Research Institute of Horticulture, Skierniewice, Poland (2011–the present). Dr. Sochacki is also the secretary of the Polish Association of Flower Bulb Growers (1994–the present). His expertise includes propagation of flower bulbs, virus detection methods and virus eradication *in vitro*, and genetic resources. Dr. Sochacki is the author of 60 refereed publications and conference papers, two cultivars of narcissus, and one cultivar of tulip.

REFERENCES

Ahmad, M., G. Zaffar, S. D. Mir, S. M. Razvi, M. A. Rather, and M. R. Mir. 2011. Saffron (*Crocus sativus* L.) strategies for enhancing productivity. *Res. J. Med. Plant* 5:630–649.

Aitken-Christie, J., T. Kozai, and S. Takayama. 1995. Automation in plant tissue culture. General introduction and overview. In *Automation and Environment Control in Plant Tissue Culture*, eds. J. Aitken-Christie, T. Kozai, and M. A. L. Smith, 1–18. Dordrecht, The Netherlands: Kluwer Acad. Publ.

Alderson, P. G., and A. G. Taeb. 1990. Effect of bulb storage on shoot regeneration from floral stems of tulip *in vitro*. *J. Hort. Sci.* 65:65–70.

Alkema, H. Y. 1975. Vegetative propagation of daffodils by double-scaling. *Acta Hort.* 47:193–199.

Anbari, S., M. Tohidfar, R. Hosseini, and R. Haddad. 2007. Somatic embryogenesis induction in *Narcissus papyraceus* cv. Shirazi. *Plant Tissue Cult. & Biotech.* 17:37–46.

Anderson, N. O. 2007. *Flower Breeding and Genetics: Issues, Challenges and Opportunities for the 21st Century*. Doordrecht, The Netherlands: Springer.

Ascough, G. D., J. E. Erwin, and J. van Staden. 2009. Micropropagation of Iridaceae—A review. *Plant Cell, Tiss. Org. Cult.* 97:1–19.

Ascough, G. D., J. van Staden, and J. E. Erwin. 2008. *In vitro* storage organ formation of ornamental geophytes. *Hort. Rev.* 34:417–445.

Bacchetta, L., P. C. Remotti, C. Bernardini, and F. Sacchardo. 2003. Adventitious shoot regeneration from leaf explants and stem nodes of *Lilium*. *Plant Cell, Tiss. Org. Cult.* 74:37–44.

Bach, A. 1992a. Micropropagation of hyacinths (*Hyacinthus orientalis* L.). In *Biotechnology in Agriculture and Forestry, Vol. 20, High-tech and Micropropagation IV*, ed. Y. P. S. Bajaj, 144–159. Berlin and Heidelberg: Springer-Verlag.

Bach, A. 1992b. Somatic embryogenesis from zygotic embryos and meristems of *Freesia hybrida*. *Acta Hort.* 325:429–433.

Bach, A., and A. Krol. 2001. Effect of light quality on somatic embryogenesis in *Hyacinthus orientalis* L. 'Delft's Blue'. *Biological Bulletin of Poznan* 38:103–107.

Bach, A., and B. Pawlowska. 2006. Effect of light qualities on cultured *in vitro* ornamental bulbous plants. In *Floriculture, Ornamental and Plant Biotechnology, Vol. II*, ed. J. A. Teixeira da Silva, 271–276, Isleworth, UK: Global Science Books.

Bach, A., and B. Pawlowska. 2009. Developmental processes and types of culture in vitro. In *Plant Biotechnology*. ed. S. Malepszy, 21–40. Warsaw: Wydawnictwo Naukowe PWN (In Polish).

Bach, A., B. Pawlowska, and K. Hura. 2010. The effect of the exogenous phenolic compound, caffeic acid on organogenesis of *Galanthus elwesii* Hook. cultured *in vitro*. *Biotechnologia* 2:139–145 (In Polish with English abstract).

Bach, A., B. Pawlowska, and K. Pulczynska. 1992. Utilization of soluble carbohydrates in shoot and bulb regeneration of *Hyacinthus orientalis* L. in vitro. *Acta Hort.* 325:487–491.

Bach, A., and A. Ptak. 2001. Somatic embryogenesis and plant regeneration from ovaries of *Tulipa gesneriana* L. in *in vitro* cultures. *Acta Hort.* 560:391–394.

Bach, A., and A. Ptak. 2005. Induction and growth of tulip 'Apeldoorn' bulblets from embryo cultures in liquid media. In *Liquid Culture Systems for in Vitro Plant Propagation*, eds. A. K. Hvoslef-Eide, and W. Preil, 359–364, Dordrecht, The Netherlands: Springer.

Bach, A., and A. Swiderski. 2000. The effect of light quality on organogenesis of *Hyacinthus orientalis* L. *in vitro. Acta Biologica Cracoviensia, series Botanica* 42/1:115–120.

Baghdadi, S. H., R. A. Shibli, M. Q. Syouf, M. A. Shatanawi, A. Arabiat, and I. M. Makhadmeh. 2010. Cryopreservation by encapsulation-vitrification of embryogenic callus of wild crocus (*Crocus hyemalis* and *Crocus moabiticus*). *Jordan J. Agric. Sci.* 6:436–442.

Berthouly, M., and H. Etienne. 2005. Temporary immersion system: A new concept for use liquid medium in mass propagation. In *Liquid Culture Systems for In Vitro Plant Propagation*, eds. A. K. Hvoslef-Eide, and W. Preil, 165–195. Dordrecht, The Netherlands: Springer.

Beruto, M., and P. Curir. 2006. Effects of agar and gel characteristics on micropropagation: *Ranunculus asiaticus*, a case study. In *Floriculture, Ornamental and Plant Biotechnology Vol. II*, ed. J. A. Teixeira da Silva, 277–284. Isleworth, UK: Global Science Books.

Blomerus, L. M., and H. A. Schreuder. 2002. Rapid propagation of *Ornithogalum* using leaf cuttings. *Acta Hort.* 570:293–296.

Borkowska, B. 2003. Photosynthetic activity as physiological marker in the evaluation of the quality of shoot cultures and obtained plantlets. *Biotechnologia* 3:30–38.

Bouman, H., and G. J. de Klerk. 1990. Cryopreservation of lily meristems. *Acta Hort.* 266:331–337.

Bryan, J. E. 2002. *Bulbs (Revised edition)*. Portland, OR: Timber Press.

Buschman, J. C. M. 2005. Globalisation–flower–flower bulbs–bulb flowers. *Acta Hort.* 673:27–33.

Chang, H. S., D. Chakrabarty, E. J. Hahn, and K. Y. Paek. 2003. Micropropagation of calla lily (*Zantedeschia albomaculata*) via *in vitro* shoot tip proliferation. *In Vitro Cell. Dev. Biol.—Plant* 39:129–134.

Cheesman, L., J. F. Finnie, and J. van Staden. 2010. *Eucomis zambesiaca* baker: Factors affecting *in vitro* bulblet induction. *S. Afr. J. Bot.* 76:543–549.

Chen, J., and M. Ziv. 2001. Ancymidol effects on oxidative stress and the regeneration potential of *Narcissus* leaves in liquid culture. *Acta Hort.* 560:299–302.

Chung, J., C. Chun, and S. Choi. 1984. Variation of totipotency from individual organ parts of *Hyacinthus orientalis* cultured *in vitro* at different growth stages. *J. Kor. Soc. Hort. Sci.* 25:297–304.

Chung, J. D., C. K. Chun, Y. K. Suh, and E. M. Rhee. 1981. *In vitro* propagation of *Hyacinthus orientalis* L. I. Effect of auxins on callus formation and bulblet differentiation from flower organ and flower stem tissue. *J. Kor. Soc. Hort. Sci.* 22:146–151.

Chung, C., Y. Chung, S. Yang, E. Ko, S. Jeong, J. Nam, G. Kim, and Y. Yi. 2006. Effects of storage temperature and sucrose on bulblet growth, starch and protein contents in *in vitro* cultures of *Hyacinthus orientalis. Biol. Plant.* 50:346–351.

Cohen, D. 1981. Micropropagation of *Zantedeschia* hybrids. *Proc. Int. Plant Prop. Soc.* 31:312–316.

Custers, J. B. M., W. Eikelboom, J. H. W. Bergervoet, and J. P. van Eijk. 1992. In ovulo embryo culture of tulip (*Tulipa* L.); effects of culture conditions on seedling and bulblet formation. *Scientia Hortic.* 51:111–122.

Davies, P. J. 1987. The plant hormones: Their nature, occurrence and functions. In *Plant Hormones and Their Role in Plant Growth and Development*, ed. P. J. Davies, 1–11. Dordrecht, The Netherlands: Martinus Nijhof Publ.

Davies, W. J., and J. M. Santamaria. 2000. Physiological markers for microplant shoot and root quality. *Acta Hort.* 530:363–369.

De Hertogh, A. A., and M. Le Nard. 1993a. Production systems for flower bulbs. In *The Physiology of Flower Bulbs*, eds. A., De Hertogh, and M. Le Nard, 45–52. Amsterdam: Elsevier.

De Hertogh, A. A., and M. Le Nard. 1993b. *Dahlia*. In *The physiology of Flower Bulbs*, eds. A. De Hertogh, and M. Le Nard, 273–283. Amsterdam: Elsevier.

De Klerk, G. J. 2007. Stress in plants cultured *in vitro. Propagation of Ornamental Plants* 7:129–137.

De Klerk, G. J. 2009. Adventitious regeneration. In *Encyclopedia of Industrial Biotechnology Bioprocess, Bioseparation and cell Technology Vol. 1*, ed. M. C. Flickinger, 72–87. Hoboken, NJ: John Wiley and Sons.

De Klerk, G. J., L. Rojas, and R. Visser. 2010. The hyperhydricity syndrome: Water logging of plant tissues as a major cause. *Propagation of Ornamental Plants* 10:169–175.

De Klerk, G. J., and R. Smulders. 2011. Epigenetics in plant tissue culture. *Plant Growth Regul.* 63:137–146.

De Klerk, G.-J., and J. ter Brugge. 2011. Micropropagation of dahlia in static liquid medium using slow-release tools of medium ingredients. *Scientia Hortic.* 127:542–547.

Doorenbos, J. 1954. Notes on the history of bulb breeding in The Netherlands. *Euphytica* 3:1–11.

Ellis, B. W. 2001. *Taylor's Guide to Bulbs. How to Select and Grow More than 400 Summer-Hardy and Tender Bulbs*. New York: Houghton Mifflin Harcourt.

Everett, T. H. 1954. *The American Gardener's Book of Bulbs*. New York: Random House.

Famelaer, I., E. Ennik, W. Eikelboom, J. M. van Tuyl, and J. Creemers-Molenar. 1996. The initiation of callus and regeneration from callus culture of *Tulipa gesneriana*. *Plant Cell, Tiss. Org. Cult.* 47:51–58.

Farrow, S. C., and R. J. N. Emery. 2008. Cytokinins in floriculture: Physiology, molecular mechanisms and impact on vegetative and reproductive trade-offs. In *Floriculture, Ornamental and Plant Biotechnology Vol. V*, ed. J. A. Teixeira da Silva, 191–205. Isleworth, UK: Global Science Books.

Ferguson, B. J., and C. A. Beveridge. 2009. Roles for auxin, cytokinin, and strigolactone in regulating shoot branching. *Plant Physiol.* 149:1229–1244.

Flint, G. J. 1982. Narcissus propagation using the chipping technique. *Annu. Rev. Kirton Experimental Hort. Station* 1981:1–9.

Franssens, J. M., and P. G. J. M. Voskens. 1992. Methods to determine abscisic acid and indole-3-acetic acid, and determinations of these hormones in tulip cv. Apeldoorn bulbs as related to the cold-treatment. *Acta Hort.* 325:267–276.

Gabryszewska, E. 2001. Effects of ABA, fluridone and light quality on growth and dormancy of tissue-cultured herbaceous peony shoots. *Acta Hort.* 560:407–410.

George, E. F. 2008. Plant tissue culture procedure—Background. In *Plant Propagation by Tissue Culture 3rd Edition. Vol. 1, The Background*, eds. E. F. George, M. A. Hall, and G.-J. de Klerk, 1–28. Heidelberg: Springer-Verlag.

Gerrits, M. M., K. S. Kim, and G. J. de Klerk. 1992. Hormonal control of dormancy in bulblets of *Lilium speciosum* cultured *in vitro*. *Acta Hort.* 325:521–527.

Gilad, Z., E. Hovav, D. Sandler-Ziv, and R. Kamenetsky. 2001. Development of *Allium aschersonianum*, an Israeli native species, as a new ornamental crop. *Acta Hort.* 552:171–177.

Grassotti, A. 1997. The effects of scale incubation temperature treatments on number and growth of regenerated bulblets, to reduce production time of commercial lily bulbs. *Acta Hort.* 430:221–226.

Gude, H., and M. H. G. E. Dijkema. 1992. The effect of light quality and cold treatment on the propagation of hyacinth bulbs. *Acta Hort.* 325:157–164.

Gude, H., and M. H. G. E. Dijkema. 1997. Somatic embryogenesis in tulip. *Acta Hort.* 430:275–280.

Hanks, G. R. 2002. Commercial production of *Narcissus* bulbs. In *Narcissus and Daffodil: The Genus Narcissus*, ed. G. R. Hanks, 53–130. London: Taylor & Francis.

Hanks, G., and S. K. Jones. 1986. Notes on the propagation of *Narcissus* by twin-scaling. *Plantsman* 8:118–127.

Hartmann, H. T., D. E. Kester, F. T. Davies, and R. L. Geneve. 2002. *Hartmann and Kester's Plant Propagation: Principles and Practices*. Upper Saddle River, NJ: Pearson Education.

Ho, C.-W., W.-T. Jian, and H.-C. Lai. 2006. Plant regeneration via somatic embryogenesis from suspension cell cultures of *Lilium × formolongi* Hort. using a bioreactor system. *In Vitro Cell. Dev. Biol.—Plant* 42:240–246.

Hol, G. M. G. M., and P. C. G. van der Linde. 1992. Reduction of contamination in bulb-explant cultures of *Narcissus* by a hot-water treatment of parent bulbs. *Plant Cell, Tiss. Org. Cult.* 31:75–79.

Hoshino, T., and J. L. Cuello. 2006. Designing the lighting environment for somatic embryogenesis. In *Floriculture, Ornamental and Plant Biotechnology, Vol. II*, ed. J. A. Teixeira da Silva, 295–298, Isleworth, UK: Global Science Books.

Huang, C. W., H. Okubo, and S. Uemoto. 1990. Importance of two scales in propagating *Hippeastrum hybridum* by twin scaling. *Scientia Hortic.* 42:141–149.

Hunter, D. C., and D. J. Burrit. 2003. Light quality: A tool to manipulate plant growth and development *in vitro*. *Research Signpost* 37/661:1–14.

Hussey, G. 1982. *In vitro* propagation of monocotyledonous bulbs and corms. In *Plant Tissue Culture*, ed. A. Fujiwara, 677–680, Tokyo: Maruzen.

Hvoslef-Eide, A. K., and C. Munster. 1999. Light quality effect on somatic embryogenesis of *Cyclamen pericum* Mill. in bioreactors. In *Reproduction of Cyclamen persicum Mill. through Somatic Embryogenesis Using Suspensor Culture Systems, A Report of the Working Group 2, COST 882*, ed. H. G. Schwenkel, 79–84. Brussels: European Commission.

Ilan, A., M. Ziv, and A. A. Halevy. 1995. Propagation and corm development of *Brodiaea* in liquid cultures. *Scientia Hortic.* 63:101–112.

Ilczuk, A., T. Winkelmann, S. Richartz, M. Witomska, and M. Serek. 2005. *In vitro* propagation of *Hippeastrum × chmielii* Chm.—influence of flurprimidol and the culture in solid or liquid medium and in temporary immersion system. *Plant Cell, Tiss. Org. Cult.* 83:339–346.

Ishimori, T., Y. Niimi, and D.-S. Han. 2007. Benzyladenine and low temperature promote phase transition from juvenile to vegetative adult in bulblets of *Lilium × formolongi* 'White Aga' cultured *in vitro*. *Plant Cell, Tiss. Org. Cult.* 88:313–318.

Ishimori, T., Y. Niimi, and D.-S. Han. 2009. *In vitro* flowering of *Lilium rubellum* Baker. *Scientia Hortic.* 120:246–249.

Jasik, J., and G. J. de Klerk. 2006. Effect of methyl jasmonate on morphology and dormancy development in lily bulblets regenerated *in vitro*. *J. Plant Growth Regul.* 25:45–51.

Jevremovic, S., A. Subotic, C. Benelli, A. de Carlo, and M. Lambardi. 2011. Cryopreservation of *Iris pumila* shoot tips by vitrification. *Acta Hort.* 908:355–359.

Kamenetsky, R. 2005. Production of flower bulbs in regions with warm climates. *Acta Hort.* 673:59–66.

Kamenetsky, R., and R. Fritsch. 2002. Ornamental alliums. In *Allium Crop Science: Recent Advances*, eds. H. D. Rabinowitch, and L. Currah, 459–492. Wallingford, UK: CAB International.

Kamenetsky, R., and H. D. Rabinowitch. 2006. The genus *Allium*: A developmental and horticultural analysis. *Hort. Rev.* 32:329–378.

Kedra, M., and A. Bach. 2005. Morphogenesis of *Lilium martagon* L. explants in callus culture. *Acta Biologica Cracoviensia Series Botanica* 47/1:65–73.

Keller, E. R. J., and A. Senula. 2010. Cryopreservation of plant germplasm. In *Plant Cell Culture: Essential Methods*, eds. M. R. Davey, and P. Anthony, 131–152. Chichester, UK: John Wiley & Sons.

Kim, K. S., E. Davelaar, and G.-J. de Klerk. 1994. Abscisic acid controls dormancy development and bulb formation in lily plantlets regenerated *in vitro*. *Physiol. Plant.* 90:59–64.

Kim, K.-W., and A. A. De Hertogh. 1997. Tissue culture of ornamental flowering bulbs (geophytes). *Hort. Rev.* 18:87–169.

Kim, Y. S., E. J. Hahn, and K. Y. Paek. 2001. A large scale production of *Lilium* bulblets through bioreactor culture. *Acta Hort.* 560:383–386.

Kim, H. H., K. Sakaguchi, and K. Ohkawa. 1997. *Zephyra elegans* D. Don., a potential new cormous crop. *Acta Hort.* 430:133–137.

Koster, J. 1993. *In vitro* propagation of tulip. Formation and bulbing of shoots on bulb-scale explants from *Tulipa gesneriana* L. PhD thesis, Rijksuniversiteit, Leiden, The Netherlands.

Krause, J. 1980. Propagation of hyacinth by leaf cuttings. Histological changes in leaf cuttings of *Hyacinthus orientalis*. *Acta Hort.* 109:271–277.

Krikorian, A. D., and D. L. Smith. 1992. Somatic embryogenesis in carrot (*Daucus carota*). In *Plant Tissue Culture Manual*, ed. K. Lindsey, 1–32. Dordrecht, The Netherlands: Kluwer Academic Publ.

Kuijpers, A.-M., and M. Langens-Gerrits. 1997. Propagation of tulip *in vitro*. *Acta Hort.* 430:321–324.

Kumar, S., M. Kashyap, and D. R. Sharma. 2005. *In vitro* regeneration and bulblet growth from lily bulbscale explants as affected by retardants, sucrose and irradiance. *Biol. Plant.* 49:629–632.

Langens-Gerrits, M., M. Albers, and G.-J. de Klerk. 1998. Hot-water treatment before tissue culture reduces initial contamination in *Lilium* and *Acer*. *Plant Cell, Tiss. Org. Cult.* 52:75–77.

Langens-Gerrits, M. M., and G. J. M. de Klerk. 1999. Micropropagation of flower bulbs. Lily and narcissus. In *Plant Cell Culture Protocols*, ed. R. D. Hall, 141–147. New York: Humana Press.

Langens-Gerrits, M., H. Lilien-Kipnis, T. Croes, W. Miller, C. Kollöffel, and G.-J. de Klerk. 1997. Bulb growth in lily regenerated *in vitro*. *Acta Hort.* 430:267–274.

Langens-Gerrits, M. M., W. B. M. Miller, A. F. Croes, and G.-J. de Klerk. 2003. Effect of low temperature on dormancy breaking and growth after planting in lily bulblets regenerated *in vitro*. *Plant Growth Regul.* 40:267–275.

Langton, A., and G. Hanks. 1993. Propagation. In *Review of Narcissus R&D*, eds. A. Langton, and G. Hanks, 1–12, Warwick, UK: Horticulture Research International—Bulb Group.

Le Nard, M., and J. Cohat. 1968. Influence des températures de conservation des bulbes sur l'élongation, la floraison et la bulbification de la tulipe (*Tulipa gesneriania* L.). *Ann. Amélior. Plantes* 18:181–215.

Le Nard, M., and A. A. De Hertogh. 1993a. *Tulipa*. In *The Physiology of Flower Bulbs*, eds. A. De Hertogh, and M. Le Nard, 617–682. Amsterdam: Elsevier.

Le Nard, M., and A. A. De Hertogh. 1993b. General chapter on spring flowering bulbs. In *The Physiology of Flower Bulbs*, eds. A. De Hertogh, and M. Le Nard, 705–739. Amsterdam: Elsevier.

Le Nard, M., C. Ducommun, G. Weber, N. Dorion, and C. Bigot. 1987. Observations sur la multiplication *in vitro* de la tulipe (*Tulipa gesneriana* L.) a partir de hampes florales preleeves chez des bulbes en cours de conservation. *Agronomie* 7:321–329.

Leshem, B., H. Lilien-Kipnis, and B. Steinitz. 1982. The effect of light and of explant orientation on the regeneration and subsequent growth of bulblets on *Lilium longiflorum* Thunb. bulb-scale sections cultured *in vitro*. *Scientia Hortic.*17:129–136.

Levin, R., R. Stav, Y. Alper, and A. A. Watad. 1997. A technique for repeated non-axenic subculture of plant tissues in a bioreactor on liquid medium containing sucrose. *Plant Tissue Culture Biotech.* 3:41–45.

Lian, M. L., D. Chakrabarty, and K. Y. Paek. 2003. Bulblet formation from bulbscale segments of *Lilium* using bioreactor system. *Biol. Plant.* 46:199–203.

Lilien-Kipnis, H., N. Azizbekova, and M. Ziv. 1994. Scaled-up proliferation and regeneration of *Nerine* in liquid cultures. Part II. Ontogeny of somatic embryos and bulblet regeneration. *Plant Cell, Tiss. Org. Cult.* 39:117–123.

Malik, M. 2008. Comparison of different liquid/solid culture systems in the production of somatic embryos from *Narcissus* L. ovary explants. *Plant Cell, Tiss. Org. Cult.* 94:337–345.

Malik, M., and A. Bach. 2010. Somatic embryogenesis induction and morphogenesis direction in narcissus culture (*Narcissus* L.) of 'Carlton' depending on the initial explants type. *Post. Nauk Rol.* 551:175–181 (In Polish with English summary).

Mansfield, T. A. 1994. Some aspects of stomatal physiology relevant to plants cultured *in vitro*. In *Physiology, Growth and Development of Plants in Culture*, eds. P. J. Lumsden, J. R. Nicholas, and W. J. Davies, 120–131. Dordrecht, The Netherlands: Kluwer Academic Publ.

Maslanka, M., and A. Bach. 2007. Effect of abscisic acid and light on the growth and development of tulip (*Tulipa gesneriana* L.) somatic embryos. *Post. Nauk Rol.* 523:155–161 (In Polish with English summary).

Matsumoto, T., A. Sakai, and K. Yamada. 1995. Cryopreservation of *in vitro*-grown apical meristems of lily by vitrification. *Plant Cell, Tiss. Org. Cult.* 41:237–241.

Matsuo, E., K. Arisumi, and H. Kawashima. 1982. Cultural practices influencing premature daughter leaf and/or shoot emergence in scale-propagated Easter lily. *HortScience* 17:196–198.

Matsuo, E., and J. M. van Tuyl. 1984. Effect of bulb storage temperature on leaf emergence and plant development during scale propagation of *Lilium longiflorum* 'White American'. *Scientia Hortic.* 24:59–66.

McKay, M. E., D. E. Byth, and J. Tommerup. 1981. The effect of corm size and division of the mother corm in gladioli. *Aus. J. Exp. Agric. Animal Husb.* 21(110):343–348.

Memon, N., M. Qasim, M. J. Jaskani, R. Ahmad, and I. Ahmad. 2009. Enhancement of corm and cormel production in gladiolus (*Gladiolus* spp.). *New Zealand J. Crop Hortic. Sci.* 37:319–325.

Minas, G. J. 2007. *In vitro* propagation of Akama tulip via adventitious organogenesis from bulb slices. *Acta Hort.* 755:313–316.

Mori, S., Y. Adachi, S. Horimoto, S. Suzuki, and M. Nakano. 2005. Callus formation and plant regeneration in various *Lilium* species and cultivars. *In Vitro Cell. Dev. Biol.—Plant* 41:783–788.

Mori, G., H. Hirai, and H. Imanishi. 1997. Vegetative propagation of *Nerine* bulbs by cross-cutting. *Acta Hort.* 430:377–381.

Murashige, T., and F. Skoog. 1962. A revised medium for rapid growth and bio assays with tobacco tissue cultures. *Physiol. Plant.* 15:473–497.

Mutafschiev, S., A. Cousson, and K. Tran Thanh Van. 1987. Modulation of cell growth and differentiation by pH and oligosaccharides. In *Advances in the Chemical Manipulation of Plant Tissue Cultures*, eds. M. B. Jackson, S. H. Mantell, and J. Blake, 29–42. London: University of London.

Naik, P. K., and S. Nayak. 2005. Different modes of plant regeneration and factors affecting *in vitro* bulblet production in *Ornithogalum virens*. *ScienceAsia* 31:409–414.

Nhut, D. T., N. T. Don, N. H. Vu, N. Q. Thien, D. T. T. Thuy, N. Duy, and J. A. T. da Silva. 2006. Advance technology in micropropagation of some important plants. In *Floriculture, Ornamental and Plant Biotechnology, Vol. II*, ed. J. A. Teixeira da Silva, 325–335. Isleworth, UK: Global Science Books.

Nhut, D. T., B. Van Le, N. T. Minh, J. T. de Silva, S. Fukai, M. Tanaka, and K. T. T. Van. 2002. Somatic embryogenesis through pseudo-bulblets transverse thin cell layer of *Lilium longiflorum*. *Plant Growth Regul.* 37:193–198.

Niederwieser, J. G., and A. M. Ndou. 2002. Review on adventitious bud formation in *Lachenalia*. *Acta Hort.* 570:135–140.

Nishiuchi, Y. 1980. Studies on vegetative propagation of tulips. IV. Regeneration of bulblets in bulb scale segments cultured *in vitro*. *J. Japan. Soc. Hort. Sci.* 49:235–240.

Nowak, J., and R. M. Rudnicki. 1993. *Hyacinthus*. In *The Physiology of Flower Bulbs*, eds. A. De Hertogh, and M. Le Nard, 335–347. Amsterdam: Elsevier.

Okubo, H., M. Chijiwa, and S. Uemoto.1988. Seasonal changes in leaf emergence from scale bulblets during scaling and endogenous plant hormone levels in Easter lily (*Lilium longiflorum* Thunb.). *J. Fac. Agr., Kyushu Univ.* 33:9–15.

Okubo, H. 1992. Dormancy in bulbous plants. *Acta Hort.* 325:35–41.

Okubo, H. 2000. Growth cycle and dormancy in plants. In *Dormancy in Plants—From Whole Plant Behaviour to Cellular Control*, eds. J.-D. Viémont, and J. Crabbé, 1–22. Wallingford, Oxon, UK: CAB International.

Olmos, E. 2006. Prevention of hyperhydricity in plant tissue culture. In *Floriculture, Ornamental and Plant Biotechnology, Vol. II*, ed. J. A. Teixeira da Silva, 285–288, Isleworth, UK: Global Science Books.

Ongaro, V., and O. Leyser. 2008. Hormonal control of shoot branching. *J. Exp. Bot.* 59:67–74.

Paek, K. Y., D. Chakrabarty, and E. J. Hahn. 2005. Application of bioreactor systems for large scale production of horticultural and medicinal plants. In *Liquid Culture Systems for In Vitro Plant Propagation*, eds. A. K. Hvoslef-Eide, and W. Preil, 95–116. Dordrecht, The Netherlands: Springer.

Paek, K. Y., C. K. Chun, and J. D. Chung. 1983. Factors influencing regeneration of bulblets and effect of chilling treatment of parent bulblets on scale segments of hyacinth *in vitro*. *J. Kor. Soc. Hort. Sci.* 24:68–75.

Paek, K. Y., and H. N. Murthy. 2002. High frequency of bulblet regeneration from bulb scale sections of *Fritillaria thunbergii*. *Plant Cell, Tiss. Org. Cult.* 68:247–252.

Park, N. B. 1996. Effect of temperature, scale position, and growth regulators on the bulblet formation and growth during scale propagation of *Lilium*. *Acta Hort.* 414:257–262.

Pawlowska, B., K. Nowak, and A. Bach. 2007. Ploidy of Martagon lily (*Lilium martagon* L.) regenerants propagated *in vitro*. In *Spontaneous and Induced Variation for the Genetic Improvement of Horticultural Crops*, ed. P. Nowaczyk, 275–281. Bydgoszcz, Poland: University Press, University of Technology and Life Science in Bydgoszcz.

Pelkonen, V.-P., and A. Kauppi. 1999. The effect of light and auxins on the regeneration of lily (*Lilium regale* Wil.) cells by somatic embryogenesis and organogensis. *Int. J. Plant Sci.* 160:483–490.

Plessner, O., and M. Ziv. 1998. *In vitro* propagation and secondary metabolite production in *Crocus sativus*. In *Saffron–Crocus sativus (Medicinal and Aromatic Plants—Industrial Profiles)*, ed. M. Negbi, 137–148. Chur, Switzerland: Hardwood Acad. Pub.

Podwyszynska, M. 2005. Somaclonal variation in micropropagated tulips based on phenotype observation. *J. Fruit Ornam. Plant Res.* 13:109–122.

Podwyszynska, M. 2006. Effect of ethylene, auxin and methyl jasmonate on bulb formation *in vitro* in tulip shoot cultures. *Zeszyty Prob. Post. Nauk Roln.* 510:461–469 (In Polish with English abstract).

Podwyszynska, M., and A. Marasek. 2003. Effect of thidiazuron and paclobutrazol on regeneration potential of tulip flower stalk explants *in vitro* and subsequent shoot multiplication. *Acta Soc. Bot. Pol.* 72:181–190.

Pospisilova, J., I. Ticha, P. Kadlecek, D. Haisel, and S. Plzakova. 1999. Acclimatization of micropropagated plants to ex vitro conditions. *Biol. Plant.* 42:481–497.

Preil, W. 2005. General introduction: A personal reflection on the use of liquid media for *in vitro* culture. In *Liquid Culture Systems for In Vitro Plant Propagation*, eds. A. K. Hvoslef-Eide, and W. Preil, 1–18. Dordrecht, The Netherlands: Springer.

Ptak, A. 2010. Somatic embryogenesis in *in vitro* culture of *Leucojum vernum* L. *Methods Mol. Biol.* 589:223–233.

Ptak, A., and A. Bach. 2007. Somatic embryogenesis in tulip (*Tulipa gesneriana* L.) flower stem cultures. *In Vitro Cell. Dev. Biol.—Plant* 43:35–39.

Rees, A. R. 1992. *Ornamental Bulbs, Corms and Tubers*. Wallingford, Oxon, UK: CAB International.

Reynolds, T. L., and R. L. Crawford. 1997. Effect of light on the accumulation of abscisic acid and expression of an early cysteine-labeled matallothionein gene in microspores of *Triticum aestivum* during induced embryogenic development. *Plant Cell Rep.* 16:458–463.

Rice, R. D., P. G. Alderson, and N. A. Wright. 1983. Induction of bulbing of tulip shoots *in vitro*. *Scientia Hortic.* 20:377–390.

Roh, S. M. 1982. Propagation of *Lilium longiflorum* Thunb. by leaf cutting. *HortScience* 17:607–609.

Roy, S. K., and A. K. M. Sayeed Hassan. 2008. Regeneration and conservation of *Gloriosa superba* L. through microtuber induction *in vitro*. In *Floriculture, Ornamental and Plant Biotechnology Vol. V*, ed. J. A. Teixeira da Silva, 253–256. Isleworth, UK: Global Science Books.

Ruffoni, B., and M. Savona. 2006. Somatic embryogenesis in floricultural crops: Experiences of massive propagation of *Lisianthus, Genista* and *Cyclamen*. In *Floriculture, Ornamental and Plant Biotechnology Vol. II*, ed. J. A. Teixeira da Silva, 305–313. Isleworth, UK: Global Science Books.

Ruffoni, B., M. Savona, S. Doveri, M. Pamato, and S. Carli. 2005. Micropropagation of *Ixia* hybrids in liquid medium. In *Liquid Culture Systems for In Vitro Plant Propagation*, eds. A. K. Hvoslef-Eide, and W. Preil, 365–372. Dordrecht, The Netherlands: Springer.

Sage, D. O. 2005. Propagation and protection of flower bulbs: Current approaches and future prospects, with special reference to *Narcissus*. *Acta Hort.* 673:107–115.

Sage, D., and N. Hammatt. 2002. Somatic embryogenesis and transformation in *Narcissus pseudonarcissus* cultivars. *Acta Hort.* 570:247–249.

Sage, D. O., J. Lynn, and N. Hammatt. 2000. Somatic embryogenesis in *Narcissus pseudonarcissus* cvs. Golden Harvest and St. Keverne. *Plant Sci.* 150:209–216.

Sandler-Ziv, D., A. Cohen, A. Ion, M. Kochba, H. Efron, and D. Amit. 1997. Improving *Hippeastrum* propagation and bulbil yield by cutting and incubation techniques. *Acta Hort.* 430:355–360.

Sandler-Ziv, D., R. Kamenetsky, Z. Gilad, M. Achiam, and G. Luria. 2011. Effective system for propagation of colored calla lily (*Zantedeschia*) by cutting and intensive growth. *Acta Hort.* 886:233–238.

Saniewski, M. 1977. Polarity of adventitious root and bulblet formation in etiolated leaves and flower buds of *Hyacinthus orientalis* L. *Bull. Acad. Polon. Sci. Ser. Biol.* 25:63–66.

Saniewski, M., and L. Kawa-Miszczak. 1992. Hormonal control of growth and development of tulips. *Acta Hort.* 325:43–54.

Santos, I., and R. Salema. 2000. Promotion by jasmonic acid of bulb formation in shoot cultures of *Narcissus triandrus* L. *Plant Growth Regul.* 30:133–138.

Selles, M., J. Viladoamt, J. Bastida, and C. Codina. 1999. Callus induction, somatic embryogenesis and organogenesis in narcissus confuses: Correlation between the state of differentiation and the content of galanthamine and related alkaloids. *Plant Cell Rep.* 18:646–651.

Sheena, A., and V. L. Sheela. 2010. Effects of the growth retardant triadimefon on the *ex vitro* establishment of gladiolus (*Gladiolus grandiflorus* L.) cv Vinks Glory. *Plant Tissue Cult. Biotech.* 20:171–178.

Shimada, Y., G. Mori, Y. Katahara, and M. Oda. 2005. The influence of BA and NAA on adventitious bud formation in leaf cuttings of *Begonia* × *tuberhybrida*. *Acta Hort.* 673:537–541.

Shimizu-Sato, S., M. Tanaka, and H. Mori. 2009. Auxin-cytokinin interaction in the control of shoot branching. *Plant Mol. Biol.* 69:429–435.

Slabbert, M. M., and J. G. Niederwieser. 1999. *In vitro* bulblet production of *Lachenalia*. *Plant Cell Rep.* 18:620–624.

Sochacki, D., and T. Orlikowska. 2000. The research on increasing of multiplication rate of narcissus using the chipping technique. *Zesz. Naukowe Inst. Sad. Kwiac.* 7:381–386 (In Polish with English abstract).

Staikidou, I., C. Selby, and G. Hanks. 2006. Stimulation of *in vitro* bulblet growth in *Galanthus* species with sucrose and activated charcoal. *Acta Hort.* 725:421–426.

Staikidou, I., S. Watson, B. M. R. Harvey, and C. Selby. 2005. *Narcissus* bulblet formation *in vitro*: Effects of carbohydrate type and osmolarity of the culture medium. *Plant Cell, Tiss. Org. Cult.* 80:313–320.

Stefaniak, B. 1994. Somatic embryogenesis and plant regeneration of gladiolus (*Gladiolus hort.*). *Plant Cell Rep.* 13:386–389.

Takahashi, S., K. Matsubara, H. Yamagata, and T. Morimoto. 1992. Micropropagation of virus free bulblets of *Liliun longiflorum* by tank culture—1. Development of liquid culture method and large scale propagation. *Acta Hort.* 319:83–88.

Takayama, S., and M. Akita. 2005. Practical aspects of bioreactor application in mass propagation of plants. In *Liquid Culture Systems for in Vitro Plant Propagation*, eds. A. K. Hvoslef-Eide and W. Preil, 61–78. Dordrecht, The Netherlands: Springer.

Takayama, S., and M. Misawa. 1980. Differentiation in *Lilium* bulb scales grown *in vitro*. Effect of activated charcoal, physiological age of bulbs and sucrose concentration on differentiation and scale leaf formation *in vitro*. *Physiol. Plant.* 48:121–125.

Takayama, S., and M. Misawa. 1982. A scheme for mass propagation of *Lilium in vitro*. *Scientia Hortic.* 18:353–362.

Takayama, S., B. Swedlund, and Y. Miwa. 1991. Automated propagation of microbulbs of lilies. In *Cell Culture and Somatic Cell Genetics of Plants: Vol. 8, Scale-Up and Automation in Plant Propagation*, ed. I. K. Vasil, 111–131. New York: Academic Press.

Tribulato, A., P. C. Remotti, H. J. M. Löffler, and J. M. van Tuyl. 1997. Somatic embryogenesis and plant regeneration in *Lilium longiflorum* Thunb. *Plant Cell Rep.* 17:113–118.

Van Aartrijk, J., and G. J. Bloem-Barnhoorn. 1984. Adventitious bud formation from bulb-scale explants of *Lilium speciosum* Thunb. *In vitro* interacting effects of NAA, TIBA, wounding, and temperature. *J. Plant Physiol.* 116:409–416.

Van Aartrijk, J., and P. C. G. van der Linde. 1986. *In vitro* propagation of flower-bulb crops. In *Tissue Culture as a Plant Production System for Horticultural Crops*, eds. R. H. Zimmerman, R. J. Griesbach, F. A. Hammerschlag, and R. H. Lawson, 317–331. Dordrecht, The Netherlands: Martinus Nijhof Publ.

Van Brenk, G., and M. Benschop. 1993. *Nerine*. In *The Physiology of Flower Bulbs*, eds. A. De Hertogh, and M. Le Nard, 559–588. Amsterdam: Elsevier.

Van der Linde, P. C. G. 2000. Certified plants from tissue culture. *Acta Hort.* 530:93–101.

Van der Linde, P. C. G., and J. A. Schipper. 1992. Micropropagation of iris with special reference to *Iris × hollandica* Tub. In *Biotechnology in Agriculture and Forestry, Vol. 20, High-Tech and Micropropagation IV*, ed. Y. P. S. Bajaj, 173–197. Dordrecht, The Netherlands: Springer.

Van Leeuwen, P. J., and J. A. van der Weijden. 1997. Propagation of specialty bulbs by chipping. *Acta Hort.* 430:351–353.

Von Arnold, S., I. Sabala, P. Bozhkov, J. Dyachok, and L. Filonova. 2002. Developmental pathways of somatic embryogenesis. *Plant Cell, Tiss. Org. Cult.* 69:233–249.

Vreeburg, P. J. M. 1986. Chipping of narcissus bulbs: A quick way to obtain large numbers of small, round bulbs. *Acta Hort.* 177:579–584.

Wang, Q., and A. Perl. 2006. Cryopreservation in floricultural plants. In *Floriculture, Ornamental and Plant Biotechnology Vol. I*, ed. J. A. Teixeira da Silva, 523–539. Isleworth, UK: Global Science Books.

Wang, S., J. Wang, J. Fan, and D. Che. 2003. Somatic embryogenesis in lily bulb scale cultures. *J. Northeast Agric. Univ.* 15(4):11–14.

Winkelmann, T. 2010. Clonal propagation of *Cyclamen persicum* via somatic embryogenesis. *Methods Mol. Biol.* 589:281–290.

Winkelmann, T., V. Mußman, and M. Serek. 2004. Cryopreservation of embryogenic suspension cultures of *Cyclamen persicum* Mill. *Plant Cell Rep.* 23:1–8.

Zaidi, N., N. H. Khan, F. Zafar, and S. I. Zafar. 2000. Bulbous and cormous monocotyledonous ornamental plants *in vitro. Science Vision* 6(1):58–73.

Ziv, M. 1990. Morphogenesis of gladiolus buds in bioreactors—Implication for scale-up propagation of geophytes. In *Current Plant Science and Biotechnology in Agriculture. Progress in Plant Cellular and Molecular Biology*, eds. H. J. J. Nijkamp, L. H. W. van der Plas, and J. van Aartrijk, 119–124. Dordrecht, The Netherlands: Kluwer Academic Publishers.

Ziv, M. 1991. Quality of micropropagated plants—Vitrification. *In Vitro Cell. Dev. Biol.—Plant* 27:64–69.

Ziv, M. 1992. The use of growth retardants for the regulation and acclimatization of *in vitro* plants. In *Progress in Plant Growth Regulation*, eds. C. M. Karssen, L. C. van Loon, and D. Vreugdenhil, 809–817. Doordrecht, The Netherlands: Kluwer Acad. Publishers.

Ziv, M. 2005. Simple bioreactors for mass propagation of plants. *Plant Cell, Tiss. Org. Cult.* 81:277–285.

Ziv, M. 2010. Silicon effects on growth acclimatization and stress tolerance of bioreactor cultured *Ornithogalum dubium* plants. *Acta Hort.* 865:29–35.

Ziv, M., and J. Chen. 2007. The anatomy and morphology of tissue cultured plants. In *Plant Propagation by Tissue Culture, the Background, Vol 1, 3rd ed.*, eds. E. F. George, M. A. Hall, and G.-J. de Klerk, 465–478. Dordrecht, The Netherlands: Springer.

Ziv, M., and H. Lilien-Kipnis. 1997. Bud cluster proliferation in bioreactor cultures of *Ornithogalun dubium. Acta Hort.* 430:307–310.

Ziv, M., and H. Lilien-Kipnis. 2000. Bud regeneration from inflorescence explants for rapid propagation of geophytes *in vitro. Plant Cell Rep.* 19:845–850.

Ziv, M., and V. Naor. 2006. Flowering of geophytes *in vitro. Propagation of Ornamental Plants* 6:3–16.

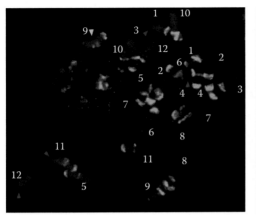

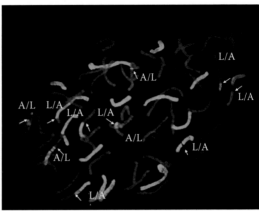

FIGURE 6.1

FIGURE 6.2a

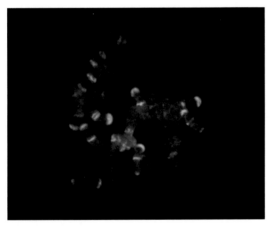

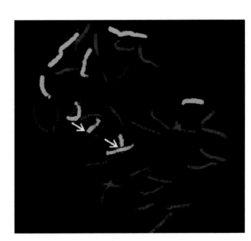

FIGURE 6.2b

FIGURE 6.3a

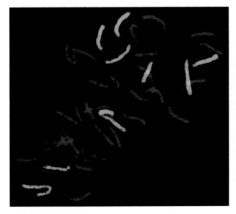

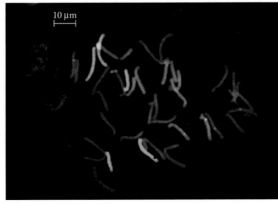

FIGURE 6.3b

FIGURE 6.4

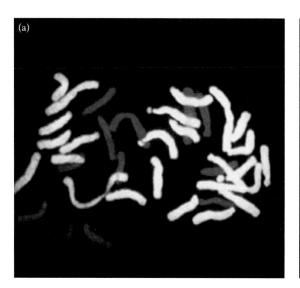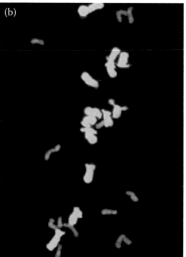

FIGURE 6.5

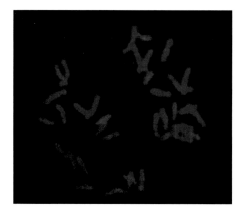

FIGURE 6.6

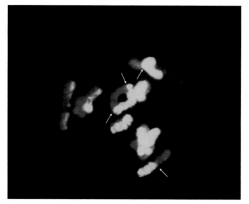

FIGURE 6.7

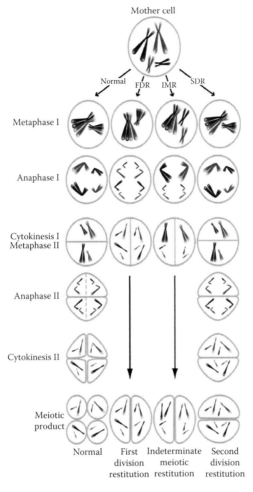

Mother cell

Normal   FDR   IMR   SDR

Metaphase I

Anaphase I

Cytokinesis I
Metaphase II

Anaphase II

Cytokinesis II

Meiotic
product

Normal   First   Indeterminate   Second
         division   meiotic   division
         restitution restitution   restitution

FIGURE 6.8

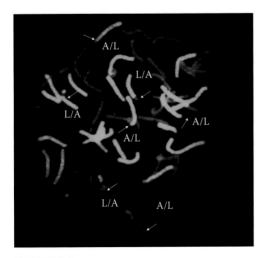

FIGURE 6.9

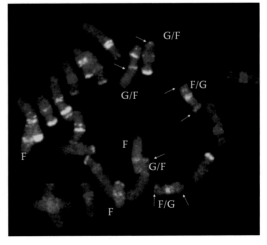

FIGURE 6.10

FIGURE 7.1

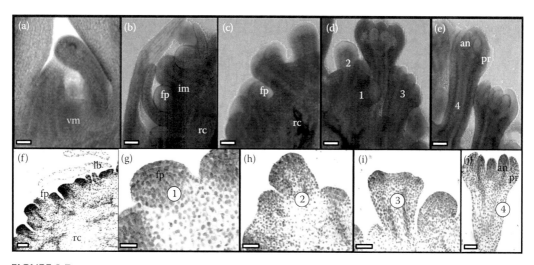

FIGURE 8.7

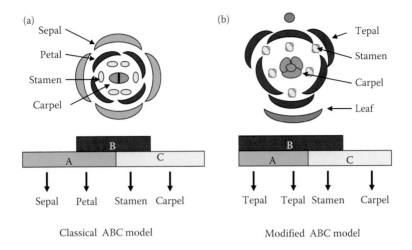

(a)
Sepal
Petal
Stamen
Carpel

B	
A	C

↓ ↓ ↓ ↓
Sepal Petal Stamen Carpel

Classical ABC model

(b)
Tepal
Stamen
Carpel
Leaf

B	
A	C

↓ ↓ ↓ ↓
Tepal Tepal Stamen Carpel

Modified ABC model

FIGURE 8.10

(a) (b)

(c) (d)

FIGURE 8.12

(a)
Whorl | 1 | 2 | 3 | 4
B
A | | C
Te | Te | St | Ca

(b)
Whorl | 1 | 2 | 3 | 4
B
A | | C
sTe | sTe | dSt | Ca

(c)
Whorl | 1 | 2 | 3 | 4
B
A | | C
Te | Te | Te | Fl*

FIGURE 8.14

FIGURE 10.1

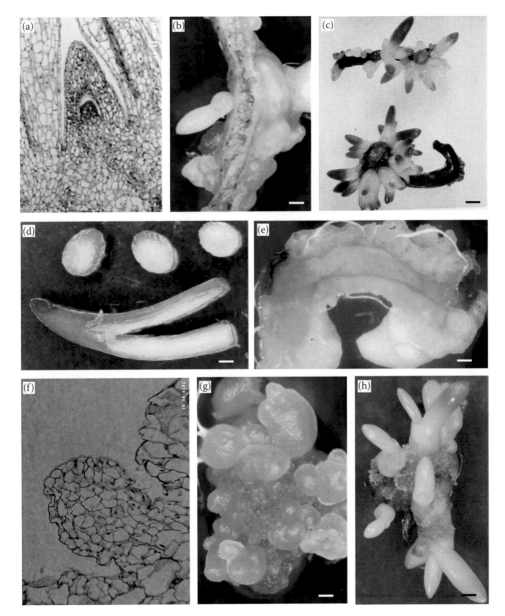

FIGURE 10.4

FIGURE 11.1

FIGURE 11.2

FIGURE 11.3

FIGURE 11.4

FIGURE 11.5

FIGURE 11.6

FIGURE 11.7

FIGURE 11.8

FIGURE 11.9

FIGURE 11.10

FIGURE 11.11

FIGURE 11.12

FIGURE 11.13

FIGURE 11.14

FIGURE 11.15

FIGURE 11.16

FIGURE 11.17

FIGURE 11.18

FIGURE 11.19

FIGURE 11.20

FIGURE 11.21

FIGURE 11.22

FIGURE 11.23

FIGURE 11.24

FIGURE 11.25

FIGURE 11.26

(a)

(b)

FIGURE 11.27

FIGURE 11.28

FIGURE 11.29

FIGURE 12.2

FIGURE 12.5

FIGURE 12.6

FIGURE 12.7

FIGURE 12.8

FIGURE 12.10

FIGURE 12.9

Control (DI water)   10 ppm CHI (24-h pulse)
Day 7   Day 7

FIGURE 12.12

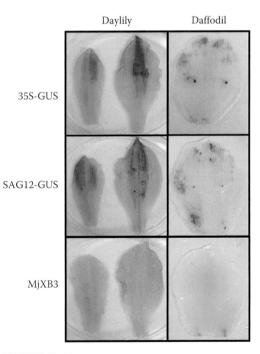

FIGURE 12.11

FIGURE 15.1

FIGURE 15.2

FIGURE 15.3

FIGURE 15.4

FIGURE 16.1

FIGURE 16.2   **FIGURE 16.3a**

FIGURE 16.3b   **FIGURE 16.4**

FIGURE 16.5

FIGURE 16.6c

FIGURE 16.6a

FIGURE 16.6b

FIGURE 16.7

FIGURE 18.1

FIGURE 18.2

FIGURE 18.3

FIGURE 18.4

FIGURE 18.5

FIGURE 18.6

FIGURE 18.7

FIGURE 18.8

FIGURE 18.9

FIGURE 18.10

FIGURE 18.11

FIGURE 18.12

FIGURE 18.13

FIGURE 18.14

FIGURE 19.1a

FIGURE 19.1b

FIGURE 19.1c

FIGURE 19.1d

FIGURE 19.2

FIGURE 19.3

11 Production Chain, Forcing Physiology, and Flower Production Systems

William B. Miller

CONTENTS

I. INTRODUCTION

Ornamental geophytes present a unique segment of horticulture because their horticultural cycle may span countries and continents, and flowering requires long-term storage that is followed by further intensive horticultural production. Unlike other commodities such as apples or bananas that are consumed after a postharvest phase, flower bulbs are again grown intensively to produce the flower or a flowering plant. Thus, the final success and quality depend on all previous conditions to which the bulb was exposed after harvest. Even when they are seemingly "dormant", bulbs are alive and metabolizing and, as De Hertogh (1974) points out, are integrating the conditions to which they are exposed and therefore essentially act as "biocomputers".

Geophytic storage organs, mainly represented in the major crops as bulbs, corms, tubers, expanded hypocotyls, rhizomes, and tuberous roots, are used in three major ways: (1) forcing by professional growers to produce cut flowers or pot/container plants, (2) providing color accent or mass ground-cover in professionally managed landscapes, and (3) used by individual gardeners in home landscapes. The progression from professional forcing to homeowner garden implies a continuum of reduced environmental control and the likelihood that bulbs will be exposed to unfavorable conditions at some point before the consumer can purchase and plant the product. This is important to be understood because the increase in planned and managed landscapes in developing countries and warmer climates implies that, ultimately, a larger share of the world's bulbs will be going into landscapes and gardens (Benschop et al. 2010). This transition will be aided by the extensive knowledge of physiology and horticulture of geophytes that has been developed for the benefit of the forcing industry.

The scientific bases of flower bulb production and forcing and the related physiology have been studied extensively and reviewed by numerous authors (Purvis 1937, 1938; Hartsema 1961; Rees 1972, 1992; De Hertogh 1974; De Hertogh and Le Nard 1993). Numerous commercial bulb production and flower forcing guides exist (Anonymous 1984; De Hertogh 1996) and many publications and resources are available online, for example, from the International Flowerbulb Centre (Anonymous 2011). A website that gives extensive information on forcing characteristics and growth regulator requirements and landscape use from a North American perspective can be found at http://www.flowerbulbs.cornell.edu/ (Miller 2011).

To some extent, the principles of successful bulb or flower production change only slowly over time. For example, the basic flowering requirements of the major bulbous crops produced in The Netherlands were worked out in Wageningen, starting in 1917 by Blaauw and coworkers; these guidelines have survived for several generations now and will continue to do so in the future. Faster to change are the horticultural systems used by bulb growers and forcers. For example, in The Netherlands, hydroponic forcing, which was a small novelty in the mid-1990s, has grown to represent ca. 75–80% of the Dutch cut tulip crop in 2011, a change that has taken only about 15 years. Also, over the last two decades, it has become commonplace to grow lily and other bulbs in the Southern Hemisphere to allow easier, higher-quality flower production for the "off-season" of the major markets in the Northern Hemisphere. Both these major technological and production shifts occurred in less than one generation. What changes even more quickly is the marketplace. Compared to the time span of horticultural knowledge, the demands of the large retail customer can change "instantaneously" and this is probably the greatest source of the instability in the industry today. Finally, the needs and wants of the ultimate "consumer" continue to be imperfectly understood. Although the consumer is well beyond the scope of this review, ultimately the entire flower bulb industry should always keep in mind the final customer.

As pointed out by Le Nard and De Hertogh (1993a), forcing of cut flowers or potted plants of bulbous species is the "system by which the potted plant or cut flower is produced from a storage organ, for example, bulb, corm, tuber". This system requires an understanding of the basic horticultural requirements of each species and cultivar, as well as an appreciation of the entire chain from bulb production to retail.

The forcing industry is based on several major concepts. The first concept is that ornamental geophytes have distinct annual growth cycles that are imposed onto the plant by temperature, rainfall, or photoperiod. These cycles lead to periods of active growth that consist of the growth of above-ground photosynthetic biomass (stems, leaves) usually followed by flowering and flower senescence, and periods of dormancy, when the plant survives unfavorable conditions as an underground geophytic organ. This natural cycle allows bulb lifting, storage under required conditions, and planting to obtain cut flowers or flowering plants.

A second concept relates to the bulb requirements for a relatively long period of active photosynthetic growth before flowering. "Spring-flowering bulbs" as a group have a minimal requirement for photosynthesis for flowering and include the major genera such as *Tulipa*, *Narcissus*, *Hyacinthus*, and *Crocus*. These genera form the vast proportion of the spring bulb industry. Crops requiring significant periods of photosynthesis for proper flowering and bulb growth are mainly summer-flowering genera such as *Allium*, *Begonia*, *Eucomis*, *Freesia*, *Gladiolus*, *Hippeastrum*, *Lilium*, *Ornithogalum*, and so on.

De Hertogh (1996) has defined five phases within the flower bulb production and utilization system as follows:

Phase 1. Bulb production. After spring or summer growth in the fields, leaf and stem senescence occurs and the bulb is ready to be lifted. After lifting, the grower has more control over the bulb's environment and can, in principle, regulate the temperature, humidity, and air flow very precisely.

Phase 2. Programming. The regulation of the bulb environment (predominantly temperature) can accelerate or retard cell division, leading to leaf, shoot, and flower formation inside of the bulb. By understanding the annual cycle of the bulb and by the application of proper temperature sequences, bulbs can be "programmed" to reliably flower earlier than normal, at the normal time, or much later than normal (referred to as delayed or retarded flowering), as needed by the marketplace.

Phase 3. Greenhouse. A properly programmed bulb (before or after planting) can be grown (forced) in a greenhouse or other structure to produce a flower or flowering plant.

Phase 4. Marketing. The product is harvested, processed as required, and then sent to the market.

Phase 5. The consumer. When all the phases have been successfully completed, there is hopefully a customer who will purchase the product and use it in her home, garden, or office.

Optimized forcing implies a remarkable degree of control over the product and is only possible with a complete understanding of the effects of temperature and other environmental factors on flower induction, initiation, and development to anthesis. This is seen in the programming and greenhouse phases of tulip forcing in The Netherlands, where the concentration of industries allows (theoretically) near-optimum handling of bulbs, operation of environmental facilities for cooling, and greenhouse forcing.

In other countries and continents, departures from the optimum are common. For example, in North America, greenhouse producers tend to plant most of the potted tulip crop in October, and hold the pots in coolers and transfer them at intervals to the greenhouse for forcing throughout the following spring. Thus, cold weeks can easily vary from ca. 14 to 22–23 weeks for forcing from Valentine's Day (February 14) to Mother's Day (the second Sunday in May). While not optimum for the crop, it is a management reality since bulbs are but one crop in the typical North American greenhouse firm, and labor and management time are needed for other crops later in the fall, for example, Christmas poinsettias. Thus, local customs and the capabilities of individual firms can impose conditions, or the need for management procedures, which do not exist or are not needed with optimum handling (e.g., the need for chemical growth regulators).

The goal of this chapter is to provide, in some detail, information on bulb and bulb-flower production systems as currently practiced in The Netherlands and North America, including bulb handling, packaging, and transport. The Netherlands is an obvious source of tradition and invention within the ornamental geophyte sector, and "Dutch fingers" are felt within most areas of bulb and bulb-flower production worldwide. Concerning North America, emphasis is placed on reviewing the selected literature on lily and tulip, published since De Hertogh and Le Nard's *The Physiology of Flower Bulbs* (1993).

II.   COMMERCIAL PRODUCTION SYSTEMS FOR ORNAMENTAL GEOPHYTES

A.   LOCATIONS

Most ornamental geophytes are produced in open fields using large-scale farming systems (Figures 11.1 and 11.2). The specific agronomic requirements vary by crop, and have been reviewed by, for example, Gould (1957), Rees (1972, 1992), Anonymous (1984), Roberts et al. (1985), and De Hertogh and Le Nard (1993). Bulb production can be, and increasingly is, successful in any location with conditions suitable for the crop of interest (see Chapter 1).

FIGURE 11.1   (**See color insert.**) A field of tulips in full flower in The Netherlands.

FIGURE 11.2   (**See color insert.**) Two Oriental hybrid lily cultivars in a field in northern Holland.

Historically, the major production center has been The Netherlands, but there have always been other countries and areas with significant production areas. Production regions develop for many reasons. One of the most common is that individuals or small groups work together to develop a local industry that grows over time, as in the bulb industry in Washington State, USA (Gould 1993) or the case of *Hippeastrum* production in South Africa. Human conflict can play a role as well, for example, the Easter lily (*Lilium longiflorum*) bulb industry in northern California and Southern Oregon got its start as a result of suspended bulb shipments from Japan after the advent of World War II. More recently, an expansion of bulb production outside The Netherlands has occurred (Benschop et al. 2010) and is driven by (1) economic and societal pressures, for example, land and labor costs and loss of bulb lands to development, (2) horticultural concerns, for example, availability of "fresh", disease-free soils, and (3) the simple realization that high-quality bulbs can be produced outside The Netherlands. A countertrend is preliminary research being done on controlled system (zero effluent) bulb production systems in Holland, which, if economically feasible, could significantly reduce the nutrient load the bulb industry puts into the environment (Gude 2011). The trend in Holland, however, has been to gradually move bulb production per se away from Holland, but with logistical, financial, and technical support originating from The Netherlands (Buschman 2005). One can conclude that the Dutch vision is to maintain as much control over this international expansion of bulb production as possible.

B. REQUIREMENTS AND PROCEDURES FOR BULB PRODUCTION

The basic requirements for bulb production are a favorable climate, quick draining, fertile soils, and the presence of knowledgeable experts in the specifics of the crop of interest. For the industry to flourish and to be competitive over time, there must be a critical mass of growers to encourage the development of the required infrastructure (availability of farming equipment, handling, sorting, cleaning and packaging machinery, agri-chemicals, etc.). As the industry grows, as quality requirements increase, and competition grows (and profit margins shrink), the sophistication of handling, information, and communication systems must necessarily increase.

Obviously, bulb farming is highly seasonal. The spring-flowering species are all planted in the fall, overwinter in the field, and flower in the spring months. Summer-flowering bulbs may be planted in the fall (e.g., *Lilium longiflorum* on the west coast of the United States), or in the spring, following controlled temperature winter storage (e.g., *Lilium* or *Dahlia* in The Netherlands). In the spring, cultivation or weed management takes place, and close inspections of crops are done to detect diseased or virus-infected plants, which are physically removed (rogued out) from the field. In The Netherlands, field inspection by the Bloembollenkeuringsdienst (BKD, or Flowerbulb Inspection Service) is important for quality certification, especially for virus infection (Knippels 2005). Finally, removal of flowers ("topping") is performed (Figure 11.3) to minimize *Botrytis* infection and promote bulb growth. Overhead irrigation may be needed, depending on the season. Usually, multiple sprays to control *Botrytis* are applied during the bulb production cycle. There has been significant research over the last 10–12 years to minimize fungicide sprays through the development of disease-forecasting systems (Van den Ende et al. 2000; see also Chapter 13 in this volume).

C. LIFTING AND INITIAL PROCESSING

In the Northern Hemisphere, most spring ornamental geophytes are harvested in June or July, while in the Southern Hemisphere, harvest occurs in December–January (Table 11.1). In all cases, aboveground stems are pulled, flailed, or mowed to remove them. Depending on the production system and soil conditions, various machines are used for lifting. On traditional sandy soil, the soil simply falls back to the ground as it is friable and breaks into clumps much smaller than the bulbs (Figure 11.4). On heavier clay soils, it is more difficult to separate soil clumps from the bulbs. Therefore,

FIGURE 11.3 **(See color insert.)** A topping or "deheading" machine, used to remove tulip flowers after field inspection.

TABLE 11.1
Production Schedules for Tulip Bulbs Grown in the Northern or Southern Hemisphere, for Northern Hemisphere Forcing

Northern Hemisphere Bulb Lifting and Programming	Southern Hemisphere Bulb Lifting and Programming
Early Forcing in the Northern Hemisphere	
Lift early June (warm climate, e.g., southern France under plastic cover)	Lift late December (Australia); January (Chile) or late January (New Zealand)
Apply 1 week 34°C to accelerate flower induction	Export to the Northern Hemisphere
Normal 17–20°C storage until G-stage	Normal 17–20°C storage until G-stage
Precooling	Plant, root at 5–9°C
Forcing for late November–early December flowering	Forcing for mid-summer (July–August) flowering
Normal Forcing in the Northern Hemisphere	
Lift late June	Lift early January (Chile); late January (New Zealand)
17–20°C storage until G-stage	Export to Northern Hemisphere
Precooling or standard forcing	Normal 17–20°C storage until G-stage
Forcing for flowering by late December–mid-January, through early May	Cool at 2°C
	Forcing for late summer (September) flowering
Very Late Forcing ("Eskimo" or "Ice" Tulips) in the Northern Hemisphere	
Lift late June	Lift early January (Chile); late January (New Zealand)
Normal 17–20°C storage until G-stage	Export to the Northern Hemisphere
Plant, root at 5–9°C until December–January	Normal 17–20°C storage until G-stage
When well rooted and after cold requirement is met, freeze at −1.5°C	Precool at 2°C
Forcing for flowering at late summer to autumn	Forcing for flowering in October–December

FIGURE 11.4   (See color insert.) Tulip lifting in late June–mid-July in The Netherlands. Note the relatively small size of the tractors and the close proximity of residential housing to the field.

several cm of soil may first be removed before lifting bulbs to minimize moving the soil to the processing area. One system involves a horizontally spinning disk with blades that fling soil off the bed. This is followed by a typical lifting machine that undercuts the bulbs and lifts them (and soil) onto a conveyor belt and ultimately to a side wagon or bulk bin (Figure 11.5).

In heavier clay soils, many bulbs are now grown in polypropylene nets that allow easier collection of the bulbs during lifting. In one system, the net is lifted via an underground blade and then laid onto the ground behind the lifting machine (Figure 11.6). A second tractor lifts the net, and passes it across a cutter to slice the net and release the bulbs. Bulbs are then conveyed to bins or a wagon in the usual manner and moved to the processing shed. The used net is wound into a roll and recycled.

FIGURE 11.5   (See color insert.) Tulip lifting on a much larger scale than in Figure 11.4. Larger tractors and large wagons are used to transport bulbs to the processing facility.

FIGURE 11.6 (**See color insert.**) "Net tulips" after lifting. The net will be lifted, slit, the bulbs swept into a box, and the net wound up and recycled.

Within the widely grown crops such as tulips and lilies, modern bulb lifting and processing methods are essentially the same throughout the world, mainly since The Netherlands is a nearly exclusive source of bulb farming equipment. Consequently, most practical issues, such as bed width and handling equipment, are standardized.

D. Washing, Peeling/Cleaning, and Sizing

If bulbs are destined for export, for phytosanitary reasons they are washed to rid them of soil (Figure 11.7). Most summer-harvested bulbs are handled in bulk bins that measure ca. $120 \times 110 \times 100$ cm, with a volume of 1.3 m^3. These bins are very sturdily made and have perforated metal or mesh bottoms that retain the bulbs but allow air to pass through. They are pressed against "air walls" or "drying

FIGURE 11.7 (**See color insert.**) Tulip bulbs after lifting and washing. These will be dried on an air wall, processed to remove roots and the old stem, peeled to remove side bulbs, sized, and then prepared for local use or export.

FIGURE 11.8 (See color insert.) An "air wall" used for the drying of flower bulbs. On the extreme right, the slots are blocked with foam rubber to direct airflow through the boxes and bulbs.

walls" that generate large volumes of air that is forced into the bins (Figure 11.8). The air wall can control the temperature, relative humidity, and air volume passed over the bulbs.

After the bulbs are dry, they are usually stored for a few days prior to "peeling" (removal of the side bulbs or offsets). Tulip bulbs usually produce one to several offsets each year, and these small bulbs are the propagation material for the future commercial crops. In earlier times, bulbs traditionally were "peeled" by hand by children of the grower or children from the village. While labor intensive, several benefits were realized, including the opportunity to visually inspect each bulb so that diseased or injured bulbs could be discarded. Since the late 1990s, however, "peeling machines" have been developed to mechanize this process. Peeling machines have a series of parallel rollers (ca. 5–6 cm diameter) that counterrotate with the effect of catching the side bulbs and rolling or pinching them off the main bulb as they pass through the rollers. A key issue is that the bulbs must be wetted before passing over the machine. While the machines work well at removing side bulbs, they do cause mechanical injury and therefore are likely to increase *Fusarium* infection. There has been scant research on this, however. Kreuk (2007) reported that traditionally hand-peeled and nonpeeled bulbs developed identical *Fusarium* incidence after incubation. If bulbs were moistened prior to hand peeling, *Fusarium* approximately doubled. Machine peeling of wet bulbs doubled *Fusarium* yet again (Table 11.2). Ultimately, machine peeling of wet bulbs increases *Fusarium*

TABLE 11.2

Change in *Fusarium* Infection in Tulip Bulb Lots as a Function of the Bulb Peeling Method

Peeling Method	Percentage of Bulbs That Became Infected with *Fusarium* after Peeling
No peeling	5
Hand peeling, dry bulbs	6
Hand peeling of "wet" bulbs	11
Machine peeling of "wet" bulbs	25

Source: Adapted from Kreuk, F. 2007. *BloembollenVisie.* 117:28–29.

incidence compared to traditional hand peeling of dry bulbs. This fact represents a major area for improvement in the mechanical handling of bulbs. After peeling, bulbs are dried and placed in ventilated storage rooms at ambient temperature.

E. STORAGE CONDITIONS, ETHYLENE, AND POTENTIAL USE OF 1-MCP IN TULIPS

After lifting and processing, tulips are stored for several weeks to months at temperatures near 17–20°C. This relatively warm temperature is needed to allow continued development and differentiation of the floral meristem within the bulb, through at least the G-stage of flower development within the bulb, which, when produced in The Netherlands, is usually complete by the 20th–30th of August.

Unfortunately, these temperatures are also highly conducive to *Fusarium* development and, consequently, to ethylene production by the fungus. In most cases, there are no commercial fungicide treatments made after lifting. Swart and Kamerbeek (1976, 1977) found that *Fusarium* isolates pathogenic to tulip produce at least 2000 times more ethylene than nonpathogenic species or forms, and that ethylene production is oxygen dependent. The rate of ethylene production by *Fusarium*-infected tulips is highly temperature dependent, and is maximal at 21°C. At this temperature, ethylene production by infected 'White Sail' tulip bulbs reached ca. 140 µL/bulb/day. This value provides the basis for industry standards for tulip ventilation requirements (de Munk 1972; de Munk and Duinveld 1986).

When tulip bulbs are exposed to ethylene in the summer or fall, injury may be nearly immediate or, alternatively, may be delayed and not visible until forcing in the spring (reviewed by Kamerbeek and de Munk 1976). There are several major ethylene-mediated disorders in tulip bulbs:

1. Gummosis. When exposed to ethylene shortly after lifting, tulip bulbs may exude polysaccharide gums, which may erupt on the outside of the bulb, fill the internal spaces between bulb scales, or be visible as "blisters" under the epidermis. Tulip cultivars vary greatly in the susceptibility to gummosis. Gummosis is seen in bulbs exposed to ethylene 2–4 weeks after lifting, but is almost nonexistent when bulbs were treated with ethylene 16 weeks after lifting (Kamerbeek et al. 1971). Therefore, the presence of gummosis in a tulip bulb shipment is a definite sign that the bulbs were exposed to ethylene after lifting, but is not a specific indication of internal damage that would inhibit normal forcing. For example, no difference was found in plant performance or flowering date in a number of 'Striped Apeldoorn' tulips with or without visible external gummosis (Table 11.3).

TABLE 11.3
Forcing of Tulips 'Striped Apeldoorn' from a Bulb Lot That Had Severe Gummosis in the Crate

Treatment	Length at Bud Color (cm)		Date of Flowering	Days in Greenhouse	Percentage Normal Plants
	Stem	Leaf			
Control	41.5	37.1	29 April	15	100
Gummosis	41.3	36.2	29 April	15	98

Source: Adapted from W. B. Miller, unpublished data.

Note: Thirty bulbs without external signs of gummosis and 180 with gummosis were forced as cut flowers. Planting and cooling began on December 3, 2001, and forcing began on April 15, 2002 (19 weeks of cold).

2. Flower bud blasting as a result of ethylene exposure during bulb storage. Kamerbeek and de Munk (1976) defined flower blasting as "phenomena in which the development of one or more floral organs is halted". Exposing bulbs to ethylene during storage can result in a range of injuries from bud necrosis to more subtle damage (e.g., reduced leaf or stem length and an increased number of side shoots). These are essentially delayed effects, as they develop as storage proceeds (even after the ethylene contamination is removed), or after planting, and are only expressed in the greenhouse during forcing. Ultimately, tulip flower blasting is expressed during forcing and may range from complete bud death (generally with petals present), anther or pistil abortion, petal deformation, and changes in leaf morphology and inhibited stem elongation (de Munk 1973). An extreme example of delayed ethylene effects is shown in Figure 11.9, where exposure of bulbs to ethylene in the fall caused 100% flower failure the following spring.

FIGURE 11.9 **(See color insert.)** An example of delayed effects of ethylene on tulip. Left, untreated control. Right, bulbs were treated with 10 µL/L ethylene for 2 weeks in the fall in October, then cooled, and forced. (a) 'Charade', an extremely sensitive cultivar. (b) 'Leen van der Mark', a cultivar that shows no obvious effect of the ethylene when forced.

3. Bud necrosis (kernrot) is a variant of bud blasting (Kamerbeek and de Munk 1976). This is a specific problem that begins with exposure of bulbs to ethylene in the presence of mites. The ethylene (mainly from *Fusarium*, but could be from any other external source) inhibits growth of primordial leaves, as compared to the bud parts, resulting in the opening of the top of the young bud inside of the bulb. At this point, the developing flower becomes accessible to mites (*Rhizoglyphus echinopus* and *Tyrophagus* sp.). Their feeding leads to a complete necrosis of the bud. When forced in the greenhouse, plants have a waxy, greasy green appearance, highly stunted stems, and an obviously necrosed stump on the stem apex (Figure 11.10). There is great variation in cultivar susceptibility to kernrot.

4. Developmental disorders caused by ethylene in the greenhouse forcing phase. Root growth can be severely inhibited by ethylene (de Munk and de Rooy 1971; Cerveny and Miller 2011), and if the greenhouse atmosphere is contaminated by ethylene, flower abortion may

(a)

(b)
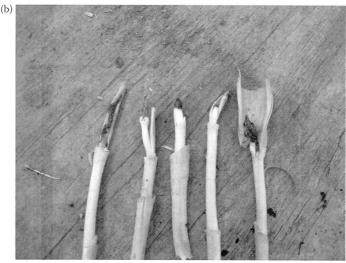

FIGURE 11.10 (**See color insert.**) Examples of "kernrot" in tulip. Greasy, waxy-green plants (a), when dissected, will exhibit a destroyed flower bud (b). Kernrot is a combination of susceptible cultivars, *Fusarium*, ethylene in storage and certain bulb mites.

FIGURE 11.11 **(See color insert.)** Damage from ethylene emanating from a *Fusarium*-infected tulip bulb, which was located in the center of the tray.

result. The influence of ethylene on tulip growth during forcing is shown in Figure 11.11, where a *Fusarium*-infected tulip bulb was planted in a forcing crate, and the resulting ethylene causes stunted growth and flower abortion in adjacent plants.

As stated above, ethylene injury in tulip bulbs can occur very quickly, as in the case of gummosis, where injury occurs 1–2 days after ethylene exposure, especially in the 2–4 weeks after lifting (Kamerbeek and de Munk 1976; de Wild et al. 2002b). Alternatively, ethylene injury may require weeks to months to manifest, as seen with flower abortion. De Munk (1973) and Liou and Miller (2011) showed that ethylene sensitivity of tulip bulbs increases as the storage progresses from July to October and later. Generally, flower bud abortion requires higher levels of ethylene (3–10 µL/L) for at least 10 days. Sensitivity to ethylene decreases as temperature falls below 20°C (de Munk 1973).

Given the significance of ethylene in tulip development, over the last decade, there has been a resurgence of ethylene research and development. In The Netherlands, emphasis has been on reengineering ventilation, with the aim of developing ways to safely reduce ventilation rates (and therefore energy costs) while maintaining a safe environment for the bulbs. Further, a reexamination of ethylene effects on tulip bulbs demonstrated that ethylene damage is a function of cultivar, time (season or developmental stage) of ethylene exposure, and ethylene concentration (de Wild et al. 2002b). Similar conclusions were reached in the United States with other cultivars and storage times (season and duration) (Liou and Miller 2011). Importantly, the work of de Wild et al. (2002b) showed that, shortly after digging, many cultivars are less sensitive to ethylene than previously thought, and that the ventilation rates needed to maintain ca. 100 nL/L (parts per billion) ethylene (as suggested by de Munk and Duineveld 1986) might be excessively high and therefore consume more energy than absolutely required to maintain bulb quality.

A second research area has been conducted at Cornell University (USA). In this work, factors affecting ethylene production during the *Fusarium*–tulip infection process have been studied. Specifically, the effect of tulip cultivar on ethylene production by *Fusarium* has been evaluated. After inoculation with *Fusarium*, an initial group of 36 tulip cultivars showed large variation in ethylene production (Miller et al. 2005). Ethylene production rates exceeding 700–800 µL/bulb/day were found in 'Furand', 'Fusor', and 'Nashville', while 'Kees Nelis', 'Pretty Woman', and 'Calgary' emitted as little as 3–5 µL/bulb/day (Miller et al. 2005). Clearly, the quantity of ethylene emitted in any lot of tulip bulbs is a function of the extent of *Fusarium* infection, and cultivar.

TABLE 11.4
Summary of Tulip Cultivar Response to Ethylene, Based on Flowering Percentage

Consistently Resistant		Consistently Susceptible	Different between Years
Blue Ribbon	Orange Princess	All Season[a]	Adamo
Bright Parrot	Pieter de Leur[a]	Angelique[a]	Agrass White
Couleur Cardinal	Pretty Woman	Annie Schilder	Fusor
Dynasty	Prominence	Apeldoorn[a]	Laura Figi
Friso	Sapporo[a]	Calgary	Louvre
Kees Nelis	Sevilla	Crème Upstar	Monte Carlo
Kikomachi[a]	The Mounties	Jan van Nes[a]	Plaisir
King's Cloak	Wirosa	Purple Flag	Strong Gold
Libretto	World's Favorite[a]		Yellow Present
Mondial	Yellow Flight		
Nashville			
Percentage of cultivars:	55%	21%	24%

Source: Adapted from S. S. Liou and W. B. Miller, unpublished.

Note: During three seasons, bulbs were exposed to 10 µL/L ethylene for 2 weeks, applied at various times from October to December at 17°C. Cultivars are indicated to be consistently resistant, consistently susceptible, or to show an inconsistent response to ethylene between years.

[a] Cultivars that were ethylene resistant or susceptible in 2 out of 3 years.

Individual tulip cultivars vary significantly in their sensitivity to applied ethylene (De Hertogh et al. 1980). Using an updated cultivar assortment, Liou and Miller (unpublished) found that many cultivars were essentially resistant to 10 µL/L ethylene given for 2 weeks during 17°C storage in October–December (Table 11.4). Three years of study demonstrated that some cultivars are consistent in susceptibility or resistance to ethylene across seasons, but for unknown reasons, many other cultivars vary a great deal in their ethylene sensitivity across years. In 2010–2011, De Waard and Miller (unpublished) reached similar conclusions in a study including 62 tulip cultivars.

Additional research on individual cultivar responses to *Fusarium* infection, or to resistance to exogenous ethylene could reveal many practical applications, including better ways to group cultivars for shipment or storage, where cultivars known to be "high ethylene producers" could be stored separately from those producing less ethylene or from cultivars that are especially ethylene sensitive. Such information should also prove useful to plant breeders in terms of developing more robust cultivars with a combination of low ethylene production when infected and a low overall sensitivity to ethylene.

A third development has been the introduction of sensors that can estimate ethylene levels in storage rooms (Hatech 2011). The Hatech "Ethyleen Analyser" has been marketed since 2006 and interfaces with storage room climate control computers and can modulate ventilation rates to maintain a specified level of ethylene. The desired result is a "safe" level of ethylene in the storage room with substantially less energy input. Taken together, tulip bulbs typically receive less ventilation in storage than in the past, with apparently good results.

The fourth and newest technology has been the use of 1-methylcyclopropene (1-MCP) as a treatment for tulips in storage. The first paper on this topic reported that pretreatment of tulips with 1-MCP gas before exposure to ethylene prevented gummosis and the respiration increase usually caused by ethylene (de Wild et al. 2002a). Subsequent work with a range of cultivars concluded that tulip bulbs can be protected from ethylene by 0.2 µL/L (200 ppb) 1-MCP for 24 h, applied at 12-day intervals (Gude and Dijkema 2005). The experiments involved a variety of assays, including evaluation of gummosis, yield of saleable bulbs, flower abortion upon forcing, and so on. Since 2009, the company AgroFresh has marketed 1-MCP to the tulip industry as a product named "FreshStart". To this point, applications have been made at ca. 12-day intervals in storage rooms or in shipping containers, before the container leaves the exporter's property. Since 1-MCP is released as a gas, the procedure requires a sealed treatment environment (storage room or shipping container). After the specified duration of MCP treatment (generally 24 h at 17–20°C), the room or container is opened, the application equipment is removed, and ventilation is reestablished. Given the 12-day "safe window" afforded by 1-MCP (Gude and Dijkema 2005), a 1-MCP treatment of tulip bulbs immediately prior to export from The Netherlands can give essentially 100% protection against ethylene in a cross-Atlantic journey, or other transport of up to ca. 12 days. For transits much longer than 12 days, for example, from The Netherlands to the West Coast of California, or to Asia, retreatment with 1-MCP would be necessary in the container. However, such a technology is not currently available.

F. Packaging/Preparation for Sale or Export

Many packing and packaging systems exist, and are highly dependent on the species (Table 11.5) and market destination (domestic vs. export). For bulbs destined for commercial forcing, extraneous packaging is neither needed nor desired. Summer-harvested bulbs are commonly exported in stackable, black plastic crates that allow excellent airflow. Two crate sizes, $60 \times 40 \times 19$ cm (the so-called "tulip crates") or a taller version (24 cm tall, "lily crate"), are used. Usually, tulips are transported in the shorter crate, with 500 bulbs of size ≥12 cm or 750 bulbs of size 11–12 cm per crate. Larger bulbs (hyacinths, daffodils) are commonly exported in the taller crates. The majority of summer-lifted bulbs need no additional packaging materials (e.g., for cushioning); in fact such materials would be undesirable from an airflow perspective. Some of the exceptions are *Muscari* and *Fritillaria meleagris* (often packed with wood shavings) and *Fritillaria imperialis* (bulbs individually wrapped in paper).

TABLE 11.5

Common Methods for Packing Bulbs for Professional Forcing

Species	Typical Packing Method	Packing Material
	Summer-Lifted	
Allium	Tulip crate	No
Crocus	Tulip crate	No
Fritillaria	Lily crate	Paper wrap
Hyacinthus	Lily crate	No
Iris (*reticulata* types)	Tulip crate	No
Muscari	Tulip crate	Wood shavings
Narcissus	Lily crate	No
Scilla	Tulip crate	No
Tulipa	Tulip crate	No
	Fall-Lifted	
Acidanthera	Lily crate	No
Begonia	Lily crate	No
Crocosmia	Lily crate	No
Dahlia	Lily crate	Plastic wrapped
Eucomis	Lily crate	No
Freesia	Lily crate	No
Gladiolus	Lily crate	No
Hippeastrum	Lily crate	No
Lilium	Lily crate	Peat moss, plastic wrapper
Ornithogalum	Lily crate	No

FIGURE 11.12   (**See color insert.**) An example of dry-sale, consumer-oriented packaging of flower bulbs.

Since export markets and relationships continue to evolve, bulbs are increasingly being shipped internationally in bulk bins (volume of ca. >1 m^3, some of which are collapsible), yielding several benefits: (1) the cost is less, (2) the materials handling logistics associated with the many export crates is eliminated, and (3) there is ease of return and reuse, especially for the collapsible boxes.

Bulbs destined for retail sales ("dry sale") generally undergo more intensive handling and packaging that is mainly oriented toward making an attractive consumer-friendly package. Such packaging must account for the basic horticultural requirements of ventilation while attempting to prompt a consumer purchase (Figure 11.12). Therefore, a large image and information on planting, flowering, time, and so on are essential. Very shortly, much of this information will migrate to the Internet and will be readily accessible by technologies such as QR codes or Microsoft tags, to make such information even more easily available.

III.   SPECIAL STORAGE SYSTEMS

A.   Ultralow Oxygen Storage at Low Temperature (Lilies)

Since the late 1970s, frozen storage (at −1°C to −2°C) has been used for longer-term storage of lily bulbs (Boontjes 1981). Frozen storage has ultimately allowed lily bulb forcing to become a year-round endeavor. Freezing is highly effective in maintaining lily bulb forcing quality, and is probably a combination of reduced respiration when frozen (as compared to 1°C storage), reduced sprout elongation, and reduced root growth and water uptake which stimulates even more sprout growth.

Commercial adoption notwithstanding, relatively few scientific reports have been published on lily freezing. Miller and Langhans (1990) stored *Lilium longiflorum* 'Nellie White' at −1°C or 4.5°C and found major changes in carbohydrates in mother and daughter scales of the bulb. After 50 days, frozen (−1°C) storage caused a ca. threefold increase in sucrose concentration, smaller but significant increases in glucose, fructose, and mannose, large increases in nonidentified oligosaccharides (probably mannose-containing oligosaccharides from glucomannan breakdown), and a 40% loss in starch. Clearly, freezing storage rapidly shifts lily bulb metabolism into a "sweetening" mode, leading to degradation of starch and glucomannan reserves and accumulation of soluble sugars. If low- (nonfrozen) temperature storage is interrupted, shoot growth can begin, with a likely rapid reduction in soluble carbohydrates. If then refrozen, these shoots are susceptible to frost (freeze) injury, resulting in "black shoot", and the bulb is useless for forcing. The concept of monitoring sugar accumulation or maintenance has been adopted into the Dutch lily industry. Through the monitoring of soluble solids (mainly sugar accumulation or maintenance) it is possible to predict when freezing may be safely employed (Gude and Kok 2005).

However, freezing storage has real limitations as illustrated by the work of Lee and Roh (2001). They forced commercially frozen bulbs (−2°C, in normal atmosphere) of Oriental hybrid lilies 'Acapulco' or 'Simplon'. As the length of storage increased, the number of flowers decreased, and the number of abnormal flowers increased, even when forced in refrigerated (16–18.5°C) greenhouses. Commercial observations also show the loss of quality and forcing ability of lily cultivars as the season progresses, and ultimately cultivars are suggested for forcing only at certain times of the year.

To improve on simple freezing storage, the Dutch lily industry, since approximately the late 1990s, has been using storage systems combining freezing temperatures with low concentrations of oxygen and/or elevated CO_2. The process has been named "ULO storage" (ultralow oxygen storage). The technical details are closely held within the industry (predominately by the bulb auction house, CNB) and have not been published. The end result, however, is the ability to store lily bulbs at −1°C to −2°C (depending on the lily group and cultivar) longer, and with less quality loss than normal freezing in peat.

B. ULTRALOW OXYGEN STORAGE AT WARM TEMPERATURE (LILIES)

In the early 2000s, Cornell University initiated a series of investigations to assess the potential of using controlled atmospheres via modified atmosphere packaging to reduce or eliminate the excessive sprouting seen in consumer "dry sale" packages at retail (Legnani et al. 2004a, b, 2006, 2010). For the consumer lily market, lilies are lifted in the fall, stored over the winter (1°C), and packaged into perforated polyethylene bags, usually with a small amount of wood shavings or peat moss. In the typical lily packaging system, the peat moss was shown to be a contributor to overall shoot elongation as it contributes moisture that stimulates sprouting (Legnani et al. 2004b). It was found that shoot growth of Asiatic hybrid lilies at warm temperatures (22–24°C) can be inhibited by reduced (1–4%) oxygen (Legnani et al. 2004a, b, 2006, 2010). Conditions that reduced or precluded plant development or flower initiation on the meristem (low oxygen levels, reduced moisture, and darkness) favored higher-quality plants after simulated storage (Legnani et al. 2004b). When bulbs were stored at 22–24°C in either atmospheric oxygen or 1% oxygen, CO_2 concentrations up to 16% had no adverse effects on plant growth, other than a slight decrease in leaf area and shoot length. While elevated (16%) CO_2 reduced shoot length when stored in atmospheric oxygen levels (21%), 16% CO_2 had no additive effect at 1% oxygen due to the much greater inhibitory effect of low oxygen. Ethylene production was increased by elevated CO_2, but had no effect on plant development.

C. STORAGE IN MODIFIED ATMOSPHERES (TULIPS)

Normally, special precooled tulip bulbs are considered to be highly perishable and rapid shipment and handling is needed to minimize injury, and the development of a reliable system to allow transportation of such bulbs without refrigeration would be beneficial. Prince et al. (1981) showed that storing precooled tulips in 3% or 5% oxygen at 17°C for 4–6 weeks reduced injury as compared with air storage or storage in 1% oxygen. Low oxygen levels also inhibited injury caused by 5 µL/L ethylene and reduced respiration. During normal dry storage, bulb respiration slowly increased from October to January when tulip bulbs were stored at 5°C or 17°C, but in January, the respiration rate at 17°C was about double the rate at 5°C (Prince et al. 1982). After January, respiration at 17°C increased rather quickly through March. Ethylene production in bulbs stored at 17°C remained low (ca. 5.5 µL/kg/day or ca. 0.2 µL/bulb/day, assuming a bulb weight of 30 g fresh weight) until mid-December, and increased abruptly through March. This is considered to be an indication of the completion of the cold requirement, but has not been tested with the use of ethylene inhibitors and other means. It is possible that the increase in ethylene could simply be due to a small and latent

Fusarium infection within one of the replicate bulbs. Ethylene levels in bulbs held at 5°C were substantially lower than those at 17°C. As cooling duration proceeded longer than 8 weeks, respiration in the two-week period *after* cold treatment increased dramatically. This was especially noticed after 12 weeks of storage, suggesting that the cold requirement had been fulfilled (Prince et al. 1982). Ultimately, the use of modified or controlled atmospheres has not been adopted for marketing precooled, ready-to-force tulips for either commercial or consumer use. Instead, precooled tulips are marketed directly from refrigerators, as in the case of tourist shops in The Netherlands.

D. ANTITRANSPIRANT COATINGS

Waxes and other edible coatings have long been used to reduce water loss in vegetables and fruit (Baldwin et al. 1995), and polyelectrolyte antitranspirant coatings can reduce *Botrytis* infection in *Lilium* leaves in lab, greenhouse, and field tests (Hsieh and Huang 1999). Coatings may contain one or more components, namely protein, lipid, carbohydrate, or resin, each with specific properties and advantages, and properties related to water retention in the product (desirable) and gas exchange (Baldwin et al. 1995), which need to be evaluated on an individual cultivar basis. For example, with apples, coatings developed for the widely grown 'Red Delicious' apple cause major postharvest problems for other cultivars and have ultimately led to a reevaluation of coating properties for use in the apple industry. Apple cultivars vary in their gas exchange properties and tolerance for anaerobiosis that can be induced with coatings especially restrictive to oxygen (Bai et al. 2002).

Antitranspirant coatings could be useful in bulbs that suffer high rates of water loss. Among the spring-flowering species, *Galanthus*, *Erythronium*, and *Fritillaria meleagris* are known to be difficult to handle, possibly due to water loss in the postharvest chain. In addition, a number of nontunicate summer-flowering ornamental geophytes could potentially benefit from such coatings (*Lilium*, *Begonia*, *Dahlia*).

Over the last 5 years, a new antitranspirant or coating product, "Liquidseal", has been marketed for lily bulbs in The Netherlands by the company P-ViAtion. The product is water soluble, and applied before bulbs are placed in cold storage. An important aspect seems to be the ability to store lily bulbs without peat moss, thereby allowing more bulbs to be packed per crate. Claimed advantages are reduced direct costs (fewer crates for a given number of bulbs and no peat moss) and, ultimately, reduced storage and shipping costs (due to reduced numbers of crates) (V. Monster 2011, pers. comm.). There is no specific information available to explain the mechanism of the material on lilies. While antitranspirant coatings could be beneficial for bulbs in the professional sector (forcing or landscape), the greatest benefit for such coatings might be in consumer packages, where environmental conditions are rarely optimum. However, there is no published research with flower bulbs in this area.

IV. BULB EXPORT: CONTAINERS AND TRANSPORTATION

The rather prolonged "dormancy" of most flower bulbs provides a predictable annual period when they can be relatively easily transported throughout the world. In fact, the history of flower bulbs is closely linked with their ease of transport, especially considering the tunicate spring-flowering species, which can tolerate long periods at room temperature. Temperature-controlled (ventilated or refrigerated) containers are a relatively recent invention, with the first refrigerated shipments occurring in the mid-1960s (Matson Navigation Company 2011). The availability of reliably controlled shipping systems has been the key to the growth in the international forcing industry and for the production of bulbs in the Southern Hemisphere.

Environmentally controlled containers allow the shipping of nearly any product around the world. As long as the bulbs are in a container and hooked to electrical power, there is no fundamental difference in sea or open road transport as regards the internal environment. However, no studies have been conducted of the effects of vibration and physical stress on bulbs.

Shipping of flower bulbs proceeds at temperature and relative humidity (ventilation) settings specified by the exporter and are dependent on the crop being shipped. Most flower bulbs are shipped in high cube containers (dimensions: 40′ long, 64 m³ volume, with internal measurements of 11.58 m length × 2.28 m width × 2.43 m height, to load line). For cargo requiring high ventilation rates (especially tulips), air exchange settings vary from 0 to 285 m³/h at 60 Hz (B. Pratt 2007. pers. comm.). Assuming that the cargo of flower bulbs occupies half of the total volume of the container, this gives approximately nine air exchanges per hour at maximum ventilation. Temperature controls are digital, and under good conditions hold within 0.5–1°C of the set point. To this point, humidity management has been mainly through ventilation. Flower bulb containers are highly engineered devices, and specific construction of the floor, the loading geometry, and clearance between the load and doors are critical for the proper functioning of environmental controls and proper airflow across the entire cargo. Modern containers can be powered by external electrical sources (at the exporter's facility, at the port waiting for transport, or on board the vessel), or by removable diesel-powered generators (over the road transport for relatively long distances).

Transportation problems do exist, however, and can be traced to both human and mechanical failure. Human failure comes from improper specification or setting of temperature or vent adjustment. In containers used for flower bulb shipping, environmental (temperature, air exchange) settings cannot be changed *en route*. Thus, all settings must be made and checked before the container is loaded on board. The bill of lading specifies settings for each container, and it is ultimately the responsibility of the shipping company to set and inspect temperature, vent, and drain settings before loading. Most exporters have developed internal checklists for shipments, including a physical inspection of the environmental settings before the container leaves the exporter's facility.

Assuming a load of tulips exported from Holland, the container is plugged into local power (at the exporter's facility) for running the onboard environmental system of the container. Once loaded with bulbs and inspected by the exporter, the container is unplugged and transported to Rotterdam where it is loaded onto a container ship. There are a fixed number of electrical slots on any container ship (in the "reefer bay"), and once physically secured, electrical connections are made. As with any mechanical device, environmental control can fail for a number of reasons within the container (motors failing, electronic controls failing), or from interrupted electrical power on board (container becoming unplugged, or a more widespread failure).

For containers destined for the east coast of North America, the journey from Rotterdam to Port Elizabeth, New Jersey, takes 8–9 days. On arrival, containers are removed by crane, placed on jitneys, and moved to locations with electrical connections and plugged in to maintain temperature. Although most flower bulbs exported to North America (and most other major export countries) have already undergone phytosanitary inspection (in The Netherlands) as part of the "preclearance program", a random agricultural inspection may or may not take place on arrival. Preclearance inspection is a vital part of the worldwide success of flower bulb transportation and logistics, as it eliminates uncontrolled and lengthy delay of the bulbs. An unscheduled inspection in the port is always a cause for concern due to disrupted temperature control during the inspection, which may require several days before the container is released. If the container is destined for a nearby location, it will proceed without further temperature control, or a generator ("genset") can be attached to provide electrical power to the container for longer journeys (if specified on the bill of lading).

Most problems during bulb transportation are related to failure of the temperature controls and can range from catastrophic, to subtle, to nonexistent. For example, a failure leading to tulips being exposed to 40°C without ventilation for 5 or 6 days will result in a total loss of the material, and freezing of the cargo is a similar loss. The difficulties come when bulbs experience out-of-specified conditions, such as several degrees of temperature change, or improper ventilation settings. Because of this, nearly all shipments are made with at least two dataloggers to provide registration and documentation of temperature conditions. The containers themselves also have environmental recording capability although, in practice, this information is extremely difficult to obtain from the shipping companies.

V. FORCING SYSTEMS FOR ORNAMENTAL GEOPHYTES

A. OVERVIEW OF RECENT ADVANCES IN FORCING SYSTEMS

In this section, key concepts for forcing systems that may be applied across genera will be presented, and perspectives on the current status of forcing of tulips and lilies in North America will be discussed, especially focusing on the period since the early 1990s. For a detailed review of the subject before the early 1990s, the reader is referred to De Hertogh and Le Nard (1993). Specific information on forcing the major crops can be found in Chapter 5 of this volume.

In recent years, there has been a rapid evolution in commercial forcing techniques and technology. These innovations have originated mainly from The Netherlands and have been driven by the need to increase efficiency and reduce cost inputs per stem or flower produced, all while maintaining or improving quality. In many cases, advances in flower bulb forcing have been made by adopting existing greenhouse mechanization technology. For example, over the last 10 years, the "Dutch Container System" for potted plant production greenhouses, which allows efficient use of greenhouse space (Figure 11.13), has been adopted in many bulb forcing facilities worldwide. The width of the benches was reduced to 60 cm to accommodate the length of plastic forcing trays and to facilitate harvesting. The forcing industry has also adopted computerized greenhouse environmental control systems and irrigation systems, all with the objective of reducing energy or input cost and maximizing quality. Development of these new technologies have allowed the largest companies in The Netherlands to force more than 75 million tulip stems per year, a quantity unheard of less than 10 years ago.

A secondary driver has been to develop more "sustainable" systems that have a lower environmental impact. Initially, elimination of solid substrates (i.e., peat moss) led to the development of hydroponic tulip systems (reviewed below). A more recent concern centers on carbon and the effects that the net output of horticultural carbon dioxide may have on global climate. At present, research is under way to evaluate energy-saving light-emitting diodes (LED) as a light source for bulb flower production (especially lilies). Studies on wavelength blend, mole quantity, duration, and time of application during crop life are being conducted. Demonstration greenhouses using multilayer technology where crops move from nearly 100% LED to full sunlight are being considered and built (Gude 2011).

FIGURE 11.13   (See color insert.) An example of "Dutch Containers" adapted for forcing tulips in black plastic crates. Transportation systems such as these allow automated movement of plants from the greenhouse to work areas for planting, harvesting, and so on. Cut lilies and other tall crops can also be grown with this system as stem supports are easily incorporated.

Over the last two decades, lily forcing has progressed from a majority of stems being grown directly in the ground within greenhouses, to the nearly universal use of black plastic crates ("bulb crates"). The tulip industry in Holland has further matured to a majority of the cut tulip crop being grown hydroponically. Although it is difficult to acquire exact statistics on hydroponic tulip production, most estimates are that in 2010–2011, 75–80% of the Dutch cut tulip crop is being produced hydroponically. To date, a commercial hydroponic system has not been introduced for cut lilies or other crops.

B. Forcing Systems for Lilies

Among bulbous crops, *Lilium* represents the most diverse genus that is forced for commercial production. Across the many species and groups, lilies are forced as pot plants (especially in North America and Europe, but increasingly in other locations as well), cut flowers (worldwide, for local use or export), and as patio-ready growing plants for summer and fall. Cut flower production is, by far, the main use worldwide, and cut flower breeding occupies the majority of the Dutch breeding effort. The variation within the genus is well known, and the genetics work of J. van Tuyl and coworkers in The Netherlands and the subsequent cultivar breeding efforts with cooperating commercial companies have led to a literal explosion of groups and cultivars in the last 20–25 years (see Chapter 6 for a detailed discussion). It is therefore likely that groups that essentially did not exist 15 years ago will become the dominant groups in the near future. Already, LA hybrids have eclipsed the acreage of Asiatic hybrids in Holland, and the OT hybrids are predicted to become the major "Oriental type" lily within the next 5–8 years (Peterse 2010; Van Tuyl and Arens 2011).

As mentioned above, lily forcing has evolved over the last ca. 20 years to make extensive use of the "lily export crate", the heavy black plastic crates ($60 \times 40 \times 24$ cm), as a mobile unit. Depending on the cultivar, season and bulb size, 8–15 bulbs are planted in the substrate in the crate, and placed in a "rooting room" for 1–3 weeks to "preroot" the bulbs. This preforcing period allows initial elongation of roots and shoots. The exact temperature regime and timing are not established, but temperatures of 5–9°C are commonly used, and crates with plants must be removed before shoots grow into the crate above. In particular with Oriental hybrids, which tend to be slow in rooting relative to the onset of shoot growth, this technique can improve stem strength and gives slightly taller plants at flowering. This procedure also reduces the greenhouse forcing time by ca. 1–2 weeks for Oriental cultivars. Crates offer convenient and uniform units that can be moved from the cooler to the greenhouse, and placed in the greenhouse, using robots designed for this purpose (Figure 11.14). To this point, most crate-grown lilies are ultimately grown on the greenhouse floor or ground, although the "Dutch container system" is increasing in use.

1. North American Perspectives

In North America, lilies are produced extensively as pot plants and as cut flowers (Miller 1992). Production statistics are available for Easter lilies (ca. 7.5–8 million pots forced per year in North America, Table 11.6), but no comparable statistics are available for hybrid lilies in pots. In 2010, the U.S. production of cut lilies totalled 94 million stems, with a wholesale value of US$61 million, with ca. 85% of this production in California (USDA-NASS 2011). To put U.S. production into perspective, this is 28% of the number of stems and 45% of the value of the 2009 Dutch crop (Anonymous 2009).

While the biology and production techniques of Easter lily (*Lilium longiflorum*) were reviewed earlier (Miller 1992, 1993), its unique position in the market, and the heavy level of research dedicated to it, led to some consideration here. Since the 1950s, the Easter lily has been forced extensively as a pot plant for the Easter holiday. The Easter lily industry in North American is unique in that there is an extremely narrow cultivar base. In the 1950s–1960s, 'Croft' was prevalent, in the 1960s–1970s, 'Ace' was predominant, and since the mid-1980s, 'Nellie White' has been essentially the only cultivar in the trade. Because the date of Easter changes yearly and the cultivar range is narrow, there has been an enormous research base built to understand vernalization responses, light, photoperiod, and temperature effects on growth rate, flowering time, and height (reviewed by Miller

FIGURE 11.14 (**See color insert.**) A robot for moving lily crates. This machine is capable of placing crates at the start of forcing, and then picking them up at the end. Here, it has a full load and is returning to the main aisle to drop them off. Note the moveable pipes that provide the "rails".

TABLE 11.6
Trends in Easter Lily (*Lilium longiflorum*) Pot Plant Production in the United States (2005–2010)

Year	Number of Growers	Total Number of Pots	Number of Pots per Grower
2005	374	6,040,000	16,200
2006	346	6,264,000	18,100
2007	317	6,597,000	20,800
2008	315	5,692,000	18,000
2009	321	6,138,000	19,100
2010	292	6,243,000	21,400

Note: These statistics do not include Canadian production, which is probably 20–25% of these numbers.

1992, 1993). There has also been a trend of fewer numbers of growers forcing Easter lilies, with each growing more units per year (Table 11.6). The major technologies routinely used in Easter lily forcing are graphical tracking (a technique to follow crop height in real time; see, for example, Fisher and Heins 2002) to integrate height and plant development and to determine optimum greenhouse temperature settings. Manipulation of day and night temperature (DIF) is a major component of this technique. Other approaches are the use of growth-retarding chemicals to restrict height growth and the use of specific gibberellins to reduce leaf senescence and promote flower longevity. These will be reviewed later.

2. Height Control of Container-Grown Lilies

The oldest height management technique is the traditional recommendation to grow lilies in maximum light (assuming winter to spring culture), with "adequate" spacing so as to avoid mutual shading and elongation (Miller 1992). Since the late 1980s, we have developed a much better understanding of the relationship of day and night temperature, and how this controls plant height. Erwin et al. (1989) described DIF as the difference between day and night temperature, where a positive DIF means day temperatures are greater than night, a zero DIF means equal day and night temperatures, and a negative DIF means that nights are warmer than days. Plants grown under positive DIF are taller than zero DIF plants, which in turn are taller than plants grown under negative DIF. Assuming a 12 h day and 12 h night, the 24 h daily average temperature is the average of the day and night temperatures. For unequal day lengths, one must multiply the day temperature by the day length in hours, add to the same product for the night, and divide by 24 h. By using a higher average daily temperature, plants will develop more quickly. By manipulating relative day and night temperatures, plant height can be controlled while maintaining a specified development rate. As mentioned above, tools such as graphical tracking (Moe and Heins 1990; Fisher and Heins 2002) incorporate these tools into easy-to-use packages.

Since 1993, another major advance in Easter lily forcing was the discovery of a "cold shock" effect, where the final stem length can be controlled by overhead irrigation with cold water. Blom et al. (2004) found that the temperature of water used in overhead irrigation has a profound effect on *L. longiflorum* growth. Within the range of 2–15°C, plant height decreased with a linear response of 1.75 cm/°C, when irrigated from emergence to flowering. When cold water was applied to the soil surface, there was no growth effect, demonstrating that the effect is specific to the shoot apical meristem.

It was also found that excluding the end-of-day far-red burst by covering plants with black cloth about 1 h before sunset reduced the final plant height of lilies (Blom et al. 1995). For greenhouses with automatic black cloth systems, this is a simple, effective, and environmentally friendly way to control lily height. Both the cold water and end-of-day black cloth techniques are currently used in commercial greenhouses.

Although the above techniques are available for use, they are not readily applicable to all situations (e.g., due to greenhouse structural limitations, making it difficult to install black cloth). As a result, North American greenhouse growers rely to some degree on chemical growth retardants for lily height control. There are many other reasons for this, among them: (1) the availability of PGR[*] that are not available elsewhere in the world; (2) an industry paradigm that emphasizes crop diversity, rather than monocultures of a single species or cultivar throughout the entire year, so other crops compete for the attention of the grower; (3) a market that, for potted plants, is highly holiday dependent and not oriented toward weekly or daily production; (4) demands of the customer; and (5) the practical fact that more 55-cm-tall plants can fit into a delivery truck than 65-cm-tall plants. At present, there is no widespread or serious concern in North America about imminent regulatory loss of PGRs on ornamental plants.

The major PGRs currently being used on lilies (ancymidol, fluprimidol, paclobutrazol, and uniconazole) are all inhibitors of gibberellin (GA) biosynthesis. Their mode of action is to reduce endogenous production of GA at different stages of biosynthesis (Rademacher 2000). As a result, plants exhibit less stem growth. Given that individual PGRs inhibit different locations of the GA pathway and that genera and species can be assumed to have slightly different pathways and/or responses to individual GAs, it is necessary to conduct empirical studies on a range of cultivars, using different concentrations and methods of application, to establish the biological and economic efficacy of a new PGR.

Lily PGR use has progressed, since the mid-1960s, from Phosphon-D through A-Rest (ancymidol), which was used as a 25–30 mg/L foliar spray or 0.25–0.5 mg/pot soil drench, to Sumagic (uniconazole), which is very effective as a low-concentration (2–5 mg/L) foliar spray (Miller 1992, 1993). This has been quite a progress in terms of usage, as the Phosphon-D dose was 2–3 grams of product per pot! Early side effects of Phosphon-D were occasionally weak and limber stems and a tendency to cause yellow leaves, problems that can occur with anti-GA PGRs to this day. Bonzi (paclobutrazol) has not proven to be a reliable product on Easter lily. Most recently, Topflor (fluprimidol) has demonstrated promise for Easter lily growth regulation, mainly as a soil drench, where it is extremely effective in the 0.25–1 mg/L range (approximately 0.03–0.12 mg/pot). It is

---

[*] Plant Growth Regulators.

not economically useful for height control as a spray, as relatively high concentrations are needed (Miller, unpublished data). Flurprimidol, paclobutrazol, and uniconazole are highly effective as preplant bulb dips for most lily species and cultivars (Krug et al. 2005a). As with many aspects of floriculture, cultivars vary a great deal in their response to PGRs, and local trials to ascertain the "optimum" rate in any greenhouse, environment, and production system will always be required. An example of cultivar differences in the effectiveness of flurprimidol, paclobutrazol, and uniconazole preplant bulb dips is shown in Figure 11.15. There is much information available online, however, and finding the starting points is much easier than it was in the past. A website by Miller (2011) (http://www.flowerbulbs.cornell.edu/forcing/pot_lilies.htm) provides an extensive listing of cultivars, data, and photographs of responses to preplant bulb dips with these chemicals.

(a)

(b)

FIGURE 11.15   (**See color insert.**) Cultivar differences in effectiveness of PGR treatments. (a) 'Acapulco', an Oriental hybrid. (b) 'Springfield', an Asiatic hybrid. Both cultivars grew from 16/18 cm bulbs in the same greenhouse. Treatments are (left to right): control, paclobutrazol at 50, 100, 200, or 300 mg/L, uniconazole at 2.5, 5, 7.5, or 10 mg/L, and flurprimidol at 10 or 50 mg/L, given as a 1 min, preplant bulb dip.

3. Reducing Leaf Chlorosis and Maximizing Flower Life

Another major advance since the early 1990s has been the widespread adoption of GA_{4+7} sprays to eliminate lower leaf senescence (yellowing) and increase flower life. This technique developed due to the need to store Easter lilies at low temperatures near the end of forcing, before flowers open. At first consideration, the need to cold store lilies suggests that they will flower before the required market date, which implies failed timing. In reality, however, there are two common reasons why this is done. The first is weather. Depending on the Easter date (which can vary by a full month) and the location of the greenhouse, very warm weather can occur in the last several weeks before Easter, making greenhouse temperature control (and therefore control of plant development and anthesis date) nearly impossible. The second reason is intentional: growers may wish to remove the lilies from the greenhouse to allow other spring crops to fill the space. The difficulty is that many lily cultivars are susceptible to rapid and dramatic lower leaf senescence (sometimes of the entire plant) within a few days of removal from cold storage. Therefore, North American market and industry circumstances have imposed the need to reliably cold-store lilies for up to 2 weeks after forcing and before sale.

Lilies have two major leaf senescence or yellowing syndromes, the rapid, postharvest disorder described above and a gradual or "greenhouse lower leaf yellowing" (Miller et al. 1995). These syndromes can be seen in all major classes of forced lilies. The gradual type occurs over many weeks during forcing and is broadly associated with suboptimal nitrogen nutrition, dense plant spacing, weakened roots, negative DIF temperature regimes, and PGR use (Miller et al. 1995; Han 2000).

There is a large body of scientific literature on rapid, postharvest leaf chlorosis. Maxie and Hasek (1974) suggested that 'Georgia' lilies could be cold-stored 2 weeks at the preflowering stage with no loss of quality. Staby and Erwin (1977), with 'Ace' and 'Nellie White', showed a loss of quality (presumably by leaf yellowing) as cold storage at 0.5°C or 6°C in boxes as storage increased to 2 weeks. Tsujita et al. (1979) found that higher doses of the antigibberellin growth regulator, ancymidol, increased leaf yellowing, as did low phosphorus levels (which were generally recommended since the common P source, super phosphate fertilizer, contains fluoride, which can cause fluoride toxicity) (Marousky and Woltz 1977). Ancymidol caused more leaf yellowing when supplied as a soil drench than as a foliar spray. Prince et al. (1987) and Prince and Cunningham (1989) found increased leaf yellowing and reduced individual flower life with increasing the duration of 2°C cold storage. Silver thiosulfate (STS) application before cold storage improved flower life, but had no effect on leaf yellowing. Phenidone, an inhibitor of lipoxygenase, had no effect on either leaf senescence or flower life (Prince et al. 1987). Ranwala et al. (2000) confirmed the cold storage and ancymidol effect, and demonstrated that postproduction cold storage greatly increased postharvest ("catastrophic") leaf yellowing.

In excised leaf experiments, GA_3 application at relatively high rates (ca. 250–1000 mg/L) was effective in reducing leaf senescence and benzyladenine had a similar effect, but at even lower rates (above 50 mg/L). Both growth regulators reduced respiration of excised leaves by 33–50% (Han 1995; Franco and Han 1997) and the results showed that in Easter lily, leaf senescence does not proceed with ethylene synthesis, nor did STS delay onset of yellowing (Franco and Han 1997). Excised leaves had a low respiration rate while in cold storage, but respiration quadrupled in the 10–14 days after removal from cold storage to 20°C. Treating leaves with 500 mg/L BA or GA_3 prevented this respiratory rise while delaying senescence, suggesting that preservation of leaf carbohydrate may be key in delaying senescence in lily. Treatment with chloramphenicol or streptomycin (inhibitors of chloroplast protein synthesis) did not delay yellowing.

However, in intact plants that were subjected to both growth regulators during greenhouse production and cold storage after production, GA_3 treatments were not effective in reducing postharvest leaf yellowing, although there was a positive effect of GA_3 in plants only treated with ancymidol, but not given cold storage (Ranwala et al. 2000).

In the late 1990s, a number of studies were confirming that GA_{4+7} was more effective than GA_3. At the time, the main form of GA_{4+7} available to agriculture was Promalin, a commercial product containing a 1:1 mixture of GA_{4+7} and the cytokinin benzyladenine (BA). It was shown that Promalin could completely prevent postharvest yellowing (Funnell and Heins 1998; Ranwala and Miller 1998; Ranwala et al. 2000; Whitman et al. 2001). Excellent effects were obtained with 25–50 mg/L foliar sprays of GA_{4+7} and BA. Using cold-stored 'Star Gazer' Oriental hybrid lilies, the GA_{4+7} was demonstrated to be the main active component (Ranwala and Miller 1998). Here, the senescence delaying effect of GA_{4+7} alone was shown to be equal to the GA_{4+7} + BA mixture. At the same concentration, BA gave no protection. However, at substantially higher concentrations (ca. 250–500 mg/L), foliar sprays of BA, alone, or as Accel (a mixture of GA_{4+7} and BA, as Promalin, only in a 1:10 GA_{4+7}:BA ratio), are also effective against leaf yellowing.

In terms of mechanism, catastrophic leaf yellowing is caused by rapid chlorophyll degradation in plants stressed by conditions that reduce the leaf carbohydrate level. One such stressor is negative DIF. Plants grown in negative DIF regimes have dramatically reduced leaf carbohydrate levels, especially in the lower leaves (Miller et al. 1993). Another stressor is cold storage. Thus, in 'Star Gazer' plants, leaf-soluble carbohydrates decreased ca. 50% during 14 days storage at 4°C (Ranwala and Miller 2000). On transfer to a warm postharvest room, leaf sugar levels were maintained in GA_{4+7}-treated leaves, but decreased in control plants, likely due to elevated respiration as shown for excised leaves by Han (1995) and Franco and Han (1997). Another piece of evidence for the importance of leaf carbohydrate is that low light levels during cold storage caused a significant increase in leaf sugar (on intact plants) during a 2-week cold storage, and such plants maintained chlorophyll after harvest (Ranwala and Miller 2000). In addition, both light- and GA_{4+7}-treated plants maintained higher levels of soluble proteins, and had less membrane degradation and increased levels of the antioxidative enzymes catalase and superoxide dismutase (Ranwala and Miller 2000). In summary, rapid leaf senescence is caused by a rapid increase in metabolic activity in lily plants with depleted reserves, when transferred from cold storage to a warm environment, and is essentially an example of chilling injury.

Gibberellins confer an additional benefit of improved flower life in Easter lilies (Kelly and Schlamp 1964; Ranwala and Miller 2000) and a wide range of Oriental, Asiatic, and LA hybrid cultivars (Ranwala and Miller 1998, 2002, 2005; Han 2001). The effects of GA_{4+7} are nothing short of spectacular, and the U.S. lily industry very quickly adopted the technique (Figure 11.16). While specific statistics are not available, it is likely that >90–95% of all Easter lilies in North America are treated with GA_{4+7}. In the mid-2000s, Valent USA released Fascination, a product specifically registered for legal use on lilies for improving postharvest quality. It is identical in active ingredients to Promalin (Table 11.7).

A number of practical problems emerged from the use of Promalin, as the GA_{4+7} can also cause unwanted stem elongation. It was shown that gibberellins do not move from leaf to leaf, and therefore uneven sprays caused irregularly green and yellow plants (Han 1997). Therefore, throughout the late 1990s, growers were encouraged to spray plants completely to get full coverage. As a result, there were examples of very tall crops, which were demonstrated to be the result of root uptake of gibberellin that dripped into soil from high-volume sprays (Ranwala et al. 2003). Interestingly, application of Fascination to Oriental hybrids is much easier owing to the fact that their stem elongation finishes relatively early in the crop, as compared with the continuing elongation of Easter, LA, or Asiatic hybrid lilies (Ranwala et al. 2002). In this case, growers can spray the entire plant within 2 weeks of harvest and obtain plants that can tolerate cold storage, have long-lasting flowers that open, and have reduced leaf yellowing without danger of stretch.

Research in postharvest physiology in lilies is complicated by the lack of uniformity of procedures, objectives, and genetic material (see also Chapter 12). Researchers have worked with excised buds and leaves (Han 1995; Franco and Han 1997; Han and Miller 2003), cut stems (Nowak and Mynett 1985; Elgar et al. 1999; Ranwala and Miller 2002), and whole plants (Ranwala and Miller 1998, 2000; Elgar et al. 1999; Ranwala et al. 2000; Celikel et al. 2002). All these studies were complicated by the use of many different cultivars and cultivar groups, and the presence or absence of cold storage. There is a need for more research in this area and for the unification of experimental procedures among researchers. Van Doorn and Han (2011) reviewed the literature on lily postharvest physiology and have reached a similar conclusion.

FIGURE 11.16   (See color insert.) Effect of cold storage and gibberellin$_{4+7}$ and benzyladenine (Fascination) treatment on catastrophic leaf yellowing on 'Ceb. Dazzle' LA hybrid lily. Plants were forced to bud stage, and then held in a dark 3°C cooler for 2 weeks. Photo taken after 1 week of 20°C postharvest evaluation. (a) Control plants (cold-stored, but no other treatment). (b) Cold-stored plants that were sprayed with 100 mg/L GA$_{4+7}$ (the main active ingredient from Fascination) before cold storage.

4. Upper Leaf Necrosis

A long-standing problem with hybrid lilies is "leaf scorch", an all-encompassing term that has also been applied to fluoride toxicity and other problems. In the early 2000s, a series of investigations demonstrated that this problem, specifically upper leaf necrosis (ULN, Figure 11.17), is caused by leaf calcium deficiency (Chang and Miller 2003, 2004, 2005). Only the upper several leaves and leaves associated with the flower buds are susceptible. In 'Star Gazer', ULN symptoms were observed ca. 35 days after planting, or 28–20 days after emergence in the greenhouse, when plants were ca. 30–35 cm tall (Chang and Miller 2005). Necrosed leaves had sixfold less calcium than healthy leaves averaging only 0.1% dry weight, while the critical Ca^{2+} level for healthy leaves was ca. 0.5–0.6% (Chang and Miller 2005). Cultivars vary greatly in susceptibility to ULN (Chang et al. 2008).

TABLE 11.7

Active Ingredients, Chemical Name, and Trade Names of Plant Growth Regulators Mentioned in the Text

Active Ingredient	Chemical Name	Trade Name(s)
Ancymidol	α-Cyclopropyl-α-(p-methoxyphenyl)-5-pyrimidinemethanol	A-Rest (also Quel, EL-531), Abide
Benzyladenine	N-(Phenylmethyl)-1H-purine-6-amine	Configure, BAP-10
Chlormequat chloride	(2-Chloroethyl) trimethylammonium chloride	Cycocel, Citadel
Ethephon	(2-Chloroethyl) phosphonic acid	Florel, Ethrel
Flurprimidol	α-(1-Methylethyl)-α-[4-trifluoromethoxy]phenyl]-5-pyrimidinemethanol	Topflor
GA$_{4+7}$ and benzyladenine (1:1 concentration ratio)	N-(Phenylmethyl)-1H-purine-6-amine plus gibberellins A$_4$A$_7$	Fascination, Fresco, Promalin
GA$_{4+7}$ and benzyladenine (1:10 concentration ratio)	N-(Phenylmethyl)-1H-purine-6-amine plus gibberellins A$_4$A$_7$	Accel
GA$_{4+7}$	Gibberellins A$_4$A$_7$	Provide
Paclobutrazol	(±)-(R*,R*)-β-[(4-Chlorophenyl)methyl]-α-(1,1-dimethylethyl)-1H-1,2,4-triazole-1-ethanol	Bonzi, Piccolo, Paczol
Uniconazole	(E)(S)-1(4-Chlorophenyl)4,4-dimethyl-2(1,2,4-triazol-1-yl)pent-1-ene-3-ol	Sumagic, Concise

During the course of growth in the greenhouse, the primordial leaves show decreasing calcium concentration (on a weight basis) until the moment they expand, allowing more calcium to enter through transpiration (Chang and Miller 2003). Normal bulbs are typically low in calcium (0.03–0.04% by dry weight). If plants were grown in a low or calcium-free nutrient regime, and the resulting bulbs were harvested, the bulbs were extremely low in calcium, and when replanted, they exhibited calcium deficiency symptoms in the middle and lower leaves (Chang and Miller 2003). The importance of transpiration of young leaves for calcium uptake was demonstrated by manually unfolding leaves away from the shoot tip. Manual leaf unfolding increased transpiration and greatly

FIGURE 11.17 (**See color insert.**) Severe symptoms of upper leaf necrosis (ULN), a calcium-deficiency disorder on 'Sunny Sulawesi' hybrid lily growing in pots. Note that the buds are unaffected.

reduced ULN incidence. Transpiration from young leaves is one of the main reasons why large bulbs are more susceptible than smaller bulbs, since their higher bud and leaf numbers lead to a situation of the youngest leaves being unable to adequately transpire due to boundary layer and other effects. Ethephon was used in this research to "chemically unfold" leaves (epinasty), as an effective way to reduce ULN and also provide a good degree of height control. Vertical air movement is also effective in reducing ULN (Chang and Miller 2004).

The light environment per se is not a causal factor in ULN development, in contrast to anecdotal industry lore that ULN is "caused" by bright days following a period of cloudy weather (Chang and Miller 2005). Foliar sprays of calcium were only effective if made daily for 14 days; five sprays at 3-day intervals were not effective in preventing ULN (Chang et al. 2004). Preplant dips into calcium solutions were also not effective. Given that plants have many leaves that pass through periods of maximal sensitivity as they grow, it is clear that any cultural method would need to be made on a frequent basis, or cause a condition (e.g., the use of ethephon to unfold leaves to increase transpiration) that is long-lasting in the 30–40 days after planting when the plants are susceptible to the injury.

C. Forcing Systems of Tulips

The basic physiology, biology, and horticulture of tulip forcing have been extensively reviewed (Le Nard and De Hertogh 1993b) and numerous production guides are available online (e.g., Anonymous 2011). Tulips are produced around the world for cut flowers and as potted plants that are used for indoor or patio decoration. A large number of bulbs are used for landscaping use by professionals and homeowners, but exact statistics on the fate of Dutch-exported bulbs (whether for forcing or landscape use) are not available.

Southern Hemisphere production of tulip bulbs now allows forcing of tulips year-round (Table 11.1). Southern-Hemisphere-grown bulbs are easily forced for the late summer to late autumn period in the Northern Hemisphere, a time when tulips were traditionally difficult to force via the "eskimo" or "ice tulip" method.

1. Traditional Forcing Systems in "Soil" or Substrate

There are two major systems for forcing tulips in "soil" or substrate. The traditional method is to plant bulbs in the fall. After planting, all or most of the required cold period to the planted bulbs is provided in a rooting room. This system is called "standard forcing" or "9°C forcing", since the initial temperature the bulbs are cooled after planting is 9°C. Temperatures are lowered as cooling proceeds to inhibit excessive rooting and to minimize stem growth during the rooting stage. Excessive rooting is not needed in tulip forcing, and may exacerbate certain diseases such as *Trichoderma*. Since most rooting rooms are very tightly packed with bulbs, it is important to keep leaf tips from growing into crates or layers above; this is mainly regulated by temperature. Standard forcing can be used for both pot and cut flower crops.

A second method for tulip forcing is "5°C forcing", where the cold is given to dry, unplanted bulbs, initially at 5°C, but potentially reduced to 2°C after 1 November. After cooling, bulbs are planted (traditionally, into the ground soil of the greenhouse which should be at a temperature of 10°C or cooler), and grown cool (10–13°C). For this method, removal of the tunic at the base of the bulb is important to allow rapid rooting. Cool temperatures during the greenhouse phase are important to balance root and shoot growth. Planting into warm soils can cause many problems if *Fusarium* is present, due to ethylene moving within the soil and causing flower abortion in adjacent plants (Figure 11.11).

Of course, hybrids of the above two strategies exist. One technique, which can yield high-quality plants, is to plant tulip bulbs into crates and at the end of the cold period, and move them into cold greenhouses, directly onto the ground. The plants root into the soil and, in combination with low growing temperatures, produce an exceptionally high crop. Dutch Iris can be forced with crates in a similar manner (Figure 11.18).

FIGURE 11.18 (**See color insert.**) Cut iris production in Ontario, Canada. Bulbs are planted into crates and then placed on the ground where they root in. Excellent crop quality.

2. Hydroponic Forcing of Tulip and Other Cut Flowers

Hydroponic forcing of cut flowers is a relatively new technique. The earliest reports on hydroponic tulip growth demonstrated a requirement for N and Ca in hydroponic production on a gravel sub-strate (De Hertogh et al. 1978), and showed calcium nitrate to be a better Ca source than the chloride (Klougart 1980) or sulfate forms (Nelson and Niedziela 1998a).

Commercial progress in hydroponic forcing has been very rapid, and as mentioned earlier, 75–80% of the Dutch cut flower crop is forced hydroponically as of 2011. For comparison, 30–35% of the crop was hydroponic in 2002 (Miller 2002). A key to successful hydroponic forcing was the realization that a "water-forced" tulip does not require the massive root system traditionally seen on pot or soil-grown cut crops. Early trials in hydroponic tulips, where the bulbs received the entire

FIGURE 11.19 (**See color insert.**) Hydroponic tulips planted into trays and in the initial rooting phase in a 9°C cooler.

cold treatment while "on water", failed due to excessive root growth, which led to anaerobic conditions and slimy roots. Hydroponic forcing has evolved whereby bulbs are given ca. 80% of their cold as dry bulbs (12–13 weeks), and then planted into the hydroponic tray filled with a dilute solution of calcium nitrate or calcium chloride (EC of ca. 1.2–1.5) (Figure 11.19). Bulbs are returned to the cold, where rooting proceeds for another 2–4 weeks, and then the bulbs are forced in a greenhouse in a normal manner (Figures 11.20 and 11.21). The time required for rooting decreases as the season progresses, and may only be 1–2 weeks late in the forcing season.

Some advantages of hydroponic tulips are: (1) a reduced requirement for cold storage facilities, as most of the cooling is done to dry bulbs that occupy less volume, (2) forcing per se is a few days faster per crop (Figure 11.22), and (3) harvesting is much cleaner, faster, and easier compared with soil culture. This is an obvious advantage for exported material where soil is forbidden.

FIGURE 11.20　(**See color insert.**) A view early in a hydroponic cut tulip crop in a Canadian greenhouse.

FIGURE 11.21　(**See color insert.**) Mechanized, moving table hydroponic tulip operation in Canada. Benches move automatically toward the worker, where they are harvested and placed on the moving belt, where they are immediately bunched and packaged.

FIGURE 11.22 (**See color insert.**) Nonformal comparison of hydroponic tulip production (left) versus traditional soil-based culture (right). This is illustrative of the 1–3 days faster forcing of the hydroponic crop.

There are several disadvantages of hydroponic tulip forcing, which include the following: (1) not all cultivars are suited for hydroponic forcing (part of this is due to difficulties in holding dry bulbs for very late plantings, but also due to cultivar susceptibility to stem topple in hydroponics) (Nelson and Niedziela 1998b), (2) very-high-quality and disease-free bulbs are needed for forcing, (3) the requirement to level and uniform benches and/or greenhouse transportation systems, (4) a need for specialized equipment such as a specific "pin tray" that holds the bulbs, (5) a need for exceptional cleanliness (automated tray and crate washers are required equipment), and (6) a slightly lowered quality of hydroponic tulips compared to soil-grown tulips. Part of the last point can be addressed by using lower temperatures in the greenhouse.

The technology for cut tulip hydroponic forcing is relatively straightforward. Automation has been developed to mechanize bulb planting. Trays with open bottom ("honey combs") have been developed as alternatives to the "pin tray". These trays, however, have a significant disadvantage, in that different trays are needed for different bulb sizes. There seems to be a transition to flowing solution hydroponic systems as this provides better solution aeration. As mentioned above, nutrition needs are mainly confined to calcium. Boron deficiency has been reported (Miller 2008), but boron addition is suggested to be not routinely required for hydroponic forcing (Nelson and Niedziela 1998a). Another problem is brown roots that are related to tannins from the tunic leaching into the forcing liquid. Current work in The Netherlands is focusing on the development of a multilayered forcing technology that, using LED technology, would take advantage of the low light requirement for tulip (Gude 2011).

There are few inherent biological limitations to producing any bulbous flower hydroponically. An experimental system based on floating hydroponics for cut iris from the early 2000s (Figure 11.23) was promising, but has apparently not been adopted commercially. Any successful system must accommodate the specific growing habit of the plant (e.g., stem roots in lilies, or the need for plant support for taller crops such as lilies or iris) and, to maximize greenhouse efficiency, allow unitization and robotic handling of plants at several stages throughout production.

D. INTEGRATED HEIGHT CONTROL OF SPRING BULB CROPS

The main spring crops (tulip, narcissus, and hyacinth) are readily grown as cut flowers. In these crops, adequate stem length for cut flower production is obtained by ensuring a sufficiently long cold treatment before forcing. Longer cold duration is associated with longer stems, hence the importance

FIGURE 11.23 (**See color insert.**) Experimental hydroponic production of Dutch iris using a float approach.

of "minimum cold-week" requirements of individual cultivars. When interpreting cultivar-specific cold-week data generated in The Netherlands, it is important to consider prevailing forcing conditions (usually, short days, cool plant temperatures, and very low light). Thus, for areas of higher light and warmer temperatures, it is almost always necessary to increase cold duration by 1–3 weeks to obtain adequately long stems.

Longer tulip stems can also be obtained by forcing under reduced light, growing cooler (at the expense of longer forcing time), or by covering the crop with light-excluding materials for several days at the start of forcing (Okubo and Uemoto 1984). A similar temporary shading will also increase stem length in narcissus and hyacinth. In the case of cut hyacinths, an important aspect for obtaining adequate stem length is maintaining 9°C for the entire cold period, to allow adequate stem growth (in storage) prior to forcing.

For pot plant production, the goal is to have a product within the "aesthetic ratio" (ratio of plant height to pot diameter of 2.6:1, Sachs et al. 1976). Or, even more likely in a market dominated by large and powerful customers, the goal is to grow a plant that meets the customer's demands. Hanks and Menhenett (1979) listed other requirements for a pot tulip, namely uniform size, no blind shoots, reasonably synchrony of flowering date, plants erect, robust, compact, and free of excessive post-flowering internode extension. One common specification for pot plants is plant height; therefore, "growth regulation" is an important issue in pot plant production. Indeed, previous authors have speculated that better regulation and control of tulip height by PGRs could be expected to rejuvenate local markets for pot tulips (Moe 1980).

There are several ways to control the spring bulb crop height for pot production. The simplest, and ultimately, most environmentally friendly, is to grow "short" (dwarf) cultivars. When the requirements for the pot plant market are recognized and as breeders respond to these needs, a greater selection of dwarf cultivars will indeed be available. Another way is to give appropriate cold duration. This is the basis of the recommended maximum cold-weeks presented by De Hertogh (1996), since the longer the cold treatment, the taller the plant will be. This is especially the case in North America, where cooling durations are routinely excessive relative to "optimum". Thus, an early-flowering cultivar with a relatively low number of cold-weeks may not be suitable for later in the season, when the bulbs likely would have many more cold-weeks. Ultimately, unlike many other floriculture crops that are highly responsive to photoperiod of day/night temperature alterations (DIF), environmental manipulations offer only small possibilities for reducing tulip height (Erwin et al. 1989). During forcing, the photoperiod has either no (Dosser and Larson 1981) or minimal (Hanks and Rees 1979, 1980) effect

TABLE 11.8

Chronological Summary of Major Reports on Chemical Plant Growth Regulation in Tulip

Chemical	Trade Name	Method of Application	Range of Concentration (Spray) or Dose (mg Active Ingredient/Pot)	Comments	Reference
Ethephon	Ethrel, Florel	In vase solution to cut flowers	1–1000 mg/L	Effective in reducing elongation in cut flowers at concentrations below 100 mg/L with few adverse effects. At higher concentrations, abscission and/or early wilting was seen. 48 mg/L ethephon was highly effective in reducing internode elongation.	Nichols (1970); Nichols and Kofranek (1982)
Ancymidol	A-Rest	In vase solution to cut flowers	3 or 25 mg/L	Ancymidol reduced top internode elongation by 31%; no effect on water uptake of flower senescence.	Einert (1971)
2,4-Dichlorobenzyl-tributylphosphonium chloride (CBBP)	n/a	In vase solution to cut flowers	6–125 mg/L	No effect at 25 mg/L or less. At 125 mg/L, CBBP caused petal browning and early flower death but no ill-effects on the leaves.	Einert (1971)
Ancymidol	EL-531 (Quel, ultimately, A-Rest)	Media drenches, foliar sprays	Drenches, up to 2 mg/pot; sprays up to 200 mg/L	Good response to drenches; sprays only effective at very high rates (>ca. 50 mg/L). Some flower delay with drenches.	Farnham and Hasek (1972)
Ancymidol	EL-531 (Quel, Reducymol, ultimately, A-Rest)	Media drenches	0.5 mg/pot	Ancymidol application at day 6 of forcing was as effective as on day 1; applications on day 17 had no effect. Ancymidol effect was reversed by GA$_{4+7}$, but not by GA$_3$. Ancymidol reduced the length of all internodes.	Shoub and De Hertogh (1974)
Ancymidol	A-Rest	Media drenches, 1–2 days of greenhouse stage	0.3–2.5 mg/pot	Generally excellent height control as rates increase. Minor adverse effects sometimes noted.	Hanks and Menhenett (1979, 1983)
Dikegulac-sodium	Atrinal	Media drenches, 1–2 days in the greenhouse	20–1000 mg/pot	Some reductions of stem length and petal length, depending on cultivar	Hanks and Menhenett (1979)

Ethephon	Ethrel	Media drenches, 1–2 days in the greenhouse	2.4–7.2 mg/pot	Relatively small effect on height; some flower blasting at higher rates	Hanks and Menhenett (1979)
Chlorphonium chloride	Phosphon	Media drenches, 1–2 days in the greenhouse	250–500 mg/pot	Ineffective at the concentrations used	Hanks and Menhenett (1979)
Dikegulac-sodium	Atrinal	Media drenches, 1–2 days of the greenhouse stage	20–1000 mg/pot	Effective for height control at higher concentrations	Hanks and Menhenett (1979)
Piproctanal bromide	Alden	Media drenches, 1–2 days in the greenhouse	250–2000 mg/pot	Ineffective at the concentrations used	Hanks and Menhenett (1979)
2,3-Dihydro-5,6-diphenyl-1,4-oxathiin	Experimental compound (UBI-P293)	Media drenches, 1–2 days in the greenhouse	100–1000 mg/pot	Excellent height control as a drench; good dose response through 1000 mg/pot	Hanks and Menhenett (1979, 1983)
Phenyl tetrazole compound	Experimental compound (PP528)	Media drenches, 1–2 days in the greenhouse	50–1000 mg/pot	While effective in height control, flower bud blasting, adverse changes in flower color, and shortened petals also occurred	Hanks and Menhenett (1979)
Morphactins	Experimental compounds (CME-74050 P, CME 10901 P, EMD 7301 W, IT 3233, IT 3456)	Media drenches, 1–2 days in the greenhouse	5–500 mg/pot	At high concentrations, most were toxic and caused flower bud blasting; at lower concentrations most reduced the first internode length, and some alterations of flower shape	Hanks and Menhenett (1979)
Ethephon	Ethrel, Florel	Media drench at ca. 10 cm shoot length	250–1000 mg/L, 50 mL drenches (12.5–50 mg/pot)	Generally excellent height control across cultivars. No ill-effects except for some bud blasting in one cultivar. Stems more responsive than leaves	Moe (1980)
Mepiquat chloride	Experimental compound (BAS 083 00W)	Media drenches, after 1 day in the greenhouse	1000–2500 mg/pot	No effect on height or flowering; some increased flower abortion in one cultivar	Hanks and Menhenett (1983)
Quaternary sulfonium carbamate compound	Experimental compound (BTS 44 584)	Media drenches, 1 day in the greenhouse	1000–2500 mg/pot	No effects on flowering or stem growth at the rates tested	Hanks and Menhenett (1983)

continued

TABLE 11.8 (continued)
Chronological Summary of Major Reports on Chemical Plant Growth Regulation in Tulip

Chemical	Trade Name	Method of Application	Range of Concentration (Spray) or Dose (mg Active Ingredient/Pot)	Comments	Reference
N-Carbamoylimidazole compound	Experimental compound (BTS 34 723)	Media drenches, 1 day in the greenhouse	1–50 mg/pot	Effective, but not commercially useful	Hanks and Menhenett (1983)
Paclobutrazol	Bonzi	Preplant bulb soaks, media drenches when shoots 5 cm	0–1 mg/pot drenches; 0–10 mg/L bulb soaks for 60 min	Good dose response to drench concentrations for scape length, 1–2 day delay in flowering. For soaks, 10 mg/L gave excessive control; better rates were 5–7.5 mg/L.	McDaniel (1990)
Ancymidol	A-Rest	Media drenches in the greenhouse	0.125–0.5 mg/pot	Extensive listing of cultivars, showing how the optimum rate tends to increase as the season (and cold duration) increases	De Hertogh (1996)
Paclobutrazol	Bonzi	Media drenches in the greenhouse	2–3 times the required A-Rest rate	From the A-Rest requirement, Bonzi rates can be estimated for a number of cultivars.	De Hertogh (1996)
Flurprimidol	Topflor	Bulb soaks, foliar sprays, media drenches in the greenhouse	0.5–1 mg/pot drench, ca. 50 mg/L preplant soaks	Excellent results as a preplant dip or media drench. Ineffective as a foliar spray	Krug et al. (2005b)
Uniconazole	Sumagic	Preplant bulb soaks	5–40 mg/L at 10 minute dips	Sumagic more effective than Bonzi	Miller (2011)

on tulip height. For pot plants, providing maximum light intensity will tend to keep plants shorter. Chemical plant growth regulators (PGRs) are another important tool for height control of potted tulips.

The history of growth regulation in tulips is quite long, and the majority of research has come from Great Britain and North America. De Hertogh and Le Nard (1993) presented a comprehensive overview of growth substances in tulips, especially in terms of physiology of stem elongation, dormancy release, and so on. In the present section, more emphasis will be placed on current and potential future use of commercially available PGRs. A listing of active ingredients, chemical names, and North American tradenames of materials used at present or experimentally in the past is given in Table 11.7, and a summary of the main findings in Table 11.8.

The main products used on tulips in North America have been ancymidol (through the mid-1990s), paclobutrazol (mid-1990s to present), and fluprimidol (mid-2000s to present). The shift away from ancymidol was based mainly on cost, as paclobutrazol and flurprimidol are less expensive on an activity basis. Tulip height control has been most successful as media drenches after plants are moved into the greenhouse. To do this, a consistent volume of material should be applied per pot, usually 60 mL per 10 cm (diameter) pot, 120 mL per 15 cm (diameter) pot, and so on. For drenches, it is important to wet the entire root zone, and thus it is recommended to irrigate plants within 24 h before application. Generally, plants are drenched within a few days of coming into the greenhouse. Later in the season, this is even more important due to the more rapid growth rate. An extensive listing of PGR drench requirements (ancymidol, paclobutrazol, and flurprimidol) for more than 150 tulip cultivars is available at the website of the Flower Bulb Research Program at Cornell University (http://www.flowerbulbs.cornell.edu/).

Unlike many other floricultural crops, tulips have rarely responded to foliar sprays of the major PGRs. This is probably due to the relative lack of foliar surface area on the plant at the time sprays would need to occur (i.e., within a few days of housing the plants). Because of labor costs, the North American forcing industry would much rather apply PGRs as foliar sprays than as media drenches. Recent research (W. B. Miller, unpublished) has suggested that it may be possible to spray appropriate PGRs onto the soil surface and follow this with a normal overhead irrigation to wash the PGR into the root zone.

Preplant bulb dips with paclobutrazol, uniconazole, or fluprimidol are a biologically and economically effective method for tulip height control (Miller 2002; Krug et al. 2005b). A range of cultivars can be effectively controlled by preplant dips, which can be surprisingly effective (Figure 11.24).

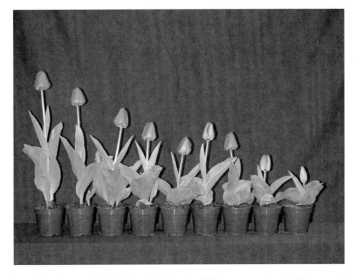

FIGURE 11.24 (**See color insert.**) 'Apeldoorn' tulips in 4″ diameter pots. Bulbs were dipped for 10 min in PGRs, planted, cooled for 16 weeks, and forced at 17°C. Left to right: the control (water), 50, 100, 200, 400 mg/L paclobutrazol, 5, 10, 20, 40 mg/L uniconazole.

Tulip bulbs can be dipped in paclobutrazol and held at 17°C at least 6 weeks before the onset of cooling with identical results to dipping immediately before cooling (W. B. Miller, unpublished). Thus, a potential exists to develop "pregrowth regulated" tulips that could be sold to customers, thereby eliminating the need for growth regulation at the forcer level.

In North America, the vast majority of daffodils and hyacinths are forced as pot plants. Since the last major review of these crops (Hanks 1993; Nowak and Rudnicki 1993), the main changes in these crops, relative to pot plant forcing, have been in more sophisticated and effective PGR usage. Historically, height control in these crops was based on foliar sprays of ethephon (Florel, Ethrel), a product that releases ethylene at cytoplasmic pH. Extensive data are available on optimal ethephon spray concentrations, by cultivar and forcing period, for a number of narcissus and hyacinth cultivars (De Hertogh 1996).

Over time, however, ethephon sprays have not always been effective on several crops (W. B. Miller, pers. observ.; Krug et al. 2006b). The possible reasons for this could be water pH or alkalinity, stage of plant development when sprayed, cultivar, or other factors. This has set the stage for evaluation of other materials, namely paclobutrazol and, more recently, flurprimidol. A key goal has been to reduce excessive stem elongation in the postharvest (consumer) phase. Krug et al. (2005c) studied 'Anna Marie' hyacinths and found that preplant bulb soaks in flurprimidol at >25 mg/L or paclobutrazol at concentrations >100 mg/L are effective in the inhibition of leaf and stem elongation. A website by Miller (http://www.flowerbulbs.cornell.edu/forcing/pot_hyacinth.

FIGURE 11.25 **(See color insert.)** Effect of 10-min preplant dips in 40 mg/L flurprimidol on hyacinth 'Blue Jacket' with 18 cold-weeks. Left to right: the control, 40 mg/L Topflor dips for 10, 30, or 60 min. (a) First flower. (b) Near end of flowering of the control.

FIGURE 11.26   (See color insert.) Effect of ethephon media drenches on the growth of narcissus 'Carlton' with 18 cold-weeks. Plants in 15 cm pots were drenched with 120 mL of ethephon at concentrations of (left to right) 0, 100, 250, and 500 mg/L.

htm) provides flurprimidiol dip recommendations (concentration and dip length) for a number of hyacinth cultivars (Figure 11.25). Media drenches with flurprimidol are somewhat less effective in controlling postharvest stretch (Krug et al. 2005c, and the website listed above). At present, flurprimidol dips prior to planting is the most effective method for controlling postharvest stem growth in hyacinth. However, a concern for dipping hyacinth bulbs is the potential spread of bacterial disease.

Krug et al. (2006b) performed similar studies on three narcissus cultivars, and found that flurprimidol soaks in the 35–40 mg/L range and paclobutrazol drenched in the 1.5–2 mg/pot range were effective for height control. A listing of the rates for flurprimidol dip and drench for a number of narcissus cultivars can be found at http://www.flowerbulbs.cornell.edu/forcing/pot_narcissus. htm. Media drenches with ethephon have yielded excellent results, with no adverse effects on postharvest life (Miller unpublished, Figure 11.26).

Ranwala et al. (2005) and Krug et al. (2006a) have shown that paclobutrazol and flurprimidol dip solutions maintain efficacy after repeated bulb dips of lily or hyacinth. Using narcissus and hyacinth as test subjects, it was recently found that flurprimidol solutions maintain efficacy over a number of weeks (W. B. Miller, unpublished). Flurprimidol solutions were used to dip tulips, narcissus, and hyacinths, and were then held in darkness for 4 or 7 weeks. Bulbs were then dipped into these aged solutions or into a freshly made solution. Results indicated that narcissus and hyacinth cultivars dipped into these solutions were all of the same height, showing that the solutions were stable over a period of at least 7 weeks (Figure 11.27).

A number of other issues remain to be solved in narcissus and hyacinth production, issues that sometimes plague other spring-forced bulbs, as well. One is "lifting", where the vigorous rooting from the bulb pushes the bulb completely out of the pot (Figure 11.28). In Europe, this problem is solved by placing ca. 6-cm-thick form rubber pads on the bulbs at the time of planting. The crate above presses down on this pad so that while rooting, the bulbs are kept pressed into the soil. These foam pads are removed (which means unstacking crates after the bulbs are well rooted). The result is a product that features the bulb (visible above the soil line) (Figure 11.29). This system is difficult, however, without specific automation and substantial investment, so it is not appropriate for many forcers where hyacinths are not a major crop.

FIGURE 11.27 (**See color insert.**) Effect of the flurprimidol dip solution age on growth of 'Pink Surprise' hyacinth (a) and 'Exception' narcissus (b). Left to right: untreated, 10 min dip into 7 week old, 4 week old, or freshly prepared 30 mg/L flurprimidol solution.

VI. CONCLUSIONS AND OUTLOOK

Geophytes represent a diverse assortment of genera, species, and cultivars, and their morphological structures and physiological traits allow relatively easy storage and transporting, facilitating international trade and enjoyment as cut flowers, pot plants, or landscape subjects in myriad environments. The successful utilization of these products is dependent on the initial production of a high-quality propagation material, safe and timely transportation of the bulbs to the place of use, and then, proper handling by the customer, whether it be a retail outlet, a professional forcer, a home gardener, or a landscaper. Modern forcing systems vary from highly sophisticated to more primitive and there are many that are successful. Local industry tradition or market chain mandates have a large effect on how bulbs are forced and how the quality of the product is defined.

For the future, it is clear that bulb production can keep increasing in those regions where it is profitable to do so. There is sufficient technical and professional information available for this to occur. The

FIGURE 11.28 **(See color insert.)** Hyacinth lifting out of the pot due to rapid and vigorous rooting. This is a persistent problem for North American forcing.

FIGURE 11.29 **(See color insert.)** Potted hyacinths ready for the European market. These were initially rooted with a soft foam pad placed on top of the bulbs, held down with another tray on top. This allows bulbs to root without lifting, and yields a final product that features the bulb itself.

main backbone of transportation is available at a reasonable cost. There is adequate greenhouse space available in many countries to force more and more flower bulbs as cut flowers or pot plants.

A different question is whether consumption will match this production. A great problem facing the worldwide flower industry is insufficient consumer demand relative to the ability to grow and transport flowers. The same holds true for geophytes and geophyte flowers. In areas where bulb plants are marketed with the bulb invisible (planted deep in the soil), perhaps a presentation featuring the bulb (Figure 11.29) and the correct promotional or educational materials could stimulate sales.

Over time, bulbs and bulb products will become more "green" and sustainable. They will last longer and present a better value to the consumer. But will these efforts in production, forcing, and research be rewarded with increased consumption? Surely, effective promotional programs that stimulate interest and curiosity about ornamental geophytes will help stimulate consumption in mature markets and are needed for the overall health of the industry.

ACKNOWLEDGMENTS

The author is grateful to the excellent students and associates who have conducted much of the work reviewed in this chapter. Special thanks are extended to Anthos, Hillegom, The Netherlands, for financial and logistical support since 1998. Other organizations providing funding that have supported students and specific aspects of this work, including Agrofresh, The American Floral Endowment, FineAmericas, Floralife, The Fred C. Gloeckner Foundation, The North American Flowerbulb Wholesalers' Association, The Pearlstein Family Foundation, SePro, The USDA Floral and Nursery Research Initiative, Valent USA, and a number of other companies, are all gratefully acknowledged.

CHAPTER CONTRIBUTOR

William B. Miller is a professor of horticulture at Cornell University (Ithaca, New York, USA). He grew up in a commercial greenhouse family that also farmed *Lilium longiflorum* bulbs in northern California. His BS is from the University of California, Davis, and his MS and PhD are from Cornell University. Dr. Miller has held faculty positions at the University of Arizona, Clemson University, and Cornell University. Since 1998, he has directed the Flower Bulb Research Program at Cornell University. His major topics of research are flower bulb physiology, greenhouse forcing, and postharvest handling. Dr. Miller has published more than 120 refereed journal articles, books or chapters, and conference papers. Dr. Miller and his PhD student, Yao-Chien (Alex) Chang, were recipients of the 2004 and 2005 Ken Post award from the American Society for Horticultural Science. From 2004–2012, Dr. Miller served as Chairman of the ISHS Working Group on Flower Bulbs and Herbaceous Perennials.

REFERENCES

Anonymous. 1984. *Bulb and Corm Production. Reference Book 62.* London: Her Majesty's Stationery Office.

Anonymous. 2009. Statiestiek CD VBN. Snijblomen Aanvoer-en prijsinformatie. FloraHolland. Naaldwijk, The Netherlands.

Anonymous. 2011. http://www.prod.bulbsonline.org (accessed February 20, 2011).

Bai, J., E. A. Baldwin, and R. H. Hagenmaier. 2002. Alternatives to shellac coatings provide comparable gloss, internal gas modification, and quality for 'Delicious' apple fruit. *HortScience* 37:559–563.

Baldwin, E. A., M. O. Nisperos-Carriedo, and R. A. Baker. 1995. Edible coatings for lightly processed fruits and vegetables. *HortScience* 30:35–38.

Benschop, M., R. Kamenetsky, M. Le Nard, H. Okubo, and A. De Hertogh. 2010. The global flower bulb industry: Production, utilization, research. *Hort. Rev.* 36:1–115.

Blom, T., D. Kerec, W. Brown, and D. Kristie. 2004. Irrigation method and temperature of water affect height of potted Easter lilies. *HortScience* 39:71–74.

Blom, T. J., M. J. Tsujita, and G. L. Roberts. 1995. Far-red at end of day and reduced irradiance affect plant height of Easter and Asiatic hybrid lilies. *HortScience* 30:1009–1012.

Boontjes, J. 1981. Can 'Rubrum' lilies also be frozen? *Bloembollencultuur* 92:364–365.

Buschman, J. C. M. 2005. Globalisation–flower–flower bulbs–bulb flowers. *Acta Hort.* 673:27–33.

Celikel, F. G., L. L. Dodge, and M. S. Reid. 2002. Efficacy of 1-MCP (1-methylcyclopropene) and Promalin for extending the post-harvest life of Oriental lilies (*Lilium* × 'Mona Lisa' and 'Stargazer'). *Scientia Hortic.* 93:149–155.

Chang, Y.-C., J. P. Albano, and W. B. Miller. 2008. Oriental hybrid lily cultivars vary in susceptibility to upper leaf necrosis. *Acta Hort.* 766:433–440.

Chang, Y.-C., K. Grace-Martin, and W. B. Miller. 2004. Efficacy of exogenous calcium applications for reducing upper leaf necrosis in *Lilium* 'Star Gazer'. *HortScience* 39:272–275.

Chang, Y.-C., and W. B. Miller. 2003. Growth and calcium partitioning in *Lilium* 'Star Gazer' in relation to leaf calcium deficiency. *J. Am. Soc. Hort. Sci.* 128:788–796.

Chang, Y.-C., and W. B. Miller. 2004. The relationship between leaf enclosure, transpiration, and upper leaf necrosis on *Lilium* 'Star Gazer'. *J. Am. Soc. Hort. Sci.* 129:128–133.

Chang, Y.-C., and W. B. Miller. 2005. The development of upper leaf necrosis in *Lilium* 'Star Gazer'. *J. Am. Soc. Hort. Sci.* 130:759–766.

De Hertogh, A. 1974. Principles for forcing tulips, hyacinths, daffodils, Easter lilies and Dutch irises. *Scientia Hortic.* 2:313–355.

De Hertogh, A. 1996. *Holland Bulb Forcer's Guide 5th Ed.* Hillegom, The Netherlands: The International Flower Bulb Centre and The Dutch Bulb Exporters Association.

De Hertogh, A. A., N. Blakeley, and J. Barrett. 1978. Fertilization of special precooled (5°C) tulips for cut-flower forcing. *Scientia Hortic.* 9:167–174.

De Hertogh, A. A., D. R. Diley, and N. Blakely. 1980. Response variation of tulip cultivars to exogenous ethylene. *Acta Hort.* 109:205–210.

De Hertogh, A., and M. Le Nard. 1993. *The Physiology of Flower Bulbs.* Amsterdam: Elsevier.

De Munk, W. J. 1972. Bud necrosis, a storage disease of tulips. III. The influence of ethylene and mites. *Neth. J. Plant Pathol.* 78:168–178.

De Munk, W. J. 1973. Flower-bud blasting in tulips caused by ethylene. *Neth. J. Plant Pathol.* 79:41–53.

De Munk, W. J., and T. L. J. Duineveld. 1986. Toleranties voor klimaatsfactoren in bewaarruimten. *Bloembollencultuur* 29:41–53.

De Wild, H. P. J., H. Gude, and H. W. Peppelenbos. 2002a. Carbon dioxide and ethylene interactions in tulip bulbs. *Physiol. Plant.* 114:320–326.

De Wild, H. P. J., H. W. Peppelenbos, M. H. G. E. Dijkema, and H. Gude. 2002b. Defining safe ethylene levels for long term storage of tulip bulbs. *Acta Hort.* 570:171–175.

Dosser, A. L., and R. A. Larson. 1981. Influence of various growth chamber environmental on growth, flowering and senescence of *Tulipa gesneriana* L. cv. Paul Richter. *J. Am. Soc. Hort. Sci.* 106:247–250.

Einert, A. E. 1971. Reduction in last internode elongation of cut tulips by growth retardants. *HortScience* 6:459–460.

Elgar, H. J., A. B. Woolf, and R. L. Bieleski. 1999. Ethylene production by three lily species and their response to ethylene exposure. *Postharvest Biol. Technol.* 16:257–267.

Erwin, J. E., R. D. Heins, and M. G. Karlsson. 1989. Thermomorphogenesis in *Lilium longiflorum. Am. J. Bot.* 76:47–52.

Farnham, D. S., and R. F. Hasek. 1972. Tulip height control with EL-531 – a progress report. *Flor. Rev.* 150(3889):22–23, 55–58.

Fisher, P. R., and R. D. Heins. 2002. *UNH FloraTrack for Poinsettia.* Durham NH: Univ. New Hampshire Cooperative Extension.

Franco, R. E., and S. S. Han. 1997. Respiratory changes associated with growth-regulator-delayed leaf yellowing in Easter lily. *J. Am. Soc. Hort. Sci.* 122:117–121.

Funnell, K. A., and R. D. Heins. 1998. Plant growth regulators reduce postproduction leaf yellowing of potted Asiflorum lilies. *HortScience* 33:1036–1037.

Gould, C. J. 1957. *Handbook on Bulb Growing and Forcing.* Mt. Vernon, Washington: Pacific Northwest Bulb Growers Association.

Gould, C. J. 1993. *History of the Flower Bulb Industry in Washington State.* Mt. Vernon, Washington: Northwest Bulb Growers Association.

Gude, H. 2011. Recent developments in the sustainable production of flower bulbs in The Netherlands. Lilytopia presentation, 23 May, 2011. Longwood Gardens, Kennett Square PA.

Gude, H., and M. Dijkema. 2005. The use of 1-MCP as an inhibitor of ethylene action in tulip bulbs under laboratory and practical conditions. *Acta Hort.* 673:243–247.

Gude, H., and J. Kok. 2005. Problem of black shoots in oriental lilies closer to a solution. Research newsletter no. 8. http://www.flowerbulbs.cornell.edu/newsletter/index.htm (accessed February 21, 2011).

Han, S. S. 1995. Growth regulators delay foliar chlorosis of Easter lily leaves. *J. Am. Soc. Hort. Sci.* 120:254–258.

Han, S. S. 1997. Preventing postproduction leaf yellowing in Easter lily. *J. Am. Soc. Hort. Sci.* 122:869–872.

Han, S. S. 2000. Growth regulators reduce leaf yellowing in Easter lilies caused by close spacing and root rot. *HortScience* 35:657–660.

Han, S. S. 2001. Benzyladenine and gibberellins improve postharvest quality of cut Asiatic and Oriental lilies. *HortScience* 36:741–745.

Han, S. S., and J. A. Miller. 2003. Role of ethylene in postharvest quality of cut oriental lily 'Stargazer'. *Plant Growth Regul.* 40:213–222.

Hanks, G. R. 1993. *Narcissus*. In *The Physiology of Flower Bulbs*, eds. A. De Hertogh, and M. Le Nard, 463–558. Amsterdam: Elsevier.

Hanks, G. R., and R. Menhenett. 1979. Response of potted tulips to new and established growth-retarding chemicals. *Scientia Hortic.* 10:237–254.

Hanks, G. R., and R. Menhenett. 1983. Responses of potted tulips to novel growth-retarding chemicals and interactions with time of forcing. *Scientia Hortic.* 21:73–83.

Hanks, G. R., and A. R. Rees. 1979. Photoperiod and tulip growth. *J. Hort. Sci.* 54:39–46.

Hanks, G. R., and A. R. Rees. 1980. Daylength and the flowering performance of tulips. *Acta Hort.* 109:177–182.

Hartsema, A. M. 1961. Influence of temperature on flower formation and flowering of bulbous and tuberous plants. In *Encyclopedia of Plant Physiology Vol. 16*, ed. W. Ruhland, 123–167. Berlin: Springer Verlag.

Hatech. 2011. http://www.hatechgas.com/Products/view/56/MacView-Ethyleen-analyser (accessed October 11, 2011).

Hsieh, T. F., and J. W. Huang. 1999. Effect of film-forming polymers on control of lily leaf blight caused by *Botrytis elliptica*. *Eur. J. Plant Pathol.* 105:501–508.

Kamerbeek, G. A., and W. J. de Munk. 1976. A review of ethylene effects in bulbous plants. *Scientia Hortic.* 4:101–115.

Kamerbeek, G. A., A. L. Verlind, and J. A. Schipper. 1971. Gummosis of tulip bulbs caused by ethylene. *Acta Hort.* 23:167–172.

Kelly, J. D., and A. L. Schlamp. 1964. Keeping quality, flower size and flowering response of three varieties of Easter lilies to gibberellic acid. *Proc. Am. Soc. Hort. Sci.* 85:631–634.

Klougart, A. 1980. Calcium uptake of tulips during forcing. *Acta Hort.* 109:89–95.

Knippels, P. J. M. 2005. The contribution of quality inspections to the improvement of the quality of the Dutch flowerbulbs and access to export markets. *Acta Hort.* 673:79–84.

Kreuk, F. 2007. Minder zuur door koudstomen en kort bevochtigen. *BloembollenVisie.* 117:28–29.

Krug, B. A., B. E. Whipker, and I. McCall. 2005a. Flurprimidol is effective at controlling height of 'Star Gazer' Oriental lily. *HortTechnology* 15:373–376.

Krug, B. A., B. E. Whipker, and I. McCall. 2006a. Hyacinth height control using preplant bulb soaks of flurprimidol. *HortTechnology* 16:370–375.

Krug, B. A., B. E. Whipker, I. McCall, and J. M. Dole. 2005b. Comparison of flurprimidol to ancymidol, paclobutrazol, and unicolazole for tulip height control. *HortTechnology* 15:370–373.

Krug, B. A., B. E. Whipker, I. McCall, and J. M. Dole. 2005c. Comparison of flurprimidol to ethephon, paclobutrazol, and unicolazole for hyacinth height control. *HortTechnology* 15:872–874.

Krug, B. A., B. E. Whipker, I. McCall, and J. M. Dole. 2006b. *Narcissus* response to plant growth regulators. *HortTechnology* 16:129–132.

Le Nard, M., and A. A. De Hertogh. 1993a. Bulb growth and development and flowering. In *The Physiology of Flower Bulbs*, eds. A. De Hertogh, and M. Le Nard, 29–43. Amsterdam: Elsevier.

Le Nard, M., and A. A. De Hertogh. 1993b. *Tulipa*. In *The Physiology of Flower Bulbs*, eds. A. De Hertogh, and M. Le Nard, 617–682. Amsterdam: Elsevier.

Lee, J. S., and M. S. Roh. 2001. Influence of frozen storage duration and forcing temperature on flowering of Oriental hybrid lilies. *HortScience* 36:1053–1056.

Legnani, G., C. B. Watkins, and W. B. Miller. 2004a. Low oxygen affects the quality of Asiatic hybrid lily bulbs during simulated dry-sale storage and subsequent forcing. *Postharvest Biol. Technol.* 32:223–233.

Legnani, G., C. B. Watkins, and W. B. Miller. 2004b. Light, moisture, and atmosphere interact to affect the quality of dry-sale lily bulbs. *Postharvest Biol. Technol.* 34:93–103.

Legnani, G., C. B. Watkins, and W. B. Miller. 2006. Tolerance of dry-sale lily bulbs to elevated carbon dioxide in both ambient and low oxygen atmospheres. *Postharvest Biol. Technol.* 41:198–207.

Legnani, G., C. B. Watkins, and W. B. Miller. 2010. Effects of hypoxic and anoxic controlled atmospheres on carbohydrates, organic acids, and fermentation products in Asiatic hybrid lily bulbs. *Postharvest Biol. Technol.* 56:85–94.

Liou, S., and W. B. Miller. 2011. Factors affecting ethylene sensitivity and 1-MCP response in tulip bulbs. *Postharvest Biol. Technol.* 59:238–244.

Marousky, F. J., and S. S. Woltz. 1977. Influence of lime, nitrogen, and phosphorus sources on the availability and relationship of soil fluoride to leaf scorch in *Lilium longiflorum* Thunb. *J. Am. Soc. Hort. Sci.* 102:799–804.

Matson Navigation Company. 2011. Refrigerated services overview. http://www.matson.com/reefer/index.html (accessed April 18, 2011).

Maxie, C. E., and R. F. Hasek. 1974. Cold storage of 'Georgia' Easter lilies, *Flr. Nsy. Rpt. Jan.* 1974, 3–4.

McDaniel, G. L. 1990. Postharvest height suppression of potted tulips with paclobutrazol. *HortScience* 25:212–214.

Miller, W. B. 1992. *Production of Easter and Hybrid Lilies.* Growers Handbook Series. Portland, OR: Timber Press.

Miller, W. B. 1993. *Lilium longiflorum.* In *The Physiology of Flower Bulbs*, eds. A. De Hertogh, and M. Le Nard, 391–422. Amsterdam: Elsevier.

Miller, W. B. 2002. A primer on hydroponic cut tulips. *Greenhouse Prod. News* 12(8):8–12.

Miller, W. B. 2008. Boron deficiency in tulip. Research newsletter no. 15. http://www.flowerbulbs.cornell.edu/newsletter/index.htm (accessed April 23, 2011).

Miller, W. B. 2011. Website for the Flower Bulb Research Program. Cornell University, Ithaca, NY, USA. http://www.flowerbulbs.cornell.edu/ (accessed February 4, 2011).

Miller, W. B., P. A. Hammer, and T. I. Kirk. 1993. Reversed greenhouse temperatures alter carbohydrate status in *Lilium longiflorum* Thunb. 'Nellie White'. *J. Am. Soc. Hort. Sci.* 118:736–740.

Miller, W. B., and R. W. Langhans. 1990. Low temperature alters carbohydrate metabolism in Easter lily bulbs. *HortScience* 25:463–465.

Miller, W. B., A. P. Ranwala, A. Hammer, T. Kirk, N. Rajapakse, and J. H. Blake. 1995. Causes and control of Easter lily leaf yellowing. *GrowerTalks* 58(9):80–88.

Miller, W. B., M. Verlouw, S. S. Liou, H. O. Cirri, C. B. Watkins, and K. Snover-Clift. 2005. Variation in *Fusarium*-induced ethylene production among tulip cultivars. *Acta Hort.* 673:229–235.

Moe, R. 1980. The use of ethephon for control of plant height in daffodils and tulips. *Acta Hort.* 109:197–204.

Moe, R., and R. Heins. 1990. Control of plant morphogenesis and flowering by light quality and temperature. *Acta Hort.* 272:81–89.

Nelson, P. V., and C. E. Niedziela Jr. 1998a. Effects of calcium source and temperature regime on calcium deficiency during hydroponic forcing of tulip. *Scientia Hortic.* 73:137–150.

Nelson, P. V., and C. E. Niedziela Jr. 1998b. Effect of ancymidol in combination with temperature regime, calcium nitrate, and cultivar selection on calcium deficiency symptoms during hydroponic forcing of tulip. *Scientia Hortic.* 74:207–218.

Nichols, R. 1970. Control of tulip stem extension. *Ann. Rept. Glasshouse Crops Res. Inst.* 1970, 74.

Nichols, R., and A. M. Kofranek. 1982. Reversal of ethylene inhibition of tulip stem elongation by silver thiosulfate. *Scientia Hortic.* 17:71–79.

Nowak, J., and K. Mynett. 1985. The effect of sucrose, silver thiosulfate and 8-hydroxyquinoline citrate on the quality of *Lilium* inflorescences cut at the bud stage and stored at low temperature. *Scientia Hortic.* 25:299–302.

Nowak, J., and R. M. Rudnicki. 1993. *Hyacinthus.* In *The Physiology of Flower Bulbs*, eds. A. De Hertogh, and M. Le Nard, 335–347. Amsterdam: Elsevier.

Okubo, H., and S. Uemoto. 1984. The application of dark treatment to cut-tulip production. *Scientia Hortic.* 24:75–81.

Peterse, A. 2010. Dutch lily breeding: Past, present and future. Presentation at Lilytopia, May 24, 2010. Longwood Gardens, Kennett Square, PA, USA.

Prince, T. A., and M. S. Cunningham. 1989. Production and storage factors influencing quality of potted Easter lilies. *HortScience* 24:992–994.

Prince, T. A., M. S. Cunningham, and J. S. Peary. 1987. Floral and foliar quality of potted Easter lilies after STS or phenidone application, refrigerated storage, and simulated shipment. *J. Am. Soc. Hort. Sci.* 112:469–473.

Prince, T. A., R. C. Herner, and A. A. De Hertogh. 1981. Low oxygen storage of special precooled 'Kees Nelis' and 'Prominence' tulips. *J. Am. Soc. Hort. Sci.* 106:747–751.

Prince, T. A., R. C. Herner, and A. A. De Hertogh. 1982. Increases in ethylene and carbon dioxide production by *Tulipa gesneriana* 'Prominence' after completion of the cold requirement. *Scientia Hortic.* 16:77–83.

Purvis, O. N. 1937. Recent Dutch research on the growth and flowering of bulbs. I. The temperature requirements of hyacinths. *Scientific Hortic.* 5:127–140.

Purvis, O. N. 1938. Recent Dutch research on the growth and flowering of bulbs. II. The temperature requirements of tulips and daffodils. *Scientific Hortic.* 6:160–177.

Rademacher, W. 2000. Growth retardants: Effects on gibberellin biosynthesis and other metabolic pathways. *Annu. Rev. Plant Physiol. Plant Molec. Biol.* 51:501–531.

Ranwala, A. P., G. Legnani, and W. B. Miller. 2003. Minimizing stem elongation during spray applications of gibberellin$_{4+7}$ and benzyladenine to prevent leaf chlorosis in Easter lilies. *HortScience* 38:1210–1213.

Ranwala, A. P., G. Legnani, M. Reitmeier, B. B. Stewart, and W. B. Miller. 2002. Efficacy of plant growth retardants as preplant bulb dips for height control in LA and Oriental hybrid lilies. *HortTechnology* 12:426–431.

Ranwala, A. P., and W. B. Miller. 1998. Gibberellin$_{4+7}$, benzyladenine, and supplemental light improve postharvest leaf and flower quality of cold-stored 'Stargazer' hybrid lilies. *J. Am. Soc. Hort. Sci.* 123:563–568.

Ranwala, A. P., and W. B. Miller. 2000. Preventive mechanisms of gibberellin$_{4+7}$ and light on low-temperature-induced leaf senescence in *Lilium* cv. Stargazer. *Postharvest Biol. Technol.* 19:85–92.

Ranwala, A. P., and W. B. Miller. 2002. Effects of gibberellin treatments on flower and leaf quality of cut hybrid lilies. *Acta Hort.* 570:205–210.

Ranwala, A. P., and W. B. Miller. 2005. Effects of cold storage on postharvest leaf and flower quality of potted Oriental-, Asiatic- and LA-hybrid lily cultivars. *Scientia Hortic.* 105:383–392.

Ranwala, A. P., W. B. Miller, T. I. Kirk, and P. A. Hammer. 2000. Ancymidol drenches, reversed greenhouse temperatures, postgreenhouse cold storage, and hormone sprays affect postharvest leaf chlorosis in Easter lily. *J. Am. Soc. Hort. Sci.* 125:248–253.

Ranwala, N. K. D., A. P. Ranwala, and W. B. Miller. 2005. Paclobutrazol and uniconazole solutions maintain efficacy after multiple lily bulb dip events. *HortTechnology* 15:551–553.

Rees, A. R. 1972. *The Growth of Bulbs. Applied Aspects of the Physiology of Ornamental Bulbous Crop Plants.* London: Academic Press.

Rees, A. R. 1992. *Ornamental Bulbs, Corms and Tubers.* Wallingford, Oxon, UK: C.A.B. International.

Roberts, A. N., J. R. Stang, Y. T. Wang, W. R. McCorkle, L. R. Riddle, and F. W. Moeller. 1985. Easter lily growth and development. *Oregon State Univ. Agric. Expt. Sta. Tech. Bull.* 148:74.

Sachs, R. M., A. M. Kofranek, and W. P. Hackett. 1976. Evaluating new pot plant species. *Flor. Rev.* 159(4116):35–36, 80–84.

Shoub, J., and A. A. De Hertogh. 1974. Effects of ancymidol and gibberellins A$_3$ and A$_{4+7}$ on *Tulipa gesneriana* L. cv. Paul Richter during development in the greenhouse. *Scientia Hortic.* 2:55–67.

Staby, G. L., and T. D. Erwin. 1977. The storage of Easter lilies. *Flor. Rev.* 161(4162):38.69.

Swart, A., and G. A. Kamerbeek. 1976. Different ethylene production *in vitro* by several species and formae speciales of *Fusarium. Neth. J. Plant Pathol.* 82:81–84.

Swart, A., and G. A. Kamerbeek. 1977. Ethylene production and mycelium growth of the tulip strain of *Fusarium oxysporum* as influenced by shaking and oxygen supply to the culture medium. *Physiol. Plant.* 39:38–44.

Tsujita, M. J., D. P. Murr, and G. L. Johnson. 1979. Leaf senescence of Easter lily as influenced by root/shoot growth, phosphorus nutrition and ancymidol. *Can. J. Plant Sci.* 59:757–761.

USDA-NASS. 2011. Floriculture Crops. 2010 Summary. http://usda.mannlib.cornell.edu/usda/current/FlorCrop/FlorCrop-04–21–2011_new_format.pdf (accessed May 1, 2011).

Van den Ende, J. E., A. T. H. J. Koster, L. J. van der Meer, M. G. Pennock-Vos, and C. Bastiaansen. 2000. BoWaS: A weather-based warning system for the control of *Botrytis* blight in lily. *Acta Hort.* 519:215–220.

Van Doorn, W. G., and S. S. Han. 2011. Postharvest quality of cut lily flowers. *Postharvest Biol. Technol.* 62:1–6.

Van Tuyl, J. M., and P. Arens. 2011. *Lilium*: Breeding history of the modern cultivar assortment. *Acta Hort.* 900:223–230.

Whitman, C. M., R. D. Heins, R. Moe, and K. A. Funnell. 2001. GA$_{4+7}$ plus benzyladenine reduce foliar chlorosis of *Lilium longiflorum*. *Scientia Hortic.* 89:143–154.

12 Postharvest
Cut Flowers and Potted Plants

Michael S. Reid and Cai-Zhong Jiang

CONTENTS

I. INTRODUCTION

Over the past 50 years, the cut flower market has changed dramatically, from a local market with growers located on city outskirts, to a global one; flowers and cut foliage sourced from around the world are sold as bunches or combined into arrangements and bouquets in the major target markets, such as North America, Japan, and the European Union. Items in a single florist arrangement often originate from countries in three or more continents. The high value of cut flowers has driven major increases in production in many developing countries. Production of cut flowers and foliage can be highly profitable in countries with an ideal growing environment (particularly those close to the equator where the environment is uniform throughout the year) and low labor costs. The costs of establishing production in the field or even in plastic houses are relatively modest, and harvest may start within a few months of planting.

Ornamental geophytes have played an important role in this change in production and marketing—herbaceous perennials with a specialized underground storage organ (whether rhizome, corm, tuber, or true bulb) are especially suited to production in low-input systems in developing countries. Dormant flower bulbs imported from The Netherlands or other producing countries, and then forced in the field of greenhouses in countries such as Colombia, Ecuador, and Kenya include popular crops—lily, tulip, iris, calla lily, freesia, and gladiolus.

A reshaping of the ornamentals market has occurred with little consideration for its postharvest consequences. Flowers that used to be obtained from local growers and retailed within days of harvest may now take as long as three weeks to arrive at the retail florist or the supermarket. The increased emphasis on holidays as the occasions for sale of cut flowers has exacerbated this trend, since the large volume of flowers required to meet the demand on the major holidays (Christmas, Valentine's Day, and Mothers' Day) has led to widespread storage.

Because of their perishability, cut flowers produced in distant growing areas traditionally are shipped by air (a transportation system whose rapidity fails to offset the disadvantages of poor temperature management and low humidity). The increasing cost of jet fuel and the volume of flowers being produced in countries such as Colombia and Kenya has led to many efforts to ship ornamentals in marine containers, further extending the time from harvest to the market. These market and transportation changes have not been accompanied by changes in postharvest technologies to offset the time/temperature effect on the life of ornamentals. The net result, especially in North America, has been a reduction in the display life of cut flowers and potted plants, disenchantment with the cut flower purchase experience, documented in many surveys, and a per capita consumption of cut flowers in the U.S. that is less than that in almost all other developed countries (Reid and Jiang 2005).

To overcome this negative impact of ornamentals, the industry needs to employ strategies to improve the display life of cut flowers and potted plants. The goal of this chapter is to describe studies that have changed our understanding of the postharvest biology of ornamental geophytes, and to indicate current optimal technologies based on the new understanding. In particular, we have focused on recent findings in relevant areas of basic plant biology and conclude with a discussion on the way in which molecular strategies are being or could be deployed in the future to extend postharvest life and reduce postharvest losses of perishable ornamental geophytes.

II. FACTORS AFFECTING THE POSTHARVEST LIFE OF ORNAMENTAL GEOPHYTES

The intersection of art, design, and horticulture represented by the ornamental plant industry has led to the use of a very wide variety of plant organs and taxa for ornamental purposes. The diversity of taxa, physiological state, and organ implies that generalizations about their biology and even technology are often misleading. In this review, we focus on the unique characteristics of the ornamental geophytes, especially on cut flowers and potted plants, since the postharvest requirements of the dormant perennating organs are discussed elsewhere (see also Chapters 9 and 11).

Most flowers of the commercial geophytes have relatively short lives. The delicate petals of flowers are easily damaged, and are often highly susceptible to disease. Even under optimum conditions, their biology leads to early wilting, abscission, or both. Foliage is longer lived, although the low light of the postharvest environment frequently leads to early leaf yellowing and, in some cases, leaf abscission. Like that of other perishable horticultural crops, the life of geophyte flowers is affected by physical, environmental, and biological factors. Quality of plant materials and preharvest factors play an important role. After harvest, temperature is of overriding importance and affects plant–water relations, growth of disease, response to physical stresses, carbohydrate status, and the interplay among endogenous and exogenous growth regulators. Much has been learnt over the last

30 years about the role of these factors and the response of ornamental geophytes to them, and some of the research findings have led to technologies that can greatly improve marketing and postharvest quality of geophyte flowers.

A. GENOTYPIC VARIABILITY

The postharvest life of geophyte flowers varies greatly, from the ephemeral flowers of taxa such as *Tigridia* to the extremely long-lived flowers of *Zantedeschia*. Less extreme, but still marked, variations are also seen within genera and even species, and certainly this variation provides a great opportunity for breeders to develop longer-lasting flowers. Unfortunately, other characteristics, such as color, form, productivity, and disease resistance, continue to be the targets of most breeding programs. This can be seen by comparing the postharvest life of different cultivars from the same breeder. In *Alstroemeria*, time to first petal fall and time to 50% leaf yellowing both showed tremendous variation (more than 100% for vase life and more than 400% for leaf yellowing) in lines released by the same breeder (Ferrante et al. 2002; Figure 12.1). Elibox and Umaharan (2008) reported vase lives of anthurium cultivars ranging from 14 to 49 days. A simple model, based on abaxial stomatal density and flower color that accurately predicted the relative vase-life ranking of different cultivars, provided an excellent tool for future breeding (Elibox and Umaharan 2008). Van der Meulen-Muisers et al. (1999) studied the genetic regulation of postharvest flower longevity in Asiatic hybrid lilies, using 10 cultivars and 45 progenies. Results from the analysis of variance for individual flower longevity revealed very significant ($p < 0.001$) variation among parents and in progenies. These studies indicate the opportunity for using a genetic approach to preventing the early flower senescence that is a common postharvest problem in geophytes used for cut flowers and potted plants.

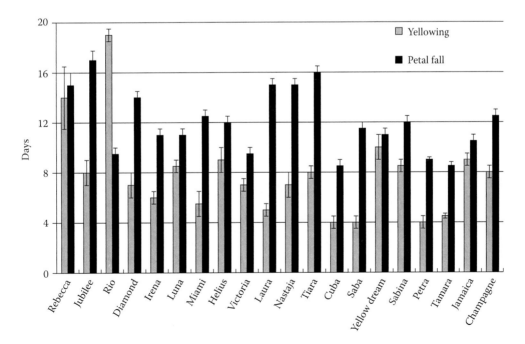

FIGURE 12.1 Variation in the postharvest performance of *Alstroemeria* cultivars. Flowers obtained from a commercial grower were placed in water at 20°C. Data show the time to first tepal fall and to yellowing of 50% of the leaves. (Adapted from Ferrante, A. et al. 2002. *Postharvest Biol. Technol.* 25:333–338.)

B. POSTHARVEST TEMPERATURE

The marked effects of temperature on the life of cut flowers were first quantified in 1973 (for carnations) by Maxie et al. (1973). Our subsequent studies have extended the findings to a wide range of other crops, including potted flowering plants (Cevallos and Reid 2000, 2001; Celikel and Reid 2002, 2005). We used a dynamic system to measure the effect of temperature on flower respiration, in which the effect of a chosen temperature was calculated on replicate single flowers (Cevallos and Reid 2000). The temperature was then increased and respiration was measured until it stabilized at the new temperature (usually within two hours). Our findings were consistent with those of Maxie and coworkers—respiration of flowers has a very high Q_{10} value, higher than that of most other perishable crops. For example, the Q_{10} for *Narcissus* was found to be nearly 7 between 0°C and 10°C (Cevallos and Reid 2000). The close link between respiration and growth and senescence in these poikilotherms means that a *Narcissus* flower held at 10°C may lose as much vase life in one day as it would if held for one week at 0°C. High (if less extreme) Q_{10} values, recorded for other geophytes, including anemone, calla lily, iris, ranunculus, and tulip, were close to 3, and were in the range of those for carnation (2–4.5 depending on cultivar) and rose (2.4–5.8) (J. Cevallos and M. Reid, unpublished).

The industry has long been aware of the importance of cool temperatures in improving long-distance marketing of ornamentals, as demonstrated by the widespread adoption of forced-air precooling, and the use of coolrooms. However, temperatures in these facilities are often well above the optimal (which is at or near 0°C for most geophyte flowers), suggesting that the true importance of temperature has not been adequately demonstrated. Simulated transport at temperatures above the freezing point results in accelerated opening and reduced vase life in many flowers, when they are subsequently held in the consumer environment (Figure 12.2). In our experiments with different flower species (Cevallos and Reid 2000, 2001; Celikel and Reid 2002, 2005), we demonstrated an extremely close relationship (Figure 12.3) between respiration during storage and residual vase life at room temperature (20°C), and were able to use such relationships to model the effects of transit temperatures on residual vase life. These mathematical models were subsequently used to program active radio-frequency identification (RFID) tags that were used in pilot studies of the value of time–temperature tags in flower marketing (Staby and Reid 2005a, 2007). Based on these results, Van Meeteren (2007) developed an equation to describe the effect of temperature on senescence rate:

FIGURE 12.2   (**See color insert.**) Effect of storage temperature on the vase life of lily flowers. Flowers were stored for 5 days at different temperatures and then returned to room temperature for evaluation of vase life. Photograph taken after 2 days at 20°C. (Photo courtesy: F. Celikel and M. Reid.)

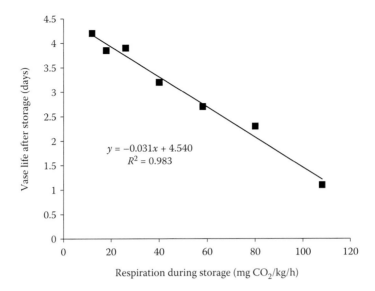

FIGURE 12.3 Relationship between vase life and respiration during storage of daffodils. Daffodil flowers were stored for 5 days at different temperatures and then held at 20°C for evaluation of vase life. (Graph courtesy: J. Cevallos and M. Reid.)

$$k_s = \frac{k_{max}}{1 + 10^{(T_{half} - T_s)\text{slope}}}$$

where k_s [day^{-1}] = rate of senescence at a temperature T_s [K], k_{max} [day^{-1}] = maximum rate of senescence, T_{half} [K] = temperature at which the senescence rate is half of k_{max}, and slope [K^{-1}] describes the steepness of the curve. This equation could potentially be used for the generalized modeling of the temperature/vase life relationships for cut geophyte flowers.

The value of time–temperature monitoring in commercial handling of cut flowers is well demonstrated by a survey conducted by Staby and Reid (2005b). Experimentally active RFID tags were included in flowers shipped by air from producers in South America and California to other US wholesalers. The data demonstrated that the temperature of the flowers varied dramatically. Some were in danger of freezing at some point during transit, while others were exposed to temperatures in excess of 35°C. Modeling the expected vase life of the flowers based on the respiration/vase life regressions suggested that vase life would be reduced in some cases by as much as 40% and that this reduction was more a matter of temperature during transit rather than of transit duration.

C. Controlled and Modified Atmospheres

The close association between flower respiration during storage and vase life after storage suggests the potential usefulness of controlled atmosphere (CA) or modified atmosphere (MA), in which the O_2 content of the storage atmosphere is reduced, sometimes with an increase in the CO_2 content. In some fruits and vegetables (particularly apples, kiwifruit, and cabbage) such atmospheres are routinely used commercially to extend the storage life. The beneficial effects are attributed to reduced respiration (resulting from low internal-oxygen concentrations) and reduced ethylene sensitivity (attributed largely to elevated CO_2 levels) (Kader et al. 1989; Kader 2003). Although such atmospheres have been frequently tested with cut flowers, the results have been disappointing (Reid 2001). Commercial trials have failed to demonstrate the benefits, and where such benefits have been claimed, the absence of valid controls has shrouded the credibility of the results. Recent studies have focused on the use of sealed packages (similar to those used in salad packages) for single flowers or small bouquets, and while the results have sometimes been promising, they only apply to specific species or even varieties of flowers, and are therefore of very little utility in general.

In an attempt to examine the reason for the lack of benefit from CA storage, respiration of flowers at different temperatures and in different concentrations of oxygen was measured (Macnish et al. 2009b). Petals held at warm temperatures (10 or 15°C) showed a fall in respiration similar to the classic curves reported by Kidd and West (1933) for apples, as O_2 partial pressure fell below 0.02% (2% by volume). However, the respiration of flowers held at low storage temperatures (0–5°C) fell only marginally, if at all. The reasons for this disparity have not been examined, but it seems possible that it may result from the difference in the surface/volume ratio of the bulky fruits and thin petals. Oxygen diffusion is likely to be a limiting factor in bulky fruits, so that the terminal oxidases are limited for O_2 at much higher external O_2 than their actual K_m would suggest. Curiously, the flower petals too failed to show the rise in respiration at very low O_2 levels (the Pasteur effect) that result from the onset of anaerobic respiration. Increased CO_2 production under anaerobic conditions has been attributed to increased glycolysis and the increased decarboxylation of phosphoenyl pyruvate. Although we have no explanation for the absence of the Pasteur effect, it is clear that low O_2 results in anaerobic respiration, because flowers stored under anaerobic conditions can smell like alcohol and may collapse shortly after placing them at room temperature in air (Macnish et al. 2009b).

A potential benefit of CA or MA that has been demonstrated but less employed is their use in controlling insects. In a study of a range of cut flowers, including several geophytes, Joyce and Reid (1985) demonstrated that high levels of CO_2 and low levels of O_2 had no positive effect on storage life of the flowers, but often had no negative effects. This implies that such atmospheres could be used in reducing disease or killing insects in cut flower shipments. The variability in the response unfortunately means that the treatment could only be applied in single-crop boxes and that may explain why it has not become popular. In a study of the potential uses of high CO_2 atmospheres, Hammer et al. (1990) found a significant reduction in *Botrytis* incidence both in naturally inoculated and in artificially inoculated rose. Unfortunately, bronzing of the leaves in response to high CO_2 atmospheres impaired the marketability of the flowers.

D. WATER RELATIONS

Adequate water relations in pot plants and cut flowers are an obvious and important element of their postharvest management. Water balance is determined by the differential between water supply and water loss, and optimal postharvest handling includes managing both sides of this relationship. The primary tool for reducing water loss is temperature control. The water content of saturated air rises in an exponential fashion (doubling for every 11°C). Depending on the humidity, therefore, water loss can rise with temperature in a similar fashion. Sealed bags or perforated polyethylene wraps can maintain a higher humidity and thus reduce water loss after harvest, but at higher temperatures the likelihood of condensation and attendant proliferation of diseases is greatly accentuated.

1. Cut Flowers

Intuitively, providing adequate water to a cut flower should be an easy matter, since the vase solution has direct access to the xylem, without the need for transport from the soil and across the tissues of the root. In practice, water uptake is frequently impeded by the desiccation that occurs during extended dry handling of the flowers, by air emboli that form when the water column in the xylem is broken, and very commonly by microbial occlusion and/or the formation of physiological plugs, tyloses, and gels (Van Doorn and Reid 1995). Differences among species and even between varieties of the same species are a function of the structure of the xylem, the size of emboli and cavitations, embolism repair ability (Brodersen et al. 2010), and the likelihood of colonization of the stem by microbes.

a. Desiccation

Although floral tissues, devoid of functional stomatas lose relatively little water to themselves, water loss can occur rapidly through the stomata of stems and leaves during postharvest handling. Surprisingly, the opening and vase life of flowers, at least in roses (Macnish et al. 2009a) and gypsophila (Rot and Friedman 2010), are not affected unless desiccation is in excess of 15% of the fresh weight. In their study, Rot and Friedman (2010) used the apoplastic fluorescent dye 8-hydroxypyrene-1,3,6-trisulfonic acid (HPTS) to measure water uptake by florets and whole stems. These dye studies verified the effects of anionic detergents (such as Triton X-100) in improving the water uptake in dehydrated flowers (Jones et al. 1993).

It has long been known that abscisic acid (ABA) plays an important role in the regulation of stomatal aperture and is therefore a key hormone in the plant's response to water stress (Radin and Ackerson 1982; Wilkinson and Davies 2002). The regulation of gas exchange in water-stressed plants involves both long-distance transport and modulation of ABA concentration in the guard cells, as well as differential sensitivity of the guard cells. Plants are thought to use the ABA signaling mechanism and other chemical signals to adjust the amount of water that is lost through the stomata in response to changes in both the root and the atmospheric environment. ABA therefore seems to be an obvious tool for reducing the water stress in cut flowers, but its other hormonal effects—stimulating ethylene synthesis or enhancing the sensitivity to ethylene (Mayak and Halevy 1972; Mayak and Dilley 1976), and accelerating petal senescence—make it a doubtful choice. Kohl and Rundle (1972), for example, demonstrated that this hormone would reduce the water use in roses, but also reduces their vase life.

b. Emboli

Formed immediately after cutting the stem, as tension in the xylem water column is released, emboli can result in a temporary reduction in water uptake that may become permanent if the rate of transpiration exceeds the water conductance of the embolized stem. In some taxa, such as *Heliconia* spp. that have very long vessels, the embolism can result in the permanent failure of conducting elements. More often, the emboli are resorbed by the xylem (apparently by water influx from the surrounding living cells, with individual droplets expanding over time, filling vessels, and forcing the dissolution of entrapped gas) (Brodersen et al. 2010). Detergent applied as brief dips or included in vase solutions, low pH, and hydrostatic pressure all overcome the emboli in the stem, as of course does recutting under water—one of the very traditional practices in the floral trade. We have recently used deep water treatments with considerable success in improving the rehydration and vase life of the recalcitrant cut flower tropical geophytes heliconia and ginger (A. Macnish, M. Reid and C.-Z. Jiang, unpublished results).

c. Microbes

A rapidly respiring and wounded stem, placed in water, quickly depletes the oxygen in the vase solution, providing perfect growing conditions for microbes (yeasts and bacteria) that benefit from the cellular contents released from the cells damaged during cutting. Occlusion by microbes and the extracellular polysaccharides that they elaborate is by far the most common cause of poor water relations in cut flowers (Macnish et al. 2008). The standard treatment for avoiding these events is the use of bactericides ($HClO_4$, $Al_2(SO_4)_3$) and quaternary ammonium compounds are among the most popular. A reduction in pH of the solution (citric acid, $Al_2(SO_4)_3$) is also helpful in reducing bacterial growth but is insufficient on its own, since acidophilic yeasts and bacteria can quickly colonize a vase or bucket solution. Not all bacteria are deleterious in the vase solution. Zagory and Reid (1986) demonstrated that some of the microbial species isolated from vase solutions have no effect on (or may even augment) the life of carnations, roses, and chrysanthemums, but the potential of a biological control system for avoiding the effects of bacteria and yeasts in the vase solution has not been explored.

2. Potted Plants

In the marketing of potted ornamental geophytes, desiccation is the most important cause of reduced quality and postharvest loss. These losses are experienced in the marketing chain (failure to water potted plants on display is a common problem in supermarkets and "club" stores), and in the consumers' homes. Bedding plants frequently fail because of wilting after planting. Produced under "luxury" conditions, where water is freely and regularly available, they have a large leaf area and are usually root-bound. Placed in the landscape, the plants quickly use all the water in the root ball, but are frequently unable to obtain sufficient water from the surrounding soil.

Postharvest water loss in potted plants starts from the moment of the last irrigation in the greenhouse and is affected by a range of environmental factors, including temperature, light, relative humidity, and air movement. However, temperature is the overriding factor affecting water loss. Cooling the plants rapidly and maintaining them at the optimal transportation temperature (typically close to 0°C) is therefore a primary approach to reducing water loss. Unfortunately, cooling potted plants is difficult and time consuming, and the industry's approach typically is simply to put plants at ambient temperature into cooled trucks for transport. This may be worse than not cooling the plants at all, because it results in condensation on the (cooled) aerial parts of the plant (given that the soil mass cools much more slowly). Such a condensation likely aggravates postharvest disease, particularly gray mold caused by *Botrytis cinerea*.

In the salad vegetable industry, difficulties at cooling products such as iceberg lettuce are overcome by the use of vacuum coolers, and these have become standard in that industry. We reasoned that vacuum cooling might be an effective strategy for cooling potted plants, since the soil water is freely available and would result in rapid cooling of the soil. We have shown that vacuum cooling is, in fact, an excellent means of cooling potted plants and that vacuum-cooled plants had improved shelf life after long-distance transportation (M. Reid, C.-Z. Jiang, J. Thompson, and S. Han, unpublished results). In this test, potted rose and campanula plants were cooled in a vacuum cooler or placed uncooled in a truck. Dataloggers included in the shipment recorded the temperatures during transport. Following transcontinental transport the vacuum-cooled plants performed as well as or better than noncooled plants.

The use of ABA to close stomata seems to have considerable potential for reducing water loss during postharvest handling of potted plants. The application of ABA to potted chrysanthemum plants significantly reduced their stomatal conductance and their rate of postharvest water loss, thereby extending the time until wilting of unwatered plants by three days (from 5 to 8 days) (Cornish et al. 1985). At that time, ABA was a prohibitively expensive chemical, so these results were of little commercial interest. As an alternative, these authors tested a postproduction drench of the soil medium with saline solutions. A solution containing 100 mM NaCl resulted in an even greater extension of unwilted life (10 days). They hypothesized that the osmotic stress following the NaCl treatment would induce ABA biosynthesis and thereby stomatal closure. However, although the stomatal conductance of the salt-treated plants was considerably lower than the controls, there was no significant difference in the ABA content of the tissues following the salt treatment. Apparently, the stomatal closure induced in salt-stressed plants was the result of some (perhaps osmotic) mechanism.

Recently, S-ABA, the biologically active form of ABA produced through microbial fermentation, has become available at a commercially viable price, and the beneficial postharvest effects of the treatment of several bedding plants with this material have recently been reported (Blanchard et al. 2007; Van Iersel et al. 2009; Waterland et al. 2010). In some species such as pansy and viola, spray applications of this chemical at the recommended rates (500–1000 mg/L) resulted in phytoxic responses, suppressed shoot elongation, and decreased flower number (Blanchard et al. 2007; Waterland et al. 2010). In experiments with roses, lavender, and impatiens, much lower concentrations of ABA provided excellent extension of shelf life without any of these negative effects (Pemberton et al. 2009). The effects of ABA on transpiration and postharvest life of ornamental geophytes remain to be explored, but the finding of Hunter et al. (2004a) that ABA accelerates tepal senescence in *Narcissus* suggests that the opportunities for the practical use of this regulator in geophyte ornamentals may be limited.

E.   ETHYLENE AND OTHER HORMONES

It has long been known that plant hormones and plant growth regulators can have dramatic effects on floral longevity. For example, the effects of pollination on orchid flowers (anthocyanin accumulation and wilting) have long been explained in terms of a response to plant hormones and the interplay among them (Arditti 1975). Ethylene is often considered the most important among the hormones affecting flower longevity, but other hormones can affect the sensitivity to ethylene, and many flowers, most of them geophytes, are insensitive to ethylene. The nature of the senescence signal in ethylene-insensitive flowers remains to be established, but there is evidence that ABA and GA may, respectively, play accelerating and retarding roles.

1.   Ethylene

Sleepiness of carnations—premature wilting of petals before the flowers even open—was known to be the result of gas leaks in greenhouses long before the active principle was shown to be ethylene (Crocker 1913), and the dramatic effect of ethylene on the senescence of flowers and abscission of flowers and flower parts was well documented in the first half of the twentieth century by researchers at the Boyce Thompson Institute (USA) and others. The role of endogenous ethylene in triggering senescence has been well documented by a range of studies reporting on the dynamics of ethylene production, changes in the activity of the biosynthetic enzymes (Bufler 1984, 1986), and up-regulation of the genes encoding these enzymes (Woodson et al. 1992). The key role of ethylene has been corroborated by studies with long-lived carnation cultivars (Wu et al. 1991) and with transgenic or VIGS constructs silencing the biosynthetic pathway (Savin et al. 1995; Bovy et al. 1999; Chen et al. 2004).

   The discovery that the action of ethylene could be inhibited by silver ions Ag$^+$ (Beyer 1976) and the subsequent development of the stable, nontoxic, yet effective silver thiosulfate complex (Veen and Van de Geijn 1978) have provided an important commercial tool, still in widespread use, for preventing ethylene-mediated senescence and abscission in cut flowers and potted plants. Other inhibitors of ethylene synthesis (aminoethoxyvinyl glycine, aminooxyacetic acid, Co^{2+}) and action (2,5-norbornadiene) were also effective to varying degrees, but none are currently being used commercially. Since many geophyte flowers are relatively insensitive to ethylene, the effects of inhibiting ethylene action have been little explored in these crops. Early results suggesting a benefit in gladiolus have not been confirmed. Using their response to exogenous ethylene, pollination, and 1-methylcyclopropene (1-MCP), flowers have been broadly classified into two groups—ethylene sensitive and ethylene insensitive. However, this classification is undoubtedly too simplistic, since some flowers show an intermediate behavior. In daffodils, for example, pollinated flowers or flowers exposed to ethylene senesce rapidly, indicating an ethylene-sensitive senescence pattern (Hunter et al. 2004b). Inhibition of ethylene action prevented this loss of longevity, suggesting that protection from ethylene would be useful when the flowers are being marketed where they are likely to be exposed to ethylene (e.g., in supermarkets). However, inhibitors of ethylene action had a minimal effect on the senescence of daffodil flowers held in ethylene-free air, indicating that natural senescence is initiated by regulators other than ethylene. This response is similar to that previously shown in *Brodiaea* (Han et al. 1991). Cyclamen show a different pattern, where pollination induces a burst of ethylene production and ethylene sensitivity, but unpollinated flowers are not affected by ethylene treatment (Halevy et al. 1984). Other geophyte flowers, traditionally considered to be ethylene insensitive, also show some responses to ethylene during floral display. In alstroemeria, overnight exposure to low concentrations of ethylene results in altered development of opening buds when the flowers were held in air (Macnish et al., in preparation). In iris, a classically ethylene-insensitive flower, a similar treatment resulted in reduced flower opening as a result of inhibition of growth of the subtending pedicel and gynoecium, as well as reduced vase life (A. Macnish, M. Reid and C.-Z. Jiang, unpublished results). There is still a considerable need for research to identify the role of ethylene and other hormones in floral senescence of ornamental geophytes.

Of particular interest in the marketing of some geophytes is the control of undesirable scape growth, and ethylene or ethephon treatments have long been known to be effective in preventing such growth. In many tulip cultivars, postharvest stem elongation results in stem bending during vase life. Van Doorn et al. (2011) noted that while stem bending could be prevented by treatment with ethylene or ethephon, these treatments resulted in poor flower opening and precocious tepal abscission. The negative effect of ethephon on flower opening was overcome by a treatment with gibberellic acid (GA$_3$), which also helped to delay leaf yellowing. Addition of benzyladenine (BA) effectively delayed leaf yellowing and also delayed tepal senescence. Browning of the bottom of the scape caused by BA was prevented by the inclusion of calcium ions in the solution. The chemical combination was successful either as a vase solution or, with altered concentrations, as a pulse treatment applied shortly after harvest.

One of the numerous olefins synthesized by E. C. Sisler, at North Carolina State University, 2,5-norbornadiene was an important tool in studies aimed at understanding the nature of ethylene binding (Sisler and Yang 1984; Sisler et al. 1984). Noting that 2,5-norbornadiene inhibited ethylene action in a competitive manner, Sisler reasoned that it would be possible to use a diazo derivative of this compound to identify the ethylene-binding site using activation tagging. He synthesized diazo-cyclopentadiene (DACP), a cyclic diolefin with an attached reactive diazo group, and found that it was very effective in inhibiting ethylene action when dissociated with UV light after being applied to the tissue (Sisler and Blankenship 1993). Curiously, the activity only required exposure to fluorescent light, not the expected shorter-wavelength UV (Sisler and Blankenship 1993), and DACP treated with fluorescent light was just as active as DACP itself (Blankenship and Sisler 1993; Sisler and Lallu 1994). An examination of the mixture of breakdown products in the irradiated DACP revealed the presence of 1-MCP, which these researchers found to be a potent inhibitor of ethylene action (Sisler and Blankenship 1996). This material has now become a standard treatment for ethylene-sensitive flowers and potted plants (Serek et al. 1994b, 1995a, b) applied either as a gas in an enclosed space or through the use of sachets that are placed in boxes prior to transportation.

2. Abscisic Acid

As noted in previous sections, ABA plays an important role in the control of stomatal aperture and thereby transpiration. There is also substantial published evidence implicating ABA in the regulation of perianth senescence. Not only have researchers shown a close association between petal senescence and increased petal ABA concentrations (Nowak and Veen 1982; Hanley and Bramlage 1989; Onoue et al. 2000), but also exogenously applied ABA has been shown to accelerate the senescence of a number of flowers (Arditti 1971; Arditti et al. 1971; Mayak and Halevy 1972; Mayak and Dilley 1976; Panavas et al. 1998b). Such an application results in many of the same physiological, biochemical, and molecular events that occur during normal senescence (Panavas et al. 1998b).

In ethylene-sensitive flowers such as carnation flowers and roses, ABA-accelerated senescence appears to be mediated through the induction of ethylene synthesis, since it is not seen in flowers that are pretreated with ethylene (Mayak and Dilley 1976; Ronen and Mayak 1981; Müller et al. 1999). This is consistent with the pattern of endogenous ABA content in rose petals, where the increase in ABA concentration occurs 2 days after the surge in ethylene production (Mayak and Halevy 1972).

Since daylilies (*Hemerocallis*) are ethylene insensitive (Lay-Yee et al. 1992), ABA presumably induces senescence independently of ethylene (Panavas et al. 1998b). The fact that ABA accumulates in daylily tepals before any increase in the activities of hydrolytic enzymes and even before the flowers have opened suggests that the hormone may coordinate early events in the transduction of the senescence signal (Panavas et al. 1998b). The application of ABA to presenescent daylily tepals resulted in a loss of differential membrane permeability, an increase in lipid peroxidation, an increase in the activities of proteases and nucleases, and the accumulation of senescence-associated mRNAs (Panavas et al. 1998b).

During senescence of daffodil flowers, Hunter et al. (2002) reported a different response to ABA. As in daylilies, ABA treatment accelerated senescence, there was no associated increase in ethylene synthesis, and treatment with inhibitors of ethylene action did not prevent exogenously applied ABA from accelerating flower senescence. However, although ABA accumulated in the tepals as they senesced, it did not appear to play a signaling role in natural senescence. The increase in ABA concentrations in the tepals occurred after the induction of senescence-associated genes. It was concluded that the increase in ABA content is most likely a consequence of the cellular stresses that occur during senescence; the hormone does not trigger senescence, but may help drive the process to completion (Hunter et al. 2002).

3. Cytokinins

The striking effects of cytokinin (CK) in delaying senescence of leaves were known (from the effects of BA) long before the first isolation of zeatin. Given the homology between leaves and petals, it is perhaps not surprising that CKs were also found to delay petal senescence (Mayak and Kofranek 1976; Eisinger 1977), an effect that was shown to be associated both with reducing the sensitivity of the corolla to ethylene (Mayak and Kofranek 1976) and with delaying the onset of ethylene biosynthesis (Mor et al. 1984). Endogenous CK content shows a pattern consistent with its putative role in delaying senescence—buds and young flowers contain high CK levels, which fall as the flower ages and senescence commences (Mayak and Halevy 1970; Van Staden and Dimalla 1980; Van Staden et al. 1990). The interplay between CK content and senescence in ethylene-sensitive flowers was elegantly demonstrated by Chang et al. (2003), who transformed petunia with a construct comprising an isopentenyltransferase gene (*IPT*) driven by the promoter of senescence-associated gene 12 (*SAG12*) designed to increase CK synthesis at the onset of senescence in leaves (Gan and Amasino 1995). CK content of corollas in the transformed plants increased after pollination, ethylene synthesis was delayed, and flower senescence was delayed 6–10 days. As in flowers treated with exogenous CKs, the flowers from the IPT-transformed plants were less sensitive to exogenous ethylene and required longer treatment times to induce endogenous ethylene production, and the symptoms of floral senescence.

Leaf senescence is also an important component of loss of quality in ornamental geophytes, particularly members of the Liliaceae, and commercial pretreatments containing CKs and/or gibberellins are recommended as a prophylaxis in sensitive genera such as *Alstroemeria* and *Lilium*. The nonmetabolized CK, thidiazuron (TDZ), has proven very useful as an amendment in tissue culture and transformation/regeneration media, and we reasoned that it might be a useful tool for preventing leaf yellowing in cut flowers. Pulse treatment of cut *Alstroemeria* stems with as little as 5 μM TDZ essentially prevented leaf yellowing in flowers of 'Diamond', where yellowing normally starts after 4–5 days (Ferrante et al. 2001). The flowers of *Alstroemeria* are relatively insensitive to ethylene, and the TDZ treatment had only a minor effect on flower life (Mutui et al. 2003). CKs have also been shown to increase the life of iris, whose natural senescence is ethylene-independent (Wang and Baker 1979). In this crop, TDZ treatment at considerably higher concentrations (200–500 μM) significantly improved flower opening (including the opening of axillary flowers, if present) and flower life (Macnish et al. 2010b). The treatment was of particular value in that it reduced the loss of vase life that results from cool storage. While control iris that were held in cool storage for two weeks had only a very short display life and did not fully open, those pretreated with TDZ had the same vase life as freshly harvested controls (Figure 12.4). Most experiments with TDZ have been conducted with flowers that are insensitive to ethylene, but in ethylene-sensitive lupins and phlox, TDZ has also been shown to improve flower opening and to reduce ethylene-mediated flower abscission and senescence (Sankhla et al. 2003, 2005). These findings indicate that TDZ acts like other CKs in decreasing ethylene sensitivity and that this potent regulator should be tested on a broader range of crops.

TDZ has proved to have a remarkable effect in improving the postharvest life of potted flowering geophytes. Leaf yellowing is a common postharvest problem with potted flowering crops, and we have found that low concentrations of TDZ are very effective in preventing this symptom in a wide

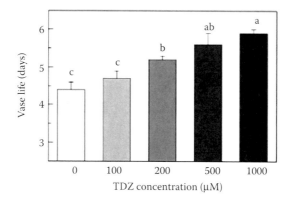

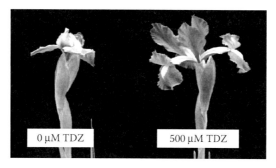

FIGURE 12.4 Effect of TDZ on the vase life of iris before and after cool storage. Iris buds were harvested at commercial maturity (pencil tip), pulsed for 24 h with different concentrations of TDZ and then (upper panel) maintained in deionized water for vase life evaluation under standard conditions or (lower panel) stored for two weeks at 0°C and then held for five days in DI at 20°C under standard conditions. (Adapted from Macnish, A. J. et al. 2010b. *Postharvest Biol. Technol.* 56:77–84.)

range of crops. The TDZ treatment appears to maintain the photosynthetic ability of the plants, since fresh and dry weights of TDZ-treated plants are much higher than those of the controls (C.-Z. Jiang and M. Reid, unpublished results). After two months, potted cyclamen plants treated with 5 μM TDZ maintained full display value, while control plants had almost ceased flowering and were showing obvious etiolation in response to the low light of the display environment (Figure 12.5). Treatment with TDZ dramatically extended the postharvest life of potted and cut freesia flowers, presumably by preventing the abortion of young buds in response to ethylene that is synthesized

FIGURE 12.5 (**See color insert.**) Effect of TDZ on the display life of potted cyclamen. Cyclamen plants at harvest maturity were sprayed with 5 μM TDZ and then held for 2 months under standard evaluation conditions. (Photo courtesy: C.-Z. Jiang and M. S. Reid.)

FIGURE 12.6 **(See color insert.)** Effect of TDZ on foliage of potted *Ornithogalum* plants. Plants were sprayed at commercial maturity with water (left) or 10 µM TDZ (right) and then held in an interior environment. The photograph was taken after 4 weeks. (Photo courtesy: C.-Z. Jiang and M. S. Reid.)

in response to low carbohydrate status (Spikman 1989). In *Ornithogalum*, the growth regulator completely inhibited leaf yellowing in potted plants held under display conditions (Figure 12.6). In tulips, a 10 µM treatment at commercial maturity resulted in reduced elongation growth, extension of flower life, and a marked inhibition of yellowing of leaves and gynoecia (Y. L. Zhang, M. Reid and C.-Z. Jiang, unpublished results, Figure 12.7).

FIGURE 12.7 **(See color insert.)** Effect of TDZ on postharvest performance of tulips. Plants were sprayed at commercial maturity (oldest bud showing color) with water or with 10 µM TDZ and then held in an interior environment. Photographs were taken 4, 12, 16, and 24 days after treatment. (Photo courtesy: Y. L. Zhang, M. Reid, and C.-Z. Jiang.)

FIGURE 12.8 (**See color insert.**) Effect of growth regulators on postharvest performance of lilies. Flowers were pulsed with STS or deionized water and then held in water or 5 mg/L GA_3. (Photo courtesy: M. Reid.)

4. Other Hormones and Regulators

Gibberellins, auxins, and other plant hormones and regulators have also been shown to have effects on floral longevity. For years, auxin was considered an important component of the pollination response in orchids and other flowers (Arditti 1975). Gibberellin treatment has been shown to improve the opening of iris flowers (Celikel and Van Doorn 1995) and also to markedly delay the natural senescence of daffodils (Hunter et al. 2004a). An important commercial use of gibberellins is in preventing postharvest leaf yellowing in monocotyledonous cut flowers, such as lilies (Han 2001; Ranwala and Miller 2002) and alstroemeria (Jordi et al. 1995; Mutui et al. 2006). Commercial preservatives containing gibberellins have been formulated for these crops. Although lily flower senescence is considered ethylene insensitive, ethylene clearly is involved in tepal abscission, since it is substantially delayed by treatment with STS (Figure 12.8). Curiously, inhibition of ethylene action results in accelerated leaf yellowing, but combining an STS pulse treatment with GA_3 in the vase solution results in excellent postharvest performance (Figure 12.8). A recent intriguing finding is that heat shock is very effective in preventing leaf yellowing and senescence in lilies (A. B. Woolf, pers. comm.). Commercial application of this observation would require the development of techniques to treat only the foliage, since flowers were damaged if they were exposed to the heat treatment.

F. Disease

Although a detailed discussion of the postharvest pathology of cut flowers and potted plants is beyond the scope of this chapter, completeness calls for a brief mention of the importance of postharvest disease in the global marketing of the ornamental geophytes. Improper temperature management, including episodes of cooling and warming in the absence of proper precooling techniques and ventilation, results in condensation and accelerated growth of pathogens on delicate petals and other floral parts. *Botrytis cinerea*, a relatively weak pathogen, is the major pathogen of these products, and a range of chemicals has been used for postharvest protection. The push for organic or sustainable production and the loss of established chemicals have led to an effort to identify alternative strategies for controlling disease. As noted above, high CO_2 levels provide effective control for species whose leaves (or petals) are not damaged by the gas. Studies with SO_2 gave the same result (Hammer et al. 1990)—good control of the pathogen, but damage to the host. Recently, Macnish et al. (2010c) reported (in roses) the efficacy of a simple dip in a solution of $NaHClO_4$, which performed as well as commercial fungicides under commercial conditions. Other strategies, including the use of ClO_2 (Macnish et al. 2008) and ozone generators, have been tested, but with inconsistent results (M. Reid, C.-Z. Jiang and A. Macnish, unpublished). Cut freesia petal specking caused by *Botrytis cinerea* infection could be suppressed by the treatment with methyl jasmonate vapor (Darras et al. 2005). These simple antimicrobial strategies remain to be tested on the majority of the geophyte ornamentals.

G. GROWTH AND GRAVITROPIC RESPONSES

As developing organs, stem elongation of many cut flowers occurs in response to environmental cues, particularly gravity, and there has been considerable research effort at understanding the mechanisms of these responses and devising strategies to prevent them. Researchers are agreed that the primary driver for gravitropic responses is the redistribution of auxin in response to its polar transport, and differential growth as a result of that redistribution. Some research has suggested a role for ethylene and/or calcium in the response. It was also reported that the gravitropic response of *Antirrhinum majus* could be avoided by a pretreatment with silver thiosulfate (Philosoph-Hadas et al. 1996). Others have not been able to reproduce these results (Woltering et al. 2005; Celikel et al. 2010). The importance of auxin redistribution in the gravitropic response is well demonstrated by the impressive effects of pretreatment with an auxin transport inhibitor naphthyl phthalamic acid (NPA) (Teas et al. 1959); it seems unfortunate that this very effective material has not been developed as a commercial pretreatment for geophyte flowers with pronounced gravitropic responses such as tulip, gladiolus, and *Kniphofia* (Figure 12.9).

H. CARBOHYDRATE SUPPLY

The rapid respiration of flowers, and the energy required for flower growth, bud opening, and floral display, necessitates substantial energy reserves in harvested cut flowers. The fact that the primary component in floral "preservatives" (sometimes termed as "fresh flower foods") is a simple sugar—fructose, glucose, or sometimes sucrose—reflects the profound effects of added carbohydrates on flower development, opening, and display life. Recent research into the effects of added carbohydrates on the life of cut flowers has focused on the potential benefits of trehalose, which has been reported to mitigate the damaging effects of ionizing radiation and to extend the life of gladiolus flowers (Otsubo and Iwaya-Inoue 2000). In a study of the mechanism of the trehalose effect, Yamada et al. (2003) found that trehalose, but not sucrose, delayed symptoms of senescence and associated programmed cell death events, including nuclear fragmentation. These data suggest that trehalose is exerting a protective effect, perhaps on membranes (Crowe et al. 1984), rather than supplying needed carbohydrate.

FIGURE 12.9   (See color insert.) Effect of naphthylphthalamic acid (NPA) on negatively geotropic curvature in *Kniphofia* (Red Hot Poker). Flowers were placed in different concentrations of NPA and then held horizontally for 24 h. (Photo courtesy: M. S. Reid.)

One of the remarkable technologies that have been successful in improving the opening and vase life of cut flowers is the provision of additional carbohydrate in high concentration or "pulse" pre-treatments in which the (usually) freshly harvested flower is placed in a solution containing a biocide and a high concentration (12–20%) sugar (Halevy and Mayak 1979). In addition to the well-known effects in gladiolus (Mayak et al. 1973), pulse or vase solution treatment with sucrose has been successful in improving the opening of bird of paradise (Halevy et al. 1978), liatris (*Liatris spicata*) (Borochov and Keren-Paz 1984), freesia (Spikman 1989), brodiaea (Han et al. 1991), and tuberose (*Polianthes tuberosa*) (Naidu and Reid 1989; Waithaka et al. 2001). In tuberose, sucrose pulsing after harvest ensures satisfactory bud opening which normally is inhibited by even brief periods of cool storage. Pulse treatments are not always an effective tool for applying carbohydrate to ornamental geophytes. In *Brodiaea*, for example, vase solutions containing up to 8% sucrose increased vase life dramatically, but an overnight pulse with 10% sucrose was not as effective, although it did allow long-term storage without reduction in vase life (Han et al. 1991). Unfortunately, treatment with sugar-containing solutions causes accelerated leaf yellowing in some flowers, notably lilies and alstroemeria. In such taxa, flower foods must be formulated with little or no sugar or amended with hormones that delay leaf yellowing.

III. BIOLOGY OF FLOWER SENESCENCE

Floral longevity is tremendously variable; ephemeral flowers may be open for only a few hours, while some flowers may remain open and receptive for many months. Even so, flowers have a very short life compared to most other plant organs, and their senescence is often precisely controlled by environmental or physiological cues. Precisely controlled senescence is likely to have an evolutionary advantage. Not only does it remove a flower from the competition for pollinators once it is pollinated (or is no longer receptive), but also it eliminates an energy sink (Ashman and Schoen 1994) and provides resources to other flowers in the inflorescence (Serek et al. 1994a).

Flower senescence has been an attractive model for studies of senescence in plants (Rogers 2006); apart from its commercial importance, it offers a range of advantages for the researcher, including a short, and often tightly controlled, time span. In addition, the onset of senescence is often readily visible (sometimes as a color change) (Macnish et al. 2010a), coordinated within a single large organ comprising relatively uniform cells, and may be manipulated by simple triggers (pollination, ethylene, photoperiod).

The signals for floral senescence are still incompletely understood. In many flowers, an increase in ethylene production, often triggered by pollination (Pech et al. 1987; Stead 1992; Van Doorn and Stead 1994; Hunter et al. 2002), clearly initiates the senescence cascade. In most geophyte ornamental species, however, ethylene appears to play no part in initiating natural senescence (Woltering and Van Doorn 1988), and despite a number of studies on model systems, including daylilies (Panavas et al. 1998a), iris (Van Doorn et al. 2003), and daffodil (Hunter et al. 2002), the signals initiating their senescence have not yet been identified.

The overall picture of floral senescence that has emerged from recent studies is one of a controlled disassembly of the cells of the corolla, probably by a mechanism homologous with apoptosis, and transport of the resulting nutrients to other parts of the inflorescence or beyond and to seeds. In agreement with this picture, increased hydrolytic activity is a common feature of floral senescence; ribonuclease and glucosidase activities increase in senescing corollas of the ephemeral morning glory (Matile and Winkenbach 1971), acid phosphatase, ribonuclease, and ATPase activities are elevated in senescing petals of carnation (Hobson and Nichols 1977) and cellulase, poly-galacturonase, and ß-galactosidase activities are greater in senescing petals of daylily (Panavas et al. 1998a).

Over the last two decades, much of the research addressing the biology of flower senescence has focused on the use of molecular tools, particularly analysis of the transcriptome, to determine the characteristic and key events of flower senescence. A number of studies have also sought to evaluate the functional importance of those changes in the senescence process.

A. Ultrastructural Changes

Ultrastructural studies suggest that autophagy is the major mechanism for large-scale degradation of macromolecules (Van Doorn and Woltering 2005, 2010). Such studies also suggest that petal cell death involves rupture of the vacuolar membrane and subsequent complete degradation of the cytoplasm, rather than a gradual increase in cell leakiness resulting from progressive degradation of the plasma membrane.

The delicacy of petal cells and their rapid collapse during senescence are a challenge to studies of ultrastructural changes during senescence. Van Doorn et al. (2003) used *Iris* as a model for examining ultrastructural and molecular changes during opening and senescence, and found dramatic changes in ultrastructure that were clearly related to eventual senescence well before any of the normal hallmarks of senescence (petal inrolling, wilting) had occurred. In particular, they noted that the plasmodesmata of mesophyll cells closed about two days before flower opening, while in the epidermis they closed concomitant with opening. Since the onset of visible senescence in the epidermal cells occurred about two days later than in mesophyll cells, it seems possible that plasmodesmatal closure may be a very early event in the senescence program.

The precise control and rapidity of floral wilting, as well as ultrastructural and biochemical observations, have led to the view that floral senescence is a process that mirrors apoptosis in animal cells. Van Doorn and Woltering (2005, 2010) have reviewed research that supports this hypothesis, including increased activity of hydrolytic enzymes, DNA laddering, and the appearance of apoptotic bodies.

In addition to characteristic DNA laddering (Eason and Bucknell 1999), ultrastructural evidence has been used to support the hypothesis that floral senescence is an apoptotic event (Van Doorn and Woltering 2008). Such evidence includes the presence of invaginations in the tonoplast and the presence of numerous vesicles in the vacuole (Matile and Winkenbach 1971; Phillips and Kende 1980; Smith et al. 1992), which is suggested to be the main site of membrane and organelle degradation. In addition, increased numbers of small vacuoles and an increase in vacuolar size have been observed in petal cells of *Ipomoea* (Matile and Winkenbach 1971), carnation (Smith et al. 1992), *Hemerocallis* (Stead and Van Doorn 1994), and *Iris* (Van Doorn et al. 2003). As senescence proceeds, cytoplasmic contents are lost; in *Iris* (Van Doorn et al. 2003) and carnation (Smith et al. 1992) the endoplasmic reticulum and attached ribosomes seem to disappear early in senescence, followed by a reduction in the number of Golgi bodies, mitochondria, and other organelles. Although the nucleus remains until late in senescence, its ultrastructure changes, with blebbing similar to that seen in apoptosis in animal cells (Serafini-Fracassini et al. 2002) and clumping of chromatin, increased fluorescence indicative of DNA condensation and sometimes a decrease in diameter (Yamada et al. 2006). Ultrastructural events during late senescence include nuclear fragmentation (Yamada et al. 2006), loss of remaining organelles, increase in vacuolar size, and eventually collapse of the tonoplast (Van Doorn and Woltering 2004). Unlike animal systems, however, a role for caspase-like enzymes or metacaspases has not yet been established in petal senescence, and there has been no clear demonstration of a role for proteins released by organelles such as the mitochondrion (Van Doorn and Woltering 2008). The fact that silencing the expression of genes encoding prohibitin accelerates floral senescence (Chen et al. 2005) suggests a possible role for the mitochondrion. Prohibitin is important for mitochondrial assembly and the maintenance of mitochondrial function, so a reduction in its synthesis could be argued to lead to impaired mitochondrial function and early release of mitochondrial proteins that initiate the senescence cascade.

B. Changes in the Transcriptome

1. Gene Expression Analysis

Researchers have worked with a range of model flowers to compile an impressive catalog of genes whose abundance changes during floral opening and senescence. Woodson and his colleagues

(Lawton et al. 1990) investigated changes in transcripts during the onset of ethylene-regulated senescence in carnations and demonstrated changes, among others, in ethylene biosynthetic genes. We used a similar approach for differential screening of cDNA libraries to identify a number of genes that were strongly up-regulated during tepal wilting in the ethylene-independent daylily (Valpuesta et al. 1995). Of particular interest was the early and massive up-regulation of a cysteine protease in this system, which might be associated with bulk protein degradation (Lay-Yee et al. 1992).

The detailed information that can be obtained from molecular studies is exemplified by the study of Hunter et al. (2002) who investigated the changes in the transcriptome of daffodil flowers using subtractive hybridization—a technique that increased the sensitivity of the differential screen. The 94 unique sequences isolated from incipiently senescent perianth tissue of daffodils selected for further analysis encoded proteins of diverse functions: from enzymes involved in protein, lipid, and nucleic acid breakdown to those involved in wall modifications, cellular signaling, and transport processes. Similar results were obtained from a study of the ephemeral flowers of *Mirabilis jalapa* (Xu et al. 2007a), and of morning glory (*Ipomoea*) (Yamada et al. 2007).

The advent of affordable microarray technology has provided an even more powerful tool to examine changes in the transcriptome. Studies in *Iris* (Van Doorn et al. 2003), in *Alstroemeria* (Breeze et al. 2004), and in wallflower (Price et al. 2008) examined changes in the abundance of transcripts of as many as a thousand genes. The results of these more sensitive analyses mirrored what had already been shown by previous differential screening studies, but did not greatly expand the list of genes showing a clear association with senescence.

2.   Functional Analysis

Identifying which of the numerous genes that are associated with floral senescence play a role in regulating the process requires functional analysis using transgenic approaches. Since transformation and regeneration is a challenge for most floricultural species and is also relatively slow and very costly, we have tested two alternative strategies—virus-induced gene silencing (VIGS) and transient expression analysis. In VIGS, plants are infected with a virus containing a fragment of a target host gene, and the phenotype of the infected plant can provide an indication of the function of the target gene. We used a purple petunia plant as our host organism and a fragment of chalcone synthase (CHS) as an indicator of silencing. Where CHS was silenced, the normally purple flowers would be white (Chen et al. 2004). Concatenating one or more host gene fragments into the viral genome allows us to test the effect of silencing those genes, and the color change due to silencing CHS allows us to select the tissues to test for changed phenotype. Using this strategy we have tested a range of candidate genes for their effect on floral senescence. A number of transcription factors identified as being associated with corolla senescence have been tested; silencing a MADS-box-containing transcription factor (MADS-1) appeared to delay senescence (C.-Z. Jiang and M. Reid, unpublished results), while silencing other transcription factors accelerated the process (Donnelly et al. 2010). VIGS has a particular advantage as a strategy for the analysis of gene function in ornamental geophytes, where a few transformation/regeneration systems have been developed, and the time from regeneration to flowering can be several years. Recently, we used this technology to successfully silence the gene that encodes phytoene desaturase (PDS) in gladiolus (A. Singh, M. Reid and C.-Z. Jiang, unpublished results, Figure 12.10).

In the transient expression assay, tissues are infected with *Agrobacterium* transformed with a target gene, and the transient phenotype induced following infection can indicate the function of the gene. Petals seem an ideal system for transient expression analysis, and we have used this strategy to test senescence-related promoters (Xu et al. 2007b). For example, by infecting petals with *Agrobacterium* containing a construct comprising a GUS reporter driven by the cauliflower mosaic 35S promoter or the senescence-specific promoter *SAG*12, we were able to demonstrate that the reporter gene was strongly expressed in daylily and daffodil tepals (Xu et al. 2007b; Figure 12.11). Curiously, the *MjXB3* (*Mirabilis jalapa* E3 ubiquitin ligase) promoter, a petal-specific senescence

FIGURE 12.10 (**See color insert.**) VIGS in *Gladiolus*. Basal leaves were inoculated with TRV–RNA2 constructs containing a fragment of the gladiolus PDS, using a 1 ml disposable syringe with a needle. Inoculated plants were maintained in the growth chamber at 18°C for 6 weeks. Typical photobleaching resulting from silencing of PDS is apparent in a newly formed leaf (right). Plants inoculated with TRV–RNA2 empty vector (left) served as controls. (Photo courtesy: A. Singh, C.-Z. Jiang, and M. S. Reid.)

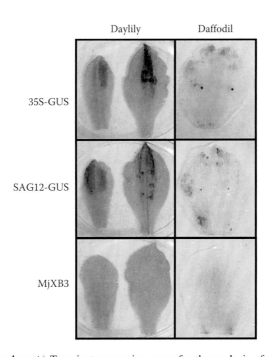

FIGURE 12.11 (**See color insert.**) Transient expression assay for the analysis of gene function during petal senescence. Tepals from opening daylilies or incipiently senescent daffodils were vacuum infiltrated with Agrobacterium transformed with a 35S-GUS construct, a *SAG*12-GUS construct, or an MjXB3-GUS construct. After 24 hours, the petals were cleared with alcohol and then stained to visualize GUS activity. (Adapted from Xu, X. et al. 2007b. *J. Exp. Bot.* 58:3623–3630.)

promoter isolated from *Mirabilis jalapa*, which drives strong up-regulation in carnations and petunia, was not active in the two geophyte taxa we tested. This technique shows promise for examining the effects of candidate genes in the senescence process, by antisense or over-expression in petals at the appropriate stage.

C. Changes in the Proteome

Although the changes described in the transcriptome suggest the proteins and enzymes that might play a key role in the regulation of senescence, interpretations based on transcript abundance are subject to the criticism that posttranscriptional, translation, and posttranslational modifications might alter the abundance or activity of the proteins that transcripts encode. Few studies have so far attempted to identify specific protein changes that might be associated with the induction of senescence. Lay-Yee et al. (1992), for example, used an *in vitro* translation technique with rabbit reticulocytes and demonstrated the synthesis of specific polypeptides during the early phases of senescence in daylily flowers. Bai et al. (2010) recently applied powerful proteomic techniques to attempt to define key changes in the proteome of senescing petunia flowers. Two-dimensional gel electrophoresis and mass spectrometry of isolated polypeptide spots were used to identify those that changed in a fashion that might suggest a role in senescence. Unfortunately, samples were made at 24, 48, and 72 h after pollination; even the earliest of these time points is known to be long after the key triggering events of pollination-induced senescence (Pech et al. 1987). In addition, proteins were applied to the gel on the basis of total protein content, potentially obscuring important changes in the well-known background of general protein degradation (Lay-Yee et al. 1992). The study identified a small number of polypeptides that appeared to be associated with senescence and were identified by mass spectrographic analysis. The authors' assertion that their data provide evidence for a disconnect between transcript abundance and translated protein products needs confirmation on the basis of a comparison using stable housekeeping proteins (actin, ubiquitin) rather than total protein content. Application of these new and powerful techniques for evaluating the changes in the proteome of geophyte flowers would help in assessing the importance of changes that have already been demonstrated in the transcriptome.

IV. TRANSGENIC STRATEGIES FOR EXTENDING FLORAL LIFE

Although floriculture crops have been a target for transgenic manipulation, the primary focus of commercial activities has been on changing the flower color, especially to produce "blue" carnations and roses. The fact that the products of these efforts are commercially available now indicates the potential for using transgenic approaches to modify other (and arguably more important) features of floral crops. Floral crops offer several advantages for commercialization of transgenic approaches. The high value of floricultural crops, the diversity of taxa to which the same transgenic approaches can be applied, and the relatively short life cycle of these crops all argue for the value of a transgenic approach to plant improvement. Since ornamentals are nonfood crops, registration of transgenic plants is much less cumbersome and expensive than for food crops, and consumer acceptance has already been demonstrated by the transgenic 'Moondust' carnations and 'Applause' blue roses. Indeed, it seems that ornamentals can be an excellent pilot program for demonstrating the value and safety of transgenic breeding in horticultural crops.

The geophyte ornamental crops with ethylene-insensitive flowers seem to be a very suitable target for deploying transgenic strategies to extend postharvest life, since ethylene inhibitors are of no value and the effect of other growth regulators are rather modest, at best. Although there is still no clear demonstration of the initiation signal for senescence in these flowers, we do know that their senescence can be delayed substantially by treating the flowers with cycloheximide, an inhibitor of protein synthesis (Jones et al. 1994; Pak and Van Doorn 2005; Figure 12.12). The use of this metabolic poison is not commercially feasible, but these results suggest a molecular strategy for extending floral life by

Control (DI water)   10 ppm CHI (24-h pulse)
Day 7   Day 7

FIGURE 12.12 **(See color insert.)** Effect of cycloheximide on senescence of tulips. Flowers were pulsed for one day as they opened, and were then held at 20°C. The photograph was taken 7 days after treatment. (Adapted from Jones, R. B. et al. 1994. *J. Am. Soc. Hort. Sci.* 119:1243–1247.)

using a molecular approach to inhibiting protein synthesis. We hypothesized that targeted expression of an antisense sequence to a protein from the ribosome should have the same effect as cycloheximide. In our laboratory we have used inducible systems that allow genes in transformed plants to be "turned on" by the application of a simple chemical regulator. For example, we have used the glucocorticoid receptor-based inducible gene expression system developed by Aoyama and Chua (1997).

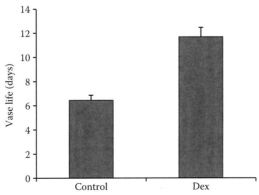

FIGURE 12.13 Effect of induced silencing of protein synthesis on the life of petunia flowers. Flowers from petunias transformed with the GVG/*Antisense RPL2* transgene were placed in water (lower flowers) or 30 μM dexamethasone (the GVG inducer), and held in ethylene-free air at 22°C. The photograph was taken after 6 days. Data are the means ± SD for flowers from four independently transformed lines. (Photo courtesy: G. Stier, M. S. Reid, and C.-Z. Jiang.)

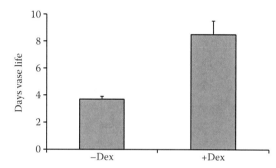

FIGURE 12.14 Effect of induced silencing of the proteasome on the longevity of petunia flowers. Flowers from petunias transformed with the GVG/*Antisense PBB2* transgene were placed in water (upper flowers) or 30 µM dexamethasone, and held in ethylene-free air at 22°C. The photograph was taken after 4 days. Data are the means ± SD for six flowers. (Photo courtesy: G. Stier, M. S. Reid, and C.-Z. Jiang.)

The system is based on the chimeric transcriptional activator GVG, consisting of three functional domains, namely the yeast Gal4 DNA-binding domain, the VP16 activation domain, and the Glucocorticoid receptor domain. Dexamethasone, a glucocorticoid, can be used to induce the expression of transformed target genes. To test the possibility of mimicking the effect of cycloheximide using a transgenic approach, we transformed petunia plants with an antisense construct for *Ribosomal Protein L2* (*RPL2*) under the control of the GVG system. The transformed plants grew normally, but when flowers were exposed to dexamethasone (the GVG system inducer) their life was considerably extended (Figure 12.13).

Most recently, we have capitalized on studies suggesting the importance of the 26S proteasome in the control of flower senescence. Pak and Van Doorn (2005) showed that inhibition of the 26S proteasome (with MG123) would extend the life of the ethylene-insensitive flowers of iris. Xu et al. (2007b) had demonstrated a very strong up-regulation, during senescence, of an E3 ubiquitin ligase that was hypothesized to be a component of the proteasome system. They also demonstrated that VIGS silencing of this gene resulted in flowers with extended longevity. We decided to test the possibility that targeted inhibition of the proteasome would extend floral longevity. In initial studies, VIGS was used to silence selected components of the proteasome in young plants. Because of the central role of the proteasome, we expected silencing to have drastic effects on plant growth and development and this proved to be the case. Silenced portions of the infected plants grew very slowly, resulting in severe malformation of leaves. The most effective silencing was seen with PBB2, the beta subunit of the 26S proteasome (Stier et al. 2010). This protein is an endopeptidase in the 20S core of the proteasome and is thought to play a key role in targeted protein degradation via the ubiquitin pathway (Sullivan et al. 2003; Smalle and Vierstra 2004). Petunia plants were transformed with the GVG system driving an antisense construct of PBB2. Transformed plants grew normally, and their flowers showed normal longevity when placed in water. However, when the flowers were treated with low concentrations of the dexamethasone inducer, their longevity was greatly increased (Stier et al. 2010) (Figure 12.14).

The beneficial effects of CKs in improving flower opening and delaying leaf senescence that have long been reported in ornamentals can be obtained by transgenic modification of CK biosynthesis. Gan and Amasino (1995) demonstrated that leaf senescence could be delayed in transgenic plants expressing IPT, an enzyme that catalyzes the rate-limiting step in CK synthesis. Chang et al. (2003) demonstrated that over-expression of IPT under the control of the *SAG12* promoter resulted in a 6–10-day delay in floral senescence, compared to wild-type flowers. Flowers from IPT- expressing plants were less sensitive to exogenous ethylene and required longer times to induce endogenous ethylene production, corolla senescence, and up-regulation of a senescence-related cysteine protease. The beneficial effects of applying CKs in the vase life of some ornamental geophytes suggest that this strategy is worth exploring when effective transformation/regeneration systems become available.

V. FUTURE PROSPECTS

Horticultural and physiological research over the past 30 years has given us a good understanding of the factors that affect the life of cut flowers, potted plants, and other ornamentals. These findings will be the key to future strategies employing transgenic and other technologies. The remarkable effects of transgenic manipulation of genes involved in petal senescence point to the potential for such strategies not only to dramatically improve the postharvest performance of ornamentals currently in the trade, but also to expand the palette of ornamentals in the trade. Many beautiful ephemeral flowers, including geophytes, are seldom seen in the vase because of their short display life, but the application of the transgenic techniques described above could enable the commercialization of spectacular flowers such as *Tigridia* (to select just one from a host of possible examples). Similarly, transgenic manipulations will be used to improve the quality and display life of potted plants by reducing their water loss, preventing leaf yellowing, and extending flower life.

The results of the few studies that have been conducted to compare the longevity of flowers of different cultivars of the same geophytes species suggest that there is also considerable potential for improving postharvest life by including this as a key trait in classical breeding programs. Marker-assisted selection can surely be used to reduce the time for the generation of flowers with reduced sensitivity to ethylene, reduced Q_{10}, improved resistance to desiccation, delayed leaf yellowing, and delayed tepal senescence and/or abscission. The characteristics of such germplasm will naturally also improve the sustainability of postharvest treatments, by reducing the need for chemical treatments, including antiethylene compounds, fungicides, and vase solution biocides.

CHAPTER CONTRIBUTORS

Michael S. Reid is a professor and extension specialist emeritus at the University of California, Davis, California, USA. He was born and educated in Auckland, New Zealand, where he obtained his BS in botany, MS in microbiology, and PhD in cell biology from the University of Auckland. After postdoctoral studies at UC Davis and the Food Research Institute, Norwich, England, and eight years as a research scientist at the New Zealand Department of Scientific and Industrial Research, he was appointed to the faculty in the Department of Environmental Horticulture at UC Davis. He retired in 2010 after 30 years working on the postharvest handling of ornamentals, including research on many geophyte crops. He has published more than 600 refereed articles, books, chapters, conference papers, and popular articles. Dr. Reid was a Fulbright Fellow, and was awarded DSc by Auckland University. He is a recipient of the Alex Laurie awards from the American Society for Horticultural Science and from the Society of American Florists and the Alan Armitage Award from the Specialty Cut Flower Growers Association, and was nominated as the Morrison Lecturer by the USDA Agricultural Research Service.

Cai-Zhong Jiang is a research plant physiologist at the USDA-ARS. He received his BS in agronomy from the Guangxi Agricultural University, China, and his MS and PhD in crop sciences from the Tokyo University of Agriculture and Technology, Japan. After postdoctoral studies at Iowa State University and the University of California, Davis, he took up a job as a senior scientist at Mendel Biotechnology, Inc., in Hayward, California. The main areas of his expertise include the functions of plant transcription factor genes as the basis for creating novel products in agriculture. Dr. Jiang has published more than 50 refereed articles, books, or chapters, conference papers, and popular articles. He has also been awarded more than 10 US patents.

REFERENCES

Aoyama, T., and N.-H. Chua. 1997. A glucocorticoid-mediated transcriptional induction system in transgenic plants. *Plant J.* 11:605–612.

Arditti, J. 1971. The place of the orchid pollen "poison" and pollenhormon in the history of plant hormones. *Am. J. Bot.* 58:480–481.

Arditti, J. 1975. Orchids, pollen poison, pollenhormon and plant hormones. *Orchid Rev.* 83:127–129.

Arditti, J., B. Flick, and D. Jeffrey. 1971. Post-pollination phenomena in orchid flowers II. Induction of symptoms by abscisic acid and its interactions with auxin, gibberellic acid and kinetin. *New Phytol.* 70:333–341.

Ashman, T.-L., and D. J. Schoen. 1994. How long should flowers live? *Nature* 371:788–791.

Bai, S., B. Willard, L. J. Chapin, M. T. Kinter, D. M. Francis, A. D. Stead, and M. L. Jones. 2010. Proteomic analysis of pollination-induced corolla senescence in petunia. *J. Exp. Bot.* 61:1089–1109.

Beyer, Jr., E. 1976. A potent inhibitor of ethylene action in plants. *Plant Physiol.* 58:268–271.

Blanchard, M. G., L. A. Newton, E. S. Runkle, D. Woolard, and C. A. Campbell. 2007. Exogenous applications of abscisic acid improved the postharvest drought tolerance of several annual bedding plants. *Acta Hort.* 755:127–132.

Blankenship, S. M., and E. C. Sisler. 1993. Response of apples to diazocyclopentadiene inhibition of ethylene binding. *Postharvest Biol. Technol.* 3:95–101.

Borochov, A., and V. Keren-Paz. 1984. Bud opening of cut liatris flowers. *Scientia Hortic.* 23:85–89.

Bovy, A. G., G. C. Angenent, H. J. M. Dons, and A.-C. van Altvorst. 1999. Heterologous expression of the *Arabidopsis etr1-1* allele inhibits the senescence of carnation flowers. *Mol. Breed.* 5:301–308.

Breeze, E., C. Wagstaff, E. Harrison, I. Bramke, H. Rogers, A. Stead, B. Thomas, and V. Buchanan-Wollaston. 2004. Gene expression patterns to define stages of post-harvest senescence in *Alstroemeria* petals. *Plant Biotechnol. J.* 2:155–168.

Brodersen, C. R., A. J. McElrone, B. Choat, M. A. Matthews, and K. A. Shackel. 2010. The dynamics of embolism repair in xylem: *In vivo* visualizations using high-resolution computed tomography. *Plant Physiol.* 154:1088–1095.

Bufler, G. 1984. Ethylene-enhanced 1-aminocyclopropane-1-carboxylic acid synthase activity in ripening apples. *Plant Physiol.* 75:192–195.

Bufler, G. 1986. Ethylene-promoted conversion of 1-aminocyclopropane-1-carboxylic acid to ethylene in peel of apple at various stages of fruit development. *Plant Physiol.* 80:539–543.

Celikel, F. G., J. C. Cevallos, and M. S. Reid. 2010. Temperature, ethylene and the postharvest performance of cut snapdragons (*Antirrhinum majus*). *Scientia Hortic.* 125:429–433.

Celikel, F. G., and M. S. Reid. 2002. Storage temperature affects the quality of cut flowers from the Asteraceae. *HortScience* 37:148–150.

Celikel, F. G., and M. S. Reid. 2005. Temperature and postharvest performance of rose (*Rosa hybrida* L. 'First Red') and gypsophila (*Gypsophila paniculata* L. 'Bristol Fairy') flowers. *Acta Hort.* 682:1789–1794.

Celikel, F. G., and W. G. van Doorn. 1995. Effects of water stress and gibberellin on flower opening in *Iris × hollandica*. *Acta Hort.* 405:246–252.

Cevallos, J. C., and M. S. Reid. 2000. Effects of temperature on the respiration and vase life of *Narcissus* flowers. *Acta Hort.* 517:335–342.

Cevallos, J.-C., and M. S. Reid. 2001. Effect of dry and wet storage at different temperatures on the vase life of cut flowers. *HortTechnology* 11:199–202.

Chang, H., M. L. Jones, G. M. Banowetz, and D. G. Clark. 2003. Overproduction of cytokinins in petunia flowers transformed with P_{SAG12}-IPT delays corolla senescence and decreases sensitivity to ethylene. *Plant Physiol.* 132:2174–2183.

Chen, J.-C., C.-Z. Jiang, T. E. Gookin, D. A. Hunter, D. G. Clark, and M. S. Reid. 2004. Chalcone synthase as a reporter in virus-induced gene silencing studies of flower senescence. *Plant Mol. Biol.* 55:521–530.

Chen, J.-C., C.-Z. Jiang, and M. S. Reid. 2005. Silencing a prohibitin alters plant development and senescence. *Plant J.* 44:16–24.

Cornish, K., A. I. King, M. S. Reid, and J. L. Paul. 1985. Role of ABA in stress-induced reduction of water loss from potted chrysanthemum plants. *Acta Hort.* 167:381–386.

Crocker, W. 1913. The effects of advancing civilization upon plants. *School Science and Mathematics* 13:277–289.

Crowe, J. H., L. M. Crowe, and D. Chapman. 1984. Preservation of membranes in anhydrobiotic organisms: The role of trehalose. *Science* 223:701–703.

Darras, A. I., L. A. Terry, and D. C. Joyce. 2005. Methyl jasmonate vapour treatment suppresses specking caused by *Botrytis cinerea* on cut *Freesia hybrida* L. flowers. *Postharvest Biol. Technol.* 38:175–182.

Donnelly, L. M., M. S. Reid, and C.-Z. Jiang. 2010. Virus-induced gene silencing of a NAC transcription factor alters flower morphology and accelerates flower senescence in petunia. *HortScience* 45:S161.

Eason, J. R., and T. T. Bucknell. 1999. DNA processing during tepal senescence of *Sandersonia aurantiaca*. *Acta Hort.* 543:143–146.

Eisinger, W. 1977. Role of cytokinins in carnation flower senescence. *Plant Physiol.* 59:707–709.

Elibox, W., and P. Umaharan. 2008. Morphophysiological characteristics associated with vase life of cut flowers of anthurium. *HortScience* 43:825–831.

Ferrante, A., D. Hunter, W. Hackett, and M. Reid. 2001. TDZ: A novel tool for preventing leaf yellowing in *Alstroemeria* flowers. *HortScience* 36:599.

Ferrante, A., D. Hunter, W. Hackett, and M. Reid. 2002. Thidiazuron-a potent inhibitor of leaf senescence in *Alstroemeria*. *Postharvest Biol. Technol.* 25:333–338.

Gan, S., and R. M. Amasino. 1995. Inhibition of leaf senescence by autoregulated production of cytokinin. *Science* 270:1986–1988.

Halevy, A. H., A. M. Kofranek, and S. T. Besemer. 1978. Postharvest handling methods for bird of paradise flowers (*Sterlitzia reginae* Ait.). *J. Am. Soc. Hort. Sci.* 103:165–169.

Halevy, A., and S. Mayak. 1979. Senescencs and postharvest physiology of cut flowers, part 1. *Hort. Rev.* 1:204–236.

Halevy, A. H., C. S. Whitehead, and A. M. Kofranek. 1984. Does pollination induce corolla abscission of cyclamen flowers by promoting ethylene production? *Plant Physiol.* 75:1090–1093.

Hammer, P. E., S. F. Yang, M. S. Reid, and J. J. Marois. 1990. Postharvest control of *Botrytis cinerea* infections on cut roses using fungistatic storage atmospheres. *J. Am. Soc. Hort. Sci.* 115:102–107.

Han, S. S. 2001. Benzyladenine and gibberellins improve postharvest quality of cut Asiatic and Oriental lilies. *HortScience* 36:741–745.

Han, S. S., A. H. Halevy, and M. S. Reid. 1991. The role of ethylene and pollination in petal senescence and ovary growth of brodiaea. *J. Am. Soc. Hort. Sci.* 116:68–72.

Hanley, K. M., and W. J. Bramlage. 1989. Endogenous levels of abscisic acid in aging carnation flower parts. *J. Plant Growth Regul.* 8:225–236.

Hobson, G. E., and R. Nichols. 1977. Enzyme changes during petal senescence in the carnation. *Ann. Appl. Biol.* 85:445–447.

Hunter, D. A., A. Ferrante, P. Vernieri, and M. S. Reid. 2004a. Role of abscisic acid in perianth senescence of daffodil (*Narcissus pseudonarcissus* 'Dutch Master'). *Physiol. Plant.* 121:313–321.

Hunter, D. A., N. E. Lange, and M. S. Reid. 2004b. Physiology of flower senescence. In *Plant Cell Death Processes*, ed. L. D. Nooden, 307–318, San Diego: Elsevier.

Hunter, D. A., B. C. Steele, and M. S. Reid. 2002. Identification of genes associated with perianth senescence in daffodil (*Narcissus pseudonarcissus* L. 'Dutch Master'). *Plant Sci.* 163:13–21.

Jones, R. B., M. Serek, C.-L. Kuo, and M. S. Reid. 1994. The effect of protein synthesis inhibition on petal senescence in cut bulb flowers. *J. Am. Soc. Hort. Sci.* 119:1243–1247.

Jones, R. B., M. Serek, and M. S. Reid. 1993. Pulsing with Triton X-100 improves hydration and vase life of cut sunflowers (*Helianthus annuus* L.). *HortScience* 28:1178–1179.

Jordi, W., G. M. Stoopen, K. Kelepouris, and W. M. van der Krieken. 1995. Gibberellin-induced delay of leaf senescence of *Alstroemeria* cut flowering stems is *not* caused by an increase in the endogenous cytokinin content. *J. Plant Growth Regul.* 14:121–127.

Joyce, D. C., and M. S. Reid. 1985. Effect of pathogen-suppressing modified atmospheres on stored cut flowers. In *Controlled Atmospheres for Storage and Transport of Perishable Agricultural Commodities*, ed. S. M. Blankenship, 185–198. Raleigh, NC: North Carolina State University.

Kader, A. A. 2003. A perspective on postharvest horticulture (1978–2003). *HortScience* 38:1004–1008.

Kader, A. A., D. Zagory, E. L. Kerbel, and C. Y. Wang. 1989. Modified atmosphere packaging of fruits and vegetables. *Crit. Rev. Food Sci. Nutr.* 28:1–30.

Kidd, F., and C. West. 1933. Effects of ethylene and of apple vapours on the ripening of fruits. *Reports, Food Investigation Board, London*, 55–58.

Kohl, H., and D. Rundle. 1972. Decreasing water loss of cut roses with abscisic acid. *HortScience.* 7:249.

Lawton, K. A., K. G. Raghothama, P. B. Goldsbrough, and W. R. Woodson. 1990. Regulation of senescence-related gene expression in carnation flower petals by ethylene. *Plant Physiol.* 93:1370–1375.

Lay-Yee, M., A. D. Stead, and M. S. Reid. 1992. Flower senescence in daylily (*Hemerocallis*). *Physiol. Plant.* 86:308–314.

Macnish, A. J., A. de Theije, M. S. Reid, and C.-Z. Jiang. 2009a. An alternative postharvest handling strategy for cut flowers—dry handling after harvest. *Acta Hort.* 847:215–221.

Macnish, A. J., C.-Z. Jiang, F. Negre-Zakharov, and M. S. Reid. 2010a. Physiological and molecular changes during opening and senescence of *Nicotiana mutabilis* flowers. *Plant Sci.* 179:267–272.

Macnish, A. J., C.-Z. Jiang, and M. S. Reid. 2010b. Treatment with thidiazuron improves opening and vase life of iris flowers. *Postharvest Biol. Technol.* 56:77–84.

Macnish, A. J., R. T. Leonard, and T. A. Nell. 2008. Treatment with chlorine dioxide extends the vase life of selected cut flowers. *Postharvest Biol. Technol.* 50:197–207.

Macnish, A. J., K. L. Morris, A. de Theije, M. G. J. Mensink, H. A. M. Boerrigter, M. S. Reid, C.-Z. Jiang, and E. J. Woltering. 2010c. Sodium hypochlorite: A promising agent for reducing *Botrytis cinerea* infection on rose flowers. *Postharvest Biol. Technol.* 58:262–267.

Macnish, A. J., M. S. Reid, and D. C. Joyce. 2009b. Ornamentals and cut flowers. In *Modified and Controlled Atmospheres for the Storage, Transportation, and Packaging of Horticultural Commodities*, ed. E. M. Yahia, 491–506. Boca Raton, FL: CRC Press.

Matile, Ph., and F. Winkenbach. 1971. Function of lysosomes and lysosomal enzymes in the senescing corolla of the morning glory (*Ipomoea purpurea*). *J. Exp. Bot.* 22:759–771.

Maxie, E. C., D. S. Farnham, F. G. Mitchell, N. F. Sommer, R. A. Parson, R. G. Snyder, and H. L. Rae. 1973. Temperature and ethylene effects on cut flowers of carnation (*Dianthus caryophyllus*). *J. Am. Soc. Hort. Sci.* 98:568–572.

Mayak, S., B. Bravdo, A. Gvilli, and A. H. Halevy. 1973. Improvement of opening of cut gladioli flowers by pretreatment with high sugar concentrations. *Scientia Hortic.* 1:357–365.

Mayak, S., and D. R. Dilley. 1976. Regulation of senescence in carnation (*Dianthus caryophyllus*). Effect of abscisic acid and carbon dioxide on ethylene production. *Plant Physiol.* 58:663–665.

Mayak, S., and A. H. Halevy. 1970. Cytokinin activity in rose petals and its relation to senescence. *Plant Physiol.* 46:497–499.

Mayak, S., and A. H. Halevy. 1972. Interrelationships of ethylene and abscisic acid in the control of rose petal senescence. *Plant Physiol.* 50:341–346.

Mayak, S., and A. M. Kofranek. 1976. Altering the sensitivity of carnation flowers (*Dianthus caryophyllus* L.) to ethylene. *J. Am. Soc. Hort. Sci.* 101:503–506.

Mor, Y., A. H. Halevy, A. M. Kofranek, and M. S. Reid. 1984. Posthavest handling of Lily of the Nile flowers. *J. Am. Soc. Hort. Sci.* 109:494–497.

Müller, R., B. M. Stummann, A. S. Andersen, and M. Serek. 1999. Involvement of ABA in postharvest life of miniature potted roses. *Plant Growth Regul.* 29:143–150.

Mutui, T. M., V. N. Emongor, and M. J. Hutchinson. 2003. Effect of benzyladenine on the vase life and keeping quality of *Alstroemeria* cut flowers. *J. Agric. Sci. Technol.* 5:91–105.

Mutui, T. M., V. E. Emongor, and M. J. Hutchinson 2006. The effects of gibberellin$_{4+7}$ on the vase life and flower quality of *Alstroemeria* cut flowers. *Plant Growth Regul.* 48:207–214.

Naidu, S. N., and M. S. Reid. 1989. Postharvest handling of tuberose (*Polianthes tuberosa* L.). *Acta Hort.* 261:313–318.

Nowak, J., and H. Veen. 1982. Effects of silver thiosulfate on abscisic acid content in cut carnations as related to flower senescence. *J. Plant Growth Regul.* 1:153–159.

Onoue, T., M. Mikami, T. Yoshioka, T. Hashiba, and S. Satoh. 2000. Characteristics of the inhibitory action of 1,1-dimethyl-4-(phenylsulfonyl)semicarbazide (DPSS) on ethylene production in carnation (*Dianthus caryophyllus* L.) flowers. *Plant Growth Regul.* 30:201–207.

Otsubo, M., and M. Iwaya-Inoue. 2000. Trehalose delays senescence in cut gladiolus spikes. *HortScience* 35:1107–1110.

Pak, C., and W. G. van Doorn. 2005. Delay of *Iris* flower senescence by protease inhibitors. *New Phytol.* 165:473–480.

Panavas, T., P. D. Reid, and B. Rubinstein. 1998a. Programmed cell death of daylily petals: Activities of wall-based enzymes and effects of heat shock. *Plant Physiol. Biochem.* 36:379–388.

Panavas, T., E. L. Walker, and B. Rubinstein. 1998b. Possible involvement of abscisic acid in senescence of daylily petals. *J. Exp. Bot.* 49:1987–1997.

Pech, J. C., A. Latche, C. Larrigaudiere, and M. S. Reid. 1987. Control of early ethylene synthesis in pollinated petunia flowers. *Plant Physiol. Biochem.* 25:431–437.

Pemberton, H. B., A. J. Macnish, M.S. Reid, C.-Z. Jiang, and W. R. Roberson. 2009. Abscisic acid: A potential treatment for reducing leaf wilting in potted bedding plants. *HortScience* 44(4):S1147.

Phillips, Jr, H. L., and H. Kende. 1980. Structural changes in flowers of *Ipomoea tricolor* during flower opening and closing. *Protoplasma* 102:199–215.

Philosoph-Hadas, S., S. Meir, I. Rosenberger, and A. H. Halevy. 1996. Regulation of the gravitorpic response and ethylene biosynthesis in gravistimulated snapdragon spikes by calcium chelators and ethylene inhibitors. *Plant Physiol.* 110:301–310.

Price, A. M., D. F. Aros Orellana, F. M. Salleh, R. Stevens, R. Acock, V. Buchanan-Wollaston, A. D. Stead, and H. J. Rogers. 2008. A comparison of leaf and petal senescence in wallflower reveals common and distinct patterns of gene expression and physiology. *Plant Physiol.* 147:1898–1912.

Radin, J. W., and R. C. Ackerson. 1982. Does abscisic acid control stomatal closure during water stress? What's New. *Plant Physiol.* 12:9–12.

Ranwala, A. P., and W. B. Miller. 2002. Effects of gibberellin treatments on flower and leaf quality of cut hybrid lilies. *Acta Hort.* 570:205–210.

Reid, M. 2001. Advances in shipping and handling of ornamentals. *Acta Hort.* 543:277–284.

Reid, M. S., and C. Z. Jiang. 2005. New strategies in transportation for floricultural crops. *Acta Hort.* 682:1667–1673.

Rogers, H. J. 2006. Programmed cell death in floral organs: How and why do flowers die? *Ann. Bot.* 97:309–315.

Ronen, M., and S. Mayak. 1981. Interrelationship between abscisic acid and ethylene in the control of senescence processes in carnation flowers. *J. Exp. Bot.* 32:759–765.

Rot, I., and H. Friedman. 2010. Desiccation-induced reduction in water uptake of gypsophila florets and its amelioration. *Postharvest Biol. Technol.* 57:189–195.

Sankhla, N., W. A. Mackay, and T. D. Davis. 2003. Reduction of flower abscission and leaf senescence in cut phlox inflorescences by thidiazuron. *Acta Hort.* 628:837–841.

Sankhla, N., W. A. Mackay, and T. D. Davis. 2005. Effect of thidiazuron on senescence of flowers in cut inflorescences of *Lupinus densiflorus* Benth. *Acta Hort.* 669:239–243.

Savin, K. W., S. C. Baudinette, M. W. Graham, M. Z. Michael, G. D. Nugent, C.-Y. Lu, S. F. Chandler, and E. C. Cornish. 1995. Antisense ACC oxidase RNA delays carnation petal senescence. *HortScience* 30:970–972.

Serafini-Fracassini, D., S. Del Duca, F. Monti, F. Poli, G. Sacchetti, A. M. Bregoli, S. Biondi, and M. Della Mea. 2002. Transglutaminase activity during senescence and programmed cell death in the corolla of tobacco (*Nicotiana tabacum*) flowers. *Cell Death Differ.* 9:309–321.

Serek, M., R. B. Jones, and M. S. Reid. 1994a. Role of ethylene in opening and senescence of *Gladiolus* sp. flowers. *J. Am. Soc. Hort. Sci.* 119:1014–1019.

Serek, M., E. C. Sisler, and M. S. Reid. 1994b. Novel gaseous ethylene binding inhibitor prevents ethylene effects in potted flowering plants. *J. Am. Soc. Hort. Sci.* 119:1230–1233.

Serek, M., E. C. Sisler, and M. S. Reid. 1995a. Effects of 1-MCP on the vase life and ethylene response of cut flowers. *Plant Growth Regul.* 16:93–97.

Serek, M., E. C. Sisler, T. Tirosh, and S. Mayak. 1995b. 1-Methylcyclopropene prevents bud, flower, and leaf abscission of Geraldton waxflower. *HortScience* 30:1310.

Sisler, E. C., and S. M. Blankenship. 1993. Diazocyclopentadiene (DACP), a light sensitive reagent for the ethylene receptor in plants. *Plant Growth Regul.* 12:125–132.

Sisler, E. C., and S. M. Blankenship. 1996. Method of counteracting an ethylene response in plants. In *US Patent 5518988*, ed. USPTO. Raleigh, NC: North Carolina State University.

Sisler, E. C., R. Goren, and M. Huberman. 1984. Effect of 2,5-norborbadiene on citrus leaf explants. *Plant Physiol.* S75:127.

Sisler, E. C., and N. Lallu. 1994. Effect of diazocyclopentadiene (DACP) on tomato fruits harvested at different ripening stages. *Postharvest Biol. Technol.* 4:245–254.

Sisler, E. C., and S. F. Yang. 1984. Anti-ethylene effects of *cis*-2-butene and cyclic olefins. *Phytochemistry* 23:2765–2768.

Smalle, J., and R. D. Vierstra. 2004. The ubiquitin 26S proteasome proteolytic pathway. *Annu. Rev. Plant Biol.* 55:555–590.

Smith, M. T., Y. Saks, and J. van Staden. 1992. Ultrastructural changes in the petals of senescing flowers of *Dianthus caryophyllus* L. *Ann. Bot.* 69:277–285.

Spikman, G. 1989. Development and ethylene production of buds and florets of cut freesia inflorescences as influenced by silver thiosulphate, aminoethoxyvinylglycine and sucrose. *Scientia Hortic.* 39:73–81.

Staby, G. L., and M. S. Reid. 2005a. *Improving the cold chain for cut flowers and potted plants, white paper, Vol. White paper.* Annapolis Maryland: WF & FSA. http://www.wffsa.org/pdf/2005/coldchainwhitepaper72505.pdf.

Staby, G. L., and M. S. Reid. 2005b. A New Technology for Floriculture—Enhanced Radio Frequency Identification (RFID) Tags. WF&FSA, Annapolis MD. http://www.wffsa.org/pdf/Robin/netWORK/RFIDtags.pdf.

Staby, G. L., and M. S. Reid. 2007. Improving the cold chain for cut flowers and potted plants, white paper II, Vol. White Paper II. WF&FSA, Annapolis MD. http://www.wffsa.org/pdf/2007/WhitePaperii FinalcorrectedApr2007.pdf.

Stead, A. D. 1992. Pollination-induced flower senescence: a review. *Plant Growth Regul.* 11:13–20.

Stead, A. D., and W. van Doorn. 1994. Strategies of flower senescence—a review. In *Molecular and Cellular Aspects of Plant Reproduction. SEB Seminar Series 55*, eds. R. J. Scott, and A. D. Stead, 215–237. Cambridge: Cambridge University Press.

Stier, G. N., P. Kumar, C.-Z. Jiang, and M. S. Reid. 2010. Silencing of a proteasome component delays floral senescence. *HortScience* 45:S147.

Sullivan, J. A., K. Shirasu, and X. W. Deng. 2003. The diverse roles of ubiquitin and the 26S proteasome in the life of plants. *Nat. Rev. Genet.* 4:948–958.

Teas, H. J., T. J. Sheehan, and T. W. Holmsen. 1959. Control of geotropic bending in snapdragon and gladiolus inflorescences. *Proc. Florida State Hortic. Soc.* 72:437–442.

Valpuesta, V., N. E. Lange, C. Guerrero, and M. S. Reid. 1995. Up-regulation of a cysteine protease accompanies the ethylene-insensitive senescence of daylily (*Hemerocallis*) flowers. *Plant Mol. Biol.* 28:575–582.

Van der Meulen-Muisers, J. J. M., J. C. van Oeveren, J. Jansen, and J. M. van Tuyl. 1999. Genetic analysis of postharvest flower longevity in Asiatic hybrid lilies. *Euphytica* 107:149–157.

Van Doorn, W. G., P. A. Balk, A. M. van Houwelingen, F. A. Hoeberichts, R. D. Hall, O. Vorst, C. van der Schoot, and M. F. van Wordragen. 2003. Gene expression during anthesis and senescence in *Iris* flowers. *Plant Mol. Biol.* 53:845–863.

Van Doorn, W. G., R. R. J. Perik, P. Abadie, and H. Harkema. 2011. A treatment to improve the vase life of cut tulips: Effects on tepal senescence, tepal abscission, leaf yellowing and stem elongation. *Postharvest Biol. Technol.* 61:56–63.

Van Doorn, W. G., and M. S. Reid. 1995. Vascular occlusion in stems of cut rose flowers exposed to air: Role of xylem anatomy and rates of transpiration. *Physiol. Plant.* 93:624–629.

Van Doorn, W., and A. Stead. 1994. The physiology of petal senescence which is not initiated by ethylene. In *Molecular and Cellular Aspects of Plant Reproduction*, eds. R. J. Scott, and A. Stead, 239–254. Cambridge: Cambridge University Press.

Van Doorn, W. G., and E. J. Woltering. 2004. Senescence and programmed cell death: substance or semantics? *J. Exp. Bot.* 55:2147–2153.

Van Doorn, W. G., and E. J. Woltering. 2005. Many ways to exit? Cell death categories in plants. *Trends Plant Sci.* 10:117–122.

Van Doorn, W. G., and E. J. Woltering. 2008. Physiology and molecular biology of petal senescence. *J. Exp. Bot.* 59:453–480.

Van Doorn, W. G., and E. J. Woltering. 2010. What about the role of autophagy in PCD? *Trends Plant Sci.* 15:361–362.

Van Iersel, M. W., K. Seader, and S. Dove. 2009. Exogenous abscisic acid application effects on stomatal closure, water use, and shelf life of hydrangea (*Hydrangea macrophylla*). *J. Environ. Hortic.* 27:234–238.

Van Meeteren, U. 2007. Why do we treat flowers the way we do? A system analysis approach of the cut flower postharvest chain. *Acta Hort.* 755:61–74.

Van Staden, J., and G. G. Dimalla. 1980. Endogenous cytokinins in *Bougainvillea* 'San Diego Red' I. Occurrence of cytokinin glucosides in the root sap. *Plant Physiol.* 65:852–854.

Van Staden, J., S. J. Upfold, A. D. Bayley, and F. E. Drewes. 1990. Cytokinins in cut carnation flowers. IX. Transport and metabolism of *iso*-pentenyladenine and the effect of its derivatives on flower longevity. *Plant Growth Regul.* 9:255–261.

Veen, H., and S. C. van de Geijn. 1978. Mobility and ionic form of silver as related to longevity of cut carnations. *Planta* 140:93–96.

Waithaka, K., M. S. Reid, and L. L. Dodge. 2001. Cold storage and flower keeping quality of cut tuberose (*Polianthes tuberosa* L.). *J. Hort. Sci. Biotech.* 76:271–275.

Wang, C. Y., and J. E. Baker. 1979. Vase life of cut flowers treated with rhizobitoxine analogs, sodium benzoate, and isopentenyl adenosine. *HortScience* 14:59–60.

Waterland, N. L., C. A. Campbell, J. J. Finer, and M. L. Jones. 2010. Abscisic acid application enhances drought stress tolerance in bedding plants. *HortScience* 45:409–413.

Wilkinson, S., and W. J. Davies. 2002. ABA-based chemical signalling: the co-ordination of responses to stress in plants. *Plant, Cell Environ.* 25:195–210.

Woltering, E. J., P. A. Balk, M. A. Nijenhuis-deVries, M. Faivre, G. Ruys, D. Somhorst, S. Philosoph-Hadas, and H. Friedman. 2005. An auxin-responsive 1-aminocyclopropane-1-carboxylate synthase is responsible for differential ethylene production in gravistimulated *Antirrhinum majus* L. flower stems. *Planta* 220:403–413.

Woltering, E. J., and W. G. van Doorn. 1988. Role of ethylene in senescence of petals - morphological and taxonomical relationships. *J. Exp. Bot.* 39:1605–1616.

Woodson, W. R., K. Y. Park, A. Drory, P. B. Larsen, and H. Wang. 1992. Expression of ethylene biosynthetic pathway transcripts in senescing carnation flowers. *Plant Physiol.* 99:526–532.

Wu, M. J., W. G. van Doorn, and M. S. Reid. 1991. Variation in the senescence of carnation (*Dianthus caryophyllus* L.) cultivars. I. Comparison of flower life, respiration and ethylene biosynthesis. *Scientia Hortic.* 48:99–107.

Xu, X., T. Gookin, C.-Z. Jiang, and M. Reid. 2007a. Genes associated with opening and senescence of *Mirabilis jalapa* flowers. *J. Exp. Bot.* 58:2193–2201.

Xu, X., C.-Z. Jiang, L. Donnelly, and M. S. Reid. 2007b. Functional analysis of a RING domain ankyrin repeat protein that is highly expressed during flower senescence. *J. Exp. Bot.* 58:3623–3630.

Yamada, T., K. Ichimura, M. Kanekatsu, and W. G. van Doorn. 2007. Gene expression in opening and senescing petals of morning glory (*Ipomoea nil*) flowers. *Plant Cell Rep.* 26:823–835.

Yamada, T., Y. Takatsu, M. Kasumi, K. Ichimura, and W. G. van Doorn. 2006. Nuclear fragmentation and DNA degradation during programmed cell death in petals of morning glory (*Ipomoea nil*). *Planta* 224:1279–1290.

Yamada, T., Y. Takatsu, T. Manabe, M. Kasumi, and W. Marubashi. 2003. Suppressive effect of trehalose on apoptotic cell death leading to petal senescence in ethylene-insensitive flowers of gladiolus. *Plant Sci.* 164:213–221.

Zagory, D., and M. S. Reid. 1986. Evaluation of the role of vase microorganisms in the postharvest life of cut flowers. *Acta Hort.* 181:207–218.

13 Sustainable Production and Integrated Management
Environmental Issues

Gary A. Chastagner, Gordon R. Hanks, Margery L. Daughtrey, Iris Yedidia, Timothy W. Miller, and Hanu R. Pappu

CONTENTS

I. INTRODUCTION

The development and implementation of cost-effective sustainable production practices that rely on an integrated approach to managing disease, pest, and weed problems are critical to the long-term field and greenhouse production of ornamental geophytes. Numerous papers have appeared in the proceedings of the 10 international symposia on flower bulbs that have been held since 1970 (Schenk 1971a; Rees and Van der Borg 1975; Rasmussen 1980; Bogers and Bergman 1986; Doss et al. 1990; Saniewski et al. 1992; Lilien-Kipnis et al. 1997; Littlejohn et al. 2002; Okubo et al. 2005; Van den Ende et al. 2011), and the crop chapters in De Hertogh and Le Nard's (1993a) book that illustrate the importance of pest management in the production of high-quality bulbs, cut flowers, and pot-grown plants. In addition to direct damage to yield and quality, the presence of quarantine pests and pathogens limits access to certain export markets.

While a variety of approaches are generally used, pesticides have played a predominant role in ornamental geophyte disease, pest, and weed management programs in the past. Increasing restrictions and the loss of widely used pesticides, such as methyl bromide, formaldehyde, and PCNB (pentachloronitrobenzene, quintozene), and problems associated with the development of resistance to pesticides have resulted in an urgent need to develop environmentally acceptable management strategies. This chapter provides a historical perspective on changes in management strategies and an overview of some recent research activities relating to disease, pest and weed management. Contributions to disease and pest management through breeding for host resistance are covered in Chapters 6 and 7. Advances in the management of postharvest diseases of cut flowers and potted plants can be found in Chapter 12.

II. DISEASES, ARTHROPOD PESTS, AND WEEDS

A number of publications are available that provide comprehensive information on specific diseases, pests, and weeds that commonly affect the production and quality of ornamental geophytes (Moore et al. 1979; Lane 1984; Chastagner and Byther 1985; Strider 1985; ADAS 1986; Rees 1992; Baker 1993; Byther and Chastagner 1993; Loebenstein et al. 1995; Van Rijn and Pfaff 1995; De Best and Zwart 2000; Van Aartrijk et al. 2000).

A. Diseases

Diseases are generally separated into two categories: those caused by infectious pathogens and those caused by noninfectious agents. Infectious or biotic diseases are often classified by the type of pathogen that causes the disease. Infectious diseases of geophytes are generally caused by fungi, oomycetes, viruses, bacteria, phytoplasmas, and nematodes. There are also a number of noninfectious or abiotic diseases.

Fungi are incapable of producing their own food but generally absorb it from living organisms (including plants) or dead organic matter. As a group, they cause a number of economically important production and postharvest diseases on bulb and cut flower crops. Soilborne fungi such as the *Fusarium* spp. that cause bulb rot and wilting diseases and *Rhizoctonia* spp., which cause bulb-, foot-, and root-rots, are among the most difficult to control. There are also a number of economically important aerially-dispersed fungi such as *Botrytis* spp. that cause leaf spots, foliage blights, bulb rots, and postharvest flower decays. If left uncontrolled, fungal diseases can have a devastating impact on the production and marketability of bulbs and cut flowers.

Oomycetes are more closely related to brown algae than true fungi. The cell walls of these fungal-like organisms are composed of cellulose rather than chitin and they produce motile spores known as zoospores that spread in water. *Pythium* and *Phytophthora* spp. are soilborne oomycetes that frequently cause root rots and/or bulb decays when geophytes are planted in poorly-drained soils or where plants are over-irrigated. Downy mildews are foliar diseases that are caused by members of the *Peronosporaceae*. *Olpidium brassicae* is a water mold that vectors *Tobacco necrosis virus* (TNV), which causes "Augusta" disease in tulips.

Viruses are infectious, submicroscopic, intracellular agents that are composed of nucleic acid (mostly RNA) and protein (Loebenstein et al. 1995; Hull 2002). Vegetatively propagated crops are very prone to the build-up of viral diseases. All viruses are obligate parasites and cannot reproduce outside of a living host cell. Virus-infected plants exhibit a range of symptoms depending on the host, virus, and environmental conditions. These can include flower break and chlorotic mosaic, streaking, or ringspot patterns on the foliage. Bulb growth is generally reduced, resulting in a significant decline in productivity of infected stocks. Allen (1975) showed that compared to infected stock, virus-free 'Enchantment' lilies were twice as tall with three times the number of leaves and bulb weight. Others have reported up to 50% increase in yields associated with the production of virus-free planting stock (Sutton et al. 1986a; Asjes 1990). In addition to the direct losses caused by viruses, there are international quarantines against several of them. Plant material with these viruses cannot be exported, resulting in severe economic damage. Viruses are transmitted from plant to plant by insects, nematodes, mites, sap (mechanical), and oomycetes (Loebenstein et al. 1995). Examples of important aphid-vectored viruses of flower bulbs include *Tulip breaking virus* (TBV), *Lily symptomless virus* (LSV), *Bean yellow mosaic virus* (BYMV), and *Cucumber mosaic virus* (CMV). Examples of viruses transmitted by other vectors include *Impatiens necrotic spot virus* (INSV) and *Tomato spotted wilt virus* (TSWV) transmitted by thrips, *Tobacco rattle virus* (TRV), which is transmitted by the nematode *Trichodorus similus*, *Tobacco necrosis virus* (TNV), which causes "Augusta" disease and is transmitted by zoospores of the soilborne water mold *Olpidium brassicae*, and *Tulip virus X* (TVX), which is probably mechanically transmitted by tulip mites.

Bacteria are single-celled prokaryotic organisms that do not have membrane-bound organelles. Bacteria have cell walls and often have flagella, which allow them to move through liquids. They infect through natural openings and wounds. A few bacterial species have the potential to cause severe damage on certain geophytes. Several *Dickeya* spp. and *Pectobacterium* spp. (formerly known as different *Erwinia* spp.) can cause serious soft rot in bulbous crops. *Xanthomonas* spp. infection leads to poor bulb quality, while infection by various *Rhodococcus* spp. leads to soft rot and growth distortion of the roots, scales, and shoots.

Phytoplasmas are organisms that lack cell walls and were formally called mycoplasma-like organisms (MLO). They are smaller than bacteria and are sometimes referred to as fastidious bacteria because they have not been cultured or can only be grown on complex, specialized media. "Grassy top" of gladiolus is caused by a phytoplasma that is transmitted by leafhoppers.

Nematodes are very small, nonsegmented roundworms that mostly feed on soil organisms and organic matter. All plant pathogenic nematodes are obligate parasites, requiring a living host for survival. They also have a sharp, hollow, needle-like feeding structure called a stylet. They have a simple life cycle, developing into a complete worm within eggs prior to hatching and then going through four larval stages before they become reproductive adults. Plant parasitic nematodes are placed into two groups: ectoparasites that can move from plant to plant and endoparasites that remain on the same plant. Within these groups are both migratory and sedentary species. Nematode infections can lead to a serious reduction in plant yield and quality. Most plant parasitic nematodes associated with the production of geophytes cause root problems. These include *Pratylenchus* spp., which cause root rot in several bulb crops. *Aphelenchoides* spp. are foliar nematodes that cause leaf spots and growth distortion. A few migratory ectoparasites are vectors of virus. The stem and bulb nematode (called the stem nematode in the United Kingdom), *Ditylenchus dipsaci*, can be a serious pest on geophytes. There are different biological races, such as the tulip race, of this pathogen. *D. dipsaci* and specific races are quarantine pests in a number of countries. In The Netherlands, infection with the tulip race leads to very stringent measures, including destruction of the infected plant material and the prohibition of flower bulb production on infected fields for at least 10 years.

Abiotic diseases There are a number of noninfectious or abiotic diseases affecting geophytes, sometimes referred to as physiological disorders (De Hertogh and Le Nard 1993b). Noninfectious diseases are generally grouped into diseases that are caused by exposure to environmental conditions such as temperature, moisture, light and nutrient availability, or exposure to harmful chemicals such as air pollutants, herbicides, salt, and so on. The causes of a number of these disorders remain unknown at this time. The upper leaf necrosis of 'Star Gazer' lily due to calcium deficiency is one abiotic disease that has been carefully studied and is well understood (Chang and Miller 2005).

B. ARTHROPOD PESTS

A number of different insect and mite pests cause damage during the production and storage of flower bulbs (Baker 1993). The most common arthropod pests include: aphids and whiteflies, especially those that vector viruses; maggots of several families of flies that are pests in the greenhouse, field and landscape; sap sucking thrips that vector important viruses such as INSV and TSWV, as well as damage bulbs, leaves, stems, and flowers; leafhoppers that vector phytoplasmas, such as the one causing "grassy top" of gladiolus; slugs; and fungus gnats, which are secondary pests that feed on roots and are associated with root rot and/or overwatering. Bulb (*Rhizoglyphus* spp.) and bulb-scale (*Steneotarsonemus laticeps*) mites are important pests of ornamental geophyte crops. For example, on lilies, the bulb mite may tunnel through roots and feed on the bud scales, thus causing stunting, distortion and reduction of flowering. The bulb mite is versatile: it can feed on fungi, nematodes, dead larvae of insects as well as plant tissue (Gerson et al. 1991), and it is hard to control with contact chemical treatments because of its ability to find hidden sites within bulbs.

C. WEEDS

In general, there are three separate and distinct potential weed problems associated with the production of ornamental geophytes. These problems are caused by perennial as well as winter and summer annual weeds. In addition to yield reduction that is caused by direct competition from weeds, weeds also serve as potential hosts for a number of pathogens, provide habitat for slugs, and can increase harvesting costs (by the clogging bulb lifting machinery, for example).

III. HISTORICAL PERSPECTIVE

Ornamental geophytes have been appreciated since antiquity. Tulips, daffodils and hyacinths were prized as early as the mid-sixteenth century and became significant commercial crops in the late-nineteenth century (Doorenbos 1954). In the early 1900s, most bulb farms were relatively small, very labor intensive, and generally relied on animal manures for fertilization, hand weeding for weed control, and pesticides such as Bordeaux mixture to minimize the impact of foliar diseases on crop yields. Organized research relating to disease, pest, and weed management on ornamental geophytes increased with the commercialization of these crops. In 1917, the UK daffodil industry was devastated by a "plague", and much like the case with potato blight and the Irish Potato Famine in the previous century, it was thought to be due to waterlogging or an "ill-defined vapor in the earth". By 1919, James Ramsbottom had identified the cause to be the stem nematode and had developed hot-water dipping as a practical method of control (Tompsett 2006a).

Early research focused on the identification and characterization of economically important plant pathogens and pests (Schenk 1971b; Gould 1993). Other than discarding diseased stocks or digging out diseased plants and drenching soils with general biocides such as formaldehyde, growers relied on crop rotation to control soilborne diseases and various types of heat treatments or fumigation with cyanide to kill pests, nematodes, and fungal and bacterial pathogens on infested planting stocks. In general, if labor and land were not limiting, planting densities were relatively low, and biodiversity was uncompromised; all factors that favored "good husbandry". Agriculture, including the production of ornamental geophytes, changed dramatically during the late 1940s and early 1950s. The introduction of mechanization in the nineteenth century and the increased use of synthetic fertilizers and pesticides and the electrification of rural areas changed farming in dramatic ways following World War II (WWII). In the United States, it was not until the early 1950s that the number of tractors on farms exceeded the number of horses and mules (Economic Research Service 2000). In addition, the increased size of tractors and the invention of the 3-point hitch and the power take-off (PTO), along with other increases in mechanization and refrigeration, influenced virtually all aspects of bulb production, including site preparation, planting, pesticide application, cultivating, harvesting, bulb cleaning, grading and storage, greenhouse production, and the shipment of bulbs and cut flowers. De Haan (1980) reported that increasing mechanization in the 1970s resulted in an increase in farm size as well as a 25–33% reduction in labor demand in The Netherlands. This pattern was evident in most major bulb production regions (Moore 1975; Gould 1990, 1993; Buschman 2005; Ohkawa 2005).

The increased availability of synthetic pesticides following WWII also changed how bulbs were grown. In the US Pacific Northwest, soil residual herbicides reduced labor costs associated with weed control by up to 90%. Bordeaux mixture was the most common fungicide used to control foliar diseases, with 35–50 applications being applied to some crops each season! The availability of new fungicides allowed for significant reductions in the volume of chemicals applied, along with more effective disease control, and eliminated phytotoxicity problems associated with the use of copper-based fungicides. In the late 1960s and early 1970s, research also showed that dipping bulbs in systemic benzimidazole fungicides improved control of *Fusarium* basal rot and reduced the viability of inoculum of pathogens such as *Botrytis tulipae* on infested bulbs, thus reducing the need for foliar applications of fungicides during the growing season (Moore 1975; Gould 1990).

The development of soil fumigants in the 1940's provided growers better control of soilborne plant pathogens, pests, and weeds. There have been a number of excellent recent reviews relating to soil fumigation (Barker and Koenning 1998; Barker 2003; Martin 2003; Schneider et al. 2003; Rosskopf et al. 2005; Zasada et al. 2010). Most of these have dealt with the impact that the loss of fumigants, particularly methyl bromide, will have on how nematodes and other soilborne pathogens and weeds are managed in the future. As pointed out by Barker (2003) and Zasada et al. (2010), it was the use of many of the fumigants that are being banned, phased-out, or heavily restricted today that allowed scientists and farmers to realize the extent to which plant-parasitic nematodes damage crops. The use of soil fumigants as research tools was also instrumental in helping to demonstrate the role of soil microbial populations in naturally occurring disease-suppressive soils (Cook 2007).

In 1947, the Chief of US Department of Agriculture Bureau of Plant Industry considered the development of soil fumigants to be one of the greatest boons to agriculture since the development of (synthetic) fertilizers (Zasada et al. 2010, p. 367). Soil fumigation, particularly with methyl bromide, became an integral part of site preparation for many high-value crops because of its consistent effectiveness in controlling nematodes, soilborne fungal pathogens and weeds. This significantly increased the health and economic viability of these crops. However, Barker (2003) and Zasada et al. (2010) pointed out that nematicides also inadvertently delayed the development of nematology as a science and minimized the need for growers to understand nematode biology because of the broad-spectrum effectiveness of fumigants.

Beginning in the 1930s, it became an accepted practice to add formaldehyde (formalin) to hot-water treatment (HWT) tanks as a biocide to augment the "kill" of stem nematode in daffodils (Chitwood et al. 1941; Chitwood and Blanton 1941) and reduce the spread of fungal pathogens (Gregory 1932; Hawker 1935). Provided that appropriate protocols were followed, long industry experience made HWT with formaldehyde an effective, crop-safe, and inexpensive treatment. Since the 1970s, concerns increased in the United Kingdom about the rising incidence of basal rot and effects on bulb exports, perhaps as a result of modern husbandry practices that favored its development and spread, and it became usual to add a fungicide (or fungicides) to the formaldehyde in the tank (Tompsett 1980). Since then, the fungicides used have changed, initially to utilize more effective or cheaper products, more recently in line with evolving pesticide legislation. A chlorpyrifos-based insecticide may also be added, if appropriate, to control large narcissus fly (Tones and Tompsett 1990). Aldrin, an organochlorine insecticide, was once used to control the large narcissus bulb fly as a spray or preplant bulb dip, but in 1989, when it contaminated the Newlyn River in southwest England, it was immediately banned—and years later an increase in the bird of prey and otter populations seemed to vindicate that decision (Tompsett 2006b). When aldrin was used as a dip in the United States, it was recommended that mercury fungicides, which were used until the mid-1960s, be added to the dip tank to control the spread of basal rot. Through such practices, HWT has therefore not only lost its "nonchemical" cachet but also faces the loss of products which has resulted in a quest for more acceptable products or reduced-rate applications.

Estimates for growing daffodils in the United Kingdom suggested that the cost of providing HWT with formaldehyde (but not fungicide) amounted to about US$885/ha (Briggs 2002; 1999 cost estimates adjusted to 2010). This cost alone is equivalent to nearly 10% of an average gross margin from over 2 years of harvesting flowers and bulbs. There will be increasing pressure on growers to use HWT tanks with effective insulation and the most energy-efficient heating, control and circulation. The field-grown fruit and vegetable sector contributes 2.5–3.0% of total UK greenhouse gas (GHG) emissions, and by implication the contribution of daffodil growing would also be low (Edwards-Jones and Plassmann 2008). Field crops grown in season without heating or protection, such as daffodils, were the least energy- and GHG-intensive crops. In a further study it was shown that the primary energy inputs for field-grown daffodil production were relatively low (37 GWh, compared with >125 GWh for field-grown vegetables but <10 GWh for orchard and soft fruit, hops and ornamental nursery stock) (Lillywhite et al. 2007). Although the greatest contribution to this footprint comprised pesticide manufacture (indicating that even a modest reduction in pesticide use would have a beneficial effect on the sustainability of the crop), the energy used in HWT and bulb drying was a further concern, again indicating the need to make saving through greater energy efficiency.

A number of key nematicides have been banned or are being phased out or severely restricted because of concerns about their safety and potential adverse effect on the environment. Barker (2003) points out that because funds and research efforts were allocated for nematicides at the expense of basic biological studies, nematology lags behind other areas of plant pathology in developing alternatives for nematode management. He goes on to say that as a result of this lack of research on basic biology and the limited developmental research on new chemistries and potentially safer nematicides, growers are faced with fewer options to manage their nematode problems. Although flooding of the soil for at least six to eight weeks has been shown to be an effective alternative to fumigation with metam-sodium for the control of soilborne nematodes, fungi, and weeds (see IPM for Weeds, Section IV.F) in the western part of The Netherlands where bulbs are grown on calcareous, coarse, sandy soils (Muller et al. 1988; De Boer 2011), some authors have pointed out that the loss of fumigants will increase the difficulty of producing "healthy" propagation stock (Schneider et al. 2003). In the short term, the loss of broad-spectrum fumigants may actually lead to the increased use of selected pesticides, including fungicides that are used to control soilborne fungal plant pathogens, such as *Pythium*, *Rhizoctonia*, and *Sclerotium*, and herbicides to control weeds.

On the other hand, several authors have noted that a positive outcome of recent restrictions on nematicides has been the deployment of IPM approaches that have contributed to the ongoing development of more comprehensive, ecology-based, and environmentally-sound strategies and systems for managing nematodes and soilborne fungal pathogens (Barker and Koenning 1998; Pouidel et al. 2001; Barker 2003; Rosskopf et al. 2005). As pointed out by these authors, the development of effective, ecologically-based management programs for nematodes, fungal pathogens, and weeds will require a multidisciplinary approach to understanding the ecology of potential biocontrol agents and antagonists in the soil. A summary of some recent research on biologically based alternatives to soil fumigation to control pests, plant pathogens, and weeds can be found in the biologically-based controls and weed management sections of this chapter.

Fungicide, insecticide, and fumigant applications have been used to provide effective protection of bulb crops for a number of decades. The fungicide PCNB, for example, was shown to provide excellent control of a number of soilborne diseases, such as those caused by various species of *Rhizoctonia*, *Sclerotium*, and *Sclerotinia*. It was first used to control soilborne diseases in the US Pacific Northwest in 1950—broadcast applications of the 20% formulation of this product were applied at the rate of 1120 kg of product per ha to protect planting stock from soilborne inoculum (Gould 1957, 1990). Later, it was shown that this material could be used as an in-furrow application at the time of planting, which reduced rates to about 73 kg/ha, or as a bulb dip treatment to reduce the spread from diseased to healthy bulbs (Gould and Russell 1965). In conjunction with other management practices, the use of PCNB has essentially eliminated *Sclerotinia bulborum* as a problem in western Washington (Gould 1990). Various chlorinated insecticides were used to control a variety of insect pests. For aphid control, the standard recommendation was once to spray DDT every 10 days during the growing season. Aldrin was used to control large bulb fly as a spray or preplant bulb dip. Growers also built rooms where bulbs could be fumigated with methyl bromide or hydrocyanic acid gas to control other pests. Some of these chemical controls of the past seem shocking today. Many of the chemicals used have been banned or discarded in favor of newer chemicals that are safer for applicators and the environment.

Even some of the "cultural" controls of the past would not be considered to be acceptable today. Early attempts in southwest England to control the large narcissus bulb fly included butterfly nets and putting out baits of syrup and arsenic (Tompsett 2006b). Daffodil growers in the US Pacific Northwest were encouraged to sort their stock prior to planting and remove any soft bulbs in order to reduce the number of large bulb flies that would emerge from the crop the following spring (Gould 1957). The recommendations went on to say how important it was to destroy the infested bulbs because if they were just left in a pile, the fly larvae would continue to develop and the adult flies would emerge from the cull pile in the spring. As part of this cultural control, it was suggested that an effective way of destroying the culls was to douse them in crankcase oil and light them on fire. It was noted that it took one gallon (3.8 l) of oil per 200 bulbs to burn them to ash.

While innovations in mechanical technology and the increased reliance on synthetic fertilizers and pesticides between the late 1940s and early 1960s resulted in significant increases in farm productivity, there were increasing problems with pest resistance to pesticides and concerns relating to the effects of some pesticides on the environment. This gave rise to the concept of Integrated Pest Management (IPM) in the late 1960s and the formation of regulatory agencies, such as the US Environmental Protection Agency (EPA) in 1970.

In the United States, concerns about the sustainability of agricultural practices were initially fostered by events such as the "Dust Bowl" in the 1930s. However, it was not until the energy crisis in the early 1970s that the concept of sustainable agriculture received serious attention. There had been a tremendous increase in mechanization and the use of energy-dependent fertilizers and pesticides since the early 1950s. Many of these changes were made possible by the availability of relatively cheap sources of energy and, along with the increase in irrigated agriculture, allowed for the average yields of crops in the United States to double in the span of a couple of decades. However, when the energy crisis hit, people began to question the sustainability of this modern production system. While the increased use of synthetic fertilizers, pesticides, bulk harvesting and handling systems, drying walls, temperature-controlled storage and greenhouses to produce cut flowers and potted plants had dramatically increased productivity and product quality, they also increased the "energy footprint" of the bulb industry.

Concerns resulting from increasing instances of pest resistance to pesticides, the effect of pesticides on the environment, and the sustainability of agricultural production systems were the impetus for a serious reformation of agricultural research by the end of the 1960s, highlighted by Rachel Carson in *Silent Spring* (1962). Initially, the focus was on the implementation of IPM programs to reduce the impact of pesticides on the environment, followed by efforts to reduce the energy consumption and cost associated with the production of bulb and cut flower crops (Rudd-Jones 1975; Van Julsingha 1980). According to Cook (2000), the goal today is economically, environmentally, and socially acceptable plant health management. The result has been a greater effort to achieve plant health while also conserving natural resources (soil, water, energy, etc.), reducing labor costs, protecting biodiversity, and making maximum use of nature's own biological systems.

IV. INTEGRATED PEST MANAGEMENT

In 1967, Smith and Van den Bosch introduced the term IPM. The recognition of the failings of the new organosynthetic insecticides, which included resistance, resurgence of primary pests, upsurges of secondary pests, and overall environmental contamination, was a major factor in the initial formulation and then the growing popularity of the integrated control concept (Kogan 1998). There are over 60 definitions of integrated control, pest management, or IPM (Bajwa and Kogan 2002). Smith and Van den Bosch (1967) promoted integrating all control strategies and the application of ecological principles to pest control in agricultural production systems. Jacobsen (1997, p. 376) points out that they also introduced the "key pest" concept, where a key pest was defined as "a serious, perennially occurring species that dominates control practices because in the absence of deliberate control by man, the pest population usually remains above the economic injury levels". Possible examples of "key pests" in terms of diseases of geophytes might include viruses that are transmitted by aphids, soilborne fungi such as *Fusarium oxysporum* and *Rhizoctonia* spp., bulb and stem nematode, bacterial soft rot, and foliage diseases, such as those caused by *Botrytis* species.

Kogan (1998) provides an excellent historical perspective on the development of IPM. IPM is a sustainable approach to managing pests that utilizes a diversity of management options to minimize health, environmental, and economic risks. These options may include biological, cultural, physical, and chemical tools. When necessary, pesticides are used judiciously and as a last resort. "Pests" may include insects and other arthropods, weeds, plant pathogens, rodents, and other animals. Success of IPM programs often has been measured by the overall reduction in the volume of pesticides used to control prevalent pests. Although reduction in pesticide usage is a potential benefit of IPM, Kogan (1998) indicates that it cannot be the only measure of success. Kogan points out that there are special circumstances in which, to maintain viable agricultural production, even under IPM guidelines it may be necessary to use more, not less, pesticides. The issue is pesticide use guided by the principles of IPM; that is, selective use after maximizing the effectiveness of natural controls.

Benbrook et al. (1996) state that IPM has provided a framework for the development of pest management programs culminating in an approach that emphasizes preventive tactics such as enhancement of natural enemies of pests, cultural methods, and plant resistance. More recently, a number of authors have focused on plant health management (PHM). According to Cook (2000), PHM is the science and practice of understanding and overcoming the succession of biotic and abiotic factors that limit plants from achieving their full genetic potential as crops, ornamentals, timber trees, or other uses. It includes and builds upon but is not a replacement for IPM.

Most key pests, pathogens, and weeds are managed by (1) integrating diverse tactics such as the use of resistant hosts, sanitation, certification, and (2) biological control via rotation, selection of planting time and site, tillage, the use of selected biological control products, and application of pesticides. The concept of a "disease triangle" is often used to illustrate the fact that disease development represents the interaction between the three "corners" of the triangle: a susceptible host, a pathogen (disease-causing organism), and a favorable environment. If all these factors are present, disease results; if one or more of the factors are not present, disease does not occur (McNew 1960). Understanding this concept helps to understand the basic methods of infectious disease control, which comprise exclusion, protection, prevention, and eradication.

It is beyond the scope of this chapter to provide specific pest, disease, and weed management recommendations for the production of field and greenhouse-grown ornamental geophytes. There are a number of excellent IPM manuals, such as the University of California's IPM for Floriculture and Nurseries (Dreistadt 2001), which contain specific IPM tactics and approaches to controlling pests, diseases, and weeds on ornamental crops. Given the regional variation in pest problems and differences in products that may be registered for use on specific crops around the world, it is important that plant health practitioners check to ensure that any products they use are registered in their region.

General IPM strategies or techniques include the monitoring of pests, weather, and site conditions; the use of thresholds (economic, action, damage, tolerance, or aesthetic); pest forecasting; accurate diagnosis; and record keeping. The concept of decision making implicitly permeates most definitions of IPM (Kogan 1998). The UC IPM manual (Dreistadt 2001) indicates that IPM methods will help reduce pesticide resistance problems, minimize potential phytotoxicity issues, reduce workforce disruptions that occur from pesticide reentry intervals, and reduce the costs of pesticides, applications, and regulatory compliance.

IPM principles are readily applied to the management of pests, diseases, and weeds. Two keys are: *monitoring* to detect pests promptly and *accurate diagnosis*. In the past, needless environmental injury resulted when applications of broad-spectrum, persistent pesticides were used to make up for poor planning. Growers using IPM are always conscious of the need to manage pests, and their thoughtful approach to pest management reduces the losses that would inevitably follow from failure to monitor or to use multiple problem-suppressive strategies such as:

Avoidance: Includes choosing planting sites that are devoid of the pathogen and/or vectors; planting times that avoid environmental conditions favoring pest and disease development or periods when susceptible host tissue is present; avoiding wounding or injury that provides entry sites for pathogens such as bacteria; and environmental manipulation.

Sanitation and Exclusion: Include quarantines that limit the spread of invasive pests, pathogens, and weeds; screening to exclude insects; the use of pest- and pathogen-free seed or stock; and certification programs to produce healthy planting stock.

Eradication: Includes rotation to nonhost crop; removal of weed hosts; destruction of infected host residues; application of some types of pesticides and disinfectants that kill target pests and pathogens, heat treatments, solarization, steam, soil fumigation, and water treatments; roguing out infected plants; destruction of vectors; and thermotherapy and/or meristem culture to produce pathogen-free propagation material.

Protection: Includes the use of biocontrol agents; application of some types of pesticides, such as some fungicides that only provide protective barriers to infection; planting resistant hosts; control of plant disease vectors; the use of antibiotics or antagonists; mandated crop-free periods to limit the spread of viruses; barrier crops; reflective mulches; cross protection; and genetically-modified (GM) resistant plants.

Finally, when pesticides are needed, growers favor the use of the least toxic material available and avoid using broad-spectrum, persistent chemicals.

A. PATHOGEN- AND PEST-FREE PLANTING STOCK

Cook (2000) states that probably the greatest collection of success stories for plant health management in the twentieth century is the number of diseases managed or all-but-eliminated by the use of healthy, high-quality seed or planting material. This is particularly important for vegetatively propagated plants, such as ornamental geophytes. A combination of approaches has been used to produce disease- and pest-free material. Improvements in pathogen detection (see Diagnosis and Detection, Section IV.H) have made it easier to determine if plant tissues are free of a number of plant pathogens. Once disease-free material is available, various approaches are used to multiply stocks and rigorous steps must be taken to protect them from becoming reinfected (see also Chapter 10).

1.   Hot and Cold Temperature Treatments

Thermal treatments are a very efficient and environmentally friendly way of controlling a number of pests and plant pathogens in plant material (Grondeau et al. 1994). The use of steam, hot air, and particularly hot-water treatments (HWT) has been intensively studied and widely applied to bulbs (Woodville 1964; Schenk 1971b; Moore et al. 1979; Byther and Chastagner 1993; Qiu et al. 1993). These treatments are used to eradicate pathogens and pests on infested and diseased planting stock, thus reducing the need for applications of pesticides during the growing season.

The history, use, and protocols of HWT were described by Gratwick and Southey (1986). Its first use, in the nineteenth century, may have been to control mites in bulbs of *Eucharis amazonica*. In the early part of the twentieth century, HWT was widely used to treat chrysanthemum stools and strawberry and mint runners to control nematodes, tarsonemid mites, aphids, and rust fungi. Gratwick and Southey (1986) reviewed the use of HWT on many ornamental geophytes—daffodil, tulip, *Allium*, begonia, *Camassia*, *Chionodoxa*, *Colchicum*, *Convallaria*, crocus, gladiolus, *Hippeastrum*, hyacinth, bulbous iris, lily, *Muscari*, *Ornithogalum*, *Oxalis*, *Puschkinia*, *Scilla*, and *Tigridia*—as well as a range of other plants and seeds.

Schenk (1971b) reported that HWT, either alone or in combination with other methods, has proven to be a powerful tool for managing diseases and pests that once threatened to wipe out the cultivation of some geophytes. Many nematodes, fungal and bacterial plant pathogens, insects, and mites carried by bulbs or corms can be controlled using this treatment. HWT is effective in eradicating fungal pathogens on cormels of gladioli. Kruyer and Boontjes (1982) also reported that populations of *Rhodococcus fascians* were reduced in *Lilium longiflorum* bulbs by immersion in 39°C water for 2 h, with complete control achieved with the addition of 0.5% formalin.

Pretreatment of bulbs and corms prior to HWT has been shown to increase the effectiveness of the HWT and also increase the thermotolerance of bulbs and corms (Hanks 1995). For example, by pretreating lily bulbs for one day at 20°C followed by one day at 20°C after the HWT, the temperature of the HWT could be safely raised to 41°C. A 2 h treatment at this temperature completely eradicated *Pratylenchus penetrans* and *Aphelenchoides fragariae* nematodes and the bulb mite *Rhizoglyphus robini* within the bulbs (Kok and Aanholt 2008, 2009). Van Leeuwen and Trompert (2011) have recently examined the effect of treatment timing on the effectiveness of HWT to control *Aphelenchoides subtenuis* on bulbs and corms of *Allium* and *Crocus*. Controlling *A. subtenuis* was effective only when a HWT of 4 h at 45°C was applied within 10–14 days after lifting the bulbs. When *Crocus* was stored at 30°C, a HWT was only effective until 10 days after lifting. When stored at 25°C, a treatment was effective until 14 days after lifting. They also found that the HWT became more effective when the bulbs were immersed in water for 24 h prior to treatment.

Daffodil production is unusual in that HWT is the primary means of controlling pests and pathogens, with bulbs typically being immersed in hot water at 43–47°C for about 2–4 h according to various protocols (Lees 1963; Hanks 1993). In daffodils, HWT is used to control the crop's most serious pest, stem nematode (*Ditylenchus dipsaci*), and gives incidental control of other pests, including bulb mites (*Rhizoglyphus* spp.), bulb-scale mite (*Steneotarsonemus laticeps*), larvae of large and small narcissus flies (*Merodon equestris* and *Eumerus* spp.), and bulb and leaf nematode (*Aphelenchoides subtenuis*) (Lane 1984). It gives a measure of control of fungal diseases such as basal rot (*Fusarium oxysporum* f. sp. *narcissi*) and leaf scorch (*Stagonospora curtisii*) (Moore et al. 1979), and possibly of smolder (*Botrytis narcissicola*) and the fungi associated with "skin diseases" (*huidziek* and *vethuidigheid*) (Van Rijn and Pfaff 1995). Less familiar targets for HWT in daffodils include dry-scale rot (*Stromatinia gladioli*) (Chastagner and Byther 1985), an *A. subtenuis*-associated rot of the base plate (Vigodsky-Hass and Lavi 1986) and black slime disease (Koster et al. 1987).

While water might be considered the ultimate environmentally friendly chemical, there are important sustainability issues connected with daffodil HWT. The prolonged heating of large volumes of water and bulbs is energy intensive, demands a specific facility that has little purpose outside the short (few weeks) window for effective treatment, and slows down the rate of processing bulbs from harvesting to sale or replanting. In addition, since the treatment of daffodil bulbs in hot water alone is insufficient to control *F. oxysporum* f. sp. *narcissi* (Van Slogteren 1931) and may not eliminate all nematodes (Hastings et al. 1952), formalin has been routinely added to the HWT tank (Chitwood et al. 1941; Chitwood and Blanton 1941), often with a fungicide (Gregory 1932; Hawker 1935; Tompsett 1980; Hanks 1993) and sometimes with an insecticide against large narcissus fly (Tones and Tompsett 1990), so there are issues of worker exposure to formalin (FormaCare 2007), large volumes of the warm dip, and disposal issues relating to spent diptank effluents. These subjects have recently been reviewed in detail elsewhere (Hanks 2002). A specific concern has been the withdrawal of formalin for agri/horticultural purposes in many countries; for example, its use was restricted in the United States to those with "third-party registrations" in the early 2000s, and it was banned in the EU in 2008, with no proven alternative being available. This illustrates the danger of reliance on a single product, especially for growers of specialty crops. Although alternative biocides, such as peroxyacetic acid/peracetic acid or glutaraldehyde, had been tested from time to time (Linfield 1991; Hanks and Linfield 1999), there had been little incentive to replace formalin. In the United States, chlorine dioxide (ClO_2) was shown to be effective in killing *Fusarium* basal rot conidia and chlamydospores and other fungal pathogens (Chastagner and Riley 2002, 2005; Copes et al. 2004, in press) and it was found to be an effective replacement for formaldehyde in HWT for the management of basal rot (Chastagner and Riley 2002). In the United Kingdom, a range of biocides was tested against *Fusarium* basal rot chlamydospores and stem nematode wool, and an iodophore biocide ("FAM 30") was selected as being both effective and crop-safe (Lole et al. 2006). "FAM 30" and ClO_2 are now undergoing farm-scale trials in the United Kingdom (G. R. Hanks 2010, pers. comm.).

The example of daffodil HWT illustrates that nonchemical pest and disease management can be very demanding on grower resources, but a recent review concluded that alternatives to HWT (e.g., microwave treatment) for controlling stem nematode would involve considerable development costs (Lole et al. 2006), while dipping bulbs in a nematicide or using nematicides in the field were largely ruled out because of a lack of approved, effective products (Hesling 1971; Damadzadeh and Hague 1979; Windrich 1986), although some foliar applications are successful and have been used in the United States (Bergeson 1955; Westerdahl et al. 1991). Lole et al. (2006) proposed that, for the short to medium term, the best option was the optimization of HWT practice through changes to the physical parameters and/or the biocides and pesticides used.

Given the extensive international trade of ornamental geophytes, the use of various thermal treatments also allows growers to meet the plant health standards to maintain access to export markets. Although high-temperature treatments are more common, chilling is sometimes employed for pest management. For example, in the absence of a suitable acaricide, infestations of greenhouse tray-grown daffodils with bulb-scale mite were traditionally dealt with by putting the crop outdoors for 24 h if conditions were freezing. This technique, however, was always inconvenient and was often ineffective (R. H. Collier 2010, pers. comm.).

Bacterial pathogens are among the most difficult to control because of the lack of effective bactericides (McManus et al. 2002). Copper-based fungicides are sometimes used to control bacterial pathogens, but they often provide inconsistent control and can be phytotoxic (Garcia-Garza et al. 2002). Hot-air treatments have been used to free hyacinth bulbs of *Xanthomonas hyacinthi*. *Rhodococcus fascians* is another bacterial pathogen that causes leafy gall, shoot proliferations, and fasciation on a number of ornamentals, including several geophytes (Lacey 1936; Putnam and Miller 2007). Soil sterilization with steam (Kruyer and Boontjes 1982) can eradicate this pathogen.

2. Roguing

Roguing to remove diseased plants and bulbs is an important historical approach that is used to maintain the health of bulb crops by reducing the development and spread of a number of plant pathogens and pests. The primary method of virus disease control is the field roguing of symptomatic plants that otherwise could serve as sources of virus inoculum for further spread by vectors (Loebenstein et al. 1995). Removal of "fire heads" caused by the infection of emerging tulip shoots by *Botrytis tulipae* and the roguing of hyacinths that exhibit symptoms of "yellow rot" caused by *Xanthomonas campestris* pv. *hyacinthi* are examples of fungal and bacterial diseases, respectively, where rouging is used to reduce disease buildup (Doornick and Bergman 1974; Moore et al. 1979; Coley-Smith et al. 1980).

B. BIOLOGICALLY BASED CONTROLS

Restrictions on the use of pesticides, particularly soil fumigants, have resulted in increased interest and efforts at using a variety of biologically-based approaches to control pests, plant pathogens, and weeds in crop production systems, including the production of geophytes (Hoitink and Boehm 1999; Cook 2000; Barker 2003; Alabouvette et al. 2009; De Boer 2011). This is especially true in relation to the management of soilborne plant pathogens and cropping systems where effective management practices based on host resistance, pesticide applications, or cultural practices are not available (Hoitink and Boehm 1999). Examples of biologically-based controls include the use of crop rotation, biofumigants and green manure crops, and biological control agents (BCA).

1. Crop Rotation

Crop rotation is a historical approach that effectively "sanitizes" the soil (Baker and Cook 1974; Cook and Baker 1983; Hoitink and Boehm 1999; Barker 2003; Agrios 2005; Cook 2007; Schumann and D'Arcy 2010) and is very beneficial in reducing populations of soilborne nematodes and fungal pathogens that cause diseases on geophytes (Gould 1957; Moore et al. 1979; Chastagner and Byther 1985; Rees 1992; Byther and Chastagner 1993). Since many foliar pathogens also survive from one season to the next in plant residues in or on the soil, crop rotation and the removal of crop residues after harvest are also beneficial in controlling foliar pathogens. With some foliar diseases, such as fire (*Botrytis tulipae*) on tulips, where emerging shoots can be infected when they come in contact with sclerotia in the soil, crop rotation is essential to disease control.

Crop rotation practices have changed during the past 50 years. In 1957, Gould recommended that in the US Pacific Northwest a 4- to 6-year-long rotation schedule was best from the standpoint of improving crop productivity and controlling insects and diseases on tulips, daffodils, and bulbous iris and the survival of "volunteer" bulbs from the previous crops. Since no one rotation would be suitable for all conditions, he suggested three potential schedules (Table 13.1).

While the benefits of crop rotation in agriculture are widely recognized, Cook (2000) points out that for economic reasons, traditional 4- and 6-year crop rotations with 2 or 3 years of pasture or hay crops were replaced with 2- and 3-year rotations or crop monoculture in the middle of the twentieth century. This coincided with the increased use of synthetic inorganic fertilizers and pesticides, particularly soil fumigants, in high value crops, and improvements in mechanical cultivation during the late 1940s to the 1960s (see Section III). Hoitink and Boehm (1999) point out that this allowed farmers to break the link between organic amendments and soil fertility, which resulted in by-products such as manures becoming solid wastes. Over time this resulted in decreased soil fertility and organic matter (OM) and a decline in soil structure, which allowed many soilborne plant pathogens to eventually develop to epidemic proportions (see Section IV.B, Cook 2000; Barker 2003; Alabouvette et al. 2009; De Boer 2011).

TABLE 13.1
Suggested Crop Rotation Schedules

Year	Schedule 1	Schedule 2	Schedule 3
1	Grass–legume mix	Cover crop	Heavily manured vegetable crop
2	Grass–legume mix	Tulip	Tulip
3	Tulips	Cover crop	Daffodil
4	Daffodils	Daffodils	Iris (only if no potatoes in rotation)
5	Iris	Cover crop	—
6	—	Iris	—

Source: Gould, C. J. 1957. *Handbook on Bulb Growing and Forcing.* Mount Vernon, Washington: Northwest Bulb Growers Association.

2. Biofumigation and Green Manure Crops

Using biofumigation crops and specific green manures to control soilborne pathogens such as nematodes would have some advantages over methods such as soil fumigation, preplanting pesticide application, or flooding (De Boer et al. 2006). For example, the nematode *Pratylenchus penetrans*, in combination with the fungus *Cylindrocarpon destructans* (*Nectria radicicola*), causes a root rot of daffodils that is referred to as "decline", "replant disease", or "soil sickness". The fungus enters through wounds on the roots caused by the nematode, spreads and results in areas of stunted, rotting plants. Reports indicated that some growers would plant montbretia (*Crocosmia × crocosmiflora*) on heavily infested patches of field, resulting in better growth of daffodils subsequently planted there. Slootweg (1956) confirmed these effects experimentally, although the benefits were considered inferior to the effect of a conventional nematicide. Again following a grower observation that daffodils did not exhibit root rot problems when grown on soil previously planted with African marigolds, *Tagetes erecta* was grown for three months before being incorporated into the soil. Following this rotation, the yield of daffodils grown on these plots was double that on control plots, with a marked reduction in the numbers of *P. penetrans* in the soil (Slootweg 1956). In later trials Conijn (1994) showed that bulb yields increased, the incidence of root rot decreased, the numbers of Trichodorid nematodes in the soil were unaffected, and the number of *P. penetrans* in the soil was reduced in daffodil and lily crops planted after the cultivation of French marigold, *T. patula*. Other relevant reports of the effects of *Tagetes* on nematodes include those by Miller and Ahrens (1969) and Paffrath and Frankenberg (2005).

In the United Kingdom, "root rot" occurs in tazetta daffodils growing on the Isles of Scilly, where it is not easy to find un-cropped land for new plantings. The use of *T. patula* 'Ground Cover' was recently investigated on two infested sites by Tompsett (2006a, 2008): the treatments were growing *Tagetes* and ploughing in before planting, soil sterilization using 1,3-dichloropropene, and a combined treatment. Bulbs of 'Royal Connection' were planted, and after two growing seasons the combined treatment gave the highest crop vigor and lowest numbers of *P. penetrans*, and the single treatments also showed improvement. Improved crop vigor was evident for 5 years at one site only, but the effects on nematode numbers were diminished.

No reports appear to be available dealing directly with the effects of *Tagetes* on daffodils infested with their major nematode pest, *Ditylenchus dipsaci*. However, the nematicidal effect of root extracts of *Helenium* and *Gaillardia* were as effective as that of *T. patula* on *D. dipsaci* in *in vitro* tests (Gommers 1972), while *T. erecta*, *T. patula*, and *Calendula officinalis* used as green manure crops reduced the numbers of *Ditylenchus* in strawberry plants to 1.4–1.6 per plant (from 19.2 per plant in controls) and the incidence of infested plants to 11–15% (from 79% in controls) (Andreeva 1983).

Conversely, biofumigant crops like 'Idagold' mustard (*Sinapis alba*) and oilseed radish (*Raphanus sativus*) gave only inconsistent control of a few weed species in the subsequent onion (*Allium cepa*) crop, and in some years, a reduced onion bulb yield (Geary et al. 2008). De Boer (2011) indicates that, given the variability seen with biofumigant crops, a broad application of this approach is still not practicable. However, she indicates that initial results from studies in The Netherlands to examine the effect of incorporation of composts or chitin and intercropping with *Tagetes patula* in conjunction with other methods show promising and long-term benefits in reducing the numbers of *P. penetrans* and enhancing the yield of lilies.

3.   Biocontrol of Plant Pathogens

Despite the increasing number of scientific papers dealing with biological control, Alabouvette et al. (2009) reported that there are still only a very limited number of commercially available products on the market. They reported that in the EU only two dozen microorganisms have been authorized for use in plant protection. Fravel (2005) listed 25 microorganisms registered by the US EPA as biological controls. Microorganisms have been incorporated into 80 formulated biocontrol products (Paulitz and Bélanger 2001) and many of these have been developed for greenhouse applications. A list published online in June 2005 of 36 biocontrol products registered in the United States included five bacterial genera (*Agrobacterium*, *Bacillus*, *Burkholderia*, *Pseudomonas*, and *Streptomyces*) and eight fungal genera (*Ampelomyces*, *Aspergillus*, *Candida*, *Coniothyrium*, *Gliocladium*, *Myrothecium* (killed), *Paecilomyces*, and *Trichoderma* spp.), as active ingredients in biocontrol products (APS Biological Control Committee 2005). The US EPA provides a current listing of biocontrol active ingredients and associated fact sheets (US EPA Biopesticide Active Ingredient Fact Sheets).

A number of reviews have been done relating to the use of biological control agents (BCA) to control plant diseases. In general, soilborne root diseases are much more difficult to control than aerial diseases (Alabouvette et al. 2005). Soils that are naturally suppressive to soilborne plant pathogens occur in a number of cropping systems. In their review, Weller et al. (2002) describe the two classical types of suppressiveness that are known to occur. The first is general suppression, which owes its activity to the total microbial biomass in soil and is not transferable between soils. The second is specific suppression, which owes its activity to the effects of individual or select groups of microorganisms and is transferable. A great diversity of microorganisms contributes to the natural development of disease-suppressive soils and composts (Weller et al. 2002; Alabouvette and Steinberg 2006; Borneman and Becker 2007; Alabouvette et al. 2009). Hoitink and Boehm (1999, p. 429) have summarized recent advances in elucidating the nature and complexity of microbial populations and substrate chemistry that have led to a better understanding of how "microbial consortia in complex or compost-amended substrates interact with one another, the pathogen, or the host to facilitate disease suppression". Borneman and Becker (2007) have described the development and utilization of a population-based approach to identifying microorganisms involved in specific pathogen suppression. As they point out, this approach has the potential to lead to the development of new strategies to create and maintain soils that are suppressive to specific plant pathogens.

Although disease-suppressive, composted pine bark-based mixes are commonly used to control diseases caused by soilborne pathogens such as *Pythium* and *Phytophthora* spp. in the production of container-grown ornamentals (Hoitink et al. 1991; Hoitink and Boehm 1999; Daughtrey and Benson 2005), introductions of single or multiple biocontrol agents have not been widely utilized by growers. Lack of adoption stems from two major problems: sometimes a BCA can control only one of several important diseases of a particular crop, and in other cases the level of control it provides has been inconsistent (Weller 1988; Hoitink and Boehm 1999). The success of biological control depends not only on plant–microbial interactions but also on the ecological fitness of the biological control agents. This has led to increased studies relating to the rhizosphere competency of BCA and the introduction of a food base with the BCA that selectively supports its activity (Harman 1992; Hoitink and Boehm 1999).

In their review, Hoitink and Boehm (1999) indicate that organic amendments such as green manures, stable manures, and composts can serve as a food base for suppressive edaphic microorganisms and have been recognized to facilitate biological control of soilborne plant pathogens ranging from *Pythium*, *Phytophthora*, and *Fusarium* spp. to *Rhizoctonia solani*. These amendments contribute to suppressive activity through the four principal mechanisms of biological control: competition, antibiosis, parasitism/predation, and systemic-induced resistance (Lockwood 1988). The concentration and availability of nutrients within the soil organic matter (OM) play a critical role in regulating these activities.

Although experimental results with biological controls have been mixed (Daughtrey and Benson 2005), products are beginning to have success in the production of ornamental crops, in part due to public pressure to accomplish plant health management without using chemical pesticides (Paulitz and Bélanger 2001). In greenhouse systems, control of powdery mildews has received a lot of attention as these pathogens are not easily regulated by environmental or cultural controls. Geophytes are not especially prone to powdery mildews with one notable exception: plants in the Ranunculaceae, including *Ranunculus asiaticus* and *Anemone coronaria*. Both these crops are vulnerable to *Erysiphe* spp., particularly *Erysiphe aquilegiae* var. *ranunculi*, which has been reported from Australia, Europe, Israel, New Zealand, South Africa, and the United Kingdom on ranunculus, and from Scandinavia, in addition, on anemones. IPM for ranunculus and anemones should certainly include management strategies for powdery mildew. Biological control experimentation has primarily been directed at greenhouse vegetables or roses since these are the most valuable of the greenhouse crops with a powdery mildew problem.

Effective biocontrols used for plant disease management today come from a limited number of organisms. Many of the commercial products developed have been used to the best effect in greenhouses because of the better cost-efficiency of application and greater control over environmental conditions found in protected cultivation. *Coniothyrium minitans* is a mycoparasite of the sclerotia of *Sclerotinia sclerotiorum* and *S. minor*. *Trichoderma virens* (= *Gliocladium virens*) produces gliovirin and gliotoxin which are toxic to other species of fungi. Several strains of *Trichoderma harzianum* have been marketed around the world for use against *Pythium* and *Rhizoctonia* spp. primarily. *Streptomyces griseoviridis* strain K61 has been used against *Fusarium* and *Pythium* species. *Gliocladium catenulatum* strain J1446 has been marketed for use against damping off, root rot and wilt pathogens. The nonpathogenic *Fusarium* strain Fo47 has been marketed in Europe for use against *Fusarium oxysporum* f. sp. *cyclaminis*, the cause of Fusarium wilt of cyclamen (Alabouvette et al. 1998). In addition to these biocontrol fungi, the spore-forming bacteria *Bacillus subtilis* and *B. subtilis* var. *amyloliquefaciens* are also used against *Pythium ultimum* and *Rhizoctonia solani*.

A number of these organisms (*Trichoderma* spp., *Streptomyces griseoviridis*, *Gliocladium catenulatum*, and *Bacillus subtilis*) used in soil have also been applied to foliage to discourage *Botrytis cinerea* and powdery mildew. *Ulocladium atrum* has been researched as a *Botrytis* control for cyclamen, with its effectiveness based on competition with *B. cinerea* for substrate (Kessel 1999). The hyperparasite *Ampelomyces quisqualis* has been studied and marketed as a control for powdery mildew (Feldman et al. 1993; Kiss et al. 2004), as has the basidiomycetous yeast, *Pseudozyma flocculosa* (Paulitz and Bélanger 2001).

For greenhouse culture of geophytes, the primary use of biocontrols would likely be as additives to the growing mix used for suppression of root rot organisms such as *Pythium* spp. and *Rhizoctonia solani*. The nature of these materials is such that they are best used preventively, rather than in response to symptoms detected by scouts. The use of growing mixes containing microorganisms competitive to pathogens may help with overall crop evenness and quality through biological control of root disease. A review article by Paulitz and Bélanger (2001) anticipates that greenhouse disease management may in the future be accomplished exclusively through the use of biological controls, pointing to the reliance on biocontrol for insect management that prevails today in some production areas.

Investigations on the biological control of *Fusarium oxysporum* f. sp. *narcissi*, which causes basal rot on daffodil, by antagonistic fungi have been reported by researchers in The Netherlands and the United Kingdom (Langerak 1977; Beale and Pitt 1990, 1995; Hiltunen et al. 1995). Nonpathogenic microorganisms (*Penicillium* and *Trichoderma* species, *Minimedusa polyspora*, and a *Streptomyces* species) were shown to inhibit the growth of the pathogen, reduce disease development, or improve the effects of using thiabendazole fungicide alone. Another instance of biocontrol synergism with chemical control was seen in the field production of iris, in which Chet et al. (1982) saw improved crown rot and dry rot control when *Trichoderma harzianum* was integrated with PCNB treatment. The reported synergism in integrated treatments of fungicides with biocontrol agents suggests that the biological control organisms are able to extend the control beyond the residual effectiveness of the fungicide. Magie (1980), for example, saw two seasons of control from the use of *Trichoderma moniliforme* against *Fusarium oxysporum* f. sp. *gladioli* on gladiolus corms.

Following these encouraging findings, fungal antagonists were field-trialed under commercial conditions at Kirton (Lincolnshire, UK). At planting, granular formulations of *Trichoderma* spp., *Minimedusa polyspora*, *Streptomyces* sp., and *Penicillium rubrum* were sprinkled over daffodil 'Golden Harvest' bulbs that had previously received hot-water treatment either with or without thiabendazole fungicide. The results were assessed 2 years later. Compared with untreated controls, some antagonist treatments resulted in a small reduction in the number of bulbs developing base rot and an accompanying small increase in yield; the combined treatment of antagonist plus thiabendazole improved the results further (Pitt 1991; Hanks and Linfield 1997). At the time these benefits were considered insufficient to warrant commercial development, and it appears that no further development was carried out. However, the environment in which the antagonists were used in the field trial—they were applied to dry soil at bulb planting and were expected to be effective over 2 years—may have been too harsh. There is a case for retesting these and similar agents using a better formulation or application method; for example, application in compost that would sustain them until invasion of the bulbs could take place.

Even when effective biocontrol agents are found, such as the use of *Pseudomonas fluorescens* strain R1SS101 to suppress *Pythium* root rot of hyacinths or the use of the mycoparasite *Verticillium biguttatum* to suppress *R. solani* AG2-2IIIB on lilies, De Boer et al. (2006) pointed out that the relatively small size of the geophyte market and resulting limited sales potential reduces the potential interest in registering these biological control agents on geophytes. Alabouvette et al. (2009) also indicated that the success of microbiological control requires a sufficient understanding of the modes of action of the BCA and also of its interactions with the host, pathogen, and the rest of the microbiota. They indicated that it takes time to achieve such an understanding and that most of the BCAs already on the market were studied for more than 20 years before registration. This echoes Barker's (2003) assessment that the success of biological control of plant nematodes has been limited by the need for information on the ecological interactions of nematodes with various soil microflora and fauna.

While significant advances have been made, Alabouvette et al. (2009) pointed out that the use of registered BCAs is limited, mostly because of the lack of efficacy and consistency of control. They indicated that ways to improve the performance of BCAs is a major question that needs additional research. Areas of future research include incorporation of rhizosphere competence as a criterion during early screening processes, elucidation of modes of action at the molecular level, characterization of the ecological fitness of the BCAs, the potential of deploying BCAs with several modes of action to improve efficacy or increase the activity spectrum, the applications of BCAs to seed or bulbs to increase the economic viability of using them to control diseases on field-grown crops (Bennett and Whipps 2008), improved production, formulation, and application processes, and the need to integrate biological control with other cultural practices (Johnson 2010), which will require taking into account all components of the agro-ecosystem.

Another area or research relates to the postharvest use of BCA to reduce losses during storage. A number of bacterial and fungal antagonists have been shown to effectively control postharvest rots of fruits and vegetables (Wilson and Wisniewski 1989; Wilson et al. 1991; Sutton 1996; Janisiewicz and Korsten 2002). Identification of effective antagonists and their commercialization on geophytes may present an alternative to the postharvest use of synthetic fungicides.

4. Biocontrol of Arthropods

Although the use of biological control agents in field-grown flower bulb crops is problematic, biocontrol of arthropods is well suited to greenhouses. Growers have in many cases found biocontrol of arthropods to be effective and have been able to reduce their use of pesticides as a result (Van Lenteren 2000a, 2007; Pilkington et al. 2010). There are several advantages to biocontrol: there is an absence of potential phytotoxicity from the treatments, workers are comfortable working around biological control organisms, there is no reentry or preharvest interval, and there is no decline of treatment effectiveness in the population due to the development of resistance, such as occurs with chemical pesticides (Van Lenteren 2000b).

Biological control is often accomplished by the repeated introduction of parasitoids or predators to the greenhouse. Although there are over 150 species available for biological control on a worldwide basis, there are 30 species that dominate the trade (Bolckmans 1999). The primary biocontrol agents marketed today are *Encarsia formosa*, *Phytoseiulus persimilis*, *Amblyseius cucumeris*, *A. swirskii*, and *Orius laevigatus* (Pilkington et al. 2010). Although biocontrol for greenhouse crops began with *Encarsia formosa* and other host-specialized parasitoids, there has now been a shift in research, such that the focus is on some of the generalist predators (Pilkington et al. 2010). IPM is itself progressing toward integration as research has begun to address the knotty question of whether parasites, predators, and parasitoids can be used compatibly for the management of an insect pest. An evaluation of the control of greenhouse whitefly found that the entomopathogenic fungus *Beauveria bassiana* could be used along with the predator *Dicyphus hesperus* and the parasitoid *Encarsia formosa* without loss of the predator and parasitoid effectiveness, at least in the short-term (Labbé et al. 2009).

Thrips management is another important concern in the greenhouse. After years of employing more specific predators, including *Neoseiulus* (= *Amblyseius*) *cucumeris*, a generalist predatory mite, *Amblyseius swirskii* is now being studied (Messelink et al. 2005; Sabelis et al. 2008). Messelink et al. (2008) showed that the population of *A. swirskii* was denser when supplied with two pests (thrips and whiteflies), which might have been an indication of the benefit of a mixed diet. *Hypoaspis* (=*Geolaelaps*) *aculeifer* has also been cited as a biocontrol for thrips (Glockemann 1992) and fungus gnats (*Bradysia* spp.) (Gillespie and Quiring 1990). Laboratory studies with a single pest–predator combination in a small area may not be sufficient for evaluating biological control effectiveness in commercial greenhouse settings.

The bothersome and pathogen-vectoring shore fly (*Scatella stagnalis*) has been controlled biologically, with limited success (Vänninen and Koskula 2003, 2004; Tilley et al. 2011). Cultural control for shore fly is given in Jacobson et al. (1999). Baker (1993) has reported that fungus gnats, which are secondary pests that feed on roots and are associated with root rot and/or overwatering, can be controlled by Gnatrol (*Bacillus thuringiensis israelensis*) in commercial flower production. The beneficial nematode, *Steinernema feltiae*, is also effective and has outperformed *B. thuringiensis israelensis* in some fungus gnat control studies (Harris et al. 1995; Jagdale et al. 2004). Efforts to control the large narcissus fly on daffodils with parasitic nematodes were reportedly unsuccessful (Conijn and Koster 1990).

The bulb mite, *Rhizoglyphus robini*, causes crop loss on *Lilium* spp. Dutch researchers seeking a biocontrol examined lily bulbs in Holland, Japan and Taiwan, and found *Hypoaspis aculeifer* frequently. *H. aculeifer* feeds on nematodes, springtails, fungus gnat larvae, pollen and fungi as well as other mites (Lesna et al. 1996). The most serious effects of bulb mites are during the 6–8 weeks of bulb propagation, but they also have an effect in storage as well as in lily cut flower production in greenhouse or field. The warm temperatures (22–23°C) during propagation allow fast generation times for the bulb mites. Treatments that are made just prior to propagation also remove natural enemies. These are usually bulb dips in fungicide mixtures containing prochloraz, captan, and carbendazim or hot-water dips at 39–41°C for 1–2 h; dips into pirimifos-methyl solution were also used in the past. Although these treatments do not entirely eliminate the bulb mites, the predatory mites are more vulnerable to these treatments and are eliminated (Lesna et al. 1996). Under ideal conditions, when predatory *H. aculeifer* mites are added to lily bulbs, they may annihilate their bulb mite prey, staying on an infested bulb until the prey is virtually eliminated, and this behavior is obviously ideal for biological control in an agricultural production system. Prey elimination was only seen at a 3:1 ratio of predator to prey, however (Lesna et al. 2000), and was more likely under greenhouse than field conditions due to various environmental factors.

Amblyseius cucumeris has been effective in controlling the bulb mites *Rhizoglyphus robini* or *R. echinopus* during the storage of hyacinth bulbs (De Boer 2011). Bulb mite-resistant cultivars are another potential solution to this problem. The population of mites on lily 'Tender' leveled off after two weeks in a trial with *Rhizoglyphus robini*, whereas populations increased for lily 'Star Gazer' or 'Le Reve' (Lesna et al. 2000).

An alternative approach, conservation biological control, improves the welfare of natural enemies by changes in cultural practices or the environment (DeBach 1974; Eilenberg 2006; Eilenberg et al. 2001). Cases of conservation biocontrol in which parasitoid wasps were fostered in greenhouses have been shown for tomato leafminer (Woets and Van der Linden 1982) and shore fly (Tilley et al. 2011). The greenhouse environment may be manipulated by the addition of alternative foods used to build up the population of a predator species, either by adding pollen or by establishing banker plants (Huang et al. 2011) amidst the crop to supply alternative prey. The banker plants provide a refuge for biological control organisms that might otherwise perish when the target pest population is low in the crop. The biocontrol organisms feed on the arthropod infestation on the banker plant—the banker plant pest is, of course, carefully chosen to be nonthreatening to the crop.

Studies of insect behavior are helpful in identifying greenhouse IPM practices that foster biological control. For example, positioning yellow sticky traps horizontally will reduce the percentage of female parasitoids caught (Tilley et al. 2011).

In their review, Zehnder et al. (2007) indicate that IPM approaches should emphasize preventive tactics such as enhancement of natural enemies of pests, cultural methods, and plant resistance. They go on to point out that due to the limited number of commercially available biological control agents, research is needed to successfully combine inundation and inoculation biological control agents with other biologically-based pest management practices. Although approved insecticides are used as a last option for the control of pests, it is likely that they will continue to play an important role, particularly in the production of high-value crops.

C.   Cultural and Environmental Controls

Cultural methods often have a very strong influence on arthropod pests and diseases (including abiotic disorders), and are effective management tools that can eliminate or reduce the need for pesticides. In general, environmental conditions that favor strong growth of the plant help reduce the impact of diseases. Stress often predisposes plants to attack by a number of weak pathogens. Since moisture is critical for the development of most fungal, bacterial, and nematode diseases, water management has a significant influence on disease development. Excessively wet soils favor root rot diseases, for example, those caused by *Pythium* and *Phytophthora* spp. Well-drained planting media must be used to minimize the potential for saturated moisture conditions in forced bulb crops (whether in the greenhouse soil or in trays of growing media). To avoid *Pythium* root rot, growers use cultural practices that avoid root injury. An appropriate fertilization technique is key to avoiding root injury; if soluble salts are allowed to accumulate in the root zone, root tips may be injured and become attractive to zoospores of *Pythium* species (Gladstone and Moorman 1987). When plants are injured by high salts, nutrients leak from the injured root tips and direct the zoospores to the compromised roots, where they encyst and subsequently infect.

Rather than relying on fungicides for control of *Botrytis cinerea* and other, more host-specific *Botrytis* species, greenhouse growers primarily rely on manipulation of the environment. Any practices that reduce condensation on plant surfaces will reduce the impact of *Botrytis* diseases, as these fungi require free moisture in order to germinate and penetrate plant tissues. Heating and ventilating the greenhouse at sunset to drive out the moisture-laden air before temperatures decline is one helpful practice. Keeping the relative humidity of the greenhouse below 80% is recommended (Hausbeck and Moorman 1996). Computer regulation of greenhouse temperature and humidity is valuable in reducing losses from *Botrytis* blight and other foliar diseases (Jewett and Jarvis 2001). During the greenhouse phase of bulb forcing, the use of fans is critical to keep air moving; adequate plant spacing is also important. Under-bench or floor heating is employed in some cases, allowing the warm air currents to rise up between plants (Hausbeck et al. 1996). Increasingly, various types of polyethylene tunnels are used for growing cut flower crops, including lilies. While offering much less protection than a modern greenhouse, tunnels provide crops with protection from rain and other weather disturbances at a much lower cost. In the United Kingdom, the structure known as the "Spanish tunnel" is favored; a relatively low-cost, easily erected structure, usually with ends and sides that can be opened or lowered. Where enhanced temperatures are not needed, the polyethylene skin may be left rolled back, producing a sturdy crop, until shortly before harvest when it is rolled out to protect the buds and blooms from rain and to facilitate harvest irrespective of weather conditions.

Minimizing periods when leaf and flower tissues are wet by using proper irrigation management is also an effective way to reduce disease foliar diseases. This usually involves some form of irrigation where water is applied directly to the planting medium or soil as opposed to overhead sprinklers. Avoiding overhead irrigation in late afternoon or evening generally reduces disease potential. Scheduling irrigation during the day when plant tissues can rapidly dry prevents disease from developing by minimizing periods of leaf wetness. However, monitoring soil moisture conditions to avoid drought-stressed plants that can be more susceptible to disease organisms is also important.

It is essential to avoid predisposing plant material to diseases by using proper handling practices and providing conditions in storage, rooting rooms, and the greenhouse that do not promote disease development. For example, cultural manipulations that reduce the mass of tulip roots at the bottom of the forcing container are beneficial for minimizing tulip leaf withering (*Trichoderma viride*). Growers in the United States have reported more difficulties with tulip leaf withering than growers in The Netherlands, presumably because of the widespread use of a sand substrate in Europe, whereas a peat-perlite mix is commonly used for forcing tulips in the United States. This disease is reduced when plants are grown in a medium with less water-holding capacity. The addition of 20% coarse, salt-free river sand to the growing medium has this effect (Miller 2004). The addition of an inch of coarse sand in the bottom of crates used for growing tulips also helps protect the roots massed at the bottom of the container from drying out. When tulips are grown in peat, a congested mass of roots often develops at the bottom of the container. When these dry out, it causes injury that helps *T. viride* to invade the root system, subsequently causing symptoms of leaf edge scorch during forcing on the greenhouse bench. Tulips may also be planted into moist soil rather than drenched thoroughly with water at planting to curb the overproduction of roots. Colder temperatures during root development have also been suggested as cultural management for tulip leaf withering. This could be achieved by reducing the temperature from 9°C to 1°C more quickly—the current practice entails a gradual temperature reduction over a 5-week period. Additional cultural controls for this disease include growing pots on open-wire mesh benching rather than on solid surfaces, and elevating bulb crates on pipes to keep the bottoms of the crates off the floor (both practices serving to keep roots from growing out of their containers).

Cultural techniques for managing Fusarium wilt diseases are not often applied during the forcing of bulb crops in crates or pots; the infected bulbs are merely discarded, a form of sanitation, once the disease is detected by cutting across the bulbs and noting discoloration. In field production systems for bulbs and other ornamentals, the pH may be raised to 6.2 or above by liming, in order to suppress Fusaria that cause vascular wilt diseases such as basal rot (Woltz and Engelhard 1973; Jones et al. 1989). Similarly, the use of calcium nitrate-based fertilizers, rather than ammonium nitrate materials, is encouraged, to capture the Fusarium-suppressive effects of the calcium nitrate (Woltz and Engelhard 1973). With daffodil and tulip, spread of basal rot is not generally an issue during the greenhouse flowering phase and the finishing time is too short for disease-suppressive cultural techniques to be of significant benefit. Management of Fusarium basal rot in tulips is extremely important in field bulb production, because ethylene production from diseased bulbs in storage and shipment can have negative effects throughout a room or container, ruining far more bulbs than were originally directly infected by the Fusarium pathogen (De Munk and Beijer 1971; De Munk 1972; see also Chapter 11).

D. REDUCTION OF HIGH-RISK PESTICIDE USE

Chemical pesticides are likely to remain an essential component of disease, pest, and weed integrated management programs for the foreseeable future. Even so, there are a number of opportunities to reduce the use of high-risk pesticides and the dosages of pesticides that are used per season. Pesticides are designated as high- or low-risk based on health and environmental impact and potential to negatively affect nontarget organisms. One approach to reducing the use of high-risk pesticides is to increase the use of reduced-risk products. For example, tulip and iris growers have used broadcast or in-furrow applications of PCNB at the time of planting to control crown rot and gray bulb rot for over 50 years (Gould and Russell 1965; Gould 1990, 1993; Chastagner 1997; Chastagner, unpublished). PCNB is typically applied at rates up to 73 kg per ha as an in-furrow application prior to planting. During the last 10–15 years, there has been increasing concern about the continued availability of PCNB due to a number of issues ranging from human health to environmental risk, including its persistence properties. The use of PCNB has recently been prohibited in the United Kingdom and the continued availability of this product in the United States is uncertain.

Research has shown that there are a number of newer reduced-risk fungicides that provide effective control of gray bulb rot and crown rot. In-furrow applications of trifloxystrobin, azoxystrobin, and flutolanil significantly reduced the development of crown rot and increased yields of bulbous iris (Chastagner et al. 1990; Chastagner 2002). Flutolanil and trifloxystrobin were also found to be the most effective materials for controlling gray bulb rot on bulbous iris. Flutolanil and fludioxonil also have the potential to provide control of gray bulb rot on tulips that is equal to that achieved with PCNB (Chastagner and DeBauw 2011b). The rates of these materials that are needed to control these diseases are significantly lower than the amount required for PCNB.

The following provides additional examples of different approaches that have been evaluated to reduce the amount of pesticides on flower bulb crops.

1. Forecasting Models

A potential approach to reducing pesticide use in a number of crops has been the development of forecasting models (decision support systems). These models can be based on reductions in the populations of pathogens and pests or the identification of critical periods relating to host susceptibility, pest life stages and activity, and infection and spread of pathogens. With monocyclic diseases, such as those caused by a number of soilborne pathogens, forecasts and subsequent treatment recommendations are often based on the direct assessment of the initial amount of inoculum at the time of planting (Adams 1981; Backman et al. 1981). In the case of polycyclic diseases, where multiple cycles of infection and inoculum production occur during the growing season, most control efforts are aimed at reducing the rate of infection and subsequent disease development. Since environmental conditions, such as temperature and periods of leaf wetness, are the primary influence on the rate of development of these disease epidemics, forecasts of polycyclic diseases are generally based on weather conditions that favor the production of inoculum, dispersal, infection, and host colonization.

a. Large Narcissus Fly

The narcissus bulb fly or the large narcissus fly (*Merodon equestris*) attacks narcissus and hyacinths (Baker 1993) and a number of other bulbs. In some years and locations it causes considerable loss of yield and can also lead to the rejection of stocks because the permitted infestation tolerances are exceeded. Although hot-water treatment (HWT) is effective in killing the larvae in the bulbs, the cost of annual lifting, treatment, and replanting of daffodils is uneconomical and they are usually grown on a 2-year or longer cycle. In UK trials only chlorpyrifos was consistently effective and is still permitted, although it gives protection for only 1 year and is not without some phytotoxic effects (Woodville 1960; Tones and Tompsett 1990; Hanks 1994). Aldrin was used in daffodil growing in the 1970s. In-furrow applications at planting provided control for 2 years (Tompsett 1974). The persistence of aldrin led to its withdrawal in 1989, since then daffodil stocks have been seen with as many as 40% of bulbs containing a larva. Formerly considered at the limit of its range in the mild southwest of England, large narcissus fly has become established in cooler parts of the country, presumably in response to increases in average temperatures over the last few decades (R. H. Collier 2010, pers. comm.). With few or no options for controlling the larvae through HWT or in-furrow treatments, attention has turned to the control of the adult fly in the field.

Regular crop sprays of omethoate were found to be effective against large narcissus fly in The Netherlands (Conijn and Koster 1990), but omethoate was not approved in the United Kingdom, and most insecticides were found to be ineffective or inconsistent (Tones et al. 1990). However, simulation models have been developed for several pest insects, including large narcissus fly (Phelps et al. 1993; Collier et al. 1995). The basis for these models is that the rates at which insects complete their life cycles are generally dependent on ambient temperatures. With this insect, one option is to use a forecasting system to target sprays between fly emergence and egg laying, ensuring that no unnecessary pesticide applications are made. Biological studies showed that adult large narcissus flies have a critical threshold for activity close to 20°C, but once mating has occurred, egg laying can take place at a lower temperature, and the base temperature for egg hatch is about 8°C (Collier and Finch 1992; Collier et al. 1995). A computer-based forecasting model incorporating all these features was developed and validated (Collier 2011).

b. Botrytis Blight on Lilies, Tulips, and Gladiolus

Significant reductions in fungicide use to control *Botrytis* blight on onions have been achieved in Canada, The Netherlands, United Kingdom, and the United States using weather-based warning systems (Jarvis 1980; Sutton et al. 1986b; Vincelli and Lorbeer 1989; De Visser 1996; Carisse et al. 2008). A weather-based forecasting system for ornamental geophytes has also been developed in The Netherlands to optimize disease control and reduce the number of fungicide sprays applied to control diseases caused by *Botrytis* spp. This system is aimed at diseases caused by *B. elliptica*, *B. tulipae*, and *B. gladiolorum* on lily, tulips, and gladiolus, respectively (Bastiaansen et al. 1997; Van den Ende et al. 2000).

These pathogens cause leaf and flower blights, stem rot, and bulb/corm rot (Price 1970; Moore et al. 1979; Trolinger and Strider 1985; Byther and Chastagner 1993). Although growers use a combination of cultural practices and fungicide bulb dips to help manage these diseases, they rely on calendar-based applications of fungicides to reduce potentially rapid disease development during the growing season. In The Netherlands, Bastiaansen et al. (1997) reported that growers spray lilies on a weekly basis from the start of the season in April until harvest in October to control fire. They estimated that this amounts to an application of approximately 25 kg of active ingredient/ha/year. In the US Pacific Northwest, Chastagner (1997) reported that growers made an average of 5.9 applications of fungicide to tulips and 14.4 applications to lilies per year. Repeated application of the same fungicide has also resulted in fungicide resistance problems (Chastagner and Riley 1990; Migheli et al. 1990; Hsiang and Chastagner 1991, 1992; Brent 1995).

Fire, caused by the fungus *Botryotinia (Botrytis) polyblastis*, can completely destroy the foliage in a field of daffodils within a few weeks under warm, moist conditions in the spring. Studies defining the period of apothecium production have identified critical spray timings (Chastagner 1983) and subsequent testing of fungicides that have been commonly used to control *Botrytis* diseases have shown that dicarboximide materials were much more effective than other products. Effective disease control could be obtained with one or two properly-timed sprays (Chastagner and DeBauw 2011a).

Using information on the relationship between temperature and leaf wetness periods (Doss et al. 1984, 1986; Van den Ende and Pennock 1997), Bastiaansen et al. (1997) developed a specific warning system for the control of *Botrytis* leaf blight in lily (BoWaS: *Botrytis* Warning System) in order to optimize spray programs. Results of research conducted on lilies grown at several research locations resulted in a 30–80% reduction of the input of active ingredients, depending on the cultivar susceptibility and year. On-farm trials at four farms in 1996 and 1997 showed that the number of calendar-based sprays ranged from 12 to 21 (avg. 16.25) and was reduced to 2–8 (avg. 5.6) for the BoWaS-based sprays. They also reported that the reduction in sprays did not have any adverse effects on quantitative and qualitative yield losses. The BoWaS forecasting system was made commercially available in The Netherlands to lily growers in 1998 and tulip and gladiolus growers in 1999 (Van den Ende et al. 2000).

c.   Smolder and White Mold on Daffodils

Smolder (*Botrytis narcissicola*) appears to attack daffodils wherever they are grown, exacting a steady yield loss. In contrast, white mold (*Ramularia vallisumbrosae*) is a particular problem in areas such as the southwest UK, where it sporadically attacks daffodil crops, developing rapidly and destroying most of the foliage in a few weeks. Smolder and white mold are rarely troublesome where daffodils are grown and lifted annually, but inoculum increases in soil and debris where it is the practice to grow them on 2-year or longer cycles.

Fungicide treatments during the bulb-handling phase give some control of fungi, and certainly of neck rot caused by *B. narcissicola* (Chastagner and DeBauw 2011c). However, the mainstay of disease management is a program of field sprays. In the United Kingdom, for example, this might include a program of four preflowering sprays, starting when shoots are 10–15 cm tall, with two further sprays after flowering, or even a three-weekly spray program (ADAS 1986).

In the late 1990s, it became apparent that the sprays used were failing to give adequate control, and disease forecasting methods were investigated with a view to achieving better disease control with fewer, but targeted, fungicide applications (Hanks et al. 2003). This work showed that it was possible to reduce the number of fungicide applications then being used from six or seven scheduled sprays to three targeted sprays if more effective active substances (such as strobilurins) were used (Hanks et al. 2003; O'Neill et al. 2004).

Smolder infection was found to be maximal at temperatures of 4–16°C when accompanied by a period of leaf wetness of at least 24 h, while at nonoptimal temperatures, longer periods of leaf wetness were needed to produce infections (O'Neill et al. 1982). White mold infection was maximal at temperatures of 5–10°C when accompanied by a period of leaf wetness of at least 12 h, while at nonoptimal temperatures longer periods of leaf wetness were again required. These data were used to derive predictive infection models for both diseases, driven by temperature and leaf-wetness duration (Hanks et al. 2003).

d.   Summary of Pesticide Use Reductions

While forecasting systems for predicting target spray dates have received support from research funders, it is clear that there are a number of unsolved problems and grower concerns that have constrained the greater use of these systems. With current fungicide-delivery systems, unless suitable weather conditions are available for spraying promptly once a spray alert is issued, the program will be compromised. In southwest England, for example, wind and rain severely limit the likely weekly numbers of "spray days", while in drier eastern England there are on average only 1.5 spray days per week during the period from daffodil emergence to early senescence (Harris and Hossell 2002). Even these few days may be diminished by the projected effects of climate change.

The BoWaS forecasting decision support system combines weather forecasts and crop sensitivity to produce a crop-specific *Botrytis* infection risk assessment and control recommendation. De Ward (2008) has reported that even though numerous demonstrations and trials have shown that the BoWaS system reduces overall fungicide use, growers have been reluctant to switch from their weekly spray schedules. Bastiaansen et al. (1997) indicated that the reasons for the grower reluctance include concern about the reliability of the weather forecasts, difficulty with the real-time implementation, and the low economic incentive. Economic benefits of using BoWaS seem to be limited, mainly as a result of high crop value and the low prices for fungicides. The potential saving in fungicide costs using BoWaS is not sufficient to balance out the perceived additional risks, which has been reported as an impediment to the utilization of other disease-forecasting systems (Campbell and Madden 1990).

The use of forecasting is also potentially impeded by the "harvest interval" and "re-entry" restrictions of the pesticides used. Although these are often being extrapolated from intervals originally developed with the safety of edible crops in mind, it is often uneconomical for the agrochemical industry to develop pesticide products specifically for use on minor ornamental crops. The economic advantage of picking flowers from field-grown crops may mean that, increasingly, pesticides will not be applied for an extended period before flowering. It follows that targeted spraying through predictive methods can be only one part of an integrated pest and disease management program. Alternative measures generally consist of cultural controls such as those for daffodil diseases, which can include the removal of crop debris between growing seasons, the removal of disease primaries, and using rotations that include the physical separation of successive years' crops (ADAS 1986), but these are measures that are often impractical or uneconomical.

The case of the large narcissus fly forecast illustrates another difficulty that may increasingly arise in incorporating forecasting models into integrated pest management programs. The implementation may be impossible because of the continued decrease in the availability of effective insecticides. However, the forecast system can still be used to guide other management practices for fly-susceptible crops. For example, it was proposed that infestation could be reduced by early foliage removal (to deter egg laying) or by early bulb lifting (before the larvae hatch and enter the bulbs (Tones et al. 2004). Consistently cool weather in spring indicates a year when the large narcissus fly damage is unlikely to be high, and no management action may be needed. Otherwise, it may be necessary to lift bulbs early to avoid infestation, in which case the curtailment of photosynthesis would lead to yield loss, especially in "early" years (Hanks 1998). Early lifting leads to marked reductions in bulb yields, more rotted bulbs, and less flower bud formation, compared with late lifting, whereas yield losses were low (<20%) when lifted at the average date (Hanks 1998). These data were integrated into a model which showed that there was potential for implementing cultural control strategies in relation to the time of fly activity, and that the yield penalties, particularly due to early lifting, might not be as great as had been supposed (Tones et al. 2004).

2. Precision Application of Pesticides

Increased technological developments relating to global positioning systems (GPS) and variable rate spray systems have fostered the development of Precision Pest Management (PPM), with the goal of achieving reduced pesticide inputs by targeting variable rates of sprays to where and when they are needed. PPM systems have been successfully developed that utilize a real-time sensing and application system to differentiate between weeds and crop plants to ensure precise application for targeted application of herbicide. Van der Zande et al. (2010) have investigated whether matching spray volume to crop canopy sizes and shapes can reduce the use of fungicides, operational costs, and environmental pollution. Working with lilies, they showed that the average spray volume of fungicide to control *Botrytis* leaf blight could be reduced by using a canopy density sprayer. This sprayer uses a sensor to quantify crop canopy combined with variable rate spray technologies. Researchers used a weekly spray scheme that started with a low rate of fungicide when the lilies were about 10 cm above the ground and raised this dosage every week by small steps, reaching the full label rate around flowering of the lilies. With this simple method, good *Botrytis* leaf blight control was achieved with at least 21% less fungicide (Van der Zande et al. 2010).

West et al. (2003) have reviewed the potential of optical canopy measurement for targeted control of field crop diseases. Unlike spray systems for targeted herbicide applications, which can utilize a real-time sensing and application system to differentiate between weeds and crop plants to ensure precise application of the herbicide, currently available technology is not able to detect certain types of diseases in real time. Barker (2003) indicated that PPM technologies have some potential in the management of soilborne nematode diseases, but strategies to overcome the high cost and other problems inherent in nematode population assessments were needed. West et al. (2003) concluded that most variable-spray systems to control diseases caused by soilborne pathogens would most likely be based upon conducting GPS-based surveys to identify the location of diseased areas to allow for targeted applications of control products.

While PPM could play an important role in reducing pesticide inputs, West et al. (2003) concluded that many technical challenges still have to be overcome before affordable practical systems operating at normal tractor speeds can be produced. In addition, they pointed out that the decision to treat should be based on the economics of expected yield or quality reductions against treatment costs. Small levels of disease near the end of the growing season may not justify the costs of treatments. Differences in the extent of damage that can be tolerated are also likely to be dependent on the end product. For example, growers producing bulbs may tolerate a higher level of infection by *Botrytis* on tulips than growers who are producing cut flowers.

3. Natural Pesticides

A number of biologically-based controls and concepts have been discussed in Section IV.B. In addition, a number of secondary plant metabolites like essential oils have been shown to provide effective control of a number of pathogens and pests (Smid et al. 1995; De Boer 2011; Regnault-Roger et al. 2011). Smid et al. (1995) evaluated the effectiveness of 15 essential oils against *Penicillium hirsutum*, which causes blue mold on tulips. Storage of tulip bulbs in atmospheres containing cuminaldehyde, perillaldehyde, salicylaldehyde, or carvone resulted in a significant reduction of the natural *Penicillium* infection. Dipping tulip bulbs in an aqueous solution of cinnamaldehyde also reduced fungal population on bulbs. They also showed that treatment of tulip bulbs with carvone, cuminaldehyde, perillaldehyde, cinnamaldehyde, or salicylaldehyde had no effect on the growth and flowering of the bulbs.

De Boer (2011) points out that several groups of essential oils that attract or repel insects or waste products of the food chain that have microbial growth-inhibiting capacities have been tested against several pests and diseases in flower bulbs. Evaporation of an essential oil from plants proved to be very effective in controlling *Thrips simplex* on corms of gladiolus in storage (Conijn and De Kogel 2008). A waste product of the citrus industry proved to be effective against *Botrytis* leaf blight in lily caused by *Botrytis elliptica*. Although it is not always as effective as standard fungicide application, De Boer (2011) concluded that it was a potentially promising, environmentally-acceptable product.

Research by Sharma and Tripathi (2008) illustrate an important concept relating to the integration of several approaches to improve disease control with natural products. They looked at the use of hot water, UV-C, and *Hyptis suaveolens* essential oil to manage postharvest losses caused by *Fusarium* rot of gladiolus corms. UV-C is shortwave germicidal ultraviolet light that is used for air, surface, and water disinfection (Hong et al. in press). Sharma and Tripathi (2008) found that integrated treatments of hot water (55°C for 30 min), UV-C (dose 4.98 kJ/m^2), and essential oil (0.8 µL/cm^3) for 2 weeks were more promising than their treatments alone, after storage for 4 and 12 weeks.

4. Gas Application of Chlorine Dioxide

Controlled-release systems to produce ClO_2 gas for a number of uses, including the horticultural industry, are being developed by a number of companies (Copes et al. in press). Recent work with ClO_2 gas has shown that gas applications can be used to sanitize tanks used for aseptic juice storage and that postharvest applications of gas to green peppers (*Capsicum annuum*) was effective in killing *Escherichia coli* on the surface of the peppers (Han et al. 1999, 2000). This technology does not require specialized equipment and could provide an opportunity to reduce the use of fungicides to control some types of postharvest diseases on bulbs and cut flowers (Chastagner and Copes 2002).

Chastagner and Riley (2005) reported that exposing *Penicillium*, *Botrytis*, and *Alternaria* spp. spores to 5 µL/L ClO_2 reduced germination by 93%, 97%, and 90%, respectively. All the spores of these pathogens were killed after a 1 h exposure at 25 µL/L ClO_2. Exposure of *Rhodococcus* cells to 5 µL/L ClO_2 reduced viability by 79%. At 25 µL/L, viability was reduced by 99.996%.

Preliminary data from two trials also indicate that exposing tulip bulbs and cut flowers to concentrations of 5–20 µL/L ClO_2 caused a significant reduction in the development of *Penicillium* blue mold on bulbs and *Botrytis* blight on the flowers. Although additional testing is needed to examine the effect of concentration and exposure period on inoculum viability on host material and to ensure that such treatments have no adverse affects on the growth of bulbs and quality of flowers, it appears that controlled-release ClO_2 gas technologies may offer the potential to control some diseases during the storage of bulbs and cut flowers. Since ClO_2 is highly active in oxidizing organic compounds, ClO_2 gas may also be beneficial in minimizing problems associated with the exposure of some types of bulbs and flowers to ethylene gas.

E. HOST RESISTANCE

Disease and insect control in agricultural production of ornamental bulbs is essential for optimal cut flower and bulb yield (Baker 1993; Byther and Chastagner 1993). Some insects and plant pathogens—mainly bacteria, phytoplasmas, viruses, and a few soilborne fungi—have no or very limited effective control measures to date. Systemic infections, mainly of viral origin, and the infestation of planting stock, which in most cases is vegetatively propagated, make the management of diseases and pests a larger problem in bulb crops than many other crops. In addition, as the use of pesticides, particularly soil fumigants like methyl bromide, is restricted or banned, growers of high-value crops such as ornamental geophytes are looking for more sustainable disease and pest management strategies.

Many cultivars that are used in the production of dry bulbs, cut flowers, and potted plants are selected for various horticultural characteristics and customer appeal, not their resistance to common biotic and abiotic diseases. For example, 'Golden Harvest', 'Carlton', and 'Dutch Master' are cultivars of daffodil that are highly susceptible to basal rot caused by *Fusarium oxysporum* f. sp. *narcissi* and yet are still widely grown. A number of disease problems could be significantly reduced by choosing to grow cultivars that are resistant.

Unfortunately, public information on the susceptibility of many commercial cultivars of geophytes to their major disease and pest problems is limited. Recent work has identified a number of commercially-available *Zantedeschia* spp. that are resistant to bacterial soft rot (Snijder and Van Tuyl 2002; Snijder et al. 2004a,b). Deng et al. (2011) have also been able to identify a number of commercial cultivars of *Caladium* that are resistant to Fusarium tuber rot, Pythium root rot, and bacterial blight (Xanthomonas leaf spot). Growers can also take advantage of the genetic variation that occurs by screening cultivars under their production system to determine their growth and disease resistance characteristics. Resistant cultivars could then be marketed to the public as an improved, "greener" product to encourage their acceptance.

Conventional breeding approaches and the increased use of biotechnology to genetically manipulate hosts for resistance to diseases and pests of ornamental geophytes are covered in Chapters 6 and 7, respectively. Over the last few years there has been increasing interest in induced resistance in ornamental geophytes as a possible mechanism to control plant pathogens in greenhouses and in field crops. The following section discusses and characterizes natural and chemical defense elicitors that have been used in bulb crops, and enumerate their activities and complexities for practical implementation in disease management of ornamental geophytes.

Plants protect themselves from pathogens through an array of constitutive and inducible defenses (Agrios 2005; Schumann and D'Arcy 2010). In susceptible plants, development of resistance may occur following elicitation with a variety of biotic and abiotic inducers. Such an induced resistance may be split broadly into two types: (1) Systemic Acquired Resistance (SAR), which may be caused by a necrotizing agent and occurs in distal plant parts, and (2) Induced Systemic Resistance (ISR),

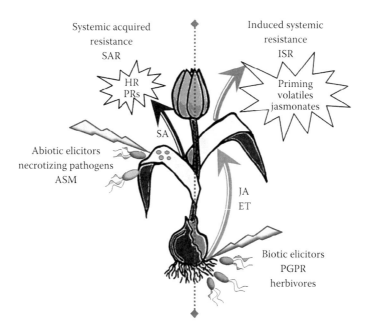

FIGURE 13.1 A scheme of the two best-characterized systemically-induced immune responses in plants, both leading to effective resistance against a broad spectrum of virulent plant pathogens. Systemic acquired resistance (SAR) is typically induced by the exposure of root or foliar tissues to abiotic or biotic elicitors such as plant activators and necrotizing pathogens. The response is dependent on the phytohormone salicylic acid (SA) and a mobile signal that travels through the plant to activate a large set of resistance genes, some encoding pathogenesis-related proteins (PRs), with antimicrobial activity. Induced systemic resistance (ISR) is typically stimulated by colonization of the root system by specific strains of beneficial microorganisms such as plant growth promoting rhizobacteria (PGPR). ISR is associated with the jasmonic acid (JA)- and ethylene (ET)-dependent signaling pathways, is independent of SA and is not associated with the accumulation of PRs. Typically ISR involves priming of the plant for accelerated JA- and ET-dependent gene expression upon subsequent pathogen attack and accumulation of secondary metabolites and volatiles. Both SAR and ISR are inducible plant defenses that are effective against a broad spectrum of virulent plant pathogens. (Original illustration by: Iris Yedidia.)

which is promoted by selected strains of nonpathogenic plant growth promoting rhizobacteria (PGPR) or beneficial fungi (Figure 13.1). Both types of resistance are considered broad-spectrum, can be long-lasting and are rarely complete, with most inducing agents providing 20–85% disease control.

The initial stages of the interaction between plants and pathogens involve elicitor molecules originating from the pathogen or the plant, such as polymers (cell wall fragments), carbohydrates, peptides, lipids, and glycopeptides. Such molecules, as well as several plant extracts and synthetic chemicals, were shown to enhance plant resistance against subsequent pathogenic attack both locally and systemically (Walters et al. 2005; Luzzatto-Knaan and Yedidia 2009; Walters 2009; Walters and Fountaine 2009).

Plant perception of pathogen-derived elicitor molecules leads to the activation of signaling pathways and consequently to the activation of defenses. The defense mechanisms include: (1) a burst of reactive oxygen species (ROS) that culminates in a programmed hypersensitive cell death at the site of pathogen invasion, (2) the production of antimicrobial secondary metabolites, such as phytoalexins, (3) accumulation of pathogenesis-related (PR) proteins, such as chitinases and glucanases, that degrade fungal and oomycete cell walls, and (4) cell wall fortification through the synthesis of callose and lignin (Hammerschmidt and Dann 1999; Pieterse et al. 2009).

Systemic Acquired Resistance (SAR) and Induced Systemic Resistance (ISR) are both preconditioned prior to elicitation and are manifested only following subsequent challenge by a pathogen or a parasite (Métraux et al. 2002; Vallad and Goodman 2004). These two forms of induced resistance may be differentiated on the basis of the elicitor and the regulatory pathways involved (Figure 13.1; Pieterse et al. 1996, 2009; Van Wees et al. 2000; Yan et al. 2002). SAR response may be elicited with plant pathogens (virulent or avirulent) or artificially with chemicals such as salicylic acid (SA) or its structural analogs 2,6-dichloroisonicotinic acid (INA) and benzo (1,2,3)-thiadiazole-7-carbo-thioic acid S-methyl ester (BTH), also recognized as acibenzolar-S-methyl (ASM) (Sticher et al. 1997; Gozzo 2003). SAR is dependent on the accumulation of salicylic acid as a signaling molecule, locally at the site of infection and often also systemically in distant tissues (Figure 13.1; Vallad and Goodman 2004; Pieterse et al. 2009). The regulatory protein NPR1 (Nonexpressor of PR Genes1) emerged as an important transducer of the SA signal, which acts upon activation as a transcriptional coactivator of pathogenesis-related (PR) gene expression (Dong 2004). Expression of PR genes and *PR-1* in particular is used as a marker for SAR induction. SAR response is usually associated with defense against biotrophic pathogens and is most commonly associated with the hypersensitive response (HR) (Walters and Fountaine 2009).

ISR response develops as a result of colonization of plant roots by certain strains of beneficial soil-borne microorganisms such as mycorrhizal fungi and plant growth-promoting rhizobacteria (PGPR) that cause no visible damage to the plant root system. ISR is mainly associated with defense against herbivores and necrotrophic pathogens and is mediated by a jasmonic acid- and ethylene-sensitive pathway (Walters 2009). Phenotypically the responses are similar; however, ISR is most often associated with priming for enhanced defense rather than direct activation of defense and the accumulation of secondary metabolites such as phytoalexins. In contrast to SAR, ISR is independent of salicylate or the accumulation of PR proteins (Pieterse et al. 1998; Yan et al. 2002; Vallad and Goodman 2004). Both responses are intertwined molecularly as shown by their reliance on NPR1, the key regulator of SA-mediated suppression of JA signaling in *Arabidopsis thaliana* (Van Wees et al. 2000; Spoel et al. 2003).

1. Induced Resistance in Monocots

There is a large amount of knowledge on inducible defense responses in dicot plants. Considering the substantial economic and environmental value of monocot geophytes, particularly ornamentals, information on induced resistance responses, and the mechanisms underlying such responses in this important group is surprisingly limited (Luzzatto et al. 2007a,b). Both types of induced resistance, SAR and ISR, have been described in monocots, as well as master regulatory components of the defense pathways such as NPR1 gene and priming (Chern et al. 2001; Conrath et al. 2002; Dong 2004; Durrant and Dong 2004; Conrath et al. 2006). The efficacy of plant defense elicitors (mainly BTH) to induce disease resistance in the monocot plants was tested mostly in Poaceae (e.g., wheat, maize, barley, and rice). BTH proved effective against several airborne fungal diseases (Vallad and Goodman 2004) and the induced resistance afforded by it was long-lasting in comparison with that seen in dicot hosts (Oostendorp et al. 2001).

2. Practical Applications of Induced Resistance in Ornamental Bulbs Using Externally Applied Biotic/Abiotic Stimulators

The use of "plant activators" to elicit natural plant defense mechanisms, also known as induced resistance, is a well-recognized approach for crop protection (Oostendorp et al. 2001; Walling 2001; Vallad and Goodman 2004; Walters et al. 2005; Walters 2009). Such resistance was proven effective against a wide spectrum of fungal, bacterial, and viral pathogens (Oostendorp et al. 2001; Conrath et al. 2002; Heil and Bostock 2002; Gozzo 2003; Durrant and Dong 2004). In the last decade, an increasing number of scientific reports deal with induced resistance against several bacterial and fungal pathogens in ornamental geophytes. These include quite a few genera, including *Cymbidium*, *Cyclamen*, *Gladiolus*, *Lilium*, *Ornithogalum*, *Tulipa*, and *Zantedeschia*. A large number of biotic and abiotic agents were reported as activators of plant resistance (Table 13.2). Plant hormones such as salicylic acid (SA), jasmonates (JA), and ethylene (ET) are well established as inducers of plant

TABLE 13.2
Effect of Natural, Synthetic, Physical, and Biotic Elicitors on Induced Resistance of Ornamental Geophytes (Summary of Major Reports)

Elicitor	Plant Species	Target Pathogen	Effect	Authors
		Natural Organic Elicitors		
Methyl jasmonate ABA	*Lilium longiflorum*	Not defined	Induction of two subclasses of PR-10 in response to both inducers	Wang et al. (1999)
SA *Botrytis elliptica*[a]	*Lilium*	*Botrytis elliptica*	Delayed symptoms; accumulation of callose in guard cells; increased expression of LsGRP1 gene	Lu and Chen (2005)
Methyl jasmonate (MJ)	*Ornithogalum dubium*	*Pectobacterium carotovorum*	Reduction of bacterial proliferation in elicited plantlets	Golan et al. (2010)
Methyl jasmonate Bion[a]	*Zantedeschia aethiopica*	*Pectobacterium carotovorum*	Priming of active glicosidic flavonoides	Luzzatto et al. (2007a)
Methyl jasmonate	*Freesia hybrida*	*Botrytis cinerea*	Reduction in disease severity, reduction in lesion numbers and lesion diameters	Darras et al. (2005)
Chitosan	*Gladiolus* 'Blanca Borrego'	*Fusarium oxysporum*	Extend vase life and overall performance of flowers	Ramos-García et al. (2009)
Chitosan	*Curcuma alismatifolia* × *Curcuma cordata*, *Curcuma* 'Laddawan'	Not defined	Extending the vase life of inflorescence	Tamala et al. (2007)
		Synthetic Organic Elicitors		
ASM	*Gladiolus*	*Fusarium oxysporum* f. sp. *gladioli*	Combinations of ASM and chemical fungicides suppressed *Fusarium* corm rot	Elmer (2006b)
ASM	*Allium cepa*	*Xanthomonas axonopodis* pv. *allii*	Reduced leaf blight disease severity by up to 50%	Lang et al. (2007)
ASM	*Cyclamen*	*Fusarium oxysporum* f. sp. *cyclaminis*	Delayed symptoms of *Fusarium* wilt	Elmer (2006a); Elmer and McGovern (2004)
ASM Methyl jasmonate[a]	*Freesia hybrida*	*Botrytis cinerea*	Suppression of postharvest infection of cut flowers	Darras et al. (2006)

Elicitor	Plant	Pathogen	Effect	Reference
BABA	*Allium cepa*	*Botrytis*	Induced resistance against *Botrytis* neck rot and priming of callose	Polyakovskii et al. 2008
BABA	*Tulipa*	*Fusarium oxysporum* f. sp. *tulipae*		Jarecka and Saniewska (2007)
BABA Bion Methyl jasmonate[a]	*Z. aethiopica*	*Pectobacterium carotovorum*	Induced resistance, reduced disease symptoms; reduced bacterial proliferation	Luzzatto et al. (2007b)
Probenazole	*Lilium* 'Star Gazer'	*Botrytis elliptica*	Protection was efficient for 14 days; stomatal closure was part of the defense mechanism	Lu et al. (2007)
Physical Elicitors				
Wounding, cell walls of yeast[a] – natural organic elicitor	*Hippeastrum* × *hortorum*	*Bacillus subtilis, B. megaterium,* larvae of *Syntomis mogadorensis*	Increase in phenylalanine ammonium lyase and peroxidase activities; enhanced synthesis of flavans and chalcone phytoalexins	Wink and Lehmann (1996)
UV irradiation	*Freesia hybrida*	*Botrytis cinerea*	Reduced lesion numbers and lesions diameter	Darras et al. (2010)
Biological Elicitors				
Bacillus cereus (isolated rhizobacteria)	*Lilium*	*Botrytis elliptica*	Up to 75% reduction in disease severity; long-lasting effect	Liu et al. (2008)
Weakly virulent strain HPF-1 (*Fusarium* sp.)	*Cymbidium* spp.	*Fusarium proliferatum*	Increase in endogenous SA; suppression of leaves, bulbs, and root diseases	Ichikawa et al. (2003)

Source: Adapted from Luzzatto-Knaan T., and I. Yedidia. 2009. *Israel J. Plant Sci.* 57:401–410.

[a] In the case of more than one elicitor for a single report, footnote 'a' highlights the second elicitor which in some cases belongs to a different category.

defense response (Walters et al. 2005; Bari and Jones 2009; Pieterse et al. 2009). Nonetheless, abscisic acid (ABA), auxin, gibberellic acid (GA), cytokinin (CK), brassinosteroids, and peptide hormones were also implicated in plant defense signaling pathways (Bari and Jones 2009). Commercial products available include: Oryzemate® (Probenazole registered in Japan, MeijiSeika Kaisha Ltd), Bion®, and Actigard® (ASM, Syngenta), Milsana® (*Reynoutria sachalinensis* extract, KHH BioScience Inc., USA), Elexa® (Chitosan, SafeScience, USA), and Messenger® (harpin protein, Eden BioScience, USA) (Lyon 2007; Walters and Fountaine 2009). Some of these inducers are pure chemical compounds (ASM, Probenazole) and some are undefined mixtures of plant origin, such as Milsana.

3. Natural Organic Elicitors

Salicylic acid is among the first discovered compounds to induce systemic resistance against biotrophic pathogens and is often associated with accumulation of PR proteins via the SAR pathway (Lyon 2007; Bari and Jones 2009). Initially, SA was postulated to be the mobile signal that moves in plants from the infected site to the systemic tissue (Métraux et al. 2002; Park et al. 2007). However, later studies identified methyl salicylate as a critical mobile signal for SAR in tobacco (*Nicotiana tabacum*). Methyl salicylate is apparently converted into SA in systemic tissues (Park et al. 2007).

Application of SA as a water drench (55 mg/L) to the potting mixture of lily plants afforded resistance against the pathogenic fungus *Botrytis elliptica*, responsible for necrosis and blight symptoms on lily leaves (Lu and Chen 2005). Results revealed up to 15 days delay in lesion development after a single application. Three successive applications with three-day intervals before inoculation provided a higher level of disease inhibition. Accumulation of callose (β-1,3-glucan) in guard cells and increased expression of *LsGRP1* (*Lilium* 'Star Gazer' Glycine Rich Protein 1) gene were also observed, sharing high homology with various plant *GRP*s (Lu and Chen 2005).

Jasmonic acid (JA) and its related compounds are often associated with wounding and insect feeding responses, indicating a role for jasmonates in intracellular signaling associated with resistance to pests and necrotrophic pathogens (Creelman and Mullet 1997a; Heil and Bostock 2002; Gozzo 2003; Rohwer and Erwin 2008). JA and ethylene are frequently implicated as signal molecules in ISR elicited by root-colonizing bacteria (Van Loon et al. 1998; Gozzo 2003). Following elicitation, the levels of these signaling compounds often remain unchanged, unless a subsequent pathogenic attack occurs, a phenomenon known as priming (Pieterse et al. 2000). ISR mediated through jasmonate and ethylene was frequently associated with the production and accumulation of plant bioactive secondary metabolites (Creelman and Mullet 1997a,b; Chen et al. 2006; Lyon 2007). This may partially explain the mechanism underlying the enhanced resistance.

Methyl jasmonate (MeJ) applied as a leaf spray 24 h prior to inoculation was shown to completely inhibit the development of disease symptoms caused by the soft-rot pathogen *Pectobacterium carotovorum* (formerly *Erwinia carotovora*) in *Zantedeschia aethiopica*. Bacterial progress was halted and the effect persisted for more than a week (Luzzatto et al. 2007a). Pathogen inhibition was accompanied by a significant accumulation of polyphenolic metabolites with antibacterial activity against *P. carotovorum* (Luzzatto et al. 2007a; Yedidia et al. 2011). MeJ (56 mg/L) applied in a similar way to *in vitro* plantlets of *Ornithogalum dubium* also conferred resistance against *P. carotovorum*. Bacterial cell proliferation in the MeJ-induced plantlets was reduced from 10^9 to 10^7 bacterial cells per g plant tissue, at seven days post-inoculation (Golan et al. 2010). Recent studies have also shown that applications of MeJ to greenhouse-grown plants, such as tomato and *Impatiens walleriana,* may induce resistance to spider mites (Rohwer and Erwin 2010).

Induction of two subclasses of *PR-10*, a stress-associated gene, in *Lilium longiflorum* anthers, floral buds, and leaves was observed 24 h after dipping of the stems in MeJ solution (Wang et al. 1999). PR proteins accumulate as part of a multicomponent defense response in many plants exposed to pathogens. MeJ and ASM treatments were tested in freesia against *B. cinerea* for suppression of postharvest infection (Darras et al. 2006). Treatments with MeJ were more effective than ASM in reducing lesion numbers and lesion diameters on cut freesia flowers. Differing efficacy between

ASM and MeJ treatments could be attributed to their differential abilities to induce the SA-mediated vs. the JA-mediated host defense pathways, respectively.

Abscisic acid (ABA) is an endogenous plant hormone involved in many aspects of plant growth and development, as well as an adaptation to environmental challenges such as freezing and water stress (Mauch-Mani and Mauch 2005). *Lilium longiflorum* anthers dipped in an ABA water solution displayed elevated levels of *PR-10* gene expression 24 h post induction (Wang et al. 1999). Recent reports suggest that ABA plays important roles in plant defense responses, mainly as a negative regulator of the signaling network activating plant defenses against some biotrophic (but not all) and necrotrophic pathogens (Peña-Cortés et al. 1995; Wang et al. 1999; Mauch-Mani and Mauch 2005; Bari and Jones 2009). ABA was shown to positively regulate defense against necrotrophs through the activation of JA biosynthesis which increases the endogenous levels of this compound, thus activating the expression of defense-related genes (Peña-Cortés et al. 1995; Wang et al. 1999; Mauch-Mani and Mauch 2005; Adie et al. 2007). ABA has been proposed to play a role in priming of callose biosynthesis after pathogen recognition, which suggests a putative mechanism for the role of ABA in defense activation (Ton and Mauch-Mani 2004). Callose (β-1,3-glucan) is normally associated with cell wall appositions (papillae) that fortify the cell after pathogen recognition, thereby inhibiting pathogen penetration.

Chitosan is a high-molecular-weight cationic polysaccharide β-1,4-glucosamine, produced by deacetylation of chitin, a natural, biodegradable, and nontoxic polymer. Chitosan is usually extracted from crustacean shell wastes and has various applications in agriculture (Ramos-García et al. 2009). It has been reported to protect seedlings against diseases, increase growth, delay ripening, prolong shelf life, and limit fungal decay of several fruits and vegetables through postharvest applications (Terry and Joyce 2004). These effects have been attributed to its direct antifungal activity, and/or induction of resistance responses such as the production of chitinases, proteinase inhibitors and phytoalexins, and lignification in plant tissues (Terry and Joyce 2004). Treatments of *Gladiolus* corms at planting with chitosan reagent or a commercial product Biorend® 1.5% water solution were found to accelerate corm germination after planting, increase the number of flowers and cormlets, and extend vase life of the flowers by 3 days (Ramos-García et al. 2009). Chitosan application to freesia corms showed an earlier emergence and shortening in the vegetative growth in comparison with control plants, while a weekly leaf spray during the growth period of *Curcuma* plants increased the growth and postharvest quality of *Curcuma alismatifolia* × *C. cordata* 'Laddawan' (Tamala et al. 2007). These findings of improved quality and vase life could be attributed to induced resistance mechanisms.

4. Synthetic Organic Elicitors

The nonprotein amino acid β-aminobutyric acid (BABA) is a potent inducer of plant resistance against microbial pathogens (Cohen 2002; Jakab et al. 2005), insects (Hodge et al. 2005), and abiotic stresses (Jakab et al. 2005). The mechanism of BABA-induced resistance in *Arabidopsis* is based mostly on priming of different pathogen-inducible defenses and includes enhanced deposition of callose containing papillae (Zimmerli et al. 2000). The pathway that controls BABA-induced resistance requires endogenous accumulation of SA and NPR1 protein, similar to SAR (Zimmerli et al. 2000, 2001). However, in contrast to SAR, BABA-treated plants displayed a clear priming effect in the induction of PR1 transcripts (Conrath et al. 2006). *Zantedeschia aethiopica* leaves induced with BABA as water drench 24 h prior to inoculation with *P. carotovorum* displayed 75% reduction in infection rate. Bacterial proliferation in leaves was reduced by two orders of magnitude, from 10^9 colony-forming units (cfu) to 10^7 cfu per g plant tissue, but the protection did not persist after 48 h (Luzzatto et al. 2007b). Treatment of onion with BABA resulted in induced resistance against neck rot caused by *Botrytis allii* and *Botrytis cinerea* and involved priming of callose depositions (Polyakovskii et al. 2008). BABA was also found to reduce bulb infection by *Fusarium oxysporum* f. sp. *tulipae* in tulip 'Apeldoorn'. Used preventively, BABA applied by soaking uncooled and cooled tulip bulbs greatly inhibited the development of *Fusarium* infection on the root system after planting. At a concentration of 200 mg/L, disease symptoms were completely halted. As

BABA did not affect the growth of the pathogen directly, induced resistance was suggested as the mechanism of *Fusarium* control (Jarecka and Saniewska 2007).

BTH is a structural analogue of salicylic acid, and perhaps the best-known synthetic elicitor. BTH was developed commercially by Novartis and is marketed by Syngenta Crop Protection (Raleigh, NC, USA) as Actigard® (acibenzolar-S-methyl, ASM) in the USA and Bion® in Europe. BTH has been successfully used to control diseases in a wide variety of plant–pathogen systems and in many cases it appeared to be as effective as classical pesticide control agents (Vallad and Goodman 2004). The role of ASM in inducing SAR against *Tomato spotted wilt virus* (TSWV) was well-studied in field crops and was found to be effective in suppressing TSWV infection in tobacco (Pappu et al. 2000; Csinos et al. 2001). This research resulted in the practical application and management of this important virus in field crops. The mechanism of the induction of SAR against TSWV was studied in detail in tobacco (Mandal et al. 2008). Since TSWV is one of the more common viruses of bulb crops, these studies could form a basis for similar investigations into the utility of SAR inducers in ornamental geophytes. BTH is considered to act differentially on monocots vs. dicots, with greater longevity of resistance elicited in monocots (Oostendorp et al. 2001; Vallad and Goodman 2004). Similar to the signaling molecule SA, resistance elicited with BTH was shown to be more effective against biotrophic pathogens than against necrotrophs (Gozzo 2003; Glazebrook 2005). In a laboratory experiment, pretreatment of *Zantedeschia aethiopica* leaves with BTH 24 h prior to inoculation with *P. carotovorum* inhibited bacterial proliferation and reduced disease symptoms by 50%. Nevertheless, following the first 48 h, resistance provided by BTH did not persist (Luzzatto et al. 2007b). Unlike the response to MJ, the defense response following BTH application and challenge with *P. carotovorum* did not involve priming (Luzzatto et al. 2007a).

BTH was also shown to protect corms of gladiolus susceptible to *Fusarium oxysporum* f. sp. *gladioli*. The treatment resulted in a 48% increase in marketable flower spikes compared to the nontreated corms (Elmer 2006b). In onion, application of BTH successfully reduced disease severity of Xanthomonas leaf blight by up to 50% (Lang et al. 2007). *Cyclamen persicum*, an ornamental dicot tuberous species, was induced using BTH against *Fusarium oxysporum* f. sp. *cyclaminis*. The major benefit of BTH treatment in cyclamen seedlings was a 3-week delay in the onset of symptoms when compared with the control, an advantage that could be integrated into a disease management program for Fusarium wilt (Elmer 2006a).

Preharvest sprays with ASM were tested for suppression of postharvest infection of cut flowers of *Freesia hybrida* by *Botrytis cinerea*. ASM treatments reduced disease symptoms; however, treatments displayed variability in reducing the incidence of *B. cinerea* disease between years, freesia varieties, incubation temperatures, and ASM concentrations (Darras et al. 2006).

Probenazole (3-allyloxy-1,2-benzisothiazole-1, 1-dioxide), the active ingredient of Oryzamate®, is capable of suppressing disease development in many plants, as demonstrated in rice, tobacco, *Arabidopsis*, and lily (Yoshioka et al. 2001; Lu and Chen 2005). Its main agricultural application is in rice for the control of rice blast disease. The mechanisms of action include alterations of plant tissues through stomatal closure, callose deposition, an increase of β-1,3-glucan polymer, in the guard cells of the foliar epidermis and the increase of expression of defense-related genes such as *PR1* in *Arabidopsis*, *PR1*, *PR2*, and *PR5* in tobacco, and *PBZ1* and *RPR1* in rice (Lu et al. 2007). *Botrytis* leaf blight caused by *B. elliptica* causes enormous loss in the lily cut-flower industry. Application of probenazole before fungal inoculation was shown to be effective in protecting Oriental lily 'Star Gazer' from infection by *B. elliptica*. The protection occurred 1 day after probenazole treatment, achieved a significant level on the second day after treatment and was maintained at a high level for 14 days (Lu et al. 2007).

5. Physical Elicitors

It is known that physical stress (e.g., wounding, CO_2 treatment, UV irradiation, ionizing irradiation, and heat treatment) can lead to induced resistance against future infection in some species (Terry and Joyce 2004).

Ultraviolet (UV) electromagnetic radiation in the visible and/or UV ranges is one of the factors that can stimulate the production of phenolic substances, for example, flavonoids known as plant photoprotective constituents. Such a response was shown using brief (1–2 min) UV exposure to simultaneously decontaminate peeled onions and enrich them in human health-enhancing phytonutrients, that is, flavonoid compounds (Rodov et al. 2010).

Wounding (e.g., by herbivores) may activate plant defense and, as a result, enhance the accumulation of defense compounds. These can either be synthesized *de novo* or modified from already present compounds that are processed to increase the concentration of biologically-active compounds (phytoalexins) (Wink and Lehmann 1996). The involvement of such compounds was demonstrated in *Hippeastrum × hortorum* bulbs which produce red pigments upon wounding. These pigments were later identified as chalcone and flavans, phytoalexins exerting biological activity on some bacteria and polyphagous larvae. Wounding also enhanced activities of phenylalanine ammonium lyase and peroxidase, enzymes that are involved in the synthesis and coupling reactions of compounds derived from the phenylpropanoid pathway (Wink and Lehmann 1996).

6. Biological-Based Elicitors

Plant growth promoting rhizobacteria (PGPR) include strains of rhizospheric bacteria that actively colonize plant roots and induce beneficial effects on plant growth and fitness (Pieterse et al. 1998; Kloepper et al. 2004; Walters and Heil 2007). In fact, PGPR strains were shown to activate indirect mechanisms acting against plant pathogens through the activation of plant defense (Pieterse et al. 1998; Kloepper et al. 2004). It was shown that PGPR-induced resistance is mediated by the ISR signaling pathway (Zehnder et al. 2001). In contrast to pathogen-induced SAR, which is SA-dependent and associated with the coordinate expression of PR proteins, ISR induced by PGPR requires jasmonic acid and ethylene signaling and does not involve PR gene expression (Pieterse et al. 2002; Solano et al. 2008). In a recent study, rhizobacteria were isolated from the rhizosphere of healthy *Lilium formosanum* plants and their capability to suppress *Botrytis* leaf blight was tested in different varieties of lilies (Liu et al. 2008). Results indicated that the application of a cell suspension of *Bacillus cereus* strain C1L as a soil drench 4 or 5 days before inoculation with *B. elliptica* reduced disease severity by 75% in *L. formosanum* seedlings. The suppression of disease symptoms persisted for at least 10 days. Such protection was also afforded by the bacteria in other *Lilium* varieties including Oriental lilies 'Star Gazer' and 'Acapulco' and thus suggesting a broader range of hosts in the genus (Liu et al. 2008). In a previous study, *LfGRP1* and *LsGRP1* transcripts were upregulated by SA in lily (Lu and Chen 2005). The application of the beneficial bacterial strain *B. cereus* C1L halted transcripts accumulation of these SA-dependent genes, implying a signal transduction pathway that is apparently independent of SA (Liu et al. 2008).

7. Cross Protection

Disease suppression by early infection with a weakly virulent strain of an otherwise virulent pathogen (bacteria, fungi, or viruses) against the same pathogen is known as cross protection (Agrios 2005). The phenomenon has recently been demonstrated in *Cymbidium* spp. where foliar inoculation with a weakly virulent strain of HPF-1 (*Fusarium* sp.) suppressed disease caused by *Fusarium* in leaves, bulbs, and roots (Ichikawa et al. 2003). The authors indicated a systemic, nonselective suppression of diseases such as yellow spot of leaves caused by *F. proliferatum* and *F. fractiflexum*, bulb and root rot caused by *F. oxysporum*, and dry rot of bulbs and roots caused by *F. solani*. After inoculation with an HPF-1 strain, an increase in the endogenous salicylic acid level was observed, suggesting that the suppressive mechanism resulted from SAR (Ichikawa et al. 2003). A similar effect was obtained in lily by means of an early infection of leaves with *B. elliptica*, which consequently suppressed blight symptoms caused by *B. elliptica* and increased the accumulation of *LsGRP1*-related transcript. The phenomenon of cross protection is considered to have a limited range in comparison with induced resistance (Hammerschmidt 2007).

A summary of the effect of natural, synthetic, physical, and biotic elicitors on induced resistance of ornamental geophytes is provided in Table 13.2.

8. Summary of Induced Resistance

Le Nard and De Hertogh (2002) emphasized the need to devote more efforts to the study of the SAR mechanism and pathways induced by elicitors in flower bulbs. Numerous studies have focused on induced resistance, considering all its aspects from fundamental to applied research. Within this field, only a small number of research papers explored the effects and the mechanisms underlying induced resistance in the highly valuable group of ornamental geophytes. Members of this group belong to several botanical taxa, although the majority of flower bulbs are monocots. Molecular data obtained in dicot model plants, such as *Arabidopsis* or tomato, do not enhance the understanding of resistance mechanisms in monocot flower bulbs, a difficulty which is further amplified by the specialized life form of bulbous plants. For instance, plant stimulators such as ABA and MeJ, which elicit resistance response in dicots, may undesirably impact morphology and dormancy in flower bulbs. Such considerations should be taken into account in the case of applying induced resistance as a mechanism to control flower bulb diseases. Overall costs and benefits must be considered carefully. In this respect, application of combined treatments integrating plant defense activating agents with chemical bactericides/fungicides may be the best answer for bulb protection. Within a few years, agricultural production of ornamental bulbs is likely to include induced resistance as an accepted alternative for chemical pesticides for effective and sustainable disease control management (Lu and Chen 2005).

F.   IPM FOR WEEDS

Weed control in ornamental bulbs is essential for the optimization of bulb yield and quality. In the US Pacific Northwest, research has shown that reductions in marketable bulb yield due to weed interference can exceed 40%, while bulb size can be reduced by over 30%. Weeds can also serve as a reservoir for pathogens, particularly viruses and nematodes, provide habitat for slugs and snails, and create problems at harvest. In addition, production of weed seed during the bulb crop can be exceedingly high if weed control is inadequate, resulting in continuing weed problems in rotationally-grown crops.

The primary weeds of concern in most ornamental bulb production regions are usually annuals, many of which are found throughout the world. When bulbs are produced where winters are mild, annual weeds that germinate in fall, winter, and early spring are most frequently problematic. These include broadleaf species such as shepherd's-purse (*Capsella bursa-pastoris*), common chickweed (*Stellaria media*), henbit (*Lamium amplexicaule*), and pineapple-weed (*Matricaria discoidea*), and the grass species, annual bluegrass (*Poa annua*). Summer annual broadleaf species include common lambsquarters (*Chenopodium album*), and smartweed/knotweed (*Polygonum* spp.), while grasses include barnyardgrass (*Echinochloa crus-galli*), the foxtails (*Setaria* spp.), and wild oat (*Avena fatua*). Many other weed species are of regional importance.

Historically, weed control in ornamental bulbs has been achieved with hand weeding, hoeing, or mechanical cultivation. In regions with mild winters, these methods are made more difficult due to wet soil in production fields at the time of weed emergence, which restricts access by weeding crews or cultivating equipment. Wet soil also reduces the effectiveness of the weeding process by slowing desiccation and allowing weeds to reestablish their roots. Weeds left to grow with the bulb crop until fields dry out can result in yield loss and make control more difficult because their larger size improves their ability to withstand efforts to control them—and this can result in damage to bulb foliage. Cultivation equipment can be used to control emerged weeds if care is taken to keep equipment away from bulb roots and foliage.

The advent of selective herbicides in the late 1940s and 1950s revolutionized weed control in practically every crop grown in agriculture and horticulture. Gould (1957, 1990) reported that labor costs associated with hand weeding were reduced by 90% with the discovery that pre-emergence applications of a dinitro-type herbicide plus propham in diesel oil were effective and nonphytotoxic on iris and narcissus. Products identified earlier as possessing preemergence selectivity in ornamental bulbs included benefin, bensulide, chlorpropham, dacthal, dinoseb, diuron, EPTC, metamitron, napropamide, oryzalin, pyrazon, and trifluralin, and many of these have been used in bulb production (Bing and Macksel 1984; Hof et al. 1987; Howard et al. 1989; Molinar 1994; Skroch et al. 1994). More recently, isoxaben, *s*-metolachlor, and pendimethalin have also been used for preemergence weed control in ornamental bulbs (Al-Khatib 1996; Miller 2010). These herbicides must be applied prior to emergence of foliage to be effective on the weeds and safe for the bulbs. Currently, the postemergence herbicides known not to injure certain ornamental bulb types are those that only control grasses; these include clethodim, fenoxyprop, fluazifop, and sethoxydim (Skroch et al. 1988; Miller 2010). Nonselective postemergence herbicides such as glyphosate or paraquat are effective after emergence of weed foliage, but before bulb foliage emerges.

There have been relatively few herbicides developed for weed management in organic ornamental bulb production systems. Acetic acid, clove/cinnamon oil extract, limonene, and pine oil extract are used for postemergence weed management: Another similar herbicide not currently registered for organic uses is pelargonic acid (Miller 2010). All are nonselective contact herbicides applied to foliage, and work best on annual broadleaf weed seedlings prior to emergence of the third or fourth true leaf and when applied at times of warm air temperatures and low relative humidity. Consequently, all are of limited usefulness in ornamental bulbs. Applications to emerged weed seedlings prior to emergence of bulb foliage have resulted in only moderate success, primarily because the timing for application corresponds to late autumn or early winter in temperate climates (T. W. Miller, unpublished data). Directed or shielded applications between ornamental bulb rows have also been only moderately successful, primarily because such applications are made in late winter to early spring or after ornamental bulb flowering when annual weeds are usually too large and are able to re-sprout from roots and crowns. Flame has also been tested for weed control in ornamental bulbs (T. W. Miller, unpublished data). Propane flame, applied either as an open flame or as heat, can kill emerged annual weed seedlings in much the same manner as a contact herbicide. As with contact herbicides, flame is nonselective, so injury is likely if bulb foliage has emerged. Seedling weeds that are too large will not be controlled, and buried seeds are also not injured by this treatment.

Two potential methods of nonchemical weed management that may be applied in ornamental bulb production fields are flooding and crop rotation (Van Aartrijk and De Rooy 1990). Flooding a production field for six weeks prior to bulb planting gave good control of Canada thistle (*Cirsium arvense*), coltsfoot (*Tussilago farfara*), and quackgrass (*Elymus repens*) (Van Zaayen et al. 1986), although the number of fields where water can be ponded for that length of time is probably limited. Crop rotation offers perhaps the widest potential for nonchemical weed control in ornamental bulbs. The most common crop rotation tool in sustainable agriculture is the use of cover crops or green manure plow-down crops. These crops are usually seeded in late summer or fall, allowed to grow until early spring, then killed using cultivation (plow-down crops) or herbicide such as glyphosate (cover crop) prior to seeding or transplanting field crops or vegetables. Among commonly tested crops are cereals such as barley (*Hordeum vulgare*), oat (*Avena sativa*), rye (*Secale cereale*), triticale (× *Triticosecale rimpaui*), and wheat (*Triticum aestivum*), legumes such as black medic (*Medicago lupulina*), clover (*Trifolium* spp.), pea (*Pisum sativum*), and vetch (*Vicia* spp.), and others such as buckwheat (*Fagopyrum esculentum*), rapeseed (*Brassica rapa*), and mustard (*Sinapis alba, Brassica* spp.) (Masiunas 1998). In areas with low winter temperatures, nonhardy cover crops may die without herbicide or cultivation (Brandsaeter and Netland 1999).

Cover crops are thought to contribute to weed control in two major ways (Masiunas 1998). First, weed seedlings germinating within the cover crop are destroyed when it is killed in early spring. Provided that the soil is not disturbed prior to seeding the subsequent crop, viable weed seed populations are depleted near the soil surface, reducing germination in the rotational crop. Second, cover crop residue left on the soil surface may retard germination of the remaining weed seed through mulching and/or allelopathic action. Lu et al. (2000) showed that tomatoes grown with hairy vetch mulch were higher yielding and more profitable than those grown with black polyethylene and no mulch system. Conversely, biofumigant crops such as yellow mustard and oilseed radish (*Raphanus sativus*) gave only inconsistent control of a few weed species in the subsequent onion crop, and in some years, reduced onion bulb yield (Geary et al. 2008).

Living mulches, a cover crop that is not killed prior to the establishment of the rotational crop, have also been tested for weed suppression in many crops (Lanini et al. 1989; Neilson and Anderson 1989; Durgy and Ashley 1993; Masiunas 1998; Valenzuela and DeFrank 1994). Living mulches often unacceptably reduce yield in the subsequent crop, particularly in areas with limited soil moisture (Elkins et al. 1983; Echtenkamp and Moomaw 1989) or show varying results depending on the species being used. Hatterman-Valenti and Hendrickson (2006) found that canola (*Brassica napus*) reduced total onion bulb yield and the yield of large-sized bulbs while barley did not reduce yield when moisture was adequate. Perhaps the majority of crop and weed suppression by living mulches is due to competition for light (Den Hollander et al. 2007). Greenland (2000) found that onion bulb size and yield was reduced if barley was allowed to grow taller than 18 cm before killing it. Living mulches have been rendered less competitive with the crop through the use of herbicides that can slow down the growth of the living mulch while not injuring the growth of the crop. For example, treating perennial ryegrass (*Lolium perenne*) with a nonlethal dose of fluazifop was promising as a living mulch with pak choi (*Brassica rapa*) (Wiles et al. 1989).

Testing of cover crops and/or living mulches specifically for weed control in ornamental bulbs has not been reported. In preliminary research in Western Washington (T. W. Miller, unpublished data), plow-down crops of white mustard, winter rapeseed, hairy vetch, and buckwheat seeded in August and plowed in mid-September did not influence weed control in the subsequent tulip, narcissus, or bulbous iris crop. Living mulches consisting of winter wheat, winter rye, winter barley, buckwheat, white mustard, or winter rapeseed planted immediately following bulb planting in early October did not achieve adequate growth to compete with weeds; consequently, acceptable weed control through May was only obtained from the diuron or glyphosate that was applied to kill the cover crops in December or January. Additional research on different living mulch or cover crop practices may generate more positive results, however, and should be encouraged.

G. IPM in the Greenhouse

The significant losses that pests and diseases can inflict on ornamental geophytes generally will be exacerbated when these crops are grown or forced under protection in greenhouses or in related structures such as polyethylene tunnels, where the environment poses additional challenges. The carefully controlled environmental conditions in a greenhouse are often very favorable for disease and arthropod pest development—hence epidemics and epiphytotics can develop explosively in greenhouse systems. Greenhouses protect plants from the outside weather, but also protect pathogens and insect and mite pests against the competition they would otherwise encounter in the outside world (Gullino et al. 1999; Pilkington et al. 2010). Compared with the outdoor environment, greenhouses will typically be warmer, more humid, more sheltered, and less well-lighted. Crop monocultures, which are typically found in greenhouses, also favor rapid buildup of pest populations. However, the ability to control the cleanliness, temperature, and moisture of the greenhouse environment, to retain predators and parasitoids within the structure, and to apply biofungicides to relatively small volumes of growing media also provides an excellent opportunity to manage certain diseases and pest problems without pesticides. Although specific pathogens and pests vary by host, some common pathogens on greenhouse-grown geophytes include species of *Botrytis*, *Fusarium*, *Pythium*, powdery mildew, foliar nematode, and a number of insect-transmitted viruses, such as *Cucumber mosaic virus* (CMV), TSWV, and *Dahlia mosaic virus* (DMV). Some of the more common arthropod pests include aphid, bulb mite (*Rhizoglyphus* etc.), broad mite, bulb-scale mite, thrips, and mealybugs.

When forcing geophytes in greenhouses, growers generally follow IPM procedures that have as their goal cost-effective pest management with minimum negative effects on workers and the environment. Because of the concentrated production that takes place in greenhouses, chemical inputs may also be concentrated. Voluntary reductions of chemical use by greenhouse growers are important for reducing the point-source pollution associated with protected cultivation. In the future, the greenhouse production of flower crops will certainly include improvements in disease forecasting and use of biological controls (see Forecasting and Precision Application of Pesticide, Sections IV.B and IV.D.2), as well as less but continued use of reduced-risk chemicals (Daughtrey and Benson 2005).

No matter whether growers are producing cut flowers or container-grown plants, an effective greenhouse IPM program should include monitoring, exclusion, sanitation, and other cultural techniques, and the selection of resistant cultivars.

1. Exclusion

Bulbs should be inspected on arrival for indications of diseases, insects, or bulb mites to exclude the inadvertent introduction of pathogens and pests into a greenhouse. Bulbs should be firm and turgid with no more than minimal symptoms of damage. Damage to the basal plate may reduce rooting capacity; this can be due to physical damage at bulb lifting or during handling, or infestation by large narcissus fly. Growers select bulbs at random and cut across them to inspect for flowers and to look for internal discoloration. In various bulbous species, internal discoloration may indicate problems with a vascular wilt fungus (*Fusarium oxysporum* f. sp.) or with *Ditylenchus dispsaci*, the bulb and stem nematode, or in the case of more subtle discoloration or feeding marks, bulb-scale mite. A sweetish or fishy odor and softened tissue might indicate contamination with bacterial soft-rot pathogens such as *Pectobacterium carotovorum* or *Dickeya zeae* (formerly *Erwinia chrysanthemi*). Superficial green mold on tulip bulbs may indicate *Trichoderma viride*, which has been implicated as the cause of tulip leaf withering, a disease that is most common when bulbs are grown in a peat-containing substrate (Muller 1986).

Sclerotia, which are hardened fungal structures that allow for long-term and long-distance survival, may be present on some bulbs, such as tulips infected with *Botrytis tulipae* or *Sclerotinia* spp. Sclerotia are not always an indication of future disease development. On narcissus, conspicuous sclerotia may only be evidence of the relatively less destructive *Botrytis cinerea*, while *B. narcissicola*, the agent of smolder, may be hidden as mycelium in the tissues at the neck of the bulb (O'Neill and Mansfield 1982). Any bulbs with conspicuous symptoms or signs of a fungal, bacterial, or nematode disease or with bulb mites should be separated from the healthy lots. As precise disease detection through the use of PCR becomes more generally available, growers will be able to have their problems more precisely identified, even in the absence of conclusive symptoms or signs.

Indication of the presence of a virus or virus infections in bulbs is not always apparent or obvious during visual inspection of bulbs. Discoloration, malformation, or misshapen bulbs may be due to virus infection but confirmation needs laboratory testing. Exclusion plays an important role in minimizing the introduction and subsequent spread of viruses into a production system, whether it is an underprotected environment or open field conditions. Introduction of virus-infected bulbs, even at a low frequency, could quickly result in its establishment and further spread should it have an insect vector already present in the area. Visual examination followed by laboratory testing to exclude introduction of virus-infected stocks is critical, especially if propagation is through cuttings. Thus, standard operating procedures to test bulb plant species against known viruses and excluding the batches/lots found to be infected could lead to reduced disease incidence at the production level.

2. Sanitation

Areas where bulbs are sorted, stored, or grown must be free of contamination from the previous season to avoid carrying arthropod pests and pathogens from one production year to the next. The storage phase may be crucial, as it may last several weeks or months, often with bulbs stored in bulk (loose or in bins) with relatively poor air flow. In structures, this sanitation step is best handled by ensuring that all organic debris is first physically removed by sweeping and applying strong jets of water ("power-washing") prior to the use of a disinfectant. The organic debris from the previous crop is an important potential source of fungal and oomycete inoculum. In particular, the root rot agents *Rhizoctonia* and *Pythium* spp, are well designed to survive in the greenhouse between crops because of their ability to produce sclerotia and oospores, respectively. Once the organic debris is removed, greenhouse growers often choose to use a quaternary ammonium-, hydrogen peroxide-, or iodophore-based disinfectants to further disinfest crates, shuttle trays, forcing boxes, benches or floors, and ceilings and walls of rooting rooms. In some cases misting may be used as an alternative to spray application, but, in any case, sufficient time should be allowed for the disinfectant to be in contact with the target organisms. Out-of-season, daffodil HWT tanks can be used to sterilize trays, etc., using a short immersion in water at 50°C. The resistant, dehydrated "wool" stage of stem and bulb nematode may need several hours contact with water before full "re-activation". It is important to realize that the substrate being disinfested will affect the efficacy of the disinfectant (Copes 2004). Production systems in the United States are now usually soil-free because of the widespread use of peat or peat–perlite mixes for containers. Growers using sand or other mineral soils as a component of their growing medium must ensure its freedom from pathogens by steam-pasteurization. Increasingly, tulips are forced hydroponically in "pin-trays", avoiding contamination via growing media but demanding good standards of hygiene for the water supply and the components used (trays, pipe-work, etc.).

H. DIAGNOSIS AND DETECTION

Accurate diagnosis is critical if disease, pest, and weed IPM management programs are going to be successful. Although diagnostic skills increase with experience, accurate identification often requires that samples be sent to a laboratory. In addition to microscopic examination of specimens, culturing of pathogens from plant or soil samples may be undertaken. Isolates are then identified based on growth characteristics, ability to grow on selective media, and spore morphology. Significant advances have been made in the use of serological tests, such as the enzyme-linked immunosorbent assay (ELISA), and nucleic acid-based tests, such as the polymerase chain reaction (PCR), to identify causal agents. In particular, these two methods have greatly enhanced the ability to conduct surveys to gain a better understanding of the incidence of a pathogen (Pahalawatta et al. 2007; Eid et al. 2009) and to rapidly test for known viruses and identify and characterize new viruses and virus strains (Pahalawatta et al. 2008a, b; Pappu et al. 2008; Miglino et al. 2010, 2011).

The development of PCR detection techniques has provided a means to also identify specific strains of pathogen and to assess infection, colonization of plant tissue, and inoculum survival in plants and soil (De Boer 2011; Van Doorn et al. 2011). The sensitivity of these detection methods will also allow the development of special certification schemes associated with the production of pathogen-free stock. This will reduce the spread of pathogens and help meet the increasing plant health regulatory requirements of countries such as the United States and China (Knippels 2011).

Management of soilborne diseases, such as *Fusarium* basal rot, could be greatly enhanced if growers were able to assay planting sites and planting stock to determine inoculum levels of these pathogens prior to planting. Fusarium basal rot is a serious problem on many different bulbs and occurs wherever bulbs are grown. It can also be a major problem during storage and when bulbs are forced to produce flower crops. There are specific taxa or formae speciales of *Fusarium oxysporum* that cause basal rot on different bulb crops (Boerema and Hamers 1988, 1989). Basal rot of daffodils is caused by *F. oxysporum* f. sp. *narcissi*. *F. oxysporum* f. sp. *tulipae* causes basal rot on tulips, while basal rot on bulbous iris and gladiolus is caused by *F. oxysporum* f. sp. *gladioli*, and *F. oxysporum* f. sp. *lilii* causes basal rot on lilies. Traditional soil dilution plate assays used to monitor soilborne pathogens cannot distinguish between the different taxa of *F. oxysporum*, thus limiting their usefulness as an assay procedure (Nash and Snyder 1962). Other external factors, such as the presence of certain microorganisms, can also conceal the presence of pathogens on agar.

The development of real-time PCR technologies provides an opportunity to develop highly specific assays that are capable of detecting and quantifying the specific taxa of *F. oxysporum* and other soilborne pathogens such as *Rhizoctonia* spp. (Chiocchetti et al. 1999; Lees et al. 2002). Unlike traditional baiting and cultural methods, real-time PCR is not affected by factors that could conceal the presence of pathogens on agar and is capable of detecting specific strains of pathogens. While standard PCR techniques have been used to detect the presence of specific pathogens in plant tissues for some time, the development of real-time PCR technologies has made it possible to rapidly quantify the levels of pathogens that are present in soils, water, air, and plant tissues (Cullen and Hirsch 1998; Nicholson et al. 1998; Bates et al. 2001; Bridge and Spooner 2001; Cullen et al. 2001, 2002; Filion et al. 2003; Okubara et al. 2005). These technologies have also made it possible to follow the colonization rate of pathogens in host tissues.

The development of real-time PCR assay protocols that can be used to quantify inoculum levels of soilborne plant pathogens such as *Fusarium* and *Rhizoctonia* in the soil and on planting stock would be very beneficial to the bulb and flower industry. Coupled with information on inoculum threshold levels necessary for disease development, this technology would lead to the development of predictive diagnostic tests to identify high-risk fields and planting stocks where the inoculum of these pathogens is above threshold values. Access to this type of information would assist growers in making management decisions relating to the application of appropriate management strategies.

PCR and real-time PCR techniques have been widely used for virus detection and diagnosis. The advantages of these methods were greatly exploited and applied for identifying, characterizing, and intercepting known and newly-discovered viruses in various bulb crops. Where sequence information is available for individual viruses to be tested, the use of group- or virus-specific primers enabled detection of single or multiple viruses in using a multiplex assay (Miglino et al. 2007).

Degenerate primers based on conserved regions across viruses within one or several genera have also been used for the rapid identification of new viruses (Miglino et al. 2010, 2011; Pappu and Druffel 2009). Molecular detection techniques are the only methods of choice for virus detection and diagnosis in situations where virus-specific antisera are not available. For example, PCR was used to determine the incidence and distribution of viruses associated with dahlia (Pahalawatta et al. 2007; Eid et al. 2009).

More recent applications of sequencing include the "deep sequencing" approach leading to discovery of new viruses in a wide range of plants. This information would be useful in developing detection tests for individual viruses that may become economically important or of quarantine significance (Studholme et al. 2011). The use of microarrays (Boonham et al. 2007) in virus detection and diagnosis has potential for application as an increasing number of viral genomes are sequenced. Chip-based, universal microarray for the detection of viruses in ornamental crops is under development (Hammond 2011).

V. OUTREACH AND EDUCATION

Implementation of integrated pest management programs that support sustainable production of ornamental geophytes is dependent in part on grower access to research-based pest management information. Vincelli (2005) points out that education is an important key to improving diagnostics and adoption of IPM strategies. Growers should be aware of the concept of the disease triangle (see Section IV) so they have the ability to identify factors that affect disease development and understand the different approaches that are used to manage monocyclic and polycyclic diseases. Growers should (1) know the pathogens, arthropod pests, and weeds that commonly occur on the crops they are growing, (2) be able to identify the symptoms and signs associated with them, (3) understand the host range of the pests and pathogens, understand how the pests and pathogens survive from one season to the next, (4) know what conditions favor inoculum survival and production of primary and secondary inoculum, and (5) understand how these pathogens are commonly spread. In short, they need to understand the disease and life cycles of the pathogens and pests, respectively, to be able develop an effective disease management program.

Jacobsen (1997) indicates that the general lack of a pesticidal "silver bullet" comparable to synthetic insecticides from the 1940s to mid-1950s has continually forced plant pathologists to emphasize nonpesticide control strategies such as genetic host resistance to pathogens, cultural controls such as rotation and tillage, and pathogen-free seed or planting stock. An exception to this has been the use of methyl-bromide fumigation to control nematodes, soilborne pathogens, and weeds. In the context of soil fumigation, Zasada et al. (2010) point out that when using practices other than soil fumigation to manage soilborne nematodes, growers will need a basic knowledge of nematode biology (including host range, life cycle, survival strategies, and longevity) as well as access to resources such as nematode identification services. In his book dealing with sustainable practices for managing plant disease, Thurston (1992, p. xvii) states that "it is quite simple to apply a pesticide or utilize a high-yielding resistant variety to manage plant diseases, but one has to know a great deal about the biology of a situation in order to use cultural management".

This underscores the importance of outreach and grower education to assist growers in implementing integrated crop protection strategies instead of using a few pesticides that are effective against several disease problems. An example is the Farming with a Future (Telen met toekomst) project in The Netherlands that uses a network approach to bring growers and stakeholders together concerning the development, testing, and implementation of Best Management Practices that are integrated into crop production strategies and contribute to a sustainable crop protection system (Brinks and De Kool 2006; De Boer 2011).

VI.   IMPEDIMENTS AND CHALLENGES TO ACHIEVING SUSTAINABLE PRACTICES

There are a number of impediments and challenges that are hampering the development of sustainable production systems for ornamental geophytes. Many of these have already been referred to in the various sections of this chapter. In general, they include a complex of issues that have to be addressed; diversity of crops, public perceptions and attitudes, grower attitudes, economics, limited availability of effective biologically-based products, a limited number of scientists, funding, and so on. The fact that ornamental geophytes represent a high value, small acreage crop is also an impediment to the registration of new control products. A product that fails to consistently provide effective control exposes the registrant to increased financial liabilities that represent a significant disincentive to support a registration.

The development of sustainable production systems requires an integration of basic and applied research that examines the whole agroecosystem. In addition, information on overall costs and benefits must be developed to provide growers with economically feasible pest, disease, and weed management options. This requires a multidisciplinary systems approach to address research needs. A major impediment to this is the continued decline in the number of scientists working on pest, disease, and weed management issues on bulb crops. For example, Gould (1993) reported that in the US Pacific Northwest, there were over 20 University and USDA scientists who had 67 projects on various aspects of bulb production in the late 1940s. This number had decreased to about a dozen scientists by the late 1970s and today there are only 3 or 4 scientists who work part-time on bulb crops. Similar reductions in research have also taken place in other countries. In England, for example, in the mid-1970s there were three substantial horticultural research institutes and a network of regional Experimental Horticulture Stations, covering the spectrum from strategic research to applied research, development, and technology transfer, all publicly funded and serving all horticultural sectors, including dedicated ornamental geophyte researchers across the major disciplines. That network now consists of one small university-based "crop centre" and the privatized remnants of one other institute and one horticulture station, and in which one scientist has a small component of daffodil research and development. Following "food security" concerns, the UK government no longer funds research on ornamental crops. This is not to decry biotechnological plant research carried out elsewhere, but that is long-term in nature, while the global crisis in farming and growing demands practical answers often short to medium-term in nature.

The need for new tools and decision aids to help farmers implement IPM systems will increase in the future (Jacobsen 1997). In his review, Jacobsen indicates that our challenge is to move beyond individual diseases and crops to an understanding of pest control in the context of the agroecosystem by building and participating in broadly-based IPM teams. He goes on to say that an equally important challenge is for administrators, grant managers, and others to find methods of rewarding individuals who contribute to interdisciplinary teams where long-term, complex research and technology transfer are the end products.

VII. CONCLUDING REMARKS

The increasing urbanization of many bulb-growing regions, increased restrictions relating to the use and disposal of pesticides, increasing costs of energy, and concern over potential environmental impacts from pesticides will continue to have a major effect on the production of ornamental geophytes. According to Poincelot (1986), a sustainable agriculture system is one in which the goal is permanence achieved through the utilization of renewable resources. Basic elements of sustainable agriculture are the conservation of energy, soil, and water. Poincelot (1986) indicates that sustainable agriculturists will have to balance resource conservation and environmental protection against maintenance of optimal productivity and minimal labor input. Return to a chemical-free, less technological agriculture would be costly in terms of increased human labor and would most likely result in decreased agricultural productivity. He goes on to indicate that agricultural chemicals and mechanization will still have a place in sustainable agriculture, but reduced use of chemical fertilizers will reduce energy demands and pollution, e.g., fertilizers would be supplemented with legumes, crop residues, green manures, and readily available local supplies of organic wastes. Their utilization would help further to reduce costs and energy demands, remove a source of potential pollution, and more importantly, renew soil organic matter.

In sustainable ornamental geophyte production, pesticides would still be used, but only as a component of IPM systems that rely primarily upon effective natural controls. Pesticides would be used only when natural controls were not effective and selection would be for the least toxic of those available. An IPM program would maintain pest control, but reduce pollution of other agricultural resources and minimize detrimental effects on environmental quality. Beneficial organisms would be less likely to be destroyed, the induction of resistance in pests would be minimal, and the elicitation of natural resistance mechanisms in the crop plants might play an important role in crop health management.

This book reviews recent advances in basic science and sustainable horticultural production of ornamental geophytes. While significant advances have been made in areas such as molecular-based diagnostic and detection systems, biotechnology, and induced resistance, it is somewhat ironic to see how little progress has been made in providing growers with practical, commercially available solutions relating to the management of diseases, pests, and weeds. This is particularly true with the intractable diseases caused by *Fusarium* and viruses. On some crops, bacterial soft rot, nematodes, and a variety of pests could be added to this list.

In the 1970s and early 1980s, there were numerous reports about efforts to produce virus-free narcissus stocks in the United Kingdom, but the cost of maintaining disease-free nuclear stock was prohibitive and this "poster child" of a virus management program was not sustainable. The development of inoculation procedures to screen material for susceptibility to various plant pathogens has allowed for the susceptibility of existing cultivars of several genera of bulbs to common pathogens to be determined. However, with the possible exception of genera that can be easily propagated *in vitro*, such as *Lilium*, reports in the literature touting the success of efforts to breed for resistance to common pathogens appear to have resulted in very few new, commercially available disease-resistant cultivars to date.

Many authors have concluded that genetic engineering will play an increasingly important role in the management of plant pathogens. However, Collinge et al. (2010) state that there is a paradox in that the enormous progress in understanding the nature of plant microbe interactions at the molecular level is yet to be translated into effective practical disease control in production systems through genetic engineering, improved plant breeding, or the development of new methods for chemical control. They go on to indicate that in addition to taxonomic, and therefore physiological, differences, there is a huge variation in "lifestyle" among bacterial and fungal pathogens, such that it has so far proved impossible to develop effective broad-spectrum disease resistance. New genes have been discovered but their efficacy has not been documented through field trials. Even though defense mechanisms effective against viruses are now relatively well understood, 20 years ago Lawson (1990) reported that the production of virus-resistant bulb stocks by the insertion of genes for production of viral coat proteins was a promising new area of research, and there have been several reports documenting the effectiveness of this approach in introducing virus resistance in some ornamental species (Hammond 2006; Hammond et al. 2006), this approach has not yet resulted in any commercially-available virus-resistant cultivars of geophytes. The expense associated with generating the required biosafety data and the environmental impact assessments for obtaining approval for commercialization from various regulatory agencies and the current climate of nonacceptance of genetically engineered crops are some of the contributing factors for the lack of commercially available virus-resistant ornamental geophytes.

During the 1980s, "Where's the beef?" was used as an advertising slogan by the Wendy's hamburger restaurant chain in North America to draw consumers' attention to the fact that a number of their competitors were selling minuscule hamburger patties in a large hamburger bun. It has become a common catch phrase questioning the substance of an idea, event, or product. Some bulb crop growers might ask "Where's the beef?" with respect to the transfer of the new scientific advances that have been made in the laboratory and greenhouse into commercially available products or management strategies for field production. As indicated above, there are a number of impediments and challenges that will have to be overcome if the advances in "basic science" are going to result in horticultural practices that allow growers to effectively and economically manage their pest, disease, and weed problems in a sustainable fashion.

CHAPTER CONTRIBUTORS

Gary A. Chastagner is a plant pathologist and extension specialist at Washington State University, Research and Extension Center, Puyallup, Washington, USA. He was born in Woodland, California, grew up in Davis, California, and earned his MS and PhD in plant pathology from the University of California, Davis. Dr. Chastagner held positions at the University of California, Davis and is currently in the Department of Plant Pathology at Washington State University. His expertise includes disease management on ornamental bulb crops and Christmas trees, management of sudden oak death (*Phytophthora ramorum*) on conifers and nursery stock, and factors that affect the postharvest quality of Christmas trees. He is the author of more than 300 peer-reviewed papers, professional papers, extension bulletins, web pages, educational videos, book chapters, technical publications, and popular press articles. He has given more than 600 presentations to regional, national, and international audiences. Dr. Chastagner received the American Phytopathological Society Excellence in Extension Award (2011), the US National Christmas Tree Association Outstanding Service Award (2010), the Pacific Northwest Christmas Tree Association (PNWCTA) Barney Douglass Research Award (2005), and the PNWCTA President's Appreciation Award (2003).

Gordon R. Hanks graduated with a BSc in botany from Royal Holloway College, University of London, UK, followed by study at the North London Polytechnic, and received an MPhil from

London University. He held research positions at the Glasshouse Crops Research Institute, West Sussex, UK and the Kirton Experimental Horticulture Station, Lincolnshire, UK. Recently, he was appointed as an Associate Fellow of the Warwick Crop Centre, University of Warwick, UK. His main areas of expertise include flower bulbs, with special interest in plant hormones, forcing and propagation. He has specialized on daffodils and has published numerous refereed and technical papers, and book chapters. G. Hanks received the Carlo Naef Trophy in recognition of his contribution to the UK flower industry.

Margery L. Daughtrey is a senior extension associate, Department of Plant Pathology and Plant-Microbe Biology, Cornell University, Riverhead, New York, USA. She was born in Charlottesville, Virginia and obtained BS from The College of William and Mary and MS from the University of Massachusetts. She has conducted an extension and research program on diseases of ornamentals at the Long Island Horticultural Research and Extension Center. The main area of her expertise is disease identification and management of floriculture crops. M. Daughtrey is the coauthor of four books on plant disease management and several book chapters and refereed journal articles, as well as hundreds of articles for the greenhouse trade press. She is the editor-in-chief, APS PRESS, and a recipient of the Society of American Florists Alex Laurie Award for Research and Education (1998).

Iris Yedidia is a researcher at the Agricultural Research Organization (ARO, The Volcani Center), Israel. She was born in Israel and obtained BSc from Haifa University and PhD from The Hebrew University of Jerusalem, Israel. After postdoctoral work at the Weizmann Institute of Science in Israel, she was appointed to a research position at ARO. Her areas of expertise include the physiology of flower bulbs with special interest in resistance mechanisms against soft rot disease. She has published numerous reviewed articles, technical papers, and reviews in the field of plant–microbe interaction.

Timothy W. Miller is extension weed scientist, Washington State University (WSU), Mount Vernon, Washington, USA. He was born in Idaho, USA, and graduated from the University of Idaho. He has held the position of extension agricultural agent and weed diagnostician at the University of Idaho prior to his arrival at WSU in 1997. His major research topics include weed management in ornamental bulbs, vegetable seed crops, small fruit, and field vegetables, as well as weed control in noncropland sites in western Washington. T. Miller has published more than 200 refereed journal articles, conference papers, and extension publications. He is the recipient of the 2007 Presidential Award of Merit from the Western Society of Weed Science, the 2007 Kenneth Morrison Excellence in Extension Award, and the 2009 Team Interdisciplinary Award from WSU.

Hanu R. Pappu is a professor and chair of the Department of Plant Pathology at Washington State University, Pullman, Washington, USA. He obtained BS from the Agricultural College, Bapatla, India, MS from the Indian Agricultural Research Institute, New Delhi, and PhD from the University of Alberta, Edmonton. Following his postdoctoral work at the University of Florida, Gainesville, he held a faculty position at the University of Georgia and a biotechnologist position at USDA-APHIS, Riverdale. His research focuses on the characterization and control of viruses of ornamentals and vegetables with an emphasis on virus diagnostics, genomics, molecular epidemiology, and the use of conventional and biotechnological approaches for virus management. Dr. Pappu has published more than 130 refereed journal articles and nine reviews and book chapters. He was a Fulbright Scholar at Alexandria University, Alexandria, Egypt in 2009 and selected as a Fulbright Specialist in 2011

REFERENCES

Adams, P. B. 1981. Forecasting onion white rot disease. *Phytopathology* 71:1178–1181.
ADAS. 1986. *Control of Diseases of Bulbs. Booklet 2524*. Alnwick, UK: MAFF (Publications).

Adie, B. A. T., J. Pérez-Pérez, M. M. Pérez-Pérez, M. Godoy, J. J. Sanchez-Serrano, E. A. Schmelz, and R. Solano. 2007. ABA is an essential signal for plant resistance to pathogens affecting JA biosynthesis and the activation of defences in *Arabidopsis*. *Plant Cell* 19:1665–1681.

Agrios, G. N. 2005. *Plant Pathology*, 5th Edition. Amsterdam: Elsevier.

Alabouvette, C., C. Olivain, F. L'Haridon, S. Aimé, and C. Steinberg. 2005. Using strains of *Fusarium oxysporum* to control wilts: Dream or reality? In *Novel Biotechnologies for Biocontrol Agent Enhancement and Management*, eds. M. Vurro, and J. Gressel, 157–177. Dordrecht, The Netherlands: Springer.

Alabouvette, C., C. Olivain, Q. Migheli, and C. Steinberg. 2009. Microbiological control of soil-borne phytopathogenic fungi with special emphasis on wilt-inducing *Fusarium oxysporum*. *New Phytol.* 184:529–544.

Alabouvette, C., B. Schippers, P. Lemanceau, and P. A. H. M. Bakker. 1998. Biological control of *Fusarium* wilts: Toward development of commercial products. In *Plant–Microbe Interactions and Biological Control*, eds. G. J. Boland, and L. D. Kuykendall, 15–36. New York: Marcel Dekker.

Alabouvette, C., and C. Steinberg. 2006. The soil as a reservoir for antagonists to plant diseases. In *An Ecological and Societal Approach to Biological Control*, eds. J. Eilenberg, and H. M. T. Hokkanen, 123–144. Dordrecht, The Netherlands: Springer.

Al-Khatib, K. 1996. Tulip (*Tulipa* spp.), daffodil (*Narcissus* spp.), and iris (*Iris* spp.) response to preemergence herbicides. *Weed Technol.* 10:710–715.

Allen, T. C. 1975. Viruses of lilies and their control. *Acta Hort.* 47:69–75.

Andreeva, V. I. 1983. Steblevye nematody sel'skokhozyaĭstvennykh kul'tur i mery bor'by s nimi [Results of the testing of some agrotechnical strategies for the control of strawberry stem nematode]. *Materialy simpoziuma, Voronezh*, URSS, 27-29 Sentyabrya 1983, 111–117 (In Russian).

APS Biological Control Committee. 2005. http://www.oardc.ohio-state.edu/apsbcc/productlist2005USA.htm. Accessed 20 October 2011.

Asjes, C. J. 1990. Production for virus freedom of some principal bulbous crops in The Netherlands. *Acta Hort.* 266:517–529.

Backman, P. A., R. Rodrigues-Kábana, M. C. Caulin, E. Beltramini, and N. Ziliani. 1981. Using the soil-tray technique to predict the incidence of *Sclerotium* rot in sugar beets. *Plant Dis.* 65:419–421.

Bajwa, W. I., and M. Kogan. 2002. *Compendium of IPM Definitions (CID)-What is IPM and How is it Defined in the Worldwide Literature? IPPC Publication No. 998.* Oregon State University, Corvallis, OR: Integrated Plant Protection Center (IPPC).

Baker, J. R. 1993. Insects. In *The Physiology of Flower Bulbs*, eds. A. De Hertogh, and M. Le Nard, 101–153. Amsterdam: Elsevier.

Baker, K. F., and R. J. Cook. 1974. Biological control of plant pathogens. In *The Biology of Plant Pathogens*, eds. A. Kelman, and L. Sequiera, 220–285. San Francisco: Freeman.

Bari, R., and J. D. G. Jones. 2009. Role of plant hormones in plant defence responses. *Plant Mol. Biol.* 69:473–488.

Barker, K. R. 2003. Perspectives on plant and soil nematology. *Annu. Rev. Phytopathol.* 41:1–25.

Barker, K. R., and S. R. Koenning. 1998. Developing sustainable systems for nematode management. *Annu. Rev. Phytopathol.* 36:165–205.

Bastiaansen, C., A. Th. J. Koster, L. J. van der Meer, D. J. E. van den Ende, I. Pennock, and F. M. P. Buurman. 1997. A disease-forecasting system of *Botrytis* blight ("fire") in lily. *Acta Hort.* 430:657–660.

Bates, J. A., E. J. A. Taylor, D. M. Kenyon, and J. E. Thomas. 2001. The application of real-time PCR to the identification, detection and quantification of *Pyrenophora* species in barley seed. *Mol. Plant Pathol.* 2:49–57.

Beale, R. E., and D. Pitt. 1990. Biological and integrated control of *Fusarium* basal rot of *Narcissus* using *Minimedusa polyspora* and other micro-organisms. *Plant Pathol.* 39:477–488.

Beale, R. E., and D. Pitt. 1995. The antifungal properties of *Minimedusa polyspora*. *Mycol. Res.* 99:337–342.

Benbrook, C. M., E. Groth, J. M. Holloran, M. K. Hansen, and S. Marquardt. 1996. *Pest Management at the Crossroads*. Yonkers, New York: Consum. Union.

Bennett, A. J., and J. M. Whipps. 2008. Beneficial microorganism survival on seed, roots and in rhizosphere soil following application to seed during drum priming. *Biol. Control* 44:349–361.

Bergeson, G. B. 1955. The use of systemic phosphates for control of *Ditylenchus dipsaci* on alfalfa and daffodils. *Plant Dis. Rep.* 39:705–709.

Bing, A., and M. Macksel. 1984. The effect of landscape herbicides on newly planted bulbs. *Proc. Northestern Weed Sci. Soc.* 38:217–220.

Boerema, G. H., and M. E. C. Hamers. 1988. Check-list for scientific names of common parasitic fungi. Series 3a: Fungi on bulbs: Liliacea. *Neth. J. Plant Pathol.* 94 Suppl. 1:1–32.

Boerema, G. H., and M. E. C. Hamers. 1989. Check-list for scientific names of common parasitic fungi. Series 3b: Fungi on bulbs: Amaryllidaceae and Iridaceae. *Neth. J. Plant Pathol.* 95 Suppl. 3:1–32.

Bogers, R. J., and B. H. H. Bergman. 1986. *Fourth International Symposium on Flower Bulbs. Acta Horticulturae 177.* Leuven, Belgium: International Society for Horticultural Science.

Bolckmans, K. J. F. 1999. Commercial aspects of biological pest control in greenhouses. In *Integrated Pest and Disease Management in Greenhouse Crops*, eds. R. Albajes, M. L. Gullino, J. C. van Lenteren, and Y. Elad, 310–318. Dordrecht, The Netherlands: Kluwer Academic Publ.

Boonham, N., J. Tomlinson, and R. Mumford. 2007. Microarrays for rapid identification of plant viruses. *Annu. Rev. Phytopathol.* 45:307–328.

Borneman, J., and J. O. Becker. 2007. Identifying microorganisms involved in specific pathogen suppression in soil. *Annu. Rev. Phytopathol.* 45:153–172.

Brandsaeter, L. O., and J. Netland. 1999. Winter annual legumes for use as cover crops in row crops in northern regions. I. Field experiments. *Crop Sci.* 39:1369–1379.

Brent, K. J. 1995. *Fungicide Resistance in Crop Pathogens: How can it be Managed? FRAC Monograph No. 1.* Brussels: GIFAP.

Bridge, P., and B. Spooner. 2001. Soil fungi: diversity and detection. *Plant and Soil* 232:147–154.

Briggs, J. B. 2002. Economics of *Narcissus* bulb production. In *Narcissus and Daffodil, the Genus Narcissus*, ed. G. R. Hanks, 131–140. London: Taylor & Francis.

Brinks, H., and S. de Kool. 2006. Farming with future: Implementation of sustainable agriculture through a network of stakeholders. In *Changing European Farming Systems for a Better Future*, eds. H. Langeveld, and N. Röling, 299–303. Wageningen, The Netherlands: Wageningen Academic Publ.

Buschman, J. C. M. 2005. Globalisation—Flower–flower bulbs–bulb flowers. *Acta Hort.* 673:27–33.

Byther, R. S., and G. A. Chastagner. 1993. Diseases. In *The Physiology of Flower Bulbs*, eds. A. De Hertogh, and M. Le Nard, 71–99. Amsterdam: Elsevier.

Campbell, C. L., and L. V. Madden. 1990. *Introduction to Plant Disease Epidemiology*. New York: John Wiley & Sons.

Carisse, O., N. McRoberts, and L. Brodeur. 2008. Comparison of monitoring- and weather-based risk indicators of *Botrytis* leaf blight of onion and determination of action thresholds. *Can. J. Plant Pathol.* 30:442–456.

Carson, R. L. 1962. *Silent Spring*. Boston: Houghton Mifflin.

Chang, Y.-C., and W. B. Miller. 2005. The development of upper leaf necrosis in *Lilium* 'Star Gazer'. *J. Am. Soc. Hort. Sci.* 130:759–766.

Chastagner, G. A. 1983. Narcissus fire: Prevalence, epidemiology and control in western Washington. *Plant Dis.* 67:1384–1386.

Chastagner, G. A. 1997. Pesticide use patterns associated with the production of ornamental bulb crops in the Pacific Northwest. *Acta Hort.* 430:661–667.

Chastagner, G. A. 2002. Potential alternatives to PCNB to control the development of crown rot and gray bulb rot on bulbous iris. *Acta Hort.* 570:301–306.

Chastagner, G. A., and R. S. Byther. 1985. Bulbs—Narcissus, tulips, and iris. In *Diseases of Floral Crops, Vol. 1*, ed. D. L. Strider, 447–506. New York: Praeger Scientific.

Chastagner, G. A., and W. E. Copes. 2002. Potential use of chlorine dioxide to control diseases in ornamental plant production systems. *Combined Proc. Intl. Plant Propagators' Soc.* 51:275–279.

Chastagner, G. A., and A. DeBauw. 2011a. Efficacy of foliar fungicides in controlling fire on daffodils. *Acta Hort.* 886:307–310.

Chastagner, G. A., and A. DeBauw. 2011b. Alternatives to PCNB for controlling gray bulb rot on tulips. *Acta Hort.* 886:311–317.

Chastagner, G., and A. DeBauw. 2011c. Effectiveness of bulb dip fungicide treatments in controlling neck rot on daffodils. *Acta Hort.* 886:319–322.

Chastagner, G. A., and K. Riley. 1990. Occurrence and control of benzimidazole and dicarboximide resistant *Botrytis* spp. on bulb crops in western Washington and Oregon. *Acta Hort.* 266:437–445.

Chastagner, G. A., and K. L. Riley. 2002. Potential use of chlorine dioxide to prevent the spread of *Fusarium* basal rot during the hot water treatment of daffodil bulbs. *Acta Hort.* 570:267–273.

Chastagner, G. A., and K. L. Riley. 2005. Sensitivity of pathogen inocula to chlorine dioxide gas. *Acta Hort.* 673:355–359.

Chastagner, G. A., J. M. Staley, and K. Riley. 1990. Control of *Sclerotium rolfsii* on bulbous iris and lilies with in-furrow fungicide applications. *Acta Hort.* 266:457–467.

Chen, H., A. D. Jones, and G. A. Howe. 2006. Constitutive activation of the jasmonate signalling pathway enhances the production of secondary metabolites in tomato. *FEBS Lett.* 580:2540–2546.

Chern, M.-S., H. A. Fitzgerald, R. C. Yadav, P. E. Canlas, X. Dong, and P. C. Ronald. 2001. Evidence for a disease-resistance pathway in rice similar to the *NPR1*-mediated signalling pathway in *Arabidopsis*. *Plant J.* 27:101–113.

Chet, I., Y. Elad, A. Kalfon, Y. Hadar, and J. Katan. 1982. Integrated control of soilborne and bulbborne pathogens in iris. *Phytoparasitica* 10:229–236.

Chiocchetti, A., I. Bernardo, M.-J. Daboussi, A. Garbaldi, M. L. Gullion, T. Langin, and Q. Migheli. 1999. Detection of *Fusarium oxysporum* f. sp. *dianthi* in carnation tissue by PCR amplification of transposon insertions. *Phytopathology* 89:1169–1175.

Chitwood, B. G., and F. S. Blanton. 1941. An evaluation of the results of treatments given narcissus bulbs for the control of the nematode *Ditylenchus dipsaci* (Kühn) Filipjev. *J. Washington Acad. Sci.* 31:296–308.

Chitwood, B. G., F. A. Haasis, and F. S. Blanton. 1941. Hot-water-formalin treatment (at 110 to 111°F) of field-grown and of forced narcissus bulbs infected with the bulb or stem nematode, *Ditylenchus dipsaci. Proc. Helminthol. Soc. Washington* 8:44–50.

Cohen, Y. R. 2002. β-Aminobutyric acid-induced resistance against plant pathogens. *Plant Dis.* 86:448–458.

Coley-Smith, J. R., K. Verhoeff, and W. R. Jarvis. 1980. *The Biology of Botrytis*. London: Academic Press.

Collier, R. H. 2011. Large narcissus fly—Forecast pattern of emergence in 2007. http://www2.warwick.ac.uk/fac/sci/lifesci/wcc/hdcpestbulletin/narcissus/lnffore.doc (accessed 10 January 2011).

Collier, R. H., and S. Finch. 1992. The effects of temperature on the development of the large narcissus fly (*Merodon equestris*). *Ann. Appl. Biol.* 120:383–390.

Collier, R. H., S. Finch, and K. Phelps. 1995. Forecasting attacks by insect pests of horticultural field crops. *British Crop Protection Conference Symposium Proceedings 63, Integrated Crop Protection: Towards Sustainability?* 423–430.

Collinge, D. B., H. J. L. Jørgensen, O. S. Lund, and M. F. Lyngkjær. 2010. Engineering pathogen resistance in crop plants: Current trends and future prospects. *Annu. Rev. Phytopathol.* 48:269–291.

Conijn, C. G. M. 1994. Cultivation of *Tagetes patula* to control rootrot in narcissus and lily caused by *Pratylenchus penetrans. Mededelingen, Faculteit Landbouwkundige en Toegepaste Biologische Wetenschappen, Universiteit Gent* 59:807–811.

Conijn, C., and W. J. de Kogel. 2008. Middel tegen gladiolentrips is er, nu de toelating nog. *Bloembollenvisie* 3 januari: 23.

Conijn, C. G. M., and A. T. J. Koster. 1990. Bestrijding van narcisvlieg. Eiafzetperiode is kritiek moment. *Bloembollencultuur* 101(18):18–19, 21.

Conrath, U., G. J. M. Beckers, V. Flors, P. Garcia-Agustin, G. Jakab, F. Mauch, M. A. Newman, C. M. Pieterse, B. Poinssot, M. J. Pozo, A. Pugin, U. Schaffrath, J. Ton, D. Wendehenne, L. Zimmerli, and B. Mauch-Mani. 2006. Priming: Getting ready for battle. *Mol. Plant–Microbe Int.* 19:1062–1071.

Conrath, U., C. M. J. Pieterse, and B. Mauch-Mani. 2002. Priming in plant–pathogen interactions. *Trends Plant Sci.* 7:210–216.

Cook, R. J. 2000. Advances in plant health management in the twentieth century. *Annu. Rev. Phytopathol.* 38:95–116.

Cook, R. J. 2007. Tell me again what it is that you do. *Annu. Rev. Phytopathol.* 45:1–23.

Cook, R. J., and K. F. Baker. 1983. *The Nature and Practice of Biological Control of Plant Pathogens*. St. Paul, MN: APS Press.

Copes, W. E. 2004. Dose curves of disinfestants applied to plant production surfaces for control of *Botrytis cinerea. Plant Dis.* 88:509–515.

Copes, W., B. Barbeau, and G. Chastagner. In press. Chlorine dioxide. In *Biology, Detection and Management of Plant Pathogens in Irrigation Water*, eds. C. Hong, G. Moorman, W. Wohanka, and C. Buettner, St. Paul, MN: APS Press.

Copes, W. E., G. A. Chastagner, and R. L. Hummel. 2004. Activity of chlorine dioxide in a solution of ions and pH against *Thielaviopsis basicola* and *Fusarium oxysporum. Plant Dis.* 88:188–194.

Creelman, R. A., and J. E. Mullet. 1997a. Biosynthesis and action of jasmonates in plants. *Annu. Rev. Plant Physiol. Plant Mol. Biol.* 48:355–381.

Creelman, R. A., and J. E. Mullet.1997b. Oligosaccharins, brassinolides, and jasmonates: Nontraditional regulators of plant growth, development, and gene expression. *Plant Cell* 9:1211–1223.

Csinos, A. S., H. R. Pappu, R. M. McPherson, and M. G. Stephenson. 2001. Management of *Tomato spotted wilt virus* in flue-cured tobacco with acibenzolar-*S*-methyl and imidacloprid. *Plant Dis.* 85:292–296.

Cullen, D. W., and P. R. Hirsch. 1998. Simple and rapid method for direct extraction of microbial DNA from soil for PCR. *Soil Biol. Biochem.* 30:983–993.

Cullen, D. W., A. K. Lees, I. K. Toth, and J. M. Duncan. 2001. Conventional PCR and real-time quantitative PCR detection of *Helminthosporium solani* in soil and potato tubers. *Eur. J. Plant Pathol.* 107:387–398.

Cullen, D. W., A. K. Lees, I. K. Toth, and J. M. Duncan. 2002. Detection of *Colletotrichum coccodes* from soil and potato tubers by conventional and quantitative real-time PCR. *Plant Pathol.* 51:281–292.

Damadzadeh, M., and N. G. M. Hague. 1979. Control of stem nematode (*Ditylenchus dipsaci*) in narcissus and tulip by organophosphate and organocarbamate pesticides. *Plant Pathol.* 28:86–90.

Darras, A. I., D. C. Joyce, and L. A. Terry. 2006. Acibenzolar-*S*-methyl and methyl jasmonate treatments of glasshouse-grown freesias suppress post-harvest petal specking caused by *Botrytis cinerea. J. Hort. Sci. Biotech.* 81:1043–1051.

Darras, A. I., D. C. Joyce, and L. A. Terry. 2010. Postharvest UV-C irradiation on cut *Freesia hybrida* L. inflorescences suppresses petal specking caused by *Botrytis cinerea. Postharvest Biol. Technol.* 55:186–188.

Darras, A. I., L. A. Terry, and D. C. Joyce. 2005. Methyl jasmonate vapour treatment suppresses specking caused by *Botrytis cinerea* on cut *Freesia hybrida* L. flowers. *Postharvest Biol. Technol.* 38:175–182.

Daughtrey, M. L., and D. M. Benson. 2005. Principles of plant health management for ornamental plants. *Annu. Rev. Phytopathol.* 43:141–169.

DeBach, P. 1974. *Biological Control by Natural Enemies*. Cambridge: Cambridge University Press.

De Best, A. L. I. C., and M. J. Zwart. 2000. *Ziekten en afwijkingen bij bolgewassen. Deel 1: Liliaceae.* Lisse, The Netherlands: Lab. Voor Bloembollenonderzoek.

De Boer, M. 2011. Producing bulbs and perennials; sustainable control of diseases, pests and weeds. *Acta Hort.* 886:59–67.

De Boer, M., G. J. van Os, V. Bijman, and J. M. Raaijmakers. 2006. Biological control of soil-borne diseases in flowerbulb cultivation in The Netherlands. *IOBC/wprs Bull.* 29(2):83–88.

De Haan, W. G. 1980. Structural developments in Dutch flowerbulb growing from 1970. *Acta Hort.* 109:311–317.

De Hertogh, A., and M. Le Nard. 1993a. *The Physiology of Flower Bulbs*. Amsterdam: Elsevier.

De Hertogh, A. A., and M. Le Nard. 1993b. Physiological disorders. In *The Physiology of Flower Bulbs*, eds. A. De Hertogh, and M. Le Nard, 155–160. Amsterdam: Elsevier.

De Munk, W. J. 1972. Bud necrosis, a storage disease of tulips. III. The influence of ethylene and mites. *Neth. J. Plant Pathol.* 78:168–178.

De Munk, W. J., and J. J. Beijer. 1971. Bud necrosis, a storage disease of tulips. I. Symptoms and the influence of storage conditions. *Neth. J. Plant Pathol.* 77:97–105.

Deng, Z., N. A. Peres, and B. K. Harbaugh. 2011. Improving disease resistance in caladium: Progress and prospects. *Acta Hort.* 886:69–76.

Den Hollander, N. G., L. Bastiaans, and M. J. Kropff. 2007. Clover as a cover crop for weed suppression in an intercropping design. II. Competitive ability of several clover species. *Eur. J. Agron.* 26:104–112.

De Visser, C. L. M. 1996. Field evaluation of a supervised control system for *Botrytis* leaf blight in spring sown onions in The Netherlands. *Eur. J. Plant Pathol.* 102:795–805.

De Ward, H. A. E. 2008. Implementation of a decision support system for control of *Botrytis* fire blight in flower bulb crops. *Book of Abstracts, 10th International Symposium on Flower Bulbs and Herbaceous Perennials. April 20-24, 2008, Lisse, The Netherlands*, 36.

Dong, X. 2004. NPR1, all things considered. *Curr. Opin. Plant Biol.* 7:547–552.

Doorenbos, J. 1954. Notes on the history of bulb breeding in The Netherlands. *Euphytica* 3:1–11.

Doornick, A. W., and B. H. H. Bergman. 1974. Infection of tulip bulbs by *Botrytis tulipae* originating from spores or contaminated soil. *J. Hort. Sci.* 49:203–207.

Doss, R. P., R. S. Byther, and G. A. Chastagner. 1990. *Fifth International Symposium on Flower Bulbs. Acta Horticulturae 266.* Leuven, Belgium: International Society for Horticultural Science.

Doss, R. P., G. A. Chastagner, and K. L. Riley. 1984. Techniques for inoculum production and inoculation of lily leaves with *Botrytis elliptica. Plant Dis.* 68:854–856.

Doss, R. P., G. A. Chastagner, and K. L. Riley. 1986. Screening ornamental lilies for resistance to *Botrytis elliptica. Scientia Hortic.* 30:237–246.

Dreistadt, S. H. 2001. *Integrated Pest Management for Floriculture and Nurseries. Publication 3402.* Oakland, CA: University of California Division of Agriculture and Natural Resources.

Durgy, R., and R. A. Ashley. 1993. Growing tomatoes in a red clover living mulch. *The Grower: Vegetable and Small Fruit Newsletter* 93(3):4.

Durrant, W. E., and X. Dong. 2004. Systemic acquired resistance. *Annu. Rev. Phytopathol.* 42:185–209.

Echtenkamp, G. W., and R. S. Moomaw. 1989. No-till corn production in a living-mulch system. *Weed Technol.* 3:261–266.

Economic Research Service. 2000. September. *A History of American Agriculture, 1607-2000.* (ERS-POST-12.) Washington, DC. http://www.agclassroom.org/gan/timeline/index.htm.

Edwards-Jones, G., and K. Plassmann. 2008. *Carbon Footprinting and UK Horticulture: Concepts and Commercial Relevance. Report on Project CP 56*. East Malling, UK: HDC.

Eid, S., K. L. Druffel, D. E. Saar, and H. R. Pappu. 2009. Incidence of multiple and distinct species of caulimoviruses in dahlia (*Dahlia variabilis*). *HortScience* 44:1498–1500.

Eilenberg, J. 2006. Concepts and visions of biological control. In *An Ecological and Societal Approach to Biological Control*, eds. J. Eilenberg, and H. M. T. Hokkanen, 1–11. Dordrecht, The Netherlands: Springer.

Eilenberg, J., A. Hajek, and C. Lomer. 2001. Suggestions for unifying the terminology in biological control. *Biocontrol* 46:387–400.

Elkins, D., D. Frederking, R. Marashi, and B. McVay. 1983. Living mulch for no-till corn and soybeans. *J. Soil Water Conserv.* 38:431–433.

Elmer, W. H. 2006a. Effects of acibenzolar-*S*-methyl on the suppression of *Fusarium* wilt of cyclamen. *Crop Prot.* 25:671–676.

Elmer, W. H. 2006b. Efficacy of preplant treatments of gladiolus corms with combinations of acibenzolar-*S*-methyl and biological or chemical fungicides for suppression of Fusarium corm rot [*Fusarium oxysporum* f. sp. *gladioli*]. *Can J. Plant Pathol.* 28:609–614.

Elmer, W. H., and R. J. McGovern. 2004. Efficacy of integrating biologicals with fungicides for the suppression of Fusarium wilt of cyclamen. *Crop Prot.* 23:909–914.

Feldman, K., M. Keren-Zur, R. Hofstein, and B. Fridlender. 1993. *Ampelomyces quisqualis*, an important component of an IPM program for the control of powdery mildew. *6th Int. Congr. Plant Pathol. Abstr. 3.2.11*, 58.

Filion, M., M. St-Arnaud, and S. H. Jabaji-Hare. 2003. Direct quantification of fungal DNA from soil substrate using real-time PCR. *J. Microbiological Methods* 53:67–76.

FormaCare. 2007. Formaldehyde toxicology—Scientific update information. Available at: http://www.formaldehyde-europe.org/fileadmin/formaldehyde/PDF/Scientific_Fact_Sheet_draft_14_09_07_ge_dp_lh.pdf

Fravel, D. R. 2005. Commercialization and implementation of biocontrol. *Annu. Rev. Phytopathol.* 43:337–359.

Garcia-Garza, J. A., T. J. Blom, W. Brown, and W. Allen. 2002. Pre- and post-applications of copper-based compounds to control *Erwinia* soft rot of calla lilies. *Can. J. Plant Pathol.* 24:274–280.

Geary, B., C. Ransom, B. Brown, D. Atkinson, and S. Hafez. 2008. Weed, disease, and nematode management in onions with biofumigants and metam sodium. *HortTechnology* 18:569–574.

Gerson, U., E. Cohen and S. Capua. 1991. Bulb mite, *Rhizoglyphus robini* (Astigmata: Acaridae) as an experimental animal. *Exp. Appl. Acarol.* 12:103–110.

Gillespie, D. R., and D. M. J. Quiring. 1990. Biological control of fungus gnats, *Bradysia* spp. (Diptera: Sciaridae) and western flower thrips, *Frankliniella occidentals* (Pergande) (Thysanoptera: Thripidae), in glasshouses using a soil-dwelling predatory mite, *Geolaelaps* sp. nr. *aculeifer* (Canestrini) (Acari: Laelapidae). *Can. Entomologist* 122:975–983.

Gladstone, L., and G. Moorman. 1987. Geranium mortality due to *Pythium* root rot associated with high levels of nutrient salts. *Phytopathology* 77:986.

Glazebrook, J. 2005. Contrasting mechanisms of defense against biotrophic and necrotrophic pathogens. *Annu. Rev. Phytopathol.* 43:205–227.

Glockemann, B. 1992. Biological control of *Frankliniella occidentalis* on ornamental plants using predatory mites. *EPPO Bulletin* 22:397–404.

Golan, A., Z. Kerem, O. M. Tun, T. Luzzatto, A. Lipsky, and I. Yedidia. 2010. Combining flow cytometry and *gfp* reporter gene for quantitative evaluation of *Pectobacterium carotovorum* ssp. *carotovorum* in *Ornithogalum dubium* plantlets. *J. Appl. Microbiol.* 108:1136–1144.

Gommers, F. J. 1972. Nematicidal principles from roots of some Compositae. *Acta Bot. Neerl.* 21:111–112.

Gould, C. J. 1957. *Handbook on Bulb Growing and Forcing*. Mount Vernon, Washington: Northwest Bulb Growers Association.

Gould, C. J. 1990. History of bulb growing in Washington State. *Acta Hort.* 266:15–23.

Gould, C. J. 1993. *History of the Flower Bulb Industry in Washington State*. Mt. Vernon, Washington: Northwest Bulb Growers Association.

Gould, C. J., and T. S. Russell. 1965. Efficiency of various methods of applying PCNB for preventing soil-borne infestations of bulbous iris by *Sclerotium rolfsii. Plant Dis. Rep.* 49:149–153.

Gozzo, F. 2003. Systemic acquired resistance in crop protection: From nature to a chemical approach. *J. Agric. Food Chem.* 51:4487–4503.

Greenland, R. G. 2000. Optimum height at which to kill barley used as a living mulch in onions. *HortScience* 35:853–855.

Gratwick, M., and J. F. Southey. 1986. *Hot-water Treatment of Plant Material. 3rd Edition, Reference Book 201*. London: Her Majesty's Stationary Office.

Gregory, P. H. 1932. The *Fusarium* bulb rot of narcissus. *Ann. Appl. Biol.* 19:475–514.

Grondeau, C., R. Samson, and D. C. Sands. 1994. A review of thermotherapy to free plant materials from pathogens, especially weeds or bacteria. *Critical Rev. Plant Sci.* 13:57–75.

Gullino, M. L., R. Albajes, and J. C. van Lenteren. 1999. Setting the stage: characteristics of protected cultivation and tools for sustainable crop protection. In *Integrated Pest and Disease Management in Greenhouse Crops*, eds. R. Albajes, M. L. Gullino, J. C. van Lenteren, and Y. Elad, 486–506. Dordrecht, The Netherlands: Kluwer Academic Publ.

Hammerschmidt, R. 2007. Introduction: Definitions and some history. In *Induced Resistance for Plant Defence: A Sustainable Approach to Crop Protection*, eds. D. Walters, A. Newton, and G. Lyon, 1–8. Oxford: Blackwell.

Hammerschmidt, R., and E. K. Dann. 1999. The role of phytoalexins in plant protection. *Novartis Found Symp.* 223:175–187.

Hammond, J. 2006. Current status of genetically modified ornamentals. *Acta Hort.* 722:117–127.

Hammond, J. 2011. Universal plant virus microarrays, broad spectrum PCR assays, and other tools for virus detection and identification. *Acta Hort.* 901:49–60.

Hammond, J., H. T. Hsu, Q. Huang, R. Jordan, K. Kamo, and M. Pooler. 2006. Transgenic approaches to disease resistance in ornamental crops. *J. Crop Improvement* 17:155–210.

Han, Y., A. M. Guentert, R. S. Smith, R. H. Linton, and P. E. Nelson. 1999. Efficacy of chlorine dioxide gas as a sanitizer for tanks used for aseptic juice storage. *Food Microbiol.* 16:53–61.

Han, Y., R. H. Linton, S. S. Nielsen, and P. E. Nelson. 2000. Inactivation of *Escherichia coli* O157:H7 on surface-uninjured and -injured green pepper (*Capsicum annuum* L.) by chlorine dioxide gas as demonstrated by confocal laser scanning microscopy. *Food Microbiol.* 17:643–655.

Hanks, G. R. 1993. *Narcissus*. In *The Physiology of Flower Bulbs*, eds. A. De Hertogh, and M. Le Nard, 463–558. Amsterdam: Elsevier.

Hanks, G. R. 1994. *Large Narcissus Fly Control: The Use of Chlorpyrifos. Report on Project BOF 24*. Petersfield, UK: Horticultural Development Council.

Hanks, G. R. 1995. Prevention of hot-water treatment damage in narcissus bulbs by pre-warming. *J. Hort. Sci. Biotech.* 70:343–355.

Hanks, G. R. 1998. *Large Narcissus Fly: The Effects of Defoliation and Lifting Date on Flower and Bulb Yield. Report on Project BOF 37 (Report on Field Trial in Lincolnshire)*. East Malling, UK: Horticultural Development Council.

Hanks, G. R. 2002. Commercial production of *Narcissus* bulbs. In *Narcissus and Daffodil, the Genus Narcissus*, ed. G. R. Hanks, 53–130. London: Taylor & Francis.

Hanks, G. R., R. Kennedy, and T. M. O'Neill. 2003. *Narcissus Leaf Diseases: Forecasting and Control of White Mould and Smoulder. Report on Project BOF 41*. East Malling, UK: Horticultural Development Council.

Hanks, G. R., and C. A. Linfield. 1997. Pest and disease control in UK narcissus growing: Some aspects of recent research. *Acta Hort.* 430:611–618.

Hanks, G. R., and C. A. Linfield. 1999. Evaluation of a peroxyacetic acid disinfectant in hot-water treatment for the control of basal rot (*Fusarium oxysporum* f. sp. *narcissi*) and stem nematode (*Ditylenchus dipsaci*) in narcissus. *J. Phytopathol.* 147:271–279.

Harman, G. E. 1992. Development and benefits of rhizosphere competent fungi for biological control of plant pathogens. *J. Plant Nutr.* 15:835–843.

Harris, D., and J. E. Hossell. 2002. Pest and disease management constraints under climatic change. *Proceedings of the BCPC Conference—Pests and Diseases 2002*. 2:635–640.

Harris, M. A., R. D. Oetting, and W. A. Gardner. 1995. Use of entomopathogenic nematodes and a new monitoring technique for control of fungus gnats *Bradysia coprophila* (Diptera: Sciaridae), in floriculture. *Biol. Control* 5:412–418.

Hastings, R. J., J. E. Bosher, and W. Newton. 1952. The revival of narcissus bulb eelworm, *Ditylenchus dipsaci* (Kühn) Filipjev, from sub lethal hot-water treatment. *Sci. Agric.* 32:333–336.

Hatterman-Valenti, H. M., and P. E. Hendrickson. 2006. Companion crop and planting configuration effect on onion. *HortTechnology* 16:12–15.

Hausbeck, M. K., and G. W. Moorman. 1996. Managing *Botrytis* in greenhouse-grown flower crops. *Plant Dis.* 80:1212–1219.

Hausbeck, M. K., S. P. Pennypacker, and R. E. Stevenson. 1996. The use of forced heated air to manage *Botrytis* stem blight of geranium stock plants in a commercial greenhouse. *Plant Dis.* 80:940–943.

Hawker, L. E. 1935. Further experiments on the *Fusarium* bulb rot of *Narcissus. Ann. Appl. Biol.* 22:684–708.

Heil, M., and R. M. Bostock. 2002. Induced systemic resistance (ISR) against pathogens in the context of induced plant defences. *Ann. Bot.* 89:503–512.

Hesling, J. J. 1971. Narcissus eelworm *Ditylenchus dipsaci*: Some aspects of its biology and control by thionazin. *Acta Hort.* 23:249–254.

Hiltunen, L. H., C. A. Linfield, and J. G. White. 1995. The potential for the biological control of basal rot of *Narcissus* by *Streptomyces* sp. *Crop Prot.* 14:539–542.

Hodge, S., G. A. Thompson, and G. Powell. 2005. Application of DL-beta-aminobutyric acid (BABA) as a root drench to legumes inhibits the growth and reproduction of the pea aphid *Acyrthosiphon pisum* (Hemiptera: Aphididae). *Bull. Entomol. Res.* 95:449–455.

Hof, N. A. A., P. A. Wolters, and P. Mantel. 1987. The effect of Goltix and Pyramin tested on humus-rich clay. *Bloembollencultuur* 98:15.

Hoitink, H. A. J., and M. J. Boehm. 1999. Biocontrol within the context of soil microbial communities: A substrate-dependent phenomenon. *Annu. Rev. Phytopathol.* 37:427–446.

Hoitink, H. A. J., Y. Inbar, and M. J. Boehm. 1991. Status of compost-amended potting mixes naturally suppressive to soilborne diseases of floricultural crops. *Plant Dis.* 75:869–873.

Hong, C., G. Moorman, W. Wohanka, and C. Buettner. In press. *Biology, Detection and Management of Plant Pathogens in Irrigation Water*. St. Paul, MN: APS Press.

Howard, S. W., E. R. Hall, and C. R. Libbey. 1989. Effects of herbicides on ornamental bulb yield. *Western Soc. Weed Sci. Res. Progress Reports*, 190–191.

Hsiang, T., and G. A. Chastagner. 1991. Growth and virulence of fungicide-resistant isolates of three species of *Botrytis. Can. J. Plant Pathol.* 13:226–231.

Hsiang, T., and G. A. Chastagner. 1992. Production and viability of sclerotia from fungicide-resistant and fungicide-sensitive isolates of *Botrytis cinerea*, *B. elliptica* and *B. tulipae*. *Plant Pathol.* 41:600–605.

Huang, N., A. Enkegaard, L. S. Osborne, P. M. J. Ramakers, G. J. Messelink, J. Pijnakker, and G. Murphy. 2011. The banker plant method in biological control. *Crit. Rev. Plant Sci.* 30:259–278.

Hull, R. 2002. *Matthews' Plant Virology, 4th ed.* New York: Academic Press.

Ichikawa, K., S. Kawasaki, C. Tanaka, and M. Tsuda. 2003. Induced resistance against Fusarium diseases of *Cymbidium* species by weakly virulent strain HPF-1 (*Fusarium* sp.). *J. General Plant Pathol.* 69:400–405.

Jacobsen, B. J. 1997. Role of plant pathology in integrated pest management. *Annu. Rev. Phytopathol.* 35:373–391.

Jacobson, R. J., P. Croft, and J. Fenlon. 1999. *Scatella stagnalis* Fallen (Diptera: Ephydridae): Towards IPM in protected lettuce crops. *IOBC/WRPS Bulletin* 22(1):117–120.

Jagdale, G. B., M. L. Casey, P. S. Grewal, and R. K. Lindquist. 2004. Application rate and timing, potting medium, and host plant effects on the efficacy of *Steinernema feltiae* against the fungus gnat, *Bradysia coprophila*, in floriculture. *Biol. Control* 29:296–305.

Jakab, G., J. Ton, V. Flors, L. Zimmerli, J.-P. Metraux, and B. Mauch-Mani. 2005. Enhancing *Arabidopsis* salt and drought stress tolerance by chemical priming for its abscisic acid responses. *Plant Physiol.* 139:267–274.

Janisiewicz, W. J., and L. Korsten. 2002. Biological control of postharvest diseases of fruits. *Annu. Rev. Phytopathol.* 40:411–441.

Jarecka, A., and A. Saniewska. 2007. The effect of D,L-β-aminobutyric acid on the growth and development of *Fusarium oxysporum* f. sp. *tulipae* (Apt.). *Acta Agrobotanica* 60(1):101–105.

Jarvis, W. R. 1980. Epidemiology. In *The Biology of Botrytis*, eds. J. R. Coley-Smith, K. Verhoeff, and W. R. Jarvis, 219–250. London: Academic Press.

Jewett, T. J., and W. R. Jarvis. 2001. Management of the greenhouse microclimate in relation to disease control: A review. *Agronomie* 21:351–366.

Johnson, K. B. 2010. Pathogen refuge: A key to understanding biological control. *Annu. Rev. Phytopathol.* 48:141–160.

Jones, J. P., A. W. Engelhard, and S. S. Woltz. 1989. Management of *Fusarium* wilt of vegetable and ornamentals by macro- and microelement nutrition. In *Soilborne Plant Pathogen: Management of Disease with Macro- and Micro-elements*, ed. A. W. Engelhard, 18–32. St. Paul, MN: APS Press.

Kessel, G. 1999. Biological control of *Botrytis* spp. by *Ulocladium atrum*, an ecological analysis. PhD thesis, Wageningen University, The Netherlands.

Kiss, L., J. C. Russell, O. Szentivanyi, X. Xu, and P. Jeffries. 2004. Biology and biological control potential *Ampelomyces mycoparasites*, natural antagonists of powdery mildew fungi. *Biocontrol Sci. Technol.* 14:635–651.

Kloepper, J. W., C.-M. Ryu, and S. Zhang. 2004. Induced systemic resistance and promotion of plant growth by *Bacillus* spp. *Phytopathology* 94:1259–1266.

Knippels, P. J. M. 2011. Recent developments in the inspection schemes of flower bulbs. *Acta Hort.* 886:147–151.

Kogan, M. 1998. Integrated pest management: Historical perspectives and contemporary developments. *Annu. Rev. Entomol.* 43:243–270.

Kok, B. J., and H. van Aanholt. 2008. Verbeterde warmwaterbehandeling met voor- en nawarmte werkt positief bij lelie. *Bloembollenvisie* 3 juli:24–25.

Kok, B. J., and H. van Aanholt. 2009. Eén dag voor/nawarmte bij wwb lelie geeft 100% mijtbestrijding. *Bloembollenvisie* 24 september:22–23.

Koster, A. T. J., P. J. M. Vreeburg, and N. A. A. Hof. 1987. Opbrengstderving niet uitsluiten. Sumisclex bestrijdt swartsnot in narcis. *Bloembollencultuur* 98:15.

Kruyer, C. J., and J. Boontjes. 1982. De warmwaterbehandeling van *Lilium longiflorum*. *Bloembollencultuur* 93:622–623.

Labbé, R. M., D. R. Gillespie, C. Cloutier, and J. Brodeur. 2009. Compatibility of an entomopathogenic fungus with a predator and a parasitoid in the biological control of greenhouse whitefly. *Biocontrol Sci. Technol.* 19:429–446.

Lacey, M. S. 1936. Studies in bacteriosis XXII: I. The isolation of a *Bacterium* associated with "fasciation" of sweet peas, "cauliflower" strawberry plants and "leafy gall" of various plants. *Ann. Appl. Biol.* 23:302–310.

Lane, A. 1984. *Bulb Pests. 7th Edition, Reference Book 51*. London: Her Majesty's Stationary Office.

Lang, J. M., D. H. Gent, and H. F. Schwartz. 2007. Management of *Xanthomonas* leaf blight of onion with bacteriophages and a plant activator. *Plant Dis.* 91:871–878.

Langerak, C. J. 1977. The role of antagonists in the chemical control of *F. oxysporum* f. sp. *narcissi*. *Neth. J. Plant Pathol.* 83 (Suppl. 1):365–381.

Lanini, W. T., D. R. Pittenger, W. L. Graves, F. Munoz, and H. S. Agamalian. 1989. Subclovers as living mulches for managing weeds in vegetables. *California Agric.* 43:25–27.

Lawson, R. H. 1990. Production and maintenance of virus-free bulbs. *Acta Hort.* 266:25–34.

Lees, P. D. 1963. Observations on hot water treatment of narcissus bulbs. *Experimental Horticulture* 8:84–89.

Lees, A. K., D. W. Cullen, L. Sullivan, and M. J. Nicolson. 2002. Development of conventional and quantitative real-time PCR assays for the detection and identification of *Rhizoctonia solani* AG-3 in potato and soil. *Plant Pathol.* 51:293–302.

Lesna, I., C. G. M. Conijn, M. W. Sabelis, and N. M. van Straalen. 2000. Biological control of the bulb mite, *Rhizoglyphus robini*, by the predatory mite, *Hypoaspis aculeifer*, on lilies: Predator–prey dynamics in the soil, under greenhouse and field conditions. *Biocont. Sci. Technol.* 10:179–193.

Lesna, I., M. Sabelis, and C. Conijn. 1996. Biological control of the bulb mite, *Rhizoglyphus robini*, by the predatory mite, *Hypoaspis aculeifer*, on lilies: Predator–prey interactions at various spatial scales. *J. Appl. Ecol.* 33:369–376.

Le Nard, M., and A. A. De Hertogh. 2002. Research needs for flower bulbs (geophytes). *Acta Hort.* 570:121–127.

Lilien-Kipnis, H., A. Borochov, and A. H. Halevy. 1997. *Proceedings of the Seventh International Symposium on Flower Bulbs. Acta Horticulturae 430*. Leuven, Belgium: International Society for Horticultural Science.

Lillywhite, R., D. Chandler, W. Grant, K. Lewis, C. Firth, C. Schmutz, and D. Halpin. 2007. *Environmental Footprint and Sustainability of Horticulture (Including Potatoes)—A Comparison with Other Agricultural Sectors*. Report to Defra.

Linfield, C. A. 1991. A comparative study of the effects of five chemicals on the survival of chlamydospores of *Fusarium oxysporum* f. sp. *narcissi*. *J. Phytopathol.* 131:297–304.

Littlejohn, G., R. Venter, and C. Lombard. 2002. *Proceedings of the Eighth International Symposium on Flowerbulbs. Acta Horticulturae 570*. Leuven, Belgium: International Society for Horticultural Science.

Liu, Y.-H., C.-J. Huang, and C.-Y. Chen. 2008. Evidence of induced systemic resistance against *Botrytis elliptica* in lily. *Phytopathology* 98:830–836.

Lockwood, J. L. 1988. Evolution of concepts associated with soilborne plant pathogens. *Annu. Rev. Phytopathol.* 26:93–121.

Loebenstein, G., R. H. Lawson, and A. A. Brunt. 1995. *Virus and Virus-like Diseases of Bulb and Flower Crops*. New York: John Wiley & Sons.

Lole, M. J., G. Hanks, and T. M. O'Neill. 2006. *Narcissus: Alternatives to Formaldehyde in Hot Water Treatment Tanks for the Control of Stem Nematode and Fusarium Basal rot. Report on Project BOF 61*. East Malling, UK: HDC.

Lu, Y.-C., K. B. Watkins, J. R. Teasdale, and A. A. Abdul-Baki. 2000. Cover crops in sustainable food production. *Food Rev. Int.* 16:121–157.

Lu, Y.-Y., and C.-Y. Chen. 2005. Molecular analysis of lily leaves in response to salicylic acid effective towards protection against *Botrytis elliptica*. *Plant Sci.* 169:1–9.

Lu, Y.-Y., Y.-H. Liu, and C.-Y. Chen. 2007. Stomatal closure, callose deposition, and increase of *LsGRP1*-corresponding transcript in probenazole-induced resistance against *Botrytis elliptica* in lily. *Plant Sci.* 172:913–919.

Luzzatto, T., A. Golan, M. Yishay, I. Bilkis, J. Ben-Ari, and I. Yedidia. 2007a. Priming of antimicrobial phenolics during induced resistance response towards *Pectobacterium carotovorum* in the ornamental monocot calla lily. *J. Agric. Food Chem.* 55:10315–10322.

Luzzatto, T., M. Yishay, A. Lipsky, A. Ion, E. Belausov, and I. Yedidia. 2007b. Efficient, long-lasting resistance against the soft rot bacterium *Pectobacterium carotovorum* in calla lily provided by the plant activator methyl jasmonate. *Plant Pathol.* 56:692–701.

Luzzatto-Knaan, T., and I. Yedidia. 2009. Induction of disease resistance in ornamental geophytes. *Israel J. Plant Sci.* 57:401–410.

Lyon, G. 2007. Agents that can elicit induced resistance. In *Induced Resistance for Plant Defence*, eds. D. Walters, A. Newton, and G. Lyon, 9–29. Oxford: Blackwell.

Magie, R. O. 1980. *Fusarium* disease of gladioli controlled by inoculation of corms with non-pathogenic *Fusaria*. *Proc. Florida State Hortic. Soc.* 93:172–175.

Mandal, B., S. Mandal, A. S. Csinos, N. Martinez, A. K. Culbreath, and H. R. Pappu. 2008. Biological and molecular analyses of the acibenzolar-S-methyl-induced systemic acquired resistance in flue-cured tobacco against *Tomato Spotted wilt Virus*. *Phytopathology* 98:196–204.

Martin, F. N. 2003. Development of alternative strategies for management of soilborne pathogens currently controlled with methyl bromide. *Annu. Rev. Phytopathol.* 41:325–350.

Masiunas, J. B. 1998. Production of vegetables using cover crop and living mulches—a review. *J. Vegetable Crop Prod.* 4:11–31.

Mauch-Mani, B., and F. Mauch. 2005. The role of abscisic acid in plant-pathogen interactions. *Curr. Opin. Plant Biol.* 8:409–414.

McManus, P. S., V. O. Stockwell, G. W. Sundin, and A. L. Jones. 2002. Antibiotic use in plant agriculture. *Annu. Rev. Phytopathol.* 40:443–465.

McNew, G. L. 1960. The nature, origin, and evolution of parasitism. In *Plant Pathology: An Advanced Treatise*, eds. J. G. Horsfall, and A. E. Dimond, 19–69. New York: Academic Press.

Messelink, G. J., R. van Maanen, S. E. F. van Steenpaal, and A. Janssen. 2008. Biological control of thrips and whiteflies by a shared predator: Two pests are better than one. *Biological Control* 44:372–379.

Messelink, G. J., S. E. F. van Steenpaal, and W. van Wensveen. 2005. *Typhlodromips swirskii* (Athias-Henriot) (Acari: Phytoseiidae): A new predator for thrips control in greenhouse cucumber. *IOBC/wprs Bull.* 28(1):183–186.

Métraux, J.-P., C. Nawrath, and T. Genoud. 2002. Systemic acquired resistance. *Euphytica* 124:237–243.

Migheli, Q., C. Aloi, and M. L. Gullino. 1990. Resistance of *Botrytis elliptica* to fungicides. *Acta Hort.* 266:429–436.

Miglino, R., K. L. Druffel, and H. R. Pappu. 2010. Identification and molecular characterization of a new potyvirus infecting *Triteleia* species. *Arch. Virol.* 155:441–443.

Miglino, R., K. L. Druffel, A. R. van Schadewijk, and H. R. Pappu. 2011. Molecular characterization of "Allium virus X", a new potexvirus in the family *Flexiviridae*, infecting ornamental allium. *Arch. Virol.* doi: 10.1007/s00705–011–1109–6.

Miglino, R., A. Jodlowska, H. R. Pappu, and T. R. van Schadewijk. 2007. A semi-automated and highly sensitive streptavidin magnetic capture-hybridization RT-PCR assay: Application to genus-wide or species-specific detection of several viruses of ornamental bulb crops. *J. Virological Methods* 146:155–164.

Miller, W. B. 2004. Trichoderma in tulips. http://www.flowerbulbs.cornell.edu/newsletter/Trichoderma%20May%202004.pdf

Miller, P. M., and J. F. Ahrens. 1969. Influence of growing marigolds, weeds, two cover crops and fumigation on subsequent populations of parasitic nematodes and plant growth. *Plant Dis. Rep.* 53:642–646.

Miller, T. 2010. Ornamental bulb, rhizome, corm, and tuber crops. Sections Q-14 through Q-16. In *Pacific Northwest Weed Management Handbook*, ed. E. Peachey, Corvallis, OR: Oregon State University.

Molinar, R. H. 1994. Weed control in field-grown flowers and bulbs. *Proc. California Weed Conf.* 46:69–73.

Moore, A. 1975. Bulb growing in England and Wales. *Acta Hort.* 47:17–23.

Moore, W. C., A. A. Brunt, D. Price, A. R. Rees (revisers), and J. S. W. Dickens (editor). 1979. *Diseases of Bulbs. 2nd edition, Reference book HPD 1*. London: Her Majesty's Stationery Office.

Muller, P. J. 1986. Leaf tip necrosis in forced tulips as a result of a root decay caused by *Trichoderma Viride*. *Acta Hort.* 177:492.

Muller, P. J., P. Vink, and A. van Zaayen. 1988. Flooding caused loss in viability and pathogenicity of sclerotia of *Rhizoctonia tuliparum*. *Eur. J. Plant Pathol.* 94:45–47.

Nash, S. M., and W. C. Snyder. 1962. Quantitative estimations by plate counts of propagules of the bean root rot *Fusarium* in field soils. *Phytopathology* 52:567–572.

Neilson, J. C., and J. L. Anderson. 1989. Competitive effects of living mulch and no-till management systems on vegetable productivity. *Western Soc. Weed Sci. Res. Progress Reports*, 148–149.

Nicholson, P., D. R. Simpson, G. Weston, H. N. Rezanoor, A. K. Lees, D. W. Parry, and D. Joyce. 1998. Detection and quantification of *Fusarium culmorum* and *Fusarium graminearum* in cereals using PCR assays. *Physiol. Mol. Plant Pathol.* 53:17–37.

Ohkawa, K. 2005. Production of flower bulbs and bulbous cut flowers in Japan—Past, present, and future. *Acta Hort.* 673:35–42.

Okubara, P. A., K. L. Schroeder, and T. C. Paulitz. 2005. Real-time polymerase chain reaction: Applications to studies of soil borne pathogens. *Can. J. Plant Pathol.* 27:300–313.

Okubo, H., W. B. Miller, and G. A. Chastagner. 2005. *Proceedings of the Ninth International Symposium on Flower Bulbs. Acta Horticulturae 673*. Leuven, Belgium: International Society for Horticultural Science.

O'Neill, T. M., G. R. Hanks, and D. W. Wilson. 2004. Control of smoulder (*Botrytis narcissicola*) in narcissus with fungicides. *Ann. Appl. Biol.* 145:129–137.

O'Neill, T. M., and J. W. Mansfield. 1982. The cause of smoulder and the infection of narcissus by species of *Botrytis*. *Plant Pathol.* 31:65–78.

O'Neill, T. M., J. W. Mansfield, and G. D. Lyon. 1982. Aspects of narcissus smoulder epidemiology. *Plant Pathol.* 31:101–118.

Oostendorp, M., W. Kunz, B. Dietrich, and T. Staub. 2001. Induced disease resistance in plants by chemicals. *Eur. J. Plant Pathol.* 107:19–28.

Paffrath, A., and A. Frankenberg. 2005. What to do against nematodes? *Gemüse* 41(3):30–31.

Pahalawatta, V., K. L. Druffel, S. D. Wyatt, K. C. Eastwell, and H. R. Pappu. 2008a. Genome structure and organization of a member of a novel and distinct species of the genus Caulimovirus associated with dahlia mosaic. *Arch. Virol.* 153:733–738.

Pahalawatta, V., K. Druffel, and H. Pappu. 2008b. A new and distinct species in the genus *Caulimovirus* exists as an endogenous plant pararetroviral sequence in its host, *Dahlia variabilis*. *Virology* 376:253–257.

Pahalawatta, V., R. Miglino, K. B. Druffel, A. Jodlowska, A. R. van Schadewijk, and H. R. Pappu. 2007. Incidence and relative prevalence of distinct cauliviruses (genus *Caulimovirus*, family *Caulimoviridae*) associated with dahlia mosaic in *Dahlia variabilis*. *Plant Dis.* 91:1194–1197.

Pappu, H. R., A. S. Csinos, R. M. McPherson, D. C. Jones, and M. G. Stephenson. 2000. Effect of acibenzolar-S-methyl and imidacloprid on suppression of tomato spotted wilt *Tospovirus* in flue-cured tobacco. *Crop Prot.* 19:349–354.

Pappu, H. R., and K. L. Druffel. 2009. Use of conserved genomic regions and degenerate primers in a PCR-based assay for the detection of members of the genus *Caulimovirus*. *J. Virol. Methods* 157:102–104.

Pappu, H. R., K. L. Druffel, R. Miglino, and A. R. van Schadewijk. 2008. Nucleotide sequence and genome organization of a member of a new and distinct Caulimovirus species from dahlia. *Arch. Virol.* 153:2145–2148.

Park, S.-W., E. Kaimoyo, D. Kumar, S. Mosher, and D. F. Klessig. 2007. Methyl salicylate is a critical mobile signal for plant systemic acquired resistance. *Science* 318:113–116.

Paulitz, T. C., and R. R. Bélanger. 2001. Biological control in greenhouse systems. *Annu. Rev. Phytopathol.* 39:103–133.

Peña-Cortés, H., J. Fisahn, and L. Willmitzer. 1995. Signals involved in wound-induced proteinase inhibitor II gene expression in tomato and potato plants. *Proc. Natl. Acad. Sci. USA* 92:4106–4113.

Phelps, K., R. H. Collier, R. J. Reader, and S. Finch. 1993. Monte Carlo simulation method for forecasting the timing of pest insect attacks. *Crop Prot.* 12:335–342.

Pieterse, C. M., A. Leon-Reyes, S. van der Ent, and S. C. van Wees. 2009. Networking by small-molecule hormones in plant immunity. *Nat. Chem. Biol.* 5:308–316.

Pieterse, C. M. J., J. A. van Pelt, J. Ton, S. Parchmann, M. J. Mueller, A. J. Buchala, J. P. Métraux, and L. C. van Loon. 2000. Rhizobacteria-mediated induced systemic resistance (ISR) in *Arabidopsis* requires sensitivity to jasmonate and ethylene but is not accompanied by an increase in their production. *Physiol. Mol. Plant Pathol.* 57:123–134.

Pieterse, C. M. J., S. C. M. van Wees, E. Hoffland, J. A. van Pelt, and L. C. van Loon. 1996. Systemic resistance in Arabidopsis induced by biocontrol bacteria is independent of salicylic acid accumulation and pathogenesis-related gene expression. *Plant Cell* 8:1225–1237.

Pieterse, C. M. J., S. C. M. van Wees, J. A. van Pelt, M. Knoester, R. Laan, H. Gerrits, P. J. Weisbeek, and L. C. van Loon. 1998. A novel signalling pathway controlling induced systemic resistance in Arabidopsis. *Plant Cell* 10:1571–1580.

Pieterse, C. M. J., S. C. M. van Wees, J. Ton, J. A. van Pelt, and L. C. van Loon. 2002. Signalling in rhizobacteria-induced systemic resistance in *Arabidopsis thaliana*. *Plant Biol.* 4:535–544.

Pilkington, L. J., G. Messelink, J. C. van Lenteren, and K. Le Mottee. 2010. "Protected Biological Control"—Biological pest management in the greenhouse industry. *Biological Control* 52:216–220.

Pitt, D. 1991. *Control of Narcissus Basal Rot by Antagonists. Report on HDC Project BOF 6*. Petersfield, UK: HDC.

Poincelot, R. P. 1986. *Toward a More Sustainable Agriculture*. Westport, CT: AVI Publ.

Polyakovskii, S. A., Zh. N. Kravchuk, and A. P. Dmitriev. 2008. Mechanism of action of the plant resistance inducer β-aminobutyric acid in *Allium cepa*. *Cytol. Genet.* 42:369–372.

Pouidel, D. D., H. Ferris, K. Klonsky, W. R. Horwath, K. M. Scow, A. H. C. van Bruggen, W. T. Lanini, J. P. Mitchell, and S. R. Temple. 2001. The sustainable farming system project in California's Sacramento Valley. *Outlook Agric.* 30:109–116.

Price, D. 1970. Tulip fire caused by *Botrytis tulipae* (lib.) Lind.; the leaf spotting phase. *J. Hort. Sci.* 45:233–238.

Putnam, M. L., and M. L. Miller. 2007. *Rhodococcus fascians* in herbaceous perennials. *Plant Dis.* 91:1064–1076.

Qiu, J., B. B. Westerdahl, D. Giraud, and C. A. Anderson. 1993. Evaluation of hot water treatments for management of *Ditylenchus dipsaci* and fungi in daffodil bulbs. *J. Nematol.* 25:686–694.

Ramos-García, M., S. Ortega-Centeno, A. N. Hernández-Lauzardoa, I. Alia-Tejacal, E. Bosquez-Molina, and S. Bautista-Baños. 2009. Response of gladiolus (*Gladiolus* spp) plants after exposure corms to chitosan and hot water treatments. *Scientia Hortic.* 121:480–484.

Rasmussen, E. 1980. *Third International Symposium on Flower Bulbs. Acta Horticulturae 109*. Leuven, Belgium: International Society for Horticultural Science.

Rees, A. R. 1992. *Ornamental Bulbs, Corms and Tubers*. Wallingford, Oxon, UK: C.A.B International.

Rees, A. R., and H. H. van der Borg. 1975. *Second International Symposium on Flower Bulbs. Acta Horticulturae 47*. Leuven, Belgium: International Society for Horticultural Science.

Regnault-Roger, C., C. Vincent, and J. T. Arnason. 2011. Essential oils in insect control: Low-risk products in a high-stakes world. *Annu. Rev. Entomol.* Online, doi: 10.1146/annurev-ento-120710–100554. Accessed 22 October 2011.

Rodov, V., Z. Tietel, Y. Vinokur, B. Horev, and D. Eshel. 2010. Ultraviolet light stimulates flavonol accumulation in peeled onions and controls microorganisms on their surface. *J. Agric. Food Chem.* 58:9071–9076.

Rohwer, C. L., and J. E. Erwin. 2008. Horticultural applications of jasmonates: A review. *J. Hort. Sci. Biotech.* 83:283–304.

Rohwer, C. L., and J. E. Erwin. 2010. Spider mites (*Tetranychus urticae*) perform poorly on and disperse from plants exposed to methyl jasmonate. *Entomol. Exp. Appl.* 137:143–152.

Rosskopf, E. N., D. O. Chellemi, N. Kokalis-Burelle, and G. T. Church. 2005. Alternatives to methyl bromide: A Florida perspective. *Plant Health Prog.* doi:10.1094/PHP-2005–1027–01-RV.

Rudd-Jones, D. 1975. Introduction. *Acta Hort.* 47:13–15.

Sabelis, M. W., A. Janssen, I. Lesna, N. S. Aratchige, M. Nomikou, and P. C. J. van Rijn. 2008. Developments in the use of predatory mites for biological pest control. *Integrated Control in Protected Crops, Temperate Climate. IOBC/WPRS Bull.* 32:187–199.

Saniewski, M., J. C. M. Beijersbergen, and W. Bogatko. 1992. *Sixth International Symposium on Flower Bulbs. Acta Horticulturae 325*. Leuven, Belgium: International Society for Horticultural Science.

Schenk, P. K. 1971a. *First International Symposium on Flowerbulbs. Acta Horticulturae 23*. Leuven, Belgium: International Society for Horticultural Science.

Schenk, P. K. 1971b. Bulbous plants in scientific research; past, present, and future. *Acta Hort.* 23:18–27.

Schneider, S. M., E. N. Rosskopf, J. G. Leesch, D. O. Chellemi, C. T. Bull, and M. Mazzola. 2003. United States Department of Agriculture-Agricultural Research Service research on alternatives to methyl bromide: Pre-plant and post-harvest. *Pest Management Sci.* 59:814–826.

Schumann, G. L., and C. J. D'Arcy. 2010. *Essential Plant Pathology, Second Edition*. St. Paul, Minnesota: APS Press.

Sharma, N., and A. Tripathi. 2008. Integrated management of postharvest Fusarium rot of gladiolus corms using hot water, UV-C and *Hyptis suaveolens* (L.) Poit. essential oil. *Postharvest Biol. Technol.* 47:246–254.

Skroch, W. A., C. J. Catanzaro, A. A. De Hertogh, and L. B. Gallitano. 1994. Preemergence herbicide evaluations on selected spring and summer flowering bulbs and perennials. *J. Environ. Hort.* 12:80–82.

Skroch, W. A., S. L. Warren, and A. A. De Hertogh. 1988. Phytotoxicity of herbicides of spring flowering bulbs. *J. Environ. Hort.* 6:109–113.

Slootweg, A. F. G. 1956. Rootrot of bulbs caused by *Pratylenchus* and *Hoplolaimus* spp. *Nematologica* 1:192–201.

Smid, E. J., Y. de Witte, and L. G. M. Gorris. 1995. Secondary plant metabolites as control agents of postharvest *Penicillium* rot on tulip bulbs. *Postharvest Biol. Technol.* 6:303–312.

Smith, R. F., and R. van den Bosch. 1967. Integrated control. In *Pest Control: Biological, Physical, and Selected Chemical Methods*, eds. W. W. Kilgore, and R. L. Doutt, 295–340. New York: Academic Press.

Snijder, R. C., H.-R. Cho, M. M. W. B. Hendriks, P. Lindhout, and J. M. van Tuyl. 2004a. Genetic variation in *Zantedeschia* spp. (Araceae) for resistance to soft rot caused by *Erwinia carotovora* subsp. *carotovora*. *Euphytica* 135:119–128.

Snijder, R. C., P. Lindhout, and J. M. van Tuyl. 2004b. Genetic control of resistance to soft rot caused by *Erwinia carotovora* subsp. *carotovora* in *Zantedeschia* spp. (Araceae), section *Aestivae*. *Euphytica* 136:319–325.

Snijder, R. C., and J. M. van Tuyl. 2002. Evaluation of tests to determine resistance of *Zantedeschia* spp. (Araceae) to soft rot caused by *Erwinia carotovora* subsp. *carotovora*. *Eur. J. Plant Pathol.* 108: 565–571.

Solano, B. R., J. B. Maicas, M. T. P. de la Iglesia, J. Domenech, and F. J. G. Mañero. 2008. Systemic disease protection elicited by plant growth promoting rhizobacteria strains: Relationship between metabolic responses, systemic disease protection, and biotic elicitors. *Phytopathology* 98:451–457.

Spoel S. H., A. Koornneef, S. M. C. Claessens, J. P. Korzelius, J. A. van Pelt, M. J. Mueller, A. J. Buchala, J. P. Métraux, R. Brown, K. Kazan, L. C. van Loon, X. Dong, and C. M. Pieterse. 2003. NPR1 modulates cross-talk between salicylate- and jasmonate-dependent defense pathways through a novel function in the cytosol. *Plant Cell* 15:760–770.

Sticher, L., B. Mauch-Mani, and J. P. Métraux. 1997. Systemic acquired resistance. *Annu. Rev. Phytopathol.* 35: 235–270.

Strider, D. L. 1985. *Diseases of Floral Crops Vol. I and Vol. II*. New York: Praeger Scientific.

Studholme, D. J., R. H. Glover, and N. Boonham. 2011. Application of high-throughput DNA sequencing in phytopathology. *Annu. Rev. Phytopathol.* 49:87–105.

Sutton, T. B. 1996. Changing options for the control of deciduous fruit tree diseases. *Annu. Rev. Phytopathol.* 34:527–547.

Sutton, M. W., G. R. Dixon, and M. Willock. 1986a. Virus-tested *Narcissus*: Progress with field evaluation in Scotland. *Acta Hort.* 177:221–226.

Sutton, J. C., T. D. W. James, and P. M. Rowell. 1986b. BOTCAST: A forecasting system to time the initial fungicide spray for managing *Botrytis* leaf blight of onions. *Agric., Ecosystems Environ.* 18:123–143.

Tamala, W., P. Jitareetat, A. Uthairatanaki, and K. Obsuwan. 2007. Effect of pre-harvest chitosan sprays on growth of Curcuma 'Laddawan' (*Curcuma alismatifolia* × *Curcuma cordata*). *Acta Hort.* 755: 387–393.

Terry, L. A., and D. C. Joyce. 2004. Elicitors of induced disease resistance in postharvest horticultural crops: A brief review. *Postharvest Biol. Technol.* 32:1–13.

Thurston, H. D. 1992. *Sustainable Practices for Plant Disease Management in Traditional Farming Systems*. Boulder, CO: Westview Press.

Tilley, L. A. N., P. Croft, and P. J. Mayhew. 2011. Control of a glasshouse pest through the conservation of its natural enemies? An evaluation of apparently naturally controlled shore fly populations. *Biol. Control* 56:22–29.

Tompsett, A. A. 1974. Bulbs. *Rosewarne Experimental Horticulture Station Annual Report 1973*, 16–86.

Tompsett, A. A. 1980. The control of narcissus basal rot (*Fusarium oxysporum* f. sp. *narcissi*). *Rosewarne and Isles of Scilly EHSs Annual Review 1979*, 13–23.

Tompsett, A. A. 2006a. *Narcissus: Overcoming the Problem of Soil Sickness with Particular Reference to Production in the Isles of Scilly. Report on HDC Project BOF 50*. East Malling, UK: Horticultural Development Council.

Tompsett, A. 2006b. Golden Harvest. The story of daffodil growing in Cornwall and the Isles of Scilly. Alison Hodge: Penzance, UK.

Tompsett, A. A. 2008. *Narcissus: Overcoming the Problem of Soil Sickness with Particular Reference to Production in the Isles of Scilly. Report on HDC Project BOF 50a*. East Malling, UK: Horticultural Development Council.

Ton, J., and B. Mauch-Mani. 2004. β-Amino-butyric acid-induced resistance against necrotrophic pathogens is based on ABA-dependent priming for callose. *Plant J.* 38:119–130.

Tones, S. J., R. W. Brown, R. L. Gwynn, R. W. Rogers, and C. J. Tavaré. 1990. Evaluation of insecticides against large narcissus fly. *Tests of Agrochemicals and Cultivars* 11 (*Ann. Appl. Biol.* 116, supplement):20–21.

Tones, S., R. Collier, and B. Parker. 2004. *Large Narcissus Fly Spatial Dynamics. Final Report on Project HH1747.* London: Defra. Available at: http://tna.europarchive.org/20050307123719/http://randd.defra.gov.uk/Document.aspx?DocumentID = 245 (accessed 10 January 2011).

Tones, S. J., and A. A. Tompsett. 1990. Tolerance of narcissus cultivars to bulb-immersion treatments with chlorpyrifos. *Tests of Agrochemicals and Cultivars 11* (*Ann. Appl. Biol.* 116, supplement):74–75.

Trolinger, J. C., and D. L. Strider. 1985. *Botrytis* diseases. In *Diseases of Floral Crops, Vol. 1*, ed. D. L. Strider, 17–101. New York: Praeger Scientific.

U. S. EPA Biopesticide Active Ingredient Fact Sheets. http://www.epa.gov/pesticides/biopesticides/ingredients/index.htm#1A. Accessed 20, October 2011.

Valenzuela, H., and J. DeFrank. 1994. Living mulches: Eggplant growth and yields in living mulch and monocultures. *Vegetable Crops Update* 4:1–7.

Vallad, G. E., and R. M. Goodman. 2004. Systemic acquired resistance and induced systemic resistance in conventional agriculture. *Crop Sci.* 44:1920–1934.

Van Aartrijk, J., and M. de Rooy. 1990. Changing views on chemical crop protection in The Netherlands and the consequences for bulb research. *Acta Hort.* 266:385–389.

Van Aartrijk, J., J. E. van den Ende, and J. M. M. Peeters. 2000. *Ziekten en afwijkingen bij bolgewassen. Deel 1: Liliaceae.* Lisse, The Netherlands: Laboratorium voor Bloembollenonderzoek.

Van den Ende, J. E., A. T. Krikke, and A. P. M. den Nijs. 2011. *Tenth International Symposium on Flower Bulbs and Herbaceous Perennials. Acta Horticulturae 886.* Leuven, Belgium: International Society for Horticultural Science.

Van den Ende, J. E., and J. G. Pennock. 1997. Influence of temperature and wetness duration on infection of lily by *Botrytis elliptica. Acta Bot. Neerl.* 46:332.

Van den Ende, J. E., M. G. Pennock-Vos, C. Bastiaansen, A. Th. J. Koster, and L. J. van der Meer. 2000. BoWaS: A weather-based warning system for the control of *Botrytis* blight in lily. *Acta Hort.* 519:215–220.

Van der Zande, J. C., V. T. J. M. Achten, H. T. A. M. Schepers, A. van der Lans, C. Kempenaar, J. M. Michielsen, H. Stallinga, and P. van Velde. 2010. Precision disease control in bed-grown crops. In *Precision Crop Protection—the Challenge and use of Heterogeneity*, eds. E.-C. Oerke, R. Gerhards, G. Menz, and R. A. Sikora, 403–415. Dordrecht, The Netherlands: Springer.

Van Doorn, J., P. J. M. Vreeburg, P. J. van Leeuwen, and R. H. L. Dees. 2011. The presence and survival of soft rot (*Erwinia*) in flower bulb production systems. *Acta Hort.* 886:365–379.

Van Julsingha, E. B. 1980. Developments in bulb research. *Acta Hort.* 109:19–24.

Van Leeuwen, P. J., and J. Trompert. 2011. Controlling *Aphelenchoides subtenuis* nematodes with a hot water treatment in *Crocus* and *Allium. Acta Hort.* 886:273–276.

Van Lenteren, J. C. 2000a. Measures of success in biological control of arthropods by augmentation of natural enemies. In *Measures of Success in Biological Control*, eds. G. Gurr, and S. Wratten, 77–103. Dordrecht, The Netherlands: Kluwer Academic Publ.

Van Lenteren, J. C. 2000b. A greenhouse without pesticides: Fact or fantasy? *Crop Prot.* 19:375–384.

Van Lenteren, J. C. 2007. Biological control for insect pests in greenhouses: An unexpected success. In *Biological Control: A Global Perspective*, eds. C. Vincent, M. S. Goettel, and G. Lazarovits, 105–117. Wallingford, UK: CAB International.

Van Loon, L. C., P. A. H. M. Bakker, and C. M. J. Pieterse. 1998. Systemic resistance induced by rhizosphere bacteria. *Annu. Rev. Phytopathol.* 36:453–483.

Van Rijn, J. F. A. T., and H. G. M. Pfaff. 1995. *Ziekten en afwijkingen bij bolgewassen. 2. Amaryllidaceae, Araceae, Begoniaceae, Cannaceae, Compositae, Iridaceae, Oxalidaceae, Ranunculaceae. 2nd edition.* Lisse, The Netherlands: Informatie en Kennis Centrum Landbouw, Laboratorium voor Bloembollenonderzoek.

Van Slogteren, E. 1931. Warm-waterbehandeling van narcissen en bolrot. 380 Weekbl. *BloembollCult.* 42: 130–145.

Van Wees, S. C. M., E. A. M. de Swart, J. A. van Pelt, L. C. van Loon, and C. M. J. Pieterse. 2000. Enhancement of induced disease resistance by simultaneous activation of salicylate- and jasmonate-dependent defense pathways in *Arabidopsis thaliana. Proc. Natl. Acad. Sci. USA* 97:8711–8716.

Van Zaayen, A., C. J. Asjes, D. Bregnan, A. Th. J. Koster, P. J. Muller, G. G. M. van der Valk, and I. Vos. 1986. Control of soil-borne diseases, nematodes and weeds in ornamental bulb cultivation by means of flooding. *Acta Hort.* 177:524.

Vänninen, I., and H. Koskula. 2003. Biological control of the shore fly (*Scatella tenuicosta*) with steinernematid nematodes and *Bacillus thuringiensis* var. *thuringiensis* in peat and rockwool. *Biocontrol Sci. Technol.* 13:47–63.

Vänninen, I., and H. Koskula. 2004. Biocontrol of the shore fly *Scatella tenuicosta* with *Hypoaspis miles* and *H. aculeifer* in peat pots. *BioControl* 49:137–152.

Vigodsky-Hass, H., and A. Lavi. 1986. Basal plate rot of narcissus bulbs and its control. *Acta Hort.* 177:485–491.

Vincelli, P. 2005. An inquiry-based approach to teaching disease cycles. *The Plant Health Instructor.* doi:10.1094/PHI-T-2005-0222–01 http://www.apsnet.org/edcenter/instcomm/TeachingArticles/Pages/InquiryBasedApproach.aspx (accessed on 12 August 2011).

Vincelli, P. C., and J. W. Lorbeer. 1989. BLIGHT-ALERT: A weather-based predictive system for timing fungicide applications on onion before infection periods of *Botrytis squamosa*. *Phytopathology* 79: 493–498.

Walling, L. L. 2001. Induced resistance: From the basic to the applied. *Trends in Plant Science* 6:445–447.

Walters, D., and M. Heil. 2007. Costs and trade-offs associated with induced resistance. *Physiol. Mol. Plant Pathol.* 71:3–17.

Walters, D., D. Walsh, A. Newton, and G. Lyon. 2005. Induced resistance for plant disease control: Maximizing the efficacy of resistance elicitors. *Phytopathology* 95:1368–1373.

Walters, D. R. 2009. Are plants in the field already induced? Implications for practical disease control. *Crop Prot.* 28:459–465.

Walters, D. R., and J. M. Fountaine. 2009. Practical application of induced resistance to plant diseases: An appraisal of effectiveness under field conditions. *J. Agri. Sci.* 147:523–535.

Wang, C.-S., J.-C. Huang, and J.-H. Hu. 1999. Characterization of two subclasses of PR-10 transcripts in lily anthers and induction of their genes through separate signal transduction pathways. *Plant Mol. Biol.* 40:807–814.

Weller, D. M. 1988. Biological control of soilborne plant pathogens in the rhizosphere with bacteria. *Annu. Rev. Phytopathol.* 26:379–407.

Weller, D. M., J. M. Raaijmakers, B. B. McSpadden Gardener, and L. S. Thomashow. 2002. Microbial populations responsible for specific soil suppressiveness to plant pathogens. *Annu. Rev. Phytopathol.* 40:309–348.

West, J. S., C. Bravo, R. Oberti, D. Lemaire, D. Moshou, and H. A. McCartney. 2003. The potential of optical canopy measurement for targeted control of field crop diseases. *Annu. Rev. Phytopathol.* 41:593–614.

Westerdahl, B. B., D. Giraud, J. D. Radewald, and C. A. Anderson. 1991. Management of *Ditylenchus dipsaci* in daffodils with foliar applications of oxamyl. *J. Nematol.* 23(supplement):706–711.

Wiles, L. J., R. D. William, G. D. Crabtree, and S. R. Radosevich. 1989. Analyzing competition between a living mulch and a vegetable crop in an interplanting system. *J. Am. Soc. Hort. Sci.* 114:1029–1034.

Wilson, C. L., and M. E. Wisniewski. 1989. Biological control of postharvest diseases of fruits and vegetables: An emerging technology. *Annu. Rev. Phytopathol.* 27:425–441.

Wilson, C. L., M. E. Wisniewski, C. L. Biles, R. McLaughlin, E. Chalutz, and S. Droby. 1991. Biological control of post-harvest diseases of fruits and vegetables: alternatives to synthetic fungicides. *Crop Prot.* 10:172–177.

Windrich, W. A. 1986. Control of stem nematode, *Ditylenchus dipsaci*, in narcissus with aldicarb. *Crop Prot.* 5:266–267.

Wink, M., and P. Lehmann. 1996. Wounding- and elicitor-induced formation of coloured chalcones and flavans as phytoalexins in *Hippeastrum × hortorum*. *Botanica Acta* 109:412–421.

Woets, J., and A. van der Linden. 1982. On the occurrence of *Opius pallipes* Wesmael and *Dacnusa sibirica* Telenga (Braconidae) in cases of natural control of the tomato leafminer *Liriomyza bryoniae* Kalt (Agromyzidae) in some large greenhouses in The Netherlands. *Med. Fac. Landbouww. Rijksuniv. Gent* 47(2):533–540.

Woltz, S. S., and A. W. Engelhard. 1973. *Fusarium* wilt of chrysanthemum: Effect of nitrogen source and lime on disease development. *Phytopathology* 63:155–157.

Woodville, H. C. 1960. Further experiments on the control of bulb fly in narcissus. *Plant Pathol.* 9:68–70.

Woodville, H. C. 1964. Lethal times and temperatures for two species of eelworm. *Experimental Horticulture* 10:90–95.

Yan, Z., M. S. Reddy, C.-M. Ryu, J. A. McInroy, M. Wilson, and J. W. Kloepper. 2002. Induced systemic protection against tomato late blight elicited by plant growth-promoting rhizobacteria. *Phytopathology* 92:1329–1333.

Yedidia, I., A. Lipsky, A. Golan, M. Yishay, A. Ion, and T. Luzzatto. 2011. Polyphenols induction in the defense response of calla lily towards *Pectobacterium carotovorum*. *Acta Hort.* 886:409–415.

Yoshioka, K., H. Nakashita, D. F. Klessig, and I. Yamaguchi. 2001. Probenazole induces systemic acquired resistance in *Arabidopsis* with a novel type of action. *Plant J.* 25:149–157.

Zasada, I. A., J. M. Halbrendt, N. Kokalis-Burelle, J. LaMondia, M. V. McKenry, and J. W. Noling. 2010. Managing nematodes without methyl bromide. *Annu. Rev. Phytopathol.* 48:311–328.

Zehnder, G., G. M. Gurr, S. Kühne, M. R. Wade, S. D. Wratten, and E. Wyss. 2007. Arthropod pest management in organic crops. *Annu. Rev. Entomol.* 52:57–80.

Zehnder, G. W., J. F. Murphy, E. J. Sikora, and J. W. Kloepper. 2001. Application of rhizobacteria for induced resistance. *Eur. J. Plant Pathol.* 107:39–50.

Zimmerli, L., G. Jakab, J.-P. Métraux, and B. Mauch-Mani. 2000. Potentiation of pathogen-specific defense mechanisms in *Arabidopsis* by β-aminobutyric acid. *Proc. Natl. Acad. Sci. USA* 97:12920–12925.

Zimmerli, L., J.-P. Metraux, and B. Mauch-Mani. 2001. β-aminobutyric acid-induced protection of *Arabidopsis* against the necrotrophic fungus *Botrytis cinerea*. *Plant Physiol.* 126:517–523.

14 Geophyte Research and Production in East and Southeast Asia

Seiichi Fukai

CONTENTS

I. INTRODUCTION

Asia is a huge continent consisting of about 50 countries with approximately 60% of the world's population. Remarkable economic progress has been achieved in East and Southeast Asia during the last two decades, and today, the countries with the second and third highest Gross Domestic Product (GDP) are located in this region. Furthermore, the economic development of China and India is expected to lead the world economy (IMF 2011). Economic development has brought about an increase in the standard of living, resulting in a rapid increase in flower production and consumption in East and Southeast Asia during the last two decades. The changes have also necessitated both further development of science and technology in flower production and expanding contacts with the world flower markets. It is assumed that flower production and consumption in this region will determine the direction of global ornamental horticulture, as well as the continuing success of the world flower market.

Commercial flower production in Asia initially developed in the temperate zones of Japan and Korea. Until now, these countries remain the center of flower production and local consumption in Asia. However, the newly emerging regions of flower production in East and Southeast Asia differ from the former production areas in both geographical and environmental conditions. The "Tropical highlands", for example, Kunming (altitude 1900 m a.s.l.; average temperature 16°C; precipitation 1000 mm) in China, Da Lat (1500 m, 16°C, 2200 mm) in Vietnam, and the Cameron highland (1500 m, 18°C, 2500 mm) in Malaysia, have recently become important centers of horticultural production. Flower production in the tropical highlands is mainly export oriented, with some advantages (e.g., mild climate and rather low labor cost) and disadvantages (e.g., insufficient technical support, infrastructures, and training and education systems). In addition, the production areas are often located far from market and consumption centers.

Also in the matter of plant genetic resources, Asia has played an important role in the development of world floriculture. Asia is characterized by a wide range of climates and ecosystems. The diverse environmental conditions allow for an enormous diversity in native plants, including geophytes with ornamental traits. In history, East Asia and Mediterranean–East Asia were the two primary centers of floriculture in ancient times (Nakao 1968). Cultivation of *Prunus mume*, *P. persica* and lotus (*Nelumbo nucifera*) as ornamental plants was mentioned in "Shi Jung", one of the oldest poetry books written in 600 B.C. (Nakao 1968). In the nineteenth century several Asian species of lilies were introduced into Europe. The introduction of these species provided a basis for the large-scale development of breeding and commercial production of this important ornamental crop. Later, the Asian species became the basis for the important groups of current lily varieties, and at present, lilies are the third largest flower crop in the global market. Unfortunately, only a few new additional ornamental geophytes have been developed from other Asian species (e.g., *Lycoris*, *Hemerocallis*), whereas many attractive geophytes have been found in this region (Zhao and Zhang 2003).

This subchapter presents an overview of plant genetic resources of ornamental geophytes in East and Southeast Asia, as well as the commercial production of geophytes in the new production areas in this region.

II. NATIVE SPECIES OF ORNAMENTAL GEOPHYTES IN ASIA

A. LILIACEAE

The genus *Lilium* consists of about 100 species distributed in a wide range of climate conditions in the Northern Hemisphere. About 60 are native to Asia. Comber (1949) classified the genus *Lilium* into seven sections, four of them (Oriental, Asiatic, Trumpet, and Dauricum) are lilies of Asian origin (see also Chapter 2). Being one of the most popular ornamental plants and highly appreciated in oriental culture, the lily species and hybrids have been described in detail by many authors (reviewed by Synge 1980; McRae 1998). Many native Asian *Lilium* species are still not used in horticulture (Zhao and Zhang 2003; Yuan et al. 2011).

The genus *Fritillaria* consists of about 100 species and is found in the Northern Hemisphere, mainly in South-Western and Himalayan Asia (Rønsted et al. 2005). *Fritillaria* species display bell-shaped flowers with a wide range of colors, including blue, pink, white, brown, and bi-color. Whereas many *Fritillaria* species in Asia have ornamental value, most have not yet been developed as commercial ornamental crops, due to their complicated biological traits and life cycle, especially in the species found in the alpine areas (Pavord 2009). Only a few species, for example *F. imperialis* and *F. persica* originating from western Asia, are currently used as commercial ornamentals.

Both the genera *Nomocharis* and *Notholirion* are closely related to the genus *Lilium* (Hayashi and Kawano 2005). *Nomocharis* consists of 14 species with flat ovoid-shaped perianth, distributed in the Himalayas and Tibet. Four species of *Notholirion* produce racemes and fragrant flowers, and are found in Central Asia. The horticultural use of both genera is still limited.

B. Amaryllidaceae

The genus *Lycoris* consists of about 20 species distributed in the temperate and subtropical regions of East Asia, from southwestern China to Japan. Based on karyotypes, *Lycoris* is divided into three groups with acrocentric, metacentric, and telocentric chromosomes (Kurita and Hsu 1998). Although the *Lycoris* species flower in late summer to autumn, based on the times of their leaf emerging, they are divided into spring and autumn types. *L. sanguinea* and *L. sprengeri* flower in late August to early September and develop leaves in spring, while *L. radiata* and *L. aurea* flower in September and early October and their leaves appear in October–November when flower stalks still exist. The inflorescence of *Lycoris* is an umbel with a wide range of petal colors: red, orange, yellow, pink, blue, and white. Many interspecific hybrids have been produced in both Japan and China (Kurita 2011); however, their commercial production and use are still limited.

The genus *Crinum* consists of about 180 species and has a pan-tropical distribution with a few species native to Asia. *C. asiaticum* is widely used as a garden plant.

C. Zingiberaceae

The genus *Curcuma* consists of about 65 species and is distributed in tropical Asia. *C. alismatifolia*, *C. harmandii*, *C. thorelii*, and some other species are used as cut flowers (Sirisawad et al. 2003). Their inflorescences show considerable morphological variation. *C. alismatifolia* produces mixed thyrsoidal inflorescences. The primary and secondary inflorescences show different developmental patterns: the main axis is indeterminate, while the secondary axis is determinate and cymose (Fukai and Udomdee 2005). The main bracts at the upper part of the inflorescence display a wide range of colors. Some other *Curcuma* species, *C. aromatica*, *C. longa*, have been used as medicinal and aromatic plants (Komatsu et al. 2007).

In addition to the well-known ornamentals, e.g., *Heliconia*, *Anthurium*, and *Strelitzia*, which are not native to Asia, several Asian Zingiberaceae genera, for example, *Alpinia*, *Globba*, *Hedychium*, and *Zingiber*, possess significant horticultural potential. Unfortunately, their commercial production is still limited, and some species, for example *Globba*, are on the list of vulnerable and endangered species because a large number of plants were collected from their native habitats for sale as ornamental plants (Branney 2005).

D. Iridaceae

The genus *Iris* consists of more than 260 species and is distributed mainly in the Northern Hemisphere. Numerous attractive *Iris* species are found in Central to East Asia. All of them are rhizomatous (Tomiyama 2004). While European *Iris* species (i.e., Dutch and German irises) have been developed into important commercial crops (see also Chapter 4), the horticultural use of the Asian *Iris* species is still limited. *Iris ensata* var. *spontanea*, *I. japonica*, *I. laevigata*, and *I. sanguinea* have been used as garden plants in Japan. Breeding of *I. ensata* var. *spontanea* was started in the middle of the nineteenth century in Japan, and about 200 cultivars were described in a gardening book at that time (Nakao 1968).

Belamcanda chinensis (syn. *Iris domestica*), a monotypic genus native to China, Japan, and India, has orange perianths with red spots. These plants have been used as both medicinal and garden plants in Japan and China. Currently, limited production of a few cultivars occurs in Israel and the United States.

E. Other Families

One of the largest and most important groups of ornamental geophytes belongs to the genus *Allium* (Alliaceae). The genus consists of about 750 species and is distributed over the entire Holarctic region, including most of Asia (Fritsch and Friesen 2002). Considerable variations in inflorescence and flower shape, flowering time, growth cycle, and cold hardiness are observed in ornamental *Allium* species (Kamenetsky and Fritsch 2002). Most of them are considered as garden plants and only a limited number of species, *A. giganteum*, *A. caeruleum*, and so on, are used as cut flowers.

The genus *Hemerocallis* (daylily) (Hemerocalidaceae) includes about 10 species distributed in China and Japan. Similar to *Lilium*, *Hemerocallis* has six perianth lobes with yellow to brawny-orange color. There are two types with regard to flowering time, namely evening-open type and daytime-open type, and both of these are one-day flowers. Numerous cultivars with colorful and large flowers, including artificially chromosome-doubled plants, have been bred over the last 60 years, but their genetic background is not clear (Tomkins et al. 2001).

Arisaema (Araceae) consists of more than 100 species, widely distributed throughout Asia, North America, and East Africa. East and Southeast Asia is the diversity center of *Arisaema* with a wide range of variations in terms of biological traits (Gusman and Gusman 2002). Only a few *Arisaema* species are currently being produced on a commercial basis as potted and garden plants.

The genus *Gloriosa* (Colchicaceae) consists of five species distributed in tropical Asia and Africa. *Gloriosa* is a climbing plant with a double-limbed tuber. The tubers, which contain colchicines, have been used as a traditional medicinal plant (Ghosh et al. 2002). *G. superba*, native to India, has been used as both a cut flower and a potted plant in the world flower market.

III. CONTRIBUTION OF ASIAN GEOPHYTES TO THE GLOBAL ORNAMENTAL HORTICULTURE

A. LILY

There is no doubt that the most important contribution of Asian geophytes to global ornamental horticulture is the genus *Lilium*. While Asian lilies were introduced in Europe in the nineteenth century, lily breeding has developed drastically over the last 20 years, using both classical and modern breeding technologies (Van Tuyl and Lim 2003, see also Chapter 6). As a result, the main groups of lily, Asiatic (A), Longiflorum (L), and Oriental (O) hybrids, have been established based on Asian lilies. Production of intersectional hybrids (e.g., LA, LO) has steadily increased and now the Asiatic and Longiflorum cultivars are being replaced by LA or LO hybrids, respectively (Van Tuyl and Arens 2011; Chapter 6). Recently, another hybrid group of Asian lily, Trumpet hybrids, has made significant contributions to the breeding of Oriental lilies, by the introduction of a yellow petal color, early bulb maturation, and resistance to leaf burn. The intersectional hybrids (OT) are also replacing the Oriental cultivars in lily production.

Since modern lily cultivars have a long breeding history, the advantages of wild Asian lily in a breeding program have already been employed, and today the use of the wild species is limited. However, resistance to virus- and soil-borne diseases is still a big issue in lily breeding, and *L. regale* and some other native Asian species are recognized as important genetic resources for these traits.

B. CURCUMA

Curcuma alismatifolia, native to northern Thailand, was introduced to the world flower market in the early 1990s. The basic physiology of flowering of *C. alismatifolia* has been studied intensively (Krasaechai 1993; Azuma and Takano 1994; Hagiladi et al. 1997; Kuehny et al. 2002) and year-round cut flower production has been achieved. Efficient methods for the propagation of *Curcuma* in tissue culture has recently been developed (Wannakrairoj 1997), and successful breeding programs of *Curcuma* have advanced in Thailand (Marcsik and Hoult 2006; Wongpiyasatid et al. 2009). Although numerous interspecific hybrids have been produced in Thailand (Wanngrairot 1997), only a few have been introduced to the world flower market because of a lack of commercial bodies to channel these flowers onto the world market.

C. Gloriosa

Gloriosa plants were introduced into Europe from Asia in the seventeenth century. Commercial production of cut flowers of *Gloriosa* started in the 1970s in The Netherlands (Cortnumme and Wehrenfenning 2006). Micropropagation (Finnie and Van Staden 1989; Custers and Bergervoet 1994) and flowering physiology (Carow 1976; Yamasaki et al. 1998; Ninomiya et al. 2008) of *Gloriosa* have been studied intensively and, as a result, year-round production of cut flowers has been achieved. Outstanding breeding of *Gloriosa* by growers was carried out in Japan. First, 'Misato Red'—a cultivar with a straight strong stem and large flowers with a bright red color—was selected. This cultivar was crossed with a genotype with short stems, resulting in the development of the 'Southern Wind', which is the leading cultivar for cut flower production. Recently, a new cultivar 'Orange Heart' was developed in collaboration between a local Japanese agricultural cooperative and research institute (Ishii et al. 2009). To expand further variation, intergeneric hybrids between *Gloriosa*, *Littonia*, and *Sandersonia* have been produced (Kuwayama et al. 2005; Nakamura et al. 2005).

IV. LOCAL PRODUCTION OF ORNAMENTAL GEOPHYTES

A. China

China is one of the largest flower-producing countries in the world. In 2010, the total cultivation area of the ornamental sector was 917,565 ha with a total production value of US\$13,790 million[*] (Forestry Bureau, China 2011). The production value is large, but 50.4% of the ornamental production is for tree nursery stocks, while cut flowers generate only 11.1% of the production value. Rapid growth in nursery production reflects the large-scale urban development in China during the last decade and the increased demand for planting trees for urban landscaping. It is expected that in the near future urban development and landscaping will increase and that this branch of ornamental horticulture will prevail for both cut flowers and ornamental pot production, including the production of ornamental geophytes.

In 2008, China produced 13,044 million stems of cut flowers with a wholesale value reaching US\$1075 million. In comparison, the wholesale value in 2008 of domestically produced cut flowers was US\$403 million in the United States (USDA 2009) and US\$2005 million in Japan (Japan Flower Promotion Center Foundation 2011). It can, therefore, be seen that China is already one of the world's leading cut flower producers.

The rose is the first and lily is the second most prevalent cut flower in China (Figure 14.1). There were 1557 million stems of cut lily flowers produced on 7484 ha in 2010 (Table 14.1). The main production areas are Kunming in Yunnan Province and Lingyuan in Liao Ning Province. A considerable number of cut lilies are exported to Hong Kong, Singapore and other Asian countries, but market data are not available.

About 1240 million bulbs were produced on 4700 ha in China in 2010 (Forestry Bureau, China 2011). The fact that China imports, from The Netherlands, only about 100 million lily bulbs annually indicates that most of the lily bulbs for cut flower production are produced domestically. About half of the lily bulb production is Oriental hybrids, 24% is Asiatic and 16% is LA hybrids from the Yunnan province (Figure 14.2).

Gladiolus is also an important cut flower in China and the annual turnover is about US\$41 million. This figure has remained relatively stable since 2004 (Table 14.1). Following the recent increase in domestic demand, tulip and hippeastrum production is also expanding (N. Kaishita 2011, pers. comm.).

The increase in the consumption and demand for quality in flowers has activated the breeding of new ornamental crops, including geophytes. Most breeding is concentrated in the public sector,

---

[*] CNY = US\$0.16, 2011.

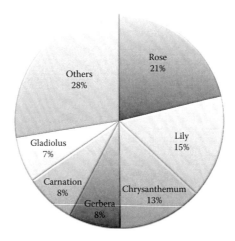

FIGURE 14.1 Production of the main cut flowers in China in 2007. Based on cultivation area. (Adapted from Forestry Bureau, China. 2011. (http://hhxh.forestry.gov.cn/portal/hhxh/s/282/content-490324.html.)

TABLE 14.1
Cut Flower Production of Gladiolus and Lily in China

	Gladiolus			Lily		
Year	Cultivated Area (ha)	No. of Stems (million)	Turnover (million CNY[a])	Cultivated Area (ha)	No. of Stems (million)	Turnover (million CNY)
2002	1738	302	336	—	—	—
2004	2001	484	299	—	—	—
2005	2523	572	313	—	—	—
2006	2140	422	396	4908	720	1568
2008	2386	484	256	5372	975	2010
2010	2891	536	256	7484	1557	3674

Source: Adapted from Forestry Bureau, China. 2011. (http://hhxh.forestry.gov.cn/portal/hhxh/s/282/content-490324.html).
[a] CNY = US$0.16, 2011.

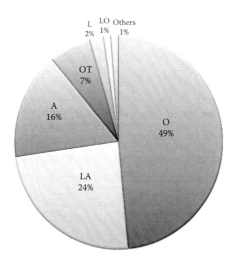

FIGURE 14.2 Bulb production of different types of lily bulbs in Yunnan Province of China in 2007. Based on cultivation area. (From N. Tsuchishita, pers. comm. 2011.)

universities and agricultural research institutes, while private sector breeding is still relatively weak in China (Yu et al. 2009). A wide range of studies, work on propagation, genetic resources evaluation, breeding and postharvest research on lily is being carried out at the China Agricultural University, Beijing Forestry University, Zheijang University, and others. New varieties of lily produced by the Flowers Research Center of the Horticultural Research Institute (Yunnan Academy of Agricultural Science) are being evaluated by the private companies in the Yunnan area.

B. SOUTH KOREA

Flower production in South Korea developed very quickly after 1980. In two decades, the greenhouse area in floriculture increased by 5.8 times, while the production value increased by 4.3 times from 1990 to 2005 (Lee et al. 2007). This great success was powered by the export of flowers to Japan—a development which has been supported and partly subsidized by the Korean government since the 1980s (Lim et al. 2007).

The main geophytes produced as cut flowers are lily, freesia, gladiolus, and calla. Cut lily production increased about 3-fold in cultivated area and 3.6 times in production value in the 1990s (Lim et al. 2007). Propagation material for lily, freesia, and gladiolus is mainly imported from The Netherlands. Domestic bulb production is rather limited, in spite of the fact that the National Horticultural Research Institute in Korea has released many new varieties of those plants (Rhee and Cho 2011). Since 2000, both the production value and area of cut lily have remained rather stable (Figure 14.3). The export of cut lily to Japan started in the early 1990s and increased very rapidly until the early 2000s (Figure 14.4). About 10 million stems of lily were exported to Japan in 2008. Other geophyte cut flowers are produced mainly for the domestic market. Only 0.4 and 0.1 million stems of gladiolus and freesia, respectively, were exported to Japan in 2008.

Research on geophytes, including propagation, flowering control, growing conditions, and postharvest physiology in lily, tulip, and iris, has been carried out actively at Donkook University (Lee et al. 2005; Suh 2011). Recent advances in the cytogenetical analysis of interspecific hybrids in *Lilium*, reported from Kyungpook National University, strongly support world lily breeding programs (Van Tuyl and Arens, 2011). Breeding of *L. ×formolongi* is also carried out at Kangwon National University (Hamid and Kim 2011).

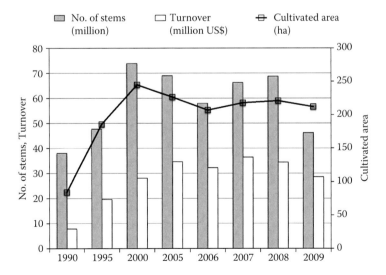

FIGURE 14.3 Cut lily production in South Korea. (Adapted from Rhee, H. K. 2007. *Horticulture in Korea*, ed. Korean Society for Horticultural Science, 276–280. Seoul: Korean Society for Horticultural Science; Ki-Byung Lim, 2011, pers. comm.)

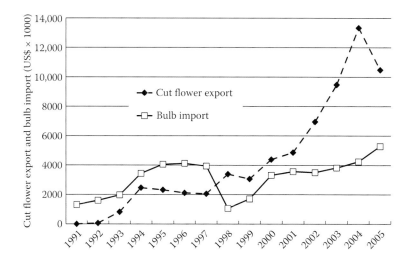

FIGURE 14.4 Cut lily flower export and lily bulb import in South Korea. (Adapted from Rhee, H. K. 2007. *Horticulture in Korea*, ed. Korean Society for Horticultural Science, 276–280. Seoul: Korean Society for Horticultural Science.)

C. VIETNAM

In the twenty-first century, flower production in Vietnam has grown rapidly (Figure 14.5). The Vietnamese government has designated several "Concentrated Production Areas" in floriculture, for example, Ha Noi, Me Linh, Ho Chi Minh, Lam Dong (Da Lat), Hai Phong, and the total area of flower production reached ca. 4500 ha in 2005 (MAFF 2010). The export of cut flowers and the import of bulbs and nursery stocks have also been increasing (Figure 14.6). The main geophytes produced as cut flowers are gladiolus, lily, dahlia, calla, and hippeastrum (Table 14.2). Five major domestic companies provide about 30 million lily bulbs per year (~80–90% of the total number of lily bulbs) to growers in Vietnam. The main production areas of cut lily and gladiolus are Da Lat and Hai Phong, respectively (MAFF 2010).

Research on the propagation and breeding of lily bulb is being carried out at the Dalat Institute of Biology (Nhut et al. 2006) and at the Hanoi University of Agriculture (Thao et al. 2008), but there is little horticulture science activity oriented to practical flower production in Vietnam.

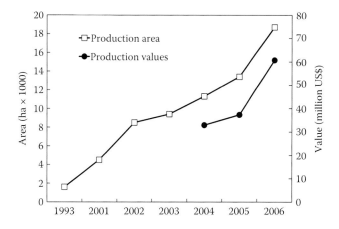

FIGURE 14.5 Production area and value of flower production in Vietnam. (From N. X. Binh, 2011, pers. comm.)

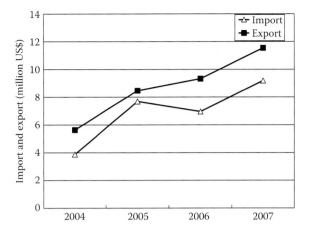

FIGURE 14.6   Flower import and export values in Vietnam. (From N. X. Binh, 2011, pers. comm.)

TABLE 14.2
Production of Cut Geophyte Flowers in Vietnam

Plant	Production Area (ha)			Production Value (×1000 US$)		
	2002	**2003**	**2004**	**2002**	**2003**	**2004**
Gladiolus	402	420	414	3800	4015	4100
Lily	57	63	80	1620	1780	2130
Dahlia	70	63	61	350	210	214
Calla	7	9	11	75	86	80
Hippeastrum	0	12	10	9	10	12

Source: From N. X. Binh, 2011, pers. comm.

D. THAILAND

Most of the temperate ornamental crops grown as cut flowers (i.e., rose, chrysanthemum, carnation, and lily) are produced in northern Thailand. The main geophytes for cut flower production are lotus, curcuma, gladiolus, and tuberose (*Polianthes tuberosa*), and most of these are produced in the low-land area of Thailand for the domestic market. Only *Curcuma* rhizomes, produced in northern Thailand, are exported to The Netherlands, Germany, and Italy in the EU and to Japan, China, and Taiwan in Asia (Figure 14.7). More than 20 million rhizomes are produced for export in Chiangmai, Chaingrai, and Lumpoon on an area of about 64 ha with a production value of about US$1.5 million[*] per year (Department of Agricultural Extension, Ministry of Agriculture, Thailand, Dr. Charelmasri, 2011, pers. comm.).

Research on *Curcuma* breeding and production management was carried out in Chiangmai University. Some of this research was carried out in collaboration with several Japanese universities (Ruamrungsri et al. 2006; Anuntalabhochai et al. 2007; Mahadtanapuk et al. 2009). Taxonomy, breeding, and postharvest studies on *Curcuma* were conducted at Kasetsart University as well as some other research institutions (Bunya-atichart et al. 2004).

[*] Thai Baht = US$0.03, 2011.

(a)
(b)

FIGURE 14.7 Propagation and production of cut flowers of *Curcuma* in Chiangmai, Thailand. (a) Cut flower production. (b) Propagation field. To avoid damage of the propagation units during harvest, the plants are grown in plastic bags with a sand and rice pod planting mixture.

E. TAIWAN

Lily as cut flowers dominates in production value among the ornamental geophytes in Taiwan. The production area of cut lily was about 340 ha and production value about US$40 million in 2009 (Council of Agriculture, Taiwan). Most of the flowers are allocated for the domestic market, but some are also exported to Hong Kong, Japan, and Singapore. Gladiolus, dahlia, and calla also belong to the popular flowers in the domestic market. Two to three million stems of gladiolus are exported to Japan annually. Although the volume of exports has recently leveled off, gladiolus remains an important export item (Council of Agriculture, Taiwan).

Breeding programs of tuberose at the National Chiayi University, including interspecific hybridization, have resulted in 19 single- or double-flowered cultivars with pink, violet, yellow, red, and purple colors and have been registered by 2009 (Huang et al. 2001). Dwarf types for potted plants have also been developed (Shen et al. 1997).

F. OTHER COUNTRIES

In many tropical and subtropical areas in Asia, gladiolus, tuberose, and lotus are produced as geophyte cut flowers, but this production is mainly aimed at the domestic markets. India has become an important flower production area in Asia, with gladiolus and tuberose as the major products of ornamental geophytes (Roy 2008). Specifically, gladiolus is the second most important cut flower crop after the rose in India (Export-Import Bank of India 2009). The country has started to produce some cut lily as an export item. Gladiolus and tuberose are also important cut flowers in Bangladesh (H. Okubo 2011, pers. comm.).

V. CONCLUSIONS

The Asian region, especially East and Southeast Asia, has a great potential for both flower production and consumption. The increasing population and the recent explosive economic growth in the region are facilitating the expansion of the flower business. It is thought that the promotion of urbanization and the improvement of the standard of living occurring in various parts of Asia will bring about an increase in flower consumption. International trade in cut flowers and propagation materials will also increase since the market is growing not only within Asia but also in neighboring countries: Russia and the former Soviet republics.

Further developments and cooperation in research on horticulture are expected in Asia. Development of growing techniques suitable for each area is required, as the expansion of areas of production and the increase in consumer demand has required quality products and longer postharvest life. The enlarged areas of production has led to the need for advanced flowering quality control based on flowering physiology, integrative pest management, environment control, and so on. Since ornamental geophytes include various genera plants, and Asia consists of various climate conditions, the development of new production techniques needs to be adapted for each crop and in each area. Advanced postharvest technology for cut flowers becomes more important as new production areas find themselves farther from consumption centers. Most cut flowers of ornamental geophytes are non-ethylene-sensitive and, so far, no effective technique exists for postponing their senescence. Since flower markets are always seeking new and improved crops, both the breeding of major popular crops for local conditions and the exploration and introduction of new crops are required. There are large numbers of unused geophytes in Asia. Development research for these species, especially Zingiberaceae, would certainly create an impact on the global flower market.

To achieve healthy and orderly growth of the market, protection of intellectual property rights is essential. Fortunately, several Asian countries have become members of the International Union for the Protection of New Varieties of Plants (UPOV), but most countries joined this organization only recently, and the establishment of registration systems working according to international rules will take time. The Chinese law of breeder's rights covers 24 genera or species, which include only a small proportion of flower resources (Yu et al. 2009). About 43% of the applications for authorization for plant breeder's rights was submitted by foreign private companies. This is because the economic benefits of cultivar protection do not cover the costs of protection due to the relatively small scale of production and consumption of the variety (Yu et al. 2009).

Research on geophytes has been developed in several Asian countries. Asian researchers have become active participants in the international scientific community through their participation in international symposia such as that on the genus *Lilium* in Korea in 1994 and the International Symposium on Flower Bulbs in Japan in 2004. In order to advance further collaboration and the exchange of knowledge in horticultural science, the International Horticultural Congress was held in South Korea in 2006 in collaboration with the Korean, Japanese, and Chinese Horticultural Societies. The establishment of an effective research network between Asian countries and the global horticultural community will certainly provide a larger base for the effective and profitable production of ornamental geophytes in the region.

ACKNOWLEDGMENTS

The author owes thanks to N. Tsuchishita, N. Kaishita, Chrysal, Japan; Dr. C. Nontaswatsri, Maejo University in Thailand; Dr. Ngo Xuan Binh, Thai Nguyen University in Vietnam, and Dr. Ting-Fang Hsieh, Floriculture Research Center in Taiwan for providing production data of Asian countries.

CHAPTER CONTRIBUTOR

Seiichi Fukai is a professor of floriculture at Kagawa University, Japan. He was born in Kanazawa, Japan, and obtained BS from Okayama University and MS and PhD from Osaka Prefectural University. He has held positions at the Osaka Prefectural Agricultural Research Center and at the Faculty of Agriculture, Kagawa University. The major topics of his research include flower bulb physiology, production, postharvest physiology, and breeding. Dr. Fukai has published more than 120 papers on various floricultural crops. He is a recipient of the Japanese Society for Horticultural Science Outstanding Achievement Award (2001).

REFERENCES

Anuntalabhochai, S., S. Sitthiphrom, and W. Thongtaksin. 2007. Hybrid detection and characterization of *Curcuma* spp. Using sequence characterized DNA markers. *Scientia Hortic.* 111:389–393.

Azuma, A., and K. Takano. 1994. Studies on the flowering control of *Curcuma alismatifolia*. II. On rest of corm, and its breaking of rest. *Bull. Kochi Agric. Res. Cent.* 3:37–45 (In Japanese).

Branney, T. M. E. 2005. *Hardy gingers*. Portland, OR: Timber Press.

Bunya-atichart, K., S. Ketsa, and W. G. van Doorn. 2004. Postharvest physiology of *Curcuma alismatifoloa* flowers. *Postharvest Biol. Technol.* 34:219–226.

Carow, B. 1976. *Gloriosa rothschildiana*: flower and tuber production. *Acta Hort.* 64:181–186.

Comber, H. F. 1949. A new classification of the genus *Lilium*. *R.H.S. Lily Year Book* 13:86–105.

Cortnumme, J., and M. Wehrenfenning. 2006. Effects of storage and greenhouse treatments on growth and flowering of Asian *Gloriosa*. *Eur. J. Hort. Sci.* 71:73–77.

Council of Agriculture, Taiwan (http://eng.coa.gov.tw/list.php?catid=8821)

Custers, J. B. M., and J. H. W. Bergervoet. 1994. Micropropagation of *Gloriosa*: Towards a practical protocol. *Scientia Hortic.* 57:323–334.

Export-Import Bank of India. 2009. *Floriculture—A Sector Study*. Mumbai, India: Export-Import Bank of India.

Finnie, J. F., and J. van Staden. 1989. In vitro propagation of *Sandersonia* and *Gloriosa*. *Plant Cell, Tiss. Org. Cult.* 19:151–158.

Forestry Bureau, China. 2011. (http://hhxh.forestry.gov.cn/portal/hhxh/s/282/content-490324.html) (In Chinese).

Fritsch, R. M., and N. Friesen. 2002. Evolution, domestication, and taxonomy. In *Allium Crop Sciencecs: Recent Advances*, eds. H. D. Rabinowitch, and L. Currah, 5–30. Wallingford, UK: CABI Publ.

Fukai, S., and W. Udomdee. 2005. Inflorescence and flower initiation and development in *Curcuma alismatifolia* Gagnep (Zingiberaceae). *Japan. J. Trop. Agric.* 49:14–20.

Ghosh, B., S. Mukherjee, T. B. Jha, and S. Jha. 2002. Enhanced colchicine production in root culture of *Gloriosa superba* by direct and indirect precursors of the biosynthetic pathway. *Biotechnol. Lett.* 24:231–234.

Gusman, G., and L. Gusman. 2002. The genus *Arisaema*. Ruggell, Lichtenstein: A. R. Gantner Verlag.

Hagiladi, A., N. Umiel, and X.-H. Yang. 1997. *Curcuma alismatifolia*. II. Effects of temperature and daylength on the development of flowers and propagules. *Acta Hort.* 430:755–761.

Hamid, B., and J. H. Kim. 2011. Cross compatibility between *Lilium × formolongi* group and *Lilium brownii*. *Afr. J. Agric. Res.* 6:968–977.

Hayashi, K., and S. Kawano. 2005. Bulbous monocots native to Japan and adjacent areas—Their habitats, life histories and phylogeny. *Acta Hort.* 673:43–58.

Huang, K.-L., I. Miyajima, H. Okubo, T.-M. Shen, and T.-S. Huang. 2001. Flower colours and pigments in hybrid tuberose (*Polianthes*). *Scientia Hortic.* 88:235–241.

IMF. 2011. Asia remains world's most dynamic region, although pockets of overheating pose risks. *Press Release No. 11/150 (April 28, 2011)*.

Ishii, Y., C. Ninomiya, and M. Matsumoto. 2009. 'Kouiku No.1', a novel cultivar of glory lily (*Gloriosa superba* L.) with reddish-orange flowers. *Bull. Kochi Agric. Res. Cent.* 18:21–24 (In Japanese).

Japan Flower Promotion Center Foundation. 2011. *Flower Data Book 2009–2010*. Tokyo: Japan Flower Promotion Center Foundation.

Kamenetsky, R., and R. M. Fritsch. 2002. Ornamental *Alliums*. In *Allium Crop Science: Recent Advances*, eds. H. D. Rabinowitch, and L. Currah, 459–491. New York: CABI Publ.

Komatsu, K., Y. Sasaki, C. Tohoda, and K. Tanaka. 2007. Origin and quality of curcuma drugs. *FFI Journal*. 212:345–356 (In Japanese).

Krasaechai, A. 1993. Day length effect on the flowering of *Curcuma sparganifolia*. *Proc. 31st Kasetsart Univ. An. Con.: Plant*, 32–36 (In Thai).

Kuehny, J. S., M. J. Sarmiento, and P. C. Branch. 2002. Cultural studies in ornamental ginger. In *Trends in New Crops and New Uses*, eds. J. Janick, and E. Whipkey, 477–482. Alexandria: ASHA Press.

Kurita, S., and P. S. Hsu. 1998. Cytological patterns in the Sino-Japanese flora; hybrid complexes in *Lycoris*, Amaryllidaceae. *Univ. Mus., Univ. Tokyo, Bull.* 37:171–180.

Kurita, S. 2011. Cultivars or artificial hybrids. http://www5e.biglobe.ne.jp/~lycoris/taxonomy-cultivar-3.html. Accessed October 2011.

Kuwayama, S., Y. Mizuta, M. Nakano, and T. Nakamura. 2005. Cross-compatibility in interspecific and intergeneric hybridization among the Colchicaceous ornamentals, *Gloriosa* spp. *Littonia modesta* and *Sandersonia aurentica*. *Acta Hort.* 673:421–427.

Lee, J.-S., J.-K Suh, K.-W. Kim, D.-H. Goo, S.-K. Chung, and K.-S. Kim. 2007. Introduction. In *Horticulture in Korea*, ed. Korean Society for Horticultural Science, 257–265. Seoul: Korean Society for Horticultural Science.

Lee, W.-H., J.-H. Kim, A.-K. Lee, J.-K. Suh, and Y.-Y. Yang. 2005. Effects of nutrient solution management and methods of storage and distribution on flowering and quality of cut iris, tulip and lily. *Acta Hort.* 673:513–518.

Lim, K.-B., Y. J. Kim, H. R. Cho, and J. K. Suh. 2007. Flower bulbs. In *Horticulture in Korea*, ed. Korean Society for Horticultural Science, 292–296. Seoul: Korean Society for Horticultural Science.

MAFF. 2010. Flower production in Vietnam. In *Vietnam, Ministry of Agriculture, Forestry and Fisheries (MAFF), Japan.* http://www.maff.go.jp/j/export/h19_zigyou/enkatu/market/vietnam/ (In Japanese).

Mahadtanapuk, S., M. Sanguansermsri, T. Handa, W. Nanakorn, and S. Anuntalahochai. 2009. Clonign of the ACC synthase gene from *Curcuma alismatifolia* Gagnep. and its use in transformation studies. *Acta Hort.* 836:277–282.

Marcsik, D., and M. Hoult. 2006. Development of *Curcuma* hybrids: An evaluation of flowering characteristics and floral features for ornamental use. *Abst. 4th International Symposium on the Family Zingiberaceae*, 59.

McRae, E. A. 1998. *Lilies: A Guide for Growers and Collectors.* Portland, OR: Timber Press.

Nakao, S. 1968. *Hana to ki no bunnkasi (Natural History of Flowers and Trees).* Tokyo: Iwanami Shoten Publ. (In Japanese).

Nakamura, T., S. Kuwayama, S. Tanaka, T. Oomiya, H. Saito, and M. Nakano. 2005. Production of intergeneric hybrid plants between *Sandersonia aurantica* and *Gloriosa rothschldiana* via ovule culture (Colchicaceae). *Euphytica* 142:283–289.

Nhut, D. T., N. T. M. Hanh, P. Q. Tuan, L. T. M. Nguyet, N. T. H. Tram, N. C. Chinh, N. H. Nguyen, and D. N. Vinh. 2006. Liquid culture as a positive condition to induce and enhance quality and quantity of somatic embryogenesis of *Lilium longiflorum. Scientia Hortic.*110:93–97.

Ninomiya, C., T. Nishiuchi, M. Hiraishi, and S. Fukai. 2008. Effects of soil temperature and temperature to enhance sprouting on node number of flower bud differentiation in *Gloriosa superba* L. *Hort. Res. (Japan)* 7:571–577 (In Japanese).

Pavord, A. 2009. *Bulb.* London: Michell Beazley.

Rhee, H. K. 2007. Lily. In *Horticulture in Korea*, ed. Korean Society for Horticultural Science, 276–280. Seoul: Korean Society for Horticultural Science.

Rhee, H. K., and H. R. Cho. 2011. Lily breeding activities at RDA, Korea. *Acta Hort.* 900:231–236.

Rønsted, N., S. Law, H. Thornton, M. F. Fay, and M. W. Chase. 2005. Molecular phylogenetic evidence for the monophyly of *Fritillaria* and *Lilium* (Liliaceae; Liliales) and the infrageneric classification of *Fritillaria. Mol. Phylog. Evol.* 35:509–527.

Roy, R. K. 2008. Floricultural boom in India. *Chronica Hort.* 48(2):14–19.

Ruamrungsri, S., N. Ohtake, K. Sueyoshi, and T. Ohyama. 2006. Determination of the uptake and utilization of nitrogen in *Curcuma alismatifolia* Gagnep. Using ^{15}N isotope. *Soil Sci. Plant Nutr.* 52:221–225.

Shen, T. M., T. S. Huang, K. L. Huang, R. S. Shen, and B. S. Du. 1997. Breeding for new flower color in *Polianthes tuberose. J. Chinese Soc. Hort. Sci.* 43:358–367.

Sirisawad, T., P. Sirirugsa, C. Suwanthada, and P. Apavatjrut. 2003. Investigation of chromosome numbers in 20 taxa of *Curcuma. Proc. 3rd Symposium on the Family Zingiberaceae,* Khon Ken, Thailand, 54–62.

Suh, J. K. 2011 Systematic strategies of lily bulb productions and forcing in Korea. *Acta Hort.* 900:71–78.

Synge, P. M. 1980. *Lilies.* London: Batsford Ltd.

Thao, N. T. P., N. T. Thuy, and N. Q. Thach. 2008. Developing an *Agrobacterium*-mediated transformation system for *Lilium × formolongi* using thin cell layer of bulb scales. *J. Sci. Devel.* 2008:123–128.

Tomiyama, M. 2004. *Wild Flowers of the World II.* Asia, Oceania, North and South America. Tokyo. Gakken (In Japanese).

Tomkins, J. P., T. C. Wood, L. S. Barnes, A. Westman, and R. A. Wing. 2001. Evaluation of genetic variation in the daylily (*Hemerocallis* spp.) using AFLP markers. *Theor. Appl. Genet.* 102:489–496.

USDA. *Floriculture Crops 2010 Summary 2009,* http://usda01.library.cornell.edu/usda/nass/FlorCrop//2000s/2009/FlorCrop-04–23–2009.pdf.

Van Tuyl, J. M., and P. Arens. 2011. *Lilium*: Breeding history of the modern cultivar assortment. *Acta Hort.* 900:223–230.

Van Tuyl, J. M., and K.-B. Lim. 2003. Interspecific hybridisation and polyploidisation as tools in ornamental plant breeding. *Acta Hort.* 612:13–22.

Wannakrairoj, S. 1997. Clonal micropropagation of patumma (*Curcuma alsimatifolia* Gagnep). *Kasetsart J. (Nat. Sci.)* 31:353–356.

Wanngrairot, S. 1997. *Curcuma, 2nd ed.* Bangkok, Thailand: Amarin Printing and Publ. (In Thai).

Wongpiyasatid, A., P. Jompuk, N. Topoonyanoon, T. Taychasinpitak, and T. Teerakathiti. 2009. Improvement of *Curcuma* hybrids by mutation induction. *Abst. 8th National Horticultural Congress*, 68.

Yamasaki, I., S. Ito, and M. Imagawa. 1998. Effect of germination temperature on growth and flowering of *Gloriosa rothschildiana* O'Brien cv. 'Rose Queen'. *Res. Bull. Aichi Agric. Res. Ctr.* 30:215–219 (In Japanese).

Yu, H., Y. Feng, and Q. Liu. 2009. Biodiversity and ornamental plant breeding in China. *Acta Hort.* 836:31–37.

Yuan, L., Q. Liu, and Q. Liu. 2011. Conservation, evaluation and enhancement of wild lily germplasm in China. *Acta Hort.* 900:53–58.

Yunnan Agricultural Science Academy (http://www.yaas.org.cn/article/ShowArticle.asp?ArticleID = 13166)

Zhao, L., and D. Zhang. 2003. Ornamental plant resources from China. *Acta Hort.* 620:365–375.

15 Geophyte Research and Production in Brazil

Antonio F. C. Tombolato, Roberta P. Uzzo,
Antonio H. Junqueira, Márcia da S. Peetz, and
Giulio C. Stancato

CONTENTS

I. INTRODUCTION

Brazil is the largest country in South America, with a geographical area of 8.5 million km² and a population of 192 million people. Including its Atlantic islands, Brazil is located between latitudes 6N and 34S and longitudes 28W and 74W. Much of the terrain lies between 200 and 800 m above sea level. The climate of Brazil is characterized by a wide range of weather conditions across a large area and varied topography, but most of the country is tropical. The different climatic conditions produce environments ranging from equatorial rainforests in the north and semiarid deserts in the northeast, to temperate coniferous forests in the south and tropical savannas in central Brazil. The country is also home to diverse wildlife, natural environments, and extensive natural resources in a variety of protected habitats.

Horticulture in Brazil is developing fast; this includes the production of not only vegetables but also fruits and flowers. The market for flowers and ornamental plants in Brazil expands annually by 8–10%. In 2010, this market generated a total value of US$ 2.2 billion (Junqueira and Peetz 2011).

The significant growth in domestic consumption of plants and flowers is caused by socioeconomic factors, the increasing rate of employment and income of the population, social mobility of major portions of the lower-income groups, inflation control, and economic stability. In addition, technological improvements, improved logistics, and expanded commercial activity in the country have resulted in the production of better and cheaper horticultural products (Junqueira and Peetz 2010a,b).

The local consumption of flowers and ornamental plants is focused on annual holidays (e.g., Mother's Day, Lover's Day (Dia dos Namorados), International Women's Day, and Christmas Eve)

FIGURE 15.1 (**See color insert.**) Veiling, the Holambra's flower auction. View of the chariots. (Cooperative Veiling Holambra: Press Photos.)

(Junqueira and Peetz 2010a), while the traditional varieties and species (roses, chrysanthemums, lilies, violets, and orchids) are enshrined in popular taste. About 90% of the standardized wholesale market of flowers and ornamental plants in Brazil is concentrated in São Paulo (based on the activities of the Society of General Warehouses of São Paulo—CEAGESP), Holambra (Cooperative Veiling Holambra and Cooperflora/Floranet, Figures 15.1 and 15.2); Campinas (Permanent Market Flowers and Ornamental Plants—CEASA Campinas) and in Mogi das Cruzes (Cooperative SP Flores) (Junqueira and Peetz 2008). From these markets, flowers and ornamental plants are distributed throughout the country, mobilizing a large workforce and a complex logistics network.

Ornamental geophytes are produced for cut flower consumption, as potted plants, and for landscaping and gardening. Species and hybrids of alstroemeria, hippeastrum, calla lily, gladiolus, and lily are among the major cut flowers and pot plants grown and marketed in the domestic market of Brazil. In 2008–2010, the production of the main ornamental geophytes increased by about 25% (Table 15.1).

In addition to significant local consumption, over the last few decades, Brazil has been prominent in the international flower market. In this respect, it imports varieties originating from leading producers abroad, especially from The Netherlands. With the recent increase in its domestic market, Brazilian import of bulbs increased from US$ 4.7 million in 2008 to US$ 8.3 million in 2009, retreating slightly to US$ 7.3 million in 2010 (Junqueira and Peetz 2011). Imported propagating material is

FIGURE 15.2 (**See color insert.**) Veiling, the Holambra's flower auction. View of the sales clocks. (Cooperative Veiling Holambra: Press Photos.)

TABLE 15.1
Trade Quantities for the Main Ornamental Bulbs Grown as Cut Flower (2008–2010)

	Alstroemeria (Package)			Calla (Dozen)			Gladiolus (Package)			Lily (Package)		
	2008	2009	2010	2008	2009	2010	2008	2009	2010	2008	2009	2010
January	58,919	101,003	106,478	5913	6540	4090	30,006	28,517	24,220	19,310	15,957	22,883
February	46,751	71,266	92,260	4747	4606	2179	30,053	23,001	28,230	17,405	15,630	13,552
March	60,253	88,968	114,724	7892	3246	2726	30,721	20,812	34,615	24,248	20,802	17,310
April	51,978	60,342	99,585	8876	4962	5052	27,880	22,946	31,417	23,855	18,380	16,803
May	55,752	60,680	111,990	13,501	6837	8351	28,925	22,277	41,785	22,478	17,819	19,714
June	58,648	72,441	121,223	17,547	10,000	9547	28,401	22,729	34,685	22,407	17,789	20,581
July	71,720	103,253	142,006	21,990	20,233	21,233	23,307	19,169	32,856	18,834	17,469	19,071
August	83,122	113,292	185,561	29,004	29,673	28,694	25,182	20,941	42,965	19,371	18,754	22,475
September	104,491	134,210	212,150	34,169	36,558	36,883	23,902	24,540	33,783	24,009	25,560	24,011
October	109,467	145,102	230,585	38,809	37,197	38,661	38,304	36,470	49,407	29,919	29,555	27,646
November	105,010	147,453	219,877	21,875	24,644	31,073	26,599	25,168	38,292	27,459	27,226	29,757
December	110,461	130,382	236,857	13,890	10,073	17,183	30,495	25,085	40,556	24,167	20,811	28,859
Total	916,571	1,228,394	1,873,297	218,215	194,571	205,672	343,774	291,653	432,811	273,462	245,752	262,661

Source: Hórtica Consultoria e Treinamento, data from CEASA—Campinas, CEAGESP, and Veiling Holambra wholesale markets.

used for the production of potted plants and cut flowers for the domestic market, as well as for bulb production for the international companies in The Netherlands, Germany, and the United States.

The comparative advantages of Brazil, such as early availability of crop due to specific ecological conditions, and competitive costs, enable the system to produce and export propagation material of gladiolus, caladium, and hippeastrum. From 1999 to 2005, Brazilian export of ornamental geophytes ranged from US$ 4–6 million annually, but recently, sales of flower bulbs to the international market has increased to US$ 13–14 million annually (Junqueira and Peetz 2010a, 2011).

II. RESEARCH AND ORGANIZATION NETWORKS

The Brazilian Society of Floriculture and Ornamental Plants (SBFPO, www.sbfpo.com.br), created in 1979, was practically the first Brazilian association dedicated to flowers. At present, it focuses mainly on academic activities, attracting a membership of researchers, teachers, and students. Every 2 years, SBFPO holds the Brazilian Congress on Floriculture and Ornamental Plants and also supports and organizes other technical and scientific events nationally and internationally. The society has produced numerous publications: annals, books, manuals, and abstracts of congresses, newsletters, books of symposia lectures, and papers presented at meetings and symposia. The *Brazilian Journal of Ornamental Horticulture* (www.sbfpo.com.br/rbho) publishes in Portuguese, Spanish, and English. It is the only technical and scientific journal in South America aimed specifically at the dissemination of research in the area of floriculture and landscape design.

The Brazilian Institute of Floriculture (Ibraflor) was established in 1994 as a forum for discussions with the goal of developing the floriculture sector. Ibraflor united various producer associations, the private sector, and members of the State's Sector Chambers. The Sector Chambers belong to the Ministry of Agriculture, Livestock and Supply, and are formed by working groups which include the invited representatives of various sectors of the supply chain. The floriculture sector has its own chamber, in which SBFPO, Ibraflor, and the Agronomic Institute (IAC, São Paulo state) participate in the discussions.

Since 1982, IAC has developed a breeding program based on the exploitation of the genetic diversity of native *Hippeastrum* species for the creation of new cultivars adapted to Brazilian conditions. The interspecific crosses in *Hippeastrum* have produced a plethora of hybrids with remarkable agronomic and commercial traits. As a result of a partnership between the IAC and the company André Boersen, three hybrids: 'IAC Neblina' (Figure 15.3), 'IAC Jaraguá', and 'IAC Itatiaia' have

FIGURE 15.3   (**See color insert.**) *Hippeastrum* 'IAC Neblina'. (Photo of A. F. C. Tombolato.)

been registered in the SNPC (National Plant Varieties Protection) for RNC (National Plant Register System). The IAC cultivars have relatively small flowers and long stems with orange-colored flowers. In the 1990s, IAC collected native species of alstroemeria to achieve interspecific crosses aimed at obtaining new cultivars adapted to tropical and subtropical climate. In addition to the breeding program, a botanical study of this genus in Brazil was carried out, which resulted in the identification of 14 new species of *Alstroemeria* (Assis 2001). Physiological studies were dedicated to the effect of winter soil temperature and summer rainfalls on flower development (Healy and Wilkins 1982; Hurka 1991).

The research groups of the Institute Biologic in São Paulo collaborate with IAC in ornamental research. However, at present, no research on geophytes is being carried out at this institute. As a service to growers, the Institute Biologic identifies viruses and diseases in ornamental geophytes.

Embrapa—Brazilian Agricultural Research Corporation—is the national governmental enterprise for agricultural research that has recently started to work on ornamental plants. The research is mainly focused on the selection of potential ornamental plants from traditional agriculture, such as rice, sunflower, pineapple, and peach tree. The scientific activity on geophytes is dedicated mainly to the rescue of native plants in endangered areas.

The Brazilian Law of Plant Variety Protection was passed in 1997, and in 1999 Brazil joined the UPOV* Convention Act of 1978. Consequently, the Ministry of Agriculture, Livestock and Supply created the National Register of Plant Varieties (RNC), in which all varieties of agricultural value must be registered in order to be marketed, and the National Plant Variety Protection (SNPC), which controls the intellectual property protection of registered plant varieties. The Brazilian Association of Protection of Plant Varieties of Flowers and Ornamental Plants (ABPCflor) was created in 1996, assembling the main national and foreign breeding companies and the IAC. The aim of ABPCflor is to accelerate the process of regulating the protection of cultivars of ornamental species.

III. MAIN CROPS OF ORNAMENTAL GEOPHYTES PRODUCED IN BRAZIL

A. *LILIUM*

Lily (*Lilium*) is the eight most popular potted plant in Brazil (Junqueira and Peetz, 2008). Within the 20–22 million bulbs planted annually in Brazil, a 60% share goes to the domestic cut flower market and the remaining 40% to the production of potted flowers. Both potted and cut flower lilies are produced mainly from the imported bulbs of various Dutch companies, which are shipped to Brazil in air-conditioned containers at a temperature of −1°C to −1.5°C.

Cut flowers are produced mainly in the mountains and plateaus of the states of São Paulo, Minas Gerais, Rio de Janeiro, and Rio Grande do Sul, at an altitude of 600–1300 m a.s.l., either in open fields or under a black or reflecting screen of 40% shading. The main production is from September to December, with the major marketing for Mother's Day, Valentine's Day, Secretary's Day, International Women's Day, Christmas, and New Year's Eve.

For cut flower production, the bulbs are planted in plastic crates and placed in storage rooms at temperatures of 12–17°C for the induction of rooting. Following rooting and sprouting, the crates are moved to a screen house and placed on bricks, but not directly on the ground. This growth technique provides better ventilation and maintenance of the plant at a lower temperature, and helps control pests and soil diseases (P. Boersen 2011, pers. comm.). Some growers prefer to produce lily cut flowers directly in the ground.

Lily cultivars produced in Brazil for cut flowers belong to six main types: Oriental, Asiatic, Longiflorum, Longiflorum × Asiatic (LA), Longiflorum × Oriental (LO), and Oriental × Trumpet (OT) hybrids (Cooperative Veiling Holambra 2010d). The most popular lily cultivars belong to the Oriental type (68%) and Asiatic type (29%). Most of the cultivars of Oriental lilies are white, while

---

* International Union for the Protection of New Varieties of Plants.

for the Asiatic lilies, market research has demonstrated a strong preference for orange flowers, followed by yellow and white. Longiflorum hybrids are also popular, for example, 'Snow Queen', 'White Elegance', 'White Europe', and 'White Heaven'. In the LA group, yellow 'Aladdin's Dazzle', cream 'Royal Fantasy', salmon 'Donau', and 'Salmon Classic' are grown in restricted numbers (L.O. Cavicchio, Terra Viva Bulbos 2011, pers. comm.).

The crop cycle for the cut flowers Asiatic lilies varies from 90 days in summer to 105 days in winter and for the Oriental lilies from 120 days in summer up to 135 days in winter. After harvesting, the cut stems are maintained at low temperatures at the producer's farm. According to quality parameters, cut flowers are classified into six classes, defined by the length of the stems (between 40 and 90 cm) and by the number of viable buds per stem, not including immature (green) buds, which will not develop into open flowers (Cooperative Veiling Holambra 2010d). In addition, cut flowers of lilies are classified into categories A1, A2, or B, based on the presence of serious defects, such as damage by disease, especially by *Botrytis*, damage by pests (slugs, aphids, caterpillars), mechanical damage, yellow leaves, and buds with malformations. Minor defects include burns on the leaves, chemical residues, soft stems, and inadequate cleaning of the bases, which may result in microbiological contamination.

After grading, the stems are arranged in packets of 10 units, wrapped in plastic and stored in cold rooms. They are packed in cardboard boxes, 50 stems per carton. In the case where the stems possess only one flower bud, they are sold in packets of 10 stems, while the flower buds are aligned in the same direction as the central stem of the plant, and the flower color should be visible. The only exception allowed is for the cultivars of yellow lilies. For the stems with several flower buds, the packets consist of five stems. The presence of open flowers is considered excessive, and the product batch is progressively less valuable, classified into categories from A1 (10% of open flowers) to B (maximum 30%). When the stems have more than 30% of open flowers, the product is no longer marketed. Prior to marketing, the producer has to declare the stage of the maturation of the plants, based on the degree of opening of flower buds, which varies between four different stages from "closed" to "mature".

The constant expansion of the Brazilian market for potted lilies has turned it into a major wholesale market with over 4 million units per year. Production is concentrated in the São Paulo region and is greatly expanded in May, focusing on marketing for Mother's Day, and in December for Christmas and New Year decorations.

In addition to the six main genetic types classified for cut flowers, potted lilies also belong to the "Rascal" group of cultivars, with shorter and firmer stems and smaller flowers. The main cultivars of this group are 'Angelique', 'Little Rainbow', 'Souvenir', and 'Diamonds'. 'Early Yellow', and 'Tarragona' from the OT group are also popular for pot production (Cooperative Veiling Holambra 2010e).

For the production of flowered lily pots, the planted bulbs are kept in a cool chamber for 2 weeks at 9–17°C for rooting induction. The cultivation is done in plastic tunnels, protected with a thermal-reflective shade net, used mainly during the warmest time of the day. According to the cultivar, the pots are ready for sale in about 8–12 weeks (L.O. Cavicchio, Terra Viva Bulbos 2011, pers. comm.). It is recommended that the height variation between plants in the pot not exceed 10 cm. Stems should be firm and self-sustained or to have individual stakes. For smaller pots with diameters of 10–14 cm, one stem for marketing is required, while for pots with diameters of 15–19 cm, three flowering stems are required. Similar to cut flowers, potted lilies are classified into categories A1, A2, and B, according to the presence of defects.

During the summer production, *Phytophthora* and *Pythium* are the major diseases, and aphid is the major pest in lily plants. The viruses detected in Brazil are *Cucumber mosaic virus* (CMV), which causes the appearance of chlorotic spots and leaf distortion, *Bean yellow mosaic virus* (BYMV), which produces symptoms of chlorotic mosaic, and *Tulip breaking virus* (TBV), which induces the appearance of necrotic lesions and streaks, mottled leaves, and lacerated petals (Alexandre et al. 2005, 2008). Recently, *Lily symptomless virus* (LSLV) was also found (Rivas 2010).

B. GLADIOLUS

The introduction of gladiolus (*Gladiolus* × *grandiflorus*) in Brazil is intertwined with the history of Dutch immigration and with the history of Holambra, a small town in the state of São Paulo, considered to be the center of the national floriculture business (ABCSEM 2010). The couple Klaas and Gemma Schoenmaker left The Netherlands in the late 1950s to immigrate to Brazil along with their 11 children. Here, they originally cultivated corn, cotton, and coffee and later devoted themselves to the cultivation of gladiolus (ABCSEM 2010; Terra Viva Bulbos n/d b). The Dutch corms of gladiolus brought by Klaas Schoenmaker won clients and consumers in Brazil, due to their long stems and varied colors. The consequent expansion of the business resulted in the formation and consolidation of Terra Viva Bulbos, currently the leading producer of gladiolus in Brazil (ABCSEM 2010).

Concurrently, the Japanese community, already established in regions close to the city of São Paulo, joined in the production of gladiolus. There are reports from the 1950s that 1000 dozens of gladiolus were sold in the São Paulo wholesale market in 30 min. During this period, the flowers were quoted at high prices and became desired by the highest-income sector of the society (Tsuboi and Tsurushima 2009).

Later, the gladiolus cut flower was gradually replaced by other ornamentals, especially with the introduction of new cultivars of lily, calla, and gerbera. The Brazilian market for gladiolus decreased over several decades, but has grown again recently following the release of new cultivars with strong and unusual colors and initiatives focused in promoting floral decoration all over Brazil.

Cultivation of gladiolus in Brazil is concentrated on 50 ha in the plateau of the north-eastern region in São Paulo state, at about 700 m a.s.l., under pivot irrigation of 100% of the planting area, as well as in some areas of the states of Rio de Janeiro, Pernambuco, Bahia, and Rio Grande do Sul. The propagation material is produced in the field. Micropropagation techniques have not yet been employed. The company Terra Viva Bulbos produces 60% (18 million) of corms for the domestic market. Another 12 million corms, mostly of white-flowered cultivars, are produced for export to The Netherlands. From The Netherlands, these corms are also sold to Portugal, Spain, and Italy.

The Brazilian consumer shows a clear preference for white flowers, with currently 40% of the market share. These are particularly used at funerals and other religious ceremonies, decorations at weddings and on New Year's Eve. There is also the demand for red flowers (25%), especially at Christmas, and finally, the other colors—yellow (12%), pink (10%), coral/cream (10%), and purple (3%)—especially for consumption on the Day of the Dead (Terra Viva Bulbos n/d b).

During corm production, crop rotation is employed in order to prevent pests and soil diseases. This rotation has occurred over the last 30 years, even though land leasing has suffered strong competition from agricultural production, especially potatoes (Tombolato 2004). In Brazil, the crop cycle is shorter than in Europe, with 65 days for the earliest cultivars. After harvest, cleaning of the corms is manual, while sorting is done mechanically.

For cut flower production, 2-year-old corms are used. In total, 200–600 thousand corms/ha are used for planting, depending on the season and variety. Planting is done in rows and in beds. The optimal season for planting is from July to September aiming to harvest in October and November (Tombolato 2004). Following harvesting, flowering stems of each cultivar are sorted and placed in buckets in water. The stems are tied into eight packs of 20 stems and are transported in an upright position to avoid damaging the stems (P. Boersen 2011, pers. comm.). According to the commercial standards, flowering stems are classified as medium (75 cm length and 0.5 cm diameter), long (90 cm length and 0.8 cm diameter), and extra long (110 cm length and 1.0 cm diameter). Inflorescence should appear in a minimum of 40% of the stem length. Quality classification is determined according to the damage caused by pests and diseases, mechanical damage, sunburn, soft stem caused by nutritional deficiency, and yellow leaves (Cooperative Veiling Holambra 2010c).

The main pests that occur in Brazil are thrips *Taeniothrips simplex* and *Thrips tabaci* that attack leaves, buds, flowers, and corm, leading to flower abortion, deformed flowers, and weakened plants, as well as virus transmission. The caterpillar *Heliothis zea*, a common pest of corn, also attacks gladiolus. The main diseases of *Gladiolus* are rust (*Uromyces transversalis*), which was first recorded in Brazil by Pitta et al. (1981), and *Stromatinia gladioli* that remains in the soil for 30 years or more, but only appears in winter crops at low temperatures. Viruses detected here are the CMV and the BYMV, which produce foliar mosaic symptoms, deformation, and flower color breaking (Alexandre et al. 2005, 2008).

C. *Hippeastrum* (Amaryllis)

Hippeastrum is a native bulb from the Americas, with 55–75 species distributed from Mexico to Argentina, of which 40 are found in Brazil in diverse ecological environments (Dahlgren et al. 1985; Dutihl 1987). Although the scientific name of the genus from South America is *Hippeastrum*, the common name "Amaryllis" continues to be used in commercial practice and trade, as well as in the name of societies devoted to the genus *Hippeastrum*. In fact, the true genus *Amaryllis* is native to Africa (Meerow et al. 1997).

Species of *Hippeastrum* from Brazil and Peru have served for the breeding of the most popular cultivars worldwide. The first commercial hybrids were obtained in Europe in the eighteenth century. Currently, this popular crop provides high market prices in the international markets, especially during the winter months. Among the 17 million bulbs of hippeastrum produced annually in Brazil, 60% are directed to international trade and 40% to the production of pot plants, cut flowers, and dry sale for gardening for domestic consumption. *Hippeastrum* bulbs are exported to The Netherlands (92%), the United States (5%), and Canada (3%).

The hippeastrum producers are located mainly in the states of São Paulo and Ceará, the latter region being totally focused on the production of bulbs for export. Four main local producers are André Boersen, Ecoflora Ornamental Plants, Bulbs Company of Ceará (CBC), and Terra Viva Bulbos.

The bulbs are produced in open fields. Experiments in bulb production in plastic tunnels proved to be economically inefficient. Hippeastrum cultivars are propagated by lateral offsets or twin-scaling. The production of commercial bulbs from offsets lasts for 2–3 years, depending on size and cultivation conditions. Twin-scaling is used in 20–25% of production volume, mainly for the introduction of new cultivars, breeding, and the renewal of cultivars with a low propagation capacity (Tombolato 2004). Micropropagation is not employed.

In general, the crop cycle for hippeastrum bulb production is shorter in Brazil than in South Africa, due to climatic and cultivation differences. Two systems are employed in bulb production:

1. Two growth cycles of 8 months in the field, planting in August/September and harvesting in April/May. The bulbs are stored at 13°C for 2–3 months or at 5–9°C for a longer period, if necessary.
2. One production cycle that involves planting in February and harvesting in May of the following year. This system is especially suitable for the cultivars 'Orange Souvereign' and 'Ferrari'.

Mechanical harvesting is done with an implement adapted for the garlic crop. Debris from the soil and old scales are removed by manual cleaning. The roots and leaves are cut, since the bulbs must be suitable for mechanical sorting, planting in pots, loose sale, or storage. For export, the harvested bulbs are transported in containers at 13°C, almost exclusively by sea.

Currently, a number of well-known cultivars are grown and marketed in Brazil, for example, red 'Red Lion', 'Ferrari', 'Bingo', orange 'Desire', white 'Intokasie', 'Matterhorn', pink 'Piquant', 'Vera', 'Apple Blossom', bicolor 'Gilmar', 'Vision', 'Minerva', and others (Terra Viva Bulbos n/d a). In 2005, the new cultivars 'IAC Neblina', 'IAC Jaraguá', and 'IAC Itatiaia' were registered by the IAC. 'Carina' is the only cultivar of the "Gracillis" group, still produced commercially, but the production volume is decreasing due to low demand in both the domestic and export markets (L. O. Cavicchio, Terra Viva Bulbos 2011, pers. comm.).

The sale of dry hippeastrum bulbs for gardening and landscaping is still incipient in Brazil. One of the innovations seen in this attractive market is the sale of bulbs packed in small plastic net bags, containing one, two, or three bulbs each. These packages are displayed in supermarkets, garden centers, and convenience stores, and even at gas stations on major highways in Brazil (Figure 15.4).

Cut flowers of hippeastrum are still considered a novelty in the Brazilian market. Marketing is done only in spring, especially during the month of September. In 2010, ca. 60,000 stems of assorted colors were sold to the wholesale domestic market by one of the producers (L. O. Cavicchio, Terra Viva Bulbos 2011, pers. comm.). The parameters for the cut flowers include stems of at least 40 cm, which are classified as thin ($\leq$7 mm stem diameter), intermediate (between 8 and 10 mm), or thick (>10 mm). The flowers are classified as small (<10 cm in diameter), medium (10–17 cm), or large (>17 cm) (Pellegrini 2007). For packaging, cardboard boxes are recommended, containing 40 stems, eight packages of five stems each.

The domestic market for flowering pots has grown, driven by increased marketing in retail stores. The red-flowered cultivars account for 70% of consumer preference. The potted flowers are marketed in containers of 13–17 cm. Peak sales are made on Mother's Day and at Christmas, but one can find potted flowers almost all year round. In terms of color, the Brazilian consumer prefers the red and orange cultivars (70%), two-tone pink and white (10%), white (10%), and two-tone red and white (10%) (Terra Viva Bulbos n/d a). Standards for potted hippeastrum for the domestic market are rather rigid and include criteria of height (23–46 cm), the number of stems per pot (1–2), stem diameter (minimum 15 mm), large bud size (8–12 cm), open flowers (or closed buds starting to open), and general performance of the plant (firm and straight stems, bulbs covered or partially covered and firmly secured in the pot). According to these characteristics, the potted plants are classified into Classes I, II, or III (Cooperative Veiling Holambra 2010b). In addition, quality categories A1, A2, and B are defined by the occurrence of defects, such as mechanical damage, damage from pests, and/or diseases and yellow leaves from burns, nutritional deficiencies, or phytotoxicity.

The main fungal disease of the crop is caused by *Stagonospora curtsii*, which attacks the leaves and bulb. The disease is more aggressive under conditions of high humidity, caused by excessive rainfall. It has also been observed that some cultivars are susceptible to *Pectobacterium* (*Erwinia*). Lately, a soil mite, facilitating fungal and bacterial infections, has been found, but it has not been identified yet. The viruses detected in Brazil are *Groundnut ringspot virus* (GRSV) and BYMV, causing chlorotic spots on the leaves, and the *Hippeastrum mosaic virus* (HiMV) associated with foliar mosaic symptoms (Alexandre et al. 2005, 2008).

FIGURE 15.4 (**See color insert.**) Plastic nets package for the retail market. (Photo of A. F. C. Tombolato.)

D. ALSTROEMERIA

Among 80 species of *Alstroemeria* found in South America, 40 are native to Brazil. They are found in diverse ecological environments, especially in the savannah, high-altitude fields, and more rarely, in the tropical forests (Assis 2001). The alstroemeria hybrids available in the international market are mainly of Andean species origin, selected and improved in The Netherlands. These flowers have fetched high market prices, especially during the winter months (Tombolato 2004).

In 2008–2010, the production of alstroemeria in Brazil has increased, and the number of cut flowers in the domestic market doubled (Table 15.1). The flowers are available from the end of July until December. More than 90% of production is currently concentrated in São Paulo state. The second region that is growing in importance is the state of Minas Gerais, with over 8% of the market. Small quantities of production are also observed in the states of Rio Grande do Sul, Paraná, and the mountain region of Rio de Janeiro (J. Zuijderwijk, Asist Consultancy 2011, pers. comm.).

The white-colored 'Fuji' is the leading cultivar in Brazil. Besides the white, the domestic market prefers pure colors, pink 'Chanel', yellow 'Firenze' and 'Shakira', and red 'Fuego'. The bicolor alstroemeria cultivars are not popular.

Planting is usually done in May/June or August, when soil temperatures of approximately 16°C promote plant development. *In vitro* propagated seedlings of alstroemeria are imported from The Netherlands. The average productivity of cut flowers is around 150–200 stems/m²/year, but it can vary, depending on the cultivar and environmental conditions. The beds remain in production for 3–4 years, depending on the cultivar (Tombolato 2004).

Harvesting is done when the flower buds begin to open and start showing color. The alstroemeria bouquets are packaged in bundles of 10–12 stems of 60–70 cm in height, which is the market standard, but the stems might range from 40 to 90 cm. Standardization is usually defined with a single color per box, which contains 10–12 packets. For small flower shops, the boxes consist of bundles of different colors (Cooperative Veiling Holambra 2010a). For further protection of the cut flowers, these bouquets are packed in cardboard boxes.

The main pests are aphids, thrips, and whitefly (species not yet identified). The main fungi are *Uromyces alstroemeriae* in the leaves (Coutinho et al. 1999) and *Botrytis* sp. in the flowers. Many viruses have been identified in this culture. The CMV produces symptoms of reticulated or mosaic-like leaf deformation, especially in young leaves, and chlorotic spots. The *Tobacco streak virus* (TSV) can induce necrotic and chlorotic spots on almost the whole leaf surface. The *Tomato spotted wilt virus* (TSWV) produces symptoms that can vary from necrotic rings and drawings, and the *Alstroemeria mosaic virus* (AlMV) inducing chlorotic streaks and mosaic. Mixed infections are frequently noticed, with the occurrence of more than one type of virus in the same plant (Duarte et al. 1999; Alexandre et al. 2005, 2008).

E. ZANTEDESCHIA

White calla lily (*Zantedeschia aethiopica*) was introduced to Brazil many years ago, and its populations spontaneously colonized wetlands and borders of streams in the plateaus of the South-Central region in the state of Minas Gerais (Almeida and Paiva 2005; Zanella 2006). The recent introduction of color cultivars (mainly hybrids of African species of *Zantedeschia*, Schoellhorn 2011) to Brazil rapidly attracted the consumers for flowers. The main production areas are concentrated in the states of São Paulo, Minas Gerais, and the highlands of Espírito Santo. Marketing is done as cut flowers, pots (13–17 cm), and as tubers in plastic net bags. The Brazilian market for calla cut flowers is currently contracting, with a decrease in the volume of production and marketing in recent years. The main reasons for the decline in consumption of white calla are their short vase life and a high susceptibility to mechanical damage in the shipping and handling of the flowers. For colored calla, the market constraints relate more to the high prices of the flowers and the irregular supply in the market, since the wholesale market experiences a clear reduction in supply during the summer

(January to March). The white callas are particularly appreciated for wedding decorations, while the green or colored ones are used in ceremonies or sophisticated environmental decorations.

Currently, 95% of the colored cultivars grown for pots are yellow, red, orange, wine, and black. Only 5% of white flower cultivars are grown as potted plants (L. O. Cavicchio, Terra Viva Bulbos 2011, pers. comm.). For cut flowers, the market preference is 40% yellow, 25% orange, 25% pink, and 10% other colors (L. O. Cavicchio, Terra Viva Bulbos 2011, pers. comm.; P. Boersen 2011, pers. comm.). The propagation material is imported from The Netherlands and is usually not used for the second cycle of flower production. Cultivation is done in well-drained soil and constant irrigation to maintain high humidity. The unfavorable weather conditions require the use of greenhouses with controlled temperature and an irrigation regime, which greatly increases production costs, without guaranteeing economic and financial results. During the summer, colored calla is susceptible to a high incidence of soft rot, caused by *Pectobacterium carotovorum*.

The growing of white calla lily is carried out almost exclusively by small growers. Greenhouse cultivation is not economically efficient, and cut flowers are produced in open fields, while the green calla (also belonging to *Z. aethiopica*) is produced in greenhouses or plastic tunnels to obtain longer stems, for flower production from May to September.

Postharvest treatment of the stems includes immersion in chlorine solutions for the prevention of soft rot (P. Boersen 2011, pers. comm.). In the market, cut flowers of the callas are classified according to the length of stems from short (40 cm) to long (80 cm), are packed in bundles of 10 units, wrapped in plastic, and stored in cold rooms. For transportation and sale, they are packed in cardboard boxes, each containing 80 flowers. The occurrence of *Dasheen mosaic virus* (DsMV), which causes chlorotic spots, necrotic rings, and the yellowing of leaves, has been detected (Alexandre et al. 2005, 2008).

F. OTHER ORNAMENTAL BULBS

Bulbs for gardening and amateurs' collections are sold via the Internet and delivered by mail. Currently, the main commercial species for dry sale are *Acidanthera bicolor*, *Tulbaghia violacea*, *Polianthes tuberosa*, *Canna indica*, *Crinum* spp., *Dahlia* spp., *Freesia* spp., *Gloriosa rothschildiana*, *Iris* spp., *Ixia flexuosa*, *Hyacinthus orientalis*, *Hymenocallis* spp., *Amaryllis belladona*, *Zephyranthes* sp., *Scadoxus multiflorus*, *Narcissus* spp., *Ornithogalum* spp., *Caladium* × *hortulanum*, and *Tulipa hybrida*. Bulbs can also be purchased in garden centers, stores for horticultural products, supermarkets, and convenience stores at gas stations located on major highways. For this service, the bulbs are packed in small quantities or singly, in plastic net bags with information cards, properly labeled and placed in specific displays.

IV. CONCLUSIONS

Ornamental bulbs have been produced in Brazil for the last two centuries, resulting mainly from human migration flows, especially from Europe and Japan. The immigrants brought not only the bulbs themselves but also appreciation for the flowers. At present, the floricultural market of Brazil is mainly domestic; however, the assortment of ornamental geophytes consists of well-known and traditional crops—alstroemeria, hippeastrum, calla lily, gladiolus, daylily, lily, and so on. This fact allows Brazil to export bulbs to other countries (e.g., hippeastrum and gladiolus are exported to Europe) and to import propagating material for the growing domestic market, especially for crops with cold requirements (e.g., *Lilium* and *Tulipa*).

One of the major obstacles to the development of the flower industry in the country is the legislation regulating the use of pesticides. Only a small number of specific pesticides are registered for use on ornamental species, which leads to either the illegal use of nonregistered specific pesticides by growers or the termination of bulb and flower production. Another concern of producers is the common occurrence of the weed *Cyperus rotundus* in the cultivation areas, control of which is very difficult and expensive.

The Brazilian flora is known worldwide as very rich and containing a large number of species with ornamental potential. Unfortunately, the process of domestication, breeding, and selection of new ornamental crops is rather slow, and is not well exploited by the Brazilian flower industry. Floricultural research in the country is still not sufficient to satisfy the demand for new cultivars and technologies. Therefore, most of the novelties are imported and introduced by the growers themselves.

The Brazilian economy, one of the fastest growing in the world, is the world's seventh largest by nominal GDP. At present, the Brazilian economy is experiencing a period of huge growth that is resulting in an increase in the standard of living of the local population. As a consequence, significant growth in the floricultural market is expected. Over the last two decades, the Brazilian floriculture sector grew at a faster pace than the economy of the country as a whole. This tendency shows that Brazilians are ready to expand their consumption of flowers, potted plants, and the flowering bulbs for gardening and landscaping.

ACKNOWLEDGMENTS

The authors would like to acknowledge the contributions of Mrs Ana Paula Sá van der Geest (Flortec—Training, Courses and Events Ltd); Mrs Ana Rita Pires Stenico (CEASA—Campinas); Mr André and Mr Peter Boersen, flower growers at Andradas—MG and Holambra—SP; Mr Carlos Godoy (Cooperative Veiling Holambra); Mr Flavio Godas (Department of Economics and Statistics, CEAGESP); Mr Graham Duncan (Kirstenbosch National Botanical Garden, South Africa); Mr Jan Zuijderwijk (Asist Consultancy); Mr Luiz Octavio Cavicchio (Terra Viva Bulbos, Holambra—SP); and Dr. Maria Amélia Vaz Alexandre (Institute Biologic of São Paulo).

CHAPTER CONTRIBUTORS

Antonio F. C. Tombolato is the director of the Botanical Garden of the Agronomic Institute (IAC) in the state of São Paulo, Brazil. He was born in Brazil, graduated from the University of São Paulo, and obtained the degree of Docteur Ingenieur from the University of Bordeaux II, France. Following postdoctoral studies at the Istituto Sperimentale per la Floricoltura in Sanremo, Italy, he has held a research position at IAC, Brazil. His main expertise includes research in the breeding of tropical flowers. Dr. Tombolato has published more than 100 refereed journal articles, books or chapters, and conference papers.

Roberta P. Uzzo is a researcher at the Institute Agronomic (IAC) in the state of São Paulo, Brazil. She was born in Brazil and obtained her BS from the Federal University of Lavras, her MS from the Institute Agronomic, and her PhD from the University of São Paulo, Brazil. Her fields of expertise include tropical flowers and ornamental palms. Dr. Uzzo has published more than 30 refereed journal articles, books or chapters, and conference papers.

Antonio H. Junqueira is a director of Junqueira and Peetz Consultants. He was born in Brazil, graduated from the University of São Paulo, and obtained his MS from the Escola Superior de Propaganda e Marketing. He is currently a PhD student at the University of São Paulo, Brazil. His expertise relates to horticultural marketing and business, and he provides advice to public companies in the agricultural and food supply business and in flower bulb marketing and consumption research. Mr Junqueira is the author of more than 100 refereed journal articles, books or chapters, and conference papers.

Márcia da S. Peetz is a director of Junqueira and Peetz Consultants. She was born in Brazil and graduated from the Faculdade de Economia São Luiz. Her expertise relates to horticultural marketing and business, and she provides advice to public companies in the agricultural and food supply business and in vegetables, fruits, flower, and ornamental plants marketing and consumption research. Mrs Peetz is the author of more than 100 refereed journal articles, books or chapters, and conference papers.

Giulio C. Stancato is a researcher at the Agronomic Institute (IAC) in the state of São Paulo, Brazil. He was born in Brazil, graduated from the University of São Paulo, and obtained his MSc and PhD from the University of Campinas, Brazil. Following his postdoctoral studies at the Istituto Sperimentale per la Floricoltura, Sanremo, Italy, he was appointed to a research position at IAC. His main area of research relates to micropropagation of tropical flowers. Dr. Stancato has published more than 50 refereed journal articles, books or chapters, and conference papers.

REFERENCES

ABCSEM—Associação Brasileira do Comércio de Sementes e Mudas, 2010. *Quatro Décadas—A Trajetória da ABCSEM*. Campinas.

Alexandre, M. A. V., L. M. L. Duarte, and A. E. C. Campos-Farinha. 2008. *Plantas Ornamentais: Doenças e Pragas Vol. 1*. São Paulo: Instituto Biológico.

Alexandre, M. A. V., E. B. Rivas, A. R. P. Tozetto, and L. M. L. Duarte. 2005. *Lista Comentada Sobre a Ocorrência Natural de Vírus em Plantas Ornamentais no Brasil*. São Paulo: Instituto Biológico.

Almeida, E. F. A., and P. D. O. Paiva. 2005. Cultivo de copo-de-leite. *Informe Agropecuário* 25(227):30–35.

Assis, M. C. 2001. *Alstroemeria L. (Alstroemeriaceae) do Brasil*. PhD thesis. Instituto de Biociências. Universidade de São Paulo, São Paulo, Brazil.

Cooperative Veiling Holambra. 2010a. Departamento de Qualidade e Pós-colheita. *Critérios de Classificação Para Alstroeméria de Corte*. Holambra, São Paulo.

Cooperative Veiling Holambra. 2010b. Departamento de Qualidade e Pós-colheita. *Critérios de Classificação Para Amaryllis em Vaso*. Holambra, São Paulo.

Cooperative Veiling Holambra. 2010c. Departamento de Qualidade e Pós-colheita. *Critérios de Classificação Para Gladíolo de Corte*. Holambra, São Paulo.

Cooperative Veiling Holambra. 2010d. Departamento de Qualidade e Pós-colheita. *Critérios de Classificação Para Lírio de Corte*. Holambra, São Paulo.

Cooperative Veiling Holambra. 2010e. Departamento de Qualidade e Pós-colheita. *Critérios de Classificação Para Lírio de Vaso*. Holambra, São Paulo.

Coutinho, L. N., O M. Russomanno, and M. B. Figueiredo, 1999. *Uromyces Alstroemeriae* uma severa e importante ferrugem da *Alstroemeria* spp. *cultivada. O Biológico* 61(1):23–26.

Dahlgren, R. M. T., H. T. Clifford, and P. F. Yeo. 1985. *The Families of Monocotyledons*. Berlin: Springer-Verlag.

Dutilh, J. H. A. 1987. *Investigações citotaxonômicas em populaces brasileiras de Hippeastrum Herb*. MSc thesis, Universidade Estadual de Campinas, Campinas, Brazil.

Duarte, L. M. L., P. V. Seabra, E. B. Rivas, S. R. Galleti, and M. A. V. Alexandre. 1999. Single and double viral infection in *Alstroemeria* sp. *Revista Brasileira de Horticultura Ornamental* 5(1):24–33.

Healy, W. E., and H. F. Wilkins. 1982. Response of *Alstroemeria* 'Regina' to temperature treatments prior to flower inducing temperatures. *Scientia Hortic.* 17:383–390.

Hurka, W. 1991. Formazione dei fiori negli ibridi di Alstroemeria. *ClamerIinforma* 1:19–24.

Junqueira, A. H., and M. S. Peetz. 2010a. Desempenho recente e tendências da floricultura brasileira. *Agrianual* 2011:291–294.

Junqueira, A. H., and M. S. Peetz. 2010b. El día de la madre 2010 en Brasil. *Plantflor* 139:2–4.

Junqueira, A. H., and M. S. Peetz. 2011. Floricultura brasileira em 2010: um balanço do comércio exterior e do mercado interno. *Cultivar Hortaliças e Frutas, 65,* 38.

Junqueira, A. H., and M. S. Peetz. 2008. Mercado interno para os produtos da floricultura brasileira: Características, tendências e importância sócio-econômica recente. *Revista Brasileira de Horticultura Ornamental* 14(1):37–52.

Meerow, A. W., J. van Scheepen, and J. H. A. Dutilh. 1997. Transfers from *Amaryllis* to *Hippeastrum* (Amaryllidaceae). *Taxon* 46:15–19.

Pellegrini, M. B. Q. 2007. *Caracterização e Seleção de Amarílis Melhorados Pelo Instituto Agronômico de Campinas—IAC para flor de corte*. MSc dissertation. Curso de Pós-Graduação em Agricultura Tropical e Subtropical. IAC, Campinas.

Pitta, G. P. B., M. B. Figueiredo, R. M. G. Cardoso, and J. F. Hennen. 1981. Ferrugem (*Uromyces Transversalis* Tuemen, Winter): uma nova doença do gladíolo (*Gladiolus* spp.) no Brasil. *O Biológico* 47(12):323–328.

Rivas, E. B. 2010. *Lily Symptomless Virus* no Brasil. *Instituto Biológico, Documento Técnico* 6:1–5.

Schoellhorn, R. 2011. Warm climate production guidelines for *Zantedeschia* (calla lily) hybrids. http://hort.ufl.edu/floriculture/pdfs/crop_production/Callas_ENHFL04-001.pdf.

Terra Viva Bulbos. n/d a. *Amaryllis: Informativo Técnico de Produção*. Holambra, SP., Brazil.

Terra Viva Bulbos. n/d b. *Gladíolo: Informativo Técnico de Produção*. Holambra, SP., Brazil.

Tombolato, A. F. C. 2004. *Cultivo comercial de plantas ornamentais*. Campinas, Brazil: Instituto Agronômico.

Tsuboi, N., and H. Tsurushima. 2009. *Introdução à história da indústria de flores e plantas ornamentais no Brasil*. Arujá, Brazil: AFLORD—Associação dos Floricultores da Região da Via Dutra.

Zanella, M. 2006. Cadeia de produção de *Zantedeschia* spp. no Estado de São Paulo. Relatório de Estágio de Conclusão de Curso de Agronomia. *Universidade Federal de Santa Catarina*, Centro de Ciências Agrárias. Florianópolis, SC, Brazil.

16 Geophyte Research and Production in Chile

Eduardo A. Olate and Flavia Schiappacasse

CONTENTS

I. INTRODUCTION

Chile is a long and narrow country on the western side of South America. It extends over 4200 km from above the Tropic of Capricorn to almost the Antarctic Circle, and therefore the climatic conditions of the country vary widely. The Andes range, with tall peaks and volcanoes, spreads linearly from north to south. The northern area of the country is the Atacama Desert—the world's driest region. Rainfall in the semiarid area leads to a predominantly Mediterranean climate in the middle of the country, which turns to colder and humid areas in the southern part and cold and dry areas in Chilean Patagonia.

Horticulture is an important sector for the Chilean economy realizing over four billion dollars in export earnings during 2010 and representing over 5% of total exports (Central Bank of Chile 2011). As a result of its geographical location and climates, Chile has no commercial production of tropical crops and exploits only a limited area of subtropical species in the far north of the country. The Central Valley of middle Chile is the main horticultural production area and the economic heart of the country. Therefore, temperate crops represent the main volume of horticultural production. The major comparative advantage of Chilean horticulture is that harvesting occurs counter season with the same crops in the Northern Hemisphere. The supply of off-season demand of fresh horticultural products has made the country a significant player in many international markets, with meaningful economic returns. During the last few decades, the country has developed a highly diversified horticultural sector and evolved a rather comprehensive and modern infrastructure including packing shed areas, refrigerated storage facilities, roads, and transportation and communications systems, all to facilitate horticultural exports (Krarup 2002).

The ornamental sector is considered the less developed sector of Chilean horticulture, but perhaps the one with the greatest potential. Chile grows around 2250 ha of cut flowers, bulbs, and other ornamental crops (ODEPA 2011). For several years the annual cut flower export value has been stagnant at around US$3 million. Currently, peonies are by far the main exported cut flower, followed by calla lilies, tulips, and lily. On the other hand, the export of flower seeds (around US$17 million in 2010) and flower bulbs (US$30 million in 2010) shows a steady increase. The advantages for Chilean floriculture development are the wide range of climates, a low incidence of pests and diseases, the unique cold winter temperatures in the central and southern part of the country, excellent soil quality, in addition to forward-looking growers and professionals and a stable political and economic system.

Lately, an increase in Chile's economy and family incomes, as well as the growing appreciation of the value of gardening, landscaping, and recreational areas, led to significant growth in the domestic demand for ornamental plants, becoming a relevant market for the ornamental industry. The annual wholesale domestic value of the ornamental industry is estimated to be between US$70 and US$80 million, with an annual increase estimated at over 10% (ODEPA 2011). Nevertheless, the relatively small population of the country (over 17 million inhabitants estimated for 2012) and the decreasing growth rate of the population (less than 1% by 2012) make the export market the main alternative for a larger and more dynamic horticulture industry in the future.

The territory of Chile can be seen as an "ecological island" with geographical barriers isolating the biological communities from the rest of the continent and, therefore, producing a high percentage of endemism. Continental Chile is home to some 5100 plant species of which about 2630 are endemic (Marticorena 1990; Marticorena and Rodríguez 1995). Such a high proportion of endemism rivals that of many islands and is one of the highest found in any region on the Earth. The temperate ecosystems of southern South America can be appreciated best in comparison with similar plant communities that exist today in South Africa, Australia, and New Zealand. Similar to these regions, the Mediterranean climate, high Andes, and desert ecosystems of Chile have evolved in isolation. The resources of the fascinating Chilean flora have still not been fully explored. Over the last two decades, native Chilean plants, including ornamental geophytes, have attracted the special attention of breeders and horticulturists. Research projects on the breeding, propagation, physiology, micropropagation, and production of flowering geophytes from Chile are currently being conducted.

II. RESEARCH AND ORGANIZATION NETWORKS

Despite the fact that the potential for development of Chilean floriculture is great, only a few universities in Chile employ floriculture researchers, that is, Pontificia Universidad Católica de Valparaíso, Universidad Austral, Universidad de Talca, Pontificia Universidad Católica de Chile, Universidad de Concepción and Universidad de Chile. Some of these research groups have developed research on Chilean geophytes. The main subjects are focused on propagation and breeding of *Alstroemeria*, *Leucocoryne*, *Chloraea*, and *Rhodophiala*. The Universidad Austral, through its Center at Trapananda in Coyhaique, and the Universidad de Magallanes in Punta Arenas have also worked on flower geophytes as production alternatives to Chilean Patagonia.

INIA is the National Institute for Agriculture Research, with several regional stations that perform applied research and extension service functions. Only one of these stations employs a floriculture researcher who has been working, along with other researchers, on production and postharvest technology of flower crops, comprising also ornamental geophytes. This support is proving to be very important for the local growers.

Different growers' associations have been formed over time consolidating various crops and/or interests. The exporters of cut flower have created an association called APEF-Chile (Asociación de Productores y Exportadores de Flores de Chile A.G.). In a similar way, peony growers and one export company specializing in peonies have gotten together to share knowledge and to solve various common problems, forming the Chilean Association of Peony Growers (Peonías de Chile A.G.).

A group of flower bulb growers has also formed the Chilean Association of Bulb Growers and Exporters A.G. (known as APEB—Chile for its acronym in Spanish) that works with local and foreign government officials to open up new markets for their products. At present, no umbrella organization unifies these various associations of growers and exporters under one roof. This is clearly a big competitive disadvantage for the country.

The Foundation for Agriculture Innovation (FIA) of the Ministry of Agriculture has supported the carrying out of innovative projects on different aspects of horticulture, including ornamental horticulture. This foundation was established in the late 1980s, and its contribution to the introduction of new species and to innovation in national floriculture has been very significant. FIA has funded horticultural field trips to various countries, the visits of international experts to Chile, and several production projects undertaken by research institutions and/or by private growers.

The Ministry of Economy through CORFO (Office for Productive Development) also supports projects that focus on new products and technological developments. In addition to the basic developments in floriculture, these projects have always had important social goals, such as for example the development of small family farms, encouraging women in agriculture and the advancement of small rural communities.

Lately, large companies in Chile have shown interest in collaborating with research organizations, and are ready to invest in breeding and other new product development. Many fruit companies (the most developed branch of horticulture in Chile) and seed production companies are interested in diversifying their businesses, and so have been exploring the ornamental industry as a possible alternative.

For many years, Chilean growers and research organizations have been interested in collaboration with the global scientific community, participating in international conferences and hosting international and regional symposia and courses, as well as bringing leading experts in floriculture and other research fields to Chile. All of these efforts show that the ornamental sector is open to incorporating new technologies and developments.

Research and funding agencies are aware that market research is a very important component, so much of applied research is market oriented. However, there is still a need to strengthen this component at all levels from research to commercialization.

III. CHILEAN GEOPHYTES: OPPORTUNITIES FOR HORTICULTURAL DEVELOPMENT

A strong effort to promote knowledge and the practical utilization of ornamental geophytes has been made by Chilean researchers. The flora of different regions of Chile has been described, and geophytes species have been identified in the Northern Andes (Squeo et al. 1993), the desert regions (Muñoz 1985; Teillier et al. 1998; Riedemann et al. 2006), the coastal area of the Central Chile region (Villagrán et al. 2007), the Central and Southern regions of the country (Riedemann and Aldunate 2001, 2003). The biological traits of Chilean geophytes, their horticultural potential, and seed and vegetative propagation have also been studied (Table 16.1). The results of all this research have laid the foundation for the development of new, attractive, and economically efficient ornamental crops.

Monocotyledonous genera represent an important percentage of the endemic flora of Chile, some of which are monospecific or consist of only a few species. Many of these plants are geophytes that can be found throughout the country, in the vast deserts of the north, in the southern humid forests, in the Patagonian pampas, along the coast and up in the mountains (Hoffmann 1989). The largest number of geophytes are found in Central Chile (latitude 28–37°S), which is characterized by a Mediterranean type of climate with short, rainy, and mild winters and long dry summers (Arriagada and Zöllner 1994).

Some of the local geophytes are already being grown and commercialized as garden plants or for land and environmental restoration by Chilean nurseries (e.g., *Alstroemeria aurea, A. ligtu, Libertia chilensis, L. sessiliflora, Oziroë biflora, Pasithea caerulea, Rhodophiala advena,* and *Sisyrinchium*

TABLE 16.1
List of Geophytes from Chilean Natural Flora, Their Breeding Status and Horticultural Potential

Family/Genus	Native and Endemic Species	Breeding and Horticultural Potential	References
		Alliaceae	
Ipheion	One native species	Garden and pot plant	
Leucocoryne	45 species, all endemic to Chile	Known as a cut flower. High horticultural potential as garden and pot plants. A breeding program in Chile resulted in registered cultivars.	Elgar et al. (2003); Mansur et al. (2002); http://www.mundoagro.cl; Riedemann and Aldunate (2001); Riedemann et al. (2006); Schiappacasse et al. (2002)
Miersia	Four endemic species	Horticultural potential is not acknowledged	
Nothoscordum	Six species, five of them are endemic	Garden and pot plant; can become invasive.	Riedemann and Aldunate (2001)
Speea	One endemic species	Horticultural potential is not acknowledged	
Tristagma	12 species, nine are endemic	Garden plant	Riedemann et al. (2008)
Zoellnerallium	One native species	Garden plant	
		Alstroemeriaceae	
Alstroemeria	37 native species, of which 32 are endemic	The genus has been extensively bred and cultivated as a cut flower, a pot plant, and a garden plant worldwide. Ongoing breeding program in Chile.	Bridgen (1992); Bridgen (1998); Molnar (1975); Olate et al. (2007); Price and Langhans (1985); Riedemann and Aldunate (2001); Riedemann and Aldunate (2003); Riedemann et al. (2006); Riedemann et al. (2008); Rojas and Müller (1987); Schiappacasse et al. (2002); Ohkawa (1994); Wilkins and Heins (1976)
Bomarea	Three endemic species	Climber plant. Recently started to be used as garden and pot plants in Chile, available in commercial nurseries. No breeding program on Chilean species.	Riedemann and Aldunate (2001); Schiappacasse et al. (2002)
Leontochir (recently merged into the genus *Bomarea*)	One species with two varieties (red and yellow)	Extraordinarily beautiful, great potential as a cut flower, a pot plant, and a garden plant.	Hofreiter (2006); Lu et al. (1995); Riedemann et al. (2006)
		Amaryllidaceae	
Phycella	Six native species, four are endemic	Potential as a cut flower, a garden plant, and a pot plant.	Schiappacasse et al. (2002)
Placea	Five endemic species	High potential as a cut flower, a garden plant, and a pot plant.	Riedemann and Aldunate (2001); Schiappacasse et al. (2002)
Rhodophiala	27 native species, 25 are endemic.	High potential as a garden and pot plants and some potential as a cut flower. Ongoing breeding program in Chile.	Riedemann and Aldunate (2001); Riedemann and Aldunate (2003); Riedemann et al. (2006); Riedemann et al. (2008); Schiappacasse et al. (2002); Schiappacasse et al. (2007)

TABLE 16.1   (continued)
List of Geophytes from Chilean Natural Flora, Their Breeding Status and Horticultural Potential

Family/ Genus	Native and Endemic Species	Breeding and Horticultural Potential	References
		Corsiaceae	
Arachnitis	One native species	Horticultural potential is not acknowledged	
		Dioscoreaceae	
Dioscorea	42 native species, 38 are endemic	Recommended as ground cover in sandy soils and in pots as hanging plant.	Riedemann and Aldunate (2001); Riedemann and Aldunate (2003); Riedemann et al. (2006)
Gilliesia	Five native species, four are endemic	Horticultural potential is not acknowledged	
Solaria	Three native species, two are endemic	Horticultural potential is not acknowledged	
		Hemerocallidaceae	
Pasithea	One endemic species	The stems are commonly picked in the wild and sold in local markets. The inflorescences are very attractive but their blue florets stain any surface below. The species has potential as a garden plant and a pot plant.	Riedemann and Aldunate (2001); Schiappacasse et al. (2002)
		Hyacinthaceae	
Oziroë	Four species, two are endemic	Garden and pot plants	Guaglianone and Arroyo-Leuenberger (2002); Riedemann and Aldunate (2001)
		Iridaceae	
Calydorea	One endemic species	Highly ephemeral, potential as garden and pot plants.	Riedemann and Aldunate (2001); Schiappacasse et al. (2002)
Herbertia	One subspecies is endemic	Ephemeral flower. Potential as garden and pot plants.	Riedemann and Aldunate (2001); Riedemann and Aldunate (2003); Schiappacasse et al. (2002)
Libertia	Six species, five are endemic	Species are currently used as garden plants in Chile and are widely available from commercial nurseries.	Riedemann and Aldunate (2001); Riedemann and Aldunate (2003); Schiappacasse et al. (2002)
Sisyrinchium	22 species, 12 are endemic	High potential as a garden plant. *S. striatum* is extensively used as a garden plant in public areas in Chile, appreciated due to its low water requirements.	Riedemann and Aldunate (2001); Riedemann and Aldunate (2003); Riedemann et al. (2008)
Solenomelus	Two species, one is endemic	Potential as a garden plant	Riedemann and Aldunate (2001)
Tigridia	One endemic species	Potential as a pot plant	

continued

TABLE 16.1 (continued)

List of Geophytes from Chilean Natural Flora, Their Breeding Status and Horticultural Potential

Family/ Genus	Native and Endemic Species	Breeding and Horticultural Potential	References
		Laxmanniaceae	
Trichopetalum	One endemic species	Potential as a pot plant	Riedemann and Aldunate (2001)
		Orchidaceae	
Aa	One endemic species	Potential as a garden plant	
Bipinnula	Four endemic species	High potential as garden and pot plants, and also as a cut flower.	Riedemann and Aldunate (2001)
Brachystele	Two native species	High potential as garden and pot plants.	Riedemann and Aldunate (2001)
Codonorchis	One native species	High potential as a garden plant.	Riedemann and Aldunate (2003)
Chloraea	30 species, 17 are endemic	High horticultural potential and value for breeding and development as a cut flower, a pot plant, and a garden plant. Currently, two ongoing breeding programs in Chile.	http://bibliotecadigital.innovacionagraria.cl Riedemann and Aldunate (2001); Riedemann and Aldunate (2003); Riedemann et al. (2008)
Gavilea	15 species, 8 are endemic	High potential as a cut flower, a pot plant, and a garden plant.	Riedemann and Aldunate (2001); Riedemann et al. (2008)
Habenaria	One native species	Potential as garden and pot plants.	
		Oxalidaceae	
Oxalis	53 species, 31 are endemic	Potential as garden plants; annuals as ground covers, perennials as pot and rock garden plants.	Riedemann and Aldunate (2001); Riedemann and Aldunate (2003); Riedemann et al. (2006); Riedemann et al. (2008)
		Ranunculaceae	
Anemone	Six species, three are endemic	Potential as garden plants. *A. multifida* has been studied as a garden plant.	Manzano et al. (2009); Riedemann and Aldunate (2003); Riedemann and Aldunate (2001)
		Tecophilaeaceae	
Conanthera	Endemic genus with five species	Ornamental potential as garden plants, pot plants and possibly as cut flowers.	Riedemann and Aldunate (2001); Schiappacasse et al. (2002); Yáñez (2001)
Tecophilaea	Endemic genus with two species	Potential as garden and pot plants.	Riedemann and Aldunate (2001); Schiappacasse et al. (2002)
Zephyra	Endemic genus with two species	Great potential as a cut flower, a garden plant, and a pot plant.	Kim et al. (1996); Kim and Ohkawa (1997); Riedemann et al. (2006)
		Tropaeolaceae	
Tropaeolum	21 species, 17 are endemic	Great potential as garden plants; some are used as climbers or in pots.	Riedemann and Aldunate (2001); Riedemann and Aldunate (2003); Riedemann et al. (2006); Riedemann et al. (2008); Schiappacasse et al. (2002)

Note: Taxonomy is updated according to Instituto Darwinion (2009).

striatum). Others have significant potential as commercial crops (Table 16.1). However, most of the Chilean geophyte species have been described only from the botanical point of view and not much work has been carried out and published on flowering physiology, propagation or methods of production. In recent years, more applied information is available on prospective species and genera, such as *Alstroemeria*, *Chloraea*, *Conanthera*, *Herbertia*, *Leucocoryne*, and *Rhodophiala*. The most promising geophyte genera from the Chilean flora are reviewed in this section.

A. *Alstroemeria*

The main center of diversity of this genus is located in South America, with 31 species found in Chile (Uphoff 1952; Bayer 1987; Aker and Healy 1990; Muñoz and Moreira 2003) and ca. 40 species in Brazil (Meerow and Tombolato 1996; Han et al. 1999; Assis 2002, 2004; Assis and Mello-Silva 2002). The different species are found in all kinds of soils in the dry desert, central valley, along the coast and coastal range, in the South, Patagonia, and the Andes (Healy and Wilkins 1986; Bayer 1987; Bridgen 1998; Bridgen and Olate 2004; Riedemann et al. 2006).

Underground rhizomes of *Alstroemeria* produce two types of shoots—vegetative and reproductive, and fibrous roots, which become thick and function as storage organs. Flowering initiation is controlled both by cold temperatures and the long-day photoperiod. Short vernalization of the rhizome has to be complete before the long-day photoperiod and temperatures between 10 and 17°C have to be used to obtain quality blooming (Bridgen 1998). New flowering stems are continuously produced after the requirements for flower initiation have been met. Once the flowering process starts, the plant will continue to flower until the soil temperature rises to above 18–21°C for extended periods (Bridgen 1998). Flowers of *Alstroemeria* are zygomorphic and the perianth consists of six uniform tepals arranged in two whorls that open simultaneously (Figure 16.1; Bridgen 1992; Riedemann et al. 2006).

Seed propagation is only used for garden mix cultivars or for breeding purposes. Seeds accumulate chemical inhibitors on the seed coat that prevent germination in the absence of water. Any soft scarification, such as hot water and overnight soaking, diluted bleach, or sulphuric acid treatments, will remove the inhibitors. Humid stratification at 15–20°C also facilitates seed germination (Bridgen 1998; Riedemann and Aldunate 2001). Clonal propagation is achieved by dividing the rhizomes in late summer or early fall, after the blooming season has finished. For a successful division the presence of undamaged fleshy storage roots and active growing rhizome is important. Plants can be divided again after two or three growing seasons. Clonal micropropagation has been widely used since the early 1980s by commercial nurseries and as a tool for mass propagation in breeding programs (Olate et al. 2007).

Commercial cultivars of *Alstroemeria*, also known as the Lily-of-the-Incas, Peruvian Lily, or Inca Lily, have been popular since the 1970s, mainly as a cut flower crop. In the Dutch and other international breeding programs, species of the Chilean group, particularly *A. aurea*, *A. ligtu*, and *A. pelegrina*, were used (Aker and Healy 1990; Kristlansen 1995), along with the Brazilian species *A. caryophyllacea* and *A. pulchella*. Breeding programs mainly base their efforts on interspecific hybridization and further embryo rescue *in vitro*, but they have also used several other biotechnology techniques to assist breeding programs, including efforts to understand the genetics behind the floral scent present in some Brazilian species (Aros et al. 2008).

In recent years, hybrid cultivars have been grown worldwide as garden flowers and potted plants. The hybrid cultivars produce beautiful, large inflorescences of many different colors including purple, lavender, red, rose, pink, yellow, peach, orange, white, and bicolors. In addition to their showy colors, the cut flowers have long postharvest vase lives of up to 2–3 weeks. When commercial *Alstroemeria* hybrids are grown properly, the plants have an everblooming performance. All these valuable characteristics have made the hybrid *Alstroemeria* very popular (Bridgen and Olate 2004). Active breeding of *Alstroemeria* is being carried out at the Universidad Católica de Chile (E. Olate).

FIGURE 16.1 (**See color insert.**) *Alstroemeria* species. (a) *A. magnifica* ssp. *magnifica*; (b) *A. aurea*. (Photographs by E. Olate.)

B. *CHLORAEA*

The genus *Chloraea* consists of 41 species in South America, 30 of which are found in Chile, with 17 species being endemic to the country (Table 16.1; Instituto Darwinion 2009). The most prominent and important species for ornamental horticulture is *C. crispa* (Figure 16.2). Its main geographical distribution ranges from Concepción to Valdivia (latitude 36–39°S). The plants are found usually in sandy sites either near the coast or in the central valley (Novoa et al. 2006).

   C. crispa is a perennial robust plant. The underground structure is a rhizome with thick and succulent roots. A basal rosette consists of lanceolate leaves, which dry during anthesis. The flowering stem reaches a height of 40–70 cm and is covered by membranaceous leaves. The inflorescence is a single spike containing a few large, white and showy flowers and bracts (Novoa et al. 2006).

   Following a juvenile period of 4–5 years, the plants flower in spring and summer, although a high percentage of the adult plants flower only every second year (G. Verdugo 2008, pers. comm.). Under natural conditions, the juvenile period can last as long as five years. However, in culture conditions, a small quantity of flowering plants can be obtained after three years of growing from seeds (Fundación para la Innovación Agraria 2007). Cut flowers have shown a long-lasting postharvest life (15–20 days) (G. Verdugo 2011, pers. comm.). After flowering, *Chloraea* plants enter summer dormancy.

FIGURE 16.2 **(See color insert.)** *Chloraea crispa* under cultivation as cut flower. (Photograph by H. Vogel.)

The seeds are very small and do not have endosperm. At the first germination stages, they need to feed through a mycorrhizae association (Mersey 2003). Seed reproduction in the field is possible if the seeds are sown next to adult plants that would provide the source of mycorrhizae. *In vitro* asymbiotic germination was attained after seven months of culture with germination rates ranging between 33 and 40% (Verdugo et al. 2007). *Chloraea* can be propagated vegetatively by rhizome division in the autumn (Riedemann and Aldunate 2001) or *in vitro* using a different type of explant (Fundación para la Innovación Agraria 2007). A breeding program at the Pontificia Universidad Católica de Valparaíso (G. Verdugo), in collaboration with the Universidad de Talca (H. Vogel), involves polyploidy induction and hybrid production. In addition, currently, research is being conducted on developing a protocol for multiplication *in vitro* using somatic embryogenesis (G. Verdugo 2011, pers. comm.).

C. CONANTHERA

Five endemic species of the genus *Conanthera* are recognized in Chile (Muñoz and Moreira 2000; Instituto Darwinion 2009). The geographical distribution of the genus in Chile ranges from latitude 30–39°S, and different species can be found in the sandy soils of the coastal areas to clay soil habitats, either in the coastal range or in the foothills of the Andes Mountains, both in the desert and in Mediterranean areas.

Conanthera possesses a small corm, covered with fibrous brown tunics. The corm is replaced annually. The above-ground parts are formed by four to seven basal linear leaves. The inflorescences are long racemes with bell-shaped flowers, ranging from light blue to purple (Figure 16.3) (Riedemann and Aldunate 2001; Schiappacasse et al. 2002).

The juvenile phase lasts about 6–7 years. During the annual cycle of the adult plant, leaves sprout in late autumn or winter, and flowering occurs in early spring (September) 5–6 weeks after flower initiation (Yáñez 2001). Under natural conditions, all Chilean species enter summer dormancy during the long dry season. Under cultivation, summer dormancy occurs even when irrigation is provided.

Seeds of *C. trimaculata* and *C. campanulata* have a very low germination rate. Mechanical scarification facilitated 90% germination of fresh seeds, but cold stratification or treatments by hot water or acid were not successful in germination (Vogel et al. 1999; Yáñez 2001). Seeds of *C. bifolia*

FIGURE 16.3   (**See color insert.**) *Conanthera* species from Chile. (a) *C. trimaculata*; (b) *C. bifolia*. (Photographs by E. Olate.)

germinated after cold stratification (Yáñez 2001). Corms of *Conanthera* do not produce cormels, and vegetative propagation can be successful by the division of the corms into sections (Vogel et al. 1999; Schiappacasse et al. 2002). At present, *Conanthera* species are not included in any breeding program or commercial production in Chile.

D. *HERBERTIA*

Herbertia consists of eight species, distributed in the United States and South America. *Herbertia lahue* subsp. *lahue*, endemic to Chile, is distributed from latitude 33°S to 39°S and is usually found in humid soils, with high levels of organic matter, close to the seashore or lakes or in humid areas near the Andes range (Morales et al. 2009).

FIGURE 16.4 (**See color insert.**) *Herbertia lahue* ssp. *lahue*. (Photograph by F. Schiappacasse.)

Bulbs of this species are ovoid and tunicated. Basal leaves are linear-lanceolate, and the flowering stem bears 1–3 purple-blue flowers (Figure 16.4). The juvenile period lasts only one year, and the adult plants flower every spring. Flower initiation and differentiation occur during plant growth (Morales 2001). In culture, flowering lasts up to 3 weeks (Morales et al. 2009).

Seeds of this plant germinate easily. After 4 days of imbibitions in water, more than 90% germination can be reached at 15°C (Schiappacasse et al. 2002). Natural vegetative propagation is limited. For artificial propagation, large bulbs can be divided into vertical sections. Following incubation for three months, they can be transplanted in soil, each one yielding 3–4 new bulbs (Morales et al. 2009).

The stems of *Herbertia lahue* subsp. *lahue* are short (5–15 cm), and anthesis lasts only one day. Therefore, the ornamental potential is limited to use as potted plant or for gardening (Schiappacasse et al. 2002). Nevertheless, its unique form and beauty make it highly regarded as a species with ornamental potential (Hoffmann 1989; Bridgen et al. 2002; Schiappacasse et al. 2002). At present, *Herbertia* is not included in any breeding program or commercial production in Chile.

E. *LEUCOCORYNE*

Leucocoryne is an endemic genus of Chile. According to different authors, the number of species varies between seven and 45 (Zöllner and Arriagada 1998; Muñoz and Moreira 2000; Zöllner 2002; Riedemann et al. 2006; Instituto Darwinion 2009). Species of *Leucocoryne* are widely distributed in Central Chile, mainly at latitudes of 30–35°S, in dry habitats from sea level up to an altitude of 1000 m a.s.l., with the largest populations found in the coastal areas north of Santiago. This area is rather dry, with scattered winter rains between May and August, and an average rainfall of 70 mm.

The relatively small bulb of *Leucocoryne* is covered with a brown dry tunic. In some cases, the bulbs form droppers, which draw the new bulb down to 40 cm below the surface to prevent drying. Plants form 2–7 leaves. Its flower stem 30–80 cm in length bears an umbel of 2–9 actinomorphic flowers (Zöllner 2002; Riedemann et al. 2006). Tepal colors range from solid white to light or deep purple, with some bicolor forms (Mansur et al. 2002; Schiappacasse et al. 2002; Zöllner 2002). Natural hybrids can also be found in native habitats (Figure 16.5).

The juvenile period lasts for about three years (Escobar et al. 2008), and temperatures of 15–16°C are favorable for bulb growth (Mansur and De la Cuadra 2004). In nature, leaves emerge during winter and flowering occurs during spring. The leaves usually dry out before flowering (Schiappacasse et al. 2002). Research projects have been dedicated to the physiology and cytogenetic of *Leucocoryne*

(Araneda et al. 2004), as well as to the horticultural traits of these species and their use as ornamental crops (Kim et al. 1998a; Kim and Ohkawa 2001; De la Cuadra et al. 2002). It was found that after leaf senescence, the bulbs of *L. coquimbensis* require a dry storage period at warm temperatures (20°C) for dormancy release, but no cold is required for flowering initiation (Kim and Ohkawa 1998; Kim et al. 1998b; Ohkawa et al. 1998).

Seed germination is achieved after 24 h water imbibition, followed by cold stratification for 3–4 weeks (*L. purpurea*), 4 weeks (*L. coquimbensis*), or 5–7 weeks (*L. ixioides*) (Schiappacasse et al. 2002). Although relatively small in size, the bulbs can be propagated vegetatively by several techniques, such as chipping and scooping. *In vitro* propagation has been successful in *L. purpurea* (Olate and Bridgen 2005; Escobar et al. 2008).

Most of the species have attracted interest as ornamental plants, either as cut flowers or garden plants (Figure 16.5), while *L. ixioides, L. coquimbensis, L. vittata,* and *L. purpurea* have especially attractive flowers (Riedemann et al. 2006). However, some ecotypes of *Leucocoryne* species produce an undesirable onion smell (Lancaster et al. 2000), which could limit their commercial desirability and, therefore, development as new ornamental crops. Two cultivars of *Leucocoryne* bred by the Universidad Católica de Valparaíso were the first Chilean ornamental plants developed in Chile

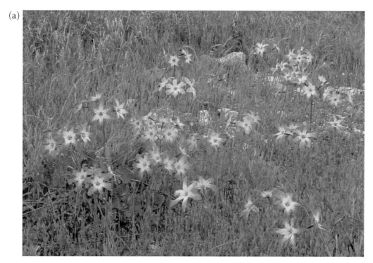

FIGURE 16.5   (**See color insert.**) *Leucocoryne* species. (a) *L. coquimbensis*; (b) *Leucocoryne* natural hybrid. (Photographs by E. Olate.)

and patented in the United States (El Mercurio de Valparaíso 2004). An active breeding program produced numerous selections and hybrids with great potential as potted or garden plants (Mansur et al. 2002).

The cut flowers of *Leucocoryne* demonstrate a long postharvest life and are produced commercially in Holland, New Zealand, and Japan. In Chile, native selections have already been exported to various foreign markets (El Mercurio de Valparaíso 2004).

F. RHODOPHIALA

The genus *Rhodophiala* is distributed throughout South America. Within 27 species identified in Chile, 25 are endemic to the country, and two species are native to Chile and Argentina. For many years, this genus was considered part of *Hippeastrum*, but *Rhodophiala* is now generally accepted as a separate genus (Meerow et al. 2000). In Chile the distribution of *Rhodophiala* spp. ranges between latitudes of 24° and 42°S, but 60% of these species are concentrated between latitudes of 30° and 38°S (Ravenna et al. 1998). They are found in the desert and semidesert zones, the coastal range, the central valley, and the planes and hills of the Andes and pre-Andes range (Schiappacasse et al. 2002). Many species have attracted great interest as ornamentals, for example, *R. advena* and *R. phycelloides*, both with red flowers, *R. bagnoldii* with a unique solid yellow color (Figure 16.6), and *R. rhodolirion* with large, attractive pink-white flowers.

The plants of *Rhodophiala* possess a perennial tunicate bulb, with linear leaves that dry out during anthesis under natural conditions (Muñoz 1985), but remain green under cultivation (Schiappacasse et al. 2002). Similar to other Amaryllidaceae species, the bulbs consist of enlarged basal tissue of the leaves and storage scales. The bulbs enlarge annually by producing independent growing units formed by scales, which protect the flower stem. Bulb offsets originate between the scales and the remains of a flower stem (Schiappacasse et al. 2002; Morales 2007). Each bulb forms two flower stems per year. The flower stem of *Rhodophiala* species reaches a length of 35–50 cm and bears an umbel with funnel-shaped flowers. The umbel holds up to six flowers, each 4–6 cm wide (Arroyo-Leuenberger and Leuenberger 1991; Riedemann and Aldunate 2001).

The juvenile stage lasts for 2–5 years depending on the species and culture conditions. *Rhodophiala* species require relatively low thermal variation and the bulbs grow faster in a growth medium heated to 20°C (Schiappacasse et al. 2006). Under natural conditions, flowering occurs during spring, summer, or even autumn, depending on the species and location. The desert species flower in the spring, both in nature and under cultivation, while most of the Andean species flower in the summer. Bulbs planted in Talca (250 km south of Santiago), at different dates in May, June, or July, flowered at about the same time in November–December, thus suggesting that flowering might respond to a specific environmental signal (Morales 2007). *Rhodophiala* species might develop symbiotic associations with some fungi species. It was shown that fibrous roots of *R. splendens*, *R. bagnoldii*, and *R. phycelloides*, collected in natural populations, have vesicular arbuscular mycorrhizae (Aguayo 2008).

In nature, the bulbs enter dormancy during fruit set, whereas under drought conditions dormancy can start even during anthesis. In cultivation, bulb dormancy is erratic, and it is not unusual to find plants in active growth next to others at rest. Thus, *R. bagnoldii* has a marked dormancy period from January to April; while slow leaf development was observed in *R. montana* from January to June, and in *R. splendens* from January to March. Plant dormancy seems to be stimulated by high temperatures, but further research is needed to confirm this assumption (Schiappacasse et al. 2006).

Seeds of *Rhodophiala* are large, flat, and shiny black. They do not have any specific requirement for germination. An exception is the seeds of *R. rhodolirion* that require 3 weeks of storage at 7°C for germination (Schiappacasse et al. 2002). It must be pointed out that *R. rhodolirion* differs from other *Rhodophiala* species in many ways (i.e., germination type, caryotype), and recent molecular studies suggest that the species belongs to a different genus (Letelier 2010; Muñoz et al. 2011).

(a)

(b)

(c)

FIGURE 16.6 (**See color insert.**) *Rhodophiala* species. (a) *R. bagnoldii*; (b) *R. phycelloides*; (c) *R. splendens*. (Photographs by E. Olate and F. Schiappacasse.)

Chipping, scoring, and twin scaling are used as traditional methods for bulb propagation or in combination with *in vitro* techniques (Basoalto 2001; Schiappacasse et al. 2002; Olate and Bridgen 2005; Jara et al. 2006; Seemann et al. 2006).

The stage for cut flower harvesting was found optimal when the florets are still protected by the bracts. In *R. bagnoldii*, pulsing with sucrose increased vase life up to 10 days, in comparison with 6–8 days in the control plants; flowers of *R. advena* last for about 7 days. Stem splitting and continued stem elongation after harvest are the main problems found in the postharvest stage (Schiappacasse et al. unpublished data).

A breeding program for *Rhodophiala* is being carried out in collaboration between Universidad de Talca and Universidad Austral de Chile. As part of the breeding program, polyploidy induction and hybrid production in *R. montana* and *R. splendens* were attempted by exposing *Rhodophiala* seeds to colchicine (Muñoz et al. 2006). Unfortunately, the flowering of polyploid clones was not achieved even after several years of cultivation (Schiappacasse et al. 2006). *Rhodophiala* species are highly compatible with the related genus *Phycella*, so the future development of intergeneric hybrids might be possible (Schiappacasse et al. 2006). Interspecific *Rhodophiala* hybrids have been developed and are currently under cultivation in Talca.

G. *ZEPHYRA*

Zephyra elegans was considered a monotypic genus and is an endemic species in Chile (Muñoz and Moreira 2000). Recently, a putative new species *Z. compacta* (Figure 16.7) was described as one separate from *Z. elegans* (Ehrhart 2001). Large populations of *Z. elegans* are only found in a very restricted region of the sandy coast of the Atacama Desert (latitude 28°–29°S).

A fibrous coated corm produces 2–3 narrow leaves and a single branched stem of 20–60 cm with a raceme of flowers, white inside and pale to bright blue on the outside (Muñoz 1985; Muñoz and Moreira 2000). The juvenile phase lasts up to four years when grown from seed (Kim et al. 1996, 1997, 1998c) or two years when propagated *in vitro* (Vidal and Niimi 2010). During the annual cycle, leaves emerge in late winter (June), and in the years with rainfall of over 100 mm the plants flower during the early spring (September–October). In seasons with no rain, a process of pupation, also described for *Freesia*, occurs and a new corm is formed on top of the existing one. The old corm crumbles, but the new corm does not form leaves. This process can take place every year until sufficient precipitation allows leaves and stem emergence, or until the storage materials in the corm deteriorate (K. Ohkawa 1999, pers. comm.). The plants do not tolerate freezing (Riedemann et al. 2006). The flowering period is during the spring, from August to October (Muñoz and Moreira 2000). When the shoot length reached 8 cm, the inflorescence was initiated in all plants observed (Kim et al. 1996). The vase life of the flower stems is about 10 days (Kim and Ohkawa 1997). The corms enter dormancy after the plant sets seeds and the foliage dries completely. Dormancy of the corms is released at the end of the warm and dry summer. Under cultivation, the corms have to be stored at 25°C for 22 weeks (Kim et al. 1996; Kim and Ohkawa 2001).

FIGURE 16.7 (**See color insert.**) *Zephyra compacta.* (Photograph by E. Olate.)

Z. elegans can be propagated by sowing the seeds by the end of the winter or in the autumn. Lifting the corms in the first season after sowing is not recommended (Riedemann et al. 2006). Vidal (2005) achieved high percentages of *in vitro* germination by treating the seeds with NaOCl (1%) and using plain agar as culture medium. There is no current breeding program on *Zephyra* in Chile.

IV.   LOCAL PRODUCTION OF ORNAMENTAL GEOPHYTES

Cut flowers are exported from Chile since 1978. Carnations were the main product during the 1980s and until the mid-1990s, when new crops were introduced into commercial production in the colder regions of the country. At present, several ornamental geophytes, such as alstroemeria, lily, liatris, tulip, calla lily, and *Allium*, are cultivated as cut flowers and also as propagating material for export.

Similar to New Zealand and South Africa, the strategic advantage of Chilean bulb production is its location in the Southern Hemisphere and the ability to supply bulbs out of season. Quality bulbs from Chile allow for early forcing of tulip and lily in Europe, Japan, or North America. In addition, the relatively low cost of labor, the availability of suitable land, and specialization of producers in Chile are also important advantages for Chilean bulb production (Wildenbeest 2008).

At present, Chile is considered the quickest-growing production area of flower bulbs outside The Netherlands. The area of production increased from ca. 100 ha in the 1990s to 500 ha in 2010. Between the years 2000 and 2004, the export of lily and tulip bulbs from Chile grew by 126%. The export value for flower bulbs was US$2.3 million in 1998, US$10.5 million in 2003 and over US$30 million in 2010 (ODEPA 2011). The Netherlands, the USA and Japan take the largest share of the exports of Chilean bulbs (Wildenbeest 2008).

Bulb production in Chile is concentrated in the Los Lagos area in the cool south. This region has favorable phyto-sanitary and climatic conditions, with cold rainy winters, which avoid the presence of virus vectors, and very light and porous soils rich in organic matter (10–20%), perfect for the growth of bulbs. Chilean bulb companies are usually joint ventures between local and Dutch investors, with local production management and a high initial investment from both parties in equipment, facilities, technology, and licenses (MundoAgro 2010). In this sense, Chilean growers are well-educated entrepreneurs willing to make large and long-term investments in the bulb production sector. The Dutch partners provide part of the initial investment, the know-how, most of the plant material, and also the market network.

Chilean production companies are managed entirely by Chilean growers. Often, they are subsidiaries of international agricultural companies, which are active in the ornamentals sector, vegetable production, and/or seed breeding. This applies, for example, to Piga Flower Bulbs Co. and to the Sone Bulb Co. Both groups have access to extensive research facilities and laboratories. This means that Chilean bulb production is well integrated into the local and regional economy. Southern Bulb Co., the largest bulb company in Chile, produces lily and tulip cultivars on 300 ha, all concentrated in the southern Lake District of the country, near the city of Osorno. A relatively new company, Sun Harvest S.A., has moved from Osorno to the warmer Los Angeles area to continue production of both lily and calla lily bulbs in more than 40 ha. Although during the last decade the number of companies involved in bulb production has not exceeded, they have all developed into large-scale business operations. All the companies are members in the APEB—Chile, the Bulb Producers and Exporters Association.

Chile bulb producers benefit from extensive free trade agreements with more than 20 countries, including EU (2003), the United States (2004), Japan, China, India, Australia, Switzerland, Turkey, South Korea, and most of the Latin American countries. Due to the recent phyto-sanitary agreement with Japan, Chile is becoming a serious competitor for New Zealand in Japan and the rest of Asia. At present, new markets, such as China and Mexico, are being developed for the Chilean bulb trade,

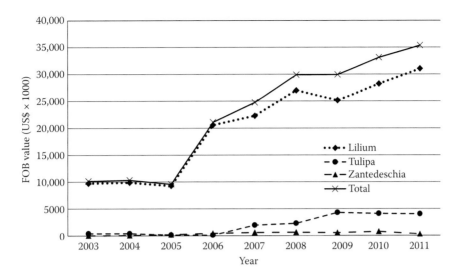

FIGURE 16.8   FOB value (US$ × 1000) of geophyte exports from Chile from 2006 to 2011. (From www. odepa.cl.)

and industry is combining efforts to open even newer markets for their products in Brazil, Colombia, and Ecuador.

Lily, tulip, and calla lily are the most important crops produced in Chile (Figure 16.8). Tulip bulbs are harvested in January and February, so that they can replace the long-stored "Ice tulips" produced in the Northern Hemisphere, and forced for the early season. Similarly, lily bulbs, harvested from June until August, are exported for forcing for Christmas and New Year, the Chinese New Year, Valentine's Day, and even for Mother's Day.

At present, Chilean companies do not conduct their own breeding programs and are, therefore, completely dependent on plant material bred overseas, acting only as a reproduction site to fill in stock gaps for some particular products and at some times of the year. Breeding is only done by university programs financed by public grants, and they all focus on native Chilean geophytes, such as *Leucocoryne*, *Rhodophiala*, *Alstroemeria*, and *Chloraea*.

V.   CONCLUSIONS

During the last two decades, Chile has become the quickest growing region of production for flower bulbs, mainly oriented for export. This fast-developing sector of ornamental horticulture is based on modern infrastructure (packing plants, refrigeration facilities, roads, transportation, and communication systems) and specialized know-how, coupled with outstanding agroecological conditions. The adoption of the recent trends in horticultural production, such as new varieties and growing systems, integrated crop management, organic production, and good agricultural practices, could lead to a more diversified and significant role for Chilean floriculture in the near future.

However, Chile's ornamental horticulture faces several challenges. Some of these are to develop more efficient marketing strategies, as well as a better understanding of the niche markets that are not covered by the large producers of the region, such as Colombia and Ecuador. Because of the great distances from the major consumers, postharvest management must be improved, including logistics and sea transportation alternatives. Efforts to legalize recognition of breeder's rights need to be strengthened; and, of course, competition from within the region should, by no means, be underestimated. The development of Chile's innovative products different from that of other Latin

American countries would seem to be a smart choice for the future of the national floriculture sector.

Rich geophyte flora can serve as a reservoir for the new and improved varieties of ornamental geophytes in Chile. Researchers, breeders, and funding agencies should strengthen breeding programs, including major geophytes crops, with a more market-oriented approach, to avoid long-term dependence on major competitors and international suppliers. However, current and future efforts might vanish if varieties are not protected and/or insufficient marketing efforts are made to introduce these new crops in the international market.

At present, a major weakness of Chilean producers of flower bulbs is their heavy reliance on Dutch knowledge and, therefore, the associated dependence on Dutch support in terms of planting material and company structure. Therefore, one of the most important tasks of the floriculture sector today is to develop a larger group of local experts to improve formal education of floriculture professionals and to strengthen applied research at the universities, agriculture schools, and private companies.

CHAPTER CONTRIBUTORS

Eduardo A. Olate is a professor at the Pontificia Universidad Catolica de Chile, Santiago, Chile. He was born and brought up in Santiago, and earned his MSc from the Pontificia Universidad Catolica de Chile, and his PhD from the University of Connecticut. He has held a faculty position at the Department of Plant Science of the Pontificia Universidad Catolica de Chile since 1994, and was elected Chair of the Department of Plant Science from 2006 until 2010. His expertise comprises the physiology, *in vitro* propagation and breeding of ornamental crops, with a special emphasis on Chilean native geophytes. Dr. Olate has published more than 20 journal articles and presented more than 30 conference papers in national and international symposia.

Flavia Schiappacasse is a professor of floriculture at the Universidad de Talca, Talca, Chile. She was born in Santiago, Chile, graduated from the Universidad Católica de Chile, and obtained her MS from Cornell University. Her main expertise comprises the physiology and breeding of ornamental geophytes and cultivation of ornamental Proteaceae. She has published more than 40 reviewed papers, technical articles, and symposia reports, and has coauthored three books on ornamental crops (in Spanish).

REFERENCES

Aguayo, E. 2008. Assesment and study of mycorriza associations on *Rhodophiala* species. Dissertation, Universidad de Talca, Talca, Chile (In Spanish).

Aker, S., and W. Healy. 1990. The phytogeography of the genus *Alstroemeria. Herbertia* 46:76–87.

Araneda, L., P. Salas, and L. Mansur. 2004. Chromosome numbers in the Chilean endemic genus *Leucocoryne* (Huilli). *J. Am. Soc. Hort. Sci.* 129:77–80.

Aros, D., H. J. Rogers, and C. Rosati. 2008. Floral scent evaluation in *Alstroemeria* through gas chromatography-mass spectrometry (GC-MS) and semiquantitative RT-PCR. *Acta Hort.* 886:19–26.

Arriagada, L., and O. Zöllner.1994. *Rhodophiala laeta* Philippi, a beautiful amaryllid from northern Chile. *Herbertia* 50:22–23.

Arroyo-Leuenberger, S. C., and B. E. Leuenberger. 1991. Notes on *Rhodophiala rhodolirion* (Amaryllidaceae) from the Andes of Mendoza, Argentina. *Herbertia* 47:80–87.

Assis, M. C. 2002. Our species of *Alstroemeria* L. (Alstroemeriaceae) from Minas Gerais, Brazil. *Revista Brasileira de Botánica* 25(2):177–182 (In Portuguese).

Assis, M. C., and R. Mello-Silva. 2002. Flora of Cipó Mountains, Minas Gerais: Alstroemeriaceae. *Boletín de Botánica de la Universidade de São Paulo* 20:49–52 (In Portuguese).

Assis, M. C. 2004. New species of *Alstroemeria* (Alstroemeriaceae) from the Brazilian savannas. *Novon* 14:17–19.

Basoalto, A. 2001. *In vitro* vegetative propagation of *Rhodophiala montana* (Phil) Traub. Dissertation, Universidad de Talca, Talca, Chile (In Spanish).

Bayer, E. 1987. The genus *Alstroemeria* in Chile. *Mitt. Bot. Staatssamml. München* 24:1–362 (In German).

Bridgen, M. P. 1992. *Alstroemeria*. In *The Physiology of Flower Bulbs*, eds. A. De Hertogh, and M. Le Nard, 201–209. Amsterdam: Elsevier.

Bridgen, M. P. 1998. Alstroemeria. In *Ball Redbook*, ed. V. Ball, 341–348. Batavia, IL: Ball Publishing.

Bridgen, M. P., E. Olate, and F. Schiappacasse. 2002. Flowering geophytes from Chile. *Acta Hort.* 570:75–80.

Bridgen, M., and E. Olate. 2004. *Alstroemeria* of Chile. *Bulbs* 6(2):9–14.

Central Bank of Chile. 2011. Macroeconomic indexes of second trimestre of 2011. Central Bank of Chile. http://www.bcentral.cl

De la Cuadra, C., L. Mansur, G. Verdugo, and L. Arriagada. 2002. Deterioration of *Leucocoryne* spp. seeds as a function of storage time. *Agric. Téc.* 62(1):46–65.

Ehrhart, C. 2001. *Zephyra compacta* (Tecophilaeaceae)—A new species from Chile. *Sendtnera* 7:47–52 (In German).

El Mercurio de Valparaíso. 2004. Two Leucocoryne varieties are patented. http://www.leucocoryne.cl/notas/nota-mercurio.htm

Elgar, H. J., T. A. Fulton, and E. F. Walton. 2003. Effect of harvest stage, storage and ethylene on the vase life of *Leucocoryne*. *Postharvest Biol. Technol.* 27:213–217.

Escobar, L. H., M. Jordan, E. Olate, L. Barrales, and M. Gebauer. 2008. Direct and indirect *in vitro* organogenesis of *Leucocoryne purpurea* (Alliaceae), a Chilean ornamental geophyte. *Propagation of Ornamental Plants* 8:59–64.

Fundación para la Innovación Agraria. 2007. Breeding of Chilean orchids of the genus *Chloraea*. Final report. http://bibliotecadigital.innovacionagraria.cl (In Spanish).

Guaglianone, E. R., and S. Arroyo-Leuenberger. 2002. The South American genus *Oziroë* (Hyacinthaceae—Oziroëoideae). *Darwiniana* 40(1–4):61–76.

Han, T.-H., M. de Jeu, H. van Eck, and E. Jacobsen. 1999. Genetic diversity of Chilean and Brazilian *Alstroemeria* species assessed by AFLP analysis. *Heredity* 84:564–569.

Healy, W. E., and H. F. Wilkins. 1986. Relationship between rhizome temperatures and shoot temperatures for floral initiation and cut flower production of *Alstroemeria* 'Regina'. *J. Am. Soc. Hort. Sci.* 111:94–97.

Hoffmann, A. E. 1989. Chilean monocotyledonous geophytes: taxonomic considerations and their state of conservation. *Herbertia* 45:13–29.

Hofreiter, A. 2006. *Leontochir*: A synonym of *Bomarea* (Alstroemeriaceae)? *Harvard Papers in Botany* 11(1):53–60.

Instituto Darwinion. 2009. Flora of Southern South America/ Flora del Cono Sur. CONICET, Argentina. http://www2.darwin.edu.ar/ (In Spanish).

Jara, G., P. Seemann, M. Muñoz, R. Riegel, F. Schiappacasse, P. Peñailillo, and A. Basoalto. 2006. Effect of meta-topoline on *in vitro* propagation of *Rhodophiala splendens* in liquid culture medium. *Agro Sur* 34(1–2):59–60.

Kim, H. H., K. Ohkawa, and K. Sakaguchi. 1996. Effects of storage temperature and duration on flower bud development, emergence and flowering of *Zephyra elegans* D. Don. *Scientia Hortic.* 67:55–63.

Kim, H.-H., and K. Ohkawa. 1997. *Zephyra elegans* D. Don as a cut flower. *Herbertia* 52:26–30.

Kim, H.-H., and K. Ohkawa. 1998. Studies on the growth cycle and flowering control of *Leucocoryne coquimbensis* F. Phil. *Herbertia* 53:64–71.

Kim, H. H., and K. Ohkawa. 2001. Introduction of two Chilean geophytes, *Leucocoryne coquimbensis* F. Phil. and *Zephyra elegans* D. Don as new ornamentals. *Acta Hort.* 552:179–184.

Kim, H. H., K. Ohkawa, and E. Nitta. 1998a. Effects of bulb weight on the growth and flowering of *Leucocoryne coquimbensis* F. Phill. *Acta Hort.* 454:341–346.

Kim, H.-H., K. Ohkawa, and E. Nitta. 1998b. Fall flowering of *Leucocoryne coquimbensis* F. Phil. after long-term bulb storage treatments. *HortScience* 33:18–20.

Kim, H. H., K. Ohkawa, and K. Sakaguchi. 1998c. Effects of corm weight on the growth and flowering of *Zephyra elegans* D. Don. *Acta Hort.* 454:335–340.

Kim, H. H., K. Sakaguchi, and K. Ohkawa. 1997. *Zephyra elegans* D. Don., a potential new cormous crop. *Acta Hort.* 430:133–137.

Krarup, C. 2002. Counter season to Mediterranean horticulture. *Chronica Hort.* 42(1):31–35.

Kristlansen, K. 1995. Interspecific hybridization of *Alstroemeria*. *Acta Hort.* 420:85–88.

Lancaster, J. E., M. L. Shaw, and E. F. Walton. 2000. *S*-Alk(en)yl-L-cysteine sulphoxides, alliinase and aroma in *Leucocoryne*. *Phytochemistry* 55:127–130.

Letelier, L. A. 2010. Phyllogeny of Amaryllidaceae genera in Chile: an approach from molecular biology. Facultad de Ciencias Agrarias, Escuela de Agronomía. Dissertation, Universidad de Talca, Talca, Chile (In Spanish).

Lu, C., Y. Ruan, and M. Bridgen. 1995. Micropropagation procedures for *Leontochir ovallei*. *Plant Cell, Tiss. Org. Cult.* 42:219–221.

Mansur, L., O. Zöellner, P. Riedemann, G. Verdugo, and C. Harrison. 2002. *Leucocoryne*, a native Chilean genus and its use as garden plant. Universidad Católica de Valparaíso, Valparaíso, Chile (In Spanish).

Mansur, L., and C. de la Cuadra. 2004. Description of the first stage of the life cycle of three genotypes of *Leucocoryne* sp.: Seed to bulb. *Agricultura técnica* (Chile) 64(2):205–212.

Manzano, E., M. Musalem, P. Seemann, and F. Schiappacasse. 2009. Flora of the Aysen Region in Patagonia: Study, propagation and management of native species with ornamental characteristics. Universidad Austral de Chile (In Spanish).

Marticorena, C., 1990. Contribution to the statistics of vascular flora of Chile. *Gayana Bot.* 47(3–4):85–113 (In Spanish).

Marticorena, C. and R. Rodríguez. 1995. *Flora of Chile: Pteridophyta-Gymnosperma. Vol. 1*. Concepción, Chile: Universidad de Concepción (In Spanish).

Meerow, A. W., C. L. Guy, Q-B. Li, and S-Y. Yang. 2000. Phylogeny of the American Amaryllidaceae based on nrDNA ITS sequences. *Syst. Bot.* 25:708–726.

Meerow, A. W., and A. F. C. Tombolato. 1996. The *Alstroemeria* of Itatiaia. *Herbertia* 51:14–21.

Mersey, L. 2003. Assesment and validation of a protocol for the germination of *Chloraea crispa* seeds inoculated with micorrizic fungi. Dissertation, Universidad Católica de Valparaíso. Chile (In Spanish).

Molnar, J. 1975. Alstroemeria—a promising new cut flower. *Ohio Florist's Assn. Bulletin* 553:1–2, 5.

Morales, P. 2001. Effect of the bulb weight on the flowering capacity and flower differentiation of *Herbertia lahue* (MOL.) Goldbl. Dissertation, Universidad de Talca, Talca, Chile (In Spanish).

Morales, M. 2007. Effect of different planting dates on the bulb growth and flowering of *Rhodophiala* sp. Dissertation, Universidad de Talca, Talca, Chile (In Spanish).

Morales, M., F. Schiappacasse, P. Peñailillo, and P. Yáñez. 2009. Effect of bulb weight on the growth and flowering of *Herbertia lahue* subsp. *lahue* (Iridaceae). *Cien. Inv. Agr.* 36(2):259–266.

Mundoagro. 2010. Revista MundoAgro Digital 12 (November): 68–73. http:// www.mundoagro.cl

Muñoz, M. 1985. *Flowers of the Small North*. Santiago, Chile: Dirección de Archivos y Museos (In Spanish).

Muñoz, M., and A. Moreira. 2000. Endemic monocotyledoneous genera of Chile. http://www.chlorischile.cl/ Monocotiledoneas/Principalbot.htm (In Spanish).

Muñoz, M., and A. Moreira. 2003. *Alstroemeria of Chile: Diversity, Distribution and Conservation*. Santiago, Chile: Taller La Era (In Spanish).

Muñoz, M., R. Riegel, P. Seemann, G. Jara, F. Schiappacasse, P. Peñailillo, and A. Basoalto. 2006. Induction and detection of polyploidy in *Rhodophiala* Presl. *Agro Sur* 34(1–2):23–25.

Muñoz, M., R. Riegel, P. Seemann, P. Peñailillo, F. Schiappacasse, and J. Núñez. 2011. Phylogenetic relationships of *Rhodolirium montanum* Phil. and related species based on ITS regions and caryotipic análisis. *Gayana Botánica* 68(1):40–48 (In Spanish).

Novoa, P., J. Espejo, M. Cisternas, M. Rubio, and E. Domínguez. 2006. *Country Guide of Chilean Orchids*. Concepción, Chile: Editorial Corporación Chilena de la Madera (In Spanish).

ODEPA. 2011. Office for Agriculture Studies and Policies, Chilean Ministry of Agriculture/ Oficina de Estudios y Políticas Agrarias, Ministerio de Agricultura, Gobierno de Chile. Available at www.odepa.cl

Ohkawa, K. 1994. *Alstroemeria*. Tokyo: Seibundo Shinkosha (In Japanese).

Ohkawa, K., H.-H. Kim, E. Nitta, and Y. Fukazawa. 1998. Storage temperature and duration affect flower bud development, shoot emergence, and flowering of *Leucocoryne coquimbensis* F. Phil. *J. Am. Soc. Hort. Sci.* 123:586–591.

Olate, E., and M. Bridgen. 2005. Techniques for the *in vitro* propagation of *Rhodophiala* and *Leucocoryne* spp. *Acta Hort.* 673:335–342.

Olate, E., C. Sepúlveda, L. Escobar, H. García, M. Gebauer, and M. Musalem. 2007. The native *Alstroemeria* plant breeding program. *Agro Sur* 35(2):17–18.

Price, G. R., and R. W. Langhans. 1985. *Alstroemeria*. Ithaca, New York: Department of Floriculture and Ornamental Horticulture. Cornell University.

Ravenna, P., S. Teillier, J. Macaya, R. Rodríguez, and O. Zöllner. 1998. Conservation status of the native bulbous plants of Chile. *Boletín del Museo Nacional de Historia Natural de Chile* 47:47–68 (In Spanish).

Riedemann, P., and G. Aldunate. 2001. *Native Flora of Ornamental Value: Identification and Propagation: Central Region*. Santiago, Chile: Editorial Andrés Bello (In Spanish).

Riedemann, P., and G. Aldunate. 2003. *Native Flora of Ornamental Value: Identification and Propagation: Southern Region.* Santiago, Chile: Editorial Andrés Bello (In Spanish).

Riedemann, P., G. Aldunate, and S. Teillier. 2006. *Native Flora of Ornamental Value: Identification and Propagation: Northern Region.* Santiago, Chile: Self-publication (In Spanish).

Riedemann, P., G. Aldunate, and S. Teillier. 2008. *Native Flora of Ornamental Value: Identification and Propagation: Andes Region.* Santiago, Chile: Self-publication (In Spanish).

Rojas, P., and C. Müller. 1987. Market analysis: Alstroemeria, new cut flower/ Análisis de mercado. Alstroemeria, nueva flor fresca de corte. *El Campesino* 118:43–64 (In Spanish).

Schiappacasse, F., P. Peñailillo, and P. Yáñez. 2002. Propagation of Chilean bulbous species. Talca, Chile: Editorial Universidad de Talca (In Spanish).

Schiappacasse, F., P. Peñailillo, A. Basoalto, P. Seemann, G. Jara, and M. Muñoz. 2006. Advances in studies of growth and flowering of *Rhodophiala* species. *Agro Sur* 34(1–2):28–29.

Schiappacasse, F., P. Peñailillo, A. Basoalto, P. Seemann, R. Riegel, M. Muñoz, and G. Jara. 2007. Biotechnological applications on plant breeding on Chilean *Rhodophiala* species: morphological and physiological studies. *Agro Sur* 35(1–2):65–67.

Seemann, P., G. Jara, M. Muñoz, R. Riegel, F. Schiappacasse, P. Peñailillo, and V. Vico. 2006. Effect of cytokinins on the *in vitro* bud regeneration of *Rhodophiala rhodolirion*. *Agro Sur* 34(1–2):65–66.

Squeo, F. A., R. Osorio, and G. Arancio. 1993. *Andean Flora of Coquimbo Region: Dona Ana Range.* La Serena, Chile: Ediciones Universidad de La Serena (In Spanish).

Teillier, S., H. Zepeda, and P. García. 1998. *Wildflowers of the Chilean Desert.* Valdivia, Chile: Marisa Cuneo Ediciones.

Uphoff, J. C. T. 1952. A review of the genus *Alstroemeria. Plant Life* 8:37–53.

Verdugo, G., J. Marchant, M. Cisternas, X. Calderón, and P. Peñaloza. 2007. Morphometric characterization of *Chloraea crispa* LINDL. (Orchidaceae) seed germination through the image analysis technique. *Gayana Bot.* 64(2):232–238.

Vidal, A. K. 2005. Creación de procotolos para la iniciación *in vitro* de semillas y cormos de *Zephyra elegans* D. Don. Master thesis, Universidad Católica de Chile. Santiago, Chile.

Vidal, A. K., and Y. Niimi. 2010. *In vitro* initiation of seeds and further corm weight gaining in *Zephyra elegans. Acta Hort.* 865:301–304.

Villagrán, C., C. Marticorena, and J. Armesto. 2007. *Flora of Vascular Species of Zapallar: Enlarged and Revised Illustrated Edition of Federico Johow's book.* Santiago, Chile: Puntángeles y Fondo Editorial U.M.C.E (In Spanish).

Vogel, H., F. Schiappacasse, M. Valenzuela, and X. Calderón. 1999. Studies of generative and vegetative propagation in *Conathera* spp. *Ciencia e Investigación Agraria* 26(1):21–26.

Wildenbeest, G. 2008. Southern threat to Holland's monopoly. FloraCulture International http://www.floraculture-international.com.

Wilkins, H. F., and R. D. Heins. 1976. Alstroemeria general culture. *Florist Review* 159(4121):78–80.

Yáñez, P. 2001. Biological studies for the ornamental use of the Chilean geophyte *Conanthera bifolia* Ruiz et Pavon (Tecophilaceae). Master thesis, Universidad de Talca, Talca, Chile (In Spanish).

Zöllner, O. 2002. Description of genus *Leucocoryne* (Alliaceae), propagation and distribution. In *Leucocoryne, un género nativo chileno y su uso como planta de jardín*, eds. L. Mansur, P. Riedemann, O. Zöllner, G. Verdugo, and C. Harrison, 21–25. Valparaíso, Chile: Universidad Católica de Valparaíso (In Spanish).

Zöllner, O., and L. Arriagada. 1998. Two species of the genus *Leucocoryne* (Alliaceae) in Chile. *Herbertia* 53:100–103.

17 Geophyte Research and Production in New Zealand

Keith A. Funnell, Ed R. Morgan, Glenn E. Clark, and Joo Bee Chuah

CONTENTS

I. INTRODUCTORY REMARKS: CHARACTERISTICS OF AND STATISTICS ON THE NEW ZEALAND FLORICULTURE INDUSTRY

A. EXPORT VERSUS DOMESTIC

The population of New Zealand (NZ) is relatively small and comprises only ca. 1.6 million households (Statistics NZ 2010). Since the average family purchases only NZ$286 per annum (US$181) of plants, cut flowers, and garden supplies, this requires the NZ floriculture industry to be export

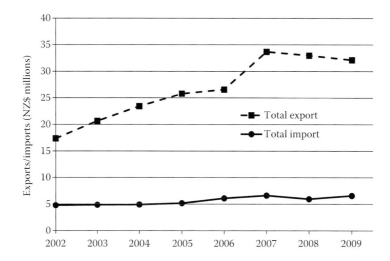

FIGURE 17.1 Annual value of NZ's exports (NZ$) and imports (NZ$) of geophyte planting material. (Adapted from Statistics New Zealand. 2010. http://www.stats.govt.nz/trade (date of download 29 April 2010).)

focused. In general, growth in exports of floriculture-related commodities began in the 1970s, stimulated by government incentives and market opportunities. Exporters promoted themselves in the Northern Hemisphere as suppliers of quality, counter-seasonally grown products.

In 1985, export values for cut flowers, foliage, and planting material of ornamental geophytes alone were NZ$8.1 million (US$4.05 million), of which cut flowers and planting material of ornamental geophytes constituted NZ$400,000 (US$200,000). By 2009, export values of cut flowers, foliage, and planting material of ornamental geophytes had grown eightfold to NZ$67 million (US$42.5 million) with the contribution of planting material of ornamental geophytes increasing nearly 80-fold to NZ$30.3 million (US$19.2 million; Figure 17.1). What drove this change? Government support had been instrumental in establishing the infrastructure and experience required, but the subsequent removal of direct government financial support and changing profitability of markets has required NZ growers to respond to changes so that the industry is currently economically sustainable. Throughout these periods of change, government support did remain in the form of investment into science and innovation, underpinning the ongoing development of this industry.

The period between 1999 and 2009 has seen exports of all ornamental plants (including ornamental geophytes), seeds, cut flowers, and foliage increase by 9% to NZ$84.9 million (US$53.9 million). However, within this category, exports of cut flower and foliage fell by 16% to NZ$38.8 million (US$24.6 million). Although exports of cut flowers from ornamental geophytes fell by 15% to NZ$29 million (US$18.4 million) between 2002 and 2009, over the same period exports and imports of plant material for ornamental geophytes increased by 85% and 37%, respectively (Figure 17.1). Hence, the commodities contributing to growth in export earnings are the planting material for ornamental geophytes rather than cut flowers. One exception to the decline in value of exported cut flowers from ornamental geophytes is the *Cymbidium* orchid, which between 1999 and 2009 actually increased by 10% to NZ$22.2 million (US$14 million).

B. ENTERPRISES, EMPLOYEE NUMBERS, AND REGIONAL DISTRIBUTION

In December 2009, NZ had 635 enterprises growing cut flowers and foliage (Statistics NZ 2010) with 11,400 people employed in the various plant-producing nurseries and cut flower and foliage enterprises (Timmins 2009). A total of 610 ha is involved in the production of ornamental geophytes in the open ground and 16.5 ha under protective cover (Statistics NZ 2010).

The region in and around NZ's largest city of Auckland has 48% of NZ's productive area for ornamental geophytes grown under protective cover. Cold-requiring geophytes such as *Lilium*, *Tulipa*, and *Paeonia* tend to be grown in more southern regions with 442 ha of ornamental geophytes grown outdoors within the South Island, while others, for example, *Zantedeschia*, are more frequently grown in more northern regions, with 168 ha of ornamental geophytes grown outdoors within the North Island.

II. CHALLENGES AND STRATEGIES OF AN ECONOMICALLY SUSTAINABLE INDUSTRY OF ORNAMENTAL GEOPHYTES

New Zealand is geographically isolated, being between 8000 and 18,000 km from significant markets in Asia, North America, or Europe. Currency exchange rates and transport costs are thus an important consideration in developing a sustainable export industry. Together with the challenges outlined in the previous section, this raises the question: what strategies should be adopted to ensure an economically sustainable export industry for ornamental geophytes in NZ?

As discussed later and subsequently illustrated by case studies, NZ pursues the development of export markets, seeking strategic advantages in

1. A novel product range
2. Timeliness of supply
3. Quality products
4. Market proximity
5. Better technologies for cost-effective production

A. NOVEL PRODUCT RANGE

Only a few of NZ's native plant species are geophytes, none of which are currently produced commercially in NZ as an ornamental. Therefore, NZ's suite of ornamental geophytes has arisen from commercializing introduced species, albeit typically less well-known ones. Additionally, the establishment of breeding programs to generate novel cultivars is seen as a key strategy for research and development. For instance, intergeneric hybrids between *Sandersonia* × *Gloriosa* (Burge et al. 2008) and *Sandersonia* × *Littonia* (Morgan et al. 2001) have resulted in a range of novel genetic combinations available for selection as cut flowers or potted plants.

As described below, *Sandersonia aurantiaca* and *Zantedeschia* spp. are examples of introduced species which NZ developed as unique export crops for both cut flowers and tubers. Both these crops have undergone dynamic changes over the last 10 years with new strategies now required to ensure their continued economic sustainability. In contrast, *Gentiana* is at the other end of the spectrum of the typical product life cycle. Export earnings for cut flowers of *Gentiana* in 1993 were NZ$50,000 (US$27,000), and by 2009 had almost doubled to NZ$75,000 (US$47,600). This expansion has to a large degree been driven by interspecific hybridization, where the novel selections are gaining market acceptance as cut flowers, potted, or landscape plants (Morgan et al. 2003; Morgan 2004).

B. TIMELINESS OF SUPPLY

Since the 1970s, NZ has utilized the counter-seasonal market advantage to supply the key Northern Hemisphere markets of Japan, Europe, and North America. *Zantedeschia*, *Paeonia*, and *Nerine sarniensis* are all ornamental geophytes grown in NZ, which supplement Northern Hemisphere production by supplying cut flowers or planting material in the counter-season market window.

Within the broad 6-month counter season, the need to target specific festivals, holidays or wedding seasons has necessitated undertaking the research and subsequent development of scheduled flowering and/or lifting of planting material for each of these geophytes (Warrington et al. 1989, 2011; Davies et al. 2002a,b; Carrillo Cornejo et al. 2003; Hall et al. 2007; Funnell 2008). The case study on *Nerine* (Section IV.C.3) illustrates this strategy.

C. Quality Product

The comparatively high costs of production and transport mean that sustained economic viability requires higher sales prices. Consequently, NZ growers of ornamental geophytes target more affluent markets focusing on superior genotypes. For example, NZ exports colored selections of *Zantedeschia* hybrids that are more sought after than the typically lower-priced *Z. aethiopica*.

In targeting the higher-priced market, NZ has the additional advantage of its naturally high solar radiation and a temperate maritime climate. This climate results in enhanced pigment expression leading to more intense flower colors and associated improved quality, for example, *Zantedeschia* (MacKay et al. 1987).

As new crops have been introduced, postharvest handling technologies have been developed to ensure that high-quality products are delivered to markets (Eason et al. 2001, 2002a,b, 2004; Funnell 2005; Chen et al. 2009).

Technologies to improve the quality and performance of the actual planting material of *Zantedeschia* and *Gentiana* are currently being developed. When integrated into NZ's production systems, these initiatives are expected to contribute toward the ongoing economic sustainability of these crops.

D. Market Proximity

As the closest large market, Asia, in particular Japan, received 51% of NZ's total exports of cut flowers and foliage in 2009. North America (primarily the United States), another large and comparatively close market, took 27% of NZ's exports in 2009. In contrast, while being a large market with counter-seasonal opportunities, Europe, in particular The Netherlands, only received 4% of NZ's 2009 exports. At more than twice the distance from NZ than Asia, a low proportion of exports going to Europe is to be expected.

Attempts to reduce NZ's dependency on the Japanese market have been successful, with the proportion of export value to this market declining from 61% to 51% between 1991 and 2009. In contrast, exports to North American have grown from 14% to 27% over the same period.

E. Better Technologies for Cost-Effective Production

As international competition reduces the profitability of growing novel crops in the counter-seasonal market window, investment in technologies that improve the economic sustainability of production becomes a vital strategy. For instance, technologies providing for a 5-fold increase in flower production with *Zantedeschia* (Funnell et al. 1992) and maintaining productivity following storage through application of gibberellins (Funnell and Go 1993) allowed NZ growers to achieve high productivity in a comparatively low-cost production system. As described in the following case studies, new technology to reduce production costs by increasing tuber quality and flower production of *Zantedeschia* is progressing (NZ Calla Council Inc. 2010). Additionally, together with industry partners new technologies are currently under development at the New Zealand Institute for Plant and Food Research. Application of this technology will enhance the efficiency of breeding of *Zantedeschia* by reducing the time before commercial release of new cultivars and should have a positive impact on the economic sustainability of this crop for NZ.

III. CASE STUDIES OF THE RESEARCH AND DEVELOPMENT OF NEW CROPS IN NEW ZEALAND

A. "COME AND GONE": *SANDERSONIA AURANTIACA*

Sandersonia was identified as a potential export crop in the early 1980s, and research led to the development of technologies that helped it develop as an important cut flower and tuber crop (reviewed by Burge et al. 2008). From the 1990s to the early 2000s, *Sandersonia* was ranked as the third largest earner amongst exported flower crops, but has since declined to the sixth largest in 2009.

1. Dynamics of Production and Export

Initially, *Sandersonia* was grown on a small scale for the local market during its natural flowering period of late November/early December. Later production systems were developed which enabled harvesting of quality stems and tubers over a 9-month period.

Limited tuber numbers meant cut flower exports grew slowly through the 1980s, reaching 120,000 stems in 1990, peaking at 4.8 million stems in 1997; then declining to 180,000 stems by 2009 (Figure 17.2a). Prices dropped from NZ$2.14 per stem (US$1.16) in 1993 to a low of NZ$0.54 (US$0.36) in 2004, but subsequently improved to NZ$0.84 (US$0.53) in 2009.

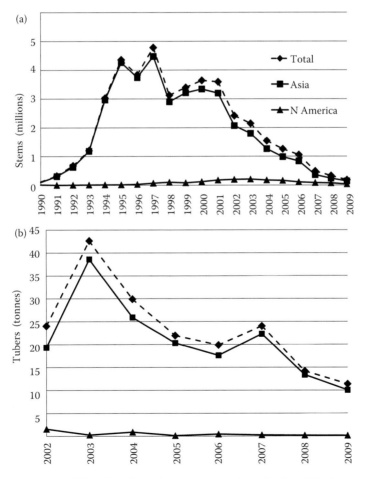

FIGURE 17.2 Exports from NZ of *Sandersonia* flower stems (a) and tubers (b). (Adapted from Statistics New Zealand. 2010. http://www.stats.govt.nz/trade (date of download 29 April 2010).)

Improved seed germination protocols resulted in increasing quantities of tubers available for export. Tuber exports peaked in 2003 and then declined to a quarter of this quantity by 2009 (Figure 17.2b). Throughout this period the major export market for S*andersonia* stems and tubers has been Asia (Figure 17.2a), primarily Japan and, to a lesser extent, Hong Kong and Taiwan. Small quantities were also exported to North America and Europe (mainly The Netherlands).

The reliance on a single primary market, that is, Japan, and a single color, has probably been a factor in the decline in sales. The crop lost its novelty value, and year-round availability of the product in the Northern Hemisphere limited the counter-seasonal opportunity for NZ growers. The decline in export returns also coincided with a downturn in the Japanese economy, and a less favorable exchange rate for NZ exporters. These factors caused many small growers to exit the industry, leaving a few larger, economically sustainable growers operating.

2. Challenges and Strategies for the Future

Increasing volumes of counter season production in Japan and increased competition from other Southern Hemisphere production countries (Chile and South Africa) mean that the development of new varieties is seen as a priority. Limited variability in the genus makes breeding and development of new varieties challenging. There have been claims of new colored forms of *Sandersonia* and one has been registered, but not commercially developed. New colors and forms are being developed using mutation breeding and intergeneric hybridization. Interspecific hybrids with *Littonia modesta* have been produced (Morgan et al. 2001). *Santonia* 'Golden Lights' (Morgan et al. 2003) was released following the development of production techniques (Clark et al. 2005a), before these techniques were sold to Japanese interests. Hybrids with *Gloriosa superba* were first developed in 1992 (Burge et al. 2008; Morgan et al. 2009) and a range of hybrids has subsequently been produced. Production studies have been carried out (Morgan et al. 2009) and the first hybrid released to commercial interests.

B. "Came, Peaked and Waning": *Zantedeschia*

New Zealand played a significant international role in the commercialization of *Zantedeschia*, with a rapid expansion of exports during the 1980s. As reviewed by Funnell (1993), research into flowering biology, postharvest, and manipulation of flower timing assisted in the development of technologies that helped it develop as an important cut flower and tuber crop. Ongoing research and a revision of strategies being implemented are currently under way so as to return this waning export industry to renewed growth.

1. Dynamics of Production and Export

Although breeding of *Zantedeschia* began in NZ in the 1930s, exports expanded following the development of an *in vitro* propagation system in the 1980s (Cohen 1981) with 250,000 stems exported in 1986. Growth in flower exports peaked at NZ$9.7 million (US$5.1 million) in the late 1990s. During the ten years between 2000 and 2009, cut flower exports declined by 51% to NZ$4.2 million (US$2.7 million; Figure 17.3). In contrast, the value of exports of planting material of *Zantedeschia* almost doubled over this same period. This change in focus from flowers to planting material (tubers and/or tissue cultured plantlets) in part reflects NZ's declining competitive advantage as an exporter of *Zantedeschia* flowers, as well as application of the improved understanding of the plant physiology underlying optimizing tuber yield and dormancy status (Funnell et al. 1998, 2002a,b; Carrillo Cornejo et al. 2003; Halligan et al. 2004) to achieve flower production all year round. If the data for 2008 and 2009 are acknowledged as being influenced by the global financial crisis, this move from flowers to plant material has resulted in total export earnings for *Zantedeschia* remaining comparatively stable over the last decade (Figure 17.3).

There were approximately 400 growers of *Zantedeschia* in NZ in the 1990s (Funnell and MacKay 1999), with 20 growers estimated to account for 60% of total exports at that time. By 2009, fewer

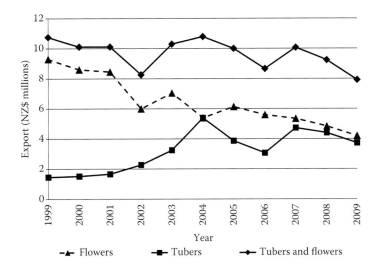

FIGURE 17.3 Value (NZ$) of exports from NZ of *Zantedeschia* flower stems and tubers. (Adapted from Statistics New Zealand. 2010. http://www.stats.govt.nz/trade (date of download 29 April 2010).)

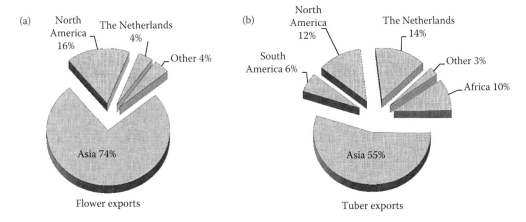

FIGURE 17.4 Distribution of NZ cut flower (a) and tuber (b) exports (NZ$ fob) of *Zantedeschia* in 2009. (Adapted from Statistics New Zealand. 2010. http://www.stats.govt.nz/trade (date of download 29 April 2010).)

than 100 growers remained, but typically were larger, more economically sustainable businesses than those operating in the late 1990s. Many of these larger businesses are more economic because they produce flowers all year round.

During 2000 to 2009, Asia (particularly Japan) imported 60–76% of NZ's *Zantedeschia* flowers, with North America being the next largest market taking 15–30% of exported value (Figure 17.4). The decline in stem numbers exported to Japan during the last 10 years has been the primary driver behind the reduction in NZ's total exports of this flower (Figure 17.3). Asia also featured as the dominant importer of planting material, with South America, North America, Africa and The Netherlands vying for the second place (Figure 17.4).

2. Challenges and Strategies for the Future

A return to growth of *Zantedeschia* exports from NZ will depend on reestablishing/maintaining an international competitive advantage. Research to extend the postharvest life of flowers by five days (Chen et al. 2009) has shown promise, but questions remain as to whether export markets are willing to pay a premium on such a product. New technology to increase tuber quality and flower production beyond that normally achieved using gibberellins is currently under commercial evaluation (NZ Calla Council Inc. 2010). While this technology is likely to improve the competitive advantage for NZ flower growers, it also offers opportunities to enhance tuber exports.

Breeding continues to offer one of the more long-term strategies for achieving growth in export earnings. Breeding for unique colors, improved yield and disease resistance remain as primary breeding objectives for the NZ industry, and is underpinned by research describing the genetic differences in spathe pigmentation (Lewis et al. 2003), flowering biology (Funnell and Go 1993), and susceptibility to bacterial softrot (Snijder et al. 2004). In addition, new technologies are currently under development to enhance the efficiency of breeding by reducing the time to cultivar release, and will impact on export earnings for NZ in the near future.

C. "Market Niche: Timeliness"—*Nerine sarniensis*

Since the late 1940s, NZ has bred and selected hybrids of *Nerine sarniensis*, but mainly as a garden plant. In the early 1980s, NZ growers began exporting the flowers targeting the counter-seasonal Northern Hemisphere market window. With the natural flowering season being short, the challenge has been both to extend the season and to target specific market windows within this extended season (Warrington et al. 1989, 2011).

1. Dynamics of Production and Export

Until 2006, export volumes of *Nerine* from NZ ranged between 1 and 1.3 million stems per annum. The temporary withdrawal of a key exporter reduced export volumes to less than 500,000 stems in 2009; however, volumes are expected to increase again by 2013. Over the last decade, over 90% of stems were exported to Asia, particularly Japan. Export of the bulbs themselves is limited to a few thousand per annum.

2. Challenges and Strategies for the Future

In NZ the natural flowering season of hybrids of *Nerine sarniensis* is limited to an 8-week period commencing mid-February. Over 70% of the stems are exported during a 2-week period, with prices typically declining to half that received during the early and late periods of these 8 weeks. The economic sustainability of this crop has, therefore, been limited by the short production window. If different populations of bulbs are either heat-treated or cool-stored, their combined periods of flower production can now be extended to cover 18 weeks (Warrington et al. 2011), resulting in a significant improvement in gross margins for growers (Lysaght et al. 1996).

While NZ's nerines comprise an extensive flower color range, there are no yellow-colored hybrids. Yellow flowers are found in other members of the Amaryllidaceae, including *Lycoris aurea*, but attempts to produce hybrids have been unsuccessful. Breeding for improved flower characteristics is also targeted, particularly for a hybrid with long stems and multiple pure-white florets.

D. "International Servicing": *Lilium* and *Tulipa* Bulb Production

Opportunities exist for NZ-based production of planting material of flowering size that provide advantages of cost, quality, and timing for flower production in the Northern Hemisphere. *Lilium*, *Tulipa*, and *Zantedeschia* are the most significant of these crops with export values increasing consistently since 1988 (Figure 17.5). The recent expansion of production for planting material coincides with large companies based in The Netherlands establishing business relationships with NZ companies. Planting material produced in NZ typically fills the gaps for early and late season supply into the Northern Hemisphere. This is because the growth season in NZ is six months out of phase with that occurring in the Northern Hemisphere and it becomes increasingly difficult to obtain

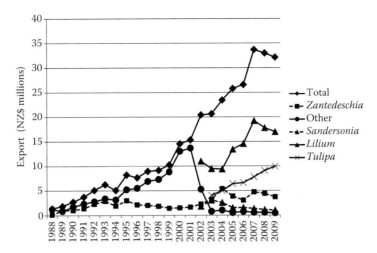

FIGURE 17.5 Export (NZ$) sales of propagation material (bulbs, tubers and corms) for major geophyte crops. "Other" includes *Tulipa* and *Lilium* until 2002/3, when subsequently listed separately. (Adapted from Statistics New Zealand. 2010. http://www.stats.govt.nz/trade (date of download 29 April 2010).)

quality fresh flowers from the bulbs produced in Europe the longer their storage approaches 12 months. Planting materials from NZ allow flower producers in Europe, Japan, and North America to obtain quality cut flowers off-season (see also Chapter 11).

Specialist bulb-growing ventures have been established, particularly in Southland and Canterbury (South Island), providing a range of economic and quality advantages. Much of the land used for bulb growing is leased, which works well with the dairy industry. The 10-year crop rotation cycle for bulb crops means that it is not viable to buy the land for this activity alone, so annual leases from dairy farmers run from March to March, and the land is returned to the farmer fertilized and reseeded in grass (MAF 2006). A further advantage is the reduced time from lifting in NZ to replanting in the Northern Hemisphere, with bulbs reaching customers in a better condition than that available domestically if stored long term.

1. *Lilium*

New Zealand has only exported small quantities of *Lilium* flower stems, with sales peaking at NZ$2.4 million (US$1.3 million) in the late 1990s, and declining to NZ$202,000 (US$128,000) in 2009. While in 1999 Asia received 79% of these stems, due to competition from other suppliers, sales to the Asian market declined to only 15% in 2009. Since 2000, the Pacific Islands (e.g., Cook Islands, French Polynesia, New Caledonia, and Samoa) have been the most consistent market for the flowers, receiving 66% of the stems in 2009.

Exports of *Lilium* bulbs have grown from NZ$11 million (US$5.1 million) in 2002 to a peak of NZ$19 million (US$14 million) in 2007. Exports in 2009 were primarily to Asia (70%) with Japan alone taking 40% (Figure 17.6a). The Netherlands (12%) is NZ's biggest market outside Asia, and ranks alongside the People's Republic of China and Taiwan in importance. Most markets have shown consistent growth, with market volatility evident with flower stem exports, while less pronounced with bulbs.

2. *Tulipa*

Tulip provides a very good example of NZ's role in the "International Servicing" side of the trade in flower bulbs. Whereas cut flowers form a minor part of the tulip export market (less than NZ$57,000 (US$26,500) per annum), over the last decade bulb exports have increased from a little over NZ$4 million (US$2.3 million) in 2003 to just under NZ$10 million (US$6.3 million) in 2009

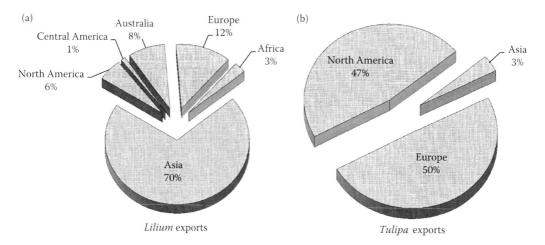

FIGURE 17.6 Distribution of *Lilium* (a) and *Tulipa* (b) bulb exports (NZ$) for the year ending December 2009. (Adapted from Statistics New Zealand. 2010. http://www.stats.govt.nz/trade (date of download 29 April 2010).)

(Figure 17.5). Similar values of tulip were exported to Europe (50%) and North America (47%) in 2009 (Figure 17.6b). Most of the bulbs sent to Europe went to The Netherlands (37%).

IV. CONCLUSIONS: A SUSTAINABLE FUTURE IN AN INCREASINGLY COMPETITIVE MARKET

New Zealand's floriculture export industry faces significant challenges, but opportunities are evident. Like other countries in the Southern Hemisphere, NZ has the challenge of distance from markets, but this cost is additionally compounded by relatively high production costs. Investment in technologies to reduce costs has enabled producers to remain competitive in the face of lower returns, with increased focus on the production of planting material providing greater opportunities for economic sustainability, in comparison with cut flower production. However, competing countries can also invest in low-cost production technologies thereby supplying competitive production into NZ's established markets at lower prices.

A common theme from exporters and producers is that they require access to both new suitable crops and new varieties. *Sandersonia* was an important export crop for about 10 years with its decline principally ascribed to a lack of new varieties. *Zantedeschia*, *Lilium*, and *Tulipa* continue to have a strong presence in international markets with continued growth being supported by the ongoing delivery of new technologies or varieties. New technologies can reduce costs, but in order to achieve economic sustainability, development and delivery of novel varieties must also occur. As achieved with *Sandersonia* and *Zantedeschia*, opportunities for NZ to introduce new crops to the global market continue to arise. It is anticipated that some new products being developed in NZ, for example, novel cultivars of *Gentiana*, and technologies, for example, for improved tuber quality in *Zantedeschia*, will provide growers with a similar advantage in the near future. Together with industry, at earlier stages of research and commercialization are other crops such as *Agapanthus* sp. (Burge et al. 2010), *Clivia* sp. (Murray et al. 2011), *Cyrtanthus* sp. (Clark et al. 2005b), and *Scadoxus* sp. (Funnell 2008). If successful, these other crops will provide growers with new opportunities on a longer-term horizon.

ACKNOWLEDGMENTS

The authors thank John Seelye and Ian Brooking of The New Zealand Institute for Plant and Food Research Ltd for a critical review of this chapter.

CHAPTER CONTRIBUTORS

Keith A. Funnell is a senior scientist at the New Zealand Institute for Plant and Food Research Limited, Palmerston North, NZ. He was born in New Zealand and obtained his BHortSci (Hons) and PhD from Massey University, Palmerston North, NZ. He held faculty positions at Massey University, and, since 2010, is a scientist with Plant and Food Research. The major topics of his research are the physiology of geophytes, production, and postharvest handling, specializing on new crops. Dr. Funnell has published more than 170 refereed journal articles, books or chapters, and conference papers. He is the chairman of the NZ Calla Council Inc. since 2005.

Ed R. Morgan is a senior scientist at the New Zealand Institute for Plant and Food Research Limited, Palmerston North, NZ. He was born in New Zealand and obtained BSc and MSc (Hons) from Massey University, Palmerston North, NZ. He held positions at Plant and Food Research and precursor organizations since 1988, and currently is leading the ornamentals research program at Plant and Food Research. His expertise comprises hybridization and breeding of geophytes, specializing on new crops. Ed Morgan has published more than 50 refereed articles in peer-reviewed journals, book chapters, and published conference proceedings.

Glenn E. Clark is a scientist at the New Zealand Institute for Plant and Food Research Limited, Pukekohe, NZ. He was born in New Zealand and earned his BSc and MSc from Waikato University, Hamilton, NZ. He held positions at Plant and Food Research and precursor organizations since 1980. The major topics of his research include the physiology and production of geophytes, specializing on new crops. Glenn has published more than 30 refereed articles in peer-reviewed journals, book chapters, and published conference proceedings.

Joo Bee Chuah was born in Malaysia and received her BBS (Accy.) from Massey University, NZ and her MSc (Information Studies) from Nanyang Technological University, Singapore. Prior to immigration to New Zealand in 2005, she was working in business-related industries in the areas of auditing, corporate finance and business information research in Singapore, and has held position of manager at PricewaterhouseCoopers (Singapore). In 2005–2011 she has held a position at the New Zealand Institute for Plant and Food Research Limited, Palmerston North, NZ.

REFERENCES

Burge, G. K., E. R. Morgan, J. R. Eason, G. E. Clark, J. L. Catley, and J. F. Seelye. 2008. *Sandersonia aurantiaca*: Domestication of a new ornamental crop. *Scientia Hortic.* 118:87–99.

Burge, G. K., E. R. Morgan, J. F. Seelye, G. E. Clark, A. McLachlan, and J. R. Eason. 2010. Prevention of floret abscission for *Agapanthus praecox* requires an adequate supply of carbohydrate to the developing florets. *S. Afr. J. Bot.* 76:30–36.

Carrillo Cornejo, C. P., K. A. Funnell, D. J. Woolley, and B. R. MacKay. 2003. Heat units may explain variation in duration of bud dormancy in *Zantedeschia*. *Acta Hort.* 618:469–475.

Chen, J., K. A. Funnell, D. J. Woolley, D. H. Lewis, and J. R. Eason. 2009. Abaxial and adaxial surfaces of spathe tissue of *Zantedeschia* differ in their pattern of re-greening. *Acta Hort.* 813:217–224.

Clark, G. E., G. K. Burge, E. R. Morgan, and C. M. Triggs. 2005a. Effects of planting date and environment on the cut flower production of Santonia Golden Lights. *Acta Hort.* 673:265–271.

Clark, G. E., G. K. Burge, and C. M. Triggs. 2005b. Effects of bulb storage, leaf and root pruning, on flower production in *Cyrtanthus elatus*. *New Zealand J. Crop Hortic. Sci.* 33:169–175.

Cohen, D. 1981. Micropropagation of *Zantedeschia* hybrids. *Proc. Int. Plant Prop. Soc.* 31:312–316.

Davies, L. J., I. R. Brooking, J. L. Catley, and E. A. Halligan. 2002a. Effects of constant temperature and irradiance on the flower stem quality of *Sandersonia aurantiaca*. *Scientia Hortic.* 93:321–332.

Davies, L. J., I. R. Brooking, J. L. Catley, and E. A. Halligan. 2002b. Effects of day/night temperature differential and irradiance on the flower stem quality of *Sandersonia aurantiaca*. *Scientia Hortic.* 95:85–98.

Eason, J. R., G. E. Clark, A. C. Mullan, and E. R. Morgan. 2002b. *Cyrtanthus*: an evaluation of cut flower performance and of treatments to maximise vase life. *New Zealand J. Crop Hortic. Sci.* 30:281–289.

Eason, J. R., E. R. Morgan, A. C. Mullan, and G. K. Burge. 2001. Postharvest characteristics of Santonia Golden Lights a new hybrid cut flower from *Sandersonia aurantiaca* × *Littonia modesta*. *Postharvest Biol. Technol.* 22:93–97.

Eason, J. R., E. R. Morgan, A. C. Mullan, and G. K. Burge. 2004. Display life of *Gentiana* flowers is cultivar specific and influenced by sucrose, gibberellin, fluoride, and postharvest storage. *New Zealand J. Crop Hortic. Sci.* 32:217–226.

Eason, J. R., T. Pinkney, J. Heyes, D. Brash, and B. Bycroft. 2002a. Effect of storage temperature and harvest bud maturity on bud opening and vase life of *Paeonia lactiflora* cultivars. *New Zealand J. Crop Hortic. Sci.* 30:61–67.

Funnell, K. A. 1993. *Zantedeschia*. In *The Physiology of Flower Bulbs*, eds. A. De Hertogh, and M. Le Nard, 683–739. Amsterdam: Elsevier.

Funnell, K. A. 2005. Nerine flower quality assurance. *Ministry of Agriculture and Forestry Sustainable Farming Fund Report*. http://www.maf.govt.nz/sff/about-projects/search/03–112/index.htm (date of download 17 June 2010).

Funnell, K. A. 2008. Growing degree-day requirements for scheduling flowering of *Scadoxus multiflorus* subsp. *katharinae* (Baker) Friis & Nordal. *HortScience* 43:166–169.

Funnell, K. A., and A. R. Go. 1993. Tuber storage, floral induction, and gibberellin in *Zantedeschia*. *Acta Hort.* 337:167–175.

Funnell, K. A., and B. R. MacKay. 1999. Directions and challenges of the New Zealand calla industry, and the use of calcium to control soft rot. In *The international symposium on development of bulbous flower industry*, eds. T.-F. Sheen, J.-J. Chen, T.-C. Yang, and M.-C. Liu, 30–44. Taichung, Taiwan: Taiwan Seed Improvement and Propagation Station.

Funnell, K. A., B. R. MacKay, and C. R. O. Lawoko. 1992. Comparative effects of Promalin® and GA$_3$ on flowering and development of *Zantedeschia* Galaxy. *Acta Hort.* 292:173–179.

Funnell, K. A., E. W. Hewett, J. Plummer, and I. J. Warrington. 2002a. Tuber dry-matter accumulation in *Zantedeschia* in response to temperature and photosynthetic photon flux. *J. Hort. Sci. Biotech.* 77:446–455.

Funnell, K. A., E. W. Hewett, J. Plummer, and I. J. Warrington. 2002b. Acclimation of photosynthetic activity of *Zantedeschia* Best Gold in response to temperature and photosynthetic photon flux. *J. Am. Soc. Hort. Sci.* 127:290–296.

Funnell, K. A., I. J. Warrington, E. W. Hewett, and J. Plummer. 1998. Leaf mass partitioning as a determinant of dry matter accumulation in *Zantedeschia*. *J. Am. Soc. Hort. Sci.* 123:973–979.

Hall, A. J., J. L. Catley, and E. F. Walton. 2007. The effect of forcing temperature on peony shoot and flower development. *Scientia Hortic.* 113:188–195.

Halligan, E. A., I. R. Brooking, K. A. Funnell, and J. L. Catley. 2004. Vegetative and floral shoot development of *Zantedeschia* Black Magic. *Scientia Hortic.* 99:55–65.

Lewis, D., S. C. Huang, K. A. Funnell, H. S. Arathoon, and E. E. Swinny. 2003. Anthocyanin and carotenoid pigments in the spathe of selected *Zantedeschia* hybrids. *Acta Hort.* 624:147–154.

Lysaght, S. D., E. A. Cameron, K. A. Funnell, and C. K. Drake. 1996. A spreadsheet model of the economics of *Nerine* production. *Proceedings of the Annual Conference of the NZ Agricultural Economics Society. Lincoln University. Agribusiness and Economics Research Unit* 144:202–208.

MacKay, B. R., K. A. Funnell, and N. W. Comber. 1987. The effect of shade and growing environment temperature on the growth and flower pigmentation development of *Zantedeschia* hybrids. *Tech. Rpt. 87/4, Dept. Hort. Sci. and N. Z. Nursery Research Centre, Massey Univ., N. Z.*, 18.

MAF. 2006. Floriculture, farm monitoring report. *Ministry of Agriculture and Forestry, NZ*. http://www.maf.govt.nz/mafnet/rural-nz/statistics-and-forecasts/farm-monitoring/2006/horticulture/horticulture-2006–06.htm (date of download 11 Jan. 2011).

Morgan, E. R. 2004. Use of *in ovulo* embryo culture to produce interspecific hybrids between *Gentiana triflora* and *Gentiana lutea*. *New Zealand J. Crop Hortic. Sci.* 32:343–347.

Morgan, E. R., G. K. Burge, J. F. Seelye et al. 2001. Wide crosses in the Colchicaceae: *Sandersonia aurantiaca* (Hook.) × *Littonia modesta* (Hook.). *Euphytica* 121:343–348.

Morgan, E. R., G. K. Burge, G. Timmerman-Vaughan, and J. Grant. 2009. Generating and delivering novelty in ornamental crops through interspecific hybridisation: some examples. *Acta Hort.* 836:97–103.

Morgan, E. R., B. L. Hofmann, and J. E. Grant. 2003. Production of tetraploid *Gentiana triflora* var. *japonica* 'Royal Blue' plants. *New Zealand J. Crop Hortic. Sci.* 31:65–68.

Murray, B. G., C. Wong, and K. R. W. Hammett. 2011. The karyotype of *Clivia mirabilis* analyzed by differential banding and fluorescence in-situ hybridization. *Plant Syst. Evol.* 293:193–196.

NZ Calla Council Inc. 2010. Technical Snippet. *CALLAnews* 20(1): 1. http://www.callas.org.nz/uploads/CallaNews20.1.pdf (date of download 24 June 2010).

Snijder, R. C., H. R. Cho, M. M. W. B. Hendriks, P. Lindhout, and J. M. van Tuyl. 2004. Genetic variation in *Zantedeschia* spp. (Araceae) for resistance to soft rot caused by *Erwinia carotovora* subsp. *carotovora. Euphytica* 135:119–128.

Statistics New Zealand. 2010. http://www.stats.govt.nz/trade (date of download 29 April 2010).

Timmins, J. 2009. *Seasonal Employment Patterns in the Horticultural Industry.* Wellington: Statistics New Zealand.

Warrington, I. J., N. G. Seager, and A. M. Armitage. 1989. Environmental requirements for flowering and bulb growth in *Nerine sarniensis. Herbertia* 45:74–80.

Warrington, I. J., I. R. Brooking, and T. A. Fulton. 2011. Lifting time and bulb storage temperature influence *Nerine sarniensis* flowering time and flower quality. *New Zealand J. Crop Hortic. Sci.* 39:107–117.

18 Geophyte Research and Production in South Africa

Graham D. Duncan

CONTENTS

I. INTRODUCTION

South Africa, in particular its winter rainfall zone, has the world's richest assemblage of native geophytes (Duncan 2006). Stretching from the Richtersveld in the far northwestern corner of the Northern Cape to Port Elizabeth in the southern part of the Eastern Cape, the area is characterized by the Mediterranean type of climate with cool rainy winters and dry summers, and is home to approximately 2100 geophytic species. Within this area, the greatest concentration falls in the southwest, in the Cape Floristic Region (CFR), extending from the Bokkeveld Mountains north of Nieuwoudtville in the Northern Cape, to Caledon in the Overberg region of the Western Cape. The vast summer rainfall parts of South Africa have relatively few geophytes, and the total number for the whole country exceeds 2500 species (Duncan 2010a).

II. NATIVE SPECIES OF ORNAMENTAL GEOPHYTES IN SOUTH AFRICA, THEIR VARIABILITY, AND BOTANICAL CHARACTERISTICS

The extraordinary variability of South African geophytes used in ornamental horticulture immediately becomes apparent when it is realized that they belong to 13 families: Agapanthaceae, Alliaceae, Amaryllidaceae, Araceae, Asphodelaceae, Colchicaceae, Haemodoraceae, Hyacinthaceae, Hypoxidaceae, Iridaceae, Oxalidaceae, Tecophilaeaceae, and Zingiberaceae (Duncan 2010a).

The Agapanthaceae has a single genus, *Agapanthus*, comprising two evergreen and four deciduous species with funnel-shaped or tubular flowers and strap-shaped leaves produced from an erect or creeping rhizome. The Alliaceae is represented by three genera in South Africa, of which only *Tulbaghia* (20 species) is of ornamental worth. It has both evergreen and deciduous members, producing umbels of small actinomorphic flowers with a characteristic corona. Its underground organs are rhizomatous, bulbous, or cormous, and most species are noted for the sharp onion- or garlic-smelling compounds given off by injured parts of the plant.

There are 18 South African genera of Amaryllidaceae which are autumn-, spring-, or summer-flowering. The most horticulturally noteworthy members include the sweet-scented, pink- or rarely white-flowered true *Amaryllis* (two species), and the universally popular *Clivia* (six species) that has trumpet-shaped or tubular flowers, and opposite rows of strap-shaped leaves. *Crinum* (23 species) has showy zygomorphic, white or pink, often striped flowers, a rosette of perennial, fleshy leaves and large bulbs with tough outer tunics. *Haemanthus* (22 species) is recognized by its congested, many-flowered white, reddish-orange or pink heads and 1–6 leathery leaves. *Nerine* (23 species) has widely flared, pink, red, or white blooms and deciduous or evergreen leaves, and *Scadoxus* (three species) has many-flowered, congested or spreading flower heads and thin-textured leaves with distinct midribs, usually carried on a distinct pseudostem. The underground storage organs are either rhizomes (e.g., *Clivia*) or bulbs (e.g., *Nerine*), noted for their poisonous amaryllid alkaloid compounds.

Zantedeschia is the only important ornamental South African member of the family Araceae. Growing from tuberous underground stems, it comprises five exclusively summer-growing species (*Z. albomaculata*, *Z. jucunda*, *Z. pentlandii* (Figure 18.1), *Z. rehmannii*, and *Z. valida*), one exclusively

FIGURE 18.1 (**See color insert.**) *Zantedeschia pentlandii* is one of five exclusively summer-growing species from South Africa. (Photo: Graham Duncan.)

winter-growing species (*Z. odorata*) and one deciduous or evergreen species (*Z. aethiopica*), including the natural, light pink color form *Z. aethiopica* 'Marsh Mallow'.

The most important ornamental geophytic members of the Asphodelaceae in South Africa are *Bulbinella* (17 winter-growing species) and *Kniphofia* (about 70 mainly evergreen species). Both have elongate racemes and grow from corm-like underground stems (*Bulbinella*) or erect or creeping rhizomes (*Kniphofia*).

The South African Haemodoraceae includes the ornamental genus *Wachendorfia* (four species) with zygomorphic, moderate-to-large yellow or brownish flowers and plicate leaves. This family is characterized by arylphenalenone compounds that impart the reddish-orange coloration to its rhizomatous or cormous rootstocks.

The Hyacinthaceae has a number of important genera including the mainly summer-growing *Eucomis* (12 species), its long-lasting inflorescences overtopped with their characteristic heads of green bracts. Hardy, summer-growing *Galtonia* (four species) has pendent white or green, bell-shaped flowers produced on long pedicels, and *Lachenalia* (more than 120 species) has very variable tubular-, bell-, or urn-shaped, often heavily scented flowers in an array of colors from white to deepest purple. *Ornithogalum* (about 120 species) produces long-lasting racemes of mostly white, yellow, or orange actinomorphic flowers and rosettes of linear, lanceolate, or ovate leaves. *Veltheimia* (two species) has tubular pink, reddish, or rarely greenish-yellow pendent blooms, rosettes of glaucous or shiny green leaves, and unique inflated capsules.

Rhodohypoxis (six species) and *Hypoxis* (about 40 species) are the two most important ornamental members within the mainly southern African family Hypoxidaceae. The storage organs are very small to medium-sized or large corms, producing rosettes of deciduous or evergreen hairy leaves, and actinomorphic star-shaped white, yellow, pink or reddish flowers with characteristic persistent tepals that remain attached to the perianth tube after the seeds have started ripening.

The extremely diverse family Iridaceae has 30 genera in South Africa that are mostly deciduous and winter-growing, and grow from corms and rhizomes, or rarely, a regenerative woody caudex. All have sheathing leaves with either unifacial or bifacial blades, and inflorescences are spicate or cymose. Its most noteworthy ornamental members include winter-growing *Chasmanthe* (three species) with its strongly curved, orange or rarely yellow perianths, and summer- or winter-growing *Crocosmia* (seven species) that has spikes of showy bright orange to red, nodding flowers and usually lance-shaped leaves with flat or plicate blades. Elegant evergreen *Dierama* (about 45 species) has erect, narrow leathery leaves and spikes of pendent or erect, bell-shaped blooms mainly in shades of pink and mauve. *Dietes* (five species) has a rhizome and is evergreen with branched inflorescences producing ephemeral regular, white or yellow blooms, often with prominent nectar guides. Fragrant *Freesia* (16 species) has spikes of funnel-shaped or irregularly star-shaped blooms and fans of lance-shaped leaves with prominent midribs. Variable *Gladiolus* (about 175 species) has simple or branched spikes of usually strongly zygomorphic flowers ranging from pure white to deepest purple. Sun-loving *Ixia* (about 65 species) has spikes of actinomorphic flowers in a wide color range including turquoise, the tepals often with contrasting inner circles. Brightly colored *Sparaxis* (15 species) has spikes of zygomorphic or regular blooms with prominent papery bracts and erect fans of lanceolate leaves, and late-flowering *Tritonia* (about 25 species) has upward- or sideward-facing orange, yellow, pink, or red flowers and unifacial leaves with prominent midribs. Deciduous or evergreen *Watsonia* (52 species), one of the more robust genera, produces simple or branched spikes of usually strongly zygomorphic and distichous flowers in shades of red, orange, pink or cream, and its leathery leaves sometimes have thickened margins.

The only noteworthy genus within the Oxalidaceae is *Oxalis*, of which all approximately 200 southern African species grow from deciduous bulbs. The family is unusual within the geophytes in that it is dicotyledonous, and furthermore, the bulbs of *Oxalis* are renewed annually and, depending on the species, produce either dry or exendospermous (green) seeds.

The genus *Cyanella* (seven species) is the only southern African member of the Tecophilaeaceae with horticultural appeal. It grows from a deep-seated corm and produces ordinary or branched racemes of white, yellow, or mauve flowers.

The ginger family Zingiberaceae has just one representative in South Africa, *Siphonochilus aethiopicus*, that grows from a series of pungent, cone-shaped rhizomes and has curious ground-dwelling, sweet-scented blooms, and attractive lance-shaped leaves (Duncan 2010a).

III. HISTORICAL OVERVIEW AND USE OF SOUTH AFRICAN NATIVE GEOPHYTE SPECIES IN GLOBAL ORNAMENTAL HORTICULTURE

South African geophytes have been cultivated in Europe and Britain for more than four centuries, a course that began with *Haemanthus coccineus*, the well-known "April Fool" lily. Bulbs of this autumn-flowering species were collected by the Belgian Gonarus de Keyser on the flats below Table Mountain in 1603 and flowered in a garden at Middelburg, The Netherlands, in 1604 (De l'Obel 1605). An ever-increasing number of Dutch visitors to the Cape during the seventeenth and eighteenth centuries resulted in a corresponding surge of geophytes and other Cape plants making their way to Europe. Despite adverse weather conditions, growers there had considerable success in bringing to flower an array of Cape geophytes, especially those with robust bulbs of the family Amaryllidaceae. During the early decades of the nineteenth century, the introduction of Cape geophytes into Europe and Britain reached a peak. Affirmation of this is seen in the number of species illustrated in horticultural and botanical works, including a sumptuous publication produced in London, *The Ladies Flower-Garden of Ornamental Bulbous Plants*, devoted largely to South African species (Loudon 1841). The two most important South African geophytes in world horticulture are undoubtedly *Gladiolus* and *Freesia*. By the 1820s, breeding in Cape *Gladiolus* species had already begun in England, and crosses between *G. cardinalis*, *G. tristis*, and *G. carneus* marked the beginning of spring-flowering plants known as the Colvillei-nanus hybrids (Barnard 1972). The later introduction of the summer-flowering *G. dalenii* (including former *G. natalensis* and *G. psittacinus*) (Figure 18.2) and *G. oppositiflorus*, and crosses between them, as well as a few other summer-grow-

FIGURE 18.2   (**See color insert.**) *Gladiolus dalenii*, one of the original parents of the modern "grandiflora" hybrids. (Photo: Graham Duncan.)

ing species, gave rise to the first of the "grandiflora" hybrids, from which the modern strains of today originate. The plant breeder's skills contributed to great achievements using only a few species (Du Plessis and Duncan 1989). By 1837, seven first-generation *Nerine* hybrids between *N. sarniensis* (Figure 18.3), *N. undulata*, and *N. humilis* had appeared in England, but the very hardy *N. bowdenii* was only introduced in the early 1900s, providing breeders with the most valuable *Nerine* species of all (Duncan 2002). By the late 1850s, the first *Clivia* hybrid between *C. nobilis* and *C. miniata* had appeared in Belgium (now known as *C.* Cyrtanthiflora Group) (Duncan 2008), and in England the first *Lachenalia* hybrid, *L.* × *nelsonii*, appeared in 1880. Breeding with *Freesia* also began in the latter part of the nineteenth century, and crosses between the white *F. alba* (now *F. leichtlinii* subsp. *alba*) (Figure 18.4), the pink form of *F. corymbosa* (Figure 18.5) and the yellow *F. leichtlinii* subsp. *leichtlinii* (Figure 18.6) provided the basis for today's modern hybrids (Hoog 1909).

The late nineteenth and early twentieth centuries saw the introduction and breeding of several more South African geophytes, the most important of which included *Agapanthus*, *Crocosmia*, *Eucomis*, and *Schizostylis coccinea* (now *Hesperantha coccinea*). A genus that has enjoyed a significantly increased interest in the late twentieth and early twenty-first centuries is *Lachenalia*, hybrids of which have been bred by the Agricultural Research Council (ARC) at Roodeplaat, South Africa (Littlejohn and Bomerus 1997).

During the 1980s and 1990s, *Sandersonia aurantiaca* became well established in cut flower production and also as a pot plant as a result of research done in New Zealand (Duncan 1999). Worldwide, a tremendous resurgence in the popularity of the cultivation, propagation, and breeding of *Clivia* began in the early 1990s and continues to the present. During the same period, cultivars of *Zantedeschia aethiopica* for cut flower use, and pot plant production of the dwarf *Z. rehmannii* and hybrids between it and other summer-growing *Zantedeschia* species, has enjoyed much popularity as a result of breeding in New Zealand. Breeding in *Eucomis* for cut flower, pot plant, and garden cultivation has been receiving renewed interest within the last decade, mainly in New Zealand.

Although a large number of the world's most popular cut flowers and flower bulbs originate from southern African genetic material, in the past, royalties were not earned by the countries of origin. This is due to the fact that the material was removed from the countries of origin centuries ago and is now in the public domain. The 1992 Convention on Biological Diversity (CBD), which covers components of biodiversity conservation, sustainable use, and benefit sharing, excludes genetic material already in the public domain (Barton 1998). Numerous potential geophyte crops from southern

FIGURE 18.3 (**See color insert.**) *Nerine sarniensis* flowering after a fire on the Cape Peninsula. (Photo: Graham Duncan.)

FIGURE 18.4 (**See color insert.**) *Freesia leichtlinii* subsp. *alba*, the white parent used in early breeding. (Photo: Graham Duncan.)

Africa could be commercialized; therefore, only the development of new cultivars from this region will result in benefit sharing (Coetzee 2002). In 1997, an important new policy was introduced in a White Paper in which the conservation and sustainable use of South Africa's biological diversity was formalized in Government Gazette no. 18163. The main thrust of this policy is that equitable sharing

FIGURE 18.5 (**See color insert.**) *Freesia corymbosa*, the pink parent used in early breeding. (Photo: Graham Duncan.)

FIGURE 18.6   **(See color insert.)** *Freesia leichtlinii* subsp. *leichtlinii*, the yellow parent used in early breeding. (Photo: Graham Duncan.)

of benefits arising from the use and development of the country's biological resources has to take place. In February 2008, the National Biodiversity Strategy and Action Plan (NBSAP) was brought about, whose main objectives include fulfilling the requirements of the CBD and translating the 1997 Biodiversity White Paper into action. Among the five strategic objectives identified (15-year targets), strategic objective no. 4 aims to "enhance human development and well-being through sustainable use of biological resources and equitable sharing of benefits". Access and benefit-sharing regulations have been developed, and are awaiting approval to publish (Lutsch 2008).

IV.   RESEARCH AND ORGANIZATION NETWORKS

Research and development in plant science and horticulture at universities and technikons (technical colleges) is supported by the National Research Foundation of South Africa (NRF), an autonomous statutory body within the government Department of Science and Technology (http://www.nrf.ac.za). Some flower bulb companies and growers conduct their own research, and formal research results are available to them through publications in national journals including the *Journal of the Southern African Society for Horticultural Science* (http://www.sashs.co.za), *South African Journal of Plant and Soil* (http://www.plantandsoil.co.za), and *South African Journal of Botany* (http://www.sabotany.com).

For many years, major research efforts in South Africa have been focused on the identification and botanical classification of the indigenous species (Du Plessis and Duncan 1989; Goldblatt and Manning 2000). This work has been conducted at the National Botanical Gardens in Kirstenbosch and Pretoria and at the KwaZulu-Natal Herbarium in Durban. In addition to contributing to this program, Duncan (1988) has written a comprehensive handbook on *Lachenalia* and described the species and some of the cultural requirements.

At the Kirstenbosch National Botanical Garden (http://www.sanbi.org), selection of superior forms of indigenous geophyte species takes place in the Bulb Nursery. Once sufficient stocks have been accumulated, material is introduced to the public via their commercial nursery outlet, as well as at the annual garden fair organized by the Botanical Society of South Africa. Public educational material is provided in articles published in the *Plant of the Week* website (www.plantzafrica.com) produced at Kirstenbosch, and in *Veld* and *Flora*, the *Journal of the Botanical Society of South Africa* (http://www.botanicalsociety.org.za) as well as in cultivation handbooks published in the *Kirstenbosch Gardening Series* (http://www.sanbi.org); titles on geophytes published in this series to date are *Grow agapanthus* (Duncan 1998), *Grow bulbs* (Duncan 2010a), *Grow clivias* (Duncan 2008), *Grow fynbos plants* (Brown and Duncan 2006) and *Grow nerines* (Duncan 2002).

The Agricultural Research Council (http://www.arc.agric.za) conducts research into the production and development of ornamentals and indigenous flora, including geophytes (Agricultural Research Council 2010). Currently, *Lachenalia* and *Ornithogalum* are its most important projects. The breeding and physiology of *Lachenalia* have been concentrated at the Vegetable and Ornamental Research Institute in Pretoria (Hancke and Coertze 1988; Niederweiser and Van Staden 1990; Coertze et al. 1992; Roodbol and Niederwieser 1998; Du Toit et al. 2002, 2004; Kleynhans 2008). The Department of Genetics at the University of the Free State (spiesjj.sci@ufs.ac.za) conducts research into cytotaxonomy and molecular systematics of geophytes with horticultural potential, especially *Lachenalia*, in order to facilitate more successful hybridization experiments in cooperation with the Agricultural Research Council.

The Research Centre for Plant Growth and Development at the University of KwaZulu-Natal in Pietermaritzburg (http://www.biology.ukzn.ac.za) has conducted research into micropropagation of numerous southern African geophytes in the past. Currently, their research is aimed mainly at geophytes with medicinal value, but this information is also beneficial to the flower bulb industry. Current projects include developing micropropagation protocols for *Agapanthus campanulatus* (medicinal purposes), *Boophone disticha* (medicinal purposes), *Dierama erectum* (horticultural and medicinal purposes), and *Tulbaghia ludwigiana* and *T. violacea* (medicinal purposes).

The Department of Horticultural Science at Stellenbosch University published a series of studies on the physiology of *Nerine bowdenii*, a species that is indigenous to South Africa and has commercial value as a landscape plant and fresh cut flower (Theron and Jacobs 1992, 1994a,b, 1996a,b).

The most widely grown bulbous crop in South Africa is the *Hippeastrum*, which originates from Central and South America (Traub 1958; Ellenbecker 1975). Hadeco is a large bulb company near Johannesburg, South Africa that produces bulb crops for the domestic markets and export. The company was established by Harry de Leeuw in 1946 (http://www.hadeco.info/about.asp) and is currently owned by the Barnhoorn family. Hadeco initiated a *Hippeastrum* breeding program in 1948 and many cultivars with unique horticultural characteristics have been released. Therefore, their products are in demand worldwide. In addition to *Hippeastrum*, the company produces over 50 different geophyte crops and many cut flower lines. Total own production from 400 ha (1000 acres) under intensive cultivation produces some 100 million bulbs and 50 million flower stems for sale per annum. The company conducts in-house research and their products are exported to over 60 countries globally. Apart from *Hippeastrum*, the flower bulb industry in South Africa produces *Gladiolus* (150 ha), *Hyacinthus* (1 ha), *Iris* (10 ha), *Lilium* (150 ha), *Narcissus* (2 ha), *Tulipa* (6 ha), and other crops (160 ha) (Stuart Baarnhoorn 2011, pers. comm.).

Multiflora is the largest flower market in the Southern Hemisphere. Today, it services 600 flower growers from South Africa and the rest of Africa, and about 400 buyers, and has a turnover in excess of US$ 37.2 million (R* 300 million) per annum.

V. RESEARCH ON NATIVE SPECIES AND PROSPECTIVES OF THEIR USE AS ORNAMENTAL CROPS

It is estimated that over 25% of the worlds' ornamental geophytes originate from South Africa (Bryan 1989). In spite of this large number of species, only *Gladiolus* and *Freesia* are grown on a large scale in the commercial trade. Although many specialty bulbs including *Agapanthus*, *Galtonia*,

---

* US$ = R 8.06, 7/12/2011

Nerine, *Ornithogalum*, *Oxalis*, and *Zantedeschia* are available in international trade, many of these are not widely grown in South Africa (Benschop et al. 2010).

A. DEVELOPMENT OF *LACHENALIA*: A STUDY CASE

The development of *Lachenalia* as a purely South African product began in 1965. The initial research involved studies of cyto-taxonomic and genetic variation using microscopic and molecular marker techniques, as well as aspects of reproduction including embryogenesis and pollination. Thirty-nine interspecific hybrid combinations were made between 10 taxa (*L. aloides* var. *aloides*, *L. bulbifera*, *L. contaminata*, *L. aloides* var. *aurea*, *L. mutabilis*, *L. orchioides*, *L. pallida*, *L. rubida*, *L. aloides* var. *quadricolor*, and *L. splendida*) of which only 18 combinations were successful and flowered (Lubbinge 1980). Selections were made and these were tested amongst several South African growers to decide whether they had commercial potential. In further breeding work, isolation barriers preventing or limiting the successful hybridization of species with different flower shapes, such as polyploidy and pollen tube length, were overcome using embryo rescue, cut-style pollination, and pollen storage techniques. It was later realized that propagation material was not available in sufficient quantities, and insufficient information regarding commercial production of the plants was available to local growers and those in The Netherlands. This fact resulted in the compilation of production protocols in order to successfully commercialize the new cultivars (Kleynhans et al. 2002). It was shown that the quality and flowering time of pot plants are influenced by the temperature regime during production. Low-temperature treatment during storage results in flowering up to eight weeks earlier and longer-lasting inflorescences; the lower the temperature, the longer the peduncle and rachis length and the higher the number of flowers per inflorescence (Du Toit et al. 2004). However, by storing bulbs at 9°C directly after harvesting, flowering time can be retarded. When stored at 20–25°C for a minimum of 18 weeks (for early-flowering cultivars) and 20 weeks (for late-flowering cultivars), optimal flower initiation and development are obtained (Kleynhans et al. 2009).

In order to stimulate vegetative growth of bulblets to obtain flowering-size bulbs, the application of four equal amounts of nitrogenous fertilizer throughout the season gave the best response in the cultivars 'Ronina' and 'Rupert', and when fertilized in this way, the mature bulbs have a low nitrogen requirement in the subsequent pot plant phase (Engelbrecht et al. 2007, 2008).

The rapid propagation of *Lachenalia* hybrids *in vitro* using leaf explants, first achieved in South Africa, is well known (Nel 1983). Virus-free material of *Lachenalia* is multiplied from leaf explants *in vitro*, then leaf cuttings are taken from mother plants and grown by the bulb production company Afriflowers near Pretoria to obtain bulblets. Mature bulbs of flowering size are produced in the Northern Cape—a different location with more suitable climatic conditions. Temperature control during the dormant period allows manipulations of flowering time, while the use of growth regulators and temperature treatments during greenhouse production contributes to high quality of pot plants. The tradename "Cape Hyacinth" was registered for commercially produced *Lachenalia* cultivars (Niederwieser et al. 1998), and so far, five cultivars have been registered for plant breeders' rights in The Netherlands. These cultivars are 'Namakwa', 'Romaud', 'Ronina', 'Rosabeth' and 'Rupert'; the blue-flowered 'Rupert' received the award for the best potted special flowerbulb at Keukenhof in 1999. It is envisaged that a further six new cultivars will be introduced to the market in 2013 (F. Hancke, Afriflowers 2011, pers. comm.). Currently, the company Afriflowers is the sole planting stock producer for the cultivars of ornamental geophytes, developed by the ARC. In 2001, growers produced an estimated 3 million marketable bulbs of which a relatively small number were sold (Kleynhans 2006), and currently only 500,000 bulbs are produced annually (F. Hancke, Afriflowers 2011, pers. comm.). Some bulbs are sold locally but most are exported to The Netherlands, where the bulbs are forced, potted, and marketed. In The Netherlands, *Lachenalia* cultivars can be forced to flower year round, provided that optimum conditions of cool growing temperature and high light intensity are maintained.

B. Research and Development of New South African Flower Bulb Crops

Efforts to increase the diversity of commercial bulbs have been made largely by the Agricultural Research Council, but due to current economic conditions and lack of state funding, the research and development of new South African flower bulb crops in the public sector are on the decline.

Ornithogalum research is being conducted by the privately owned company Frontier Laboratories in the Western Cape and at the ARC, Roodeplaat. At Frontier Laboratories, research is aimed mainly at obtaining long-stemmed cultivars of *O. dubium* and *O. maculatum* for cut flower production and short cultivars of these species for pot plant production. At Roodeplaat, several cultivars of *O. dubium* and *O. thyrsoides* have been developed by means of controlled pollination, using embryo rescue and tissue culture multiplication and pot plant trials were performed to determine nutrient requirements (Reinten and Coetzee 2002). The cultivars 'Namib Sunrise', 'Namib Star', 'Oranjezicht', and 'Tipper' are being distributed in The Netherlands by Zuidgeest Nurseries (F. Hancke, Afriflowers 2011, pers. comm.). *Ornithogalum* breeding at Roodeplaat has also been done between ecotypes of the orange-flowered *O. dubium* and *O. maculatum*, and research into two genetically modified lines of *Ornithogalum* are being tested for virus resistance (F. Hancke, Afriflowers 2011, pers. comm.).

Selection of superior yellow and pink forms of *Veltheimia bracteata* and the blue-flowered *Merwilla plumbea* (= *Scilla natalensis*) is also being conducted. Frontier Laboratories specializes in the production of tissue-cultured material and exports 70% of its products, mainly to The Netherlands and Australia (A. Hackland, Frontier Laboratories 2011, pers. comm.).

Research into flowering physiology and inflorescence development in *Cyrtanthus elatus* was conducted to determine the optimum period for flower forcing. It was shown that due to the presence of two to four inflorescences at different developmental stages within the bulb at any time, flower forcing could be problematic, as forcing of one inflorescence might affect the next season's bloom (Slabbert 1997).

Currently, research programs on *Agapanthus*, *Clivia*, *Eucomis*, *Lachenalia*, and *Ornithogalum* are ongoing. *Agapanthus* research is centered at the privately owned nursery Black Dog Plants in the Western Cape. The aim of this program is twofold: the selection of superior forms of native species for garden and pot plants and the production of hybrids between evergreen and deciduous species with a view to combining the evergreen habit with the tubular, pendulous flower shape of certain deciduous species. Once hybrids and selections have been made, the crop is developed by mass production in tissue culture, then hardened-off in polyethylene tunnels and finally multiplied in open fields or planted into containers for sale to garden centers. Material is not exported and a number of products have already been introduced to the South African market, including first-generation crosses between the evergreen *A. praecox* and the deciduous *A. inapertus*, such as the breeder's lines *A.* 'Ballerina', *A.* 'Petticoat Lane', and *A.* 'Purple Cloud'.

Breeding of improved pastel forms of *Clivia miniata* and interspecific hybrids is being undertaken mainly by amateurs in many parts of the country. Hybrids and selections are multiplied either vegetatively or, once true-breeding strains have been obtained, from seed. A number of products have already been introduced in limited numbers to the "specialist" market in South Africa and abroad, one of the best being *C. miniata* 'Vico Peach', a cross between *C. miniata* 'Vico Yellow' and *C. miniata* 'Chubb's Peach' (Duncan 2008).

Eucomis research is centered at the privately owned Shosholoza Nursery near Mooi River in the midlands of KwaZulu-Natal and at the Agricultural Research Council (ARC) at Roodeplaat, Pretoria. At Shosholoza Nursery, emphasis is placed on the development of dwarf, pink-flowered cultivars for pot plant production and robust, pink-flowered cultivars for garden cultivation. Products are obtained through hybridization and selection and then multiplied vegetatively in open fields. In addition to eucomis, this nursery also exports dieramas, kniphofias, and tulips (for forcing) to countries of the European Union. At Roodeplaat, research is being conducted into flower formation, especially in *E. comosa* types. It has been found that when bulbs of *E. comosa* types are harvested from open ground and stored, they fail to flower the following season, whereas in bulbs that remain in ground during the dormant phase, flower formation takes place normally (F. Hancke, Afriflowers 2011, pers. comm.).

Valuable research into tissue culture protocols for various genera resulted in the development of *in vitro* propagation technologies for five *Cyrtanthus* species (McAlister et al. 1998b) and *Tulbaghia simmleri* (Zschocke and Van Staden 2000). Speeding up the conventional multiplication process of *Babiana* species was achieved in tissue culture using seeds sown aseptically, then cut into hypocotyl explants and placed on medium. The use of seeds is advantageous in that they are easy to sterilize and germinate readily, providing a sterile explant source, whereas corms are difficult to decontaminate (McAlister et al. 1998a). Similarly, *Gladiolus carneus* can be multiplied rapidly from hypocotyl explants that produce multiple shoots after three to four weeks, and the shoots readily form corms (Jager et al. 1998). The bright pink *Lapeirousia silenoides* has potential as a spring-flowering pot plant, but the seeds are erratic in germination and the corms do not form offsets. However, good *in vitro* proliferation of a section of the corm containing a growing point can be achieved. Exposure to cytokinin enhances shoot proliferation, and a 100% survival rate can be obtained when rooted *in vitro* plantlets are transplanted into soil (Louw 1989).

With respect to *Veltheimia bracteata*, *in vitro* bud formation can be initiated using leaf explants of the pink forms and the cultivar 'Lemon Flame' (Taylor and Van Staden 1997). *Ornithogalum maculatum* can also be rapidly propagated *in vitro* from leaf tissue, provided that the leaves are cut from donor plants of which the first flowers have opened. It has also been found that the formation of bulbs on *in vitro*-formed shoots is stimulated by the presence of growth regulators and the sucrose concentration (Van Rensburg et al. 1989).

VI. UNDERUTILIZED SPECIES

A. *BULBINELLA*

Two winter-growing, half-hardy *Bulbinella* taxa deserve attention. *B. gracilis* is an elegant, free-flowering dwarf species 20–30 cm high with narrow yellow racemes. It is suited to 20–25 cm diameter pots and flowers in winter. *B. latifolia* subsp. *doleritica* is the only species with bright orange flowers. When grown in 25 cm diameter pots, it grows to 70 cm high, and can also be grown in garden beds in areas with dry summers (Duncan 2010a).

B. *EUCOMIS*

Three dwarf, summer-growing *Eucomis* species are suited to cultivation in 15–20 cm diameter pots. *E. schijffii* has beautiful blue-green leaves and interesting purplish-maroon inflorescences 12–15 cm high (Figure 18.7), and *E. vandermerwei* (10–24 cm high) and *E. zambesiaca* (15–30 cm high) make excellent, long-lasting pot plants. *E. vandermerwei* has cryptically marked maroonish leaves and striking brownish maroon flowers; the only disadvantage of this species and *E. schijffii* is that their fetid scent is apparent at close quarters (Duncan 2011). *E. zambesiaca* has short, bright green leaves and sweet-scented creamy-white flowers. All three species are frost hardy (Duncan 2007, 2010a).

C. *FREESIA GRANDIFLORA*

This species has spikes of large deep red flowers and is suited to pot plant production in 15 cm diameter containers, in slightly shaded or bright light conditions, or as a subject for woodland gardens in mild climates (Figure 18.8). In the wild it is a summer grower but in cultivation it is versatile and easily adapts to a winter-growing cycle if required. The plants, which are easily raised from fresh seeds, can flower within eight months under appropriate conditions (Duncan 2010b), and can also be propagated from cormlets produced on subterranean scaly rhizomes (Duncan 2010a,b).

FIGURE 18.7   (**See color insert.**) The dwarf *Eucomis schijffii* has beautiful blue-green leaves and interesting purplish maroon inflorescences. (Photo: Graham Duncan.)

FIGURE 18.8   (**See color insert.**) *Freesia grandiflora*, an underutilized summer- or winter-growing species. (Photo: Graham Duncan.)

D. *HAEMANTHUS*

Haemanthus humilis subsp. *hirsutus* is a deciduous, spring- or summer-flowering plant with rounded heads of white or light-pink flowers and well-exerted stamens, with two hairy leaves. It is an excellent subject for 20 cm diameter pots and needs lightly shaded conditions (Figure 18.9). *Haemanthus deformis* is an evergreen autumn-flowering plant with a large white inflorescence produced on a short, hairy scape, and has two broad, leathery leaves. It requires deeply shaded conditions and

FIGURE 18.9 (**See color insert.**) The spring- or summer-flowering *Haemanthus humilis* subsp. *hirsutus* performs well in containers. (Photo: Graham Duncan.)

performs well as an indoor pot plant in 20 cm diameter pots or in deeply shaded rock garden pockets in mild climates (Duncan 2010a).

E. *ORNITHOGALUM MACULATUM*

This winter-growing, spring-flowering plant is ideally suited to pot cultivation. It is a variable species, growing 0.8–50 cm high, depending on the form; the robust, large-flowered, deep orange form with long stems is best suited to cultivation (Figure 18.10). The bulbs require a very well-drained,

FIGURE 18.10 (**See color insert.**) A robust, large-flowered form of *Ornithogalum maculatum*. (Photo: Graham Duncan.)

sandy growing medium and only occasional heavy drenching during the growing period. The flowers need bright light and warm conditions to open fully, but even in cool weather, the closed flowers remain attractive. The bulbs have to be kept absolutely dry during the summer dormant period (Duncan 2010a).

F. SCADOXUS MEMBRANACEUS

The evergreen *S. membranaceus* is a long-lived, usually dwarf, species 12–50 cm high with orange flower heads. It is an ideal indoor plant for 15–25 cm diameter pots (depending on the particular form) or for shady rock garden pockets in mild climates. It requires low-light levels, minimal watering in winter, and only occasional watering in summer. The orange flower heads are followed by attractive bright-red berries (Duncan 2010a).

G. SIPHONOCHILUS AETHIOPICUS

The "Natal Ginger" is a deciduous, summer-growing plant 0.3–1.0 m high with attractive lance-shaped leaves borne alternately on an erect pseudostem. This is an attractive summer foliage plant for positions in filtered light, either grown indoors in 25 cm diameter pots or in shaded gardens where temperatures do not fall below freezing. When grown in full sun, the plant produces large mauve, sweetly scented blooms at ground level (Duncan 2010a).

VII.   LOCAL PRODUCTION OF ORNAMENTAL GEOPHYTES

A. INDIGENOUS GENERA

The main indigenous crops (in alphabetical order) being produced in South Africa include hybrids and/or cultivars of *Agapanthus*, *Clivia*, *Eucomis*, *Freesia*, *Gladiolus*, *Kniphofia*, *Lachenalia* (Figure 18.11), and *Zantedeschia* (including the naturally occurring pink- and green-flushed forms 'Marsh Mallow' and 'Green Goddess', respectively). In addition, five wild South African species are grown in large numbers: *Dietes bicolor*, *D. grandiflora*, *Ornithogalum thyrsoides* (including the cultivar 'Mt Fuji'), and *Tulbaghia violacea*. With the exception of *Eucomis*, *Lachenalia*, *Ornithogalum*, and

FIGURE 18.11   (**See color insert.**) *Lachenalia* cultivar production at "Afriflowers" near Cullinan, Gauteng, South Africa. (Photo: Graham Duncan.)

FIGURE 18.12 **(See color insert.)** *Watsonia* hybrid in cultivation at the Kirstenbosch National Botanical Garden, Cape Town. (Photo: Graham Duncan.)

certain *Agapanthus* cultivars that are initially multiplied by tissue culture, most ornamental geophytes are propagated vegetatively in open fields. *Zantedeschia* pots and cut flowers of *Freesia*, *Ixia*, *Ornithogalum thyrsoides* 'Mt Fuji', *Gladiolus*, and *Zantedeschia* are delivered to the local market. Cut flowers of *Ornithogalum thyrsoides* 'Mt Fuji' and *Zantedeschia* are exported.

Smaller indigenous (specialty) crops (in alphabetical order) include hybrids and/or cultivars of *Crocosmia, Cyrtanthus, Dierama, Ixia, Lachenalia, Sparaxis, Tritonia,* and *Watsonia* (Figure 18.12).

In addition, numerous pure species are produced, of which the most important are: *Crinum bulbispermum, Cyrtanthus elatus, Galtonia candicans, Gloriosa superba, Nerine bowdenii, Ornithogalum dubium, Sandersonia aurantiaca, Scadoxus multiflorus* (Figure 18.13), *Scadoxus puniceus,* and *Veltheimia bracteata* (Figure 18.14). There are also several small, ultra-specialist nurseries that supply limited quantities of rare indigenous geophytic species for the local and overseas "specialist-collector" market, such as those belonging to *Babiana, Cyrtanthus, Gethyllis, Haemanthus,* and *Nerine.*

B. Introduced Genera

Hybrid cultivars of the genus *Hippeastrum* (also known in commercial production and trade as Amaryllis) are the biggest flower bulb crop in South Africa. Material is multiplied in tissue culture before hardening-off in greenhouses, and then grown-on to maturity in open fields. Other major introduced crops (in alphabetical order) being produced (mainly by Hadeco) include hybrids and/or cultivars of *Alstroemeria, Canna, Dahlia, Hemerocallis, Iris* (Dutch and German), *Lilium* (Asiatic), *L. longiflorum, Narcissus, Ranunculus,* and *Tulipa* (Hadeco Bulbs 2011).

Specialty crops (in alphabetical order) comprise hybrids and/or cultivars of *Anemone, Begonia* (tuberous), *Brodiaea, Crocosmia, Cyclamen, Hyacinthus, Muscari, Polianthes,* and *Tigridia,* and the species *Allium cowanii, Chlidanthus fragrans, Liatris spicata, Leucojum vernum, Scilla peruviana,* and *Sprekelia formosissima.*

FIGURE 18.13 (**See color insert.**) *Scadoxus multiflorus* subsp. *katherinae* flowers in low-light conditions. (Photo: Graham Duncan.)

The major bulb production areas are in the northern, summer rainfall parts of South Africa. They are located mainly in and around Johannesburg and Pretoria in the province of Gauteng, as well as in the provinces of North West and Mpumalanga, located to the west and east of Gauteng, respectively. Additional major production areas are located near Mooi River in the midlands of KwaZulu-Natal, at Clocolan in the eastern Free State, and at Elgin and Riversdale in the southwestern and

FIGURE 18.14 (**See color insert.**) The winter-growing *Veltheimia bracteata* is an ideal pot plant. (Photo: Graham Duncan.)

southern parts of the Western Cape, respectively. Different climatic zones are used to match the growing requirements of certain geophytes such as tulips that are grown for forcing in the Northern Hemisphere by Hadeco at Belfast in the Mpumalanga highlands, one of the coldest areas in South Africa, and by Shosholoza Nursery at Mooi River. Some specialty crops are also produced in Nieuwoudtville in the northwestern Cape, Malmesbury, Somerset West and Brackenfell in the southwestern Cape, and Napier and George in the southern Cape.

VIII.   CONCLUSIONS: PROBLEMS AND HOPES IN THE PRESENT AND FUTURE

The flower bulb industry in South Africa is somewhat fragmented, with most growers operating independently. A closer working association between growers and the Agricultural Research Council needs to be developed to foster greater cooperation and increase production. International collaboration already exists with a few countries including The Netherlands and Germany, but greater efforts should be made to market existing products and collaborate with a wider marketing and consumer audience.

Although much potential exists for the development of new geophyte crops in South Africa, this is being hampered by insufficient funding for research and development, and loss of skilled researchers to the private sector. In current economic conditions, the greatest potential for development, job creation, and foreign exchange earnings in the flower bulb industry rests with strong cooperation between the private and public sectors.

ACKNOWLEDGMENTS

The author thanks Fransie Hancke of Afriflowers, Richard Jamieson of Black Dog Plants, Allan Tait, and Mandy Fick and Dr. Luise Ehrich of New Plant Nursery for fruitful discussions.

CHAPTER CONTRIBUTOR

Graham D. Duncan is a specialist horticulturist for geophytes, the Kirstenbosch National Botanical Garden, South African National Biodiversity Institute, Cape Town. He grew up amongst the wild geophytes of the Cape west coast of South Africa, and obtained his MSc from the University of KwaZulu-Natal, Pietermaritzburg, South Africa. His expertise comprises the cultivation, propagation, and biology of South African geophytes, and the classification of *Lachenalia* (Hyacinthaceae). G. Duncan has published 13 books, 64 refereed journal articles, 4 book chapters, and more than 130 popular articles. He is a recipient of the International Bulb Society's Herbert Medal (2001) and the Recht Malan Prize (1989) for non-fiction awarded by the Nasionale Boekhandel for the book *Bulbous Plants of Southern Africa* (Tafelberg Publishers).

REFERENCES

Agricultural Research Council. 2010. *Lachenalia*. http://www.arc.agric.za

Barnard, T. T. 1972. On hybrids and hybridization. In *Gladiolus, a Revision of the South African Species. Journal of South African Botany, Supplementary Vol. 10*, eds. G. J. Lewis, and A. A. Obermeyer, 304–310. Cape Town: Purnell.

Barton, J. H. 1998. Acquiring protection for improved germplasm and inbred lines. In *Intellectual Property Rights in Agricultural Biotechnology*, eds. F. H. Erbisch, and K. M. Maredia, 19–30. Wallingford, Oxon, UK: CAB International.

Benschop, M., R. Kamenetsky, M. Le Nard, H. Okubo, and A. De Hertogh. 2010. The global flower bulb industry: Production, utilization, research. *Hort. Rev.* 36:1–115.

Brown, N. A. C., and G. D. Duncan. 2006. *Grow Fynbos Plants. Kirstenbosch Gardening Series*. Cape Town: South African National Biodiversity Institute.

Bryan, J. 1989. *Bulbs*. London: Christopher Helm Publ.

Coertze, A. F., F. L. Hancke, E. Louw, J. G. Niederwieser, and P. J. Klesser. 1992. A review of hybridization and other research on *Lachenalia* in South Africa. *Acta Hort.* 325:605–609.

Coetzee, J. H. 2002. Benefit sharing from flowering bulb—Is it still possible? *Acta Hort.* 570:21–27.

De l'Obel, M. 1605. *Animadversiones in Rondelet.* London: Thomas Purfoot.

Duncan, G. D. 1988. *The Lachenalia Handbook. Annals of Kirstenbosch Botanic Gardens, Vol. 17.* Cape Town: National Botanic Gardens.

Duncan, G. D. 1998. *Grow Agapanthus. Kirstenbosch Gardening Series.* Cape Town: National Botanical Institute.

Duncan, G. D. 1999. Christmas Bells—The cultivation and propagation of *Sandersonia aurantiaca. Veld & Flora* 85:178–180.

Duncan, G. D. 2002. *Grow Nerines. Kirstenbosch Gardening Series.* Cape Town: National Botanic Institute.

Duncan, G. D. 2006. Bulbous wealth at the Cape. *The Alpine Gardener* 74:296–315.

Duncan, G. D. 2007. Lesser-known *Eucomis. Plantsman New Series* 6:98–103.

Duncan, G. D. 2008. *Grow Clivias. Kirstenbosch Gardening Series.* Cape Town: South African National Biodiversity Institute.

Duncan, G. D. 2010a. *Grow Bulbs. Kirstenbosch Gardening Series.* Cape Town: South African National Biodiversity Institute.

Duncan, G. D. 2010b. Cultivation and propagation of *Freesia* species. In *Botany and Horticulture of the Genus Freesia (Iridaceae). Strelitzia 27*, eds. J. Manning, and P. Goldblatt, 96–103. Pretoria: South African National Biodiversity Institute.

Duncan, G. D. 2011. *Eucomis vandermerwei. Curtis's Botanical Magazine* 28(3):176–189.

Du Plessis, N., and G. Duncan. 1989. *Bulbous Plants of S. Africa—A Guide to their Cultivation and Propagation.* Cape Town: Tafelberg.

Du Toit, E. S., P. J. Robbertse, and J. G. Niederwieser. 2002. Effects of growth and storage temperature on *Lachenalia* cv. Ronina bulb morphology. *Scientia Hortic.* 94:117–123.

Du Toit, E. S., P. J. Robbertse, and J. G. Niederwieser. 2004. Temperature regime during bulb production affects foliage and flower quality of *Lachenalia* cv. Ronina pot pants. *Scientia Hortic.* 102:441–448.

Ellenbecker, M. 1975. Geographical distribution of the Amaryllidaceae. *Plant Life* 31:37–49.

Engelbrecht, G. M., C. C. Du Preez, and J. J. Spies. 2007. Response of *Lachenalia* growing in soil to nitrogen fertilization during the nursery phase. *S. Afr. J. Plant Soil* 24(4):220–227.

Engelbrecht, G. M., C. C. Du Preez, and J. J. Spies. 2008. Response of *Lachenalia* growing in soil to nitrogen fertilization during the pot plant phase. *S. Afr. J. Plant Soil* 25(2):92–98.

Goldblatt, P., and J. Manning. 2000. *Wildflowers of the Fairest Cape.* Cape Town: ABC Press.

Hadeco Bulbs. 2011. *Export.* http://www.hadecobulbs.com.

Hancke, F. L., and A. F. Coertze. 1988. Four new *Lachenalia* cultivars with yellow flowers. *HortScience* 23:923–924.

Hoog, T. 1909. Die neuen Freesienhybriden in der handelsgartnerei der firma C. G. van Tubergen jun., Haarlem, Holland. *Gartenweld Berlin* 13:199–201.

Jager, A. K., B. G. McAlister, and J. van Staden. 1998. *In vitro* culture of *Gladiolus carneus. S. Afr. J. Bot.* 64(2):146–149.

Kleynhans, R. 2006. *Lachenalia.* In *Flower Breeding and Genetics: Issues, Challenges and Opportunities*, ed. N. O. Anderson, 491–516. Dordrecht, The Netherlands: Springer.

Kleynhans, R. 2008. Potential new lines in the Hyacinthaceae. *Book of Abstracts 10th Intl. Symp. Flower Bulbs and Herbaceous Perennials, Lisse, The Netherlands*, 21.

Kleynhans, R., J. G. Niederwieser, and F. L. Hancke. 2002. Development and commercialisation of a new flower crop. *Acta Hort.* 570:81–86.

Kleynhans, R., J. G. Niederwieser, and E. Louw. 2009. Temperature requirements for good quality *Lachenalia* pot plants. *Acta Hort.* 813:641–648.

Littlejohn, G. M., and L. M. Blomerus. 1997. Breeding with indigenous South African *Ornithogalum* species. *Acta Hort.* 325:549–553.

Loudon, J. W. 1841. *The Ladies Flower Garden of Ornamental Bulbous Plants.* London: W. Smith.

Louw, E. 1989. *In vitro* propagation of *Lapeirousia silenoides. S. Afr. J. Bot.* 55(3):369–371.

Lubbinge, J. 1980. *Lachenalia* breeding. 1. Introduction. *Acta Hort.* 109:289–295.

Lutsch, W. 2008. *National Biodiversity Strategy and Action Plan.* Department of Environmental Affairs, Pretoria. http://www.environment.gov.za

McAlister, B. G., A. K. Jager, and J. van Staden. 1998a. Micropropagation of *Babiana* spp. *S. Afr. J. Bot.* 64(1):88–90.

McAlister, B. G., A. Strydom, and J. van Staden. 1998b. *In vitro* propagation of some *Cyrtanthus* species. *S. Afr. J. Bot.* 64(3):229–231.

Nel, D. 1983. Rapid propagation of *Lachenalia* hybrids *in vitro*. *S. Afr. J. Bot.* 2(3):245–246.

Niederwieser, J. G., P. Anandajayasekeram, M. Coetzee, D. Martella, B. Pieterse, and C. Marasas. 1998. Research impact assessment as a management tool: *Lachenalia* research at ARC-Roodeplaat as a case study. *J. S. Afr. Hort. Sci.* 8:80–84.

Niederwieser, J. G., and J. van Staden. 1990. The relationship between genotype, tissue age and endogenous cytokinin levels on adventitious bud formation on leaves of *Lachenalia*. *Plant Cell, Tiss. Org. Cult.* 22:223–228.

Reinten, E., and J. H. Coetzee. 2002. Commercialization of South African indigenous crops: aspects of research and cultivation of products. In *Trends in New Crops and New Uses*, eds. J. Janick, and A. Whipkey, 76–80. Alexandria, VA: ASHS Press.

Roodbol, F., and J. G. Niederwieser. 1998. Initiation, growth and development of bulbs of *Lachenalia aloides* Romelia (Hyacinthaceae). *J. S. Afr. Soc. Hort. Sci.* 8:18–20.

Slabbert, M. M. 1997. Inflorescence initiation and development in *Cyrtanthus elatus*. *Acta Hort.* 430:139.

Taylor, J. L. S., and J. van Staden. 1997. *In vitro* propagation of *Veltheimia bracteata* and *V. bracteata* Lemon Flame. *S. Afr. J. Bot.* 63(3):158–161.

Theron, K. I., and G. Jacobs. 1992. Inflorescence abortion in *Nerine bowdenii* W. Wats. *Acta Hort.* 325:97–103.

Theron, K. I., and G. Jacobs. 1994a. Comparative growth and development of *Nerine bowdenii* W. Watson: Bulbs *in situ* versus replanted. *HortScience* 29:1493–1496.

Theron, K. I., and G. Jacobs. 1994b. Periodicity of inflorescence initiation and development in *Nerine bowdenii* W. Watson (Amaryllidaceae). *J. Am. Soc. Hort. Sci.* 119:1121–1126.

Theron, K. I., and G. Jacobs. 1996a. Changes in carbohydrate composition of the different bulb components of *Nerine bowdenii* W. Watson (Amaryllidacae). *J. Am. Soc. Hort. Sci.* 121:343–346.

Theron, K. I., and G. Jacobs. 1996b. The effect of irradiance, defoliation, and bulb size on flowering of *Nerine bowdenii* W. Watson (Amaryllidacae). *J. Am. Soc. Hort. Sci.* 121:115–122.

Traub, H. P. 1958. *The Amaryllis Manual*. New York: Macmillan.

Van Rensburg, J. G. J., B. M. Vcelar, and P. A. Landby. 1989. Micropropagation of *Ornithogalum maculatum*. *S. Afr. J. Bot.* 55(1):137–139.

Zschocke, S., and J. van Staden. 2000. *In vitro* propagation of *Tulbaghia simmleri*. *S. Afr. J. Bot.* 66(1):86–89.

19 Geophyte Research and Production in Turkey

Ibrahim Baktir

CONTENTS

I. INTRODUCTION

Geographically, Turkey is located in two continents, Asia and Europe, forming a natural bridge between them. The country extends more than 1600 km from east to west and about 800 km from north to south, and its total area covers about 783,562 km² of which 756,816 km² is in Anatolia and the rest in Europe. The Anatolian peninsula is the westernmost point of Asia and is also known as Asia Minor. The European part is called Thrace and is located on the Balkan Peninsula. Half of the country is higher than 1000 m and two-third is higher than 800 m above sea level, with 129 peaks exceeding 3000 m. The land surface has rough, broken, and deep gorges in many regions (www. allaboutturkey.com). The diverse nature of the landscape and topography results in significant differences in climatic conditions in the country. Annual precipitation ranges from about 300 mm in the Salt Lake (Tuz Gölü) district of Konya Province to nearly 2200 mm in Rize Province in the Northeastern corner of the country. The Aegean and Mediterranean coasts have a typical Mediterranean climate with mild rainy winters and hot dry summers. Annual precipitation in the Mediterranean and Black Sea regions varies from 500 to 1300 mm. The mountain chains located parallel to the northern and southern coasts create barriers for the penetration of rain clouds to the inland and cause abundant rainfall on the mountain slopes facing the coasts. In contrast, the mountain ranges located perpendicular to the coast are divided by broad valleys such as Menders and Gediz, which allow the maritime climate to prevail several kilometers inland in the Aegean Region. The Anatolian Plateaus in the Central Eastern and Southeastern parts of the country have a typical continental climate with cold winters and dry hot summers. In these regions, a minimum temperature of –30°C or even lower can often be experienced. Temperature differences between day and night and between summer and winter are very sharp, rain is infrequent and the climate diverse due to topography along with unique geographical location of these regions.

The Turkish flora is rich both in the number of species and in its level of endemism. Three phytogeographical regions of the country—the Mediterranean, Euro Siberian, and Irano-Turanian—contain more than 10,000 native vascular species, which is almost equal to the entire European Flora and also equal to nearly half of the total species of the whole Mediterranean region. The level of endemism of the Turkish flora is 34.4%. Every year about 60 new species are added to the flora

(Davis 1965–1988; Ekim et al. 2000; Blamey and Grey-Wilson 2004). Most of the endemic species of Turkey are restricted to the higher mountains and deep gorges that provide optimal conditions for their development. However, many species have specific habitat requirements. For example, the sea daffodil *Pancratium maritimum* is found only in coastal sandy habitats, whereas snowflakes *Leucojum aestivum* prefer seasonal lake shores and wet areas (Göktürk et al. 2009). The local flora is very rich in flowering geophytes. More than 1000 (Özhatay et al. 2005; Kaya 2009) native geophyte species belong mainly to the Liliaceae, Alliaceae, Amaryllidaceae, Iridaceae, and Orchidaceae families. Many of these geophytes have good potential in terms of ornamental features (Kaya 2009).

Turkey is a fast-developing country and is one of the world's most important producers with respect to quantity and diversity in horticulture. Globally, Turkey is ranked in the 9th place in horticultural production, leading in apricot, fig, filbert, and sweet cherry production. Open field vegetable culture has been a traditional practice for years in the country. In recent years, Turkey has become an important vegetable producer in greenhouses (60,000 ha), while nearly 50% of the greenhouses are located in Antalya and the surrounding regions (ATIM 2011). Since the 1980s, the country exports flowers and ornamental plants. The annual export value of the ornamental industry in 2009 is US$48.60 million with spray carnation taking the major share (FTMT 2010). Floriculture in Turkey is growing rapidly. In ten years, the area of cut flower production increased by 42% from 855 ha in 2000 to 1213 ha in 2009, and the export value of cut flowers has reached US$24.38 million (ATIM 2011). The demand for Turkish carnations and gerbera in international markets is quite high, not only in Eastern and Western Europe but also in Japan.

Ornamental geophytes have a significant share in Turkish floriculture. The area of commercial bulb production increased from 13 ha in 2000 to 65 ha in 2009 (ATIM 2011). The demand for wild collected and commercially propagated flower bulbs such as snowdrops, windflower, winter aconite, snowflake and cyclamen is great in Western Europe. At present, five Turkish companies export flower bulbs, four of them have been active in the market for more than 20 years.

Although government support to floriculture is not as strong as it is to the other horticultural sectors, during the last few decades infrastructural problems were significantly improved and financial support for research and the development of ornamental crops has increased. At the same time, the high cost of energy for greenhouse flower production, transportation costs, and postharvest handling remain the main problems in the production of ornamental geophytes.

II. USE OF THE NATIVE GEOPHYTES SPECIES IN GLOBAL HORTICULTURE

The use of and trade in flower bulbs in Turkey date back to the 15th and 16th centuries. Hyacinth, narcissus, and tulip were commonly grown in the gardens of the Ottoman Empire. It is well known that tulips were introduced into Europe from Turkey in the sixteenth century. During the reign of Ottoman Sultan Ahmed III in the eighteenth century, a list of 1323 tulip hybrids was compiled (Bryan 1989). The beginning of the eighteenth century was known as the "Tulip Era" of the Ottoman Dynasty, and upper class society of this period developed an immense fondness for tulips. Tulip motifs were commonly used in Turkish art and folklore. Many embroidery and textile clothing, handmade by women, as well as carpets, tiles, miniatures, and drawings had tulip designs and shapes. The presence of tulips in textiles was considered by museum experts to be almost a guarantee that the fabric was of Turkish origin (Synge 1961; Crockett 1971). Saffron *Crocus sativus* was another very important flowering bulb of the Ottoman Empire (Arslan et al. 2009). This species was defined as a spice rather than for its ornamental uses, although it was also used in gardens. Saffron was widely cultivated in the northern town of Safranbolu. The area is still known for its annual saffron-harvesting festivals.

The interest in flower bulbs decreased during the stagnation and periods of decline of the Ottoman Empire, although bulb exports from Izmir were recorded in the late 1890s (Arslan et al. 2006, 2009). Exports of flower bulbs were reestablished in the late 1960s from the Aegean region, around Izmir, and later from other regions, in particular the Taurus Mountain Range (Baktir and

Ekim 2008). At present, several models of production of ornamental geophytes for export are employed in Turkey: (1) collecting from native populations, (2) commercial production in specialized farms, and (3) semi-commercial production in special plots in the areas of natural distribution.

During 1960–1980, only bulbs collected in their natural habitats were exported. The trade steadily increased until the quantities of bulbs exported annually amounted to millions. A total of 18 species from 14 genera have been exported (Ekim et al. 1997). The most popular species, for years collected in the wild populations for export, were *Galanthus elwesii* var. *elwesii*, *Cyclamen* spp., *Leucojum aestivum*, *Anemone blanda*, and *Eranthis hyemalis*. The late 1970s and early 1980s were dramatic years for native bulb trade, with exports increasing by a huge amount and reaching over 80 million bulbs only in *Galanthus elwesii*. Turkey became the largest global exporter of wild bulbs (Entwistle et al. 2002; MARA 2009, 2010; Table 19.1). The extensive and unsustainable collection of flower bulbs in the wild jeopardized the natural resources of the country. Botanists, horticulturalists, and other concerned organizations and people began to examine and strongly criticize this situation at almost every level. As a result of this activity, wild flower bulb collecting in Turkey was banned from 1985 for five years. Consequently, serious measures were proposed in order to preserve Turkish flora (Ekim et al. 1997, 2000; Baktir 2010). In the 1990s, the quota system for bulb collecting in nature was established by the Turkish Government (Tables 19.2 and 19.3). These strict precautions were taken even before Turkey became an active member of CITES[*]. Legislation regulating bulb collection and export was first enacted in 1989 and revised in 1991 and 1995. Advisory and Technical Committees were established in 1989. They consisted of scientists, representatives of the Ministry

TABLE 19.1

Export of the Five Most Important Ornamental Geophytes from Turkey during 1972–2010 (in bulb number)[a]

	Species					
Year	Anemone	Cyclamen	Eranthis	Galanthus	Leucojum	Total
1972	2,967,000	1,366,985	2,734,500	15,197,200	1,794,000	24,059,685
1975	1,176,000	324,000	5,205,000	7,333,000	1,515,000	15,553,000
1979	6,849,500	1,099,725	11,073,887	18,636,824	5,028,500	42,688,436
1981	9,334,000	1,290,870	12,118,000	25,173,950	4,750,350	52,667,170
1983	5,085,100	3,573,445	12,418,000	36,874,150	6,029,770	63,980,465
1984	16,975,000	6,632,000	33,082,500	82,808,500	14,525,500	154,023,500
1989	11,103,800	1,811,168	11,650,000	34,925,290	7,742,880	67,233,138
1993	6,641,900	1,487,630	7,406,400	13,561,250	4,490,800	33,587,980
1996	9,000,000	400,000	9,000,000	7,000,000	4,000,000	29,400,000
1997	9,000,000	2,010,000	8,000,000	8,000,000	5,000,000	32,010,000
2007	6,000,000	2,800,000	3,500,000	8,100,000	4,000,000	24,400,000
2008	6,000,000	2,800,000	3,500,000	8,100,000	4,000,000	24,400,000
2009	6,000,000	2,800,000	3,500,000	8,100,000	4,000,000	24,400,000
2010	6,000,000	2,800,000	3,500,000	8,100,000	4,000,000	24,400,000

[a] MARA: Ministry of Agriculture and Rural Affairs, Ankara, Turkey, 2010.

[*] CITES (the Convention on International Trade in Endangered Species of Wild Fauna and Flora, 1973) is an international agreement between governments. Its aim is to ensure that international trade in specimens of wild animals and plants does not threaten their survival. CITES was drafted as a result of a resolution adopted in 1963 at a meeting of members of IUCN (The World Conservation Union). The text of the Convention was agreed to at a meeting of representatives of 80 countries in Washington DC., USA, in 1973, and on 1 July 1975 CITES came into force.

TABLE 19.2

List of Turkish Geophytes Species, Forbidden for Collecting in Natural Habitats for Export

1. *Allium* (all species)
2. *Arum creticum*
3. *Biarum* (all species)
4. *Crocus* (all species)
5. *Cyclamen* (all species, except for *C. coum*, *C. cilicicum*, and *C. hederifolium*)
6. *Eminium* (all species)
7. *Fritillaria* (all species, except for *F. persica* and *F. imperialis*)
8. *Galanthus* (all species, except for *G. elwesii* and *G. woronowii*)
9. *Gentiana lutea*
10. *Hyacinthus orientalis*
11. *Iris* (all species)
12. *Lilium* (all species, except for *L. candidum*)
13. *Muscari* (all species)
14. Nympheaceae (all species)
15. Orchidaceae (all species)
16. *Pancratium maritimum*
17. *Sternbergia* (all species, except for *S. lutea*)
18. *Tulipa* (all species)
19. Other geophytes

Note: The list appears in Annex I of the Ministry of Agriculture and Rural Affairs of Turkey (2009).

TABLE 19.3

List of Turkish Geophytes Species, Allowed for Export According to Quota

Species	Collecting in Native Populations	Production	Minimum Size (cm)[a]
1. *Anemone blanda*	7,000,000		4.0
2. *Arum italicum*	118,000		6.0
3. *A. dioscorides*	78,930		6.0
4. *Cyclamen cilicicum*	250,000		8.0
5. *C. coum*	300,000		8.0
6. *C. hederifolium*		1,800,000	10.0
7. *Dracunculus vulgaris*	250,000		10.0
8. *Eranthis hyemalis*	4,000,000		3.5
9. *Galanthus elwesii*	4,500,000	350,000	4.0
10. *G. woronowii*	1,500,000		4.0
11. *Leucojum aestivum*		2,000,000	7.5
12. *Urginea maritima*	5000		20.0
13. *Ornithogalum nutans*	150,000		7.0
14. *Geranium tuberosum*	750,000		5.0
15. *Fritillaria persica*		250,000	10.0+
16. *F. imperialis*		200,000	18.0+
17. *Lilium martagon*		1000	10.0+
18. *L. ciliatum*		2000	14.0+

Note: The list appears in the Annex II of the Ministry of Agriculture and Rural Affairs of Turkey (2009).

[a] Circumference (cm).

of Agricultural and Rural Affairs, nongovernmental organizations, and customs and trade officials. In 1995 and 2004, national legislation adopted CITES terminology and a national scientific CITES committee was appointed. An important step in bulb export control and monitoring was the establishment of interim storage warehouses in the Taurus mountains. Following bulb collection and cleaning, the bulbs were stockpiled in the warehouses for inspection by independent committees. Finally, an education program was developed for the "collecting" villages, to help optimize the collection process and to encourage nature conservation (Ekim et al. 1997; Baktir 2010).

The quota of wild-grown bulbs for export is set by a Technical Committee of the Ministry of Agriculture and Rural Affairs following a field inspection by scientists and other teams of experts. The status of the geophytes populations determines the policy of collecting. Thus, collecting *Galanthus elwesii* and *G. ikariae* from nature was completely prohibited during 1995–1996. In contrast, the export quota for *Anemone blanda* and *Eranthis hyemalis* was increased to 10 million bulbs per species when a professional survey of wild populations indicated that their condition was good (Ekim et al. 1997). Beginning in 2009, the collecting of numerous wild species has been completely banned and the export quota of these species is now totally met by commercial production in Thrace and the Black Sea Regions (Tables 19.2 and 19.3).

Currently, about one-third of the exported bulbs are being produced on farms (Table 19.3). According to the geophytes production programs of the bulb companies (data received from the Ministry of Agriculture and Rural Affairs of Turkey (MARA 2010), flower bulbs are produced on ca. 30 ha in three locations: the Balikesir, Izmir, and Yalova provinces. The production areas are regularly visited by the members of both Turkish Scientific Committee in Flowering Geophytes and Technical Committee of the Ministry of Agriculture and Rural Affairs of Turkey and appointed local agricultural experts. *Leucojum aestivum* and *Cyclamen hederifolium* are the most important crops in commercial bulb production, with 7.0 and 2.5 ha of commercial production, respectively. *Fritillaria imperialis* and *F. persica* (Figure 19.1) are grown in the Eastern and South-Eastern regions on 4.0 ha in total (Alp 2006). In addition, *Leucojum aestivum, Sternbergia lutea,* and *Lilium candidum* are commercially grown for foreign markets on a small scale on local farms, in home gardens, and in small plots in the Aegean, Black Sea, and Marmara regions.

An example of the semi-commercial production of ornamental geophytes in the areas of their natural distribution is the cultivation of giant snowdrop *Galanthus elwesii* bulbs, carried out in the western part of the Taurus Mountains. This unusual species (Figure 19.2) is found in two different habitats at altitudes below or above 1000 m. At lower altitudes, it is usually grown on the northern slopes, under canopies of deciduous trees, where they get enough sunlight during the winter period, while relatively wet soil covered by leaves protects the bulbs from sun burn during the hot summer season. On the other hand, at altitudes above 1000 m giant snowdrops bulbs are found among rocks, and under native shrubs and juniper trees (Davis 2000; Baktir 2010). Small plots for bulb propagation were recently scattered in the Taurus Mountains, while semi-commercial propagation is employed on various species on about 10 ha in Izmir. The bulbs must be harvested before June in order to prevent bulb damage under the hot summer conditions in Izmir.

Propagation material for commercial production is initially gathered from wild habitats, since it is prohibited to use export size bulbs for propagation. According to the latest regulations, the same rule applies to geophyte production under semi-commercial conditions.

Micropropagation of flowering geophytes is very limited. Only one tissue culture laboratory, located in the Free Zone of Antalya, is actively propagating a few geophytes such as daylily (*Hemerocallis* spp.), bleeding heart (*Dicentra* spp.), flamingo flower (*Anthurium andraeanum*), and *Geranium tuberosum*. All the plantlets propagated in the laboratory are exported to The Netherlands.

The annual export value of Turkish geophytes ranged between US$2 and 3 million (MARA 2010). However, the amount and export value of geophytes has decreased in recent years due to the global economic crises (MARA 2010). The attainment of the export ratio from the given quota in 2009 varies from 100% for *Galanthus elwesii* var. *elwesii* to 0% for *Scilla bifolia* and *Lilium ciliatum*. The amount of exported *Anemone blanda* and *Eranthis hyemalis* also decreased in

FIGURE 19.1 (**See color insert.**) Native *Fritillaria* species. (a) Wild population of *F. imperialis* in Siverek, Şanlıurfa; (b) commercial production of *F. imperialis* in open field; (c) yellow form of *F. imperialis*, (d) *F. persica*. (Photo courtesy Ş. Alp.)

FIGURE 19.2 (**See color insert.**) Giant snowdrop (*Galanthus elwesii*) in natural population in Akseki, Antalya. (Photo courtesy R. Kamenetsky.)

comparison with that of previous years. For some species (e.g., *Geranium tuberosum, Arum italicum, Dracunculus vulgaris,* and *Urginea maritima*) annual quotas have never been 100% fulfilled. These species are considered to be of minor importance and are rarely collected from the wild since their commercial production fulfills the export demand.

III. PRODUCTION OF ORNAMENTAL GEOPHYTES FOR THE DOMESTIC MARKET

In contrast to exports, the domestic market does not utilize the native flower bulbs, noted in the CITES list. At the same time, in 2009, Turkey imported cultivars of major exotic ornamental geophytes (anemone, tulip, gladiolus, hyacinth, narcissus, lilies, pot cyclamen, calla, freesia, and iris) in the amount of US$6.36 million (MARA 2009). Anemone, gladiolus, and lilies are cultivated in greenhouses for domestic consumption of cut flowers, whereas narcissus and hyacinths are used either in home gardens or for pot flower production.

The imported tulip bulbs are used as stock materials for local bulb production. In 2003, a project called "Tulips are Returning to Their Motherland" was initiated, which resulted in increased commercial production of Dutch cultivars to 70 ha in Balikesir, Istanbul, Konya, and Sakarya Provinces (MARA 2010). The produced bulbs are used by major municipalities for tulip exhibitions in spring. It should be noted that tulips are grown in Turkey only for bulb production and are used for urban landscaping, but not as cut flowers.

Commercial production of cut flowers in an open field was common in a few regions of Turkey even before the 1970s, when the Madonna lily *Lilium candidum* was grown for the perfume industry in Balikesir Province. Nowadays, cut flowers of *Narcissus tazetta* are produced commercially in Karaburun/ Izmir on 133.5 ha in open fields, where 20% of them are irrigated and 80% are not. The pleasant scent of this flower is highly appreciated by people in the coastal areas in the very early winter (ITIM 2006). *Narcissi* are unique flowers as far as their marketing is concerned; they are mainly sold in the neighborhoods' open bazaars in many cities. In the early 1990s, local *Allium neapolitanum* was considered an important cut flower in Izmir, but this is no longer so.

The greenhouse production of flowering geophytes as cut flower is still very limited, with a total area of about 41 ha. Production is located in Antalya, Izmir, and Yalova Provinces (MARA 2010). Plants are grown in a soilless medium, especially in volcanic ash, which is easily available in the country at a low cost. The main crops produced in the greenhouses are *Anemone blanda, Lilium longiflorum, L. candidum,* and *Gladiolus* hybrids. In the winter, *Strelitzia reginae, Ranunculus asiaticus, Anthurium andraeanum, Freesia, Iris,* and *Alstroemeria* are also grown in the greenhouses covering small acreage. Although a few attempts have been made for the propagation of exotic geophytes in open fields in different parts of the country, the results have been successful only for gladiolus. Currently, propagation material for lilies, irises, and freesia are mostly imported from The Netherlands. Anemone and calla tubers for cut flower production are imported from Israel. All cut flower production is consumed in the domestic market.

Traditionally, some of the Turkish geophytes are used as medicinal or vegetable plants. For example, the cooked leaves of *Arum dioscorides* and *Eremurus spectabilis* are commonly used as vegetables by local people. Bulbs of some *Crocus* species and tubers of *Geranium tuberosum* are consumed fresh. *Allium* species are used as green vegetables. Terrestrial orchids are traditionally used for ice cream making, and these plants are currently facing extinction due to heavy harvesting in native populations. It is well known that a number of geophytes have been used for centuries as folk medicine or alternative drugs, especially in old world countries. For example, *Crocus, Ornithogalum,* and *Paeonia* species are known as sedatives, while *Allium, Colchicum, Iris, Muscari,* and *Urginea maritima* are used as diuretics. Some of these species have been collected, dried, and preserved for medical purposes (Baytop 1999). In recent years, fresh leaves of summer snowflakes *Leucojum aestivum* have been exported to Bulgaria for the extraction of galanthamine, used for the treatment of Alzheimer's disease. An increasing demand for this particular species will lead, in the future, to the enlargement of its plantations in suitable regions.

IV. RESEARCH AND DEVELOPMENT NETWORK

National agricultural research and development in Turkey is carried on in a large number of universities and research, academic and technical institutions, overseen by several government ministries.

One of the main priorities in plant research is the identification and characterization of natural flora, with special emphasis on the endemic species of Turkey. The richness and diversity of local flora were described in nine volumes (1966–1988) of *Flora of Turkey and East Aegean Islands*, edited by P. H. Davis (Royal Botanic Garden, Edinburgh, UK). The last supplement was published in 2001 and was edited by the leading Turkish botanists A. Güner, N. Özhatay, T. Ekim, K. Hüsnü, and C. Başer. A recent series of field research (Özhatay et al. 2005, the Istanbul University) identified 122 important plant areas, covering 11 million ha to conserve 10,000 native species, 34.4% of which are endemic. The *Red Data Book* of Turkey indicates the status of rare and threatened plants of Turkey according to the International Union for Conservation of Nature and Natural Resources (IUCN) categories (Ekim et al. 2000; Bulut and Yilmaz 2010). A seed collection project of Turkish endemic species has been coordinated by T. Ekim (Gazi University in Ankara and the Istanbul University) under the joint sponsorship of the Turkish Scientific and Technical Research Council (TUBITAK) and the State Planning Organization (DPT). The project involved the work of numerous botanists from 12 universities and several seed banks. The collected seeds were deposited at seed banks, particularly in Menemen-Izmir, in Turkey. The project resulted in the collection of many rare species, and field observations provided the most recent and reliable information on the status, population richness, and distribution of many endemic species.

Besides the Flora of Turkey and the Aegean Islands, colorful catalogues and books were published on flowering geophytes of Turkey and Antalya (Göktürk et al. 2009; Kaya 2009).

Since the early 1980s, a number of research projects have been undertaken to provide the basis for the commercial production of native geophytes. Target species with commercial value were defined, their ecological requirements were determined, and cultivation practices were developed (Altan et al. 1984; Ekim et al. 1991). Most of the recent projects have concentrated on breeding, physiological aspects, and cultural practices of geophytes. Recently, breeding programs were initiated on tulips, fritillaries, irises, and lilies in some universities and research stations of the Ministry of Agriculture in different regions of the country. Akdeniz University in Antalya, Abant Izzet Baysal University in Bolu, and Ege University in Izmir are actively involved in this research (ICARDA 1999). In addition, Atatürk Central Horticultural Research Institute in Yalova, near Istanbul, has specialized in floricultural crops.

The project "Alternatives for the bulb trade from Turkey. A case study of indigenous bulb propagation" was initiated in the early 1990s. This 10-year project was carried out in the Göksu River Valley on the Western Taurus Mountain, and resulted in an alternative source of plant material for the international trade in bulbous plants from Turkey. Over 250 villagers from three villages were involved in commercial bulb propagation. The project demonstrated that native bulbs species can be produced for the international market using farming technologies and therefore to meet CITES criteria for artificial propagation (Entwistle et al. 2002). Today the villagers in these regions are successfully continuing commercial production of snowdrop.

The comprehensive two-step project is dedicated to the domestication of native species and introduction of new species and cultivars into the floricultural sector. The first part of the project was completed in 2009 and resulted in the publication of "The Catalogue of Native Turkish Flowers" (Kaya 2009). The research group collected 49 species of *Colchicum*, 38 *Fritillaria*, 2 *Hyacinthus*, 47 *Iris*, 11 *Lilium*, 3 *Nectaroscordum*, 8 *Polygonatum*, and 16 *Tulipa* species, most of which have commercial potential. Seven research institutes in different regions, 8 universities and 18 private companies were involved in the first part of the project with excellent collaboration. According to the results, a number of species were selected for further research (Table 19.4). The second part of the project started in 2010. The main purposes of the project are the breeding of promising genera

TABLE 19.4

List of Geophytes Species from Native Flora of Turkey, Selected for Future Introduction and Development as New Ornamental Crops

Genus (References)	Species
Fritillaria (Aslay et al. 2009)	*imperialis*
	persica
	armena
	aurea
	caucasica
	zagrica
Lilium (Kaya 2009)	*ciliatum*
	akkusianum
	martagon
	ponticum ssp. *artvinense*
Iris (Erken 2009)	*galatica*
	nectarifera var. *mardinensis*
	purpureobractea
	schachtii
	xanthospuria
Paeonia (Kaya 2009)	*arietina*
	daurica
	kesrouanensis
	mascula ssp. *arasicola*
Sternbergia (Baytop 1999)	*candida*
Tulipa (Saraç 2009)	*agenensis*
	armena var. *armena*
	sintenisii
	julia
	praecox

and phytochemical analysis of the species. Phytochemical analysis of the species is being carried out at the Faculty of Pharmacy at Ankara University, molecular studies are being conducted at TUBITAK Research Institute in Gebze, and breeding studies are being conducted at seven different research institutes. For instance, tulip breeding is performed at the Black Sea Research Station in Samsun, fritillaria breeding at the East Anatolian Research Station in Erzincan, crocus breeding in Menemen Research Institute in Izmir. A peony breeding program has been going on at the Yalova Atatürk Central Research Station since 2001 (Figure 19.3), and a number of peony cultivars ('Tombak', 'Eful', 'Alev Topu', 'Kaya', 'Kan Çiçeği', and 'Bocur') have been registered by E. Kaya and his coworkers. The framework of the project also includes the establishment of a small botanical garden for geophyte species in Yalova.

Financial support for these projects has been provided by the Scientific and Technological Research Council of Turkey (TUBITAK), the World Wildlife Fund (WWF), the Turkish Ministry of Agriculture and Rural Affairs, the Ministry of Forestry, and the Ministry of Environment, Universities and Municipalities. The Turkish Association of Growers and Exporters of Botanical Flowerbulbs actively supports research on indigenous species, mainly giant snowdrop *Galanthus elwesii*, cyclamen species *Cyclamen cilicicum*, *C. coum*, and *C. hederifolium* (Baktir 1996), and

FIGURE 19.3 (**See color insert.**) *Paeonia* as new ornamental crops. (a) *Paeonia* × *kayae*—natural hybrid; (b) Peony 'Tombak'. (Photo courtesy E. Kaya.)

Fritillaria (Alp 2006). In addition, grower organizations and unions, such as Antalya Cut Flower Exporters Union, have been active in flower propagation, distribution, and exporting networks.

Agricultural extension services in Antalya, Balikesir, Istanbul, Izmir, and Yalova are quite active in the development of geophyte production, but this practice is not common in most regions of the country. The key issue in geophyte production in Turkey is education and training in geophyte biology, agrotechniques, postharvest handling, and marketing. A graduate course on flowering geophytes is given at some universities, including Akdeniz University in Antalya. In addition, short professional courses on commercial and semi-commercial production of indigenous species have been periodically organized in the mountainous regions for local people.

V. NATIVE ORNAMENTAL GEOPHYTES: OPPORTUNITIES FOR HORTICULTURAL DEVELOPMENT

In-depth research on flowering geophytes, native to Turkey, has shown significant potential of numerous species for their commercial production as new ornamental crops (Kaya 2009; Table 19.4). However, in addition to biological traits, other aspects of possible commercial production should be considered, for example, climatic and cultivation requirements, socioeconomic structure of the local population, and the possible location of commercial production. For instance, it was shown that the South-Eastern and Southern parts of Eastern Turkey have suitable ecological conditions for the cultivation of *Fritillaria imperialis* and *F. persica*, which are native to the regions (Alp 2006). Although the growers in these regions still need to improve their skills and technologies in order to get higher yields and good bulb quality and to handle the bulbs after harvesting, CITES authorities urge the bulb traders in the country to obtain their quotas for fritillaria bulbs from these regions. Six *Fritillaria* species (out of 38 species native to Turkey) have ornamental potential and need to be studied and introduced into the culture (Aslay et al. 2009).

Tulips have been regaining their popularity in recent years, and a number of commercial companies are planning to invest a significant amount of money in tulip production and trading. The plateaus of the Central and Eastern Regions, Istanbul, and its surroundings have appropriate conditions for tulip bulb production. Certainly, there is a growing trend to use tulips as garden plants in big cities and as pot plants. Cut flower tulip production is very limited and is not likely to be implemented in the near future, although the use of tulip cut flowers is becoming common. Breeding programs are being undertaken in different regions in order to develop disease-resistant cultivars of good quality. At least five native species are promising for future use and breeding activities (Saraç 2009).

In comparison with other native geophytes, *Sternbergia lutea* is an easily grown species in backyards and its native areas. Being an autumn flowering and disease-resistant species, it has good potential as a garden flower. When the demand for this particular species in the market develops, large numbers of bulbs can be produced for marketing. However, research on flowering physiology of *Sternbergia* is needed, in order to prolong its flowering period. The endemic and endangered species *Sternbergia candida* with white flowers and spring bloom can also be a suitable garden plant when bulb production is increased in the South-Western region of the country. Currently, there is very limited production of this species in home gardens (MARA 2010).

The natural flora of Turkey is one of the richest sources of cyclamen. Within 10 local species, three are highly appreciated in international markets (Tables 19.2 and 19.3). The Turkish cyclamen species might provide a basis for the development of new cultivars, as was the case for *Cyclamen persicum*, and they can also be used as a source for the resistance of current cultivars against diseases. Even though hybridization between cyclamen species is difficult due to the differences in chromosome numbers and incompatibility mechanisms, some unique leaf characteristics of *Cyclamen graecum* ssp. *anatolicum* can be transferred to forthcoming hybrids by using conventional hybridization techniques (Grey-Wilson and Matthews 1988; Mathew and Özhatay 2001).

Iris can be produced as an ornamental crop or be used in hybridization (Erken 2009). Nearly half of the 47 native species are endemic to Turkey. *Iris nectarifera* with its brownish veined pale yellow flowers is endemic to Mardin Province in the South-Eastern part of the country, while *I. xanthospuria* with its vivid yellow flowers is endemic to Muğla Province in the Western Mediterranean Region. Both species are totally different in their ecological requirements.

Lilies of the Turkish flora include 11 taxa, some of which have good potential as flower crops (Table 19.4). *L. ciliatum* is adapted to rainy and mild ocean climate and is traded in international markets, although the demand for this species has been dramatically reduced in recent years. *L. akkusianum*, *L. martagon*, and *L. ponticum* ssp. *artvinense* are very attractive species with different flower colors and ecological adaptation to the mild Mediterranean climate. The collecting of these species in their native populations is forbidden.

Paeonia is a very picturesque genus of the local flora. All 12 species in this genus are grown at higher altitudes, above 1000 m. Four native species (Table 19.4) are most important for the further development of new cultivars and also as a gene pool for future trials (Kaya 2009). *Nectaroscordum* and *Polygonatum* are two other important genera with ornamental potential (Kaya 2009). Musky grape hyacinth *Muscari muscarimi*, *Hyacinthus orientalis*, and *Lilium candidum* are promising species as garden and pot flowers as well as perfume plants and have good chances of being cultivated in large-scale areas similar to *Narcissus tazetta* (Özzambak et al. 2006).

VI. CONCLUSIONS

The natural climatic conditions and geographic location of Turkey provide the most favorable conditions for the horticultural production of ornamental geophytes. Local flora includes many species which have already been used in modern floriculture or possess great potential for future breeding and development.

During the last few decades, the export of native Turkish flower bulbs to the international markets has undergone a transition from uncontrolled plant collecting in nature to planned and systematic commercial production in various regions of Turkey. This process obviously has had to be supported by significant research activity, legislation, quota regulation and education. At the same time, increasing productivity and product quality is essential for meeting the market demands and to take advantage of the emerging agro-industries and export opportunities. Such objectives can be achieved only through a sustainable and efficient use of improved agricultural technologies. Thus, agricultural research and development are of vital importance to the floral industry. The research results, including production practices, Integrated Crop Management, and postharvest technologies, will all be transferred to the producers.

On the other hand, the development of a modern infrastructure (packing plants, refrigerated facilities, roads, transportation and communication systems, etc.) is essential. The future challenges could be met through a significant increase in financial resources, public investment, and nongovernmental sources. Private companies, domestic or foreign, must be involved in flowering geophytes, similar to the seed companies that provide large financial support for horticultural research in Turkey. This process will not only improve the bulb industry of Turkey, but will also have a positive impact on the socioeconomic development of the country, especially in agricultural regions.

In parallel with the development of commercial bulb production, precautions are being taken to prevent the over-collection of bulbs from their natural habitats. Currently, collecting from the wild is allowed only in a given period of time (less than a month), with rotation in three or four years' cycles for the same areas. The latest status of the natural resources of Turkish geophytes, a system developed by the authorities, and its application were accepted as an excellent model by CITES members in 2000 (Baktir and Ekim 2008; Baktir 2010). It was proposed that the areas with a large number of geophyte species (e.g., Ibradi County in the Western Taurus Mountains) should be declared conservation areas by the state. In addition, as a result of the development of resorts and infrastructures in the region of Antalya, the well-known tourist destination, populations of valuable geophytes are increasingly fragmented and isolated. Overgrazing is another problem in many parts of the country. Necessary legislations and environmental control should prevent further destruction of these valuable populations.

Owing to the global economical crises, the number of exported bulbs and their export value have clearly been reduced in recent years (MARA 2010). Therefore, the development of new and existing markets for flower bulbs and cut flowers is an essential goal of the floricultural sector in Turkey.

CHAPTER CONTRIBUTOR

Ibrahim Baktir is a professor of horticulture at Akdeniz University, Antalya, Turkey. He was born in Turkey and obtained BS from Atatürk University, Erzurum, Turkey and MS and PhD from West Virginia University, USA. He held a research position at the Turkish Scientific and Technical Research Council in Ankara, Çukurova University, Adana, and was dean at the European University of Lefke (Cyprus). His expertise comprises the ecology, physiology, and production of flower bulbs. Dr. Baktir has published numerous papers, chapters, and books, mostly in Turkish. He is a recipient of Alpha Zeta, ISHS medals and a number of honor plates in Turkey.

REFERENCES

Alp, Ş. 2006. *Native flower bulbs: Protection and propagation of Fritillaria No. 2.* Yalova, Turkey: Doğal Çiçek Soğancilari Derneği, Yayın.

Altan, T., G. Uzun, S. Altan, I. Baktir, C. Özsoy, E. Tanrısever, M. F. Altunkasa, and M. Yücel. 1984. A research on determination ecological requirements, distribution areas and effects of digging for sell of some geophytes which grow naturally in the Coastal Mediterranean Region of Turkey. *Doğa* 13:478–486.

Arslan, N., B. Gürbüz, A. Ipek, S. Özcan, E. Sarıhan, A. M. Daeshian, and M. S. Moghadassi. 2006. The effect of corm size and different harvesting time on saffron (*Crocus sativus* L.) regeneration. *Acta Hort.* 739:113–117.

Arslan, N., A. S. Özer, and R. Akdemir. 2009. Cultivation of saffron (*Crocus sativus* L.) and effects of organic fertilizers to the flower yield. *Acta Hort.* 826:237–240.

Aslay, M., U. Rastgeldi, and A. Kesici. 2009. *Fritillaria.* In *Native Flower Catalogue of Turkey, No. 91*, ed. E. Kaya, 81–119. Yalova, Turkey: Atatürk Bahçe Kültürleri Merkez Araştırma Enstitüsü, Tasarım Matbaacılık Hizmetleri.

ATIM. 2011. Antalya Agricultural Department-Antalya Tarım Il Müdürlüğü, Antalya.

Baktir, I. 1996. A research on *in situ* propagation of giant snowdrop *Galanthus elwesii* Hooker f. *Akdeniz Üniversitesi, Ziraat Fakültesi Dergisi* 9:342–346.

Baktir, I. 2010. Sustainable snowdrop production in Turkey. *Bull. UASVM Hortic.* 67(1):292–297.

Baktir, I., and T. Ekim. 2008. Latest regulations on exported geophytes in Turkey. *Book of Abstract 10th International Symposium on Flower Bulbs and Herbaceous Perennials, Lisse, The Netherlands*, 2.

Baytop, T. 1999. *Cure with Plants: Old Time and Now.* Çapa, Istanbul: Nobel Tıp Kitapevleri Ltd.

Blamey, M., and C. Grey-Wilson. 2004. *Wild Flowers of the Mediterranean.* London: A&C Black.

Bryan, J. E. 1989. *Bulbs, Vol. 1.* Portland, OR: Timber Press.

Bulut, Z., and H. Yilmaz. 2010. The current situation of threatened endemic flora in Turkey: Kemaliye (Erzincan) case. *Pakistan J. Bot.* 42:711–719.

Crockett, J. U. 1971. *The Time-Life Encyclopedia of Gardening: Bulbs.* New York: Time-Life Books.

Davis, A. P. 2000. *The Genus Galanthus: A Botanical Magazine Monograph.* Portland, OR: The Royal Botanic Gardens in association with Timber Press.

Davis, P. H. 1965–1988. *Flora of Turkey and East Aegean Islands. Vols. 1–8.* Edinburgh: Edinburgh University Press.

Ekim, T., N. Arslan, and M. Koyuncu. 1997. Developments in conservation and propagation of flowerbulbs native to Turkey. *Acta Hort.* 430:773–777.

Ekim, T., M. Koyuncu, H. Duman, M. Vural, Z. Aytaç, and N. Adigüzel. 2000. *Red Data Book of Turkish plants (Pteridophta and Spermatophyta).* Ankara: Turkish Association for the Conservation of Nature & Van Centennial University, Barışcan Ofset.

Ekim, T., M. Koyuncu, A. Güner, S. Erik, B. Yıldız, and M. Vural. 1991. *Taxonomic and Ecologic Investigations on Economic Geophytes of Turkey. Sıra No. 669, Seri No. 65.* Ankara, T.C. Tarım, Orman ve Köyişleri Bakanlığı, Orman Genel Müdürlüğü.

Entwistle, A., S. Atay, A. Byfield, and S. Oldfield. 2002. Alternatives for the bulb trade from Turkey: A case study of indigenous bulb propagation. *Fauna and Flora International* 36(4):333–341.

Erken, K. 2009. *Iris.* In *Native flower catalogue of Turkey No. 91*, ed. E. Kaya, 123–167. Yalova, Turkey: Atatürk Bahçe Kültürleri Merkez Araştırma Enstitüsü, Tasarım Matbaacılık Hizmetleri.

FTMT. 2010. Annual report of native flower bulbs, 2010. *Foreign Trade Ministry of Turkey Ankara*, 123–167.

Göktürk, R. S., Z. K. Elinç, I. Baktir, and B. Takmer. 2009. *Geophytes of Antalya.* Antalya, Turkey: Akdeniz Üniversitesi.

Grey-Wilson, C., and V. Matthews. 1988. *The Genus Cyclamen.* Portland, OR: The Royal Botanic Gardens, Timber Press.

ITIM. 2006. *Annual Reports on Flowers.* Izmir Tarim Il Müdürlüğü, Izmir.

Kaya, E. 2009. *Native Flower Catalogue of Turkey, No. 91.* Yalova, Turkey: Atatürk Bahçe Kültürleri Merkez Araştırma Enstitüsü, Tasarım Matbaacılık Hizmetleri.

MARA. 2009. *Annual Reports on Native Flower Bulbs.* Ankara: Ministry of Agriculture and Rural Affairs of Turkey.

MARA. 2010. *Annual Reports on Native Flower Bulbs.* Ankara: Ministry of Agriculture and Rural Affairs of Turkey.

Mathew, B., and N. Özhatay. 2001. *The Turkish Cyclamen: Propagation Guide of Native Cyclamen Species in Turkey.* Istanbul: WWF. Doğal Hayatı Koruma Derneği Yayınları.

Özhatay, N., A. Byfield, and S. Atay. 2005. *122 Important Plant Areas of Turkey*. Istanbul: WWF. Türkiye Doğal Hayatı Koruma Vakfı Yayınları.

Özhatay, N., A. Ilçin, and T. Ok. 2009. *Silent beauties of Ahır Mountains: 200 Wild Flowers*. Kahramanmaraş: T. C. Kahramanmaraş Valiliği Yayınları, Il Çevre ve Orman Müdürlüğü.

Özzambak, M. E., E. Zeybekoğlu, Ö. Kahraman, and N. Acarsoy. 2006. Revitalization and development of daffodil production in Mordoğan through a sustainable agriculture approach. *III. Ulusal Süs Bitkileri Kongresi, İzmir*, 9–13.

Saraç, Y. 2009. *Tulipa*. In *Native Flower Catalogue of Turkey, No. 91*, ed. E. Kaya, 195–210. Yalova, Turkey: Atatürk Bahçe Kültürleri Merkez Araştırma Enstitüsü, Tasarım Matbaacılık Hizmetleri.

Synge, P. M. 1961. *Collins Guide to Bulbs*. London: R&R Clark. www.allaboutturkey.com (Accessed on September 5, 2010).

20 Conclusions and Future Research

Rina Kamenetsky, Gary A. Chastagner, and Hiroshi Okubo

CONTENTS

Ornamental geophytes present one of the most fascinating and challenging sectors in the global flower industry. Traditionally, the commercial production of these crops prevailed in temperate-climate regions. However, over the last two decades, globalization and increased market competition have led to the development of new production centers in the Southern Hemisphere, Africa and Asia. In addition, novel approaches to address environmentally friendly, sustainable production and integrated management have stimulated new research directions, and innovative biochemical and molecular methods have opened new avenues in geophytes' taxonomy, breeding, and plant protection.

The most recent comprehensive treatise *Physiology of Flower Bulbs* by A. A. De Hertogh and M. Le Nard was published in 1993. Over the last 20 years, the general approach to plant research and horticultural production changed in rather dramatic ways. Therefore, a collective effort to analyze the current status in geophytes science, technology, and production systems was imperative.

Even within the last decade, the primary goal of most geophytes research was to provide the flower bulb industry with the technologies that enable them to produce crops with highly efficient growth cycles and meet high-quality market standards. These goals are still essential, and a great deal of the current research is dedicated to plant development, quality, and reproduction. However, at present the global bulb industry faces new challenges, which were difficult to foresee even a few years ago. Environmental issues, regulations, and globalization will likely continue to shape future research goals and directions. Scientists are challenged not only to study plant life, support industry, or to develop certain products and technologies, but also to serve society in general. This goal necessitates combining comprehensive knowledge of plants with ecological vision, market prospective, and great responsibilities for the well-being of future generations.

Unfortunately, in a continuously changing world, it is difficult to predict future needs and advancements. Therefore, this chapter will provide a short review of the up-to-date trends and research challenges in geophyte science, technologies for sustainable production and conservation of natural resources.

I. EMPLOYMENT OF NEW SCIENTIFIC TOOLS AND TECHNIQUES

Tremendous advancement in molecular science and technology in the last couple of decades has brought geophyte research to an entirely new level: genes can be analyzed at relatively low costs, proteomics and metabolomics have become valuable research tools, and gene transfer and genome sequencing, using the most complicated technologies, have been achieved in several species.

Molecular technologies have increased our ability to detect and identify pests and pathogens, which is a major benefit for clean plant programs. Recent advances in taxonomy and phylogeny also owe greatly to the employment of new methods of molecular and biochemical analyses.

The identification and introduction of genes, which control and regulate flower development, storage organ formation and dormancy *in vivo* and *in vitro*, await further progress. Another innovative area of biotechnology, which may have great potential for application in geophytes, is the genetic control of the process of senescence to prolong flower shelf life.

To date, most of the genetic studies on ornamental geophytes have been carried out on commercially important crops, such as lilies, narcissi, or tulips. To advance the investigation of geophyte biology, a model plant which the scientific community has agreed upon, should be chosen for genetic research. Such a model species has to meet several criteria, for example, a short juvenile period, easy pollination and seed germination, well-known morphological and physiological traits, an established transformation system, and a rapid regeneration ability. Several species have been proposed as possible candidates for the "geophyte *Arabidopsis*" (e.g., *Iris*, *Ornithogalum*, *Nelumbo*, *Gladiolus*, and *Lilium formosanum*), but the scientific community has still not reached consensus in this area. In addition to scientists, the new model species should be proposed also to funding agencies and industry, so that research teams are able to obtain financial support and to unify their efforts at understanding the most important developmental mechanisms in geophytes.

The acquired knowledge about genetic mechanisms and the control of the expression of the genes regulating the major developmental process will, of course, have a major impact on the breeding of new cultivars with the desired traits. Modern breeding goals include variety diversity, resistance to pests and diseases, year-round flowering ability, tolerance to abiotic stress factors, vigorous growth and bulb production, and better postharvest life. For ornamental geophytes, forcing ability and short production cycle are also important breeding goals.

In spite of the great benefits of bioengineering, classic breeding remains the major force for the development of new cultivars. Recently, new methods have enlarged the breeders' toolbox, enabling interspecific hybridization and ploidy level manipulations that have enhanced breeding for new traits. A remarkable advancement has been achieved in the breeding of several major crops (e.g., lilies and tulips) using state-of-the-art methods and approaches. In many other species and genera the potential for creating new variation is evident, but still awaits application.

In addition to aesthetic traits and physiological characteristics, classical breeding supported by modern biotechnical tools (e.g., marker-assisted selection, *in vitro* technologies or the introduction of functional genes) should prioritize novel key traits. For instance, reduced sensitivity to ethylene, postharvest longevity, resistance to desiccation, and pest and pathogen resistance of future crops will improve the sustainability of production and postharvest treatments by reducing the need for chemicals and energy.

II.   SUSTAINABLE PRODUCTION SYSTEMS

A sustainable agriculture system encompasses conservation of energy, soil, and water, as well as environmental protection. Social demand for sustainable agriculture calls for the development of economically feasible and environmentally friendly methods and technologies associated with the use of pesticides, chemicals, and nutrients. In bulb production, development of effective pest and disease control systems with reduced pesticide inputs is one of the vital questions, which will require additional research. Although experts predict that we will not be able to completely discontinue the use of pesticides, in the sustainable production of ornamental geophytes pesticides will be used as a component of Integrated Pest Management (IPM) systems. Therefore, new technologies for the optimization of pesticide applications need further development. In order to reduce pollution and minimize the damaging effects on the environment, it is necessary to utilize natural controls and the least toxic pesticides. In-depth studies of pest/pathogen–plant interactions are of special importance, and should be brought to a higher level. In the past, the availability of highly effective, broad-spectrum pesticides minimized the urgency to understand parasite biology, which limited comprehensive research in this field. At present, however, the cessation in the use of fumigants, particularly methyl bromide, affects the future management of nematodes and other soilborne pathogens and weeds. Consequently, there is an immediate need to develop new knowledge about pest/pathogen–plant interactions, as well as new technologies of chemical and biological control.

The most promising approach for the future sustainable production of geophytes are the intentional breeding for host resistance, the elicitation of natural resistance mechanisms in the crop plants, and the development of new methods for induction of systemic acquired resistance in ornamental crops.

III. APPLIED RESEARCH, KNOWLEDGE TRANSFER, AND RESEARCH–INDUSTRY INTERACTIONS

In spite of significant advances in biotechnology and bioengineering, molecular-based diagnostic for plant pathogens, and detection of numerous genes, relatively little progress has been made in providing growers with practical, commercially available solutions for pest management, shorter production cycle, sustainable production systems, and so on. The efforts to breed for resistance to common pathogens have so far resulted in a very few new commercially available disease-resistant cultivars. Thus, more emphasis needs to be placed on the transfer of the remarkable progress in understanding the intimate processes of plant development and plant–pathogen interaction to effective products and practical technologies.

Product quality and value can greatly benefit from the recent development of new technologies for plant production, forcing, transportation, and postharvest display. Some examples are: bulb storage at ultra-low oxygen and storage of cut flowers in a controlled or modified atmosphere, antitranspirant coatings, height control by chemical or cultural means, hydroponic production of cut flowers, integrated control of flowering timing and stem elongation, desiccation prevention during product marketing, and so on. Precise knowledge of flower development and its efficient regulation will facilitate and shorten the production cycle and, therefore, maximize the use of greenhouse space and decrease energy consumption for climate control.

At present, transfer of new scientific advances from the laboratory and greenhouse into commercially available products or management strategies is one of the most crucial needs for the flower bulb industry. The advances in basic and applied research should finally result in horticultural practices that will allow growers to effectively and economically manage their production cycles, pathogen problems, and postharvest treatments in a sustainable way. Unfortunately, over the last 20 years, university and government extension services have been reduced, eliminated, or privatized in many countries. As a result, many successful and helpful research results never reach the growers, and can only be found in scientific reports.

IV. GLOBALIZATION AND MARKET-ORIENTED RESEARCH

The process of interaction and integration between diverse cultures, people, and economic activity, resulting from international trade and investments, has strongly affected global production and consumption of horticultural products. Globalization has contributed to the economic growth of different countries through increased specialization based on the principle of comparative advantage. In addition to the traditional bulb producers (The Netherlands, the United Kingdom, France, Japan, and the United States), new players are rapidly emerging in regions with favorable agroecological conditions (e.g., Chile and Turkey), relatively cheap labor (e.g., Brazil and South Africa), or new commercial niches (e.g., New Zealand). As a result, we can foresee remarkable changes in global crop production, marketing, and consumption.

One can assume that the new production regions can certainly take advantage of modern technologies and solutions that have been developed to address problems in traditional geophyte production regions, and to use the best cultivation and management practices possible. However, adaptation of new crops/varieties to local conditions and optimization of the assortment, production, storage, and transportation techniques will be required for each location. This fact necessitates significant research activity in the new production regions, as well as an international effort at collaboration and knowledge transfer. On the other hand, the increasing competition in international markets is resulting in a growing volume of confidential and corporate research, which is not available to large scientific audiences. This contradiction sometimes leads to lose–lose situations, when all parties must develop new management protocols or production systems in parallel, without any opportunity for exchanging information and accelerating the development of quality products.

In general, the objectives of research and development of ornamental geophytes in new production regions include: (1) the breeding and improvement of existing crops, suitable for the local climate; (2) research on potentially useful new species and their development as new commercial crops and novel products; (3) knowledge transfer and development of the scientific and technical basis for progress toward new, sustainable and economically viable systems. The development of quality products has to fulfill the expectations of consumers regarding quality and environmentally sustainable production, in combination with reasonable prices and a suitable market niche.

The introduction of new and underutilized crops is one of the most vital issues in geophyte research and industry. Long-term programs are running in numerous centers of global floriculture. However, this is a lengthy and expensive process. The inability to rapidly build up commercial quantities of new cultivars limits the potential benefits of breeding and/or relegates them to only minor roles in the future. In addition, considerable skills required for the registration and marketing of the novel crops are not always available in the new production regions.

V. ECOLOGICAL VISION: HARMONY BETWEEN NATURAL RESOURCES AND COMMERCIAL PRODUCTION

The collecting of ornamental geophytes from various regions of the world is an important instrument for germplasm enhancement and breeding. This process facilitates the search for new or useful traits for major crops, or the introduction and development of new plant assortments. Numerous examples of successful gene "hunting" in nature are known (e.g., high variability of tulip and lily cultivars or new cultivars of alstroemeria). In spite of the aspiration to achieve maximum genetic diversity between and within populations and to collect the most useful material from the wild, a balance should be kept between profitable commercial production and natural resource exploitation. Therefore, one of the most important research aims and responsibilities for the future is to ensure the conservation and preservation of geophyte biodiversity in their natural habitats.

Protecting biodiversity and the prevention of genetic erosion are in our self-interest. Collecting genetic resources and sampling strategies must be conducted according to national and international laws and for the benefit of future generations. Efforts at plant conservation *in situ* and *ex situ*, genebanks, natural reserves, and support for ecological tourism—these actions will eventually not only help preserve natural resources, but promote beautiful flowers and commercially produced rare species or underutilized crops.

VI. EDUCATION AND RESEARCH NETWORK

In spite of increasing interest in plant sciences and huge advancements in understanding the intimate genetic regulation of the plant organism, the number of scientists working on applied research in horticulture is continuously declining. For example, in the 1970s three large horticultural research institutes and a network of regional Experimental Horticulture Stations were publically funded in England. The research goals of these organizations covered a large spectrum of basic and applied research, development, and technology transfer in all horticultural sectors, including flower bulb production. At present, the government of the United Kingdom has reduced funding for research on ornamental crops and the research network consists of one small university-based "crop centre" and the privatized remnants of two other organizations. Likewise, the activity of the International Flower Bulb Centre in The Netherlands (IBC) was terminated in December 2011. The responsibilities of this center included market research, public relations, advertising, technical advice, and product information, promotion, and education.

The majority of organizations that have supported research and educational programs in the past belonged to government or semi-government agencies and industry. Unfortunately, in most countries, financial support has been discontinued or reduced and links between industry and research have weakened. However, further creation and production of superior products in the competitive global market will require access to up-to-date knowledge, and this will necessitate continuous and substantial amounts of financial support for research and education.

At present, government agencies set regulations and industry has to develop products and systems that satisfy the regulations and/or needs. In spite of the fact that corporate market-oriented research, sponsored by industry, directly solves practical problems, the results are often confidential. Therefore, significant portions of the general areas of geophyte science (e.g., plant development, propagation technologies, sustainable production, and nature preservation) must remain in the public domain for the benefit of the entire scientific and horticultural community.

Education in practical horticulture has declined during the last two decades, but at present it is being reintegrated into classes at some universities. The new generation of educated horticulturalists will be fortified with modern technologies, general knowledge about molecular and biochemical regulation of developmental processes and ecological vision. Hopefully, renewed efforts at the training of a new generation of scientists will strengthen and contribute to the continuity of applied horticultural research.

Obviously, international research networks and collaborative efforts are needed for the further development of geophyte research. International and national symposia and conferences, many of them under the aegis of the International Society of Horticultural Science (ISHS), allow for a constant exchange of new ideas and research results. In addition, multidisciplinary research teams are largely supported by the scientific programs of the US government and the European Community. Unfortunately, ornamental horticulture is not included in the list of priorities for the majority of these programs.

The authors of this book have provided detailed reviews of the recent advances and future perspectives for research and development in geophyte science, technologies for sustainable production, and the conservation of natural resources. A Sustainable Scientific Community dealing with an integrated assortment of basic and applied research, constant knowledge exchange, graduate student education, creative and cooperative industry, marketing and consumer-focused outreach educational programs is key to addressing the numerous challenges of an economically viable sustainable industry of ornamental geophytes.

Appendix: Botanical Names Mentioned in This Book

Index

Note: The botanical terms and page numbers are in the Appendix.

Printed and bound by CPI Group (UK) Ltd, Croydon, CR0 4YY

21/10/2024

01777112-0019